New Perspectives and Applications of Modern Control Theory

Julio B. Clempner · Wen Yu
Editors

New Perspectives and Applications of Modern Control Theory

In Honor of Alexander S. Poznyak

Springer

Editors
Julio B. Clempner
Centro de Investigaciones Económicas, Administrativas y Sociales
Instituto Politécnico Nacional
Mexico city
Mexico

Wen Yu
Department of Automatic Control
Center for Research and Advanced Studies
Mexico city
Mexico

ISBN 978-3-319-62463-1 ISBN 978-3-319-62464-8 (eBook)
https://doi.org/10.1007/978-3-319-62464-8

Library of Congress Control Number: 2017949137

Printed on acid-free paper

This Springer imprint is published by Springer Nature
The registered company is Springer International Publishing AG
The registered company address is: Gewerbestrasse 11, 6330 Cham, Switzerland

To Dr. Alexander S. Poznyak on the occasion of his seventieth birthday. He is an excellent colleague, mentor and friend. We greatly benefited from his persistent advice, criticism, clear-thinking, constant availability, prompt feedback and rigorous standards.

Preface

With enormous pleasure, we bring you this book of Control Theory to honor Prof. Dr. Alexander S. Poznyak. Since Dr. Poznyak turned seventy this year, it is a good occasion to celebrate his accomplishments in control theory. We requested chapters from researches around the world. Dr. Poznyak is a relevant researcher and it is fitting that we are able to bring out this book to honor him.

Dr. Poznyak interest in control theory stated from the beginning in his carrier and he realized important and significant contributions. Dr. Poznyak was graduated from Moscow Physical Technical Institute (MPhTI) in 1970. He earned Ph.D. and Doctor Degrees from the Institute of Control Sciences of Russian Academy of Sciences in 1978 and 1989, respectively. From 1973 to 1993, he served this institute as researcher and leading researcher, and in 1993 he accepted a post of full Professor (actually 3-F) at CINVESTAV of IPN in Mexico. He is Regular Member of Mexican Academy of Sciences and System of National Investigators (SNI-Investigador Nacional Emerito from 2014). During 8 years, he was the Head of the Automatic Control Department. He is the Director of 42 Ph.D. thesis (37 in Mexico). It is impossible to summarize in this shot foreword all of his accomplishments. He has published 10 books, more than 220 papers in scientific journals and numerous papers for national and international conference proceedings. He has delivered several invited speeches at various universities across the world as a well as at many national and international conferences. He also serves as an Associate Editor of diverse scientific journals, member of various national and international scientific committees. He has also served as General Chairman program and Chairman program of numerous scientific international conferences. Dr. Poznyak was contemporary to very important people in the control arena, some of them were among founders of control theory, the others made contributions in its chapters: Ya. Z. Tzypkin, V.A. Yakubovich, V.G. Boltyanski, M.A. Krasnoselskii, V.I. Utkin, R. Sh. Liptser, K. Furuta, V.G. Sragovich, V.L. Kharitonov, G.A. Leonov, A.L. Fradkov, B.T. Polyak, V. Razvan, K. Najim, V.B. Kolmanovski, Yu.B. Shtessel, H. Sira-Ramirez, S.K. Spurgeon, Yu. V. Orlov, A. Levant, A. Polyakov, L.M. Fridman and many others.

Dr. Poznyak personal qualities of commitment, integrity, leadership and initiate leave lasting impression on his colleagues, students and friends, including those who are honouring him with the following chapters. Much of the research presented for the contributors in this book draw extensively too in their colleagueship and friendship over the years with Dr. Poznyak, for which they remain deeply grateful. It is in this spirit that we offer this book in honour of Prof. Dr. Alexander on his seventieth birthday.

We wish Dr. Poznyak the best of health, wealth, happiness, prosperity and success in all his endeavours, and we count on his contributions to control theory for many years to come.

Mexico city, Mexico
May 2017

Dr. Julio B. Clempner
Dr. Wen Yu

Acknowledgements

The act of completing a book is in fact a hard work. However, to be satisfied with the result is still harder. For the success and the completion of this project, we are grateful to many people.

We are most grateful to all the contributors and reviewers for their patience and efficiency with meeting the requirements of a rigorous and time-constrained review process. We are proud to have put together such a collection of chapters of excellent quality and variety in the field of control theory. We believe that this book offers a unique and wonderful glimpse in this classical but lively field of research.

A special gratitude to Alin Cârsteanu and César Solis who helped us in solving editorial and technical problems. We would like to thank the Department of Automatic Control at CINVESTAV-IPN for their continuous support. Julio B. Clempner also would like to acknowledge the financial support of the project 20170245 of the Secretaría de Investigación y Posgrado del Instituto Politécnico Nacional. Finally, to all those who we forget to mention please accept our apologies and thanks.

Contents

Contributors

Norelys Aguila-Camacho Department of Electrical Engineering and AMTC, University of Chile, Santiago, Chile

M.A. Aguilar-Orduña Departamento de Ingeniería Eléctrica, Sección de Mecatrónica, CINVESTAV-IPN, Ciudad de México, Mexico

Fathi M. Aldukali Department of Electrical and Computer Engineering, The University of Alabama in Huntsville, Huntsville, AL, USA

Yazan M. Alsmadi Department of Electrical Engineering, Jordan University of Science and Technology, Irbid, Jordan

Yannick Aoustin LS2N, UMR CNRS 6004, École Centrale, University of Nantes, Nantes Cedex 3, France

Christine Chevallereau LS2N, UMR CNRS 6004, École Centrale, University of Nantes, Nantes Cedex 3, France

Julio B. Clempner Centro de Investigaciones Económicas, Administrativas y Sociales (Center for Economics, Management and Social Research), Instituto Politécnico Nacional (National Polytechnic Institute), Ciudad de México, México

Michele Cucuzzella Dipartimento di Ingegneria Industriale e dell'Informazione, University of Pavia, Pavia, Italy

José Paulo V.S. Cunha State University of Rio de Janeiro, Rio de Janeiro, Brazil

Manuel A. Duarte-Mermoud Department of Electrical Engineering and AMTC, University of Chile, Santiago, Chile

Antonella Ferrara Dipartimento di Ingegneria Industriale e dell'Informazione, University of Pavia, Pavia, Italy

Leonid Fridman Facultad de Ingeniería, Universidad Nacional Autonoma de México, Mexico City, Mexico

Katsuhisa Furuta Tokyo Denki University, Tokyo, Japan

Javier A. Gallegos Department of Electrical Engineering, University of Chile, Santiago, Chile

Rosalba Galván-Guerra Unidad Profesional Interdisciplinaria de Ingeniería Campus Hidalgo, Instituto Politécnico Nacional, Hidalgo, Mexico

E. Hernández-Flores Departamento de Ingeniería Eléctrica, Sección de Mecatrónica, CINVESTAV-IPN, Ciudad de México, Mexico

Liu Hsu COPPE/Federal University of Rio de Janeiro, Rio de Janeiro, Brazil

Rafael Iriarte Facultad de Ingeniería, Universidad Nacional Autonoma de México, Mexico City, Mexico

M. Joaquin Collado Automatic Control Department, CINVESTAV, Mexico City, Mexico

Maria Kiseleva St. Petersburg State University, Saint Peterburg, Russia

Gabriel Tavares Melo State University of Rio de Janeiro, Rio de Janeiro, Brazil

Oscar Montano CICESE, Ensenada, Baja California, Mexico

Jaime A. Moreno Instituto de Ingeniería, Universidad Nacional Autónoma de Mexico (UNAM), Coyoacán, Ciudad de Mexico, Mexico

Tiago Roux Oliveira State University of Rio de Janeiro, Rio de Janeiro, Brazil

Yury Orlov CICESE, Ensenada, Baja California, Mexico

Yaodong Pan Honeywell Aerospace, Mississauga, ON, Canada

Frank Plestan Ecole Centrale de Nantes, Institut de Recherche en Communications et Cybernétique de Nantes - IRCCyN, Nantes, France

Andrey Polyakov Inria Lille, Villenueve d'Ascq, France

Alexander S. Poznyak Department of Control Automático, CINVESTAV-IPN, Ciudad de México, México

Yuri B. Shtessel Department of Electrical and Computer Engineering, The University of Alabama in Huntsville, Huntsville, AL, USA

H. Sira-Ramírez Departamento de Ingeniería Eléctrica, Sección de Mecatrónica, CINVESTAV-IPN, Ciudad de México, Mexico

Cesar U. Solis Departamento de Control Automático, CINVESTAV-IPN, Ciudad de México, México

Sarah K. Spurgeon Department of Electronic and Electrical Engineering, University College London, Torrington Place, UK

Juan C. Travieso-Torres Department of Industrial Technologies, University of Santiago de Chile, Santiago, Chile

Kristal K. Trejo Department of Automatic Control, Center for Research and Advanced Studies, Mexico, Mexico

Vadim Utkin Department of Electrical and Computer Engineering, The Ohio State University, Columbus, OH, USA

Juan-Eduardo Velázquez-Velázquez Unidad Profesional Interdisciplinaria de Ingeniería Campus Hidalgo, Instituto Politécnico Nacional, Hidalgo, Mexico

Longya Xu Department of Electrical and Computer Engineering, The Ohio State University, Columbus, OH, USA

Xinggang Yan School of Engineering and Digital Arts, University of Kent, Canterbury, UK

Wen Yu Departamento de Control Automatico, CINVESTAV-IPN, Mexico City, Mexico

Xinghuo Yu RMIT University, Melbourne, VIC, Australia

Dongya Zhao Department of Chemical Equipment and Control Engineering, China University of Petroleum, Huangdao District, Qingdao, China

Jiaming Zhu Yangzhou University, Yangzhou, China

E.W. Zurita-Bustamante Departamento de Ingeniería Eléctrica, Sección de Mecatrónica, CINVESTAV-IPN, Ciudad de México, Mexico

Chapter 1
Dr. Alexander Semionovich Poznyak Gorbatch: Biography

Alexander S. Poznyak

1.1 Current Position

Fig. 1.1 Dr. Alexander Semionovich Poznyak Gorbatch

Prof. Titular 3-F CINVESTAV-IPN.
Dept. Control Automático.
Institution: CINVESTAV-IPN, AP-14-740, México, Ciudad de México.
Tel: (+52 55) 5747-3741
FAX: (+52 55) 5747-7089
e-mail: apoznyak@ctrl.cinvestav.mx (Fig. 1.1)

A.S. Poznyak (✉)
Department of Control Automático, CINVESTAV-IPN, AP-14-740, Ciudad de México, México
e-mail: apoznyak@ctrl.cinvestav.mx

J.B. Clempner and W. Yu (eds.), *New Perspectives and Applications of Modern Control Theory*, https://doi.org/10.1007/978-3-319-62464-8_1

1.2 Resume

Alexander S. Poznyak (Alexander Semion Pozniak Gorbatch) was born on December 6 ,1946 in Moscow and graduated from Moscow Physical Technical Institute (MPhTI) in 1970. He earned Ph.D. and Doctor Degrees from the Institute of Control Sciences of Russian Academy of Sciences in 1978 and 1989, respectively. From 1973 up to 1993, he served this institute as researcher and leading researcher, and in 1993 he accepted a post of full professor (3-F) at CINVESTAV of IPN in Mexico.

For 8 years he was the head of the Automatic Control Department. He is the director of 41 PhD thesis' (37 in Mexico). He has published more than 220 papers in different international journals and 13 books.

He is Regular Member of Mexican Academy of Sciences and System of National Investigators (SNI-Emerito from 2014). He is Fellow of IMA (Institute of Mathematics and Its Applications, Essex UK) and Associated Editor of Oxford-IMA Journal on Mathematical Control and Information, of Kybernetika (Chech Republic), Nonlinear Analysis: Hybrid systems (IFAC) as well as Iberamerican Int. Journal on Computations and Systems.

He was also Associated Editor of CDC, ACC, and Member of Editorial Board of IEEE CSS. He is a member of the Evaluation Committee of SNI (Ministry of Science and Technology) responsible for Engineering Science and Technology Foundation in Mexico, and a member of Award Committee of Premium of Mexico on Science and Technology. In 2014, he was invited by the USA Government to serve as the member of NSF committee on Neuro Sciences and Artificial Intelligence.

1.3 Photos from the Personal Archive of Prof. A.Poznyak

See Figs. 1.2, 1.3, 1.4, 1.5, 1.6 and 1.7.

(a) Alexander in 1949.

(b) Seminar on Mechanics in MPhTI in 1980.

(c) July 2001: Lab.7 of Institute of Control Sciences of Russian Academy of Sciences.

Fig. 1.2 Photo album 1

(a) G.Arkhipova, A.Poznyak, B.Polyak, Ya.Tzypkin and S.Faina, Moscow, 1995.

(b) Z. and V. Kantorovich, V.Boltyanski and V.Kharitonov, Mexico, 1998.

Fig. 1.3 Photo album 2

(a) A.Poznyak with his teacher Academic Ya. Z. Tzypkin, February 1993.

(b) A. and T. Poznyak with V .Razvan. Bukharest, Rumania, October 2010.

(c) V. Razvan, A. Poznyak, V. Kolmanovski and A. Rodkina. Atlanta, 2002.

(d) Ya. Z. Tzypkin and V. A. Yakubovich, Moscow 1995.

Fig. 1.4 Photo album 3

(a) Lecture in Yale University, May 2013.

(b) A. Poznyak with the lecture " My life in Control", CINVESTAV, Mexico, December 2016.

(c) Academic A. Kurzhanskii, Prof. Yu B. Stessel, Prof. Leonid Fridman and Prof. A.Poznyak with his students, CINVESTAV, Mexico, February 2006.

(d) West Verginia, USA, March 2008.

Fig. 1.5 Photo album 4

(a) Huntsville, Alabama, USA. February 2004.

(b) Plenary lecture at VSS, Villanova, Spain, 2004.

(c) L. Fridman and Yu B. Stessel, Mexico, March 2006.

Fig. 1.6 Photo album 5

(a) A. S. Poznyak and V. G. Boltyanski, Ixtapa, December 2002.

(b) Alex Poznyak, Galina and Alexander Nazin, Versalle, France, April, 2006.

(c) A. Poznyak and V. Utkin. Bohum, Germany, 2012.

Fig. 1.7 Photo album 6

Books, Articles and Conferences

1. Aguilar, R., Martinez-Guerra, R., Poznyak, A.S.: Nonlinear PID controller for the regulation of fixed bed bioreactors. In: Proceedings of the 41st IEEE Conference on Decision and Control, vol. 4, pp. 4126–4131 (2002). https://doi.org/10.1109/CDC.2002.1185014
2. Alazki, H., Ordaz, P., Poznyak, A.S.: Robust bounded control for the flexible arm robot. In: Proceedings of the 52nd IEEE Conference on Decision and Control, pp. 3061–3066 (2013). https://doi.org/10.1109/CDC.2013.6760349
3. Alazki, H., Poznyak, A.S.: Output linear feedback tracking for discrete-time stochastic model using robust attractive ellipsoid method with LMI application. In: Proceedings of the 2009 6th International Conference on Electrical Engineering, Computing Science and Automatic Control (CCE), pp. 1–6 (2009). https://doi.org/10.1109/ICEEE.2009.5393429
4. Alazki, H., Poznyak, A.S.: Constraint robust stochastic discrete-time tracking: attractive ellipsoids technique. In: Proceedings of the 7th International Conference on Electrical Engineering Computing Science and Automatic Control, pp. 99–104 (2010). https://doi.org/10.1109/ICEEE.2010.5608567
5. Alazki, H., Poznyak, A.S.: Probabilistic analysis of robust attractive ellipsoids for quasi-linear discrete-time models. In: Proceedings of the 49th IEEE Conference on Decision and Control (CDC), pp. 579–584 (2010). https://doi.org/10.1109/CDC.2010.5717662
6. Alazki, H., Poznyak, A.S.: Averaged attractive ellipsoid technique applied to inventory projectional control with uncertain stochastic demands. In: Proceedings of the 50th IEEE Conference on Decision and Control and European Control Conference, pp. 2082–2087 (2011). https://doi.org/10.1109/CDC.2011.6160847
7. Alazki, H., Poznyak, A.S.: Robust stochastic tracking for discrete-time models: designing of ellipsoid where random trajectories converge with probability one. Int. J. Syst. Sci. **43**(8), 1519–1533 (2012). https://doi.org/10.1080/00207721.2010.547664
8. Alazki, H., Poznyak, A.S.: A class of robust bounded controllers tracking a nonlinear discrete-time stochastic system: attractive ellipsoid technique application. J. Frankl. Inst. Eng. Appl. Math. **350**(5), 1008–1029 (2013). https://doi.org/10.1016/j.jfranklin.2013.02.001
9. Alazki, H., Poznyak, A.S.: Robust output stabilization for a class of nonlinear uncertain stochastic systems under multiplicative and additive noises: the attractive ellipsoid method. J. Ind. Manag. Optim. **12**(1), 169–186 (2016). https://doi.org/10.3934/jimo.2016.12.169
10. Alazki, H.S., Poznyak Gorbatch, A.S.: Inventory constraint control with uncertain stochastic demands: attractive ellipsoid technique application. IMA J. Math. Control Inf. **29**(3), 399–425 (2012). https://doi.org/10.1093/imamci/dnr038
11. Alvarez, I., Poznyak, A.S.: Game theory applied to urban traffic control problem. Proc. ICCAS **2010**, 2164–2169 (2010). https://doi.org/10.1109/ICCAS.2010.5670234
12. Alvarez, I., Poznyak, A.S., Malo, A.: Urban traffic control problem via a game theory application. In: Proceedings of the 46th IEEE Conference on Decision and Control, pp. 2957–2961 (2007). https://doi.org/10.1109/CDC.2007.4434820
13. Alvarez, I., Poznyak, A.S., Malo, A.: Urban traffic control problem a game theory approach. In: Proceedings of the 47th IEEE Conference on Decision and Control, pp. 2168–2172 (2008). https://doi.org/10.1109/CDC.2008.4739461
14. Azhmyakov, V., Poznyak, A.S.: A variational characterization of the sliding mode control processes. In: Proceedings of the American Control Conference (ACC), pp. 5383–5388 (2012). https://doi.org/10.1109/ACC.2012.6315542
15. Azhmyakov, V., Boltyanski, V., Poznyak, A.S.: The dynamic programming approach to multimodel robust optimization. Nonlinear Anal. Theory, Methods Appl. Int. Multidiscip. J. **72**(2), 1110–1119 (2010). https://doi.org/10.1016/j.na.2009.07.050
16. Azhmyakov, V., Boltyanski, V.G., Poznyak, A.S.: First order optimization techniques for impulsive hybrid dynamical systems. In: Proceedings of International Workshop on Variable Structure Systems, pp. 173–178 (2008). https://doi.org/10.1109/VSS.2008.4570703

17. Azhmyakov, V., Boltyanski, V.G., Poznyak, A.S.: On the dynamic programming approach to multi-model robust optimal control problems. In: Proceedings of the American Control Conference, pp. 4468–4473 (2008). https://doi.org/10.1109/ACC.2008.4587199
18. Azhmyakov, V., Boltyanski, V.G., Poznyak, A.S.: Optimal control of impulsive hybrid systems. Nonlinear Anal. Hybrid Syst. **2**(4), 1089–1097 (2008). https://doi.org/10.1016/j.nahs.2008.09.003
19. Azhmyakov, V., Cabrera Martinez, J., Poznyak, A.S.: Optimal fixed-levels control for nonlinear systems with quadratic cost-functionals. Optim. Control Appl. Methods **37**(5), 1035–1055 (2016). https://doi.org/10.1002/oca.2223
20. Azhmyakov, V., Egerstedt, M., Fridman, L., Poznyak, A.S.: Approximability of nonlinear affine control systems. Nonlinear Anal. Hybrid Syst. **5**(2), 275–288 (2011). https://doi.org/10.1016/j.nahs.2010.07.005
21. Azhmyakov, V., Galvan-Guerra, R., Poznyak, A.S.: On the hybrid LQ-based control design for linear networked systems. J. Frankl. Inst. Eng. Appl. Math. **347**(7), 1214–1226 (2010). https://doi.org/10.1016/j.jfranklin.2010.05.012
22. Azhmyakov, V., Martinez, J.C., Poznyak, A.S., Serrezuela, R.R.: Optimization of a class of nonlinear switched systems with fixed-levels control inputs. In: Proceedings of the American control Conference (ACC), pp. 1770–1775 (2015). https://doi.org/10.1109/ACC.2015.7170989
23. Azhmyakov, V., Polyakov, A., Poznyak, A.S.: Consistent approximations and variational description of some classes of sliding mode control processes. J. Frankl. Inst. Eng. Appl. Math. **351**(4), 1964–1981 (2014). https://doi.org/10.1016/j.jfranklin.2013.01.011
24. Azhmyakov, V., Poznyak, A.S., Gonzalez, O.: On the robust control design for a class of nonlinearly affine control systems: the attractive ellipsoid approach. J. Ind. Manag. Optim. **9**(3), 579–593 (2013). https://doi.org/10.3934/jimo.2013.9.579
25. Azhmyakov, V., Poznyak, A.S., Juárez, R.: On the practical stability of control processes governed by implicit differential equations: the invariant ellipsoid based approach. J. Frankl. Inst. Eng. Appl. Math. **350**(8), 2229–2243 (2013). https://doi.org/10.1016/j.jfranklin.2013.04.016
26. Baev, S., Shkolnikov, I., Shtessel, Y., Poznyak, A.S.: Parameter identification of non-linear system using traditional and high order sliding modes. In: Proceedings of the American Control Conference, p. 6 (2006). https://doi.org/10.1109/ACC.2006.1656620
27. Baev, S., Shkolnikov, I.A., Shtessel, Y.B., Poznyak, A.S.: Sliding mode parameter identification of systems with measurement noise. Int. J. Syst. Sci. **38**(11), 871–878 (2007). https://doi.org/10.1080/00207720701622809
28. Bejarano, F.J., Fridman, L., Poznyak, A.S.: Output integral sliding mode with application to the LQ - optimal control. In: Proceedings of the International Workshop on Variable Structure Systems VSS'06, pp. 68–73 (2006). https://doi.org/10.1109/VSS.2006.1644495
29. Bejarano, F.J., Fridman, L., Poznyak, A.S.: Estimation of unknown inputs, with application to fault detection, via partial hierarchical observation. In: Proceedings of the European Control Conference (ECC), pp. 5154–5161 (2007)
30. Bejarano, F.J., Fridman, L., Poznyak, A.S.: Exact state estimation for linear systems with unknown inputs based on hierarchical super-twisting algorithm. Int. J. Robust Nonlinear Control **17**(18), 1734–1753 (2007). https://doi.org/10.1002/rnc.1190
31. Bejarano, F.J., Fridman, L., Poznyak, A.S.: Hierarchical observer for strongly detectable systems via second order sliding mode. In: Proceedings of the 46th IEEE Conference on Decision and Control, pp. 3709–3714 (2007). https://doi.org/10.1109/CDC.2007.4434968
32. Bejarano, F.J., Fridman, L.M., Poznyak, A.S.: Output integral sliding mode control based on algebraic hierarchical observer. Int. J. Control **80**(3), 443–453 (2007). https://doi.org/10.1080/00207170601080205
33. Bejarano, F.J., Fridman, L.M., Poznyak, A.S.: Output integral sliding mode for min-max optimization of multi-plant linear uncertain systems. IEEE Trans. Autom. Control **54**(11), 2611–2620 (2009). https://doi.org/10.1109/TAC.2009.2031718

34. Bejarano, F.J., Fridman, L.M., Poznyak, A.S.: Unknown input and state estimation for unobservable systems. SIAM J. Control Optim. **48**(2), 1155–1178 (2009). https://doi.org/10.1137/070700322
35. Bejarano, F.J., Poznyak, A.S., Fridman, L.: Hierarchical second-order sliding-mode observer for linear time invariant systems with unknown inputs. Int. J. Syst. Sci. Princ. Appl. Syst. Integr. **38**(10), 793–802 (2007). https://doi.org/10.1080/00207720701409280
36. Bejarano, F.J., Poznyak, A.S., Fridman, L.: Min-max output integral sliding mode control for multiplant linear uncertain systems. In: Proceedings of the American Control Conference, pp. 5875–5880 (2007). https://doi.org/10.1109/ACC.2007.4282716
37. Bejarano, F.J., Poznyak, A.S., Fridman, L.M.: Observation of linear systems with unknown inputs via high-order sliding-modes. Int. J. Syst. Sci. **38**(10), 773–791 (2007). https://doi.org/10.1080/00207720701409538
38. Boltyanski, V.G., Poznyak, A.S.: Robust maximum principle for minimax mayer problem with uncertainty from a compact measured set. In: Proceedings of the American Control Conference (IEEE Cat. No.CH37301), vol. 1, pp. 310–315 (2002). https://doi.org/10.1109/ACC.2002.1024822
39. Boltyanski, V.G., Poznyak, A.S.: A compact uncertainty set. The Robust Maximum Principle (2012). https://doi.org/10.1007/978-0-8176-8152-4_17
40. Boltyanski, V.G., Poznyak, A.S.: Dynamic programming for robust optimization. The Robust Maximum Principle (2012). https://doi.org/10.1007/978-0-8176-8152-4_12
41. Boltyanski, V.G., Poznyak, A.S.: Extremal problems in banach spaces. The Robust Maximum Principle (2012). https://doi.org/10.1007/978-0-8176-8152-4_7
42. Boltyanski, V.G., Poznyak, A.S.: Finite collection of dynamic systems. The Robust Maximum Principle (2012). https://doi.org/10.1007/978-0-8176-8152-4_8
43. Boltyanski, V.G., Poznyak, A.S.: Introduction. The Robust Maximum Principle (2012). https://doi.org/10.1007/978-0-8176-8152-4_1
44. Boltyanski, V.G., Poznyak, A.S.: Linear multimodel time optimization. The Robust Maximum Principle (2012). https://doi.org/10.1007/978-0-8176-8152-4_10
45. Boltyanski, V.G., Poznyak, A.S.: Linear quadratic optimal control. The Robust Maximum Principle (2012). https://doi.org/10.1007/978-0-8176-8152-4_4
46. Boltyanski, V.G., Poznyak, A.S.: LQ-stochastic multimodel control. The Robust Maximum Principle (2012). https://doi.org/10.1007/978-0-8176-8152-4_16
47. Boltyanski, V.G., Poznyak, A.S.: The maximum principle. The Robust Maximum Principle (2012). https://doi.org/10.1007/978-0-8176-8152-4_2
48. Boltyanski, V.G., Poznyak, A.S.: A measurable space as uncertainty set. The Robust Maximum Principle (2012). https://doi.org/10.1007/978-0-8176-8152-4_11
49. Boltyanski, V.G., Poznyak, A.S.: Min-max sliding-mode control. The Robust Maximum Principle (2012). https://doi.org/10.1007/978-0-8176-8152-4_13
50. Boltyanski, V.G., Poznyak, A.S.: Multimodel Bolza and LQ problem. The Robust Maximum Principle (2012). https://doi.org/10.1007/978-0-8176-8152-4_9
51. Boltyanski, V.G., Poznyak, A.S.: Multimodel differential games. The Robust Maximum Principle (2012). https://doi.org/10.1007/978-0-8176-8152-4_14
52. Boltyanski, V.G., Poznyak, A.S.: Multiplant robust control. The Robust Maximum Principle (2012). https://doi.org/10.1007/978-0-8176-8152-4_15
53. Boltyanski, V.G., Poznyak, A.S.: The Robust Maximum Principle. Foundations and Applications. Birkhauser, New York, Systems and Control (2012)
54. Boltyanski, V.G., Poznyak, A.S.: The tent method in finite-dimensional spaces. The Robust Maximum Principle (2012). https://doi.org/10.1007/978-0-8176-8152-4_6
55. Boltyanski, V.G., Poznyak, A.S.: Time-optimization problem. The Robust Maximum Principle (2012). https://doi.org/10.1007/978-0-8176-8152-4_5
56. Bregeault, V., Brgeault, V., Plestan, F., Shtessel, Y., Poznyak, A.S.: Adaptive sliding mode control for an electropneumatic actuator. In: Proceedings of the 11th International Workshop on Variable Structure Systems (VSS), pp. 260–265 (2010). https://doi.org/10.1109/VSS.2010.5544714

57. Cabrera, A., Poznyak, A.S., Poznyak, T., Aranda, J.: Some experiments on identification of a fed-batch culture via differential neural networks. In: Proceedings of the IEEE International Conference on Control Applications (CCA '01), pp. 152–156 (2001). https://doi.org/10.1109/CCA.2001.973855
58. Carrillo, L., Escobar, J.A., Clempner, J.B., Poznyak, A.S.: Optimization problems in chemical reactions using continuous-time Markov chains. J. Math. Chem. **54**(6), 1233 (2016). https://doi.org/10.1007/s10910-016-0620-0
59. Castillo, R.G., Clempner, J.B., Poznyak, A.S.: Solving the multi-traffic signal-control problem for a class of continuous-time Markov games. In: Proceedings of the 12th International Conference on Electrical Engineering, Computing Science and Automatic Control (CCE) 2015, pp. 1–5 (2015). https://doi.org/10.1109/ICEEE.2015.7357932
60. Chairez, I., Fuentes, R., Poznyak, A.S., Poznyak, T.: Robust identification of uncertain Schrödinger type complex partial differential equations. In: Proceedings of the 7th International Conference on Electrical Engineering, Computing Science and Automatic Control, pp. 170–175 (2010). https://doi.org/10.1109/ICEEE.2010.5608635
61. Chairez, I., Fuentes, R., Poznyak, A.S., Poznyak, T.: Robust identification of uncertain Schrödinger type complex partial differential equations. In: Proceedings of the 7th International Conference on Electrical Engineering, Computing Science and Automatic Control, CCE 2010 (Formerly known as ICEEE) IEEE, Tuxtla Gutierrez, Mexico, 8–10 Sept 2010, pp. 170–175 (2010). https://doi.org/10.1109/ICEEE.2010.5608635
62. Chairez, I., Fuentes, R., Poznyak, A.S., Poznyak, T., Escudero, M., Viana, L.: Neural network identification of uncertain 2D partial differential equations. In: Proceedings of the 6th International Conference on Electrical Engineering, Computing Science and Automatic Control (CCE) 2009, pp. 1–6 (2009). https://doi.org/10.1109/ICEEE.2009.5393456
63. Chairez, I., Fuentes, R., Poznyak, A.S., Poznyak, T., Escudero, M., Viana, L.: DNN-state identification of 2D distributed parameter systems. Int. J. Syst. Sci. **43**(2), 296–307 (2012). https://doi.org/10.1080/00207721.2010.495187
64. Chairez, I., Garca, A., Poznyak, A.S., Poznyak, T.: Model predictive control by differential neural networks approach. In: Proceedings of the International Joint Conference on Neural Networks (IJCNN), pp. 1–8 (2010). https://doi.org/10.1109/IJCNN.2010.5596521
65. Chairez, I., Poznyak, A.S., Poznyak, T.: Dynamic neural observer with sliding mode learning. In: Proceedings of the 3rd International IEEE Conference on Intelligent Systems, pp. 600–605 (2006). https://doi.org/10.1109/IS.2006.348487
66. Chairez, I., Poznyak, A.S., Poznyak, T.: New sliding-mode learning law for dynamic neural network observer. IEEE Trans. Circuits Syst. II: Express Briefs **53**(12), 1338–1342 (2006). https://doi.org/10.1109/TCSII.2006.883096
67. Chairez, I., Poznyak, A.S., Poznyak, T.: High order dynamic neuro observer: application for ozone generator. In: Proceedings of the International Workshop on Variable Structure Systems, pp. 291–295 (2008). https://doi.org/10.1109/VSS.2008.4570723
68. Chairez, I., Poznyak, A.S., Poznyak, T.: High order sliding mode neurocontrol for uncertain nonlinear SISO systems: theory and applications. Modern Sliding Mode Control Theory (2008). https://doi.org/10.1007/978-3-540-79016-7_9
69. Chairez, I., Poznyak, A.S., Poznyak, T.: Stable weights dynamics for a class of differential neural network observer. IET Control Theory Appl. **3**(10), 1437–1447 (2009). https://doi.org/10.1049/iet-cta.2008.0261
70. Clempner, J.B., Poznyak, A.S.: Convergence method, properties and computational complexity for Lyapunov games. Appl. Math. Comput. Sci. **21**(2), 349–361 (2011). https://doi.org/10.2478/v10006-011-0026-x
71. Clempner, J.B., Poznyak, A.S.: Analysis of best-reply strategies in repeated finite Markov chains games. In: Proceedings of the 52nd IEEE Conference on Decision and Control, pp. 568–573 (2013). https://doi.org/10.1109/CDC.2013.6759942
72. Clempner, J.B., Poznyak, A.S.: Simple computing of the customer lifetime value: a fixed local-optimal policy approach. J. Syst. Sci. Syst. Eng. **23**(4), 439 (2014). https://doi.org/10.1007/s11518-014-5260-y

73. Clempner, J.B., Poznyak, A.S.: Computing the strong Nash equilibrium for Markov chains games. Appl. Math. Comput. **265**, 911–927 (2015). https://doi.org/10.1016/j.amc.2015.06.005
74. Clempner, J.B., Poznyak, A.S.: Modeling the multi-traffic signal-control synchronization: a Markov chains game theory approach. Eng. Appl. Artif. Intell. **43**, 147–156 (2015). https://doi.org/10.1016/j.engappai.2015.04.009
75. Clempner, J.B., Poznyak, A.S.: Stackelberg security games: computing the shortest-path equilibrium. Expert Syst. Appl. **42**(8), 3967–3979 (2015). https://doi.org/10.1016/j.eswa.2014.12.034
76. Clempner, J.B., Poznyak, A.S.: Analyzing an optimistic attitude for the leader firm in duopoly models: a strong Stackelberg equilibrium based on a Lyapunov game theory approach. Econ. Comput. Econ. Cybern. Stud. Res. **4**(50), 41–60 (2016)
77. Clempner, J.B., Poznyak, A.S.: Conforming coalitions in Markov Stackelberg security games: setting max cooperative defenders vs. non-cooperative attackers. Appl. Soft Comput. **47**, 1–11 (2016). https://doi.org/10.1016/j.asoc.2016.05.037
78. Clempner, J.B., Poznyak, A.S.: Constructing the Pareto front for multi-objective Markov chains handling a strong Pareto policy approach. Comput. Appl. Math. 1 (2016). https://doi.org/10.1007/s40314-016-0360-6
79. Clempner, J.B., Poznyak, A.S.: Convergence analysis for pure stationary strategies in repeated potential games: Nash, Lyapunov and correlated equilibria. Expert Syst. Appl. **46**, 474–484 (2016). https://doi.org/10.1016/j.eswa.2015.11.006
80. Clempner, J.B., Poznyak, A.S.: Solving the Pareto front for multiobjective Markov chains using the minimum Euclidean distance gradient-based optimization method. Math. Comput. Simul. **119**, 142–160 (2016). https://doi.org/10.1016/j.matcom.2015.08.004
81. Clempner, J.B., Poznyak, A.S.: Multiobjective Markov chains optimization problem with strong Pareto frontier: principles of decision making. Expert Syst. Appl. **68**, 123–135 (2017). https://doi.org/10.1016/j.eswa.2016.10.027
82. Clempner, J.B., Poznyak, A.S.: Using Manhattan distance for computing the multiobjective Markov chains problem. Int. J. Comput. Math. (2017) (To be published)
83. Clempner, J.B., Poznyak, A.S.: Using the extraproximal method for computing the shortest-path mixed Lyapunov equilibrium in Stackelberg security games. Math. Comput. Simul. **138**, 14–30, (2017). https://doi.org/10.1016/j.matcom.2016.12.010. (To be published)
84. Clempner, J.B., Poznyak, A.S.: A Tikhonov regularized penalty function approach for solving polylinear programming problems. J. Comput. Appl. Math. **328**, 267–286 (2018)
85. Clempner, J. B., Poznyak, A.: Negotiating The Transfer Pricing Using The Nash Bargaining Solution. *Int J Appl Math Comput Sci.* (To be published)
86. Clempner, J.B., Poznyak, A.: A Tikhonov Regularization Parameter Approach For Solving Lagrange Constrained Optimization Problems. *Eng Optimiz.* (To be published)
87. Davila, J., Poznyak, A.S.: Sliding modes parameter adjustment in the presence of fast actuators using invariant ellipsoids method. In: Proceedings of the 6th International Conference on Electrical Engineering, Computing Science and Automatic Control (CCE) 2009, pp. 1–6 (2009). https://doi.org/10.1109/ICEEE.2009.5393474
88. Davila, J., Poznyak, A.S.: Attracting ellipsoid method application to designing of sliding mode controllers. In: Proceedings of the 11th International Workshop on Variable Structure Systems (VSS), pp. 83–88 (2010). https://doi.org/10.1109/VSS.2010.5544627
89. Davila, J., Poznyak, A.S.: Design of sliding mode controllers with actuators using attracting ellipsoid method. In: Proceedings of the 49th IEEE Conference on Decision and Control (CDC), pp. 72–77 (2010). https://doi.org/10.1109/CDC.2010.5717774
90. Davila, J., Poznyak, A.S.: Dynamic sliding mode control design using attracting ellipsoid method. Automatica **47**(7), 1467–1472 (2011). https://doi.org/10.1016/j.automatica.2011.02.023
91. Davila, J., Poznyak, A.S.: Sliding mode parameter adjustment for perturbed linear systems with actuators via invariant ellipsoid method. Int. J. Robust Nonlinear Control **21**(5), 473–487 (2011). https://doi.org/10.1002/rnc.1599

92. Davila, J., Fridman, L., Poznyak, A.S.: Observation and identification of mechanical systems via second order sliding modes. Int. J. Control **79**(10), 1251–1262 (2006). https://doi.org/10.1080/00207170600801635
93. Escobar, J., Poznyak, A.S.: Continuous-time identification using LS-method under colored noise perturbations. In: Proceedings of the 46th IEEE Conference on Decision and Control, pp. 5516–5521 (2007). https://doi.org/10.1109/CDC.2007.4434168
94. Escobar, J., Poznyak, A.S.: Robust continuous-time matrix estimation under dependent noise perturbations: sliding modes filtering and LSM with forgetting. CSSP **28**(2), 257–282 (2009). https://doi.org/10.1007/s00034-008-9080-5
95. Escobar, J., Poznyak, A.S.: Time-varying parameter estimation in continuous-time under colored perturbations using "equivalent control concept" and LSM with forgetting factor. In: Proceedings of the 11th International Workshop on Variable Structure Systems (VSS), pp. 209–214 (2010). https://doi.org/10.1109/VSS.2010.5544662
96. Escobar, J., Poznyak, A.S.: Time-varying matrix estimation in stochastic continuous-time models under coloured noise using LSM with forgetting factor. Int. J. Syst. Sci. **42**(12), 2009–2020 (2011). https://doi.org/10.1080/00207721003706852
97. Escobar, J., Poznyak, A.S.: Benefits of variable structure techniques for parameter estimation in stochastic systems using least squares method and instrumental variables. Int. J. Adapt. Control Signal Process. **29**(8), 1038–1054 (2015). https://doi.org/10.1002/acs.2521
98. Fridman, L., Poznyak, A.S., Bejarano, F.: Decomposition of the min-max multi-model problem via integral sliding mode. Int. J. Robust Nonlinear Control **15**(13), 559–574 (2005). https://doi.org/10.1002/rnc.1009
99. Fridman, L., Poznyak, A.S., Bejarano, F.J.: Decomposition of the mini-max multimodel optimal problem via integral sliding mode control. Proceedings of the American Control Conference **1**, 620–625 (2004)
100. Fridman, L., Poznyak, A.S., Bejarano, F.J.: Hierarchical second-order sliding-mode observer for linear systems with unknown inputs. In: Proceedings of the 45th IEEE Conference on Decision and Control, pp. 5561–5566 (2006). https://doi.org/10.1109/CDC.2006.377463
101. Fridman, L., Poznyak, A.S., Bejarano, F.J.: Fault detection. Robust Output LQ Optimal Control via Integral Sliding Modes (2014). https://doi.org/10.1007/978-0-8176-4962-3_8
102. Fridman, L., Poznyak, A.S., Bejarano, F.J.: Integral sliding mode control. Robust Output LQ Optimal Control via Integral Sliding Modes (2014). https://doi.org/10.1007/978-0-8176-4962-3_2
103. Fridman, L., Poznyak, A.S., Bejarano, F.J.: Introduction. Robust Output LQ Optimal Control via Integral Sliding Modes (2014). https://doi.org/10.1007/978-0-8176-4962-3_1
104. Fridman, L., Poznyak, A.S., Bejarano, F.J.: Magnetic bearing. Robust Output LQ Optimal Control via Integral Sliding Modes (2014). https://doi.org/10.1007/978-0-8176-4962-3_10
105. Fridman, L., Poznyak, A.S., Bejarano, F.J.: Multimodel and ISM control. Robust Output LQ Optimal Control via Integral Sliding Modes (2014). https://doi.org/10.1007/978-0-8176-4962-3_6
106. Fridman, L., Poznyak, A.S., Bejarano, F.J.: Multiplant and ISM output control. Robust Output LQ Optimal Control via Integral Sliding Modes (2014). https://doi.org/10.1007/978-0-8176-4962-3_7
107. Fridman, L., Poznyak, A.S., Bejarano, F.J.: Observer based on ISM. Robust Output LQ Optimal Control via Integral Sliding Modes (2014). https://doi.org/10.1007/978-0-8176-4962-3_3
108. Fridman, L., Poznyak, A.S., Bejarano, F.J.: Output integral sliding mode based control. Robust Output LQ Optimal Control via Integral Sliding Modes (2014). https://doi.org/10.1007/978-0-8176-4962-3_4
109. Fridman, L., Poznyak, A.S., Bejarano, F.J.: Stewart platform. Robust Output LQ Optimal Control via Integral Sliding Modes (2014). https://doi.org/10.1007/978-0-8176-4962-3_9
110. Fridman, L., Poznyak, A.S., Shtessel, Y., Bejarano, F.J.: Sliding mode multimodel control. In: Advances in Variable Structure and Sliding Mode Control, Lecture Notes in Control and Information Sciences, vol. 334, pp. 247–267. Springer, Berlin (2006). https://doi.org/10.1007/11612735_12

111. Fuentes, R., Poznyak, A.S., Chairez, I., Poznyak, T.: Neural numerical modeling for uncertain distributed parameter systems. In: Proceedings of the International Joint Conference on Neural Networks, pp. 909–916 (2009). https://doi.org/10.1109/IJCNN.2009.5178909
112. Fuentes, R.Q., Chairez, I., Poznyak, A.S.: Neuro-observer based on backstepping technique for distributed parameters systems. In: Proceedings of the 9th International Conference on Electrical Engineering, Computing Science and Automatic Control (CCE) 2012, pp. 1–6 (2012). https://doi.org/10.1109/ICEEE.2012.6421213
113. Fuentes, R.Q., Chairez, I., Poznyak, A.S., Poznyak, T.: 3D nonparametric neural identification. J. Control Sci. Eng. (2012). https://doi.org/10.1155/2012/618403
114. Fuentes, R.Q., Poznyak, A.S., Figueroa, I., Garcia, A., Chairez, I.: Continuous neural networks and finite element application for the tissue deformation reconstruction dynamic. In: Proceedings of the VI Andean Region International Conference, pp. 157–160 (2012). https://doi.org/10.1109/Andescon.2012.44
115. Garca, A., Chairez, I., Poznyak, A.S.: Hybrid differential neural network identifier for partially uncertain hybrid systems. Recent Advances in Intelligent Control Systems (2009). https://doi.org/10.1007/978-1-84882-548-2_7
116. Garcia, A., Chairez, I., Poznyak, A.S., Poznyak, T.: Robust identification of uncertain nonlinear systems with state constrains by differential neural networks. In: Proceedings of the International Joint Conference on Neural Networks, pp. 917–924 (2009). https://doi.org/10.1109/IJCNN.2009.5178825
117. Garcia, A., Poznyak, A.S., Chairez, I., Poznyak, T.: Projectional differential neural network observer with stable adaptation weights. In: Proceedings of the 47th IEEE Conference on Decision and Control, pp. 3652–3657 (2008). https://doi.org/10.1109/CDC.2008.4738950
118. Garcia, A., Poznyak, A.S., Oria, I.C., Poznyak, T.: Projectional differential neural network observer with stable adaptation weights. In: Proceedings of the 47th IEEE Conference on Decision and Control, CDC 2008, 9–11 Dec 2008, Cancún, México, pp. 3652–3657 (2008). https://doi.org/10.1109/CDC.2008.4738950
119. García, P., Poznyak, A.S.: Multi-model LQ-constrained min-max control. Optim. Control Appl. Methods **37**(2), 359–380 (2016). https://doi.org/10.1002/oca.2173
120. Gmez-Ramrez, E., Najim, K., Poznyak, A.S.: Saddle-point calculation for constrained finite Markov chains. J. Econ. Dyn. Control **27**(10), 1833–1853 (2003)
121. Godoy, M., Ramrez, E.G., Poznyak, A.S., Najim, K.: Noncooperative constrained finite games: alternate linear programming approach. In: Proceedings of the 41st IEEE Conference on Decision and Control, vol. 4, pp. 3958–3963 (2002). https://doi.org/10.1109/CDC.2002.1184985
122. Godoy-Alcántar, M., Poznyak, A.S., Gómez-Ramírez, E.: Generalization of the Mangasarian-Stone theorem for Markov chain finite N-person games: LPM-approach. Dyn. Syst. Appl. **12**(3–4), 489–508 (2003)
123. Gomez-Ramirez, E., Godoy-Alcantar, M., Poznyak, A.S.: Genetic algorithm for static games with N players. Nonlinear Stud. Int. J. **14**(1), 5–19 (2007)
124. Gonsales-Garsiya, S., Polyakov, A.E., Poznyak, A.S.: Application of the method of invariant ellipsoids for the robust linear output stabilization of a spacecraft. Rossiĭskaya Akademiya Nauk. Avtomatika i Telemekhanika **3**, 81–97 (2011). https://doi.org/10.1134/S0005117911030064
125. Gonzalez, O., Poznyak, A.S., Azhmyakov, V.: On the robust control design for a class of nonlinear affine control systems: the invariant ellipsoid approach. In: Proceedings of the 6th Int. Conf. Electrical Engineering, Computing Science and Automatic Control (CCE) 2009, pp. 1–6 (2009). https://doi.org/10.1109/ICEEE.2009.5393387
126. Gonzalez-Garcia, S., Polyakov, A., Poznyak, A.S.: Output linear feedback for a class of nonlinear systems based on the invariant ellipsoid method. In: Proceedings of the 5th International Conference on Electrical Engineering, Computing Science and Automatic Control 2008, pp. 7–12 (2008). https://doi.org/10.1109/ICEEE.2008.4723431
127. Gonzalez-Garcia, S., Polyakov, A., Poznyak, A.S.: Robust stabilization of a spacecraft with flexible elements using invariant ellipsoid technique. In: Proceedings of the 5th International Conference on Computing Science and Automatic Control 2008, pp. 97–101 (2008). https://doi.org/10.1109/ICEEE.2008.4723429

128. Gonzalez-Garcia, S., Polyakov, A., Poznyak, A.S.: Linear feedback spacecraft stabilization using the method of invariant ellipsoids. In: Proceedings of the 41st Southeastern Symposium on System Theory, pp. 195–198 (2009). https://doi.org/10.1109/SSST.2009.4806834
129. Gonzalez-Garcia, S., Polyakov, A., Poznyak, A.S.: Output linear controller for a class of nonlinear systems using the invariant ellipsoid technique. In: Proceedings of the American Control Conference, pp. 1160–1165 (2009). https://doi.org/10.1109/ACC.2009.5160434
130. Guerra, R.M., Poznyak, A.S., Leon, V.D.D.: Robustness property of high-gain observers for closed-loop nonlinear systems: theoretical study and robotics control application. Int. J. Syst. Sci. **31**(12), 1519–1529 (2000). https://doi.org/10.1080/00207720050217296
131. Hernández-Santamaría, V., de Teresa, L., Poznyak, A.S.: Hierarchic control for a coupled parabolic system. Portugaliae Mathematica. J. Port. Math. Soc. **73**(2), 115–137 (2016). https://doi.org/10.4171/PM/1979
132. Hernández-Santamaría, V., de Teresa, L., Poznyak, A.S.: Hierarchic control for a coupled parabolic system. Portugaliae Mathematica. Nova Série **73**(2), 115–137 (2016). https://doi.org/10.4171/PM/1979
133. Jimenez, M., Poznyak, A.S.: ϵ-equilibrium in LQ differential games with bounded uncertain disturbances: robustness of standard strategies and new strategies with adaptation. Int. J. Control **79**(7), 786–797 (2006). https://doi.org/10.1080/00207170600690624
134. Jimenez, M., Poznyak, A.S.: Robust and adaptive strategies with pre-identification via sliding mode technique in LQ differential games. In: Proceedings of the American Control Conference, pp. 6 (2006). https://doi.org/10.1109/ACC.2006.1657475
135. Jimenez-Lizarraga, M., Poznyak, A.S.: Near-Nash equilibrium strategies for LQ differential games with inaccurate state information. Math. Probl. Eng. Art. ID **21509**, 24 (2006). https://doi.org/10.1155/MPE/2006/21509
136. Jimenez-Lizarraga, M., Poznyak, A.S.: Equilibrium in LQ differential games with multiple scenarios. In: Proceedings of the 46th IEEE Conference on Decision and Control, pp. 4081–4086 (2007). https://doi.org/10.1109/CDC.2007.4434346
137. Jimenez-Lizarraga, M., Poznyak, A.S.: Quasi-equilibrium in LQ differential games with bounded uncertain disturbances: robust and adaptive strategies with pre-identification via sliding mode technique. Int. J. Syst. Sci. Princ. Appl. Syst. Integr. **38**(7), 585–599 (2007). https://doi.org/10.1080/00207720701431938
138. Jiménez-Lizárraga, M., Poznyak, A.S.: Robust Nash equilibrium in multi-model LQ differential games: analysis and extraproximal numerical procedure. Optim. Control Appl. Methods **28**(2), 117–141 (2007). https://doi.org/10.1002/oca.795
139. Jiménez-Lizárraga, M., Poznyak, A.S.: Necessary conditions for robust Stackelberg equilibrium in a multi-model differential game. Optim. Control Appl. Methods **33**(5), 595–613 (2012). https://doi.org/10.1002/oca.1018
140. Jimenez-Lizarraga, M., Poznyak, A.S., Alcorta, M.A.: Leader-follower strategies for a multi-plant differential game. In: Proceedings of the American Control Conference, pp. 3839–3844 (2008). https://doi.org/10.1109/ACC.2008.4587092
141. Jiménez-Lizárraga, M.A., Poznyak, A.S.: e-equilibrium strategies for LQ differential games with output measurement. In: H.R. Arabnia, G.A. Gravvanis (eds.) Proceedings of The 2005 International Conference on Scientific Computing, CSC 2005, Las Vegas, Nevada, USA, 20–23 June 2005, pp. 191–197, CSREA Press (2005)
142. Jiménez-Lizárraga, M.A., Poznyak, A.S.: Quasi-equilibrium in LQ differential games with bounded uncertain disturbances: robust and adaptive strategies with pre-identification via sliding mode technique. Int. J. Syst. Sci. **38**(7), 585–599 (2007). https://doi.org/10.1080/00207720701431938
143. Juárez, R., Azhmyakov, V., Poznyak, A.S.: Practical stability of control processes governed by semiexplicit DAEs. Math. Probl. Eng. Art. ID **675408**, 7 (2013)
144. Juarez, R., Poznyak, A.S., Azhmyakov, V.: On applications of attractive ellipsoid method to dynamic processes governed by implicit differential equations. In: Proceedings of the 8th International Conference on Electrical Engineering, Computing Science and Automatic Control 2011, pp. 1–6 (2011). https://doi.org/10.1109/ICEEE.2011.6106585

145. Keshtkar, S., Poznyak, A.S.: Adaptive sliding mode controller based on super-twist observer for tethered satellite system. Int. J. Control **89**(9), 1904–1915 (2016). https://doi.org/10.1080/00207179.2016.1185669
146. Keshtkar, S., Poznyak, A.S.: Tethered space orientation via adaptive sliding mode. Int. J. Robust Nonlinear Control **26**(8), 1632–1646 (2016). https://doi.org/10.1002/rnc.3371
147. Keshtkar, S., Hernandez, E.E., Oropeza, A., Poznyak, A.S.: Orientation of radio-telescope secondary mirror via adaptive sliding mode control. Neurocomputing **233**, 43–51 (2017). https://doi.org/10.1016/j.neucom.2016.08.116
148. Keshtkar, S., Keshtkar, J., Poznyak, A.S.: Adaptive sliding mode control for solar tracker orientation. In: Proceedings of the American Control Conference (ACC), pp. 6543–6548 (2016). https://doi.org/10.1109/ACC.2016.7526700
149. Keshtkar, S., Poznyak, A.S., Hernandez, E., Oropeza, A.: Orientation of radio-telescope secondary mirror via parallel platform. In: Proceedings of the 12th International Conference on Electrical Engineering, Computing Science and Automatic Control (CCE), pp. 1–5 (2015). https://doi.org/10.1109/ICEEE.2015.7357899
150. Keshtkar, S., Poznyak, A.S., Keshtkar, N.: Magnetic control of tethered cube-satellite stabilized by rotating. In: Proceedings of the 11th International Conference on Electrical Engineering, Computing Science and Automatic Control (CCE), pp. 1–5 (2014). https://doi.org/10.1109/ICEEE.2014.6978291
151. León, J.A., Lozada-Castillo, N.B., Poznyak, A.S.: Estabilidad de ecuaciones diferenciales estocásticas lineales anticipativas. Matemáticas: Enseñanza Universitaria (Nueva Serie) **15**(2), 51–64 (2007)
152. León, J.A., Lozada Castillo, N.B., Poznyak, A.S.: Stability of anticipating linear stochastic differential equations. Matemáticas. Enseñanza Universitaria **15**(2), 51–64 (2007)
153. Lozada-Castillo, N., Poznyak, A.S., Chairez, I.: Control of multiplicative noise stochastic gene regulation systems by the attractive ellipsoid technique. Int. J. Control Autom Syst. **12**(5), 1018 (2014). https://doi.org/10.1007/s12555-013-0226-2
154. Lozada-Castillo, N.B., Alazki, H., Poznyak, A.S.: Robust stabilization of linear stochastic differential models with additive and multiplicative diffusion via attractive ellipsoid techniques. In: Proceedings of the 8th International Conference on Electrical Engineering, Computing Science and Automatic Control, pp. 1–6 (2011). https://doi.org/10.1109/ICEEE.2011.6106685
155. Lozada-Castillo, N.B., Alazki, H., Poznyak, A.S.: Robust control design through the attractive ellipsoid technique for a class of linear stochastic models with multiplicative and additive noises. IMA J. Math. Control Inf. **30**(1), 1–19 (2013). https://doi.org/10.1093/imamci/dns008
156. Martinez-Guerra, R., Aguilar, R., Poznyak, A.S.: Estimation for HIV transmission using a reduced order uncertainty observer. In: Proceedings of the American Control Conference (Cat. No. 01CH37148), vol. 6, pp. 4603–4604 (2001). https://doi.org/10.1109/ACC.2001.945705
157. Martínez-Guerra, R., Poznyak, A.S., Leon, V.D.D.: Robustness property of high-gain observers for closed-loop nonlinear systems: theoretical study and robotics control application. Int. J. Syst. Sci. **31**(12), 1519–1529 (2000). https://doi.org/10.1080/00207720050217296
158. Medel, J.d.J., Poznyak, A.S.: Adaptive tracking for dc-derivate motor based on matrix forgetting. Computación y Sistemas **4**(3), 205–212 (2001). http://cys.cic.ipn.mx/ojs/index.php/CyS/article/view/945/1041
159. Mera, M., Castaños, F., Poznyak, A.S.: Quantised and sampled output feedback for nonlinear systems. Int. J. Control **87**(12), 2475–2487 (2014). https://doi.org/10.1080/00207179.2014.928948
160. Mera, M., Poznyak, A.S., Azhmyakov, V., Fridman, E.: Robust control for a class of continuous-time dynamical systems with sample-data outputs. In: Proceedings of the 6th International Conference on Electrical Engineering, Computing Science and Automatic Control (CCE), pp. 1–7 (2009). https://doi.org/10.1109/ICEEE.2009.5393420
161. Mera, M., Poznyak, A.S., Azhmyakov, V., Polyakov, A.: A robust dynamic controller for a class of nonlinear systems with sample-data outputs. In: Proceedings of the 9th International Conference on Electrical Engineering, Computing Science and Automatic Control (CCE), pp. 1–7 (2012). https://doi.org/10.1109/ICEEE.2012.6421198

162. Miranda, F.A., Castaños, F., Poznyak, A.S.: Min-max piecewise constant optimal control for multi-model linear systems. IMA J. Math. Control Inf. **33**(4), 1157–1176 (2016). https://doi.org/10.1093/imamci/dnv030
163. Moya, S., Poznyak, A.S.: Numerical method for finding a static Stackelberg-Nash equilibrium: the case of favorable followers. In: Proceedings of the 46th IEEE Conference on Decision and Control, pp. 145–149 (2007). https://doi.org/10.1109/CDC.2007.4434769
164. Moya, S., Poznyak, A.S.: Numerical methods for Stackelberg-Nash equilibrium calculation with favorable and unfavorable followers. In: Proceedings of the 5th International Conference on Electrical Engineering, Computing Science and Automatic Control, pp. 125–130 (2008). https://doi.org/10.1109/ICEEE.2008.4723369
165. Moya, S., Poznyak, A.S.: Stackelberg-nash concept applied to the traffic control problem with a dominating intersection. In: Proceedings of the 5th International Conference on Electrical Engineering, Computing Science and Automatic Control, pp. 137–142 (2008). https://doi.org/10.1109/ICEEE.2008.4723420
166. Moya, S., Poznyak, A.S.: Extraproximal method application for a Stackelberg -Nash equilibrium calculation in static hierarchical games. Part B (Cybernetics). IEEE Trans. Syst. Man Cybern. **39**(6), 1493–1504 (2009). https://doi.org/10.1109/TSMCB.2009.2019827
167. Moya, S., Poznyak, A.S.: Extraproximal method application for a Stackelberg-Nash equilibrium calculation in static hierarchical games. IEEE Trans. Syst. man Cybern. Part B Cybern. Publ. IEEE Syst Man Cybern. Soc. **39**, 1493–1504 (2009). https://doi.org/10.1109/TSMCB.2009.2019827
168. Murano, D.A., Poznyak, A.S.: Adaptive stochastic tracking: DNN-approach. In: Proceedings of the 41st IEEE Conference on Decision and Control, vol. 2, pp. 2202–2207 (2002). https://doi.org/10.1109/CDC.2002.1184858
169. Najim, K., Poznyak, A.S.: Learning Automata, 1 edn. Pergamon, Oxford (1994). Literaturangaben S. 206–214
170. Najim, K., Poznyak, A.S., Gomez, E.: Adaptive policy for two finite Markov chains zero-sum stochastic game with unknown transition matrices and average payoffs. Automatica **37**(7), 1007–1018 (2001). https://doi.org/10.1016/S0005-1098(01)00050-4
171. Najim, K., Poznyak, A.S., Ikonen, E.: Optimization based on a team of automata with binary outputs. Automatica **40**(8), 1349–1359 (2004). https://doi.org/10.1016/j.automatica.2004.03.013
172. Ordaz, P., Poznyak, A.S.: The Furuta's pendulum stabilization without the use of a mathematical model: attractive ellipsoid method with KL-adaptation. In: Proceedings of the 51st IEEE Conference on Decision and Control (CDC), pp. 7285–7290 (2012). https://doi.org/10.1109/CDC.2012.6426722
173. Ordaz, P., Poznyak, A.S.: 'KL'-gain adaptation for attractive ellipsoid method. IMA J. Math. Control Inf. **32**(3), 447–469 (2015). https://doi.org/10.1093/imamci/dnt046
174. Ordaz, P., Alazki, H., Poznyak, A.S.: A sample-time adjusted feedback for robust bounded output stabilization. Kybernetika **49**(6), 911–934 (2013). http://www.kybernetika.cz/content/2013/6/911
175. Oria, I.C., Poznyak, A.S., Poznyak, T.: Practical stability analysis for DNN observation. In: Proceedings of the 47th IEEE Conference on Decision and Control, CDC 2008, 9–11 Dec 2008, Cancún, México, pp. 2551–2556. IEEE (2008). https://doi.org/10.1109/CDC.2008.4738995
176. Perez, C., Azhmyakov, V., Poznyak, A.S.: Practical stabilization of a class of switched systems: dwell-time approach. IMA J. Math. Control Inf. **32**(4), 689–702 (2015). https://doi.org/10.1093/imamci/dnu011
177. Perez, C., Poznyak, A.S., Azhmyakov, V.: On the practical stability for a class of switched system. In: Proceedings of the 9th International Conference on Electrical Engineering, Computing Science and Automatic Control (CCE), pp. 1–6 (2012). https://doi.org/10.1109/ICEEE.2012.6421209
178. Perez-Cruz, J.H., Poznyak, A.S.: Estimation of the precursor power and internal reactivity in a nuclear reactor by a neural observer. In: Proceedings of the 4th International Conference on

Electrical and Electronics Engineering, pp. 310–313 (2007). https://doi.org/10.1109/ICEEE.2007.4345030
179. Perez-Cruz, J.H., Poznyak, A.S.: Identification of measurable dynamics of a nuclear research reactor using differential neural networks. In: Proceedings of the IEEE International Conference on Control Applications, pp. 473–478 (2007). https://doi.org/10.1109/CCA.2007.4389276
180. Perez-Cruz, J.H., Poznyak, A.S.: Neural control for power ascent of a TRIGA reactor. In: Proceedings of the American Control Conference, pp. 2190–2195 (2008). https://doi.org/10.1109/ACC.2008.4586817
181. Perez-Cruz, J.H., Poznyak, A.S.: Trajectory tracking based on differential neural networks for a class of nonlinear systems. In: Proceedings of the American Control Conference, pp. 2940–2945 (2009). https://doi.org/10.1109/ACC.2009.5160014
182. Pérez-Cruz, J.H., Poznyak, A.S.: Control of nuclear research reactors based on a generalized hopfield neural network. Intell. Autom. Soft Comput. **16**(1), 39–60 (2010). https://doi.org/10.1080/10798587.2010.10643062
183. Plestan, F., Shtessel, Y., Brégeault, V., Poznyak, A.S.: New methodologies for adaptive sliding mode control. Int. J. Control **83**(9), 1907–1919 (2010). https://doi.org/10.1080/00207179.2010.501385
184. Plestan, F., Shtessel, Y.B., Brégeault, V., Poznyak, A.S.: New methodologies for adaptive sliding mode control. Int. J. Control **83**(9), 1907–1919 (2010). https://doi.org/10.1080/00207179.2010.501385
185. Polyakov, A., Poznyak, A.S.: Lyapunov function design for finite-time convergence analysis of "twisting" and "super-twisting" second order sliding mode controllers. In: Proceedings of the International Workshop on Variable Structure Systems, pp. 153–158 (2008). https://doi.org/10.1109/VSS.2008.4570699
186. Polyakov, A., Poznyak, A.S.: Lyapunov function design for finite-time convergence analysis: "twisting" controller for second-order sliding mode realization. Automatica **45**(2), 444–448 (2009). https://doi.org/10.1016/j.automatica.2008.07.013
187. Polyakov, A., Poznyak, A.S.: Minimization of the unmatched disturbances in the sliding mode control systems via invariant ellipsoid method. In: Proceedings of the 2009 IEEE Control Applications, (CCA) and Intelligent Control (ISIC), pp. 1122–1127 (2009). https://doi.org/10.1109/CCA.2009.5280842
188. Polyakov, A., Poznyak, A.S.: Reaching time estimation for "super-twisting" second order sliding mode controller via Lyapunov function designing. IEEE Trans. Autom. Control **54**(8), 1951–1955 (2009). https://doi.org/10.1109/TAC.2009.2023781
189. Polyakov, A., Poznyak, A.S.: Invariant ellipsoid method for minimization of unmatched disturbances effects in sliding mode control. Automatica **47**(7), 1450–1454 (2011). https://doi.org/10.1016/j.automatica.2011.02.013
190. Polyakov, A., Poznyak, A.S.: Unified Lyapunov function for a finite-time stability analysis of relay second-order sliding mode control systems. IMA J. Math. Control Inf. **29**(4), 529–550 (2012). https://doi.org/10.1093/imamci/dns007
191. Polyakov, A., Poznyak, A.S., Richard, J.P.: Robust output stabilization of time-varying input delay systems using attractive ellipsoid method. In: Proceedings of the 52nd IEEE Conference on Decision and Control, pp. 934–939 (2013). https://doi.org/10.1109/CDC.2013.6760002
192. Polyakov, A.E., Poznyak, A.S.: The method of Lyapunov functions for systems with higher-order sliding modes. Rossiĭskaya Akademiya Nauk. Avtomatika i Telemekhanika **5**, 47–68 (2011). https://doi.org/10.1134/S0005117911050043
193. Poznyak, A.S.: Matrix forgetting factor. Int. J. Syst. Sci. **30**(2), 165–174 (1999). https://doi.org/10.1080/002077299292515
194. Poznyak, A.S.: Differential neural networks for robust nonlinear control (2001). http://www.worldscientific.com/worldscibooks/10.1142/4703#t=toc (Includes bibliographical references and index)
195. Poznyak, A.S.: Stochastic output noise effects in sliding mode state estimation. Int. J. Control **76**(9–10), 986–999 (2003). https://doi.org/10.1080/0020717031000099001 (Dedicated to Vadim Utkin on the occasion of his 65th birthday)

196. Poznyak, A.S.: Stochastic output noise effects in sliding mode observation. In: Variable Structure Systems: From Principles to Implementation, IEE Control Eng. Ser., vol. 66, pp. 81–97. IEE, London (2004)
197. Poznyak, A.S.: Kaddour najim, enso ikonen ait-kadi daoud (eds.), Stochastic Processes: Estimation, Optimization and Analysis, kogan page science, London and Sterling, va, ISBN: 1-903996-55-4, 2004 (332 pp.). Automatica **42**(7), 1237–1240 (2006). https://doi.org/10.1016/j.automatica.2006.03.002
198. Poznyak, A.S.: Robust maximum principle: multi-model dynamic optimization. Int. J. Tomogr. Stat. **5**(W07), 2–19 (2007)
199. Poznyak, A.S.: Advanced mathematical tools for automatic control engineers. Elsevier, Amsterdam [u.a.] (2008)
200. Poznyak, A.S.: Least squares method for dynamic systems identification. In: Models in statistics and probability theory (Spanish), Aportaciones Mat. Comun., vol. 39, pp. 107–150. Soc. Mat. Mexicana, México (2008)
201. Poznyak, A.S.: Least squares method for dynamic systems identification. In: Modelos en estadística y probabilidad, pp. 107–150. México: Sociedad Matemática Mexicana; México: Universidad Nacional Autónoma de México (2008)
202. Poznyak, A.S.: Advanced Mathematical Tools for Automatic Control Engineers (2009). http://www.sciencedirect.com/science/book/9780080446738 (Includes bibliographical references (pp. 529-533) and index)
203. Poznyak, A.S.: Attractive Ellipsoids in Robust Control (2014). https://doi.org/10.1007/978-3-319-09210-2
204. Poznyak, A.S.: Non-smooth missiles guidance: interceptor-defender scenario with uncertainties. In: Proceedings of the 13th International Workshop on Variable Structure Systems (VSS), pp. 1–6 (2014). https://doi.org/10.1109/VSS.2014.6881125
205. Poznyak, A.S.: Robust feedback design for stabilization of nonlinear systems with sampled-data and quantized output: attractive ellipsoid method. Autom. Remote Control **76**(5), 834–846 (2015). https://doi.org/10.1134/S0005117915050094 (Translation of Avtomat. i Telemekh. 2015, no. 5, 130-144)
206. Poznyak, A.S.: Stochastic sliding mode control: what is this? In: Proceedings of the 14th International Workshop on Variable Structure Systems (VSS), pp. 328–333 (2016). https://doi.org/10.1109/VSS.2016.7506939
207. Poznyak, A.S.: Stochastic super-twist dynamics. In: Proceedings of the 14th International Workshop on Variable Structure Systems (VSS), pp. 334–339 (2016). https://doi.org/10.1109/VSS.2016.7506940
208. Poznyak, A.S.: Sliding mode control in stochastic continuous-time systems: μ-zone ms-convergence. IEEE Trans. Autom. Control **62**(2), 863–868 (2017). https://doi.org/10.1109/TAC.2016.2557759
209. Poznyak, A.S., Azhmyakov, V., Mera, M.: Practical output feedback stabilisation for a class of continuous-time dynamic systems under sample-data outputs. Int. J. Control **84**(8), 1408–1416 (2011). https://doi.org/10.1080/00207179.2011.603097
210. Poznyak, A.S., Bejarano, F.J., Fridman, L.: Numerical method for weights adjustment in minimax multi-model LQ-control. Optim. Control Appl. Methods **28**(4), 289–300 (2007). https://doi.org/10.1002/oca.805
211. Poznyak, A.S., Chairez, I., Poznyak, T.: Sliding mode neurocontrol with applications. In: Proceedings of the International Workshop on Variable Structure Systems VSS'06, pp. 5–10 (2006). https://doi.org/10.1109/VSS.2006.1644484
212. Poznyak, A.S., Chairez, I., Poznyak, T.: Sliding mode neurocontrol for the class of dynamic uncertain non-linear systems. Int. J. Control **81**(1), 74–88 (2008). https://doi.org/10.1080/00207170701278303
213. Poznyak, A.S., Escobar, J., Shtessel, Y.B.: Sliding modes time varying matrix identification for stochastic system. Int. J. Syst. Sci. **38**(11), 847–859 (2007). https://doi.org/10.1080/00207720701620142

214. Poznyak, A.S., Escobar, J.A., Shtessel, Y.B.: Stochastic sliding modes identification. In: Proceedings of the International Workshop on Variable Structure Systems VSS'06, pp. 226–231 (2006). https://doi.org/10.1109/VSS.2006.1644522
215. Poznyak, A.S., Fridman, L., Bejarano, F.J.: Mini-max integral sliding-mode control for multimodel linear uncertain systems. IEEE Trans. Autom. Control **49**(1), 97–102 (2004). https://doi.org/10.1109/TAC.2003.821412
216. Poznyak, A.S., Gallegos, C.J.: Multimodel prey-predator LQ differential games. In: Proceedings of the American Control Conference, vol. 6, pp. 5369–5374 (2003). https://doi.org/10.1109/ACC.2003.1242582
217. Poznyak, A.S., Godoy-Alcantar, M., Gomez-Ramirez, E.: Learning for repeated constrained games in counter-coalition space. In: Proceedings of the 42nd IEEE International Conference on Decision and Control (IEEE Cat. No.03CH37475), vol. 5, pp. 4410–4415 (2003). https://doi.org/10.1109/CDC.2003.1272207
218. Poznyak, A.S., Ljung, L.: On-line identification and adaptive trajectory tracking for nonlinear stochastic continuous time systems using differential neural networks. Automatica **37**(8), 1257–1268 (2001). https://doi.org/10.1016/S0005-1098(01)00067-X
219. Poznyak, A.S., Najim, K.: Learning Automata and Stochastic Optimization. No. 225 in Lecture Notes in Control and Information Sciences. Springer, London [u.a.] (1997) Literaturangaben
220. Poznyak, A.S., Najim, K.: Adaptive control of constrained finite Markov chains. Automatica **35**(5), 777–789 (1999). https://doi.org/10.1016/S0005-1098(98)00219-2
221. Poznyak, A.S., Najim, K.: Adaptive zero-sum stochastic game for two finite Markov chains. In: Proceedings of the 39th IEEE Conference on Decision and Control (Cat. No.00CH37187), vol. 1, pp. 717–722 (2000). https://doi.org/10.1109/CDC.2000.912852
222. Poznyak, A.S., Najim, K.: Bush-mosteller learning for a zero-sum repeated game with random pay-offs. Int. J. Syst. Sci. **32**(10), 1251–1260 (2001). https://doi.org/10.1080/00207720110042347
223. Poznyak, A.S., Najim, K.: Learning through reinforcement for n-person repeated constrained games. IEEE Trans. Syst. Man Cybern. Part B **32**(6), 759–771 (2002). https://doi.org/10.1109/TSMCB.2002.1049610
224. Poznyak, A.S., Najim, K., Chtourou, M.: Analysis of the behaviour of multilevel hierarchical systems of learning automata and their application for multimodal functions optimization. Int. J. Syst. Sci. **27**(1), 97–112 (1996). https://doi.org/10.1080/00207729608929192
225. Poznyak, A.S., Najim, K., Chtourou, M.: Learning automata with continuous inputs and their application for multimodal functions optimization. Int. J. Syst. Sci. **27**(1), 87–95 (1996). https://doi.org/10.1080/00207729608929191
226. Poznyak, A.S., Najim, K., Gmez-Ramrez, E., Najim, K., Poznyak, A.S.: Self-learning Control of Finite Markov Chains. No. 4 in Control engineering. Marcel Dekker, New York [u.a.] (2000)
227. Poznyak, A.S., Nazin, A., Murano, D.: Suboptimal robust asymptotic observer for stochastic continuous time nonlinear system: numerical procedure and convergence analysis. In: Proceedings of the 41st IEEE Conference on Decision and Control, vol. 3, pp. 2588–2593 (2002). https://doi.org/10.1109/CDC.2002.1184228
228. Poznyak, A.S., Nazin, A., Murano, D.: Observer matrix gain optimization for stochastic continuous time nonlinear systems. Syst. Control Lett. **52**(5), 377–385 (2004). https://doi.org/10.1016/j.sysconle.2004.02.013
229. Poznyak, A.s., Polyakov, A., Azhmyakov, V.: Attractive ellipsoid method with adaptation. Attractive Ellipsoids in Robust Control (2014). https://doi.org/10.1007/978-3-319-09210-2_12
230. Poznyak, A.S., Polyakov, A., Azhmyakov, V.: Attractive ellipsoids in sliding mode control. Attractive Ellipsoids in Robust Control (2014). https://doi.org/10.1007/978-3-319-09210-2_8
231. Poznyak, A.S., Polyakov, A., Azhmyakov, V.: Bounded robust control. Attractive Ellipsoids in Robust Control (2014). https://doi.org/10.1007/978-3-319-09210-2_11

232. Poznyak, A.S., Polyakov, A., Azhmyakov, V.: Control with sample-data measurements. Attractive Ellipsoids in Robust Control (2014). https://doi.org/10.1007/978-3-319-09210-2_5
233. Poznyak, A.S., Polyakov, A., Azhmyakov, V.: Introduction. Attractive Ellipsoids in Robust Control (2014). https://doi.org/10.1007/978-3-319-09210-2_1
234. Poznyak, A.S., Polyakov, A., Azhmyakov, V.: Mathematical background. Attractive Ellipsoids in Robust Control (2014). https://doi.org/10.1007/978-3-319-09210-2_2
235. Poznyak, A.S., Polyakov, A., Azhmyakov, V.: Robust control of implicit systems. Attractive Ellipsoids in Robust Control (2014). https://doi.org/10.1007/978-3-319-09210-2_7
236. Poznyak, A.S., Polyakov, A., Azhmyakov, V.: Robust control of switched systems. Attractive Ellipsoids in Robust Control (2014). https://doi.org/10.1007/978-3-319-09210-2_10
237. Poznyak, A.S., Polyakov, A., Azhmyakov, V.: Robust output feedback control. Attractive Ellipsoids in Robust Control (2014). https://doi.org/10.1007/978-3-319-09210-2_4
238. Poznyak, A.S., Polyakov, A., Azhmyakov, V.: Robust stabilization of time-delay systems. Attractive Ellipsoids in Robust Control (2014). https://doi.org/10.1007/978-3-319-09210-2_9
239. Poznyak, A.S., Polyakov, A., Azhmyakov, V.: Robust state feedback control. Attractive Ellipsoids in Robust Control (2014). https://doi.org/10.1007/978-3-319-09210-2_3
240. Poznyak, A.S., Polyakov, A., Azhmyakov, V.: Sample data and quantifying output control. Attractive Ellipsoids in Robust Control (2014). https://doi.org/10.1007/978-3-319-09210-2_6
241. Poznyak, A.S., Polyakov, A.E., Strygin, V.V.: An analysis of finite-time convergence by the method of Lyapunov functions in systems with second-order sliding modes. Rossiĭskaya Akademiya Nauk. Prikladnaya Matematika i Mekhanika **75**(3), 410–429 (2011). https://doi.org/10.1016/j.jappmathmech.2011.07.006
242. Poznyak, A.S., Poznyak, T., Chaĭrez, I.: Dynamic neural observers and their application for the identification of an ozone water treatment process. Rossiĭskaya Akademiya Nauk. Avtomatika i Telemekhanika **6**, 61–74 (2006). https://doi.org/10.1134/S0005117906060051
243. Poznyak, A.S., Sánchez, E.N.: Non-linear system on-line identification using dynamic neural networks. Intell. Autom. Soft Comput. **5**(3), 201–209 (1999). https://doi.org/10.1080/10798587.1999.10750761
244. Poznyak, A.S., Shtessel, Y., Fridman, L., Davila, J., Escobar, J.: Identification of parameters in dynamic systems via sliding-mode techniques. Advances in Variable Structure and Sliding Mode Control. Lecture Notes in Control and Information Sciences, pp. 313–347. Springer, Berlin (2006). https://doi.org/10.1007/11612735_15
245. Poznyak, A.S., Shtessel, Y.B.: Minimax sliding mode control with minimal-time reaching phase. In: Proceedings of the American Control Conference, vol. 1, pp. 174–179 (2003). https://doi.org/10.1109/ACC.2003.1238933
246. Poznyak, A.S., Shtessel, Y.B., Gallegos, C.J.: Min-max sliding-mode control for multimodel linear time varying systems. IEEE Trans. Autom. Control **48**(12), 2141–2150 (2003). https://doi.org/10.1109/TAC.2003.820068
247. Poznyak, A.S., Yu, W.: Robust identification by dynamic neural networks using sliding mode learning. Appl. Math. Comput. Sci. **8**(1), 135–144 (1998)
248. Poznyak, A.S., Yu, W.: Robust asymptotic neuro-observer with time delay term. Int. J. Robust Nonlinear Control **10**(7), 535–559 (2000). https://doi.org/10.1002/1099-1239(200006)10:7<535::AID-RNC492>3.0.CO;2-6
249. Poznyak, A.S., Yu, W., Poznyak, T.I., Najim, K.: Simultaneous states and parameters estimation of an ozonation reactor based on dynamic neural network. Differ. Equ. Dyn. Syst. Int. J. Theory Appl. Comput. Simul. **12**(1–2), 195–221 (2004)
250. Poznyak, A.S., Yu, W., Sánchez, E.N., Pérez, J.P.: Nonlinear adaptive trajectory tracking using dynamic neural networks. IEEE Trans. Neural Netw. **10**(6), 1402–1411 (1999). https://doi.org/10.1109/72.809085
251. Poznyak, A.S., Zuniga, R.S.: Information inequalities and limiting possibilities of adaptive control strategies in ARX models with a general quadratic criterion. J. Math. Syst. Estim. Control **7**(1), 107–110 (1997)

252. Salgado, I., Moreno, J., Fridman, L., Poznyak, A.S., Chairez, I.: Design of variable gain super-twisting observer for nonlinear systems with sampled output. In: Proceedings of the 7th International Conference on Electrical Engineering, Computing Science and Automatic Control, pp. 153–157 (2010). https://doi.org/10.1109/ICEEE.2010.5608643
253. Sánchez, E.M., Clempner, J.B., Poznyak, A.S.: A priori-knowledge/actor-critic reinforcement learning architecture for computing the mean-variance customer portfolio: the case of bank marketing campaigns. Eng. Appl. Artif. Intell. **46**, 82–92 (2015). https://doi.org/10.1016/j.engappai.2015.08.011
254. Sánchez, E.M., Clempner, J.B., Poznyak, A.S.: Solving the mean-variance customer portfolio in markov chains using iterated quadratic/lagrange programming: a credit-card customer limits approach. Expert Syst. Appl. **42**(12), 5315–5327 (2015). https://doi.org/10.1016/j.eswa.2015.02.018
255. Shtessel, Y.B., Moreno, J.A., Plestan, F., Fridman, L.M., Poznyak, A.S.: Super-twisting adaptive sliding mode control: a Lyapunov design. In: Proceedings of the 49th IEEE Conference on Decision and Control (CDC), pp. 5109–5113 (2010). https://doi.org/10.1109/CDC.2010.5717908
256. Shtessel, Y.B., Poznyak, A.S.: Parameter identification of affine time varying systems using traditional and high order sliding modes. In: Proceedings of the 2005, American Control Conference, vol. 4, pp. 2433–2438 (2005). https://doi.org/10.1109/ACC.2005.1470331
257. Solis, C.U., Clempner, J.B., Poznyak, A.: Fast terminal sliding mode control with integral filter applied to a van der Pol oscillator. IEEE Trans. Ind. Electron. **99**(1) (2017). https://doi.org/10.1109/TIE.2017.2677299
258. Solis, C.U., Clempner, J.B., Poznyak, A.S.: Designing a terminal optimal control with an integral sliding mode component using a saddle point method approach: a cartesian 3D-crane application. Nonlinear Dyn. **86**(2), 911 (2016). https://doi.org/10.1007/s11071-016-2932-9
259. Solis, C.U., Clempner, J.B., Poznyak, A.S.: Modeling multileader-follower noncooperative Stackelberg games. Cybern. Syst. **47**(8), 650–673 (2016). https://doi.org/10.1080/01969722.2016.1232121
260. Trejo, K.K., Clempner, J.B., Poznyak, A.S.: Computing the lp-strong Nash equilibrium looking for cooperative stability in multiple agents Markov games. In: Proceedings of the 12th International Conference on Electrical Engineering, Computing Science and Automatic Control (CCE) 2015, pp. 1–6 (2015). https://doi.org/10.1109/ICEEE.2015.7357926
261. Trejo, K.K., Clempner, J.B., Poznyak, A.S.: Computing the Stackelberg/Nash equilibria using the extraproximal method: convergence analysis and implementation details for Markov chains games. Int.l. J Appl. Math. Comput. Sci. **25**(2), 337–351 (2015)
262. Trejo, K.K., Clempner, J.B., Poznyak, A.S.: A Stackelberg security game with random strategies based on the extraproximal theoretic approach. Eng. Appl. Artif. Intell. **37**, 145–153 (2015). https://doi.org/10.1016/j.engappai.2014.09.002
263. Trejo, K.K., Clempner, J.B., Poznyak, A.S.: Adapting strategies to dynamic environments in controllable Stackelberg security games. In: Proceedings of the 55th IEEE Conference on Decision and Control, CDC 2016, Las Vegas, NV, USA, 12–14 Dec 2016, pp. 5484–5489. IEEE (2016). https://doi.org/10.1109/CDC.2016.7799111
264. Trejo, K.K., Clempner, J.B., Poznyak, A.S.: An optimal strong equilibrium solution for cooperative multi-leader-follower Stackelberg Markov chains games. Kybernetika **52**(2), 258–279 (2016). https://doi.org/10.14736/kyb-2016-2-0258
265. Trejo, K.K., Clempner, J.B., Poznyak, A.S.: Computing the strong L_p-Nash equilibrium for Markov chains games: convergence and uniqueness. Appl. Math. Model. Simul. Comput. Eng. Environ. Syst. **41**, 399–418 (2017). https://doi.org/10.1016/j.apm.2016.09.001
266. Trejo, K.K., Clempner, J.B., Poznyak, A.S.: Nash bargaining equilibria for controllable Markov chains games. In: 20th IFAC World Congress. Toulouse, France (2017) (To be published)
267. Utkin, V.I., Poznyak, A.S.: Adaptive sliding mode control. Advances in Sliding Mode Control (2013). https://doi.org/10.1007/978-3-642-36986-5_2

268. Utkin, V.I., Poznyak, A.S.: Adaptive sliding mode control. Advances in sliding mode control. Lecture Notes in Control and Information Sciences, pp. 21–53. Springer, Heidelberg (2013). https://doi.org/10.1007/978-3-642-36986-5_2
269. Utkin, V.I., Poznyak, A.S.: Adaptive sliding mode control with application to super-twist algorithm: equivalent control method. Automatica **49**(1), 39–47 (2013). https://doi.org/10.1016/j.automatica.2012.09.008
270. Utkin, V.I., Poznyak, A.S., Ordaz, P.: Adaptive super-twist control with minimal chattering effect. In: Proceedings of the 50th IEEE Conference on Decision and Control and European Control Conference, pp. 7009–7014 (2011). https://doi.org/10.1109/CDC.2011.6160720
271. Villafuerte, R., Mondie, S., Poznyak, A.S.: Practical stability of neutral type time delay systems: LMI's approach. In: Proceedings of the 5th International Conference on Electrical Engineering, Computing Science and Automatic Control 2008, pp. 75–79 (2008). https://doi.org/10.1109/ICEEE.2008.4723435
272. Villafuerte, R., Mondie, S., Poznyak, A.S.: Practical stability of time delay systems: LMI's approach.. In: Proceedings of the 47th IEEE Conference on Decision and Control, pp. 4807–4812 (2008). https://doi.org/10.1109/CDC.2008.4738801
273. Yu, W., Poznyak, A.S., Li, X.: Multilayer dynamic neural networks for nonlinear system on-line identification. Int. J. Control **74**(18), 1858–1864 (2001). https://doi.org/10.1080/00207170110089816
274. Yu, W., Poznyak, A.S., Sanchez, E.N.: Neural adaptive control of two-link manipulator with sliding mode compensation. In: Proceedings of the IEEE International Conference on Robotics and Automation (Cat. No.99CH36288C), vol. 4, pp. 3122–3127 (1999). https://doi.org/10.1109/ROBOT.1999.774073
275. Zhermolenko, V.N., Poznyak, A.S.: Criteria of robust stability for time-varying 2D Wang–Mitchel differential systems: integral funnel method. Int. J. Control **89**(11), 2297–2310 (2016). https://doi.org/10.1080/00207179.2016.1155752
276. Zhermolenko, V.N., Poznyak, A.S.: Necessary and sufficient conditions for stabilizability of planar parametrically perturbed control systems. IMA J. Math. Control Inf. **33**(1), 53–68 (2016). https://doi.org/10.1093/imamci/dnu028

Chapter 2
Luenberger Observer Design for Uncertainty Nonlinear Systems

Wen Yu

Abstract Most of the nonlinear observers require the nonlinear systems to be known. If the systems are partly unknown, model-free observers such as high-gain observers, sliding mode observers, and neural observers, can be applied. However, the performances of these observers are not satisfactory, for example, they are sensitive to measurement noise and they can only estimate the derivative of the output. In this chapter, we use the structure of Luenberger observers for partially unknown nonlinear systems. Using a Riccati differential equation, we design a time-varying observer gain such that the observer error is robust with respect to bounded uncertainties. Compared with the other robust nonlinear observers, this observer is simple and effective with respect to the uncertainties in the nonlinear systems.

2.1 Introduction

The state observation problem is one of the essential points in modern control theory. If the systems are linear, the well-known observers can be the Kalman filters (stochastic noise cases) and Luenberger observers (deterministic cases) [11]. Since the early 80s, many implementations in theory and practice focus on the observers for nonlinear systems. Many nonlinear observers have been developed, such as Lie-algebra-based observers [24], Luenberger-like observers [12], high-gain observers [10], optimization-based observers [6], linearization approaches [25], stochastic systems [26], and reduced-order nonlinear observers [7], adaptive high-gain observers [2], etc. A basic requirement of the above observers is the nonlinear systems that are completely known [14]. The observers duplicate the nonlinear dynamics. There are no internal and external uncertainties.

If the uncertainties in the nonlinear systems can be parameterized, the nonlinear adaptive observers [20] can obtain the states. H_∞ and L_2 techniques are applied to

W. Yu (✉)
Departamento de Control Automatico, CINVESTAV-IPN, Mexico City, Mexico
e-mail: yuw@ctrl.cinvestav.mx

J.B. Clempner and W. Yu (eds.), *New Perspectives and Applications of Modern Control Theory*, https://doi.org/10.1007/978-3-319-62464-8_2

construct the robust observers for linear systems in [3]. The results are extended to the Lipschitz nonlinear systems in [15]. These robust observers are very complex, and require the uncertainties to satisfy desired properties. If the uncertainties are structural, i.e., the uncertainties and disturbances are assumed to be bounded, nonlinear model-free observers are needed [13]. High-gain observers are the most popular model-free nonlinear observers, they are robust against model uncertainty and disturbances [17]. However, they can only estimate the derivative of the output and are sensitive to measurement noises. Sliding mode observers do not require estimate the derivative of the output [27]. However, the measurement noise and disturbances case chattering in the nonlinear observers [18]. Another model-free nonlinear observer uses neural networks to approximate the unknown part and then to design an observer [4]. However, the observation error is large, because the output has to be used for neural approximation and observer estimation [16].

In this chapter, the nonlinear observer is also for unknown or partially unknown nonlinear system. We use a Luenberger-like structure. Compared with the other robust nonlinear observers, high-gain observer, sliding mode observer, and neural observer, our Luenberger-like nonlinear observer is more simple. It can estimate not only the derivative of the output. It is not sensitive to measurement noises. It does not necessarily model the nonlinear system. Compared with the other Luenberger-like nonlinear observers [1, 19], our observer do not need to copy the complete nonlinear system for the observer design. They assume the nonlinear systems do not have any uncertainties.

We assume both external disturbances (noises) and internal uncertainties (unmodeled dynamics) are bounded. The assumptions on the partly known nonlinear systems are standard, such as Lipschitz and uniformly observability The gain of Luenberger observer is specially selected to guarantee the property of robustness using a Riccati differential equation. The stability of this robust nonlinear observer is proven. An example demonstrates the effectiveness of this observer for the system containing complex uncertain nonlinearities.

2.2 Robust Nonlinear Observer

We discuss a class of single input and single output nonlinear systems given by

$$\begin{aligned} \dot{x} &= f(x) + g(x)u_t + \xi_{1,t} \\ y_t &= Cx + \xi_{2,t} \end{aligned} \tag{2.1}$$

where $f(x) = \begin{pmatrix} x_2 \\ \vdots \\ x_n \\ \varphi_0(x) + \Delta\varphi(x) \end{pmatrix}$, $g(x,t) = \begin{pmatrix} g_1(x_1) + \Delta g_1(x_1) \\ g_2(x_1, x_2) + \Delta g_2(x_1, x_2) \\ \vdots \\ g_n(x_1 \cdots x_n) + \Delta g_n(x_1 \cdots x_n) \end{pmatrix}$, $x \in \Re^n$ is the state vector of the system at time $t \in R^+ := \{t : t \geq 0\}$, $u_t \in \Re$ is a given control action, it is assumed to be bounded as $|u_t|^2 \leq \overline{u}$, $y_t \in \Re$ is the output vector that is suggested to be measurable at each time t, $f(\cdot)$ and $g(\cdot) : \Re^{n+1} \to \Re^n$, are nonlinear functions describing the dynamic operator of the given system, $C = [1, 0 \cdots 0]$, $\xi_{1,t}$ and $\xi_{2,t}$ are the vector functions presenting the external state and output perturbations.

We assume that the nominal parts of nonlinear system $F(x)$ and $G(x)$ are known,

$$\begin{aligned} f(x) &= F(x) + \Delta f(x) \\ g(x) &= G(x) + \Delta g(x) \end{aligned} \tag{2.2}$$

where $F(x) = [x_2, \ldots x_n, \varphi_0(x)]$, $G(x) = [g_1(x_1), \ldots g_n(x_1 \cdots x_n)]$, $\Delta f(x) = [0, \ldots 0, \Delta\varphi(x)]$, and $\Delta g(x) = [\Delta g_1(x_1), \ldots \Delta g_n(x_1 \cdots x_n)]$.

In this chapter, we use the following four assumptions to design a robust nonlinear observer for the system (2.1).

A1. The perturbations $\xi_{1,t}$ and $\xi_{2,t}$ satisfy the following "bounded power" condition

$$\limsup_{T\to\infty} \frac{1}{T} \int_0^T \left\|\xi_{i,t}\right\|^2_{\Lambda_{\xi_i}} dt = \Upsilon_i < \infty \quad , i = 1, 2 \tag{2.3}$$

where $\Lambda_{\xi_i} = \Lambda_{\xi_i}^T > 0$, $\left\|\xi_{i,t}\right\|^2_{\Lambda_{\xi_i}} = \xi_{i,t}^T \Lambda_{\xi_i} \xi_{i,t}$.

A2. The normal parts of the nonlinear system $F(x, t)$ and $G(x, t)$ are observable and Lipschitz in a compact set U

$$\begin{aligned} \left\|F^T(x_1) - F^T(x_2)\right\|_{\Lambda_f} &\leq L_f \left\|x_1 - x_2\right\|_{\Lambda_{fx}} \\ \left\|G^T(x_1) - G^T(x_2)\right\|_{\Lambda_g} &\leq L_g \left\|x_1 - x_2\right\|_{\Lambda_{gx}} \end{aligned} \tag{2.4}$$

where $\Lambda_f \in \Re^{n\times n}$, $\Lambda_{fx} \in \Re^{n\times n}$, $\Lambda_g \in \Re^{m\times m}$, $\Lambda_{gx} \in \Re^{n\times n}$ are strictly positive-definite matrices, or Lipschitz constant matrices. L_f and L_g are Lipschitz constants.

A3. The unmodeled dynamics Δf and Δg in (2.2) satisfy the "strip bounded" conditions

$$\begin{aligned} \Delta f^T(x)\Lambda_{\Delta f}\Delta f(x) &\leq C_{\Delta f} + x^T D_{\Delta f} x \\ \Delta g^T(x)\Lambda_{\Delta g}\Delta g(x) &\leq C_{\Delta g} + x^T D_{\Delta g} x \end{aligned} \tag{2.5}$$

where $0 < \Lambda_{\Delta f}^T = \Lambda_{\Delta f} \in \Re^{n\times n}$, $0 < D_{\Delta f}^T = D_{\Delta f} \in \Re^{n\times n}$, $0 < D_{\Delta g}^T = D_{\Delta g} \in \Re^{n\times n}$ are known constant matrices, $C_{\Delta f}$ and $C_{\Delta g}$ are known positive constants characterizing the behavior of the corresponding unmodeled dynamic mappings at the point $x = 0$.

A4. There exist a stable matrix A_0, strictly positive-defined matrices Q_0 and R_t, such that the following matrix differential Riccati equation

$$\dot{P}_t + P_t A_0 + A_0^T P_t + P_t R_t P_t + Q_0 = 0 \tag{2.6}$$

has strictly positive solution P_t.

The assumption **A1** requires the noises or unknown disturbances are bounded. This is a basic assumption for the analysis of nonlinear systems [9]. Otherwise, the nonlinear systems are flooded with the noises.

The assumption **A2** comes from the fact that a nonlinear system is uniformly observable for any input if and only if it can be presented as the observer form (2.1), see Theorem 2 of [10]. The nominal parts are also Lipschitz, as

$$\begin{aligned} F(x + \Delta x) &= F(x) + \nabla^T F(x)\Delta x + v_f \\ G(x + \Delta x) &= G(x) + \nabla^T G(x)\Delta x + v_g \end{aligned} \tag{2.7}$$

where the vector v_g can be estimated as $\left\|v_f\right\| \leq 2L_f \left\|\Delta x\right\|, \left\|v_g\right\| \leq 2L_g \left\|\Delta x\right\|.$

The "strip bounded" condition **A3** is more general than the bounded unmodeled dynamic [10] and the sector condition [5]. If the unmodeled dynamics are bounded, $D_{\Delta f} = D_{\Delta g} = 0$. If the unmodeled dynamics satisfy sector conditions, $C_{\Delta f} = C_{\Delta g} = 0$.

A special case of **A4** is the following differential Riccati equation with constant A_0, R_0, and Q_0

$$\dot{P}_t' + P_t' A_0 + A_0^T P_t' + P_t' R_0 P_t' + Q_0 = 0 \tag{2.8}$$

Equation (2.8) has a positive solution P_t', if the pair $(A_0, R_0^{1/2})$ is controllable and the pair $(Q_0^{1/2}, A_0)$ is observable. The detailed proof of it can be found in [22]. This is an extend of the matrix Riccati equation

$$A^T P + PA + PRP + Q = 0 \tag{2.9}$$

Equation (2.9) has a positive solution P, if the pair $(A, R^{1/2})$ is controllable, the pair $(Q^{1/2}, A)$ is observable. The controllability and observability conditions for A, R, and Q are equivalent to the following local frequency condition [22]

$$A^T R^{-1} A - Q \geq 0 \tag{2.10}$$

Lemma 2.1 in the next section shows that if the uncertainties are not very large, i.e., the condition (2.16) is satisfied, we can use condition (2.10) for (2.6). This means that it is not difficult to select a Hurwitz matrix A and Q such that (2.10) is established. So (2.10) and (2.16) are the conditions for **A4.**

The robust observer is selected as Luenberger's form [11],

$$\frac{d}{dt}\widehat{x} = F(\widehat{x}) + G(\widehat{x})u_t + K_t \left[y_t - C\widehat{x}\right] \tag{2.11}$$

where $K_t \in \Re^{n\times m}$ is the gain matrix. The observer uses only the available information: the nominal nonlinear mappings $F(\cdot)$, $G(.)$, and C, and the online measurement y_t.

The observer error is defined as

$$\Delta_t = \widehat{x} - x$$

The object of robust observer is to find a K_t such that following performance index

$$J(K_t) = \limsup_{T\to\infty} \frac{1}{T}\int_0^T \Delta_t^T Q \Delta_t dt \tag{2.12}$$

is minimized. $J(K_t)$ is the average of the observer error, $\|\Delta_t\|_Q^2 = \Delta_t^T Q \Delta_t$. Here the strictly positive constant matrix Q is a given to present different physical nature.

If K_t in the observer (2.11) is a constant matrix, (2.11) becomes the well-known high-gain observer [10]. It has the following property

$$\|\Delta_t\| \le k(\theta)\exp\left(-\frac{\theta}{3}t\right)\|x_0 - \widehat{x}_0\|, \tag{2.13}$$

where x_0 and $\widehat{x}_0$ are initial states, θ is big enough positive constants, $k(\theta)$ is a positive function related to observer gain. It means that the observation error is asymptotically stable uniformly on initial conditions. This property requires the perturbations in **A1** be zero, and the unmodeled dynamics $\triangle f$ and $\triangle g$ in **A3** also be zero. In order to deal with these uncertainties, in this chapter, we use a time-variant gain K_t, which is designed in the next section.

2.3 Stability of the Robust Nonlinear Observer

The following lemma guarantees that the assumption **A4** is correct.

Lemma 2.1 *Consider the following two matrix differential Riccati equations: Riccati differential equation with time-varying parameter and algebraic Riccati equation*

$$\begin{array}{l} \dot{P}_1(t) + A_1^T(t)P_1(t) + P_1(t)A_1(t) - P_1(t)R_1(t)P_1(t) + Q_1(t) = 0 \\ A_2^T P_2 + P_2 A_2 - P_2 R_2 P_2 + Q_2 = 0 \end{array} \tag{2.14}$$

with initial condition

$$P_1(0) > P_2 \tag{2.15}$$

the corresponding Hamiltonians are given by

$$H_{1,t} = \begin{bmatrix} Q_1(t) & A_1(t)^T \\ A_1(t) & -R_1(t) \end{bmatrix} \quad H_2 = \begin{bmatrix} Q_2 & A_2^T \\ A_2 & -R_2 \end{bmatrix}$$

If

$$H_2 \geq H_{1,t} \geq 0 \tag{2.16}$$

and the pair (A_2, R_2) *are stable, then*

$$P_1(t) \geq P_2 \geq 0, \forall t > 0 \tag{2.17}$$

Proof Let us define $e_t = P_1(t) - P_2$. Using condition (2.16) and the definition $H = \begin{bmatrix} H_{11} & H_{12} \\ H_{21} & H_{22} \end{bmatrix}$, the Riccati equation (2.14) in the Hamiltonian form is

$$-\dot{P}_1(t) = \begin{bmatrix} I & P_1(t) \end{bmatrix} H_{1,t} \begin{bmatrix} I \\ P_1(t) \end{bmatrix}$$
$$0 = \begin{bmatrix} I & P_2 \end{bmatrix} H_2 \begin{bmatrix} I \\ P_2 \end{bmatrix}$$

we derive:

$$\begin{aligned} -\dot{e}_t &= \begin{bmatrix} I & e_t + P_2 \end{bmatrix} H_{1,t} \begin{bmatrix} I \\ e_t + P_2 \end{bmatrix} \\ &\leq \begin{bmatrix} I & e_t + 0 & P_2(t) \end{bmatrix} H_2 \begin{bmatrix} I \\ e_t \end{bmatrix} + \begin{bmatrix} I \\ P_2(t) \end{bmatrix} \\ &= \left(A_2^T + P_2 R_2\right) e_t + e_t \left(A_2 - R_2 P_2\right) - e_t R_2 e_t + Q_t - Q_t = L_t - Q_t \end{aligned}$$

where

$$L_t = \left(A_2^T + P_2 R_2\right) e_t + e_t \left(A_2 - R_2 P_2\right) - e_t R_2 e_t + Q_t$$

Based on Theorem 3 of [23], the term $\left(A_2^T - P_2 R_2\right)$ is stable when (A_2, R_2) is stable. From (2.15), $e_{t=0} > 0$. By Lemma 1 of [23]), there exists $Q_{t=0} > 0$ such that $L_{t=0} = 0$. This leads to

$$\dot{e}_t \geq Q_t > 0.$$

The solution of the Riccati differential equation is a continuous function, if its time-varying parameters are continuous. So we conclude that for time $t > 0$, there exists a $\varepsilon > 0$ such that

$$Q_\tau > 0 \; \forall \tau \in [t, t+\varepsilon].$$

As a result,

$$e_{t+\varepsilon} = e_t + \int_t^{t+\varepsilon} \dot{e}_\tau d\tau \geq e_t + \int_t^{t+\varepsilon} Q_\tau d\tau \geq e_t + \inf_\tau [\lambda_{\min}(Q_t)] I\varepsilon > 0$$

that leads to

$$P_1(\tau) > P_2 \; \forall \tau \in [0, 0+\varepsilon]$$

Iterating this procedure for the next time interval $[\varepsilon, 2\varepsilon]$, we obtain the final result (2.17).

This lemma shows that if we let $A_0 = A_1(t)$, $Q_0 = Q_1(t)$, $R(t) = R_1(t)$, then (2.6) is (2.14). By Lemma 2.1, the solution of (2.6) $P(t)$ is not less than the solution of

$$A_2^T P + P A_2 + P R_2 P + Q_2 = 0 \tag{2.18}$$

providing than the initial condition of (2.6) is larger than that of (2.18). This means the existence condition for **A4** is always satisfied.

Now, we define R_t and Q_0 in (2.6) as

$$\begin{aligned} R_t &= R_0 + \beta_t \left(C^+ \Lambda (C^+)^T\right)^{\frac{1}{2}} \times \left(I + \Pi^{-1}\right)\left(C^+ \Lambda (C^+)^T\right)^{\frac{1}{2}} \beta_t^T \\ Q_0 &= \left(2\Lambda_{fx} + K_1 \lambda_{\max}(\Lambda_{\partial f}) I\right) + \left(2\overline{u}\Lambda_{gx} + \overline{u} K_2 \lambda_{\max}(\Lambda_{\partial g}) I\right) + Q \end{aligned} \tag{2.19}$$

where $R_0 = \Lambda_f^{-1} + \Lambda_g^{-1} + \Lambda_{\xi_1}^{-1} + \Lambda_{\Delta f}^{-1} + \Lambda_{\Delta g}^{-1} + \Lambda_{\partial f}^{-1} + + \Lambda_{\partial g}^{-1}$, $\Lambda = \Lambda_{\xi_2}^{-1}$, K_1 and K_2 are positive constants, $\overline{u}$ is the upper bound of control u_t, the matrix $\beta_t \in \Re^{n \times n}$ is defined as

$$\beta_t = \frac{\partial}{\partial x} F^T(\widehat{x}) + \frac{\partial}{\partial x} G^T(\widehat{x}) u_t - A_0 \tag{2.20}$$

C^+ is the pseudoinverse matrix of C in Moor–Penrose sense. A_0 is a Hurwitz matrix.

The following theorem formulates the main result of the chapter. It also provides the upper bound of the performance index (2.12).

Theorem 2.1 *Given a class of nonlinear system $\mathcal{H}$ satisfies the assumptions* ***A1–A4****, for any gain matrix K_t, the following upper bound of the performance index (2.12) holds,*

$$J(K_t) \leq J^+(K_t) = \overline{C} + D + \Upsilon_1 + \Upsilon_2 + \phi(K_t) \tag{2.21}$$

*where the constants Υ_1, Υ_2 are defined by **A1**, $\overline{C} = C_{\triangle f} + C_{\triangle g}$,*

$$
\begin{array}{l}
D = \limsup\limits_{T\to\infty} \frac{1}{T} \int\limits_0^T x^T \left(D_{\triangle f} + \overline{u} D_{\triangle g}\right) x dt \\
\phi\left(K_t\right) = \limsup\limits_{T\to\infty} \frac{1}{T} \int\limits_0^T \Delta_t^T \left(X_t \widehat{\Omega}_t^{\frac{1}{2}} + \widehat{\Omega}_t^{-\frac{1}{2}}\right) \left(X_t \widehat{\Omega}_t^{\frac{1}{2}} + \widehat{\Omega}_t^{-\frac{1}{2}}\right)^T \Delta_t dt
\end{array}
\tag{2.22}
$$

$X_t = P_t \left(\beta_t - \widehat{K}_t C\right)$, $K_t = \widehat{K}_t C C^+$, $\widehat{\Omega}_t = C^+ \Lambda^{\frac{1}{2}} \left(I + \Pi\right) \Lambda^{\frac{1}{2}} (C^+)^T + \delta I$, $\delta > 0$.

Proof To start the proof, we need to derive the error dynamic. Taking into account (2.1) and (2.11),

$$
\begin{array}{l}
\dot{\Delta}_t = \dot{\widehat{x}} - \dot{x} \\
= F\left(\widehat{x}\right) + G(\widehat{x}) u_t + K_t \left[y_t - C\widehat{x}\right] \\
-F(x) - G(x) u_t - \triangle f(x) - \triangle g(x) u_t - \xi_{1,t}
\end{array}
\tag{2.23}
$$

Let us denote

$$
\begin{array}{l}
F_t = F(x, \Delta_t, u_t \mid K_t) \\
= F(x + \Delta_t) - F(x) + G(x + \Delta_t) u_t - G(x) u_t - K_t C \Delta_t, \\
\triangle H_t = \triangle H(\xi_{1,t}, \xi_{2,t}, \triangle f \mid K_t) = K_t \xi_{2,t} - \triangle f(\cdot) - \triangle g(\cdot) u_t - \xi_{1,t}.
\end{array}
\tag{2.24}
$$

The vector function F_t describes the dynamic of a nominal model and the function $\triangle H_t$ corresponds to unmodeled dynamics and external disturbances. So

$$
\dot{\Delta}_t = F_t + \triangle H_t
$$

Define a Lyapunov like function as

$$
V_t = \Delta_t^T P_t \Delta_t, \quad P_t^T = P_t > 0
\tag{2.25}
$$

Along with the trajectories of the differential equation (2.23), we derive

$$
\frac{dV_t}{dt} = \frac{\partial V_t}{\partial P_t} + \left(\frac{\partial V_t}{\partial \Delta_t}, \dot{\Delta}_t\right) = \Delta_t^T \dot{P}_t \Delta_t + 2\Delta_t^T P_t \left[F_t + \triangle H_t\right]
\tag{2.26}
$$

Using (2.4),

$$
\begin{array}{l}
F(x + \Delta_t) - F(x) = \frac{\partial}{\partial x} F^T(x) \Delta_t + v_f \\
G(x + \Delta_t) - G(x) = \frac{\partial}{\partial x} G^T(x) \Delta_t + v_g
\end{array}
\tag{2.27}
$$

According to (2.7),

$$
\left\| v_f \right\|_{\Lambda_f} \leq 2 \left\| \Delta \right\|_{\Lambda_{fx}}, \ \left\| v_g \right\|_{\Lambda_g} \leq 2 \left\| \Delta \right\|_{\Lambda_{gx}}
\tag{2.28}
$$

Substitute (2.27) into (2.24),

$$
\begin{aligned}
F_t &= \tfrac{\partial F^T(x)}{\partial x}\Delta_t + v_f + \tfrac{\partial G^T(x)}{\partial x}u_t\Delta_t + u_t v_g - K_t C\Delta_t \\
&= \left[\tfrac{\partial F^T(x)}{\partial x} + \tfrac{\partial G^T(x)}{\partial x}u_t - K_t C\right]\Delta_t + v_f + u_t v_g
\end{aligned}
\tag{2.29}
$$

Because $\Delta_t^T P_t v_f$ is scalar, applying (2.28) and matrix inequality [16]

$$
X^T Y + Y^T X \le X^T \Lambda X + Y^T \Lambda^{-1} Y, \tag{2.30}
$$

where $X \in \Re^{n\times k}$, $Y \in \Re^{n\times k}$, $\Lambda = \Lambda^T > 0$, $\Lambda \in \Re^{n\times n}$ are any matrices. We obtain the inequalities:

$$
\begin{aligned}
2\Delta_t^T P_t v_f &\le \Delta_t^T\left(P_t\Lambda_f^{-1}P_t + 2\Lambda_{fx}\right)\Delta_t, \\
2\Delta_t^T P_t v_g u_t &\le \Delta_t^T\left(P_t\Lambda_g^{-1}P_t + 2\overline{u}\Lambda_{gx}\right)\Delta_t,
\end{aligned}
\tag{2.31}
$$

Using the assumptions **A1** ,we have

$$
\begin{aligned}
-2\Delta_t^T P_t \xi_{1,t} &\le \Delta_t^T P_t \Lambda_{\xi_1}^{-1} P_t \Delta_t + \xi_{1,t}^T, \Lambda_{\xi_1}\xi_{1,t}, \\
2\Delta_t^T P_t K_t \xi_{2,t} &\le \Delta_t^T P_t K_t \Lambda_{\xi 2}^{-1} K_t^T P_t \Delta_t + \xi_{2,t}^T, \Lambda_{\xi_2}\xi_{2,t},
\end{aligned}
\tag{2.32}
$$

Using the assumptions **A3**

$$
\begin{aligned}
-2\Delta_t^T P_t \triangle f &\le \Delta_t^T P_t \Lambda_{\triangle f}^{-1} P_t \Delta_t + C_{\triangle f} + x^T D_{\triangle f} x \\
-2\Delta_t^T P_t \triangle g u_t &\le \Delta_t^T P_t \Lambda_{\triangle g}^{-1} P_t \Delta_t + \overline{u} C_{\triangle g} + \overline{u} x^T D_{\triangle g} x
\end{aligned}
\tag{2.33}
$$

Let us denote

$$
A_t = \frac{\partial F^T(x)}{\partial x} + \frac{\partial G^T(x)}{\partial x}u_t - K_t C. \tag{2.34}
$$

Using the identity $2\Delta_t^T A_t \Delta_t = \Delta_t^T A_t \Delta_t + \Delta_t^T A_t^T \Delta_t$ and adding and subtracting the term $\Delta_t^T Q \Delta_t$ $\left(Q = Q^T > 0\right)$ in the right hand side of (2.26), we obtain

$$
\frac{dV_t}{dt} \le \Delta_t^T L_t \Delta_t + \left(C_{\triangle f} + \overline{u} C_{\triangle g}\right) + \Upsilon_t + x^T\left(D_{\triangle f} + \overline{u} D_{\triangle g}\right)x - \Delta_t^T Q \Delta_t \tag{2.35}
$$

where $\Upsilon_t = \xi_{1,t}^T \Lambda_{\xi_1}\xi_{1,t} + \xi_{2,t}^T \Lambda_{\xi_2}\xi_{2,t}$,

$$
\begin{aligned}
L_t &= P_t\left(\Lambda_f^{-1} + \Lambda_g^{-1} + \Lambda_{\xi_1}^{-1} + \Lambda_{\triangle g}^{-1}\right)P_t \\
&+ \dot{P}_t + P_t A_t + A_t^T P_t + P_t K_t \Lambda_{\xi_2}^{-1} K_t^T P_t + 2\left(\Lambda_{fx} + \overline{u}\Lambda_{gx}\right) + Q,
\end{aligned}
\tag{2.36}
$$

Choosing any Hurwitz matrix A_0, we can rewrite (2.34) as follows:

$$\begin{array}{l} A_t = A_0 + (\frac{\partial}{\partial x}F^T(\widehat{x}) + \frac{\partial}{\partial x}G^T(\widehat{x})u_t \\ -K_tC - A_0) + \left(\frac{\partial}{\partial x}F^T(x) - \frac{\partial}{\partial x}F^T(\widehat{x})\right) + \left(\frac{\partial}{\partial x}G^T(x) - \frac{\partial}{\partial x}G^T(\widehat{x})\right)u_t \end{array} \tag{2.37}$$

Denote

$$\begin{aligned} \widehat{A}_t &= \tfrac{\partial}{\partial x}F^T(\widehat{x}) + \tfrac{\partial}{\partial x}G^T(\widehat{x})u_t - K_tC - A_0 \\ \partial f_t &= \tfrac{\partial}{\partial x}F^T(x) - \tfrac{\partial}{\partial x}F^T(\widehat{x}) \\ \partial g_t &= \tfrac{\partial}{\partial x}G^T(x)u_t - \tfrac{\partial}{\partial x}G^T(\widehat{x})u_t. \end{aligned}$$

From (2.27), we know

$$\partial f_t = (F(x + \Delta_t) - F(x)) - (F(x + 2\Delta_t) - F(x + \Delta_t))$$

Similar to (2.32), we deduce

$$\begin{aligned} 2\Delta_t^T P_t \partial f_t &\le \Delta_t^T \left(P_t \Lambda_{\partial f}^{-1} P_t + \partial f_t \Lambda_{\partial f} \partial f_t^T\right) \Delta_t \le \Delta_t^T \left(P_t \Lambda_{\partial f}^{-1} P_t + K_1 \lambda_{\max}(\Lambda_{\partial f}) I\right) \Delta_t, \\ 2\Delta_t^T P_t \partial g_t &\le \Delta_t^T \left(P_t \Lambda_{\partial g}^{-1} P_t + \overline{u} K_2 \lambda_{\max}(\Lambda_{\partial g}) I\right) \Delta_t, \end{aligned}$$

where K_1 and K_2 are positive constants. Substituting these inequalities into (2.36) and using (2.37), we obtain

$$L_t \le \left(\dot{P}_t + P_t A_0 + A_0^T P_t + P_t R_0 P_t + Q_0\right) + \left(P_t K_t \Lambda K_t^T P_t + P_t \widehat{A}_t + \widehat{A}_t^T P_t\right) \tag{2.38}$$

where the matrices R_0, Q_0, and Λ are defined in (2.19). If we select K_t have a special structure $K_t = \widehat{K}_t C C^+$, using the pseudoinverse property $CC^+C = C$, we can present

$$\begin{array}{l} P_t K_t \Lambda K_t^T P_t + P_t \widehat{A}_t + \widehat{A}_t^T P_t \\ = P_t \left(\beta_t - \widehat{K}_t C\right) + \left(\beta_t - \widehat{K}_t C\right)^T P_t + P_t \widehat{K}_t C G C^T \widehat{K}_t P_t \end{array} \tag{2.39}$$

where

$$G = C^+ \Lambda (C^+)^T. \tag{2.40}$$

Using the definition of (2.22), the last term in (2.39) can be estimated as follows:

$$\Delta_t^T P_t \widehat{K}_t C G C^T \widehat{K}_t^T P_t \Delta_t = \left\| G_t^{\frac{1}{2}} \left(\widehat{K}_t C\right)^T P_t \Delta_t \right\|^2$$
$$= \left\| G_t^{\frac{1}{2}} x \Delta_t - G_t^{\frac{1}{2}} \beta_t P_t \Delta_t \right\|^2$$
$$\leq \Delta_t^T x G_t^{\frac{1}{2}} (I + \Pi) G_t^{\frac{1}{2}} x^T \Delta_t + \Delta_t^T P_t \beta_t G_t^{\frac{1}{2}} \left(I + \Pi^{-1}\right) G_t^{\frac{1}{2}} \beta_t^T P_t \Delta_t$$

where Π is any positive matrix. Define

$$\Phi_t = \beta_t G_t^{\frac{1}{2}} \left(I + \Pi^{-1}\right) G_t^{\frac{1}{2}} \beta_t^T,$$

The first term of (2.39) can be rewritten in the following equivalent form:

$$P_t K_t \Lambda K_t^T P_t \leq x \Omega_t x^T + P_t \Phi_t P_t$$

Since $\Omega_t \in \Re^{n \times n}$ and $rank\,(\Omega_t) = min\,\{n, m\}$, when $m < n$ the inverse matrix of Ω_t does not exist. We introduce the matrix

$$\widehat{\Omega}_t = \Omega_t + \delta I, \ \ \delta > 0$$

Equation (2.39) can be written as

$$P_t K_t \Lambda K_t^T P_t + P_t \widehat{A}_t + \widehat{A}_t^T P_t \leq X_t + X_t^T + X_t \Omega_t X_t^T + P_t \Phi_t P_t$$
$$\leq \left(x \widehat{\Omega}_t^{\frac{1}{2}} + \widehat{\Omega}_t^{-\frac{1}{2}}\right) \left(x \widehat{\Omega}_t^{\frac{1}{2}} + \widehat{\Omega}_t^{-\frac{1}{2}}\right)^T - \widehat{\Omega}_t^{-1} - \delta X_t X_t^T + P_t \Phi_t P_t$$
$$\leq \left(x \widehat{\Omega}_t^{\frac{1}{2}} + \widehat{\Omega}_t^{-\frac{1}{2}}\right) \left(x \widehat{\Omega}_t^{\frac{1}{2}} + \widehat{\Omega}_t^{-\frac{1}{2}}\right)^T + P_t \Phi_t P_t$$

If we define R_t as in (2.19), using the definition (2.22), we transform (2.35) as follows:

$$\frac{dV_t}{dt} \leq \Delta_t^T L_t' \Delta_t + \overline{C} + \Upsilon_t + D_t - \Delta_t^T Q \Delta_t + \Delta_t^T \widehat{\gamma}_t \Delta_t \tag{2.41}$$

where

$$L_t' = \dot{P}_t + P_t A_0 + A_0^T P_t + P_t R_t P_t + Q_0$$
$$\widehat{\gamma}_t = \left(x \widehat{\Omega}_t^{\frac{1}{2}} + \widehat{\Omega}_t^{-\frac{1}{2}}\right) \left(x \widehat{\Omega}_t^{\frac{1}{2}} + \widehat{\Omega}_t^{-\frac{1}{2}}\right)^T$$

According to the assumption **A4**, we conclude that $\widehat{L}_t = 0$. Integrating (2.41) within the interval $t \in [0, T]$ and dividing both sides on T, we obtain:

$$\frac{1}{T}\int_0^T \Delta_t^T Q\Delta_t dt \leq \overline{C} + \frac{1}{T}\int_0^T \Upsilon_t dt + \frac{1}{T}\int_0^T D_t dt + \frac{1}{T}\int_0^T \Delta_t^T \widehat{\gamma}_t \Delta_t - \frac{1}{T}(V_T - V_0)$$
$$\leq \overline{C} + \frac{1}{T}\int_0^T \Upsilon_t dt + +\frac{1}{T}\int_0^T D_t dt + \frac{1}{T}\int_0^T \Delta_t^T \widehat{\gamma}_t \Delta_t - \frac{1}{T}V_0.$$

Taking the limit of $T \to \infty$, we finally obtain (2.21), and

$$\phi(K_t) = \limsup_{T\to\infty} \frac{1}{T}\int_0^T \Delta_t^T \widehat{\gamma}_t \Delta_t dt$$

The following theorem provides the design process of the robust observer.

Theorem 2.2 *If $\widehat{K}_t$ satisfies*

$$K_t = \left\{ P_t^{-1}\left[C^+ \Lambda^{\frac{1}{2}}(I + \Pi)\Lambda^{\frac{1}{2}}(C^+)^T + \delta I \right]^{-1} + \beta_t \right\} C^+ \tag{2.42}$$

the upper bound (2.12) is minimum

$$\min_{K_t} J = \overline{C} + D + \Upsilon_1 + \Upsilon_2$$

Proof Since $\overline{C}$, D, Υ_1, Υ_2, and $\phi\left(\widehat{K}_t\right)$ are positive, to minimize (2.21), we must choose

$$\phi(K_t) = 0, \quad X_t = -\widehat{\Omega}_t^{-1} \tag{2.43}$$

that leads to

$$P_t(\beta_t - \widehat{K}_t C) = -(\Omega_t + \delta I)^{-1}$$

In equivalent form,

$$\widehat{K}_t C = \beta_t + P_t^{-1}(\Omega_t + \delta I)^{-1} \tag{2.44}$$

Using $K_t = \widehat{K}_t C C^+$, (2.42) is established.

If there are no any unmodeled dynamics $\left(\overline{C} = D = 0\right)$ and no any external disturbances $(\Upsilon_1 = \Upsilon_2 = 0)$, the robust observer (2.11) with the optimal matrix gain given by (2.42) guarantees "the stability in average",

$$\limsup_{T\to\infty} \frac{1}{T}\int_0^T \Delta_t^T Q\Delta_t dt = 0.$$

It is equivalent to the fact that

$$\lim_{t\to\infty} \Delta_t = 0$$

The robust nonlinear observer (2.11) has Luenberger form. It is very simple. The time-varying gain K_t is calculated by (2.42). Here C, Π, and Λ are constant matrices, β_t is calculated by the normal parts (2.20), P_t is obtained from the Riccati differential equation (RDE) (2.6). This RDE has a time-varying matrix R_t. It is not easy to discuss the existence conditions for $P(t)$. Lemma 2.1 shows how to use a time-invariant algebraic Riccati equation to decide the solution of a RDE with time-varying parameters.

Now, we compare our observer (2.11) with the others. If the nonlinear system (2.1) is known completely, and the perturbations $\xi_{1,t}$ and $\xi_{2,t}$ are zero, the model-based nonlinear observer in Luenberger form is

$$\frac{d}{dt}\widehat{x} = f(x) + g(x)u_t + K_t (y_t - C\widehat{x})$$

If $f(x)$ and $g(x)$ are unknown or partly known as $F(\widehat{x})$ and $G(\widehat{x})$, the model-based methods [2, 6, 7, 10, 12, 24–26] cannot be used.

The high-gain observer require the nonlinear system (2.1) can be transformed to

$$\begin{aligned} \dot{x}_i &= x_{i-1}, \quad i = 1 \cdots n-1 \\ \dot{x}_n &= F(x) + G(x)u \\ y &= x_1 \end{aligned} \tag{2.45}$$

by a local diffeomorphism, here $F(x)$ and $G(x)$ are unknown. The high-gain observer is [13, 17]

$$\begin{aligned} &\tfrac{d}{dt}\widehat{x}_i = \widehat{x}_{i-1} + \tfrac{\alpha_i}{\varepsilon^i}(y - \widehat{y}) \quad i = 1 \cdots n-1 \\ &\tfrac{d}{dt}\widehat{x}_n = \tfrac{\alpha_n}{\varepsilon^n}(y - \widehat{y}) \\ &\widehat{y} = \widehat{x}_1 \end{aligned} \tag{2.46}$$

where $\alpha_i > 0$ are constant design parameters, ε is a small positive constant. Obviously, the observed states $\widehat{x}_i$ are derivative of the output in different degrees. From (2.1), we can see that the high gain $\frac{\alpha_n}{\varepsilon^n}$ enlarges the measurement noise $\xi_{2,t}$ directly. However, our Luenberger observer (2.11) uses time-varying gain K_t which only use the upper bound $\Lambda = \Lambda_{\xi_2}^{-1}$ of the measurement noise in (2.19). The measurement noise $\xi_{2,t}$ is filtered in the Luenberger observer (2.11).

If the nonlinear system (2.1) cannot be transformed into (2.45), or we do not want to estimate the derivative of the output, the sliding mode observer can be applied [8, 18]

$$\frac{d}{dt}\widehat{x} = F(\widehat{x}) + G(\widehat{x})u_t - KC^T \operatorname{sign}(y_t - C\widehat{x}) \tag{2.47}$$

where $K > 0$, is a big gain matrix. The measurement noises are included inside the sign function, it causes chattering.

A smooth version model-free nonlinear observer is the neural observer [16]

$$\frac{d}{dt}\widehat{x} = F(\widehat{x}) + G(\widehat{x})u_t + W_{1,t}\sigma(V_{1,t}\widehat{x}) + W_{2,t}\phi(V_{2,t}\widehat{x})u_t + K\left(y_t - C\widehat{x}\right) \quad (2.48)$$

where $W_{1,t}\sigma(V_{1,t}\widehat{x})$ and $W_{2,t}\phi(V_{2,t}\widehat{x})$ are neural networks to approximate the unknown functions $f(x)$ and $g(x)$. The output error $(y_t - C\widehat{x})$ has to be used for the observer (2.48) and the weights training as

$$\dot{W}_{1,t} = K_1 C^+ \left(y_t - C\widehat{x}\right)\sigma^T$$

So the observer accuracy is low, compared with the other model-free observers.

2.4 Simulations

We consider a single-link robot rotating in a horizon plane. It is described as

$$\begin{aligned} \begin{bmatrix} \dot{x}_1 \\ \dot{x}_2 \end{bmatrix} &= \begin{bmatrix} x_2 \\ -\sin(x_1) + u \end{bmatrix} - 0.05 \begin{bmatrix} x_1 \cos(x_1) \\ x_2 \sin(x_2) \end{bmatrix} + \begin{bmatrix} w_1 \\ w_2 \end{bmatrix} \\ y_t &= x_1 + w_3 \end{aligned} \quad (2.49)$$

where w_1, w_2 are external state perturbations modeling square wave and saw-tooth functions; w_3 is an output perturbation with the "white noise" nature. We use a PD control to force the system following the reference signal $x_{1,d} = \sin t$, $x_{2,d} = \cos t$, $u = -5\left(x_1 - x_{1,d}\right) - 10\left(x_2 - x_{2,d}\right)$. The unmodeled dynamics are given by

$$\Delta f(\cdot)^T = -0.05\left[x_1\cos(x_1);\, x_2\sin(x_2)\right]$$

This example is similar to [21], but we consider a general case, i.e., there are bounded unmodeled dynamics and external disturbances. Obviously the assumptions **A1–A3** are satisfied. (2.49) is already the observer form (2.1). We construct the following robust observer

$$\begin{aligned} \begin{bmatrix} \frac{d}{dt}\widehat{x}_1 \\ \frac{d}{dt}\widehat{x}_2 \end{bmatrix} &= \begin{bmatrix} \widehat{x}_2 \\ -\sin(\widehat{x}_1) + u \end{bmatrix} + K_t\left(x_1 - \widehat{x}_1\right) \\ K_t &= \left\{P_t^{-1}\left[C^+\Lambda^{\frac{1}{2}}\left(I + \Pi\right)\Lambda^{\frac{1}{2}}(C^+)^T + \delta I\right]^{-1} + \beta_t\right\}C^+ \end{aligned} \quad (2.50)$$

where the gain matrix K_t is computed according to (2.42), $C = [1, 0]$, $\beta_t = \begin{bmatrix} 0 & 1 \\ -\cos(\widehat{x}_1) & 0 \end{bmatrix} + \begin{bmatrix} 0 & 0 \\ 0 & 1 \end{bmatrix} u_t - A_0$. In order to satisfy **A4,** we select $A_0 = \begin{bmatrix} -2 & 0 \\ 0 & -2 \end{bmatrix}$,

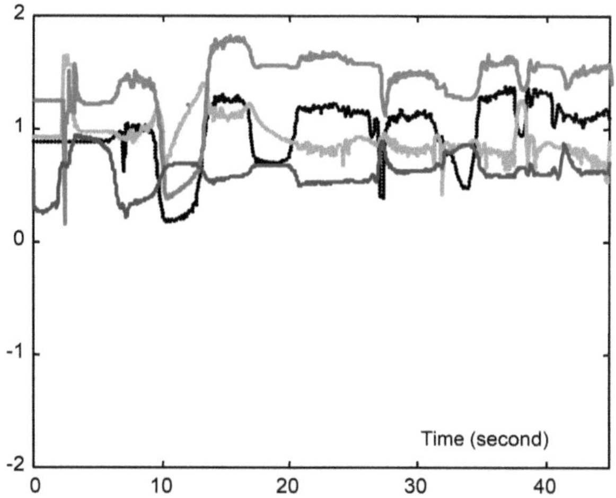

Fig. 2.1 Time evolution of P_t

$Q = \begin{bmatrix} 0.2 & 0 \\ 0 & 0.2 \end{bmatrix}$, $\Lambda = \Pi = I$. So $Q_0 = \begin{bmatrix} 1.2 & 0 \\ 0 & 1.2 \end{bmatrix}$, $R_0 = \begin{bmatrix} 2.5 & 0 \\ 0 & 2.5 \end{bmatrix}$. To obtain the solution of the differential Riccati Equation (2.6), we start from the initial conditions $P_0 = \begin{bmatrix} 0.3 & -0.01 \\ -0.01 & 0.3 \end{bmatrix}$ which is the solution of the corresponding algebraic Riccati equation ($\dot{P}_t = 0$). The time evolution for the elements P_t is shown in Fig. 2.1.

To illustrate the effectiveness of our observer, we compare the observer (RO) (2.50) with the other four nonlinear observers: (1) mode-based nonlinear observer in Luenberger form (LO) [12]; (2) model-free high-gain observer (HO) [17]; (3) model-free sliding mode observer (SO) [8]; (4) model-free neural observer (NO) [16]. They are

$$\text{LO:} \begin{bmatrix} \frac{d}{dt}\widehat{x}_1 \\ \frac{d}{dt}\widehat{x}_2 \end{bmatrix} = \begin{bmatrix} \widehat{x}_2 \\ -\sin(\widehat{x}_1) + u \end{bmatrix} + 20\left(x_1 - \widehat{x}_1\right)$$

$$\text{HO:} \begin{bmatrix} \frac{d}{dt}\widehat{x}_1 \\ \frac{d}{dt}\widehat{x}_2 \end{bmatrix} = \begin{bmatrix} \widehat{x}_2 \\ \frac{-1}{0.001}\left(x_1 - \widehat{x}_1\right) \end{bmatrix}$$

$$\text{SO:} \begin{bmatrix} \frac{d}{dt}\widehat{x}_1 \\ \frac{d}{dt}\widehat{x}_2 \end{bmatrix} = \begin{bmatrix} \widehat{x}_2 \\ -\sin(\widehat{x}_1) + u \end{bmatrix} + \begin{bmatrix} 0 \\ -100\text{sign}\left(x_1 - \widehat{x}_1\right) \end{bmatrix}$$

$$\text{NO:} \begin{bmatrix} \frac{d}{dt}\widehat{x}_1 \\ \frac{d}{dt}\widehat{x}_2 \end{bmatrix} = \begin{bmatrix} \widehat{x}_2 \\ -\sin(\widehat{x}_1) + u \end{bmatrix} + W_{1,t}\sigma(\widehat{x}) + W_{2,t}\phi(\widehat{x})u_t + 15\left(x_1 - \widehat{x}_1\right)$$

$$\dot{W}_{1,t} = 10C^{+}\left(x_1 - \widehat{x}_1\right)\sigma^T$$

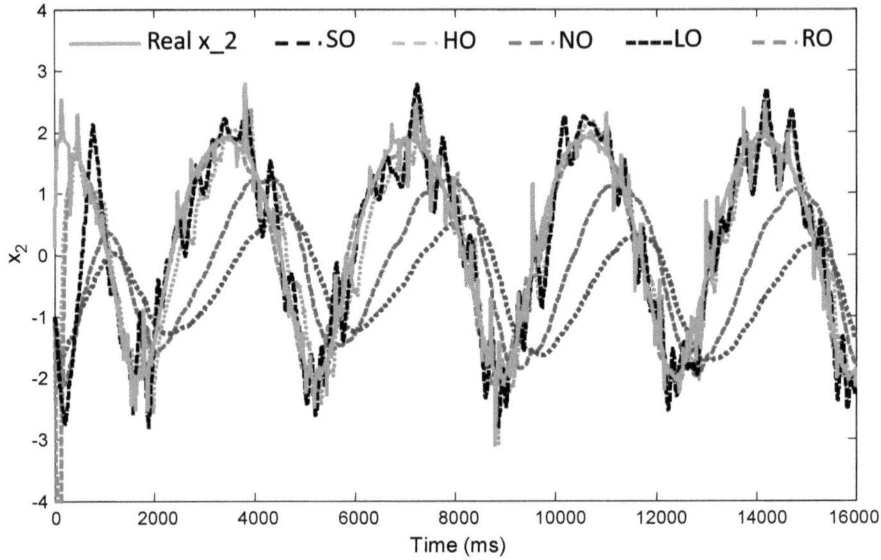

Fig. 2.2 The comparison results for x_2

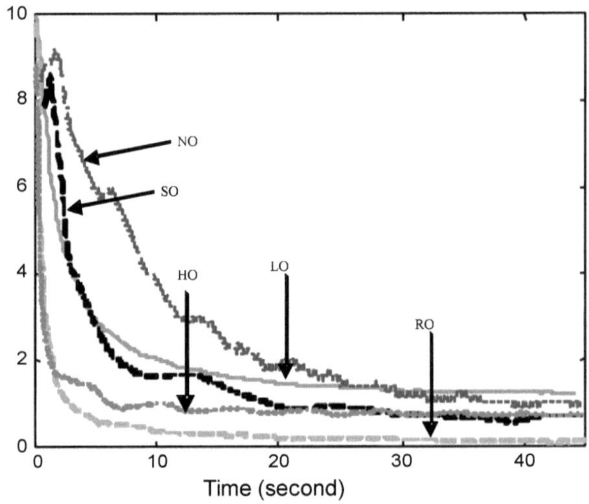

Fig. 2.3 Average of observation errors

The initial conditions are $x_1(0) = 2$, $x_2(0) = 1$, $\widehat{x}_1(0) = 1$, $\widehat{x}_2(0) = 1$. The comparison results are shown in Fig. 2.2. Defining performance indexes $J_T = \frac{1}{T}\int_0^T (x_2 - \widehat{x}_2)dt$, the average observation errors are shown in Fig. 2.3.

From the simulations of these five nonlinear observers, we obtain the following conclusions:

(1) In Fig. 2.2, our robust observers based on the Riccati differential equation (RO) has similar instant observation error as the high-gain observer (HO) [17] and the sliding mode observer (SO) [8]. These three observers are better than the mode-based nonlinear Luenberger observer (LO) [12] and the neural observer (NO) [16], although LO and NO are more smooth. The reason is LO and NO do not consider uncertainties.

(2) In Fig. 2.3, for the average of observation error after a long time, RO is better than HO and SO, because HO and SO are sensitive to the measurement noises w_1 and w_2.

(3) The structure of our observer RO is more simple than NO, and more complex than HO and SO. However, our observer is much better than NO, HO, and SO.

2.5 Conclusion

In this chapter, we propose a novel robust observer for nonlinear uncertainty systems with unmodeled dynamics and external perturbations. This observer has a simple structure, being a Luenberger-like observer. Using a Riccati differential equation and a time-varying observer gain, the stability of the observer is proven. The observer proposed in this chapter can be considered as an alternative approach to the nonlinear robust feedback control. Future work is to extend the results to multi-input and multi-output unknown nonlinear systems.

References

1. Andrieu, V., Praly, L.: On the existence of a Kazantzis-Kravaris/Luenberger observer. SIAM (2006). doi:10.1137/040617066
2. Andrieu, V., Praly, L., Astolfi, A.: High gain observers with updated gain and homogeneous correction terms. Automatica **45**(2), 422–428 (2009)
3. Bernhard, P.: H∞-Optimal Control and Related Minimax Design Problems: A Dynamic Game Approach. Birkhäuser, Basel (1991) Please check the edits made in References [3, 6, 15, 20], and correct if necessary
4. Chen, M., Ge, S.S.: Direct adaptive neural control for a class of uncertain nonaffine nonlinear systems based on disturbance observer. IEEE Trans. Cybern. **43**(4), 1213–1225 (2013). doi:10.1109/TSMCB.2012.2226577
5. Desoer, C.A., Vidyasagar, M.: Feedback systems: input-output properties. SIAM (2009)
6. Dong, Y., Wang, H., Wang, Y.: Design of observers for nonlinear systems with H∞ performance analysis. Math. Methods Appl. Sci. **37**(5), 718–725 (2014). doi:10.1002/mma.2830
7. Karagiannis, D., Carnevale, D., Astolfi, A.: Invariant manifold based reduced-order observer design for nonlinear systems. IEEE Trans. Autom. Control **53**(11), 2602–2615 (2008). doi:10.1109/TAC.2008.2007045
8. Kazraji, S.M., Soflayi, R.B., Sharifian, M.B.B.: Sliding-mode observer for speed and position sensorless control of linear-PMSM. Electr. Control Commun. Eng. **5**(1), 20–26 (2014)
9. Khalil, H.K.: Nonlinear Systems. Prentice-Hall, New Jersey (1996)
10. Khalil, H.K., Praly, L.: High-gain observers in nonlinear feedback control. Int. J. Robust Nonlinear Control **24**(6), 993–995 (2014)

11. Luenberger, D.G.: Observing the state of linear system. IEEE Trans. Mil. Electron **8**(2), 74–80 (1964)
12. Marconi, L., Praly, L., Isidori, A.: Output stabilization via nonlinear Luenberger observers. SIAM J. Control Optim. **45**(6), 2277–2298 (2007)
13. Naifar, O., Boukettaya, G., Oualha, A., Ouali, A.: A comparative study between a high-gain interconnected observer and an adaptive observer applied to IM-based WECS. Eur. Phys. J. Plus **130**(5), 1–13 (2015)
14. Periasamy, V., Tade, M.: An adaptive non-linear observer for the estimation of temperature distribution in the planar solid oxide fuel cell. Journal (2013)
15. Pertew, A.M., Marquezz, H.J., Zhao, Q.: H∞ observer design for Lipschitz nonlinear systems. IEEE Trans. Autom. Control **51**(7), 1211–1216 (2006). doi:10.1109/TAC.2006.878784
16. Poznyak, A.S., Sanchez, E., Palma, O., Yu, W.: Robust asymptotic neuro observer with time delay term. In: Proceedings of the 2000 IEEE International Symposium on Intelligent Control, 2000, pp. 19–24. IEEE (2000)
17. Prasov, A.A., Khalil, H.K.: A nonlinear high-gain observer for systems with measurement noise in a feedback control framework. IEEE Trans. Autom. Control **58**(3), 569–580 (2013). doi:10.1109/TAC.2012.2218063
18. Resendiz, J., Yu, W., Fridman, L.: Two-stage neural observer for mechanical systems. IEEE Trans. Circuits Syst. II Express Br. **55**(10), 1076–1080 (2008). doi:10.1109/TCSII.2008.2001962
19. Solsona, J.A., Valla, M.I.: Disturbance and nonlinear Luenberger observers for estimating mechanical variables in permanent magnet synchronous motors under mechanical parameters uncertainties. IEEE Trans. Ind. Electron. **50**(4), 717–725 (2003). doi:10.1109/TIE.2003.814866
20. Stamnes, Ø.N., Aamo, O.M., Kaasa, G.O.: Adaptive redesign of nonlinear observers. IEEE Trans. Autom. Control **56**(5), 1152–1157 (2011). doi:10.1109/TAC.2011.2107090
21. Tyukina, I.Y., Steurb, E., Nijmeijerc, H., Van Leeuwen, C.: Adaptive observers and parameter estimation for a class of systems nonlinear in the parameters. Automatica **49**(8), 2409–2423 (2013)
22. Willems, J.C.: Least squares optimal control and algebraic Riccati equations. IEEE Trans. Autom. Control **16**(6), 621–634 (1971). doi:10.1109/TAC.1971.1099831
23. Wimmer, H.K.: Monotonicity of maximal solutions of algebraic Riccati equations. Syst. Control Lett. **5**(5), 317–319 (1985)
24. Xingling, S., Honglum, W.: Trajectory linearization control based output tracking method for nonlinear uncertain system using linear extended state observer. Asian J. Control **18**(1), 316–327 (2016)
25. Zeitz, M.: The extended Luenberger observer for nonlinear systems. Syst. Control Lett. **9**(2), 149–156 (1987)
26. Zhao, C.R., Xie, X.J.: Output feedback stabilization using small-gain method and reduced-order observer for stochastic nonlinear systems. IEEE Trans. Autom. Control **58**(2), 523–529 (2013). doi:10.1109/TAC.2012.2208313
27. Zhu, F., Xu, J., Chen, M.: The combination of high-gain sliding mode observers used as receivers in secure communication. IEEE Trans. Circuits Syst. I Regul. Pap. **59**(11), 2702–2712 (2012). doi:10.1109/TCSI.2012.2190570

Chapter 3
Hill Equation: From 1 to 2 Degrees of Freedom

M. Joaquin Collado

Abstract After the introduction, in the first part of the chapter, we review some properties of the scalar Hill equation, a second-order linear ordinary differential equation with periodic coefficients. In the second part, we extend and compare the vectorial Hill equation; most of the results are confined to the case of two degrees of freedom (DOF). In both cases, we describe the equations with parameters (α, β), the zones of instability in the $\alpha - \beta$ plane are called Arnold Tongues. We graphically illustrate the properties wherever it is possible with the aid of the Arnold Tongues.

3.1 Introduction

Hill equation (3.10) was introduced by George Hill in the 1870s, but it was not published until 1886 in [28]. It is a linear second-order ordinary differential equation with a periodic function, originally an even function, to describe the variations in the lunar orbit. It matched so well with the data available in those days, that of immediately gained wide diffusion. In the lapse between the results that Hill got and the date of publication, the Floquet Theorem [20] was published. Nowadays, any study of Hill equation is based on Floquet's result. More than half a century later, the Nobel prize winner Piotr Kapitsa [32] used the newer Theory of Perturbations to find a condition in which the upper equilibrium point of a pendulum[1] may be stabilized varying periodically its suspension point. In detail, if the suspension point of a pendulum of mass M and length L, variates periodically as $z = A\cos\omega t$, then the upper equilibrium point becomes stable if $A\omega > \sqrt{2gL}$, where $g = 9.81\,\text{m/s}^2$, is the acceleration of the gravity. The pendulum with periodic variation of its suspension point is called *Kapitsa's Pendulum*. After this result was published, some authors reported that Stephenson [46] had obtained earlier a similar result, in opinion of the

[1] When the pendulum is assumed a mass M hanging of rigid massless rod.

M. Joaquin Collado (✉)
Automatic Control Department, CINVESTAV, 07360 Mexico City, Mexico
e-mail: jcollado@ctrl.cinvestav.mx

J.B. Clempner and W. Yu (eds.), *New Perspectives and Applications of Modern Control Theory*, https://doi.org/10.1007/978-3-319-62464-8_3

author this is partially true, because Stephenson's paper claims that it is possible, but he did not express any condition. In 1928, van der Pol and Strutt[2] [47] published the first Arnold Tongues,[3] they claimed belong to the Mathieu equation (Eq. 3.10, when $q(t) = \cos t$), but actually they reproduced the Arnold Tongues diagram for the Meissner Equation (Eq. 3.10, when $q(t) = sign(\cos t)$).

Even more interesting is that there exist a seven-century tradition at the Cathedral of Santiago of Compostela, since then they have experienced a Kapitsa's Pendulum with a large censer (O Botafumeiro), which reaches approximately $\pm 82°$, in 17 cycles and it takes approximately 80 s to achieve the maximum excursion [42]. This effect is contained in Kapitsa's result, because when the condition $A\omega > \sqrt{2gL}$ is satisfied, simultaneously the upper equilibrium becomes stable and the lower equilibrium becomes unstable, as in the Botafumeiro. Around 1940 the Romantic era for Hill equation ended. Late 1940s until 1960s, two prominent Russian academicians Krein and Yakubovich, established the foundation of linear Hamiltonian with periodic coefficients; we mention two celebrated references, [34, 48]. Other important contributions were made by Gelfand–Lidskii [24], Starzhinskii [45], Bolotin [6], Atkinson [4], and Eastham [19]; above all of them, it was Lyapunov himself, who contributed approximately half of his Ph.D. Thesis to the problem of stability of Linear Periodic Systems [35]. Relations, only for the scalar case with the Sturm–Liouville Theory, appear in Atkinson [4], Eastham [19], Yakubovich and Starzhinskii [48], and the excellent book of Marchenko [37]; a recent reference is [10]. A recent application of parametric resonance in the roll effect of ships appeared in [21]. Two excellent surveys are Champneys [11] or Seyranian [43]; encyclopedic and very deep results related to the spectrum of Hill's equation were presented by McKean and Moerbeke [38].

The *Direct Problem* refers to: *Given a Hill equation, find the spectral bands or Arnold Tongues associated*; this paper deals *only* with this case. The *Inverse Problem*, consists in: *Given the spectral data, to recover the equation which has the given spectrum*. The inverse problem of the Sturm–Liouville problem related to the scalar Hill equation was solved in the 1960s by Gelfand and Levitan [23] and others. But, it was Borg [7] who defined the problem and gave the first key results. Atkinson [4] and Eastham [19] gave interesting results. In the opinion of the author, the inverse problem associated with the vectorial Hill equation is far from being solved, because it is not equivalent to a vectorial Sturm–Liouville (SL) Problem, although some early attempts are found in [26] and more recently in [5]. In physics literature, they name to the Hill equation, the 1-dimensional Schrödinger equation with periodic potential.

This chapter is organized as follows: the first section is an introduction and historical overview, in Sect. 3.2 we present mathematical preliminaries particularly concerning matrices, Sect. 3.3 is dedicated to survey the results for scalar Hill equation, in

[2]More well known as Lord Rayleigh, more correctly Baron Rayleigh because Baron is a higher novelty title than Lord.

[3]The name *Arnold Tongues* was introduced after [2].

Sect. 3.4 we present the 2-DOF Hill equation or vectorial Hill equation, the objective of Sect. 3.5 gives a set of open problems and different possible generalizations, finally in Sect. 3.6 we present some conclusions.

3.2 Preliminaries

In this section, we present the main background required subsequently, namely Floquet Theory, which gives us the basic property of the solutions of a linear ordinary differential equations with periodic coefficients. Then we review the Hamiltonian systems, and the associated Hamiltonian and symplectic matrices with their main properties.

3.2.1 *Floquet Theory*

Given a linear system described as a set of first-order linear ordinary differential equations with periodic coefficients, as:

$$\dot{x} = A(t)x \tag{3.1}$$

where $A(t)$ is an $n \times n$ matrix whose components are piecewise continuous, and periodic with minimum period T; i.e., $A(t+T) = A(t)$ for all t; for the sake of brevity we will say that $A(t)$ is T-*periodic*. The solutions of (3.1) may be expressed in terms of the *state transition matrix*[4] $\Phi(t, t_0)$, which has the following basic properties, see [8] or [12]:

(a) $\Phi(t, t) = I_n \ \forall t \in \mathbb{R}$

(b) $[\Phi(t, t_0)]^{-1} = \Phi(t_0, t)$

(c) $\Phi(t_2, t_0) = \Phi(t_2, t_1)\, \Phi(t_1, t_0) \ \forall t_0, t_1, t_2 \in \mathbb{R}$,

(d) $\frac{\partial}{\partial t}\Phi(t, t_0) = A(t)\,\Phi(t, t_0)$, and

(e) $\forall\, x(t_0) = x_0 \in \mathbb{R}^n$, the solution of (3.1) is $x(t) = \Phi(t, t_0)\, x_0$.

Using the state transition matrix previously reviewed, Floquet Theory [8] asserts:

Theorem 3.1 (Floquet) *Given the periodic linear system (3.1), its state transition matrix satisfies:*

$$\Phi(t, t_0) = P^{-1}(t)\, e^{R(t-t_0)} P(t_0), \tag{3.2}$$

[4] *Matriciant* in the russian literature [1]. Also denominated as *Cauchy Matrix* or *Normalized Fundamental Matrix*.

where $P(t+T) = P(t)$ is an $n \times n$ periodic matrix of the same period T of the system (3.1), and R is a constant $n \times n$ matrix, not necessarily real even if (3.1) is real.[5]

If we make $t_0 = 0$ in (3.2) and by property (a), we get $P^{-1}(0) = I_n$, then we get the most well-known version:

Corollary 3.1 (Floquet Theorem) *Given the system (3.1), for $t_0 = 0$ its state transition matrix satisfies:*

$$\Phi(t,0) = P^{-1}(t)\, e^{Rt} \tag{3.3}$$

where $P(t+T) = P(t)$ is an $n \times n$ periodic matrix of the same period as the system (3.1), and R is a constant $n \times n$ matrix.

Now if we evaluate (3.3) at $t = T$, taking into account that $P(t)$ is T-periodic, $P(T) = I_n$, then

$$M = \Phi(T,0) = e^{RT}. \tag{3.4}$$

The last constant matrix is particularly important, it is called *Monodromy Matrix* and will be designated by M.

Remark 3.1 The Monodromy matrix defined by (3.4) is dependent of the initial time t_0; but not its spectrum. Let us designate $M_{t_0} = \Phi(T+t_0, t_0)$, then using (3.2) for $t = T + t_0$, $\Phi(T+t_0, t_0) = P^{-1}(T+t_0)\, e^{RT} P(t_0) = P^{-1}(t_0)\, e^{RT} P(t_0) = P^{-1}(t_0)\, M P(t_0)$. This shows that $\Phi(T+t_0, t_0) = M_{t_0}$ and M are similar. As long as our use of the Monodromy matrix is reduced to its spectrum, there is no difference to use M or M_{t_0}.

Two consequences of the Floquet Theorem are of great importance: *Reducibility and Stability*.

I.- Reducibility

Given a system $\dot{x} = A(t)x$, if we make the following change of coordinates $z(t) = T(t)x(t)$, where the square $n \times n$ matrix $T(t)$ satisfies:

(i) $T(t)$ is differentiable and invertible $\forall t$, and

(ii) The matrices $T(t)$, $\dot{T}(t)$, and $T^{-1}(t)$ are all bounded

Then the Transformation matrix $T(t)$ is called a *Lyapunov Transformation*.[6]

Roughly speaking, the system in coordinates x or z, keep their stability property if, $T(t)$ the matrix which relates x and z is a Lyapunov Transformation. For properties of Lyapunov Transformations see [1, 8, 22].

---

[5] The necessary and sufficient condition for R to be real is that the real negative eigenvalues of $\Phi(T,0)$, be of algebraic multiplicity even [1].

[6] This transformation was introduced by Lyapunov himself [35], other reference is [8].

Definition 3.1 A time-varying linear system (not necessarily periodic) $\dot{x} = A(t)x$, is said to be 'reducible,' if there exists a linear time-varying Lyapunov Transformation $T(t)$ such that $z(t) = T(t)x(t)$

$$\dot{z} = \left[T^{-1}(t)A(t)T(t) + T^{-1}(t)\dot{T}(t)\right]z \tag{3.5}$$

where $\left[T^{-1}(t)A(t)T(t) + T^{-1}(t)\dot{T}(t)\right] = R$ a constant matrix

Any system (3.1) T-periodic is reducible, the result is expressed formally in the next theorem. All the symbols refer to the factorization given in (3.3).

Theorem 3.2 *Given a T-periodic linear system $\dot{x} = A(t)x$, the change of coordinates $z(t) = P^{-1}(t)x(t)$ transforms the system into a linear time-invariant system:*

$$\dot{z} = Rz. \tag{3.6}$$

Remark 3.2 It follows that for linear periodic systems, there is a linear periodic transformation, which transforms the original periodic time-varying system into a linear time-invariant system. Unfortunately this result, while very useful for analysis, it is not so for synthesis; because one requires the solution in order to perform this change of coordinates.

II.- Stability

Recall the stability definition in the sense of Lyapunov [35] (or contemporary reference [33]):

Definition 3.2 The zero solution of $\dot{x} = A(t)x$

(a) Stable, if $\forall\,\varepsilon > 0, \exists\,\delta > 0$ such that $\|x(t_0)\| < \delta \Longrightarrow \|x(t)\| < \varepsilon\ \forall\, t \geq t_0$
(b) Asymptotically stable if the zero solution is stable and $\lim_{t\to\infty} x(t) = \mathbf{0}$.

In our system (3.1) $\dot{x} = A(t)x$, for $t \geq 0$, t may be expressed as: $t = kT + \tau$, k a non-negative integer and $\tau \in [0, T)$; then the solution satisfies $t_0 = 0$ and $x(0) = x_0$:

$$\begin{aligned}
x(t) &= \Phi(t,0)x_0 \\
&= \Phi(kT+\tau,0)x_0 \\
&= \Phi(kT+\tau,kT)\Phi\left(kT,(k-1)T\right)\Phi\left(kT,(k-1)T\right)\cdots\Phi\left(T,0\right)x_0 \\
&= \Phi(\tau,0)\underbrace{\Phi(T,0)\,\Phi(T,0)\cdots\Phi(T,0)}_{k\ times}x_0 \\
&= \Phi(\tau,0)\,M^k x_0
\end{aligned}$$

from the last step, we can conclude that

(a) *Asymptotic Stability*: $x(t) \to \mathbf{0}$ if only if $\sigma(M) \subset \overset{\circ}{D}_1 \triangleq \{z \in \mathbb{C} : |z| < 1\}$
(b) *Stability*: $x(t)$ remains bounded $\forall\, t \geq 0$ iff $\sigma(M) \subset \overline{D}_1 \triangleq \{z \in \mathbb{C} : |z| \leq 1\}$ and if $\lambda \in \sigma(M)$ and $|\lambda| = 1$, λ is a simple root of the minimal polynomial of M.

Remark 3.3 Notice that both properties, reducibility, and stability, for linear T -periodic systems could be obtained thanks to the Floquet factorization (3.3).

3.2.2 Hamiltonian Systems

Given a differentiable function $\mathcal{H}(q, p)$, called a *Hamiltonian function*, which depends on vectors q and p, which satisfies the equations:

$$\begin{aligned} \dot{q} &= \left(\frac{\partial \mathcal{H}(q,p)}{\partial p}\right)^T \\ \dot{p} &= -\left(\frac{\partial \mathcal{H}(q,p)}{\partial q}\right)^T \end{aligned} \tag{3.7}$$

is called a *Hamiltonian System*, also called in Russian literature *Canonical System*. The Hamiltonian function represents the energy of the system and for the case in which this function $\mathcal{H}(q, p)$ does not depend explicitly of time, this quantity being preserved along the solutions of (3.7); when this property holds the system is called *Conservative*. This guarantees that the system (3.7) has a first integral, [3, 14, 39]. The Hamiltonian systems (3.7) are always of even order, say $2n$ if $q, p \in \mathbb{R}^n$. For further properties see [3, 39].

We shall consider Hamiltonian functions that are also function of time, i.e., $\mathcal{H}(t, q, p)$, in this case the Hamiltonian system is no longer conservative. Also we shall only regard linear Hamiltonian systems, then the Hamiltonian function is a quadratic homogeneous form, i.e.,

$$\mathcal{H}(t, q, p) = \begin{bmatrix} q \\ p \end{bmatrix}^T H(t) \begin{bmatrix} q \\ p \end{bmatrix} \tag{3.8}$$

where $H(t)$ is a $2n \times 2n$ symmetric matrix, in this case the Hamiltonian System (3.7) may be expressed as:

$$\frac{d}{dt}\begin{bmatrix} q \\ p \end{bmatrix} = JH(t)\begin{bmatrix} q \\ p \end{bmatrix} \tag{3.9}$$

where $J = \begin{bmatrix} 0 & I_n \\ -I_n & 0 \end{bmatrix}$. Notice that $J^{-1} = J^T = -J$ and $J^2 = -I_{2n}$.

Finally, if the linear Hamiltonian system is T-periodic, then $H(t+T) = H(t) = H^T(t)$. We are going to assume this relation to hold from now on.

Definition 3.3 ([39]) An even-order matrix $A \in \mathbb{R}^{2n\times 2n}$ is called Hamiltonian Matrix, if $A = JH$, where H is a symmetric matrix; equivalently $JA + A^T J = 0$

From $JA + A^T J = 0$, we get $A = J^{-1}\left(-A^T\right)J$, i.e., A is similar to $-A^T$ therefore they have the same spectrum:

$$\sigma(A) = \sigma\left(-A^T\right) = \sigma(-A).$$

We have proven the key property of constant Hamiltonian matrices, that is, its spectrum is symmetric with respect to the imaginary axis.

Theorem 3.3 *Let $A \in \mathbb{R}^{2n\times 2n}$ be a Hamiltonian matrix, then if $\lambda \in \sigma(A) \Longrightarrow -\lambda \in \sigma(A)$.*[7] *Equivalently, the characteristic polynomial of a Hamiltonian matrix has only even powers or it is an even polynomial.*

Remark 3.4 Notice also that, the trace of a Hamiltonian matrix is always zero.

Hamiltonian matrices are closely related to another kind of matrices, the symplectic ones.

Definition 3.4 ([39]) An even-order real matrix $M \in \mathbb{R}^{2n\times 2n}$ is called a Symplectic Matrix, if $M^T JM = J$.

The determinant of a symplectic matrix is $+1$, moreover the set of symplectic matrices of a given order form a Group [39]. The key property of constant symplectic matrices is that its spectrum is symmetric with respect to the unit circle, it may be easily proven, from the definition and the fact that a symplectic matrix is always invertible, then $M^T = JM^{-1}J^{-1}$, i.e., $\sigma\left(M^T\right) = \sigma\left(M^{-1}\right) = \sigma(M) \Longrightarrow$ if $\lambda \in \sigma(M)$ then $\lambda^{-1} \in \sigma(M)$. Let us express this fact formally in the next theorem:

Theorem 3.4 *Let $M \in \mathbb{R}^{2n\times 2n}$ be a symplectic matrix, then if $\lambda \in \sigma(A) \Longrightarrow \lambda^{-1} \in \sigma(A)$. Equivalently, the characteristic polynomial of a Symplectic matrix is self-reciprocal [39] or palindromic [31], i.e., $p_M(\lambda) = \lambda^{2n} p_M\left(\lambda^{-1}\right)$.*

The property that relates Hamiltonian matrices with symplectic ones in a given Hamiltonian system is:

Theorem 3.5 *Let $\frac{d}{dt}\begin{bmatrix} q \\ p \end{bmatrix} = JH(t)\begin{bmatrix} q \\ p \end{bmatrix}$ for some $H(t) = H^T(t)$ be a linear time-varying*[8] *Hamiltonian system, then its state transition matrix is a symplectic matrix.*

---

[7]Given a square matrix A, by $\sigma(A)$ we denote its spectrum, i.e., the set of all the eigenvalues.

[8]Not necessarily periodic.

Remark 3.5 A linear time-invariant Hamiltonian system can not be asymptotically stable, because of the symmetry of its eigenvalues; this property goes to all the Hamiltonian Systems time-invariant or not; and linear or not.

Remark 3.6 Hamiltonian systems enjoy another important property: For arbitrary $2n$-dimensional system $2n-1$ independent first integrals are required in order to arrive at a first-order 1-dimensional ODE, which may be integrated by quadratures to finally integrate the whole system. The Liouville Theorem ensures that a $2n$-dimensional Hamiltonian system is integrated if we know *only* n independent first integrals [3, 39]; only half of the work!

3.3 Hill Equation: The Scalar Case

In this section, we are going to present the main properties of scalar Hill's equation, namely

$$\ddot{y} + [\alpha + \beta q(t)] y = 0 \tag{3.10}$$

where $q(t)$ is T-periodic.[9] The parameter α represents the square of the natural frequency for $\beta = 0$; the parameter β is the *amplitude* of the parametric excitation, and the periodic function $q(t)$ is called the *excitation function*. For comparison reasons, we are going to use three different excitation functions: (a) $q(t) = \cos t$, in this case the equation is called *Mathieu equation*; (b) $q(t) = sign(\cos t)$, in this case the equation is called *Meissner equation,* and (c) $q(t) = \cos t + \cos 2t$, which was used originally by Lyapunov [35].

Notice that a linear second-order differential equation with periodic coefficients:

$$\ddot{z} + a(t)\dot{z} + b(t)z = 0 \tag{3.11}$$

where $a(t)$ and $b(t)$ are T-periodic, may be always transformed with $y = e^{-\frac{1}{2}\int a(\tau d\tau)}z$, into (3.10), therefore there is no loss of generality to consider with respect to (3.10). Note also that this is *not* a Lyapunov Transformation in general [27].

If we define the 2-dimensional vector $x \triangleq \begin{bmatrix} y \\ \dot{y} \end{bmatrix}$, the Eq. (3.10) may be rewritten as

$$\dot{x} = \begin{bmatrix} 0 & 1 \\ -\alpha - \beta q(t) & 0 \end{bmatrix} x = \left\{ \underbrace{\begin{bmatrix} 0 & 1 \\ -1 & 0 \end{bmatrix}}_{J} \underbrace{\begin{bmatrix} \alpha + \beta q(t) & 0 \\ 0 & 1 \end{bmatrix}}_{H(t)} \right\} x \tag{3.12}$$

[9]We will assume through the paper that $q(t)$ is piecewise continuous, integrable in $[0, T]$ and of zero average, i.e., $\int_0^T q(t)\,dt = 0$.

where J is as in Eq. (3.9) for $n = 1$, and $H(t+T) = H(t) = H^T(t)$, satisfies the condition for linear Hamiltonian systems. Then the state transition matrix of Hill's equation in the format (3.12) is a Symplectic matrix for all t. Therefore its Monodromy matrix M is also a symplectic matrix. The characteristic polynomial $p_M(\mu)$ of the Monodromy Matrix M is of the form:

$$p_M(\mu) = \mu^2 - tr(M)\mu + 1 \tag{3.13}$$

Definition 3.5 The eigenvalues of the Monodromy matrix M, equivalently the roots of its characteristic polynomial $p_M(\mu)$, are called multipliers of (3.12) or (3.10), denoted by μ. For Hamiltonian systems are symmetric with respect to the unit circle.

Definition 3.6 Associated to every multiplier μ, there exist (an infinite) numbers called *characteristic exponents* λ related to a multiplier by $\mu = e^{\lambda T}$.

The roots of $p_M(\mu)$ or the multipliers of (3.10) are:

$$\mu_{1,2} = \left[tr(M) \pm \sqrt{tr^2(M) - 4}\right]/2 \tag{3.14}$$

- If $tr^2(M) < 4$ the multipliers are complex conjugates and their modulus is $\left|\mu_{1,2}\right|^2 = \dfrac{tr^2(M)}{4} + \dfrac{4 - tr^2(M)}{4} = 1$. The two eigenvalues are different, which implies that the minimal and the characteristic polynomials of M are the same. This case corresponds to a stable system.
- If $tr^2(M) > 4$, the multipliers are real and reciprocal, $\mu_1 = \left[tr(M) + \sqrt{tr^2(M) - 4}\right]/2$ and $\mu_2 = \left[tr(M) - \sqrt{tr^2(M) - 4}\right]/2$. Obviously $\mu_1 + \mu_2 = tr(M)$ and $\mu_1\mu_2 = 1$, so $\mu_2 = \mu_1^{-1}$. If one of the multipliers, say $\mu_1 > 1$, then this case corresponds to instability.
- If $tr^2(M) = 4$ the multipliers are real and equal to $+1$ if $tr(M) = +2$, or the multipliers are equal to -1 if $tr(M) = -2$. In this case, Hill equation is stable if only if M is a diagonal matrix or *scalar matrix,* otherwise the Hill equation is unstable.[10]

The boundaries between stability-instability correspond to this last case, i.e., when $|tr(M)| = 2$. It is clear that M depends on the parameters α, β. It is customary to define [36][11] $\phi(\alpha, \beta) \triangleq tr(M)$. Hochstadt [29] was the first to recognize the important properties of $\phi(\alpha, \beta)$.

---

[10]When the multipliers are ± 1 and the Monodromy matrix is diagonal, and we say that there is a point of *Coexistence*, because there are two linearly independent periodic solutions of Hill equation; T-periodic for multipliers $+1$, and $2T$-periodic for the multipliers equal to -1.

[11]In Magnus [36] the function that we call $\phi(\alpha, \beta)$, is denoted as $\Delta(\lambda)$, because λ is used instead of our α, and the parameter β is not used in the cited work.

Theorem 3.6 (Hochstadt) *The function $\phi(\alpha, \beta)$ for any β constant, is an entire function of order $\frac{1}{2}$. The functions $\phi(\alpha, \beta) \pm 2 = 0$ have an infinite number of roots. For any β_0, and for α_0 sufficiently negative, $\phi(\alpha_0, \beta_0)$ is positive, therefore increasing α appears the first root for the equation $\phi(\alpha, \beta) - 2 = 0$, which corresponds to a double multiplier at $+1$, and from there appear two roots (not necessarily different) at -1, then two roots $+1$, up to infinity.*

Due to the Hochstadt Theorem, there are two infinite sequences:

$$\begin{gathered} \lambda_0, \lambda_1, \lambda_2, \lambda_3, \lambda_4, \lambda_5, \ldots \\ \overline{\lambda}_1, \overline{\lambda}_2, \overline{\lambda}_3, \overline{\lambda}_4, \overline{\lambda}_5, \ldots \end{gathered} \tag{3.15}$$

The first sequence corresponds to roots of $\phi(\alpha, \beta) + 2 = 0$, and the second sequence corresponds to $\phi(\alpha, \beta) - 2 = 0$. Moreover they interlace as:

$$\lambda_0, \overline{\lambda}_1, \overline{\lambda}_2, \lambda_1, \lambda_2, \overline{\lambda}_3, \overline{\lambda}_4, \lambda_3, \lambda_4, \overline{\lambda}_5, \overline{\lambda}_6, \ldots \tag{3.16}$$

This fact is illustrated in Fig. 3.1

Remark 3.7 Notice that for the values in which $\phi(\alpha, \beta_0) \in [-2, 2]$ the multipliers lie on the unit circle, and for the values in which $|\phi(\alpha, \beta_0)| > 2$ the multipliers are both positive or both negative, and one the reciprocal of the other. Also if for some value of $\alpha = \alpha_1$ both multipliers lie on -1, and increasing this value up to the point $\alpha = \alpha_2$ for which both multipliers lie on $+1$; the path of the multiplier from the point -1 to $+1$ should be through arcs on the unit circle, they can not go from -1 to $+1$

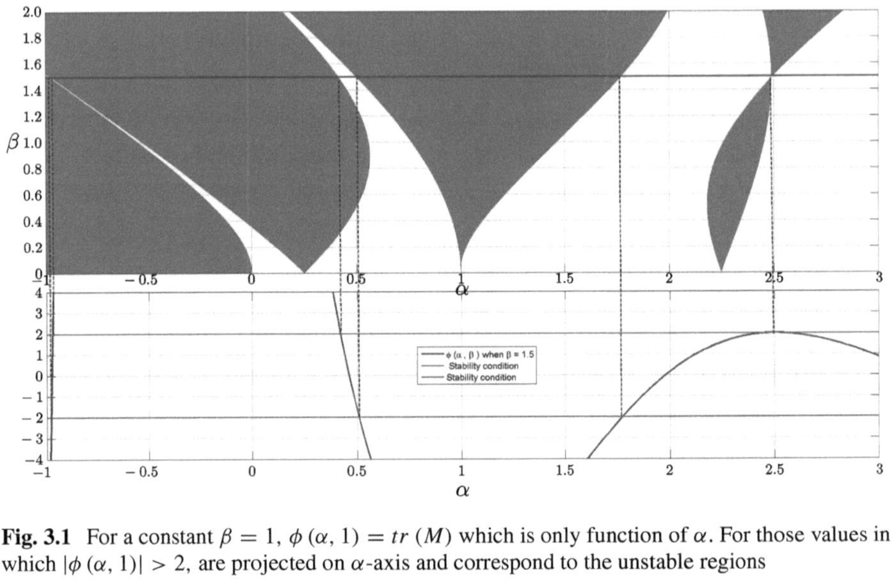

Fig. 3.1 For a constant $\beta = 1$, $\phi(\alpha, 1) = tr(M)$ which is only function of α. For those values in which $|\phi(\alpha, 1)| > 2$, are projected on α-axis and correspond to the unstable regions

on the real line, because at 0 the non-singularity of the Monodromy matrix would be violated. This property goes to any degree of freedom as long as the system is Hamiltonian. Moreover, even in the non-Hamiltonian case the Monodromy matrix is always nonsingular, because it is the state transition matrix, evaluated at the end of a period.

3.3.1 Multipliers of Hamiltonian Systems

Given in general a linear Hamiltonian system:

$$\dot{x} = JH(t)\,x \tag{3.17}$$

where J was defined in (3.9) and $H(t) \in \mathbb{R}^{2n\times 2n}$ is a real symmetric matrix. Due to the fact that a Hamiltonian system cannot exhibit asymptotic stability, then the accepted definition for *weak stability* of a Hamiltonian system is that all the solutions be bounded in $(-\infty, +\infty)$. The following definition is also required:

Definition 3.7 The Hamiltonian system (3.17) is *strongly stable* if it is stable (bounded) and there exists an $\varepsilon > 0$, such that for all $\widetilde{H}(t)$ $2n \times 2n$ symmetric matrices, while $\left\|\widetilde{H}(t) - H(t)\right\| < \varepsilon$, all the Hamiltonian systems

$$\dot{x} = J\widetilde{H}(t)\,x$$

are stable.

The condition of strong stability for Hamiltonian Systems was formulated more than 50 years ago, the sufficiency by Krein [34], and the necessity by Gelfand and Lidskii [24]; another definition is required for an *indefinite inner product* associated to the symplectic geometry of the Hamiltonian System [48].

Definition 3.8 Given an even-order real vector space of dimension $2n$, and the standard inner product $(x, y) \triangleq y^T x$, and given any Hermitian nonsingular matrix $H \in R^{2n\times 2n}$, it is possible to define the *indefinite inner product* as $\langle x, y\rangle \triangleq (Gx, y)$. We are going to use $H = iJ$.[12]

For any multiplier on the unit circle μ, its associated eigenvector v_μ is such that $\langle v_\mu, v_\mu\rangle \neq 0$, if $\langle v_\mu, v_\mu\rangle > 0$, μ is called *Multiplier of the First Kind*; if $\langle v_\mu, v_\mu\rangle < 0$, μ is called *Multiplier of the Second Kind* [48]. If $|\mu| \neq 1$, $\langle v_\mu, v_\mu\rangle = 0$, but if we extend the definition of Multiplier of first kind for $\mu : |\mu| < 1$; and Multiplier of the second kind for those $\mu : |\mu| > 1$. Then all the multipliers are of first or second kind,

[12] Recall that given any skew-hermitian matrix J, then (iJ) is an hermitian matrix. [22, 48].

and moreover, for a Hamiltonian system of dimensions $2n$, n multipliers are of first kind and the remaining n multipliers are of second kind.[13]

Remark 3.8 The *key property* due to Krein is that the multipliers including their kind are continuous functions with respect to variations in the Hamiltonian functions, in our case the symmetric Matrix $H(t)$, [34, 48].

Due to the last remark, if two multipliers coincide on the unit circle and both are of the same kind, they cannot leave the unit circle, because they would violate continuity of the kind of multipliers.

Finally, to formulate the Gelfand–Lidskii–Krein Theorem, we require this last definition:

Definition 3.9 A multiplier μ with algebraic multiplicity r, is said to be definite of first or second kind, if $\langle q, q \rangle$ is of the same sign for all q in the eigenspace associated to μ.

Theorem 3.7 (Krein–Gelfand–Lidskii) *The linear periodic Hamiltonian system* $\dot{x} = JH(t)x$ *is strongly stable iff all the multipliers lie on the unit circle and those with algebraic multiplicity greater than one are definite or all are of the same kind.*

3.3.2 Arnold's Tongues

If we mark in the $\alpha - \beta$ plane the points of instability, which correspond to $|tr(M)| > 2$ with some color, and leave blank the points of stability which correspond to $|tr(M)| < 2$; this diagram is called *Ince-Strutt diagram*. Figure 3.2 shows the Ince-Strutt diagram for the Mathieu equation.

Remark 3.9 We have to emphasize that the Ince-Strutt diagram was obtained numerically, i.e., gridding 1000 points in each of the chosen intervals for $\alpha \in [-1, 10]$ and $\beta \in [0, 10]$. Then integrating the differential equation in the time interval $[0, 2\pi]$ with the initial conditions $\begin{bmatrix} 1 & 0 \end{bmatrix}^T$, we get the solution $\mathbf{x}_1(t)$, similarly for initial condition $\begin{bmatrix} 0 & 1 \end{bmatrix}^T$ we get another solution $\mathbf{x}_2(t)$ on *each* one of the grid points; finally the Monodromy matrix is $M = \begin{bmatrix} \mathbf{x}_1(2\pi) & \mathbf{x}_2(2\pi) \end{bmatrix}$.

[13]Equivalently, if we increase the Hamiltonian, i.e., $\widetilde{H}(t) - H(t) > 0$, and μ was an isolated multiplier on the unit circle associated to $H(t)$, when $H(t)$ is increased to $\widetilde{H}(t)$, μ moves on the unit circle to $\widetilde{\mu}$; if $\arg\widetilde{\mu} > \arg\mu$, the multiplier μ is said to be a *Multiplier of the First Kind*, contrarily, i.e., $\arg\widetilde{\mu} < \arg\mu$, the multiplier μ is a *Multiplier of the Second Kind* [48].

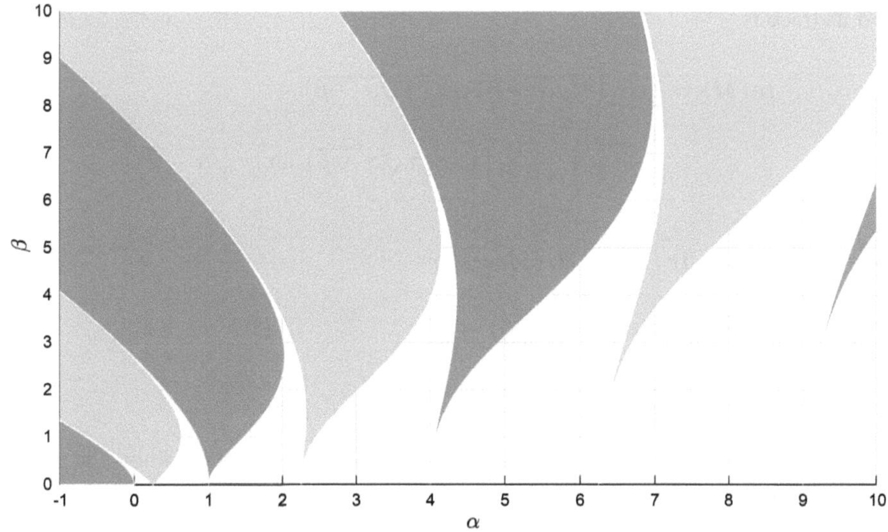

Fig. 3.2 Arnold Tongues for the Mathieu equation. Boundaries of the *blue zones* correspond to a 2π-periodic solution, *red zones* correspond to 4π -periodic solution

3.3.3 Meissner Equation

The exceptional cases in which an analytic solution of the scalar Hill equations may be obtained [40] are: (a) $q(t)$ a train of impulses, (b) $q(t)$ piecewise constant, (c) $q(t)$ piecewise linear and d) $q(t)$ elliptic functions[14]. The case b) for $q(t) = sign(\cos t)$, which corresponds to the Meissner Equation, is particularly simple. It is easy to get the Monodromy matrix analytically, full details are in [44, pp. 276–278]. For $\alpha > \beta \geq 0$, we have

$$M = \begin{pmatrix} \cos\left(\sqrt{\alpha-\beta}\pi\right) & \frac{1}{\left(\sqrt{\alpha-\beta}\right)}\sin\left(\sqrt{\alpha-\beta}\pi\right) \\ -\left(\sqrt{\alpha-\beta}\right)\sin\left(\sqrt{\alpha-\beta}\pi\right) & \cos\left(\sqrt{\alpha-\beta}\pi\right) \end{pmatrix} \bullet$$

$$\bullet \begin{pmatrix} \cos\left(\sqrt{\alpha+\beta}\pi\right) & \frac{1}{\left(\sqrt{\alpha+\beta}\right)}\sin\left(\sqrt{\alpha+\beta}\pi\right) \\ -\left(\sqrt{\alpha+\beta}\right)\sin\left(\sqrt{\alpha+\beta}\pi\right) & \cos\left(\sqrt{\alpha+\beta}\pi\right) \end{pmatrix} =$$

$$\begin{pmatrix} \cos\pi\sqrt{\alpha+\beta}\cos\pi\sqrt{\alpha-\beta} - \left(\sin\pi\sqrt{\alpha+\beta}\sin\pi\sqrt{\alpha-\beta}\right)\frac{\sqrt{\alpha+\beta}}{\sqrt{\alpha-\beta}} & \frac{1}{\sqrt{\alpha+\beta}}\left(\sin\pi\sqrt{\alpha+\beta}\right)\left(\cos\pi\sqrt{\alpha-\beta}\right) + \frac{1}{\sqrt{\alpha-\beta}}\left(\cos\pi\sqrt{\alpha+\beta}\right)\left(\sin\pi\sqrt{\alpha-\beta}\right) \\ -\left(\sin\pi\sqrt{\alpha+\beta}\cos\pi\sqrt{\alpha-\beta}\right)\sqrt{\alpha+\beta} - \left(\cos\pi\sqrt{\alpha+\beta}\sin\pi\sqrt{\alpha-\beta}\right)\sqrt{\alpha-\beta} & \left(\cos\pi\sqrt{\alpha+\beta}\right)\left(\cos\pi\sqrt{\alpha-\beta}\right) - \frac{\sqrt{\alpha-\beta}}{\sqrt{\alpha+\beta}}\left(\sin\pi\sqrt{\alpha+\beta}\right)\left(\sin\pi\sqrt{\alpha-\beta}\right) \end{pmatrix}$$

---

[14] In the case that the periodic function $q(t)$ is an elliptic function, called Lamé Equation.

and its trace is

$$tr\,(M) = 2\cos\left(\pi\sqrt{\alpha+\beta}\right)\cos\left(\pi\sqrt{\alpha-\beta}\right) - \left[\tfrac{\sqrt{\alpha-\beta}}{\sqrt{\alpha+\beta}} + \tfrac{\sqrt{\alpha+\beta}}{\sqrt{\alpha-\beta}}\right]\left(\sin\left(\pi\sqrt{\alpha+\beta}\right)\sin\left(\pi\sqrt{\alpha-\beta}\right)\right)$$

then the condition $|tr\,(M)| = 2$, reduces to:

$$\left|2\cos\left(\pi\sqrt{\alpha+\beta}\right)\cos\left(\pi\sqrt{\alpha-\beta}\right) - \left[\tfrac{\sqrt{\alpha-\beta}}{\sqrt{\alpha+\beta}} + \tfrac{\sqrt{\alpha+\beta}}{\sqrt{\alpha-\beta}}\right]\left(\sin\left(\pi\sqrt{\alpha+\beta}\right)\sin\left(\pi\sqrt{\alpha-\beta}\right)\right)\right| = 2$$

If we make $\beta = 0$ in this last expression $|tr\,(M)| = 2$, in order to know the points at which the Arnold Tongues are born, we get

$$\begin{aligned}\left|2\left[\cos\left(\pi\sqrt{\alpha}\right)\cos\left(\pi\sqrt{\alpha}\right)\right] - 2\left[\sin\left(\pi\sqrt{\alpha}\right)\sin\left(\pi\sqrt{\alpha}\right)\right]\right| &= 2\\ &\Updownarrow\\ \left|2\cos\left(2\pi\sqrt{\alpha}\right)\right| &= 2\\ &\Updownarrow\\ 2\pi\sqrt{\alpha} &= k\pi\end{aligned}$$

which finally leads us to $\boxed{\alpha = \dfrac{k^2}{4} \text{ for } k = 0, 1, 2, \ldots}$

It is customary to assign a number to each Arnold Tongue according to the rule: kth Arnold Tongue touches the α-axis at $\dfrac{k^2}{4}$. We may also say that in the boundaries of even-order Arnold's Tongues there is at least one T-periodic solution, similarly, in the boundaries of odd-order Arnold's Tongues there is at least one $2T$-periodic solution.

Figure 3.3 shows the Meissner equation, i.e., the Hill equation for $q\,(t) = sign\,(\cos t)$.

Remark 3.10 Notice in the Ince-Strutt diagram for the Meissner equation, starting from the 3rd Arnold Tongue, the appearance of zero-length intervals in the α directions; these points are called *Coexistence*, and correspond a parameters in which all the solutions are T-periodic if they belong to an even-order tongue, or $2T$-periodic if they belong to an odd-order tongue. Notice also that coexistence points are exceptional ones.[15]

[15] Chulaevsky [13] justifies the fact that coexistence points are exceptional ones, because: ... 'From a topological point of view the scalar matrices, which correspond to coexistence points, form a subvariety in the variety of 2×2 Jordan Cells.'

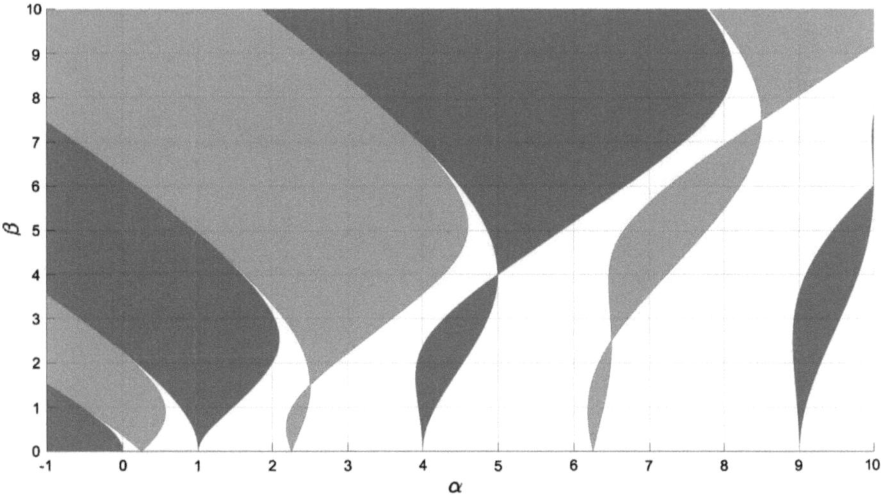

Fig. 3.3 Arnold Tongues for the Meissner equation. Notice the coexistence point approx at $\alpha \approx 2.5$ and $\beta \approx 1.5$

If we introduce the notation

$Tongue\,(i) \triangleq \{(\alpha, \beta) : (\alpha, \beta)$ belongs to the ith Arnold Tongue$\}$,
in the above notation we include their boundaries. We may express compactly the next fundamental property:

Remark 3.11 (Non-intersecting) All the Arnold Tongues are non-intersecting, i.e., $Tongue\,(i) \cap Tongue\,(j) = \phi, \forall\, i \neq j.$

3.3.4 *Critical Lines*

The following question arises: *What happen when we analyze in large intervals* [9] *of* (α, β)? In Fig. 3.4 we show the same diagram for Meissner equation, but now in the intervals $\alpha \in [0, 120]$ and $\beta \in [0, 120]$.

We may observe from Fig. 3.4 that below 45° the region is 'essentially stable' and above this line is 'essentially unstable': this line was designated by Broer [9] as the '*critical line*', and it is independent of the function $q\,(t)$ used, as long as $q\,(t)$ is of zero average and $(\|q\,(t)\|_2 \triangleq \left[\int_0^T |q\,(t)|^2\, dt\right]^{1/2} = \|\cos t\|_2$.

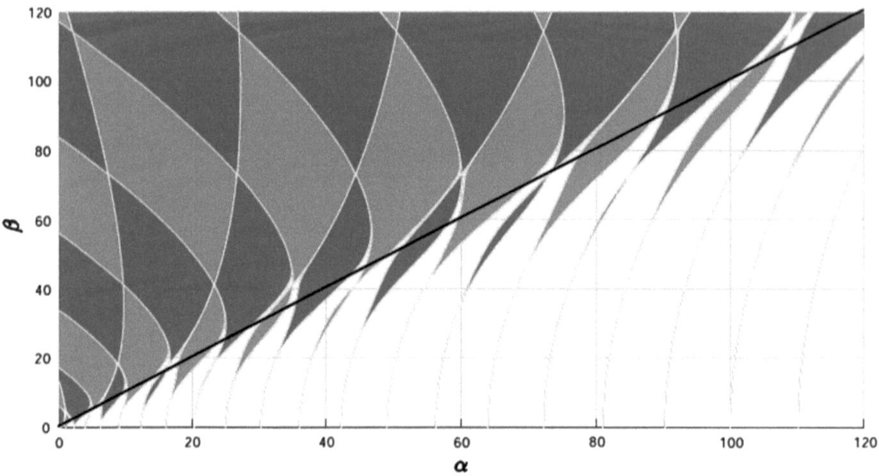

Fig. 3.4 Arnold Tongues for the Meissner equation at a larger scale, notice that above 45° almost everything is unstable

3.3.5 *Forced Hill Equation*

In [41] the T-periodically forced Hill equation was analyzed, i.e.,

$$\ddot{y} + [\alpha + \beta q(t)]\, y = f(t), \text{ where } f(t+T) = f(t). \tag{3.18}$$

It is known [36], that in the stable regions there exists kT-periodic solutions, for $k \geq 3$, of the homogeneous equation (3.18 with $f(t) = 0$), for these values of (α, β) there are two independent kT-periodic solutions. Figure 3.5 shows these kT*-periodic lines* for $k = 3, 5, 9$, and 14.

When we apply a forced periodic term $f(t+T) = f(t)$, of the same period T as the exciting function $q(t)$. In [41], we prove that if $f(t)$ contains a kT-periodic harmonic, then the corresponding kT-periodic line becomes unstable, due to linear resonance.

3.3.6 *Open-Loop Stabilization of Hill Equation*

The last point considered for the scalar Hill equations is: Given a Hill equation for some set of parameters (α_0, β_0) : $\ddot{y} + [\alpha_0 + \beta_0 q(t)]\, y = 0$, where $q(t)$ is a T -periodic function. If the equation for these parameters is unstable, the following problem is posed (Fig. 3.6):

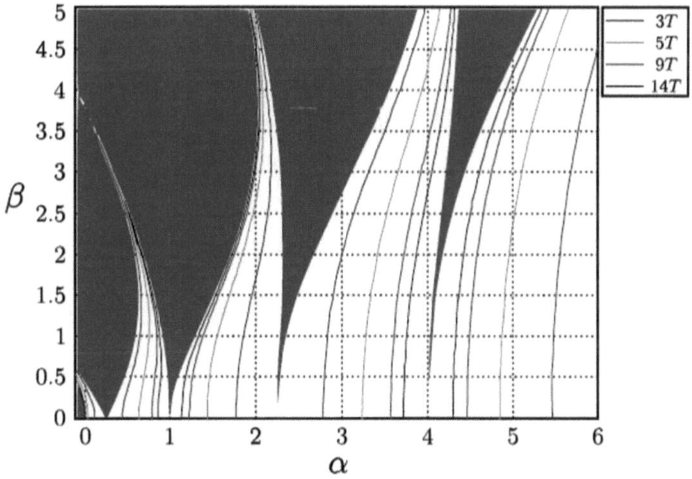

Fig. 3.5 *Colored lines* represent kT-periodic solutions in the homogeneous case, but also *Linear Resonance* for the forced case

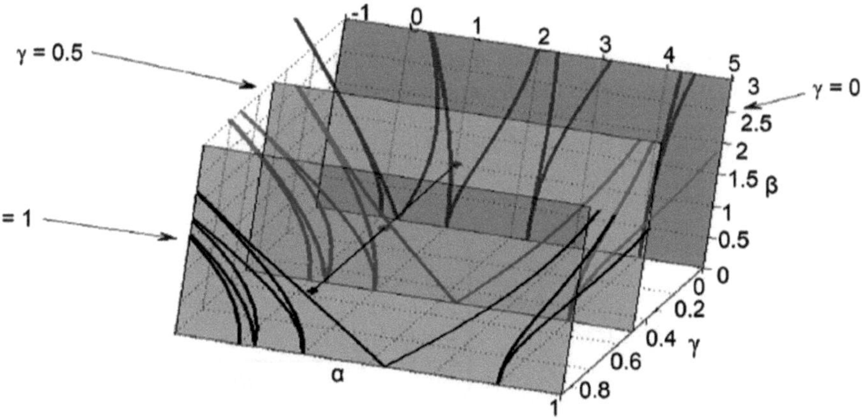

Fig. 3.6 Illustrates graphically the solution proposed to the problem of stabilization of a Hill equation adding to $q(t)$ another T-periodic function

Problem 3.1 There exists another T-periodic function $r(t)$ such that the new Hill equation $\ddot{y} + \left[\alpha_0 + \beta_0 \left(q(t) + \gamma r(t)\right)\right] y = 0$ is stable for the same set of parameters (α_0, β_0)? [16].

Solution 3.1 Suppose (α_0, β_0) is unstable, equivalently $(\alpha_0, \beta_0) \in Tongue(i)$ for some $i \geq 1$, add a T-periodic function $r(t)$ to $q(t)$ such that $(\alpha_0, \beta_0) \in Tongue$

$(i+1)$. This guarantees that if $tr\,[M(\alpha_0, \beta_0, q(t))] > 2$, then $tr\,[M(\alpha_0, \beta_0, q(t) + r(t))] < -2$.[16]

Due to the continuity of $tr\,[M(\alpha_0, \beta_0, q(t))]$, if we perform the convex combination of $q(t)$ and $r(t)$, i.e., $q(t) \to q(t) + \gamma r(t)$, for some $\gamma \in [0, 1]$,

$tr\left[M(\alpha_0, \beta_0, q(t) + \gamma r(t))\right]\big|_{\gamma=0} > 2$ and similarly
$tr\left[M(\alpha_0, \beta_0, q(t) + \gamma r(t))\right]\big|_{\gamma=1} < -2, \Longrightarrow$
$\exists\, \gamma_0 \in (0, 1) : tr\left[M(\alpha_0, \beta_0, q(t) + \gamma r(t))\right]\big|_{\gamma=\gamma_0} = 0$, which corresponds to a stable system.

Notice that the previous solution rests heavily on Remark 11 (Non-Intersecting).

3.4 Hill Equation: Two Degrees of Freedom Case

In the 2 degrees of freedom case, $y(t) \in \mathbb{R}^2$

$$\ddot{y} + [\alpha A + \beta B q(t)]\, y = 0. \tag{3.19}$$

Notice, that we have included matrices $A, B \in \mathbb{R}^{2\times 2}$, and we keep our two-parameter (α, β) in order to make some comparisons with the 1-DOF case.

Similarly to the 1-DOF case, if we define $x = \begin{bmatrix} y \\ \dot{y} \end{bmatrix} \in \mathbb{R}^4$, we may express (3.19) in state space as:

$$\dot{x} = \begin{bmatrix} 0 & I_2 \\ -\alpha A - \beta B q(t) & 0 \end{bmatrix} x = \left\{ \underbrace{\begin{bmatrix} 0 & I_2 \\ -I_2 & 0 \end{bmatrix}}_{J} \underbrace{\begin{bmatrix} \alpha A + \beta B q(t) & 0 \\ 0 & I_2 \end{bmatrix}}_{H(t)} \right\} x. \tag{3.20}$$

In order to the system described by (3.20) be a Hamiltonian $(H(t) = H^T(t))$, the restrictions: $A = A^T$ and $B = B^T$ should be satisfied.

Without loss of generality, we may assume matrix A diagonal with positive entries, which represents the square of the two natural frequencies of the system without parametric excitation. An early publication appears in [26], where the author analyzes a pair of Mathieu equations coupled.

Now there are four multipliers, eigenvalues of the Monodromy matrix, they have symmetry with respect to the real axis because we are treating real matrices, and there is a symmetry with respect to the unit circle because the state transition matrix is symplectic. Now there are three possibilities for multipliers to abandon the unit

---

16 Here $tr\left[M(\alpha_0, \beta_0, q(t) + \gamma r(t))\right]$ refers to the trace of the Monodromy Matrix associated to $\ddot{y} + \left[\alpha_0 + \beta_0 \left(q(t) + \gamma r(t)\right)\right] y = 0$.

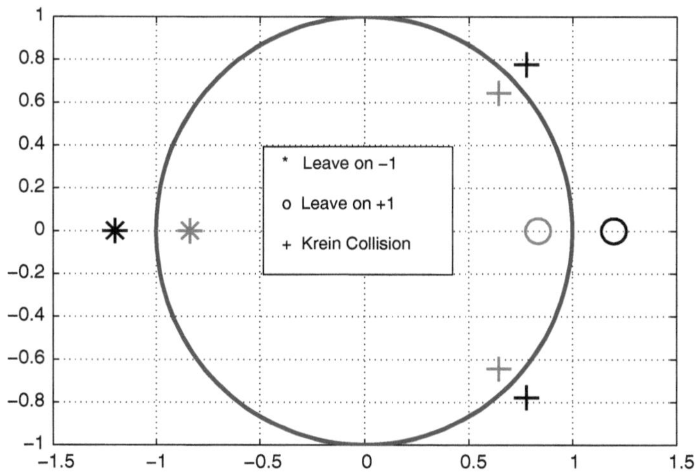

Fig. 3.7 Points where multipliers for a 2-DOF Hamiltonian system may leave the unit circle. Note that for leaving the unit circle at the points ± 1, only two multipliers are required; but to leave the unit circle at $1\measuredangle\theta$, for $\theta \neq 0$ or π, the four multipliers should satisfy the configuration shown

circle, namely: (a) a pair of multipliers leaving at the point $+1$, (b) a pair leaving at the point -1, and (c) two conjugate pairs leaving the unit circle at any point $1\measuredangle\theta$, $\theta \in (0, \pi)$.[17] The cases (a) and (b) already appear in the 1-DOF case; but (c) is a new case for systems having at least 2-DOFs, and it is called *Krein Collision* of the multipliers. Figure 3.7 represents the three case above.

3.4.1 Reduction of the Characteristic Polynomial

Because of the symmetry of the characteristic polynomial of the Monodromy Matrix, $p_M(\mu) = \mu^4 - A\mu^3 + B\mu^2 - A\mu + 1$, is a self-reciprocal polynomial, Howard and MacKay [30] introduced a new variable $\rho = \mu + \mu^{-1}$, in this variable the characteristic polynomial of M reduces to degree 2, and is given by:

$$Q(\rho) = \rho^2 - A\rho + B - 2 \tag{3.21}$$

their corresponding eigenvalues are:

$$\rho_{1,2} = \tfrac{1}{2}\left[A \pm \left(A^2 - 4B + 8\right)^{1/2}\right] \tag{3.22}$$

[17] We use $r\measuredangle\theta$ to represent a complex number with modulus r, and argument θ.

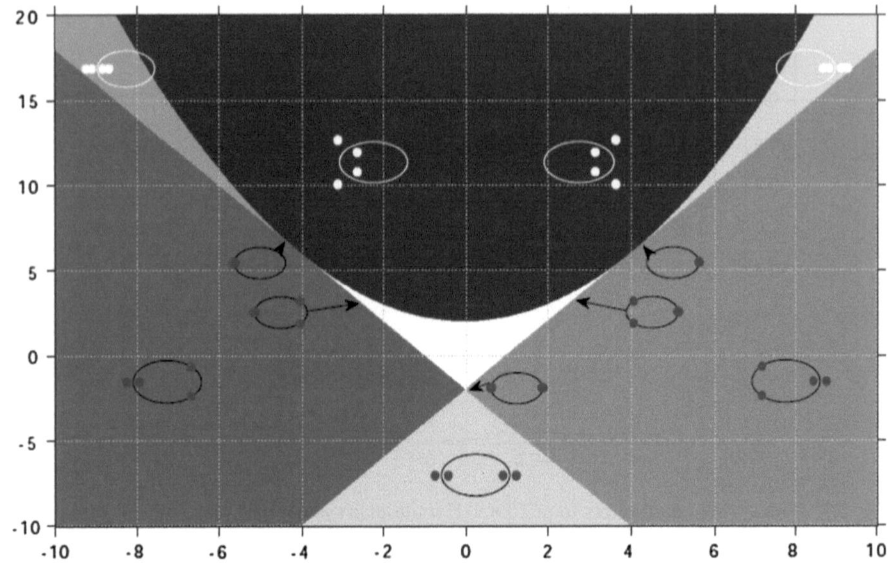

Fig. 3.8 Regions of stability for the reduced polynomial in white. *Red* for some $\mu < -1$; *Green* for some $\mu > 1$; *Yellow* for some $\mu < -1$ and another $\widetilde{\mu} > 1$; *Pink* for two multipliers < 1; *Cyan* for two multipliers > 1. *Blue* for two multipliers not real outside the unit disk

and the eigenvalues of $p_M(\mu)$ are recovery from:

$$\mu = \frac{1}{2}\left[\rho \pm i\left(4-\rho^2\right)^{1/2}\right]. \tag{3.23}$$

Remark 3.12 The symmetry property inherited by the Hamiltonian nature allows to reduce the order in the analysis to one half[18]

The transition boundaries defined when a multiplier leave the unit circle or equivalently using (3.22) are given by two lines and a parabola:

$$\begin{array}{lll} (a) & \mu = +1 & B = +2A - 2 \\ (b) & \mu = -1 & B = -2A - 2 \\ (c) & Krein\ collision & B = A^2/4 + 2. \end{array} \tag{3.24}$$

Figure 3.8 shows the relationships given in (3.24) indicating the typical multiplier positions. The reduced polynomial $Q(\rho) = \rho^2 - A\rho + B - 2$, the white zone represents parameters A, B which produces multipliers of $p_M(\mu) = \mu^4 - A\mu^3 + B\mu^2 - A\mu + 1$ on the unit circle through formula (3.23). Colored regions correspond

[18]In [17] this property is extended to Hamiltonian systems with dissipation, strictly speaking this class of systems is not longer Hamiltonian.

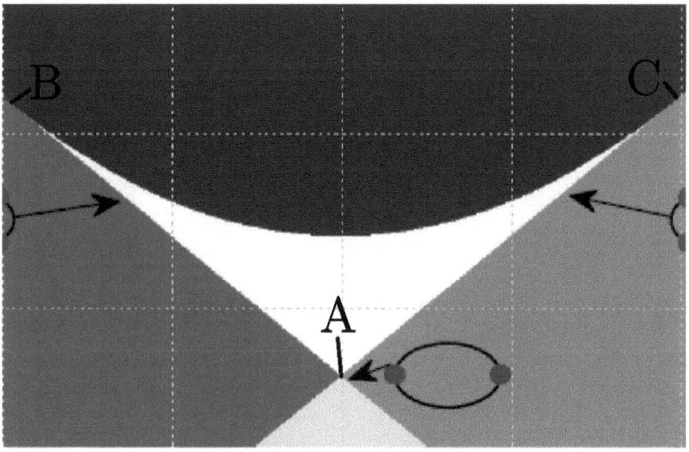

Fig. 3.9 Zoom of the Fig. 3.8 for the point **A** there is at least one T-periodic and at least one $2T$ -periodic solutions; for the point **B** there are at least two linearly independent $2T$-periodic solutions; and for the point **C** there are at least two linearly independent T-periodic solutions

to unstable zones, in the boundary $B = +2A - 2$ there is at least one T-periodic solution; in the boundary $B = -2A - 2$ there is at least one $2T$-periodic solution; in the parabola boundary $B = A^2/4 + 2$ there are a couple of multipliers at some point of the unit circle except ± 1, and have two periodic solutions of any frequency in general (Fig. 3.9).

Remark 3.13 Figures 3.10 and 3.11 use this same colors code.

Using this code of colors, Fig. 3.10 shows the Arnold Tongues for a 2-DOFs Mathieu equation, and Fig. 3.11 shows the Arnold Tongues for a 2-DOFs Meissner equation. For comparison reasons we chose the same matrices A and B.

For the Figs. 3.10 and 3.11 we use the following equation:

$$\dot{x} = \begin{bmatrix} 0 & I_2 \\ -\alpha A - \beta B q\,(t) & 0 \end{bmatrix} x, \tag{3.25}$$

with $A = \begin{bmatrix} 1 & 0 \\ 0 & 2 \end{bmatrix}$ and $B = \begin{bmatrix} 1 & -1 \\ -1 & 2 \end{bmatrix}$ and we have used $q\,(t) = \cos t$ for the Fig. 3.10; and $q\,(t) = sign\,(\cos t)$ for the Fig. 3.11.

Zones of instability occur when some pair of multipliers coincide in the point $+1$ or -1 and after that they leave the unit circle, as in the scalar case, but there are multipliers associated to each of the natural frequencies of the subsystems; therefore there are two possible ways to leave at each of the points ± 1, each one associated with the two subsystems, these Tongues are called *Principal*. But the true new characteristic is that the multiplier now may leave the unit circle at any point $1\measuredangle\theta$ for some $\theta \in (0, \pi)$, these Krein Collision give unstable zones called *Combination*

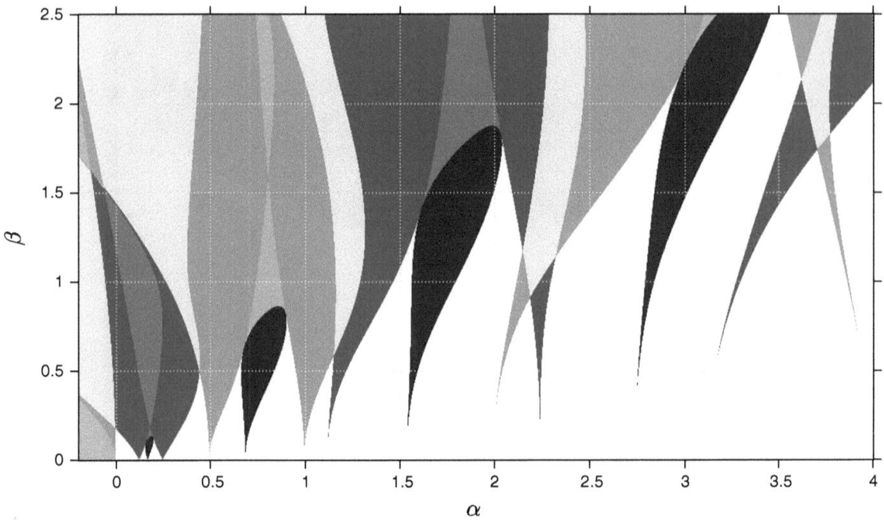

Fig. 3.10 Arnold Tongues for a 2-DOF Mathieu equation (3.25) with $q(t) = \cos t$. *Blue zones* correspond to *Combination Arnold Tongues*

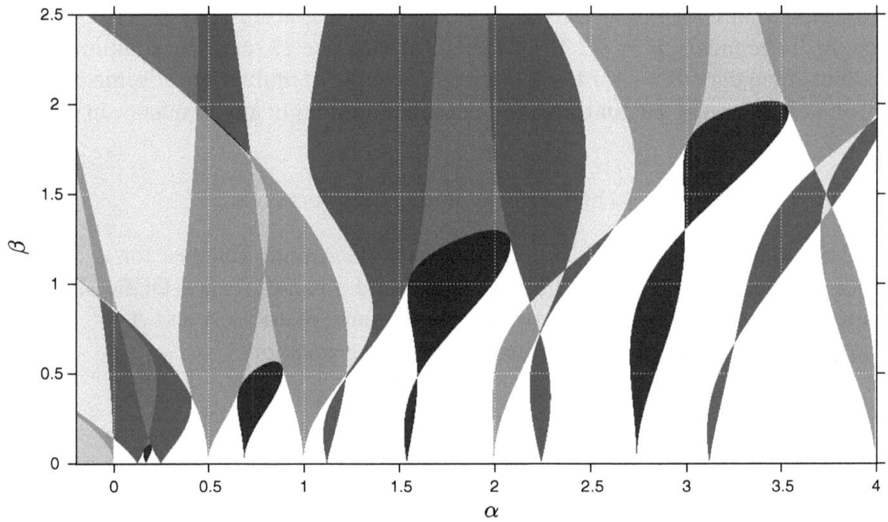

Fig. 3.11 Arnold Tongues for a 2-DOF Meissner equation (3.25) with $q(t) = sign(\cos t)$. *Blue zones* correspond to *Combination Arnold Tongues*

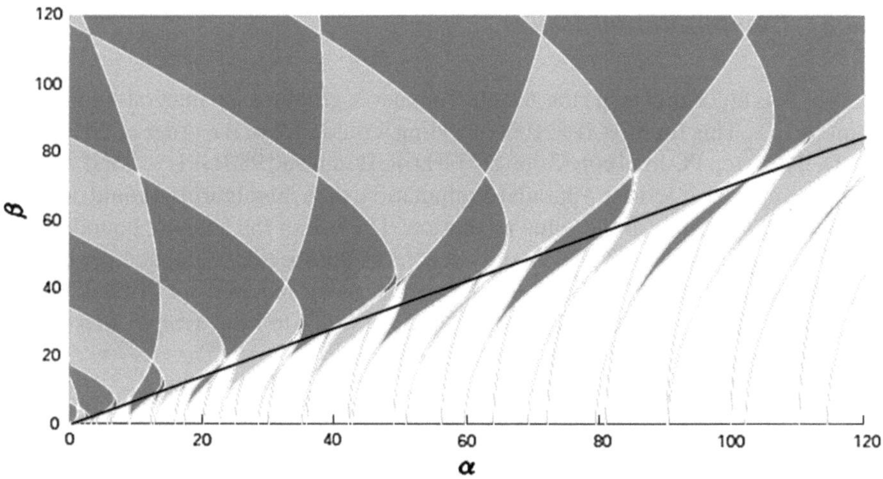

Fig. 3.12 This figure critical line for a 2-DOFs system, now the critical line is 33.6°, arctan(2/3)

Arnold Tongues. There are two kinds of Combination Arnold Tongues: summing or difference, see [48] for further information.

Remark 3.14 In the 2-DOF case, some comments are in order. First, there are Arnold Tongues related to each of the natural frequencies of each subsystem. And of course, the Arnold Tongues associated to each subsystem are non-overlapping. Nevertheless, generically the Arnold Tongues for some rationally independent natural frequencies are intersecting.

Remark 3.15 (Critical lines for 2-DOF) With respect to the critical line for Hill equation of at least 2-DOFs, we claim that generically this critical line now form an angle with the horizontal axis lower than 45°, see Fig. 3.12 for the Lyapunov like equation, i.e., $q(t) = \cos t + \cos 2t$ [15].

Remark 3.16 (Forced 2-DOF Hill Equation) There is no chance to extend the property obtained to simultaneously have parametric instability (Arnold Tongues) and linear resonance in the same Ince-Strutt diagram as in Fig. 3.5.

Remark 3.17 (Open-loop stabilization) In general, it is not possible for 2-DOF systems to take advantage of the non-intersecting property because this property does not exist in n-DOF for $n \geq 2$. Nevertheless for very small values of β, we could develop this idea, see Fig. 3.6.

3.4.2 Computational Issues

The obvious algorithm to get the Arnold Tongues is gridding and integrating for every point (α, β). This for a 1000×1000 gridding, could take of the order of 20 h to run, in a Dell desktop PC Intel core 2 duo 2.8 GHz 4GB ram and 980 GPU's. So, if we keep this naive approach and use a parallel computation, then this algorithm could decrease its speed up to 2 min for the same resolution. The use of the analytic boundaries of the reduced polynomial (3.21) not only decreases the speed of computation, but also gives better precision. For large scales such as those required for critical lines, it is required to use symplectic integrators [25] in order to keep *symplecticity* of the Monodromy matrix, which guarantees good precision.

3.5 Future Work

We are going to enumerate the possible extensions of the scalar and 2-DOF Hill equation.

3.5.1 Generalizations of the Scalar Case

Given the scalar equation:

$$\ddot{y} + [\alpha + \beta q(t)]\, y = 0,$$

it always represents a Hamiltonian System, therefore the only generalization possible is:

- **Scalar I**.- The function $q(t)$ is no longer T-periodic, could be Quasi-Periodic or Almost Periodic. Notice that in this case there is not a Floquet Theorem![19]

A function $q(t)$ is *Quasi-Periodic*, denoted $q(t) \in QP$, if it is the sum of a finite number of Periodic functions of frequency not rationally related; for instance $q(t) = \sin t + \sin \pi t$.

Recall that a function $q(t)$ is *Periodic* if it admits a Convergent *Fourier Series* of the form:

$$q(t) = \sum_{k=-\infty}^{\infty} \rho_k e^{j(k\omega_0)t}$$

[19]There is a reduced for of a Floquet theorem, no factorization is possible, but there is a reducibility part.

Remark 3.18 Notice that there exists a fundamental frequency ω_0 and its harmonics $(k\omega_0)$, rationally related.

A function $q(t)$ is *Almost-Periodic*, denoted $q(t) \in AP$, if it admits a Convergent *Generalized Fourier Series* of the form:

$$q(t) = \sum_{k=-\infty}^{\infty} \rho_k e^{j\varpi_k t}$$

where the sequence $\{\ldots, \rho_k, \rho_{k+1}, \ldots\} \in \ell_2$[20], this condition guarantees the convergence.

3.5.2 Generalizations of the 2-DOFs

Given the 2-DOFs Hill equation expressed in state variables:

$$\dot{x} = \begin{bmatrix} 0 & I_2 \\ -\alpha A - \beta B q(t) & 0 \end{bmatrix} x.$$

The following possible open problems are listed in increasing level of complexity:

- **2-DOFs I**.- If we keep the function $q(t)$ T-periodic, but matrix B is no longer symmetric. It is possible to solve the problem, because we still may apply the Stability consequence of the Floquet Theorem. But the Monodromy Matrix is no longer symplectic. So the same condition for stability, all the multiplier lie on the unit circle, there is no result for strong stability. It is unknown how the multiplier leaves the unit circle.
- **2-DOFs II**.- The function $q(t)$ is no longer T-periodic, then $q(t) \in QP$ or $q(t) \in AP$. Again, there is not a Floquet Theorem, there is no Monodromy Matrix, therefore no analytic condition of the stability based on the multipliers, etc.
- **2-DOFs III**.- The function $q(t)$ is no longer T-periodic, but $q(t) \in QP$ or $q(t) \in AP$; and B is no longer symmetric. The resulting system is no longer Hamiltonian, neither T-Periodic. None of the tools used are valid. Very scarce results exist in this area.
- **IV**.- This item does not belong to n-DOFs Hamiltonian systems, if the dimension of the state equation is odd (never is a Hamiltonian system), i.e.,

$$\dot{x} = A(t)x, \; A(t) = A(t+T), \quad x(t) \in \mathbb{R}^{2n+1}$$

[20]A sequence $\{\ldots, x_k, x_{k+1} \ldots\}$ double infinite belongs to ℓ_2 if $\sum_{k=-\infty}^{\infty} |x_k|^2 = M < \infty$. See for instance [18].

the system is still periodic, we may apply the Floquet Theorem, but for stable or bounded systems there is always a real multiplier at $+1$ or -1, it is unknown how to leave the unit circle, etc.

A final comments about the relationship between Hill equation [36] and Sturm–Liouville Theory [14]. In the scalar case if we write the standard SL Problem

$$\text{Hill Equation rewritten as} \quad \ddot{y} + \beta q(t)\, y = -\alpha y$$

$$\text{with boundary conditions:} \quad \text{(a) } y(0) = y(T) \ \& \ \dot{y}(0) = \dot{y}(T)$$

and the spectral parameter α, we recovery the T-Periodic boundaries of the Arnold Tongues with the above *Boundary Value Problem.*

If we replace the boundary conditions by: (b) $y(0) = -y(T)$ & $\dot{y}(0) = -\dot{y}(T)$, we get as a solution the $2T$-periodic boundaries of the Arnold Tongue [19] or [36].

Completely different is the 2-DOF case, because with the above boundaries conditions (a) and (b) we recovery the boundaries of the principal Arnold Tongues, but the boundaries of the Combination Arnold Tongues do not correspond to some specific boundary condition.

3.6 Conclusions

We may summarize the differences explained in the previous exposition, in the following table:

Property	1-DOF	2-DOF
Multiplier leaving the unit circle	only at $+1$ or -1	any $1\measuredangle\theta,\ \theta \in [0, \pi]$
Arnold Tongues	Not interesting	for high excitation β generically intersecting
Boundaries of the Arnold Tongues	$\exists T$-periodic or $2T$-periodic sols	a) May have T-periodic sol b) May have $2T$-periodic sol c) May have T & $2T$-periodic sol d) A periodic solution noncommensurable with T
Combination Tongues	NO	YES
Critical Lines	45°	less than 45°
Equivalent with SL Problem	YES	NO

Acknowledgements The author wishes to thank to A. Rodriguez, C. Franco and M. Ramirez for a detailed revision of the first draft and also to G. Rodriguez for the computational work.

Long life to Alex!
С наилучшими пожеланиями к Alex!
Salud por el Poznyako!

References

1. Adrianova, L.Y.: Introduction to Linear Systems of Differential Equations. American Mathematical Society, Providence (1995)
2. Arnol'd, V.I.: Remarks on the perturbation theory for problems of Mathieu type. Russ. Math. Surv. **38**(4), 215–233 (1983)
3. Arnold, V.I.: Mathematical Methods of Classical Mechanics. Translated from the 1974 Russian original by K. Vogtmann and A. Weinstein. Corrected reprint of the second (1989) edition. Graduate Texts in Mathematics, vol. 60 (1989)
4. Atkinson, F.V.: Discrete and Continuous Boundary Problems. Academic Press, New York (1964)
5. Belokolos, E.D., Gesztesy, F., Makarov, K.A., Sakhnovich, L.A.: Matrix-valued generalizations of the theorems of Borg and Hochstadt. Lect. Notes Pure Appl. Math. **234**, 1–34 (2003)
6. Bolotin, V.V.: The Dynamic Stability of Elastic Systems. Holden- Day lnc, San Francisco (1964)
7. Borg, G.: Eine umkehrung der Sturm-Liouvilleschen eigenwertaufgabe. An inverse problem of the self-adjoint Sturm-Liouville equation. Acta Math. **78**(1), 1–96 (1946) (In German)
8. Brockett, R.: Finite Dimensional Linear Systems. Wiley, New York (1969)
9. Broer, H., Levi, M., Simo, C.: Large scale radial stability density of Hill's equation. Nonlinearity **26**(2), 565–589 (2013)
10. Brown, B.M., Eastham, M.S., Schmidt, K.M.: Periodic Differential Operators, vol. 228. Springer Science & Business Media, New York (2012)
11. Champneys, A.: Dynamics of parametric excitation. Mathematics of Complexity and Dynamical Systems, pp. 183–204. Springer, New York (2012)
12. Chen, C.T.: Linear System Theory and Design. Oxford University Press, Oxford (1998)
13. Chulaevsky, V.A.: Almost Periodic Operators and Related Nonlinear Integrable Systems. Manchester University Press, Manchester (1989)
14. Coddington, E.A., Levinson, N.: Theory of Ordinary Differential Equations. Tata McGraw-Hill Education, New York (1955)
15. Collado, J.: Critical lines in vectorial Hill's equations. Under preparation (2017)
16. Collado, J., Jardón-Kojakhmetov, H.: Vibrational stabilization by reshaping Arnold Tongues: a numerical approach. Appl. Math. **7**, 2005–2020 (2016)
17. Collado, J., Ramirez, M., Franco, C.: Hamiltonian systems with dissipation and its application to attenuate vibrations. Under preparation (2017)
18. Corduneaunu, C.: Almost Periodic Oscillations and Waves. Springer, Berlin (2009)
19. Eastham, M.S.P.: The Spectral Theory of Periodic Differential Equations. Scottish Academic Press, London (1973)
20. Floquet, G.: Sur les équations différentielles linéaires à coefficients périodiques. Annales scientifiques de l'École normale supérieure **12**, 47–88 (1883)
21. Fossen, T., Nijmeijer, H. (eds.): Parametric Resonance in Dynamical Systems. Springer Science & Business Media, New York (2011)
22. Gantmacher, F.R.: The Theory of Matrices, vol. 2. Chelsea, Providence (1959)

23. Gelfand, I.M., Levitan, B.M.: On the determination of a differential equation from its spectral function. Izvestiya Rossiiskoi Akademii Nauk. Seriya Matematicheskaya **15**(4), 309–360. English Trans. Am. Math. Soc. Trans. **1**(2), 253–304, 1955 (1951)
24. Gelfand, I.M., Lidskii, V.B.: On the structure of the regions of stability of linear canonical systems of differential equations with periodic coefficients. Am. Math. Soc. Transl. **2**(8), 143–181 (1958)
25. Hairer, E., Lubich, C., Wanner, G.: Geometric Numerical Integration: Structure-Preserving Algorithms for Ordinary Differential Equations, vol. 31. Springer Science & Business Media, New York (2006)
26. Hansen, J.: Stability diagrams for coupled Mathieu-equations. Ingenieur-Archiv **55**(6), 463–473 (1985)
27. Hayashi, C.: Forced Oscillations in Non-Linear Systems. Nippon Printing and Publishing Co, Osaka (1953)
28. Hill, G.W.: On the part of the motion of the lunar perigee which is a function of the mean motions of the sun and moon. Acta Math. **8**(1), 1–36 (1886)
29. Hochstadt, H.: Function theoretic properties of the discriminant of Hill's equation. Mathematische Zeitschrift **82**(3), 237–242 (1963)
30. Howard, J.E., MacKay, R.S.: Linear stability of symplectic maps. J. Math. Phys. **28**(5), 1036–1051 (1987)
31. Kalman, D.: Uncommon Mathematical Excursions: Polynomial and Related Realms. Dolciani Mathematical Expositions, vol. 35. Mathematical Association of America (2009)
32. Kapitsa, P.L.: Dynamic stability of the pendulum when the point of suspension is oscillating. Sov. Phys. JETP **21**, 588 (In Russian). English Translation: In Collected works of P. Kapitsa (1951)
33. Khalil, H.K.: Nonlinear Systems. Prentice-Hall, Upper Saddle River (2001)
34. Krein, M.G.: Fundamental aspects of the theory of $\lambda-$ zones of stability of a canonical system of linear differential equations with periodic coefficients. To the memory of AA Andronov [in Russian], Izd. Akad. Nauk SSSR, Moscow, 414–498. English Translation: Am. Math. Soc. Transl. **120**(2), 1–70 (1955)
35. Lyapunov, A.M.: The general problem of the stability of motion. Int. J. Control **55**(3), 531–534. Also: Taylor and Francis (1992). From the original Ph.D. thesis submitted in Kharkov (1882)
36. Magnus, W., Winkler, S.: Hill's Equation. Dover Phoenix Editions. Originally published by Wiley, New York 1966 and 1979 (2004)
37. Marchenko, V.A.: Sturm-Liouville operators and their applications. Kiev Izdatel Naukova Dumka. Engl. Transl. (2011). Sturm-Liouville Operators and Applications, vol. 373. American Mathematical Society (1977)
38. McKean, H.P., Van Moerbeke, P.: The spectrum of Hill's equation. Inventiones mathematicae **30**(3), 217–274 (1975)
39. Meyer, K., Hall, G., Offin, D.: Introduction to Hamiltonian Dynamical Systems and the N-Body Problem, vol. 90. Springer Science & Business Media, New York (2008)
40. Richards, J.A.: Modeling parametric processes–a tutorial review. Proc. IEEE **65**(11), 1549–1557 (1977)
41. Rodriguez, A., Collado, J.: Periodically forced Kapitza's pendulum. In: American Control Conference (ACC), pp. 2790–2794. IEEE (2016)
42. Sanmartín Losada, J.R.: O Botafumeiro: parametric pumping in the middle age. Am. J. Phys. **52**, 937–945 (1984)
43. Seyranian, A.P.: Parametric resonance in mechanics: classical problems and new results. J. Syst. Des. Dyn. **2**(3), 664–683 (2008)
44. Seyranian, A.P., Mailybaev, A.A.: Multiparameter Stability Theory with Mechanical Applications, vol. 13. World Scientific, Singapore (2003)
45. Staržhinskiĭ, V.M.: Survey of works on conditions of stability of the trivial solution of a system of linear differential equations with periodic coefficients. Prik. Mat. Meh. **18**, 469–510 (Russian). Engl. Transl. Am. Math. Soc. Transl. Ser. **1**(2) (1955)

46. Stephenson, A.: On induced stability. Lond. Edinb. Dublin Philos. Mag. J. Sci. **15**(86), 233–236 (1908)
47. van der Pol, B., Strutt, M.J.O.: On the stability of the solutions of Mathieu's equation. Lond. Edinb. Dublin Philos. Mag. J. Sci. **5**(27), 18–38 (1928)
48. Yakubovich, V.A., Starzhinskii, V.M.: Linear Differential Equations with Periodic Coefficients, vol. 1 & 2. Halsted, Jerusalem (1975)

Chapter 4
Sliding Mode Control Devoid of State Measurements

H. Sira-Ramírez, E.W. Zurita-Bustamante, M.A. Aguilar-Orduña and E. Hernández-Flores

Abstract An input-output approach is presented for sliding mode control of linear and nonlinear switched systems of the differentially flat type. Two sliding mode control design options are presented: (1) a Delta-Sigma modulation implementation of a robust continuous output feedback controller design, such as the Active Disturbance Rejection Control scheme and (2) An Integral reconstructor-based approach, involving a suitable linear combinations of iterated integrals of the available system input and the measured output of the system. These reconstructors provide the synthesis of a suitably compensated stabilizing sliding surface coordinate function for stabilization or trajectory tracking problems. The relation of the second approach with Delta-Sigma modulation is also established. Two experimental case studies are presented including nonlinear mechanical plants: An under-actuated convey crane and a DC-motor-Single-link manipulator system.

4.1 Introduction

Classical sliding mode control fundamentally relies on the need to measure, for sliding surface synthesis, the entire state vector of the system (See Utkin [26], Utkin et al. [27], Slotine and Li [25], Edwards and Spurgeon [2]). For this task, asymptotic state observers were originally proposed. In spite of the lack of a clear cut separation principle, in nonlinear systems, the closed-loop stability of an observer-based sliding mode controlled system can be established without major difficulties [2] (See also Khalil et al. [13]). The need to free sliding mode control from the state space formulation may point toward new research areas involving systems described by integral equations, distributed parameter systems, and fractional order systems.

H. Sira-Ramírez (✉) · E.W. Zurita-Bustamante · M.A. Aguilar-Orduña · E. Hernández-Flores
Departamento de Ingeniería Eléctrica, Sección de Mecatrónica,
CINVESTAV-IPN, Av. IPN No. 2508, Col. San Pedro Zacatenco AP 14750,
07300 Ciudad de México, Mexico
e-mail: hsira@cinvestav.mx

E.W. Zurita-Bustamante
e-mail: wzurita@cinvestav.mx

J.B. Clempner and W. Yu (eds.), *New Perspectives and Applications of Modern Control Theory*, https://doi.org/10.1007/978-3-319-62464-8_4

Well-established research trends, known as Active Disturbance Rejection (ADRC), as well as more recent developments, known as Model Free control (MFC) (See Fliess and Join [3], Fliess et al. [4]) have been explored and developed, including numerous theoretical results and interesting practical applications. The main assumption is that all the acting disturbances can be lumped into a simplified, time-varying, uniformly absolutely bounded total disturbance signal affecting the input-output model. Once this simplifying feature is accepted, the estimation of the disturbance proceeds via either powerful algebraic estimation techniques (Sira-Ramírez et al. [21]), or, alternatively, via disturbance observers of extended nature (Han [12], Gao et al. [6, 7], Zheng et al. [29, 30], Zhao and Gao [28]). In this chapter, we will be largely relying in the simplifying philosophy of ADRC and MFC by lumping the description of the effects of exogenous and endogenous additive disturbances and uncertainties into a single additive unstructured term in the input-output model of the system. This is much in the same underlying spirit of the paradigm of scalar sliding mode control, for the sliding surface coordinate function controlled evolution, in using sufficiently large bounded switching amplitudes in front of the lumped, unstructured, yet bounded, uncertainties.

In this chapter, we explore two input-output sliding mode based controller design approaches for differentially flat switched systems. The first one is based on Delta-Sigma modulation implementation of an average designed (output) feedback controller. We explore the use of the robust ADRC technique as an appropriate option. ADRC constitutes a functionally adaptive output feedback controller built on the basis of an Extended State Observer (ESO), online estimating the additive disturbances, and an appropriate canceling controller which is based on the flat output estimated phase variables. The ESO online estimates the, so called total disturbance affecting the input-output system model (exogenous disturbances, neglected state-dependent nonlinearities and the effects of unmodeled dynamics are comprised into a single, additive, time-varying signal). The ADRC controller is designed as if the switched-controlled system input was of a continuous nature. This designed control is taken as an average feedback control which is to be implemented through a Delta-Sigma modulator on the actual switched system. Delta-Sigma modulation is a classical communications dynamical analog signal encoding technique, suitably modified for sliding mode control. The modulator allows continuous control inputs to be converted into equivalent binary switched signals which, based on the equivalent control concept, achieve a qualitatively similar behavior as the average design. An experimental illustrative application of this approach in a convey crane system is here presented.

The second approach here explored is an integral reconstructor-based sliding mode control for the simplified input-to-flat output model of the given nonlinear system. Its relation with Delta-Sigma modulation (Sira-Ramírez [19]) is clearly established. Integral reconstructors were introduced by Fliess et al. [5] in the context of Generalized Proportional Integral Control for the specific case of linear systems. GPI control is a model-based technique which relies on producing structural estimates of the states via finite linear combinations of iterated integrations of input and output variables. The state estimates differ from the actual values in terms describable as "classical disturbances" (constant, ramp, parabolic time-polynomials, etc.). Such errors are easily compensated via linear combinations of a sufficient number of iterated output tracking errors or output stabilization errors. A second experimental illustrative application of this approach in a DC-motor-Single-link manipulator system is presented.

This chapter is organized as follows: In Sect. 4.2 we present Delta-Sigma modulation as a means of translating an average, smooth, feedback controller design into a binary-valued (switched) control input. The switchings emerge from the creation of a sliding regime in the state space of a one-dimensional subsystem which is exogenous to the plant. The switched control input reproduces, in an average (equivalent control) sense, the features of the original smooth, average, designed input signal. ADRC is presented as a viable robust option for an average output feedback controller implementable through a Delta-Sigma modulator for the switched system. Section 4.3 introduces, in a tutorial manner, integral reconstructor-based sliding mode control for pure integration, linear, plants. Section 4.3 also develops the manner in which integral reconstructor-based sliding control is related to flat SISO systems naturally reducible to perturbed pure integration systems. Section 4.4 presents two experimental case studies. An under-actuated convey crane controlled via an ADRC output feedback controller implemented through a Delta-Sigma modulator. An integral reconstructor-based sliding mode controller is developed for a single-link manipulator including the complete DC motor dynamics. Section 4.5 is devoted to the conclusions of this chapter.

4.2 Delta-Sigma Modulation

In this section, we provide the basics of Delta-Sigma modulation. These modulators constitute analog-to-binary conversion tools, specifically tailored for sliding mode control. They translate a continuous average feedback control law design, possibly dynamic in nature, into a switched (i.e., binary-valued) signal, whose average behavior precisely recovers the features of the continuous closed-loop behavior in an equivalent control sense. The modulators achieve such a translation on the basis of the creation of a sliding regime on a state space which is exogenous to the state space of the plant system and of minimal dimension.

Consider we are given a sufficiently smooth scalar signal, $\mu(t)$, which is bounded by the closed interval $(-W, W) \subset \mathbb{R}$. i.e., $\sup_t |\mu(t)| < W$. Suppose it is desired to produce a binary-valued signal $u(t) \in \{-W, W\}$ so that the average value of $u(t)$, in a sliding mode control sense, coincides with $\mu(t)$ (Fig. 4.1).

Consider the Delta-Sigma modulation encoding circuit, shown in Fig. 4.2. The following relations describe the mathematical model of such a system

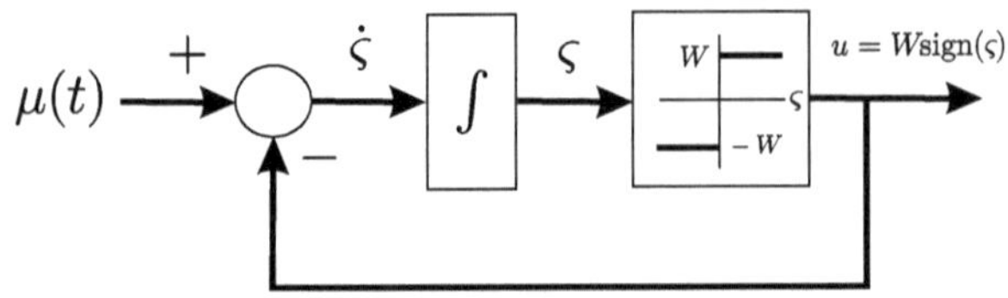

Fig. 4.1 Delta-Sigma modulation of input signal μ

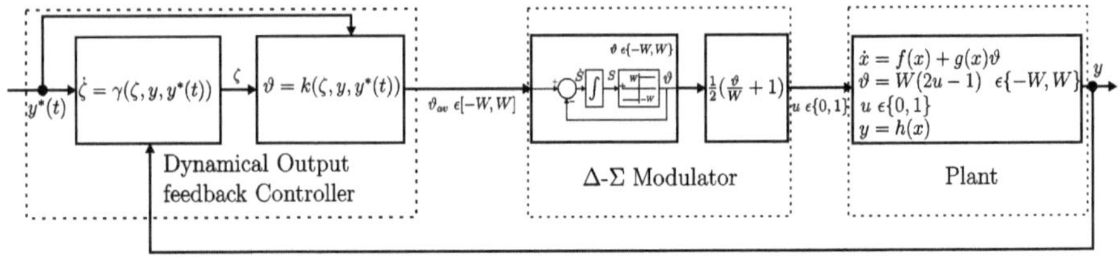

Fig. 4.2 Delta-Sigma modulation implementation of average output feedback controller design

$$\begin{aligned} \dot{\varsigma} &= \mu(t) - u(t) = \mu(t) - W\text{sign}(\varsigma) \\ \sup_t |\mu(t)| &< W, \quad u(t) \in \{-W, W\} \ \forall \ t \end{aligned} \tag{4.1}$$

The signal ς is known as the encoding error. The switching amplitudes $\{-W, W\}$ are addressed as the encoding limits, the signal $u(t)$ is the locally coded signal.

A sliding regime globally exists on the set, $\varsigma = 0$, whose corresponding ideal sliding dynamics is described by $\mu(t) = u_{eq}(t)$.

The proof of this result is quite simple. Consider the product $\varsigma\dot{\varsigma}$:

$$\varsigma\dot{\varsigma} = \varsigma(\mu(t) - u(t)) = \varsigma\mu(t) - \varsigma u(t) \tag{4.2}$$

A sliding regime globally exists on $\varsigma = 0$ if and only if $\varsigma\dot{\varsigma} < 0$ in the open hemispaces delimited by the sliding manifold $\varsigma = 0$. Indeed, the choice $u = W\text{sign}(\varsigma)$ yields $\varsigma\dot{\varsigma} < 0$. On the manifold, $\varsigma = 0$, the ideal sliding dynamics is obtained from the invariance condition: $\dot{\varsigma} = 0$, yielding $u_{eq}(t) = \mu(t)$ for all t. The initial conditions of ς are irrelevant in the previous analysis and the sliding motion, on $\varsigma = 0$, takes place globally with respect to the possible initial conditions.

The use of Delta-Sigma modulators in the control of dynamical systems, subject to binary-valued control inputs, is immediate by producing an average feedback controller design as if the control input to the system were of a smooth nature. After establishing the uniform hard bounds of the designed feedback controlled input signal, this signal is to be injected to a Delta-Sigma modulator with suitable switching limits (encoding limits) which are, at least, as large as the obtained bounds.

4.2.1 Use of Delta-Sigma Modulation in Average Controller Design Implementation for Switched Systems

For simplicity, we consider that the given nonlinear plant is a Single-Input Single-Output (SISO) nonlinear system and remark that the results are extendable to Multi-Input Multi-Output (MIMO) systems and infinite dimensional systems characterized by inputs with finite (known) delays.

Suppose we are given the following smooth nonlinear (SISO) system,

$$\begin{aligned} \dot{x} &= f(x) + g(x)\vartheta, \quad x \in \mathbb{R}^n, \vartheta \in \mathbb{R} \\ y &= h(x), \quad y \in \mathbb{R} \end{aligned} \tag{4.3}$$

for which it is known that the dynamic scalar output feedback control law

$$u = k(\zeta, y, y^*(t)), \quad \dot{\zeta} = \gamma(\zeta, y, y^*(t)), \quad \zeta \in \mathbb{R}^{\mu}, \quad \mu \leq n \tag{4.4}$$

globally accomplishes a smooth, possibly in an exponentially asymptotic manner, output reference trajectory tracking task, i.e., $\lim_{t\to\infty} y(t) = y^*(t)$. Suppose, furthermore, that from a sufficiently large set of initial conditions, the control input values, $\vartheta(t)$, remain uniformly bounded by the compact interval of the real line, $[-W, W]$, for some suitable but fixed amplitude $W > 0$.

One of the questions that concern us here is: How can we take advantage of the above knowledge if at the outset, for some valid reason, the actual plant control input signal, $u(t)$, were restricted to take values on the binary-valued set: $\{-W, W\}$, for all times? In other words, is there a sliding surface coordinate function s on which a sliding regime exists, such that its corresponding equivalent control $u_{eq}(t)$ values coincide with the obtained $u(t)$ for which the ideal sliding dynamics is none other than the continuously achieved closed-loop behavior?

The answer to this question cannot be found in the realm of the composite state space $(x, \zeta) \in \mathbb{R}^{n+\mu}$ since any forced algebraic restriction of the extended state variables, (x, ζ), disrupts the very nature of the average feedback control law, thus producing a different closed-loop behavior.

Consider the following simple exogenous relations for, respectively, some switched input signal, ϑ, and a sufficiently continuous average signal, ϑ_{av}, satisfying: $\vartheta(t) \in \{-W, W\}$, $\forall t$, and, $\vartheta_{av}(t) \in [-W, W]\ \forall t$,

$$\dot{s} = \vartheta_{av} - \vartheta, \quad \vartheta = W\text{sign(s)} \tag{4.5}$$

Clearly, a sliding regime exists on $s = 0$ provided $\vartheta_{av} \in (-W, W)$. Moreover, the invariance conditions, $s = 0$, $\dot{s} = 0$, imply $\vartheta_{eq} = \vartheta_{av}$. Notice that s is an exogenous variable, foreign, to the considered plant system, when u is taken to be the actual (switched) control input and ϑ_{av} is the control signal resulting from the continuous feedback design. The equivalent control of the exogenous sliding regime, on $s = 0$, coincides with the continuous values of the designed average feedback control input.

The above result is the basis of Delta-Sigma modulation based control strategies. It just makes available to the designer a means of translating the outcome of his favorite methodology, for synthesizing smooth feedback controllers, into an implementable control strategy with binary-valued control input restrictions which equivalently generate a good average approximation to the designed (and possibly tested) smooth responses.

The following figure depicts the Delta-Sigma modulation scheme, acting on an average dynamic output feedback controller design for a nonlinear system.

4.2.2 Active Disturbance Rejection as an Average Controller for a Switched System

ADRC constitutes a mature research area, with a wealth of theoretical and applications results [10, 11, 14, 29]. Typically, a given nonlinear feedback linearizable plant description, of the form,

$$\dot{x} = f(x) + g(x)\vartheta, \quad x \in \mathbb{R}^n, \quad y = h(x), \quad \vartheta \in \mathbb{R}, \quad y \in \mathbb{R} \tag{4.6}$$

which is monovariable in nature $u \in \mathbb{R}$, is replaced by a simpler input-to-flat output, perturbed, pure integration model. This is carried out to facilitate the controller design. The preferred type of model used is centered around the class of input-output models with a simple structure, such as in Ramírez–Neria et al. [18]. The additive state-dependent terms, ignored in the simplification process, are lumped into a total disturbance term. Generally speaking, the class of input-to-flat output models treated is of the form

$$y^{(n)} = \eta(y, \dot{y}, \cdots, y^{(n-1)}) + \gamma(y, \dot{y}, \cdots, y^{(n-1)})\vartheta \tag{4.7}$$

We assume that the input gain function is either constant or perfectly known. In traditional ADRC, as well as in MFC, a phenomenological model is used to replace the plant in all the design considerations by a simpler plant. In this case, the system may be replaced by the following system,

$$y^{(n)} = \gamma(\cdot)\vartheta + \xi(t) \tag{4.8}$$

where $\gamma(\cdot)$ is either a known function of a known constant gain, replacing the actual gain $\gamma(y, \dot{y}, \cdots, y^{(n-1)})$ and $\xi(t)$ represents the total disturbance including the non-linear state-dependent term $L_f^n y(x)$.

The observer-based average ADRC output feedback controller is prescribed as,

$$\begin{aligned}
\vartheta_{av} &= \frac{1}{\gamma(\cdot)}\left[(y^*(t))^{(n)} - z - \sum_{i=0}^{n-1}\kappa_i\left(y_i - [y^*(t)]^{(i)}\right)\right] \\
\dot{y}_0 &= y_1 + \lambda_n(y - y_0) \\
\dot{y}_1 &= y_2 + \lambda_{n-1}(y - y_0) \\
&\vdots \\
\dot{y}_{n-1} &= \gamma\vartheta_{av} + z + \lambda_1(y - y_0) \\
\dot{z} &= \lambda_0(y - y_0)
\end{aligned} \tag{4.9}$$

The last $n + 1$ differential equations, correspond with a total disturbance observer known as the Extended State Observer (ESO). This output feedback controller, whose validity has been rigorously demonstrated from the theoretical viewpoint in numerous articles [10, 11, 14, 29], as well as in many laboratory case studies (see Ramírez-Neria et al.[16–18, 24], J. Linares et al. [22, 23], Cortés-Romero et al.[1].) is known as the ESO based ADRC controller.

The robust average ADRC design, shown above, may be implemented for the trajectory tracking of a binary-valued switched system through a Delta-Sigma modulator, provided the smooth average control input $\vartheta_{av}(t) \in [-W, +W]$ for all t, for some sufficiently large real number W.

An actual implementation of this controller for a binary-valued control input is discussed in the second experimental case study presented in this chapter.

We remark that the area of MFC exhibits many similar philosophical features with ADRC (See Fliess and Join [3], Fliess et al. [4]). In MFC a more radical simplification of the plant is introduced by systematically reducing the order of the uncertain plant to either 1 or 2, with a well engineered constant gain and the use of algebraic estimation techniques for the online estimation of the total disturbance and, possibly, the input gain. However, in output feedback ADRC of flat systems, the order n of the plant system is preserved. If the control input gain $L_g L_f^{n-1} y(x)$ is unknown, or nonconstant, then it is replaced by a constant gain and the disturbance is estimated by means of a linear extended controller, otherwise, the known gain is used.

4.3 Integral Reconstructors and Sliding Mode Control for Pure Integration Systems

Here we review, in a tutorial fashion, the rôle of integral reconstructors (See Fliess et al. [5]) in an elementary example, constituted by a third order, pure integration, plant and its natural connections with Delta-Sigma modulation. Consider the third order integration plant,

$$y^{(3)} = \vartheta, \quad \vartheta \in \{-W, W\} \tag{4.10}$$

Suppose it is desired to track a smooth output reference trajectory signal $y^*(t)$ with the available, binary-valued, control input. The corresponding nominal control input is, according to the inversion of the model, $\vartheta^*(t) = [y^*(t)]^{(3)}$

The tracking error dynamics for e_y satisfies,

$$e_y^{(3)} = e_\vartheta, \quad e_y = y - y^*(t), \quad e_\vartheta = \vartheta - \vartheta^*(t) \tag{4.11}$$

Integral reconstructors for the second and first order time derivative of y are readily devised as,

$$\widehat{\ddot{e}}_y = \int_0^t e_\vartheta(\sigma)d\sigma, \quad \widehat{\dot{e}}_y = \int_0^t \int_0^\sigma e_\vartheta(\sigma_1)d\sigma_1 d\sigma \tag{4.12}$$

where, the actual values of the output time derivatives satisfy:

$$\begin{aligned} \ddot{e}_y &= \widehat{\ddot{e}}_y + \ddot{e}_y(0), \\ \dot{y} &= \widehat{\dot{e}}_y + \ddot{e}_y(0)t + \dot{e}_y(0) \end{aligned} \tag{4.13}$$

The presence of constant and time polynomial ramp errors in the structural estimates, or integral reconstructors, of the time derivatives of y, prompts the use of integral output tracking error compensation. A suitably compensated sliding surface coordinate function is proposed as

$$\begin{aligned} s &= \widehat{\ddot{e}}_y + k_3\widehat{\dot{e}}_y + k_2 e_y + k_1 \int_0^t e_y(\sigma)d\sigma + k_0 \int_0^t \int_0^\sigma e_y(\sigma_1)d\sigma_1 d\sigma \\ &= \int_0^t e_\vartheta(\sigma)d\sigma + k_3 \int_0^t \int_0^\sigma e_\vartheta(\sigma_1)d\sigma_1 d\sigma + k_2 e_y \\ &\qquad + k_1 \int_0^t e_y(\sigma)d\sigma + k_0 \int_0^t \int_0^\sigma e_y(\sigma_1)d\sigma_1 d\sigma \\ &= k_2 e_y + \int_0^t \left[e_\vartheta(\sigma) + k_1 e_y(\sigma) + \int_0^\sigma \left(k_3 e_\vartheta(\sigma_1) + k_0 e_y(\sigma_1)\right) d\sigma_1 \right] d\sigma \end{aligned} \tag{4.14}$$

An expression for $\dot{s}$, to be used later, is readily obtained as:

$$\begin{aligned} \dot{s} &= e_\vartheta + k_3 \int_0^t e_\vartheta(\sigma)d\sigma + k_2\dot{e}_y + k_1 e_y + k_0 \int_0^t e_y(\sigma)d\sigma \\ &= (\vartheta - \vartheta^*(t)) + k_3(\ddot{e}_y - \ddot{e}_y(0)) + k_2\dot{e}_y + k_1 e_y + k_0 \int_0^t e_y(\sigma)d\sigma \\ &= (\vartheta - \vartheta^*(t)) + k_3\ddot{e}_y + k_2\dot{e}_y + k_1 e_y + k_0\rho \end{aligned}$$

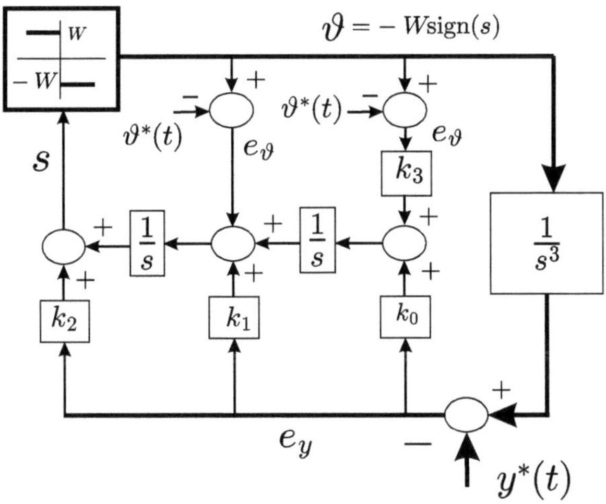

Fig. 4.3 Integral reconstructors-based sliding mode control scheme for third order integration plant

$$\dot{\rho} = e_y, \quad \rho(0) = -\frac{k_3}{k_0}\ddot{e}_y \tag{4.15}$$

A switching policy aiming at creating a sliding regime on $s = 0$ is obtained by exercising all the available control possibilities in the perusal of the condition $s\dot{s} < 0$. The switching policy is clearly given by

$$\vartheta = -W\text{sign}(s). \tag{4.16}$$

This switching policy, in combination with the last expression found for the sliding surface coordinate function s, yields the interpretation of the sliding mode control scheme of Fig. 4.3.

4.3.1 Relation with Delta-Sigma Modulation

Using simple block diagram algebra manipulation, a reinterpretation of the proposed controller scheme, in terms of an average controller passing through a Delta-Sigma modulator, is presented in Fig. 4.4.

Define,

$$\rho_1 = \int_0^t e_y(\sigma)d\sigma + \frac{k_0}{k_1}\int_0^t \rho_2(\sigma)d\sigma + \frac{\ddot{e}_y(0) + k_3\dot{e}_y(0)}{k_1}$$

$$\rho_2 = \int_0^t y(\sigma)d\sigma + \frac{k_3}{k_1}\ddot{e}_y(0) \tag{4.17}$$

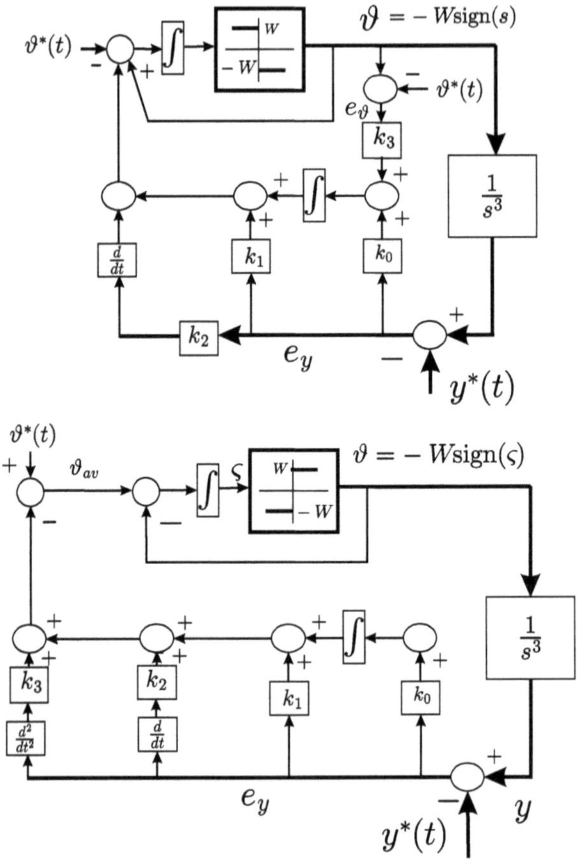

Fig. 4.4 Reinterpretation of integral reconstructor-based sliding mode control scheme for third order integration plant, in terms of an average state feedback controller and a Delta-Sigma modulator

It follows that on $\sigma = 0$ the dynamics of the tracking error is governed by,

$$\begin{aligned}
\ddot{e}_y &= -k_3\dot{e}_y - k_2 e_y - k_1\rho_1 \\
\dot{\rho}_1 &= y + \frac{k_0}{k_1}\rho_2, \quad \rho_1(0) = \frac{\ddot{e}_y(0) + k_3\dot{e}_y(0)}{k_1} \\
\dot{\rho}_2 &= y, \quad \rho_2(0) = \frac{k_3}{k_1}\ddot{e}_y(0)
\end{aligned} \tag{4.18}$$

The characteristic equation of the linear system is obtained as:

$$p(\frac{d}{dt})e_y = \left[\left(\frac{d}{dt}\right)^4 + k_3\left(\frac{d}{dt}\right)^3 + k_2\left(\frac{d}{dt}\right)^2 + k_1\left(\frac{d}{dt}\right) + k_0\right]e_y = 0 \tag{4.19}$$

The set of coefficients $\{k_3, k_2, k_1, k_0\}$ is chosen as a Hurwitz set so as to guarantee global exponential asymptotic stability of e_y to the origin of the tracking error coordinates.

The equivalent control is implicitly obtained from the invariance conditions $s = \dot{s} = 0$. This yields:

$$\begin{aligned} e_{\vartheta,eq} &= -k_3 \int_0^t e_\vartheta(\sigma)d\sigma - k_2\dot{e}_y - k_1 e_y - k_0 \int_0^t e_y(\sigma)d\sigma \\ &= -k_3\ddot{e}_y - k_2\dot{e}_y - k_1 e_y - k_0\rho \\ \dot{\rho} &= e_y, \quad \rho(0) = -\frac{k_3}{k_0}\ddot{e}_y(0) \end{aligned} \tag{4.20}$$

The ideal sliding dynamics for the output tracking error, $e_y^{(3)} = e_{\vartheta,eq}$, is then given by,

$$\begin{aligned} e_y^{(3)} &= -k_3\ddot{e}_y - k_2\dot{e}_y - k_1 e_y - k_0\rho \\ \dot{\rho} &= e_y, \quad \rho(0) = -\frac{k_3}{k_0}\ddot{e}_y(0) \end{aligned} \tag{4.21}$$

which has the same characteristic equation as the tracking error dynamics previously found for the condition $s = 0$.

4.3.2 Sliding Mode Control of Differentially Flat Systems

Consider the smooth nonlinear system

$$\dot{x} = \tilde{f}(x) + \tilde{g}(x)\left(W[2u - 1]\right) = f(x) + g(x)u, \quad x \in \mathbb{R}^n, u \in \{0, 1\} \tag{4.22}$$

Clearly $f(x) = \tilde{f}(x) - \tilde{g}(x)W$, $g(x) = 2W\tilde{g}(x)$. Alternatively, the results below also applied without major modifications to systems controlled by a binary-valued control input signal, ϑ, taking values in the set $\{-W, W\}$. The system would read as

$$\dot{x} = f(x) + g(x)\vartheta, \quad x \in \mathbb{R}^n, \vartheta \in \{-W, W\} \tag{4.23}$$

Independently of the nature of the control signal, assume the smooth vector fields f and g, locally, satisfy the exact feedback linearization conditions:
(a) The set of vector fields: $[g, ad_f g, \cdots, ad_f^{n-2} g]$ span an involutive distribution
(b) $[g, ad_f g, \cdots ad_f^{n-2} g\; ad_f^{n-1} g]$ is a full rank matrix.

The set of vector fields $\{f, g\}$ are said to conform an exactly feedback linearizable pair. The system will be said to be flat. The flat output may be (non-uniquely) found via its gradient. The following result is well-known (See Sira-Ramírez and Agrawal [20]). We call the matrix $K = [g, ad_f g, \cdots ad_f^{n-2} g\; ad_f^{n} g]$, the Kalman matrix.

Proposition 4.1 *If conditions (a) and (b) are locally satisfied, the gradient of the flat output, y, is given by*

$$\frac{\partial y(x)}{\partial x^T} = [0, 0, ..., 0, 1]\left[g, ad_f g, \cdots, ad_f^{n-2} g, ad_f^{n-1} g\right]^{-1} \tag{4.24}$$

i.e., the row gradient of the scalar flat output is given by the last row of the inverse of the Kalman matrix.

In general, from the previous statement, the flat output is not uniquely defined and there are, indeed, infinitely many scalar outputs which qualify as a flat output.

The map

$$\underline{y}(x) = \begin{bmatrix} y(x) \\ \dot{y}(x) \\ \vdots \\ y^{(n-1)}(x) \end{bmatrix} = \Phi(x) = \begin{bmatrix} y(x) \\ L_f y(x) \\ \vdots \\ L_f^{n-1} y(x) \end{bmatrix} \tag{4.25}$$

is a local diffeomorphism on $\mathbb{R}^n$.

The nonlinear system is equivalent to the following integration system:

$$y^{(n)} = L_f^n y(\Phi^{-1}(\underline{y})) + L_g L_f^{(n-1)} y(\Phi^{-1}(\underline{y}))\vartheta \tag{4.26}$$

where $L_\phi^j y(x)$ denotes the j times iterated directional derivative of the scalar function $y(x)$ along the vector field $\phi(x)$. Clearly, $L_g L_f^{(n-1)} y(\Phi^{-1}(\underline{y}))$ is locally assumed to be nonzero.

Customarily, the flat output is naturally provided with a key physical meaning, which eases the understanding of the underlying physical system, it trivializes the controller design and it is enormously helpful in off-line planning of physically sound, meaningful, trajectories. We assume that the flat output y is a measured variable of the system (See the Case Studies section). Moreover, thanks to the differential parametrization property, all states are differentially parameterizable in terms of the flat output and a finite number of its time derivatives. This off-line adverts of possible nominal state and input restrictions violations for the desired nominal flat output trajectory.

The system, describing the dynamics of the scalar flat output, y, may be locally written as:

$$y^{(n)} = \eta(y, \dot{y}, \cdots, y^{(n-1)}) + \gamma(y, \dot{y}, \cdots, y^{(n-1)})u, \quad \gamma(y, \dot{y}, \cdots, y^{(n-1)}) \neq 0 \tag{4.27}$$

If the gain function $\gamma(y, \dot{y}, \cdots, y^{(n-1)})$ is known, then the input coordinate transformation $\gamma(y, \dot{y}, \cdots, y^{(n-1)})u = \vartheta$ yields a system of the form:

$$y^{(n)} = \vartheta + \eta(y, \dot{y}, \cdots, y^{(n-1)}) \tag{4.28}$$

In line with the MFC and ADRC philosophy, we would consider such a system as a pure integration system, with all exogenous disturbances and the nonlinearities in the term: $\eta(y, \dot{y}, \cdots, y^{(n-1)})$, playing the rôle of an unknown, unstructured, time-varying total disturbance, $\xi(t)$. i.e., sliding mode considerations will be based on the simplified perturbed pure integration model

$$y^{(n)} = \vartheta + \xi(t) \tag{4.29}$$

A rigorous justification of this simplification procedure, from a powerful combination of differential algebra and functional analysis, can be found in Fliess and Join [3] (See also: Han [12], and Guo and Jin, [8, 9] for similar approaches). If the nonlinear input gain function, $\gamma(y, \dot{y}, \cdots, y^{(n-1)})$, is not known, the ADRC procedure indicates to go ahead and replace the unknown gain function by an appropriate constant based on engineering judgment. In the MFC approach, algebraic identification is always a viable, and quite effective, online option (See Morales et al. [15]). In our two illustrative experimental case studies, presented below, the gains are known and constant and, hence, we will have no need to use online algebraic identification techniques nor of the input gain replacement option.

4.4 Experimental Case Studies

4.4.1 An Inverted Pendulum on a Car System

Consider an inverted pendulum on a car system shown in Fig. 4.5. The mass of the car is M, while L is the length of the pendulum with concentrated mass m, θ denotes the angle between the pendulum rod and the vertical direction.

4.4.1.1 Mathematical Model of the Inverted Pendulum on a Car

The model of the system is derived from the Euler–Lagrange equations, and it is:

$$(M+m)\ddot{x} - mL\ddot{\theta}\cos(\theta) + mL\dot{\theta}^2\sin(\theta) = F \tag{4.30}$$
$$-mL\cos(\theta)\ddot{x} + mL^2\ddot{\theta} - mgL\sin(\theta) = 0 \tag{4.31}$$

Writing the model in matrix form, the equations of the inverted pendulum on a car are:

$$M(q)\ddot{q} + C(q, \dot{q}) + G(q) = u(F) \tag{4.32}$$

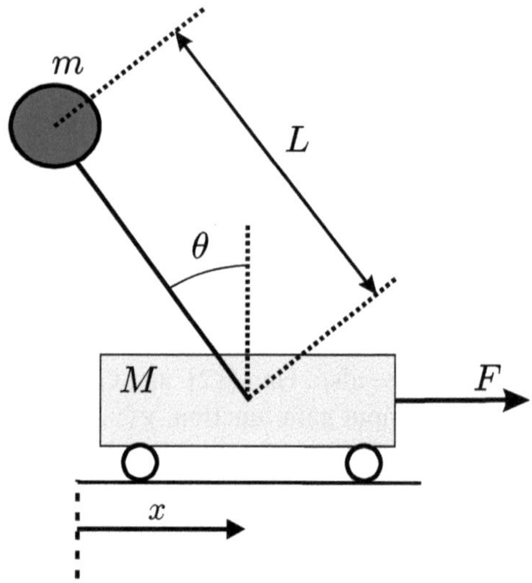

Fig. 4.5 An inverted pendulum on a car system

where

$$q = \begin{bmatrix} x \\ \theta \end{bmatrix} \quad M(q) = \begin{bmatrix} M+m & -mL\cos(\theta) \\ -mL\cos(\theta) & mL^2 \end{bmatrix} \quad C(q,\dot{q}) = \begin{bmatrix} 0 & mL\sin(\theta)\dot{\theta} \\ 0 & mL^2 \end{bmatrix} \tag{4.33}$$

$$G(q) = \begin{bmatrix} 0 \\ -mgL\sin(\theta) \end{bmatrix} \quad u(F) = \begin{bmatrix} F \\ 0 \end{bmatrix} \tag{4.34}$$

The input force (F) is obtained from a DC motor. The control input torque can be expressed as a function of the motor voltage through the relation: $F = \frac{k_t}{rR}V(t)$ where R is the armature resistance of the motor, k_t is the motor torque constant and r is the radius of the mesh that is used to move the car.

Now, we linearize the system around the stable equilibrium point

$$\bar{x} = 0; \quad \bar{\dot{x}} = 0; \quad \bar{\theta} = 0; \quad \bar{\dot{\theta}} = 0; \quad \bar{V} = 0; \tag{4.35}$$

In this case, the linearization is around the origin, all incremental variables coincide, then, with the actual variables. The linearized system is written as,

$$(M+m)\ddot{x}_\delta - mL\ddot{\theta}_\delta = \frac{k_t}{Rr}V_\delta(t) \tag{4.36}$$

$$-mL\ddot{x}_\delta + mL^2\ddot{\theta}_\delta = mgL\theta_\delta \tag{4.37}$$

where $x_\delta = x - \bar{x}$, $\theta_\delta = \theta - \bar{\theta}$ and $V_\delta = V - \bar{V}$

The linearized system in (4.36) is flat [20], then is observable from the incremental flat output and the system is, clearly, controllable. The flat output is obtained as the last row of the inverse of the controllability matrix multiplied by the state vector. The flat output is found to be the incremental value of the linearized projection of the pendulum tip position onto the horizontal line (i.e., the "shadow" of the tip on the horizontal.):

$$y = -x_\delta + L\theta_\delta \tag{4.38}$$

The consecutive time derivatives of the flat output are computed through the observability matrix of the linearized system. This results in:

$$\begin{aligned} y &= -x_\delta + L\theta_\delta; \quad \dot{y} = -\dot{x}_\delta + L\dot{\theta}_\delta; \quad \ddot{y} = g\theta_\delta; \quad y^{(3)} = g\dot{\theta}_\delta \\ y^{(4)} &= \frac{gk_t}{rRLM}V_\delta + \frac{g^2(M+m)}{LM}\theta_\delta \end{aligned} \tag{4.39}$$

The second order time derivative of the flat output is seen to be proportional to the measured angular position θ. Therefore, it is not necessary to estimate the flat output acceleration since this is conformed from the measured angular position. On the other hand, the first and third order time derivatives of the flat output are combinations of the velocities of the measured positions.

The inverse relation yields the following differential parametrization of the all system variables:

$$x_\delta = -y + \frac{L}{g}\ddot{y}; \quad \dot{x}_\delta = -\dot{y} + \frac{L}{g}y^{(3)}, \quad \theta_\delta = \frac{1}{g}\ddot{y}; \quad \dot{\theta}_\delta = \frac{1}{g}y^{(3)}$$

The control input, in terms of the flat output, is just:

$$V_\delta = \frac{rRLM}{gk_t}y^{(4)} + \frac{rR(M+m)}{k_t}\ddot{y} \tag{4.40}$$

The input-output dynamics of the linearized system is, thus, given by

$$y^{(4)} = \frac{gk_t}{rRLM}V_\delta + \frac{g(M+m)}{LM}\ddot{y} \tag{4.41}$$

The simplified input-output dynamics which is quite useful in simplifying the design and implementation of an ADRC and sliding mode controller is:

$$y^{(4)} = \beta V_\delta + \xi(t), \quad \beta = \frac{gk_t}{rRLM}, \quad \xi(t) = -\frac{g(M+m)}{LM}\ddot{y} \tag{4.42}$$

4.4.1.2 An Active Disturbance Rejection Control

Suppose it is desired to have the output y track a given trajectory $y^*(t)$, oriented by the need to stabilize the car and the pendulum on a given constant position without remaining oscillations of the pendulum. Define $e_y = y - y^*(t)$, $e_V = (V - V^*(t))$ where $V^*(t)$ is the nominal control obtained from (4.40). Similarly, we let $e_\theta = \theta_\delta - \theta_\delta^*$. The incremental state variables nominal trajectories, $\theta^*(t)$, is easily obtained from the flatness property which differentially parametrizes all system state variables in terms of the flat output and a finite number of its time derivatives. In our case, the tangent linearized system enjoys such a special property.

Consider the following relations satisfied by the even order time derivatives of the incremental flat output trajectory tracking errors:

$$\ddot{e}_y = \ddot{y} - \ddot{y}^* = g e_\theta; \quad e_y^{(6)} = \beta e_V \tag{4.43}$$

These relations may be viewed as dynamical sub-systems with input given by linear combinations of the generalized position tracking error variables. The first relation is used to build an ESO observers for the unmeasured, first-time derivatives of the incremental flat output tracking error.

Let $\hat{e}_{y0}$ denote the redundant estimate of the measured flat output tracking error $e_y = y - y^*(t)$. The first observer is devised as follows:

$$\begin{aligned} \frac{d}{dt}\hat{e}_{y0} &= \hat{e}_{y1} + k_1(e_y - \hat{e}_{y0}) \\ \frac{d}{dt}\hat{e}_{y1} &= g e_\theta + k_0(e_y - \hat{e}_{y0}) \end{aligned} \tag{4.44}$$

This observer produces an estimate $\hat{e}_{y1}$ of the first order time derivative of e_y.

The second relation from (4.43) prompts an ESO that simultaneously estimates the third order time derivative of the flat output tracking error, here denoted by $\hat{e}_y^{(3)}$, and the possible unknown disturbances, summarized in the extension variable z, due to neglected nonlinearities as well as exogenous perturbations (unmodeled dynamics, force disturbances and so on). We have,

$$\begin{aligned} \frac{d}{dt}\hat{e}_{y2} &= \hat{e}_{y3} + k_4(g e_\theta - \hat{e}_{y2}) \\ \frac{d}{dt}\hat{e}_{y3} &= \hat{z} + \beta e_V + k_3(g e_\theta - \hat{e}_{y2}) \\ \frac{d}{dt}\hat{z} &= k_2(g e_\theta - \hat{e}_{y2}) \end{aligned} \tag{4.45}$$

The output observation error, $e_0 = e_y - \hat{e}_{y0}$, and its time derivatives $e_1 = g e_\theta - \hat{e}_{y2}$ generate the following set of reconstruction error dynamics,

$$\begin{aligned} \ddot{e}_0 + k_1\dot{e}_0 + k_0 e_0 &= 0 \\ e_1^{(3)} - k_4\ddot{e}_1 - k_3\dot{e}_1 - k_2 e_1 &= 0 \end{aligned}$$

An appropriate choice of coefficients ($k_4,\ldots, k_0$) guarantee exponentially decreasing estimation errors, e_0 and e_1. Using, respectively, a second and third order Hurwitz polynomial of the form: $(s^2 + 2\zeta_1\omega_1 s + \omega_1^2)$ and $(s^2 + 2\zeta_2\omega_2 s + \omega_2^2)(s + p)$, one obtains:

$$k_1 = 2\zeta_1\omega_1; \quad k_0 = \omega_1^2; \quad k_4 = 2\zeta_2\omega_2 + p; \quad k_3 = 2p\zeta_2\omega_2 + \omega_2^2; \quad k_2 = p\omega_2^2$$

The tracking controller is synthesized with a canceling strategy in mind based on the estimate of the total disturbance input function, $\xi(t)$. The estimation of $\xi(t)$ is denoted by $\hat{z}$. The output feedback control $e_u = u - u^*(t)$ is given by,

$$e_u = \frac{1}{\beta}\left[\hat{z} - \lambda_3(\hat{e}_{y3} - y^{*(3)}) - \lambda_2(ge_\theta - \ddot{y}^*) - \lambda_1(\hat{e}_{y1} - \dot{y}^*) - \lambda_0(e_y - y^*)\right] \tag{4.46}$$

The linear controller gains $\{\lambda_3,\ldots,\lambda_0\}$, were set by using the coefficients of a desired fourth Hurwitz polynomial of the form, $(s^2 + 2\zeta\omega s + \omega^2)$. We obtained:

$$\lambda_3 = 4\zeta\omega; \quad \lambda_2 = (4\zeta^2\omega^2 + 2\omega^2); \quad \lambda_1 = 4\omega^3\zeta; \quad \lambda_0 = \omega^4$$

In Fig. 4.6 a block diagram of the ADRC scheme for the inverted pendulum on a car system is shown. From (4.46), we observe that $e_u = V(t) - V^*(t)$, where $e_u = \upsilon_{av}$, represents the average input control of the plant. If we use a Σ-Δ modulator we have that the corresponding sliding surface s correspond with the average input control. Figure 4.7 shows the corresponding block diagram of the ADRC and Σ-Δ modulator for the system, where $u(t)$ represents the discontinuous control obtained trough the modulator.

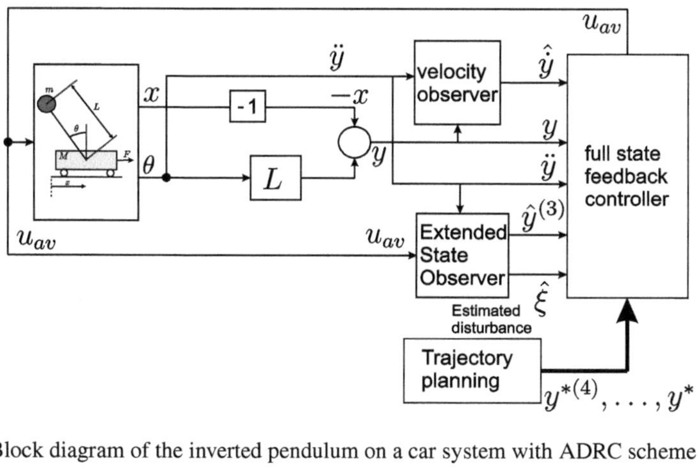

Fig. 4.6 Block diagram of the inverted pendulum on a car system with ADRC scheme

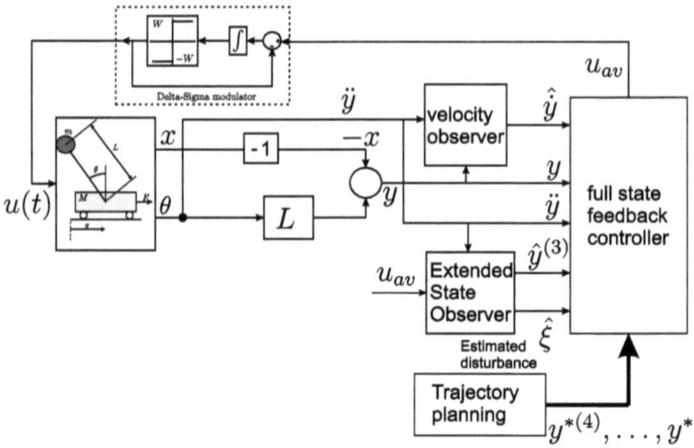

Fig. 4.7 Block diagram of the inverted pendulum on a car system with ADRC scheme and Σ-Δ modulator

4.4.1.3 Description of the Experimental Platform and Experimental Results.

The experimental platform consists of two rails, which allowed an aluminum car to move along them by using a rack and a pinion driven by a DC motor with gearbox of transmission ratio, 30:1. The primary radius of the pinion is 0.03 [m] and the motor supply voltage is 12V. Both the rail and the rack are, respectively, located in the bottom and top of the car. These are fixed to a rigid structure made with a PTR profile and a steel plate. On the car, two incremental rotary encoders, of 2000 pulses per revolution, are found. The pinion is mounted on the car shaft and the hanging pendulum, made of aluminum with a length of 0.56 [m] and a mass of 0.2 [kg], can freely rotate on its hinge, as shown in Fig. 4.8.

The angular position of the pendulum is measured using one of the incremental rotary encoders and the position of the car is measured by the second encoder. The DC motor, powered by a DC voltage supplied by a "Monster motor shield" H-bridge,

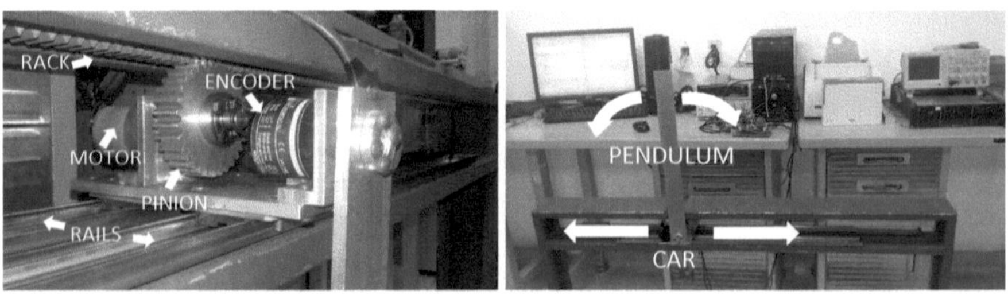

Fig. 4.8 Experimental platform of the inverted pendulum on a car system

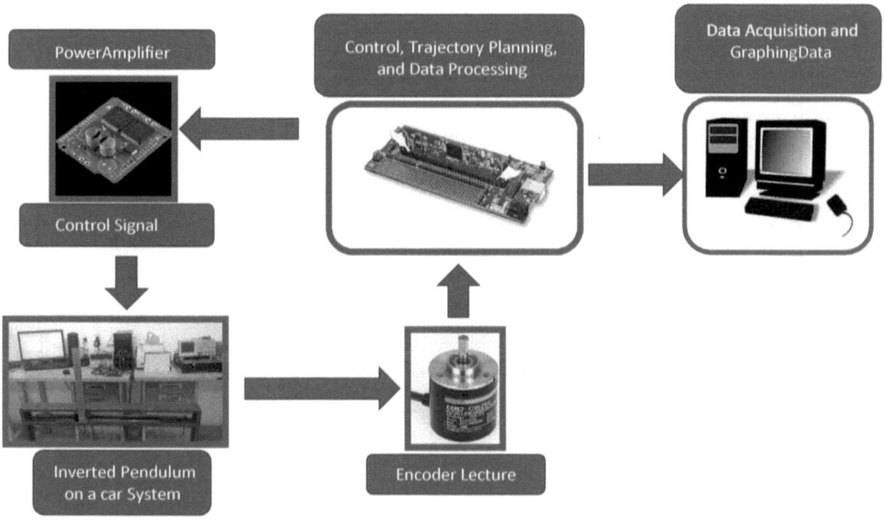

Fig. 4.9 Schematics of the implementation of the controller

is controlled by means of a target "C2000 Delfino TMS320F28335" card. This card reads the signal from the incremental rotary encoders and sends it to the computer where the data is plotted. This system is shown in Fig. 4.9

The initial conditions of the pendulum and car position are given by $[x = 0, \theta = 0]$. The reference trajectory is specified via a "rest-to-rest" maneuver synthesized by means of a Bézier polynomial, interpolating between 0 and 0.5 [m] in 3 s. The parameters of the dominating the closed-loop characteristic polynomial in the linear controller:

$$p_1(s) = s^4 + \lambda_3 s^3 + \lambda_2 s^2 + \lambda_1 s + \lambda_0 \tag{4.47}$$

are obtained by equating this polynomial to a desired Hurtwitz polynomial of the form $(s^2 + 2\zeta_1\omega_1 + \omega_1^2)^2$. The parameters were adjusted to $\zeta_1 = 1$, $\omega_1 = 9$.

An appropriate choice of coefficients $(k_4, \ldots, k_0)$ guarantee exponentially decreasing estimation errors, e_0 and e_1. Using, respectively, a second and third order polynomial of the form:

$$p_2(s) = s^2 + k_1 s + k_0 \tag{4.48}$$

$$p_3(s) = s^3 + k_4 s^2 + k_3 s + k_2 \tag{4.49}$$

Equating this polynomials to a desired Hurtwitz polynomials of the form $(s^2 + 2\zeta_2\omega_2 s + \omega_2^2)$ and $(s^2 + 2\zeta_3\omega_3 s + \omega_3^2)(s + p)$ the desired parameters are obtained. The parameters were adjusted to $\zeta_2 = 1$, $\omega_2 = 120$,$\zeta_3 = 1$, $\omega_3 = 120$ and $p = 120$.

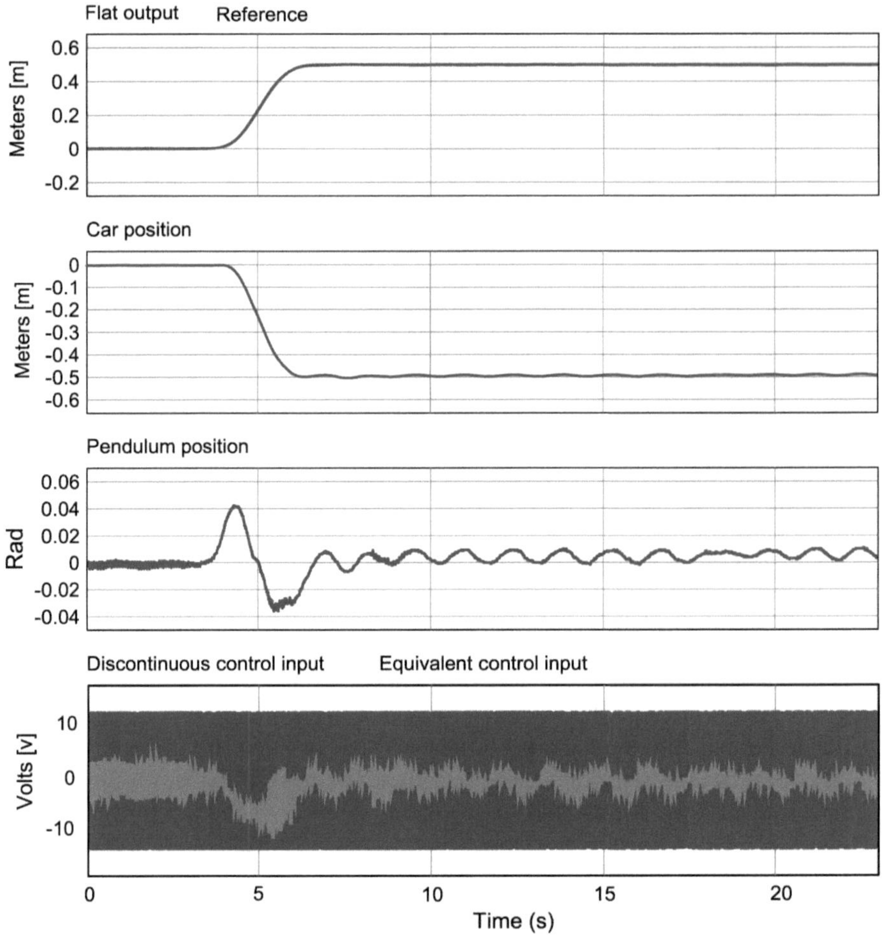

Fig. 4.10 Sliding mode controlled trajectory tracking features for the inverted pendulum on a car variables and applied control input, compared with the corresponding equivalent control

In Fig. 4.10 the closed-loop system response is depicted, when the desired trajectory initially coincides with the flat output reference trajectory. The desired trajectory is followed with good precision and reaches the desired value in the required time. The sliding mode control input signal is plotted along with the computed equivalent control. This is contained within the expected sliding mode existence range.

Figure 4.11 shows the response of the closed-loop system when it is disturbed by an instantaneous external force at the tip of the pendulum once the system is in equilibrium (approximately, at $t = 11$ [s] and $t = 17$ [s]). The reaction of the controller allows to bring the flat output to the desired reference equilibrium while avoiding the input effects of the external disturbance. This demonstrates the remarkable robustness of the proposed controller.

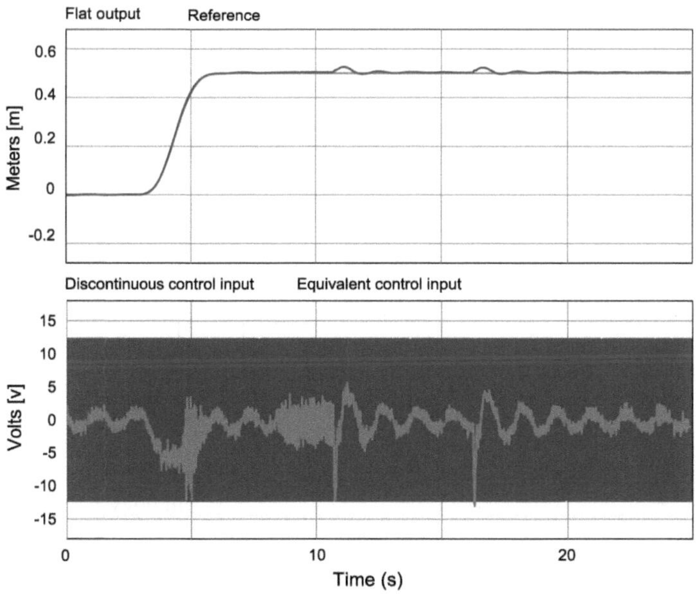

Fig. 4.11 Tracking, control input voltage

Table 4.1 Experimental platform parameters

Parameter	Value
Armature resistance R	2.4 Ω
Torque constant k_t	0.35 Nm/A
Pendulum mass	0.278 kg
Car mass	1.672 kg
Pendulum length	0.56 m
Primitive radius of the pinion	0.03 m

Experimental platform parameters, used in this experimental test, are presented in Table 4.1.

4.4.2 A Nonlinear Manipulator-DC Motor System

Consider the single-link robot manipulator shown in Fig. 4.12. It is desired to track a pre-specified angular position trajectory, given by $\theta^*(t)$. Suppose that only the angular position of the pendulum, $\theta(t)$, is accessible for measurement.

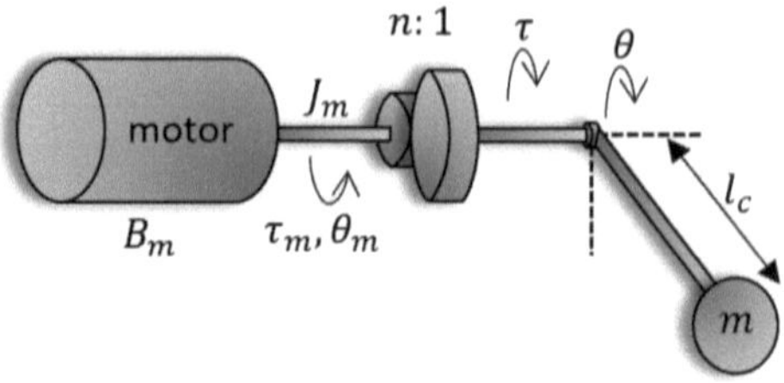

Fig. 4.12 Single-link manipulator: simple link manipulator with DC motor

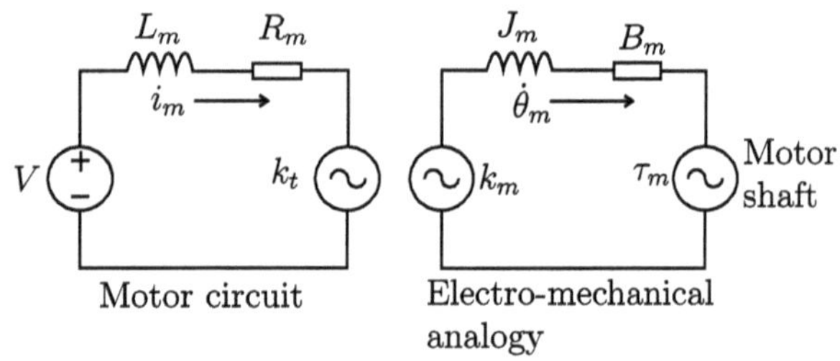

Fig. 4.13 Motor scheme with electro-mechanical analogies

Using the Euler–Lagrange formalism, the single-link manipulator model is obtained as:

$$\left(J + ml_c^2\right)\ddot{\theta} + mgl_c\sin\theta = \tau \tag{4.50}$$

where m is the mass of the link, J is the moment of inertia of the link, l_c is the distance from the pivot to the center of gravity of the beam and θ describes the angular position of the link with respect to the vertical.

Considering the model of the DC motor, shown in Fig. 4.13, this model is given by the following equations:

$$L_m\frac{d}{dt}i_m + R_m i_m + k_m\dot{\theta}_m = V \tag{4.51}$$

$$J_m\ddot{\theta}_m = k_t i_m - \tau_m - B_m\dot{\theta}_m \tag{4.52}$$

where L_m is the armature inductance, R_m is the armature resistance, i_m is the armature current, k_m the electromotive force (EMF) constant, V is the applied voltage, J_m is the motor's rotor inertia, B_m is the coefficient of viscous friction of the motor, k_t is the torque constant, θ_m the angular position of the rotor, and τ_m the load torque on the motor shaft.

Since the motor and the pendulum are coupled by means of a gear-head box (with a gear relation n:1) and a rigid coupling, we can say that the motor's torque and angle are related with the pendulum's torque and angle via the following relations:

$$n\tau_m = \tau; \quad \frac{1}{n}\theta_m = \theta$$

Using the relationships above, we can combine the pendulum's equation with the motor equations. In terms of the pendulum's angular displacement, and considering that $k_m = k_t = K$. Thus, measured in the International System of Units the model of the full system is:

$$L_m \frac{d}{dt} i_m + R_m i_m + K n\dot{\theta} = V \tag{4.53}$$

$$\left(J + m l_c^2 + J_m n^2\right) \ddot{\theta} + B_m n^2 \dot{\theta} + m g l_c \sin\theta = K i_m n \tag{4.54}$$

The composite system is flat. The flat output can be taken to be the angular position of the pendulum $F = \theta$. This flat output induces the following differential parametrization of the system variables and of the control input:

$$\begin{aligned} i_m &= \left[\left(J + m l_c^2 + J_m n^2\right) \ddot{F} + B_m n^2 \dot{F} + m g l_c \sin(F)\right] / K n \\ \theta &= F \\ \omega &= \dot{F} \end{aligned} \tag{4.55}$$

$$\begin{aligned} v = V = &\frac{L_m}{K n} \left(F^{(3)} \left(J + m l_c^2 + J_m n^2\right) + B_m n^2 \ddot{F} + m g l_c \dot{F} \cos(F)\right) \\ &+ \frac{R_m}{K n} \left(\ddot{F} \left(J + m l_c^2 + J_m n^2\right) + B_m n^2 \dot{F} + m g l_c \sin(F)\right) + K n \dot{F} \end{aligned} \tag{4.56}$$

The input-to-flat output dynamics of the system is given by:

$$\begin{aligned} F^{(3)} = &-\frac{\beta}{n k_t} \left[L_m B_m n^2 + R_m \left(J_m n^2 + J + m l_c^2\right)\right] \ddot{F} \\ &- \frac{\beta}{n k_t} \left(L_m m g l_c \cos F + k_m k_t n^2 + R_m B_m n^2\right) \dot{F} \\ &- \alpha \sin F + \beta V \end{aligned} \tag{4.57}$$

where α and β are defined as:

$$\alpha = \frac{R_m m g l_c}{L_m \left(J + m l_c^2 + J_m n^2\right)}; \quad \beta = \frac{n K}{L_m \left(J + m l_c^2 + J_m n^2\right)}$$

Let the constant gain β be known. Given the nonlinear system (4.57), with the control input represented by the motor's armature circuit voltage V. Define $u_{av} = \beta V$. Consistent with the methodology advocated by the ADRC methodology, consider the following simplified third order system dynamics,

$$F^{(3)} = \theta^{(3)} = \beta V + \xi = u_{av} + \xi \tag{4.58}$$

where, generally speaking, ξ represents the systems nonlinearities and possible exogenous and endogenous disturbances as well as unmodeled uncertainties affecting the plant behavior. Here, we have,

$$\xi = -\frac{\beta}{nK}\left[L_m B_m n^2 + R_m\left(J + ml_c^2 + J_m n^2\right)\right]\ddot{F}$$

$$-\frac{\beta}{nK}(L_m mgl_c \cos F + K^2 n^2 + R_m B_m n^2)\dot{F} - \alpha \sin F \tag{4.59}$$

ξ represents the state-dependent *total disturbance* including the nonlinear and linear expressions affecting the dynamics of the angular position third order time derivative. This disturbance term also includes all the exogenous disturbances, modeling errors, friction terms, and unmodeled dynamics affecting the system behavior.

The output tracking error $e_\theta = \theta - \theta^*(t)$ and the auxiliary input error $e_u = u_{av} - u^*(t)$, with $u^*(t) = [y^*]^{(3)}$. The simplified description of the tracking error dynamics, treated in the previous section, except for the disturbance term, is given by:

$$e_\theta^{(3)} = e_u + \xi \tag{4.60}$$

We propose the following integrally reconstructed sliding surface (no integration limits are shown for simplicity):

$$\widehat{\sigma} = \widehat{\ddot{e}_\theta} + k_5\widehat{\dot{e}_\theta} + k_4 e_\theta + k_3 \int e_\theta(\lambda)\, d\lambda + k_2 \iint e_\theta(\lambda_1)\, d\lambda_1 d\lambda$$

$$+ k_1 \iiint e_\theta(\lambda_2)\, d\lambda_2 d\lambda_1 d\lambda k_0 \iiiint e_\theta(\lambda_3)\, d\lambda_3 d\lambda_2 d\lambda_1 d\lambda \tag{4.61}$$

where, as before, the estimated phase variables of the plant are computed as integrals of the input error without taking into account the disturbance. The estimates are then structural reconstructors which we know beforehand they are in error. The only variation we are introducing is the addition of two extra iterated integrals of third and fourth order. This provision is taken just to have one more degree of integral compensation in the actual physical system operation. The effects of the (matched) phase variables dependent disturbance will be handled in part by the integral compensation and in part by the intrinsic robustness of the sliding regime under matching conditions. The parameters $k_0, k_1, k_2, k_3, k_4, k_5$, are design constants to be determined so that the closed-loop characteristic polynomial of the unperturbed system is a Hurwitz polynomial. As before, a suitable switching policy is given by:

$$\beta V = u_{av} = -W \text{sign}\widehat{\sigma} \tag{4.62}$$

Under ideal sliding conditions we have that the following invariance conditions are satisfied: $\widehat{\sigma} = 0$, $\frac{d}{dt}\widehat{\sigma} = 0$. From these two conditions, the equivalent control is readily determined as shown in the next Eq. (4.63):

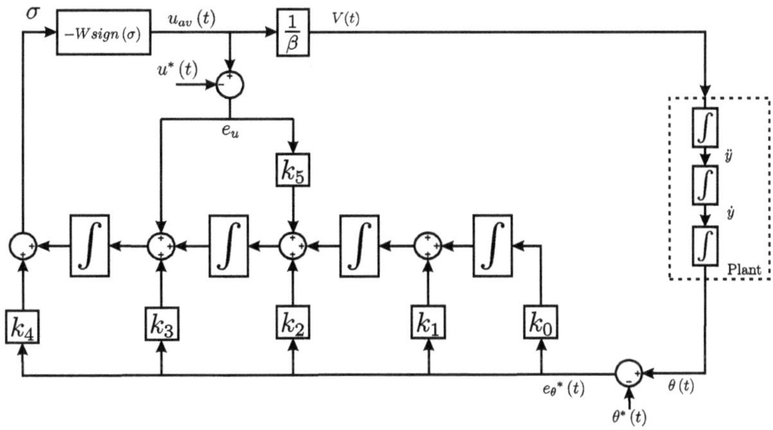

Fig. 4.14 Block diagram for the sliding mode control with integral reconstructor and two extra integrators for a third order plant

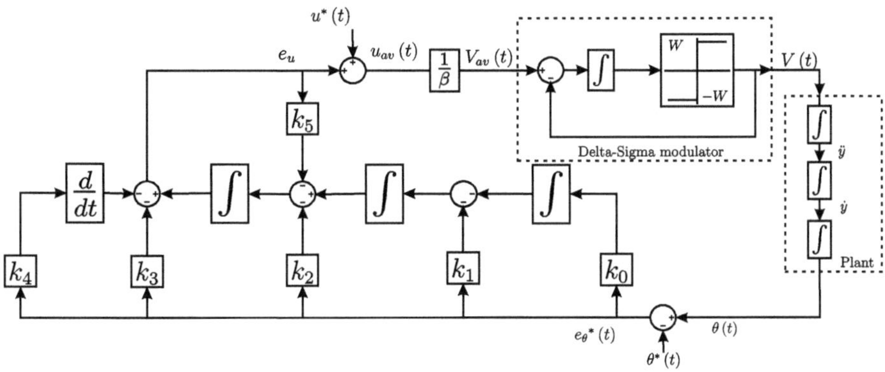

Fig. 4.15 Sliding mode controller equivalence with a sigma delta modulator based scheme

$$
\begin{aligned}
e_{ueq} &= -k_5\ddot{e}_\theta - k_4\dot{e}_\theta - k_3 e_\theta - k_2 \int e_\theta - k_1 \iint e_\theta - k_0 \iiint e_\theta \\
&= -k_5 \int e_u - k_4 \iint e_u - k_3 e_\theta - k_2 \int e_\theta - k_1 \iint e_\theta - k_0 \iiint e_\theta \quad (4.63)
\end{aligned}
$$

The diagram for the third order sliding mode control scheme, based on integral reconstructors, is presented in Fig. 4.14.

As explained before, this diagram can be reinterpreted in the form of a Delta-Sigma modulator based scheme, as shown in Fig. 4.15.

Also taking advantage of the integral reconstructors to estimate $\dot{e}_\theta$, we finally arrived to the scheme on Fig. 4.16 which corresponds to a GPI controller with an extra integration term and a sigma delta modulator.

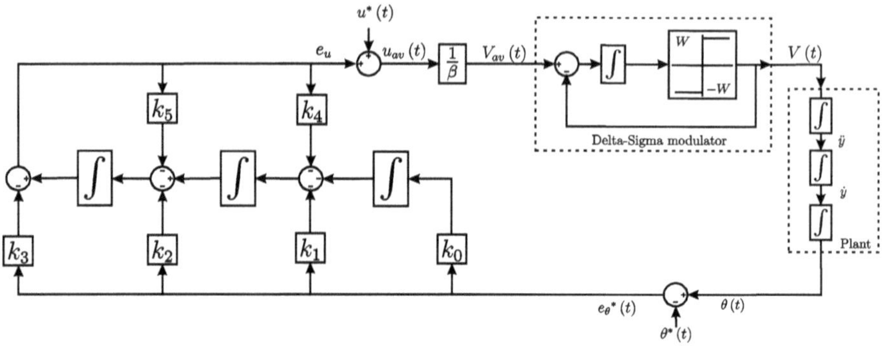

Fig. 4.16 GPI and Delta-Sigma modulator controller scheme

Fig. 4.17 Experimental platform of a single-link manipulator

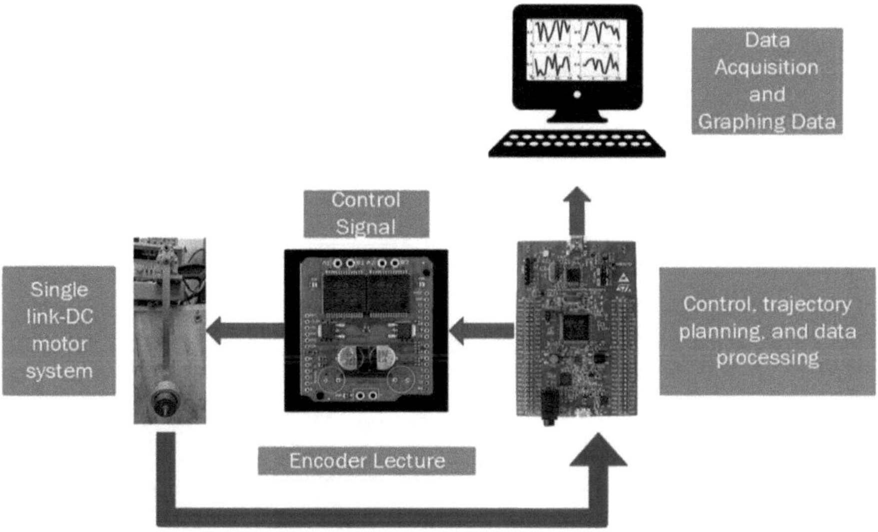

Fig. 4.18 Experimental platform diagram

4.4.2.1 Description of Platform and Experimental Results

An experimental platform of the nonlinear manipulator-DC motor system was built, consisting of a pendulum with a 250 [g] mass attached to a 0.27 [m] thin aluminum bar. The link was bounded by a rigid coupling to the axis of a 12V DC motor with a 131:1 gear ratio and with 64 counts per revolution. A two-channel Hall effect encoder is used to sense the rotation of a magnetic disk on a rear protrusion of the motor shaft (see Fig. 4.17).

The DC motor is excited by a DC voltage source through a "Monster motor shield" H-bridge, controlled with the STM32F407 discovery kit, which also read the motor's encoder and is able to send the data to a computer where the data is graphed, as shown in Fig. 4.18.

The output reference trajectory tracking experiment considered a Bézier polynomial, interpolating between 0 and $\frac{3}{2}\pi$ [rad] in 2 seconds. The closed-loop system is dominated by the ideal linear dynamics whose characteristic polynomial is:

$$p(s) = s^6 + k_5 s^5 + k_4 s^4 + k_3 s^3 + k_2 s^2 + k_1 s + k_0 \tag{4.64}$$

whose coefficients are obtained from a corresponding desired Hurwitz polynomial of the form: $(s^2 + 2\zeta\omega_n + \omega_n^2)^3$. For the experimental results, the values of the parameters were set to: $W = 12$, $\zeta = 0.5$ and $\omega_n = 27$.

In Fig. 4.19, the response of the closed-loop system is depicted when the initial condition of the desired reference trajectory coincides with the initial angular position of the link. The position maneuver from 0 to $3\pi/2$ [rad] is achieved in a significantly

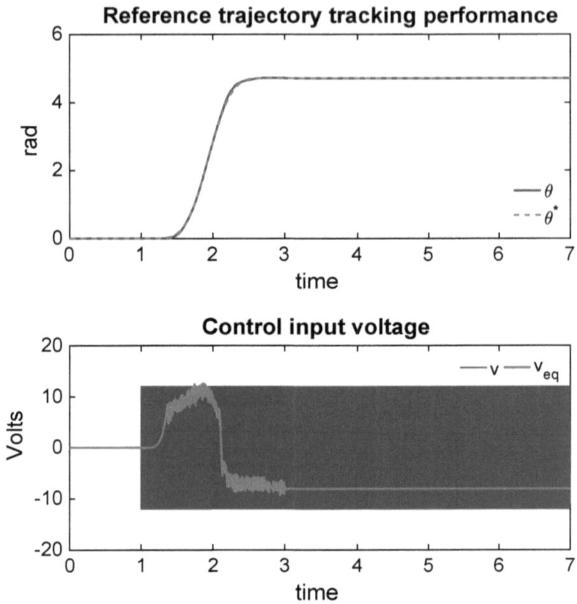

Fig. 4.19 Tracking, control input voltage

Table 4.2 Motor parameters

Lm	Jm	Km	Kt	Rm
2 mH	5.5942e-4 kgm^2	10.065e-3 V*s/rad	10.065e-3 Nm/A	1.8 Ω

short period of time (2 [s]) and the desired reference trajectory is tracked with good accuracy. The GPI control signal is a bit noisy during the transition from one rest to an other, but when the reference trajectory is reached the noise disappears. The motor parameters, used in this experimental test, are presented in Table 4.2.

4.5 Conclusions

In this chapter, we have introduced an input-output approach for sliding mode control of switched linear and (a large class of) nonlinear systems (namely, the class of feedback linearizable systems, or flat systems). Although the developments were centered around the single-input single-output case, the results may be generalized for the multivariable case, provided dynamic feedback linearization is allowed and appropriately handled. One of the explored options is represented by the switched implementation of a sound smooth average output feedback controller design for the plant, via a Delta-Sigma modulator. We took for such a prototypical controller the Active Disturbance Rejection Control (ADRC) scheme which is a rather robust output feedback controller. The implementation module is represented by a Delta-Sigma modulator, taking the average designed bounded control input signal as the modulator input. The modulator produces, at the output, the switched, or binary-valued, version of the input. The modulator thus provides a command signal for the switched plant. Delta-Sigma modulators absorb the sliding regime creation problem into its exogenous, one-dimensional, state space, foreign to the given plant. It provides, in turn, a switching (binary-valued) output whose average value, in an equivalent control sense, coincides with the input signal to the modulator. Here, in this chapter, we illustrated the case of an average ADRC output reference trajectory tracking design for an under-actuated switched nonlinear inverted pendulum on a car system using its controllable tangent linearization. The desired output reference trajectory takes the system variables arbitrarily far from the operating equilibrium used for the approximate linearization. The second option, for the input-output approach to sliding mode control design, is based on the possibilities of structural phase variables reconstruction, or integral reconstruction in pure integration systems. Using this approach, a sliding surface coordinate function, free of phase variables measurements, is synthesized using a variation of the Generalized Proportional Integral (GPI) control idea. The class of feedback linearizable, or flat, systems has been shown here to benefit from the integral reconstruction procedure in sliding mode schemes. The connection of integral reconstructor-based sliding regimes with Delta-Sigma modulators was also clearly established. We have provided here two case studies, including laboratory prototypes, as an illustrative implementation of the foregoing developments.

References

1. Cortés-Romero, J., Rojas-Cubides, H., Coral-Enriquez, H., Sira-Ramírez, H., Luviano-Juárez, A.: Active disturbance rejection approach for robust fault-tolerant control via observer assisted sliding mode control. Math. Probl. Eng. **2013** (2013)
2. Edwards, C., Spurgeon, S.: Sliding Mode Control: Theory and Applications. CRC Press, Boca Raton (1998)
3. Fliess, M., Join, C.: Model-free control. Int. J. Control **86**(12), 2228–2252 (2013)
4. Fliess, M., Join, C., Mboup, M., Sira-Ramirez, H.: Vers une commande multivariable sans modele. In: Actes Conf. Francoph. Automat.(CIFA 2006), Bordeaux (2006)
5. Fliess, M., Marquez, R., Delaleau, E., Sira-Ramírez, H.: Correcteurs proportionnels-intégraux généralisés. ESAIM: Control Optim. Calc. Var. **7**, 23–41 (2002)
6. Gao, Z.: Active disturbance rejection control: a paradigm shift in feedback control system design. In: 2006 American Control Conference, p. 7 (2006)
7. Gao, Z., Huang, Y., Han, J.: An alternative paradigm for control system design. In: Proceedings of the 40th IEEE Conference on Decision and Control, 2001, vol. 5, pp. 4578–4585 (2001)
8. Guo, B., Jin, F.: The active disturbance rejection and sliding mode control approach to the stabilization of the Euler–Bernoulli beam equation with boundary input disturbance. Automatica **49**(9), 2911–2918 (2013)
9. Guo, B., Jin, F.: Sliding mode and active disturbance rejection control to stabilization of one-dimensional anti-stable wave equations subject to disturbance in boundary input. IEEE Trans. Autom. Control **58**(5), 1269–1274 (2013)
10. Guo, B.Z., Zhao, Z.L.: On the convergence of an extended state observer for nonlinear systems with uncertainty. Syst. Control Lett. **60**(6), 420–430 (2011)

11. Guo, B.Z., Zhao, Z.L.: On convergence of the nonlinear active disturbance rejection control for MIMO systems. SIAM J. Control Optim. **51**(2), 1727–1757 (2013)
12. Han, J.: From PID to active disturbance rejection control. IEEE Trans. Ind. Electron. **56**(3), 900–906 (2009)
13. Khalil, H.K.: A note on the robustness of high-gain-observer-based controllers to unmodeled actuator and sensor dynamics. Automatica **41**(10), 1821–1824 (2005)
14. Madoński, R., Herman, P.: Survey on methods of increasing the efficiency of extended state disturbance observers. ISA Trans. **56**, 18–27 (2015)
15. Morales, R., Sira-Ramírez, H., Feliu, V.: Adaptive control based on fast online algebraic identification and GPI control for magnetic levitation systems with time-varying input gain. Int. J. Control **87**, 1604–1621 (2014)
16. Ramirez-Neria, M., Garcia-Antonio, J., Sira-Ramirez, H., Velasco-Villa, M., Castro-Linares, R.: An active disturbance rejection control of leader-follower Thomson's jumping rings. Control Theory Appl. **30**(12), 1564–1572 (2013)
17. Ramírez-Neria, M., Sira-Ramírez, H., Garrido-Moctezuma, R., Luviano-Juarez, A.: Linear active disturbance rejection control of underactuated systems: the case of the Furuta pendulum. ISA Trans. **53**(4), 920–928 (2014)
18. Ramírez-Neria, M., Sira-Ramírez, H., Garrido-Moctezuma, R., Luviano-Juárez, A.: Active disturbance rejection control of singular differentially flat systems. In: 54th Annual Conference of the Society of Instrument and Control Engineers of Japan (SICE), 2015, pp. 554–559 (2015)
19. Sira-Ramírez, H.: Sliding Mode Control: The Delta-Sigma Modulation Approach. Birkhäuser, Boston (2015)
20. Sira-Ramírez, H., Agrawal, S.K.: Differentially Flat Systems. Marcel Dekker, New York (2004)
21. Sira-Ramírez, H., García-Rodríguez, C., Cortés-Romero, J., Luviano-Juárez, A.: Algebraic Identification and Estimation Methods in Feedback Control Systems. Wiley, New York (2014)
22. Sira-Ramirez, H., Linares-Flores, J., Garcia-Rodriguez, C., Contreras-Ordaz, M.: On the control of the permanent magnet synchronous motor: an active disturbance rejection control approach. IEEE Trans. Control Syst. Technol. **22**(5), 2056–2063 (2014). doi:10.1109/TCST.2014.2298238
23. Sira-Ramírez, H., Linares-Flores, J., Luviano-Juarez, A., Cortés-Romero, J.: Ultramodelos globales y el control por rechazo activo de perturbaciones en sistemas no lineales diferencialmente planos. Revista Iberoamericana de Automática e Informática Industrial RIAI **12**(2), 133–144 (2015)
24. Sira-Ramírez, H., Ramírez-Neria, M., Rodríguez-Angeles, A.: On the linear control of nonlinear mechanical systems. In: 49th IEEE Conference on Decision and Control, pp. 1999–2004. Atlanta, USA (2010)
25. Slotine, J.J.E., Li, W., et al.: Applied Nonlinear Control. Prentice-Hall, Englewood Cliffs (1991)
26. Utkin, V.: Sliding Modes in Optimization and Control Problems. Springer, New York (1992)
27. Utkin, V., Guldner, J., Shi, J.: Sliding Mode Control in Electro-Mechanical Systems. Taylor and Francis, London (2009)
28. Zhao, S., Gao, Z.: An active disturbance rejection based approach to vibration suppression in two-inertia systems. Asian J. Control **15**(2), 350–362 (2013)
29. Zheng, Q., Gao, L., Gao, Z.: On stability analysis of active disturbance rejection control for nonlinear time-varying plants with unknown dynamics. In: Proceedings of the IEEE Conference on Decision and Control, New Orleans, LA, pp. 12–14, Dec 2007
30. Zheng, Q., Gao, Z.: On practical applications of active disturbance rejection control. In: Proceedings of the 29th Chinese Control Conference, pp. 6095–6100 (2010)

Chapter 5
Design of Asymptotic Second-Order Sliding Mode Control System

Yaodong Pan and Katsuhisa Furuta

Abstract A chattering-free sliding mode (SM) control system can be realized by a second-order sliding mode (2nd-SM) control based on the derivative model of the original system. In this case, the derivative of a switching function, which may be unavailable for the control implementation, is required for the finite time convergence to a 2nd-SM. In this chapter, a new asymptotic SM control algorithm, without using the derivative of the switching function, is proposed for a class of nonlinear systems, to ensure the asymptotically convergence to a 2nd-SM. The locally and asymptotically stability is guaranteed by a Lyapunov function.

5.1 Introduction

In a variable-structure control system with a sliding mode (SM), a switching function and an SM control law should be designed such that the system reaches the SM in finite time and stays there after that, and the reduced-order system on the SM is asymptotically stable [6, 16]. To keep the system state on the SM, the SM control input has to be switched in a high frequency, which results in chattering phenomena. As one of effective methods to deal with the chattering problem, second-order SM (2nd-SM) control algorithms have been proposed and have been implemented in many mechanical control systems [2–5, 10, 12].

To design a chattering-free SM control system by a 2nd-SM control algorithm, the derivative of the control input may be considered as an alternative control input and an SM control law is designed for the alternative control input based on the derivative model of the system. Thus a smooth control input as the integration of an 2nd-SM control law is obtained. To guarantee the finite time convergence to a 2nd-

Y. Pan (✉)
Honeywell Aerospace, Mississauga, ON, Canada
e-mail: yaodong.pan@yahoo.com

K. Furuta
Tokyo Denki University, Tokyo, Japan
e-mail: kfuruta@jcom.home.ne.jp

J.B. Clempner and W. Yu (eds.), *New Perspectives and Applications of Modern Control Theory*, https://doi.org/10.1007/978-3-319-62464-8_5

SM with a 2nd-SM control law, both the switching function and its derivative (or its derivative-related information) are required for the 2nd-SM control implementation [11]. In practical control systems, however, the derivatives of the switching function may be unmeasurable or unavailable for the control implementation, although it can be estimated by an observer.

An asymptotic 2nd-SM control algorithm without using the derivative has been proposed [1, 7, 15] such that a 2nd-SM is reached, asymptotically with a first-order SM (1st-SM) control law, i.e., a relay control law. The asymptotic convergence to the 2nd-SM can also be observed when an SM control is implemented to a system with fast dynamic actuators [8] or inertial sensors [9]. It has been proved in [14] for a linear time-invariant system that a high order SM can be reached locally and asymptotically with a reduced-order SM control law and the asymptotic reduced-order SM control system is locally and asymptotically stable if the sum of system poles is less than the sum of system zeros. A similar necessary condition for the convergence to the 2nd-SM with a relay control law is also required for nonlinear time-varying cases [7]. Therefore, the asymptotic 2nd-SM control algorithm can only be implemented to those systems satisfying the necessary condition [7, 14]. In [13], the condition has been represented by linear matrix inequalities (LMI).

In this chapter, a new chattering-free asymptotic 2nd-SM control algorithm is proposed for a class of nonlinear systems, without using the derivative of the switching function. A 1st-SM control law is implemented such that a 2nd-SM is reached, locally and asymptotically together with a pre-feedback control term letting the necessary condition mentioned above hold. A Lyapunov function is provided to show the local and asymptotic stability of the proposed asymptotic SM control system.

This chapter is organized as follows: Sect. 5.2 describes the problem to be considered in this chapter; Sect. 5.3 proposes the asymptotic SM control algorithm and shows its local and asymptotic stability; Sect. 5.4 presents simulation results; and Sect. 5.5 concludes the chapter.

5.2 Problem Description: Nonlinear Model

Consider a single-input single-output nonlinear system, described by

$$y^{(n)}(t) = f(y(t), \dot{y}(t), \ldots, y^{(n-1)}(t), u(t)), \tag{5.1}$$

where $y(t) \in R$ and $u(t) \in R^1$ are the output and the input variables, respectively, and $f(y, \dot{y}, \ldots, y^{(n-1)}, u) \in C^k$ $(k \geq 1)$ is a continuously differentiable nonlinear function. The control input $u(t)$ is essentially bounded in a compact subset $U \subset R$.

Define a state vector $x \in R^n$ as

$$x(t) = \begin{bmatrix} x_1(t) \\ x_2(t) \\ \vdots \\ x_n(t) \end{bmatrix}$$

where $x_i(t) = y^{(i-1)}(t)$ $(i = 1, 2, \ldots, n)$. It is assumed that the state $x(t)$ is considered in a bounded compact subset $X \subset R^n$, i.e., $x(t) \in X \subset R^n$. Denoting the nonlinear function $f(y, \dot{y}, \ldots, y^{(n-1)}, u)$ as $f(x, u)$, the system (5.1) can be described as

$$\dot{x}(t) = \begin{bmatrix} x_2(t) \\ x_3(t) \\ \vdots \\ x_n \\ f(x, u) \end{bmatrix}. \tag{5.2}$$

Define a switching function as

$$\sigma(t) = Cx(t) \tag{5.3}$$

where $C = \begin{bmatrix} c_0 \; c_1 \; \cdots \; c_{n-2} \; 1 \end{bmatrix} \in R^{1\times n}$. It is assumed that the switching function $\sigma(t)$ is designed such that all roots of the polynomial equation

$$s^{n-1} + c_{n-2}s^{n-2} + \cdots + c_1 s + c_0 = 0 \tag{5.4}$$

have negative real parts.

Taking the derivative of both sides of the system dynamical equation (5.2) yields

$$\ddot{x}(t) = A(x, u)\dot{x}(t) + B(x, u)\dot{u}(t), \tag{5.5}$$

where $A(x, u) \in R^{n\times n}$ and $B(x, u) \in R^{n\times 1}$ are determined by

$$A(x, u) = \left[\begin{array}{c|c} O_{(n-1)\times 1} & I_{n-1} \\ \hline -a_0(x, u) & -a_1(x, u) \; \ldots \; -a_{n-1}(x, u) \end{array} \right]$$

$$B(x, u) = \left[\begin{array}{c} O_{1\times(n-1)} \\ \hline b(x, u) \end{array} \right].$$

Here, $a_i(x, u)$ $(i = 0, 1, \ldots, n-1)$ and $b(x, u)$ are defined as

$$a_i(x, u) = -\frac{\partial}{\partial x_{i+1}} f(x, u), \quad (i = 0, 1, \ldots, n-1) \tag{5.6}$$

$$b(x, u) = \frac{\partial}{\partial u} f(x, u). \tag{5.7}$$

In this chapter, I_i and $O_{j \times k}$ are the $i \times i$ ($i = 2, 3, \ldots$) identity matrix and the $j \times k$ ($j = 1, 2, \ldots; k = 1, 2, \ldots$) zero matrix, respectively.

Assumption 1 There exist two positive constants $\underline{b}$ and $\bar{b}$ such that $b(x, u)$ is bounded by

$$0 < \underline{b} \leq b(x, u) \leq \bar{b}, \quad \forall x \in X \subset R^n, \forall u \in U \subset R.$$

Define an extended state vector $\bar{x} \in R^{n+1}$ as

$$\bar{x}(t) = \begin{bmatrix} y(t) \\ \dot{y}(t) \\ \vdots \\ y^{(n)}(t) \end{bmatrix} = \begin{bmatrix} x(t) \\ \dot{x}_n(t) \end{bmatrix} = \begin{bmatrix} x_1(t) \\ \dot{x}(t) \end{bmatrix}.$$

Then the extended system dynamics of (5.5) can be described as

$$\dot{\bar{x}}(t) = \bar{A}(x, u)\bar{x}(t) + \bar{B}(x, u)\dot{u}(t) \tag{5.8}$$

where $\bar{A}(x, u) \in R^{(n+1)\times(n+1)}$ and $\bar{B}(x, u) \in R^{(n+1)\times 1}$ are defined as

$$\bar{A}(x, u) = \left[\begin{array}{c|c} 0 & 1 \; O_{1\times(n-1)} \\ \hline O_{n\times 1} & A(x, u) \end{array}\right]$$
$$\bar{B}(x, u) = \left[\begin{array}{c} 0 \\ \hline B(x, u) \end{array}\right].$$

Then the switching function (5.3) and its derivative can be written as

$$\sigma(t) = Cx(t) = \begin{bmatrix} C \; 0 \end{bmatrix} \bar{x}(t) = \begin{bmatrix} c_0 \; c_1 \; \cdots \; c_{n-2} \; 1 \; 0 \end{bmatrix} \bar{x}(t)$$
$$\dot{\sigma}(t) = C\dot{x}(t) = \begin{bmatrix} 0 \; C \end{bmatrix} \bar{x}(t) = \begin{bmatrix} 0 \; c_0 \; \cdots \; c_{n-3} \; c_{n-2} \; 1 \end{bmatrix} \bar{x}(t)$$

Define a new state vector as

$$w(t) = \begin{bmatrix} \sigma(t) \\ \dot{\sigma}(t) \\ z(t) \end{bmatrix} \in R^{n+1},$$

where $z(t) \in R^{n-1}$ is the internal state vector defined as

$$z(t) = \begin{bmatrix} x_1(t) \\ x_2(t) \\ \vdots \\ x_{n-1}(t) \end{bmatrix} \in R^{n-1}.$$

A state transformation matrix $T \in R^{(n+1)\times(n+1)}$

$$T = \left[\begin{array}{cccc|cc} c_0 & c_1 & \cdots & c_{n-2} & 1 & 0 \\ 0 & c_0 & \cdots & c_{n-3} & c_{n-2} & 1 \\ \hline & & I_{n-1} & & O_{(n-1)\times 2} & \end{array}\right]$$

transforms the state from $\bar{x}(t)$ to $w(t)$ as

$$w(t) = T\bar{x}(t).$$

Then the system (5.8) is transformed with the transformation matrix T as

$$\dot{w}(t) = \tilde{A}(x,u)w(t) + \tilde{B}(x,u)u(t) \tag{5.9}$$

with $\tilde{A}(x,u) = T\bar{A}(x,u)T^{-1}$ and $\tilde{B}(x,u) = T\bar{B}(x,u)$ being determined by

$$\tilde{A}(x,u) = \begin{bmatrix} 0 & 1 & O_{1\times(n-1)} \\ -\alpha(x,u) & -\beta(x,u) & C_z(x,u) \\ B_z & O_{(n-1)\times 1} & A_z \end{bmatrix}$$

$$\tilde{B}(x,u) = \begin{bmatrix} 0 \\ b(x,u) \\ O_{(n-1)\times 1} \end{bmatrix},$$

where $A_z \in R^{(n-1)\times(n-1)}$, $B_z \in R^{(n-1)\times 1}$, $C_z(x,u) \in R^{1\times(n-1)}$, $\alpha(x,u) \in R$, and $\beta(x,u) \in R$ are given by

$$A_z = \left[\begin{array}{c|ccc} O_{(n-2)\times 1} & & I_{n-2} & \\ \hline -c_0 & -c_1 & \cdots & -c_{n-2} \end{array}\right] \tag{5.10}$$

$$B_z = \left[\begin{array}{c} O_{(n-2)\times 1} \\ \hline 1 \end{array}\right]$$

$$\begin{aligned} C_z(x,u) &= \begin{bmatrix} b_0(x,u) & b_1(x,u) & \cdots & b_{n-2}(x,u) \end{bmatrix} \\ \alpha(x,u) &= a_{n-2}(x,u) + c_{n-2}^2 - c_{n-3} - c_{n-2}a_{n-1}(x,u) \\ \beta(x,u) &= a_{n-1}(x,u) - c_{n-2}. \end{aligned} \tag{5.11}$$

Here $b_i(x, u)$ $(i = 0, 1, \ldots, n-2)$ is defined as

$$\begin{aligned} b_i(x,u) = {} & c_{i-2} - a_{i-1}(x,u) - c_i(c_{n-3} - a_{n-2}(x,u)) \\ & + (c_{n-2}c_i - c_{i-1})(c_{n-2} - a_{n-1}(x,u)) \end{aligned}$$

with $a_{-1}(x, u) = c_{-1} = c_{-2} = 0$.

Therefore, the system (5.9) can be represented as two subsystems:

$$\ddot{\sigma}(t) = -\beta(x,u)\dot{\sigma}(t) - \alpha(x,u)\sigma(t) + C_z(x,u)z(t) + b(x,u)\dot{u}(t), \quad (5.12)$$

$$\dot{z}(t) = A_z z(t) + B_z \sigma(t). \quad (5.13)$$

The control objective is to stabilize the system (5.1) asymptotically by a chattering-free asymptotic 2nd-SM control law without using $\dot{\sigma}(t)$, the derivative of $\sigma(t)$.

5.3 Design of Asymptotic SM Control System

Based on the dynamical equations (5.12) and (5.13), most of 2nd-SM control algorithms can realize a chattering-free SM control but the derivative of $\sigma(t)$ is required for the control implementation. For example, using the twisting 2nd-SM control [12] designed as

$$\dot{u}(t) = -r_1 \operatorname{sign}(\dot{\sigma}(t)) - r_2 \operatorname{sign}(\sigma(t)), \quad (5.14)$$

a chattering-free SM control law is given by

$$u(t) = -r_1 \int_0^t \operatorname{sign}(\dot{\sigma}(t))dt - r_2 \int_0^t \operatorname{sign}(\sigma(t))dt,$$

where r_1 and r_2 are positive constants.

To avoid using the derivative of $\sigma(t)$, it is possible to stabilize the system (5.1) by the asymptotic 2nd-SM control with a 1st-SM control law [1, 7, 14, 15]

$$\dot{u}(t) = -k_2 \operatorname{sign}(\sigma(t)). \quad (k_2 > 0)$$

In this case, the 2nd-SM can be reached with the 1st-SM control law locally and asymptotically if the time-varying coefficient $\beta(x, u)$ in (5.12) is positive. For practical systems, however, $\beta(x, u)$ may be unknown and may be negative. In this case, a feedback control term is added to the 1st-SM control law as

$$\dot{u}(t) = -k_1\dot{\sigma}(t) - k_2 \operatorname{sign}(\sigma(t))$$

such that $b(x,u)k_1 + \beta(x,u)$ is positive with a sufficiently large positive constant k_1 for a bounded $\beta(x,u)$. Thus, a new chattering-free asymptotic 2nd-SM control law, without using the derivative of $\sigma(t)$, is proposed in this chapter as

$$u(t) = -k_1\sigma(t) - k_2 \int_0^t \text{sign}(\sigma(t))dt. \tag{5.15}$$

Here, k_1 and k_2 are positive constants to be designed such that the system (5.1) is stabilized locally and asymptotically.

5.3.1 A Stable Sliding Mode

Lemma 5.1 *Consider the system (5.1) with the switching function $\sigma(t)$ defined by (5.3). The reduced-order system on the 2nd-SM $\sigma(t) = \dot{\sigma}(t) = 0$ is asymptotically stable.*

Proof On the 2nd-SM, $\sigma(t) = \dot{\sigma}(t) = 0$ holds. Thus, the stability of the reduced-order system on the 2nd-SM $\sigma(t) = \dot{\sigma}(t) = 0$ is totally determined by the dynamics (5.13) of the internal state $z(t)$. The characteristic equation of the system matrix A_z defined by (5.10), of the subsystem (5.13) is given by the polynomial equation (5.4). As the switching function $\sigma(t)$ defined by (5.3) is designed such that all roots of the polynomial equation (5.4) have negative real parts, all eigenvalues of the system matrix A_z have negative real parts, too. Thus, the reduced-order system on the 2nd-SM is asymptotically stable.

Lemma 5.2 *Consider the system (5.1) with the switching function $\sigma(t)$ defined by (5.3). A necessary condition for the 2nd-SM to be reached asymptotically with the chattering-free asymptotic 2nd-SM control law (5.15) is that the inequality*

$$k_1 b(x,u) + \beta(x,u) = k_1 b(x,u) + a_{n-1}(x,u) - c_{n-2} > 0 \tag{5.16}$$

holds with a sufficiently large gain k_1 of the control law (5.15), where $\beta(x,u)$, $a_{n-1}(x,u)$, and $b(x,u)$ are defined by (5.6), (5.7), and (5.11) respectively, and c_{n-2} is one of coefficients of the switching function $\sigma(t)$ defined by (5.3).

Proof With the SM control law (5.15), the subsystem (5.12) can be rewritten as

$$\begin{aligned}\ddot{\sigma}(t) = &-\lambda(x,u)\dot{\sigma}(t) - \alpha(x,u)\sigma(t)\\ &+C_z(x,u)z(t) - b(x,u)k_2\,\text{sign}(\sigma(t))),\end{aligned} \tag{5.17}$$

where $\lambda(x,u) = b(x,u)k_1 + \beta(x,u)$.

It has been proven in [1, 7] that one of the necessary conditions for the system to converge asymptotically to the 2nd-SM $\sigma(t) = \dot{\sigma}(t) = 0$ is that the coefficient

$\lambda(x,u)$ in (5.17) to be positive. Thus, the following inequality is a necessary condition for the 2nd-SM to be reached asymptotically:

$$\lambda(x,u) = k_1 b(x,u) + a_{n-1}(x,u) - c_{n-2} > 0.$$

Remark 5.1 If the inequality (5.16) holds for the state $x \in X \subset R^n$ and the control input $u \in U \subset R$, then only the local and asymptotic convergence to the 2nd-SM can be guaranteed.

5.3.2 Convergence to Sliding Mode

Theorem 5.1 *Consider the description (5.12) and (5.13) of the system (5.1) with the switching function $\sigma(t)$ defined in (5.3). Then the 2nd-SM $\sigma(t) = \dot{\sigma}(t) = 0$ can be reached locally and asymptotically with the chattering-free asymptotic 2nd-SM control law (5.15) if*

1. *the system state vector $w(t) = \left[\delta(t)\ \dot{\delta}(t)\ z^T(t)\right]^T$ is considered in a bounded compact subset Ω of R^{n+1}, i.e., $w(t) \in \Omega \subset R^{n+1}, \forall t \in [0,+\infty)$ and the following holds:*

$$\begin{cases} x(t) \in X \in R^n \\ u(t) \in U \subset R \end{cases} \Leftrightarrow \begin{cases} w(t) \in \Omega \subset R^{n+1} \\ u(t) \in U \subset R \end{cases};$$

2. *Assumption 1 holds for $w(t) \in \Omega \subset R^n$;*
3. *the positive coefficient k_1 is chosen to be large enough such that the following conditions hold for $x(t) \in X$ ($w(t) \in \Omega$) and $u(t) \in U$:*

$$\lambda(x,u) = k_1 b(x,u) + \beta(x,u) > 0, \tag{5.18}$$
$$2\lambda(x,u) b(x,u) + \dot{b}(x,u) > 0 \tag{5.19}$$

   and

4. *the following inequality holds with a sufficiently large positive coefficient k_2 for $w(t) \in \Omega$ and $u(t) \in U$:*

$$k_2 > \max\{k_{21}(w,u), k_{22}(w,u)\}, \tag{5.20}$$

   where $k_{21}(w,u)$ and $k_{22}(w,u)$ are respectively determined by

$$k_{21}(w,u) = \frac{\mu_1 + |\alpha(x,u)\sigma(t) - C_z(x,u)z(t)| + \lambda(x,u)|\dot{\sigma}(t)|}{b(x,u)}$$
$$k_{22}(w,u) = \frac{\mu_2 + |h_1(x,u)\sigma(t) + h_2(x,u)\dot{\sigma}(t) + h_3(x,u)z(t)|}{2\lambda(x,u)b(x,u) + \dot{b}(x,u)}.$$

Here, μ_1 and μ_2 are positive constants and $h_1(x,u)$ $(i = 1, 2, 3)$ is defined as

$$\begin{aligned} h_1(x,u) &= 2\lambda\alpha(x,u) + \dot{\alpha}(x,u) - C_z(x,u)B_z \\ h_2(x,u) &= 2\alpha(x,u) \\ h_3(x,u) &= -\dot{C}_z(x,u) - C_z(x,u)(A_z + 2\lambda I_{n-1}) \end{aligned}$$

Proof Rewrite the dynamical equation (5.12) with the SM control law (5.15) as

$$\ddot{\sigma}(t) = -\lambda(x,u)\dot{\sigma}(t) - \eta(t)\,\mathrm{sign}(\sigma(t)), \tag{5.21}$$

where $\lambda(x,u) = k_1 b(x,u) + \beta(x,u)$ is positive according to Lemma 5.2 when the inequality (5.18) holds, and $\eta(t) \in R$ is defined as

$$\eta(t) = (\alpha(x,u)\sigma(t) - C_z(x,u)z(t))\,\mathrm{sign}(\sigma) + b(x,u)k_2. \tag{5.22}$$

Using the inequality (5.20), it can be confirmed that

$$\begin{aligned} \eta(t) &\geq -|\alpha(x,u)\sigma(t) - C_z(x,u)z(t)| + b(x,u)k_2 \\ &> -|\alpha(x,u)\sigma(t) - C_z(x,u)z(t)| + b(x,u)k_{21}(w,u) \\ &> \lambda(x,u)|\dot{\sigma}(t)| + \mu_1 \\ &\geq \mu_1 > 0. \end{aligned}$$

Choose four positive constants γ, ϵ, ν_1, and ν_2 $(\forall w(t) \in \Omega, \forall u(t) \in U)$ satisfying

$$\begin{cases} \epsilon \leq \lambda(x,u)\left(\frac{\mu_2}{2\eta^2(t)} - \nu_1\right) \\ \eta(t) \geq |\gamma - \lambda(x,u)||\dot{\sigma}(t)| + \nu_2 \\ \epsilon < \frac{\bar{\lambda}^2}{\eta(t)} \\ \nu_1 < \frac{\mu_2}{2\eta^2(t)} \end{cases}. \tag{5.23}$$

Then, a Lyapunov function candidate is defined as

$$\begin{aligned} V_\sigma(t) &= \frac{1}{2}\begin{bmatrix}\sigma(t) & \dot{\sigma}(t)\end{bmatrix}\begin{bmatrix}\frac{1}{\eta(t)} & \frac{\epsilon}{\gamma} \\ \frac{\epsilon}{\gamma} & \epsilon\end{bmatrix}\begin{bmatrix}\sigma(t) \\ \dot{\sigma}(t)\end{bmatrix} + |\sigma(t)| \\ &= \frac{\dot{\sigma}^2(t)}{2\eta(t)} + \frac{\epsilon}{2}\sigma^2(t) + \frac{\epsilon}{\gamma}\sigma(t)\dot{\sigma}(t) + |\sigma(t)|, \end{aligned} \tag{5.24}$$

which is continuously differentiable when $\sigma(t) \neq 0$. As ϵ and $\frac{1}{\eta(t)}$ are positive and $\frac{\epsilon}{\eta(t)} - \frac{\epsilon^2}{\gamma^2}$ is also positive according to the third inequality in (5.23), the 2×2 symmetric matrix

$$\begin{bmatrix} \frac{1}{\eta(t)} & \frac{\epsilon}{\gamma} \\ \frac{\epsilon}{\gamma} & \epsilon \end{bmatrix}$$

is positive definite. Therefore, the Lyapunov function candidate $V_\sigma(t)$ is positive definite for all $w(t) \in \Omega$ and for all $u(t) \in U$.

The derivative of $V_\sigma(t)$ for $\sigma(t) \neq 0$ along the trajectories of (5.17) is

$$\begin{aligned}
\dot{V}_\sigma(t) &= \frac{\dot{\sigma}(t)\ddot{\sigma}(t)}{\eta(t)} - \frac{\dot{\sigma}^2(t)}{2\eta^2(t)}\dot{\eta}(t) + \epsilon\sigma(t)\dot{\sigma}(t) \\
&\quad + \frac{\epsilon}{\gamma}(\sigma(t)\ddot{\sigma}(t) + \dot{\sigma}^2(t)) + \operatorname{sign}(\sigma)\dot{\sigma}(t) \\
&= \frac{\dot{\sigma}(t)}{\eta(t)}\left(\ddot{\sigma}(t) - \frac{\dot{\sigma}(t)}{2\eta(t)}\dot{\eta}(t) + \eta(t)\operatorname{sign}(\sigma)\right) \\
&\quad + \frac{\epsilon}{\gamma}\sigma(t)(\gamma\dot{\sigma}(t) + \ddot{\sigma}(t)) + \frac{\epsilon}{\gamma}\dot{\sigma}^2(t) \\
&= \frac{\dot{\sigma}(t)}{\eta(t)}\left(-\lambda(x,u)\dot{\sigma}(t) - \frac{\dot{\sigma}(t)}{2\eta(t)}\dot{\eta}(t)\right) - \frac{\epsilon}{\gamma}\eta(t)|\sigma(t)| \\
&\quad + \frac{\epsilon(\gamma - \lambda(x,u))}{\gamma}\sigma(t)\dot{\sigma}(t) + \frac{\epsilon}{\gamma}\dot{\sigma}^2(t) \\
&= -\frac{\dot{\sigma}^2(t)}{2\eta^2(t)}(2\lambda(x,u)\eta(t) + \dot{\eta}(t)) - \frac{\epsilon}{\gamma}\eta(t)|\sigma(t)| \\
&\quad + \frac{\epsilon(\gamma - \lambda(x,u))}{\gamma}\sigma(t)\dot{\sigma}(t) + \frac{\epsilon}{\gamma}\dot{\sigma}^2(t).
\end{aligned} \tag{5.25}$$

It follows from the inequality (5.20), i.e., $k_2 > \max\{k_{21}(w,u), k_{22}(w,u)\} \geq k_{22}(w,u)$ that

$$\begin{aligned}
2\lambda(x,u)\eta(t) + \dot{\eta}(t) &= (2\lambda(x,u)b(x,u) + \dot{b}(x,u))k_2 + (h_1(x,u)\sigma(t) \\
&\quad + h_2(x,u)\dot{\sigma}(t) + h_3(x,u)z(t))\operatorname{sign}(\sigma) \\
&> (2\lambda(x,u)b(x,u) + \dot{b}(x,u))k_{22}(w,u) \\
&\quad - |h_1(x,u)\sigma(t) + h_2(x,u)\dot{\sigma}(t) + h_3(x,u)z(t)| \\
&\geq \mu_2.
\end{aligned}$$

Substituting the above inequality for (5.25) and using the first two inequalities in (5.23) yield

$$\begin{aligned}
\dot{V}_\sigma(t) &< -\frac{\sigma}{2\eta^2(t)}\dot{\sigma}^2(t) - \frac{\epsilon}{\gamma}\eta(t)|\sigma(t)| + \frac{\epsilon(\gamma - \lambda(x,u))}{\gamma}\sigma(t)\dot{\sigma}(t) + \frac{\epsilon}{\gamma}\dot{\sigma}^2(t) \\
&= -\left(\frac{\mu_2}{2\eta^2(t)} - \frac{\epsilon}{\gamma}\right)\dot{\sigma}^2(t) - \frac{\epsilon}{\gamma}\left(\eta(t) - (\gamma - \lambda(x,u))\operatorname{sign}(\sigma)\dot{\sigma}(t)\right)|\sigma(t)|
\end{aligned}$$

$$
\begin{aligned}
&\leq -\nu_1\dot{\sigma}^2(t) - \frac{\epsilon}{\gamma}\nu_2|\sigma(t)| \\
&\leq 0, \quad (\dot{V}_\sigma(t) = 0 \Leftrightarrow \sigma(t) = \dot{\sigma}(t) = 0)
\end{aligned}
$$

which means that $V_\sigma(t)$ keeps decreasing until both $\sigma(t)$ and $\dot{\sigma}(t)$ converge to zero. Therefore, it is proven that the 2nd-SM $\sigma(t) = \dot{\sigma}(t) = 0$ is reached locally and asymptotically with the SM control law (5.15) for $w(t) \in \Omega$ and $u(t) \in U$.

5.3.3 Asymptotic Stability

When the switching function $\sigma(t)$ in (5.3) is designed such that the reduced-order system (5.13) on the 2nd-SM is asymptotically stable, there exists a symmetric positive definite matrix $P_z \in R^{(n-1)\times(n-1)}$ satisfying

$$P_z A_z + A_z^T P_z = -I_{n-2}.$$

Define a Lyapunov function candidate

$$V_z(t) = z^T(t) P_z z(t). \tag{5.26}$$

Then its derivative along the trajectories of (5.13) is given by

$$
\begin{aligned}
\dot{V}_z(t) &= z^T(t)(P_z A_z + A_z^T P_z)z(t) + 2z^T(t)P_z B_z \sigma(t) \\
&= -z^T(t)z(t) + 2z^T(t)P_z B_z \sigma(t) \\
&= -||z(t)||^2 + 2z^T(t)P_z B_z \sigma(t) \\
&\leq -||z(t)||^2 + 2||z(t)|| \times ||P_z B_z|| \times |\sigma(t)| \\
&= -||z(t)|| \left(||z(t)|| - 2||P_z B_z|| \times |\sigma(t)|\right).
\end{aligned}
$$

Therefore, the Lyapunov function candidate $V_z(t)$ keeps decreasing as long as $||z(t)|| > 2||P_z B_z|| \times |\sigma(t)|$, which means that the internal state $z(t)$ converges to a neighborhood determined by

$$||z(t)|| \leq 2||P_z B_z|| \times |\sigma(t)|. \tag{5.27}$$

Thus, $z(t)$ is bounded for any bounded $\sigma(t)$. Furthermore, the asymptotic stability of the system (5.1) with the chattering-free asymptotic 2nd-SM control law (5.15) is shown in the following theorem.

Theorem 5.2 *Consider the system (5.1) with the switching function $\sigma(t)$ defined in (5.3). The system (5.1) with the chattering-free asymptotic 2nd-SM control law (5.15) is locally and asymptotically stable if*

1. *all conditions given in Theorem 5.1 are satisfied;*
2. *the switching function $\sigma(t)$ in (5.3) is designed such that the reduced-order system (5.13) on the 2nd-SM is asymptotically stable;*
3. *the following inequality holds with a sufficiently large positive coefficient k_2 for $w(t) \in \Omega$ and $u(t) \in U$:*

$$k_2 \geq \max\{k_{23}(w, u), k_{22}(w, u)\}, \tag{5.28}$$

where Ω and $k_{22}(w, u)$ are defined in Theorem 5.1, and $k_{23}(w, u)$ is defined as $k_{23}(w, u) = \frac{k_{23}^n(w,u)}{b(x,u)}$ with

$$k_{23}^n(w, u) = \mu_1 + |\alpha(x, u)\sigma(t) - C_z(x, u)z(t)| + 2|B_z^T P_z z(t)| + \lambda(x, u)|\dot{\sigma}(t)|.$$

Here, μ_1 is the positive constant defined in Theorem 5.1.

Proof Define a Lyapunov function candidate as

$$V(t) = V_\sigma(t) + \frac{\epsilon}{\gamma} V_z(t), \tag{5.29}$$

where the positive constants ϵ and γ satisfy the inequalities in (5.23), and $V_\sigma(t)$ and $V_z(t)$ are defined by (5.24) and (5.26), respectively. Corresponding to the change of the condition for the positive coefficient k_2, the second condition in (5.23) is modified as

$$\eta(t) \geq |\gamma - \lambda(x, u)||\dot{\sigma}(t)| + 2|B_z^T P_z z(t)| + \nu_2. \tag{5.30}$$

Substituting (5.24) and (5.26) for (5.29) yields

$$\begin{aligned} V(t) &= \frac{1}{2} \begin{bmatrix} \sigma(t) & \dot{\sigma}(t) \end{bmatrix} \begin{bmatrix} \frac{1}{\eta(t)} & \frac{\epsilon}{\gamma} \\ \frac{\epsilon}{\gamma} & \epsilon \end{bmatrix} \begin{bmatrix} \sigma(t) \\ \dot{\sigma}(t) \end{bmatrix} + |\sigma(t)| + \frac{\epsilon}{\gamma} z^T(t) P_z z(t) \\ &= \frac{\dot{\sigma}^2(t)}{2\eta(t)} + \frac{\epsilon}{2}\sigma^2(t) + \frac{\epsilon}{\gamma}\sigma(t)\dot{\sigma}(t) + |\sigma(t)| + \frac{\epsilon}{\gamma} z^T(t) P_z z(t). \end{aligned}$$

Thus, $V(t)$ is positive definite and is continuous differentiable for $\sigma(t) \neq 0$ if the reduced-order system on the SM is stable and all conditions given in Theorem 5.1 are satisfied.

It follows from the definition of $\eta(t)$ in (5.22) and the inequality (5.28) given in this theorem that

$$\begin{aligned} \eta(t) &= (\alpha(x, u)\sigma(t) - C_z(x, u)z(t)) \operatorname{sign}(y) + b(x, u)k_2 \\ &\geq -|\alpha(x, u)\sigma(t) - C_z(x, u)z(t)| + b(x, u)k_2 \\ &\geq -|\alpha(x, u)\sigma(t) - C_z(x, u)z(t)| + b(x, u)k_{23}(w, u) \\ &= \lambda(x, u)|\dot{\sigma}(t)| + 2|B_z^T P_z z(t)| + \mu_1 > 0, \quad \forall w(t) \in \Omega, \forall u(t) \in U. \end{aligned}$$

With this relation, the modification of the second condition in (5.23)–(5.30) is possible. Furthermore, it can be confirmed that the derivative of $V(t)$ for $\sigma(t) \neq 0$ along the trajectories of (5.12) and (5.13) is negative as shown below.

$$
\begin{aligned}
\dot{V}(t) &= \dot{V}_\sigma(t) + \frac{\epsilon}{\gamma}\dot{V}_z(t) \\
&\leq -\nu_1\dot{\sigma}^2(t) - \frac{\epsilon}{\gamma}z^T(t)z(t) + \frac{\epsilon}{\gamma}2z^T(t)P_zB_z\sigma(t) \\
&\quad -\frac{\epsilon}{\gamma}\left(\eta(t) - (\gamma - \lambda(x,u))\,\mathrm{sign}(\sigma)\dot{\sigma}(t)\right)|\sigma(t)| \\
&= -\nu_1\dot{\sigma}^2(t) - \frac{\epsilon}{\gamma}z^T(t)z(t) - 2z^T(t)P_zB_z\,\mathrm{sign}(\sigma)) \\
&\quad -\frac{\epsilon}{\gamma}|\sigma(t)|(\eta(t) - (\gamma - \lambda(x,u))\,\mathrm{sign}(\sigma)\dot{\sigma}(t) \\
&\leq -\nu_1\dot{\sigma}^2(t) - \frac{\epsilon}{\gamma}z^T(t)z(t) - \frac{\epsilon\nu_2}{\gamma}|\sigma(t)| \\
&\leq 0, \quad (\dot{V}(t) = 0 \Leftrightarrow w(t) = 0).
\end{aligned}
$$

Therefore, the system (5.1) with the chattering-free asymptotic 2nd-SM control law (5.15) is locally and asymptotically stable for $w(t) \in \Omega$.

Remark 5.2 It is clear that $k_{23}(w,u)$ is larger than $k_{21}(w,u)$. Thus, the following holds:

$$
\max\{k_{23}(w,u), k_{22}(w,u)\} \geq \max\{k_{21}(w,u), k_{22}(w,u)\}.
$$

According to the gain conditions (5.20) and (5.28) given in Theorems 5.1 and 5.2, respectively, the gain determined in Theorem 5.1 is larger than or equal to the one given in Theorem 5.2. Therefore, the gain k which ensures the asymptotic stability of the system (5.1) with the chattering-free asymptotic 2nd-SM control law (5.15) should be larger than or equal to the one guaranteeing the asymptotic convergence to the 2nd-SM $\sigma(t) = \dot{\sigma}(t) = 0$. In other words, with a gain k chosen to ensure the asymptotic convergence to the 2nd-SM, the asymptotic stability of the system (5.1) with the chattering-free asymptotic 2nd-SM control law (5.15) may not be guaranteed.

5.4 Simulation Results

In the simulation, a second-order system

$$
\begin{aligned}
\ddot{y}(t) &= y(t)\dot{y}(t) + 2y(t) + 3\dot{y}^2(t) \\
&\quad + sin(5\dot{y}(t)) + (2 + sin(3y(t)) + y^2(t))u(t)
\end{aligned}
\tag{5.31}
$$

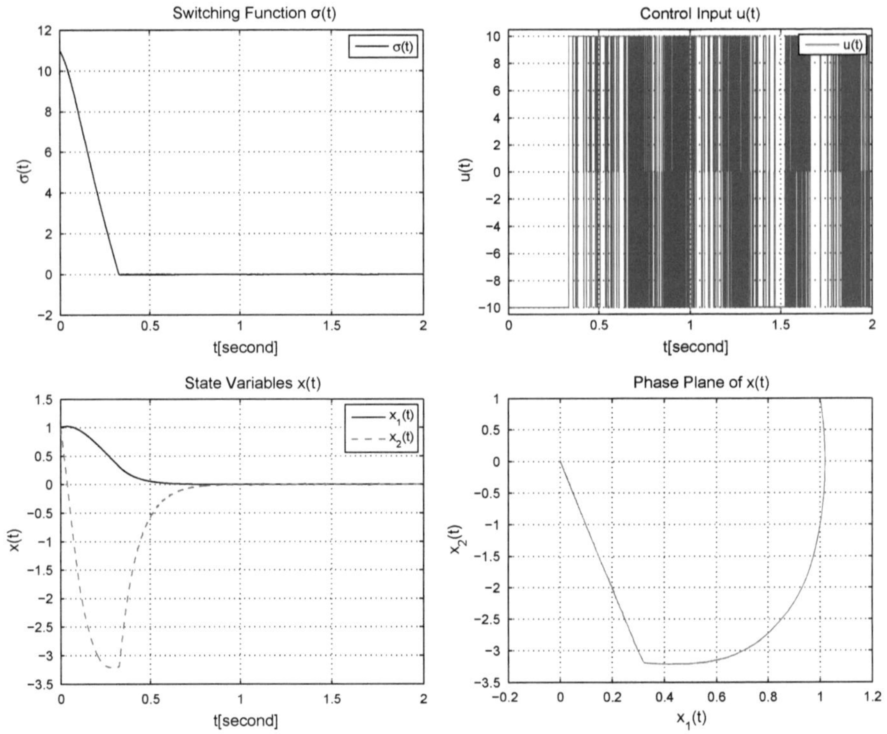

Fig. 5.1 Simulation result with 1st-SM control

is considered as an example, which can be rewritten as the state equation description

$$\dot{x}(t) = \begin{bmatrix} x_2(t) \\ f(x(t), u(t)) \end{bmatrix}$$

with $x(t) = \begin{bmatrix} x_1(t) \; x_2(t) \end{bmatrix}^T$, $x_1(t) = y(t)$ and $x_2(t) = \dot{y}(t)$, and

$$\begin{aligned} f(x(t), u(t)) &= x_1(t)x_2(t) + 2x_1(t) + 3x_2^2(t) + sin(5x_2(t)) \\ &\quad +(2 + sin(3x_1(t)(t)) + x_1^2(t))u(t) \end{aligned}$$

In the simulation, it is assumed that the initial condition x_0 is equal to $\begin{bmatrix} 1 \; 1 \end{bmatrix}^T$ and the control input $u(t)$ is bounded with

$$|u(t)| < 10.$$

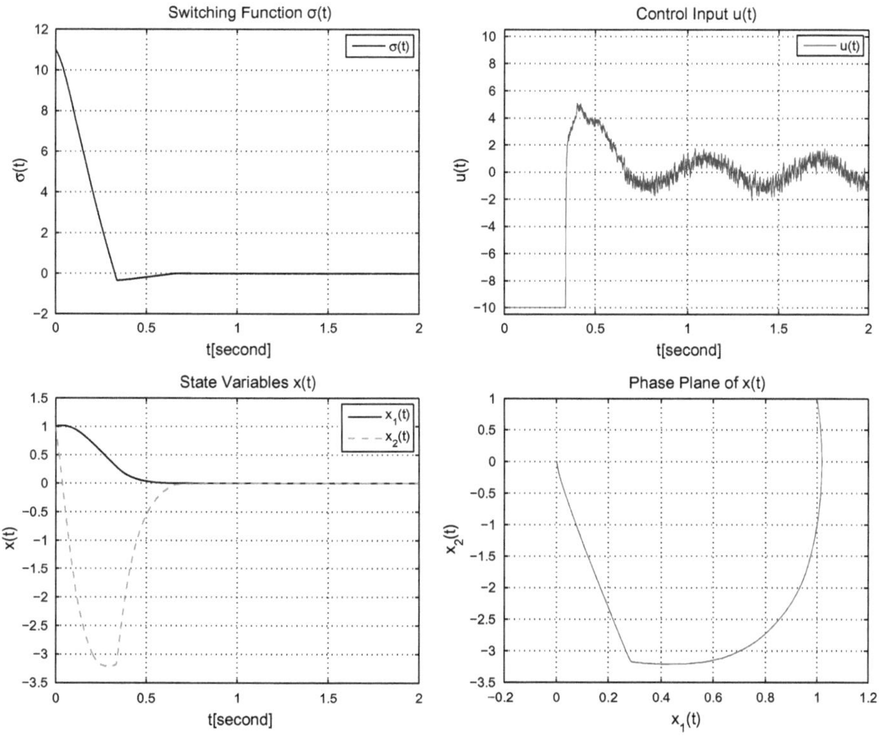

Fig. 5.2 Simulation result with chattering-free asymptotic 2nd-SM control

A switching function is designed as

$$\sigma(t) = 10x_1(t) + x_2(t) = 10y(t) + \dot{y}(t), \tag{5.32}$$

such that the reduced-order system on the sliding mode $\sigma(t) = 0$ is asymptotically stable.

At first, a 1st-SM control law

$$u(t) = -10\,\mathrm{sign}(\sigma(t))$$

is implemented in the simulation. The results given in Fig. 5.1 show that the sliding mode $\sigma(t) = 0$ is reached in finite time with the 1st-SM control law and the reduced-order system on the sliding mode $\sigma(t) = 0$ designed above is asymptotically stable. In this case, however, the chattering phenomena can be observed as the 1st-SM control input $u(t)$ is switched between ± 10 in a high frequency.

The simulation results with the proposed SM control law (5.15) are shown in Fig. 5.2, where the control parameters are chosen as $k_1 = 190$ and $k_2 = 200$. It is clear that the SM $\sigma(t) = 0$ is reached asymptotically, the state variable $x(t)$ converges to

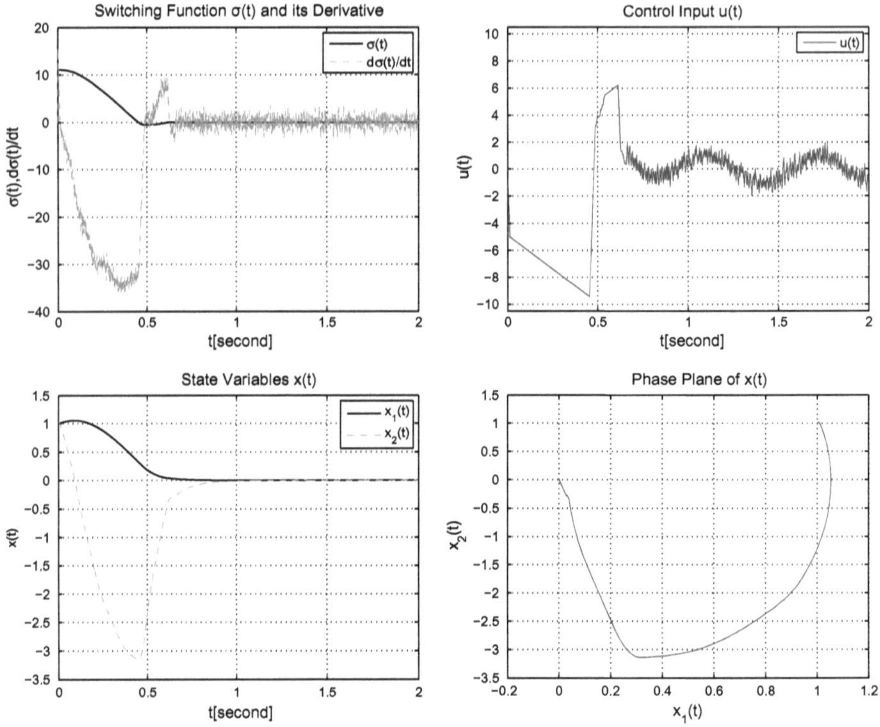

Fig. 5.3 Simulation result with 2nd-SM twisting control

the origin asymptotically, and $u(t)$ is smooth in comparison with the 1st-SM control law shown in Fig. 5.1. Thus, the system (5.31) with the chattering-free asymptotic 2nd-SM control law (5.15) is asymptotically stable without chattering. It is also confirmed in the simulation that the system cannot be stabilized by the integrated 1st-SM control law, i.e., the proposed SM control law (5.15) with $k_1 = 0$ and $k_2 = 200$.

For comparison, the twisting 2nd-SM control (5.14) is also simulated to control the system (5.31) with the same control parameters $r_1 = k_1 = 190$ and $r_2 = k_2 = 200$ used for the chattering-free asymptotic 2nd-SM control simulation. The simulation results are shown in Fig. 5.3. Obviously, the 2nd-SM control algorithm using the derivative of the switching function $\sigma(t)$ ensure the finite time convergence to the 2nd-SM $\sigma(t) = \dot{\sigma}(t) = 0$. There is, however, no remarkable improvement for the convergence of the system state $x(t)$ to the origin as it is mainly determined by the dynamics of the reduced-order system on the SM.

5.5 Conclusion

This chapter proposed a new chattering-free asymptotic 2nd-SM control algorithm for a class of nonlinear systems, without using the derivative of the switching function. The SM can be reached, locally and asymptotically with the proposed 2nd-SM control law. The pre-feedback control term included in the control law ensures that the necessary condition for the SM to be reached locally and asymptotically is satisfied. The locally and asymptotical stability is proven by a Lyapunov function.

References

1. Anosov, D.V.: On stability of equilibrium points of relay systems. Autom. Remote Control **2**(1), 135–149 (1959)
2. Bartolini, G., Ferrara, A., Pisano, A., Usai, E.: On the convergence properties of a 2-sliding control algorithm for nonlinear uncertain systems. Int. J. Control **74**(7), 718–731 (2001)
3. Bartolini, G., Pisano, A., Punta, E., Usai, E.: A survey of applications of second-order sliding mode control to mechanical systems. Int. J. Control **76**(9–10), 875–892 (2003)
4. Boiko, I., Fridman, L., Castellanos, M.I.: Analysis of second order sliding mode algorithms in the frequency domain. IEEE Trans. Autom. Control **49**(6), 946–950 (2004)
5. Damiano, A., Gatto, G., Marongiu, I., Pisano, A.: Second-order sliding-mode control of DC drives. IEEE Trans. Indust. Electron. **51**(2), 364–373 (2004)
6. Edwards, C., Spurgeon, S.: Sliding Mode Control: Theory and Applications. Taylor and Francis, London (1998)
7. Fridman, L., Levant, A.: Higher order sliding modes as a natural phenomenon in control theory. Lect. Notes Control Inf. Sci. Robust Control Var. Struct. Lyapunov Tech. **217**, 107–133 (1996)
8. Fridman, L.M.: Stability of motions in singularly perturbed discontinuous control systems. In: Proceedings of IFAC World Conference, pp. 367–370. Sydney (1993)
9. Fridman, L.M.: Chattering analysis in sliding mode systems with inertial sensors. Int. J. Control **76**(9–10), 906–912 (2003)
10. Levant, A.: Sliding order and sliding accuracy in sliding mode control. Int. J. Control **58**(6), 1247–263 (1993)
11. Levant, A.: Higher-order sliding modes, differentiation and output-feedback control. Int. J. Control **76**, 924–941 (2003)
12. Levant, A.: Principles of 2-sliding mode design. Automatica **43**(4), 576–586 (2007)
13. Marquez, R., Tapia, A., Bernal, M., Fridman, L.: Lmi-based second-order sliding set design using reduced order of derivatives. Int. J. Robust Nonlinear Control **25**, 3763–3779 (2015)
14. Pan, Y., Kumar, K., Liu, G.: Reduced-order design of high-order sliding mode control system. Int. J. Robust Nonlinear Control **21**(18), 2064–2078 (2011)
15. Shtessel, Y.B., Krupp, D.R., Shkolnikov, I.A.: 2-sliding-mode control for nonlinear plants with parametric andd dynamic uncertainties. In: Proceedings of AIAA Guidance, Navigation, and Control Conference, pp. 1–9. Denver, CO (2000)
16. Utkin, U.: Sliding Modes in Control and Optimization. Springer, Berlin (1992)

Chapter 6
Fractional-Order Model Reference Adaptive Controllers for First-Order Integer Plants

Manuel A. Duarte-Mermoud, Norelys Aguila-Camacho, Javier A. Gallegos and Juan C. Travieso-Torres

Abstract In this chapter, we extend the ideas of the model reference adaptive control (MRAC), developed for integer-order plants with integer-order adaptive laws, to the case of integer-order plants but with fractional-order adaptive laws. Two cases are analyzed in detail; the direct MRAC (DMRAC) and the combined MRAC (CMRAC). In both cases, boundedness of all the signals in the resultant adaptive scheme is theoretically proved and a discussion on the error, and parameter convergence is provided in each case. The study is performed for scalar first-order time-invariant plants, since extensions to the vector case are currently under investigation.

6.1 Introduction

Model Reference Adaptive Control (MRAC) is one of the most popular adaptive control techniques, whose origins can be traced back to the late fifties [27]. Although in its origin MRAC was conceived for deterministic continuous-time systems in the context of the so-called direct control (control strategy where no estimation of the plant parameters is attempted), later it was generalized and used for stochastic discrete-time systems in the context of indirect control (control strategy whose first

M.A. Duarte-Mermoud (✉) · N. Aguila-Camacho
Department of Electrical Engineering and AMTC, University of Chile,
Av. Tupper, 2007 Santiago, Chile
e-mail: mduartem@ing.uchile.cl

N. Aguila-Camacho
e-mail: naguila@ing.uchile.cl

J.A. Gallegos
Department of Electrical Engineering, University of Chile, Av. Tupper,
2007 Santiago, Chile
e-mail: jgallego@ing.uchile.cl

J.C. Travieso-Torres
Department of Industrial Technologies, University of Santiago de Chile,
Av. El Belloto, 3735 Santiago, Chile
e-mail: juancarlos.travieso@usach.cl

J.B. Clempner and W. Yu (eds.), *New Perspectives and Applications of Modern Control Theory*, https://doi.org/10.1007/978-3-319-62464-8_6

stage is to perform a plant parameter estimation and then the controller is designed based on those estimates). In its early stages, it have been recognized four schools to have started research in adaptive control and even today they continue doing research on adaptive control, namely Ioan Landau et al. at Laboratoire d'Automatique de Grenoble, Ecole Nationale Supèrieure d'Ingénieria Electriciens de Grenoble; Karl Aström et al. at Lund Institute of Technology, Department of Automatic Control; Graham Goodwin et al. at the University of Newcastle, Dept. of Electrical Engineering and Computer Science; and Kumpati Narendra et al. and Stephen Morse et al. at Yale University, Dept. of Electrical Engineering, each working independently.

Perhaps one of the cornerstones in adaptive control was set in 1980, when the complete adaptive control problem for the ideal case (time-invariant linear systems under no perturbations) was resolved in four different general cases and published in 1980 in the memorable June issue of the IEEE Transactions on Automatic Control, all solutions in the same issue, so that all the authors could share this privilege [14, 16, 20, 21].

Soon after the ideal case was solved, researchers switched their attention to relaxing the four basic hypotheses under which the original solution was obtained (namely, stable zeros of the plant, sign of the high frequency gain known, order of the plant known or an upper bound thereof, and knowledge of the plant relative degree) and also to developing solutions for the case of plants with time-varying parameters and under external perturbations, giving rise to the robust adaptive control theory. In 1988, a generalization of the MRAC was presented in [19], called combined MRAC (CMRAC), where the basic idea was to combine the identification and the control processes being performed on a plant, such that the information obtained from each one could be shared to achieve the control objective, i.e., the output of the plant follows asymptotically the output of the model reference. Since more information is being used to control the plant, the CMRAC is supposed to exhibit a better behavior as compared with the direct control and the indirect control when used separately. In this combined scheme, the control adaptive laws as well as the parameter estimation updating laws make use of the control error, the identification error and the so-called closed-loop estimation errors [8, 9].

Another interesting characteristic of the CMRAC is that it can be viewed as a generalized MRAC scheme from which direct MRAC (DMRAC) and dynamical indirect MRAC (DIMRAC) can be obtained as particular cases. It is important to point out that the DIMRAC uses the same prior information as the algebraic indirect MRAC (AIMRAC), and it proves that both control schemes (direct and indirect) are equivalent, as far as the prior information needed for the solution is concerned. Remember that AIMRAC requires, besides the sign of the high frequency gain, the knowledge of an upper bound on the high frequency gain (if it is negative) or of a lower bound on the high frequency gain (if it is positive). This is to avoid division by zero of the high frequency gain estimate in the AIMRAC when solving the certainty equivalence principle on which the AIMRAC is based [18].

In this chapter, we extend the ideas of the DMRAC and the CMRAC, developed for integer-order plants with integer-order adaptive laws, to the case when the adaptive laws for control and estimation are updated using derivatives of fractional order. The study is performed for scalar first-order time-invariant plants, although the extensions to the vector case are currently under investigation.

The use of fractional adaptive laws in the DMRAC for integer-order plants has been reported at simulation level (see for instance [1, 24, 26]), however, as far as the authors know, analytical results are not available in the technical literature for this particular scheme at this moment. Some advances have been made using the error model approach for the analysis, but in this chapter, the analysis is detailed specifically for the DMRAC. In the case of the combined approach, no results are available in the technical literature, either analytical or at simulation level, thus the results presented in this chapter are novel, even being developed for a simple scalar plant.

The chapter is organized as follows. After the basic concepts are introduced in Sect. 6.2, the control problem is presented in Sect. 6.3. The direct approach is introduced in Sect. 6.4, together with the analytical proof of boundedness for all the signals in the fractional adaptive scheme, and also some conclusions about the evolution of the errors. The corresponding results for the combined approach are presented in Sect. 6.5. Simulations comparing the behavior of both approaches are presented in Sect. 6.6, where some general comments are stated on their properties.

6.2 Preliminaries

The Riemann–Liouville fractional integral is one of the main concepts of fractional calculus, and is presented in Definition 6.1.

Definition 6.1 (*Riemann–Liouville fractional integral* [15]) The Riemann–Liouville fractional integral of order $\alpha > 0$ of a function $f(t) \in \mathbb{R}$ is defined as

$$I_{t_0}^{\alpha} f(t) = \frac{1}{\Gamma(\alpha)} \int_{t_0}^{t} \frac{f(\tau)}{(t-\tau)^{1-\alpha}} d\tau, \quad t > t_0, \tag{6.1}$$

where $\Gamma(\alpha)$ corresponds to the Gamma Function [15], given by

$$\Gamma(\alpha) = \int_{0}^{\infty} t^{\alpha-1} e^{-t} dt.$$

Regarding the fractional derivative of order $\alpha > 0$ of a function, there exist several definitions. The results presented in this work use the Caputo definition given in Definition 6.2, which has been used in literature for modeling systems and also as part of controllers.

Definition 6.2 (*Caputo Fractional Derivative*) Let $\alpha \geq 0$ and $n = [\alpha]$. According to [6, Definition 3.1], the Caputo fractional derivative of order α is defined as follows

$$ {}_C D_{t_0}^{\alpha} f(t) = \frac{1}{\Gamma(n-\alpha)} \int_{t_0}^{t} \frac{f^{(n)}(\tau)}{(t-\tau)^{\alpha-n+1}} d\tau \tag{6.2} $$

whenever $f^{(n)} \in L_1[t_0, t]$.

Lemma 6.1 ([7]) *Let $x(t) \in \mathbb{R}^n$ be a vector of differentiable functions. Then, for all $t \geq t_0$, the following relationship holds*

$$ \frac{1}{2} {}_C D_{t_0}^{\alpha} \left(x^T(t) P x(t) \right) \leq x^T(t) P \, {}_C D_{t_0}^{\alpha} x(t), \tag{6.3} $$

where $\alpha \in (0, 1]$ and $P \in \mathbb{R}^{n \times n}$ is a constant, square, symmetric, and positive definite matrix.

Particular cases of relation (6.3) were proposed in [3, 4].

Lemma 6.2 ([2]) *Let $x(\cdot) : \mathbb{R}^+ \to \mathbb{R}$ be a bounded nonnegative function. If there exists some $\alpha \in (0, 1]$ such that*

$$ \frac{1}{\Gamma(\alpha)} \int_{t_0}^{t} \frac{x(\tau)}{(t-\tau)^{1-\alpha}} d\tau < M, \quad \forall t \geq t_0, \quad with \ M \in (0, \infty) \tag{6.4} $$

then

$$ \lim_{t \to \infty} \left[t^{\alpha-\varepsilon} \frac{\int_{t_0}^{t} x(\tau) d\tau}{t} \right] = 0, \qquad \forall \varepsilon > 0 \tag{6.5} $$

6.3 The Control Problem

Let an integer-order plant with input–output pair $\{u(\cdot), x_p(\cdot)\}$ be described by the scalar integer-order differential equation (IODE)

$$ \dot{x}_p(t) = a_p x_p(t) + k_p u(t), \tag{6.6} $$

where a_p, k_p are unknown plant parameters, but the sgn (k_p) is assumed to be known.

A reference model is described by the first-order IODE

$$ \dot{x}_m(t) = a_m x_m(t) + k_m r(t), \tag{6.7} $$

where $a_m < 0$, a_m and k_m are known constants and r is a piecewise continuous bounded function of time. It is assumed that a_m, k_m, and r have been chosen so that

$x_m(t)$ represents the desired output of the plant at time t. The aim here is to determine a bounded control input $u(t)$ so that all the signals in the adaptive scheme remain bounded and

$$\lim_{t\to\infty} |x_p(t) - x_m(t)| = 0. \tag{6.8}$$

If the plant parameters were known, the plant behavior could be modified using a control signal $u(t)$ that includes feedforward and feedback gains, having the form

$$u(t) = \theta^* x_p(t) + k^* r(t), \tag{6.9}$$

where

$$\theta^* \triangleq \frac{a_m - a_p}{k_p} \quad \text{and} \quad k^* \triangleq \frac{k_m}{k_p}. \tag{6.10}$$

It can be seen that applying the control signal (6.9) in (6.6), the transfer function of the plant together with the control will be the same as that of the reference model (6.7). In that way, control objective (6.8) can be achieved.

However, control signal (6.9) cannot be practically implemented, since the plant parameters a_p and k_p are unknown. Thus, feedforward and feedback gains θ^*, k^* given by (6.10) cannot be computed.

One possible solution to this problem is to use adjustable parameters $\theta(t)$ and $k(t)$, which estimate the unknown parameters θ^* and k^*, respectively. Thus, the control signal to be applied to the plant will be given by

$$u(t) = \theta(t) x_p(t) + k(t) r(t). \tag{6.11}$$

How to obtain the estimated parameters $\theta(t)$ and $k(t)$ is a question that has, in fact, several answers. For instance, three approaches can be found in [18], namely direct, indirect algebraic, and indirect dynamic MRAC. One more can be found in [8], namely the combined MRAC. All these four approaches use the same control signal (6.11), the main difference among them being the procedure for calculating $\theta(t)$ and $k(t)$. Nevertheless, one common point among the four approaches is that they use integer-order differential equations to adaptively estimate the controller parameters and/or the plant parameters, e.g., integer-order adaptive laws.

In this work, fractional adaptive laws are introduced to estimate the parameters adaptively, instead of using the integer-order adaptive laws proposed in the above-mentioned works. Since IODE can be seen as particular cases of fractional-order differential equations (FODE), the fractional adaptive laws proposed here can be seen as a generalization of the classic integer-order adaptive laws found in [8, 18]. Eventhough there exist four possible cases, in this work, we will focus only on two of them; the direct fractional-order MRAC (DFOMRAC) and the combined fractional-order MRAC (CFOMRAC). Analysis of signal boundedness in these two schemes is detailed in Sects. 6.4 and 6.5, respectively, together with some results regarding the convergence of the errors.

6.4 The Direct Approach

Although the real desired control parameters θ^* and k^* depend on the plant parameters a_p, k_p, no attempt to identify them is made within the direct approach. Instead, the controller parameters $\theta(t), k(t)$ are directly adjusted and these estimates used in (6.11) to compute the control signal $u(t)$.

Using (6.11) in (6.6), the following expression results for plant + control signal

$$\dot{x}_p(t) = \left(a_p + k_p\theta(t)\right) x_p(t) + k_p k(t) r(t). \tag{6.12}$$

Defining the control error as

$$e_c(t) \triangleq x_p(t) - x_m(t) \tag{6.13}$$

and the parameter errors as

$$\phi_\theta(t) \triangleq \theta(t) - \theta^* \quad \text{and} \quad \phi_k(t) \triangleq k(t) - k^* \tag{6.14}$$

we can obtain the error equation by subtracting (6.7) from (6.12), which yields

$$\dot{e}_c(t) = a_m e_c(t) + k_p\phi_\theta(t) x_p(t) + k_p\phi_k(t) r(t), \quad a_m < 0. \tag{6.15}$$

In this approach, the controller parameters $\theta(t), k(t)$ are directly adjusted, using the following fractional adaptive laws

$$\begin{aligned} {}_C D_{t_0}^{\alpha}\theta(t) &= -\gamma_1 \mathrm{sgn}\left(k_p\right) e_c(t) x_p(t), \\ {}_C D_{t_0}^{\alpha}k(t) &= -\gamma_2 \mathrm{sgn}\left(k_p\right) e_c(t) r(t), \end{aligned} \tag{6.16}$$

where $\alpha \in (0, 1]$ is the order of the adaptive laws and $\gamma_1, \gamma_2 \in \mathbb{R}^+$ are the adaptive gains, used to handle the convergence speed.

Summarizing, Table 6.1 contains all the details of the adaptive scheme proposed in this section, which will be referred as DFOMRAC. Figure 6.1, on the other hand,

Table 6.1 DFOMRAC implementation details

Plant	$\dot{x}_p(t) = a_p x_p(t) + k_p u(t)$
Reference model	$\dot{x}_m(t) = a_m x_m(t) + k_m r(t), \quad a_m < 0$
Control law	$u(t) = \theta(t) x_p(t) + k(t) r(t)$
Control error	$e_c(t) = x_p(t) - x_m(t)$
Adaptive laws	${}_C D_{t_0}^{\alpha}\theta(t) = -\gamma_1 \mathrm{sgn}\left(k_p\right) e_c(t) x_p(t)$
	${}_C D_{t_0}^{\alpha}k(t) = -\gamma_2 \mathrm{sgn}\left(k_p\right) e_c(t) r(t)$, $\alpha \in (0, 1]$

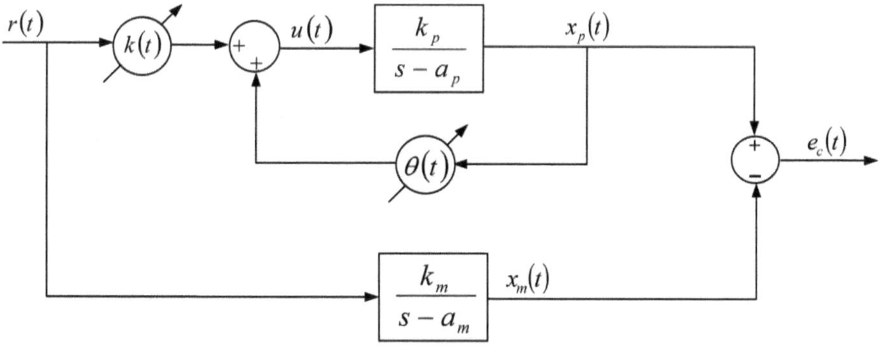

Fig. 6.1 Block diagram for the DFOMRAC scheme

shows the corresponding block diagram, which helps to understand the structure of the controller.

6.4.1 Boundedness of the Signals in the DFOMRAC Scheme

As it was stated above, the evolution of the control error $e_c(t)$ is given by (6.15). In the case of the parameter errors $\phi_\theta(t)$ and $\phi_k(t)$, given by (6.14), their Caputo fractional derivatives coincide with the Caputo fractional derivatives of the estimated parameters (6.16). Thus, the behavior of the DFOMRAC scheme is represented by the following set of differential equations

$$
\begin{aligned}
\dot{e}_c(t) &= a_m e_c(t) + k_p \phi_\theta(t) x_p(t) + k_p \phi_k(t) r(t), \quad a_m < 0, \quad e_c(t_0) = e_{c0} \\
{}_C D_{t_0}^{\alpha} \phi_\theta(t) &= -\gamma_1 \mathrm{sgn}(k_p) e_c(t) x_p(t), \quad \phi_\theta(t_0) = \phi_{\theta_0} \\
{}_C D_{t_0}^{\alpha} \phi_k(t) &= -\gamma_2 \mathrm{sgn}(k_p) e_c(t) r(t), \quad \phi_k(t_0) = \phi_{k_0}
\end{aligned}
\tag{6.17}
$$

To prove boundedness of signals in the adaptive scheme (6.17), the following hypotheses are needed.

1. The parameter errors ϕ_θ and ϕ_k are differentiable and uniformly continuous functions.
2. The control error e_c is a uniformly continuous function.

The differentiability of the parameter errors is required in order to use Lemma 6.1 in the proof. For the differential system (6.17), where the order of the output error differential equation is different from the order of the parameter errors differential equations (mixed order), differentiability of ϕ_θ and ϕ_k has not been analytically proved yet. Some important advances have been recently made in this direction [13], however, the complete proof of differentiability for this particular problem is not yet available.

In the case of the uniform continuity of the parameter errors and the control error, it has not been proved analytically neither, although research is being conducted in this direction as well.

Nevertheless, it should be mentioned that extensive simulation studies performed show that for several bounded piecewise continuous reference signals $r(t)$ tested in scheme (6.17), the parameter errors result in differentiable and uniformly continuous functions and the control error was uniformly continuous as well.

Proof Since $\phi_\theta(t)$ and $\phi_k(t)$ are differentiable functions, Lemma 6.1 can be applied to obtain the following expression

$$\frac{1}{2}\frac{\mathrm{d}}{\mathrm{dt}}e_c^2(t) + \frac{|k_p|}{2\gamma_1}{}_C D_{t_0}^{\alpha}\left\{\phi_\theta^2\right\}(t) + \frac{|k_p|}{2\gamma_2}{}_C D_{t_0}^{\alpha}\left\{\phi_k^2\right\}(t) \le \begin{bmatrix} e_c(t)\dot{e}_c(t) + \\ \frac{|k_p|}{\gamma_1}\phi_\theta(t)\,{}_C D_{t_0}^{\alpha}\phi_\theta(t) + \\ \frac{|k_p|}{\gamma_2}\phi_k(t)\,{}_C D_{t_0}^{\alpha}\phi_k(t) \end{bmatrix}. \tag{6.18}$$

Using (6.17) in the right-hand side of (6.18), it follows that

$$\frac{1}{2}\frac{\mathrm{d}}{\mathrm{dt}}e_c^2(t) + \frac{|k_p|}{2\gamma_1}{}_C D_{t_0}^{\alpha}\left\{\phi_\theta^2\right\}(t) + \frac{|k_p|}{2\gamma_2}{}_C D_{t_0}^{\alpha}\left\{\phi_k^2\right\}(t) \le a_m e_c^2(t). \tag{6.19}$$

If the integer-order integral is applied to expression (6.19), then it results

$$\frac{1}{2}e_c^2(t) - \frac{1}{2}e_{c0}^2 + \frac{|k_p|}{2\gamma_1}\int_{t_0}^{t}{}_C D_{t_0}^{\alpha}\left\{\phi_\theta^2\right\}(\tau)\,d\tau + \frac{|k_p|}{2\gamma_2}\int_{t_0}^{t}{}_C D_{t_0}^{\alpha}\left\{\phi_k^2\right\}(\tau)\,d\tau \le a_m\int_{t_0}^{t}e_c^2(\tau)\,d\tau. \tag{6.20}$$

Using properties of the fractional integrals and derivatives given in [6, Definition 3.1, Corollary 2.3], it follows that

$$\frac{|k_p|}{2\gamma_1}\int_{t_0}^{t}{}_C D_{t_0}^{\alpha}\left\{\phi_\theta^2\right\}(\tau)\,d\tau = \frac{|k_p|}{2\gamma_1}I_{t_0}^{1-\alpha}\left\{\phi_\theta^2 - \phi_{\theta_0}^2\right\}(t), \tag{6.21}$$

$$\frac{|k_p|}{2\gamma_2}\int_{t_0}^{t}{}_C D_{t_0}^{\alpha}\left\{\phi_k^2\right\}(\tau)\,d\tau = \frac{|k_p|}{2\gamma_2}I_{t_0}^{1-\alpha}\left\{\phi_k^2 - \phi_{k_0}^2\right\}(t). \tag{6.22}$$

Thus, if the following notation is used

$$\phi(t) \triangleq \left[\frac{\phi_\theta(t)}{\sqrt{\gamma_1}} \quad \frac{\phi_k(t)}{\sqrt{\gamma_2}}\right]^T \quad \text{and} \quad \phi_0 \triangleq \left[\frac{\phi_{\theta_0}}{\sqrt{\gamma_1}} \quad \frac{\phi_{k_0}}{\sqrt{\gamma_2}}\right]^T,$$

then substituting (6.21) and (6.22) in (6.20) and rearranging terms it follows that

$$\frac{1}{2}e_c^2(t) - \frac{1}{2}e_{c_0}^2 - a_m \int_{t_0}^{t} e_c^2(\tau)\,d\tau + \frac{|k_p|}{2} I_{t_0}^{1-\alpha}\left\{\phi^T\phi - \phi_0^T\phi_0\right\}(t) \le 0. \tag{6.23}$$

We will now prove boundedness of all the signals, by contradiction, using expression (6.23).

Assume that $\phi(t)$ is not bounded. Since $\phi(t)$ is uniformly continuous, then intervals of increasing length will exist where $\phi^T(t)\phi(t) > L > \phi_0^T\phi_0$. Thus, the fractional integral given by $I^{1-\alpha}\left\{\phi^T\phi - \phi_0^T\phi_0\right\}(t)$ in (6.23) will diverge. This contradicts the fact that expression (6.23) is always nonpositive, because the terms $e_c^2(t)$ and $-a_m \int_{t_0}^{t} e_c^2(\tau)\,d\tau$ are always non negative and the nonpositive term $-e_{c_0}^2$ is bounded, contradicting the hypothesis that $\phi(t)$ is unbounded. Thus, $\phi(t)$ remains bounded $\forall t \ge t_0$.

The lowest value that the fractional integral in (6.23) can achieve would be for the case when $\phi^T\phi = 0\ \forall t \ge 0$. Thus, from (6.23) we can write the following

$$\frac{1}{2}e_c^2(t) - \frac{1}{2}e_{c_0}^2 - a_m \int_{t_0}^{t} e_c^2(\tau)\,d\tau \le \frac{|k_p|\phi_0^T\phi_0}{2\Gamma(1-\alpha)}(t-t_0)^{1-\alpha}. \tag{6.24}$$

Assume now that $e_c(t)$ is not bounded. Since $e_c(t)$ is uniformly continuous, then intervals of increasing length will exist where $e_c^2(t) > C > 0$. Thus, the integral given by $-a_m \int_{t_0}^{t} e_c^2(\tau)\,d\tau$ in (6.24) will diverge as $t \to \infty$. Since it is an integer-order integral, its grow rate will be higher than $t^{1-\alpha}$, which establishes a contradiction in (6.24) as $t \to \infty$. Thus, it can be concluded that $e_c(t)$ remains bounded $\forall t \ge t_0$.

Since $e_c(t) = x_p(t) - x_m(t)$ is bounded and $x_m(t)$ is the output of an asymptotically stable system with a bounded input $r(t)$, then it implies that $x_p(t)$ remains bounded as well.

Given that the estimated parameters $\theta(t), k(t)$ remain bounded, this result also implies that the control signal $u(t)$ (6.11) will remain bounded.

Moreover, using all these conclusions in (6.17) it follows that $\dot{e}_c(t)$, ${}_C D_{t_0}^{\alpha}\phi_\theta(t)$ and ${}_C D_{t_0}^{\alpha}\phi_k(t)$ remain bounded as well.

Thus, all the signals in the DFOMRAC scheme remain bounded, and this concludes the proof. □

6.4.2 Convergence of the Parameter Errors in the DFOMRAC Scheme

Although the convergence of parameter errors $\phi_\theta(t)$, $\phi_k(t)$ to zero is not an objective in this adaptive control scheme, but only the convergence of the control error to zero, it is interesting to analyze under which conditions it can be achieved.

In the integer-order case, the concept of Persistent Excitation (PE) [17] is used to state what properties of the input signal $r(t)$ must hold in order to guarantee the parameter error convergence to zero in the control scheme.

The concept of PE in the integer-order case arose naturally in the identification problems around the 1960s, but in 1966 it was formally stated in [5], related to identification problems. In the case of the adaptive control, the concept of PE was also addressed in [17].

In the case of fractional adaptive control, like the problem addressed in this chapter, the concept of PE has not been fully developed so far. However, some advances have been made, which can be applied to the DFOMRAC proposed in Sect. 6.4, for the particular case when $k_p = \gamma_1 = \gamma_2 = 1$. The analysis for the rest of the cases is currently under investigation. In what follows, the available results are briefly introduced.

Equations in (6.17) can be compactly rewritten as

$$\begin{aligned}\dot{e}_c(t) &= a_m e_c(t) + \phi^T(t)\,\omega(t)\\ {}_C D_{t_0}^{\alpha}\phi(t) &= -e_c(t)\,\omega(t)\end{aligned} \tag{6.25}$$

where $\phi(t) \triangleq [\phi_\theta(t) \quad \phi_k(t)]^T$ and $\omega(t) \triangleq \left[x_p(t) \quad r(t)\right]^T$. The type of mixed order fractional differential equations like (6.25), has been studied in [11].

Let us analyze the vector of functions defined as

$$\omega_m(t) \triangleq [x_m(t) \quad r(t)]^T \tag{6.26}$$

which can be written in Laplace domain as

$$\omega_m(s) = [H_m(s) \quad 1]^T r, \tag{6.27}$$

where $H_m(s) = k_m/(s - a_m)$ is the transfer function of the reference model (6.7), which is asymptotically stable ($a_m < 0$ by definition).

From arguments in the proofs of [23, Theorem 2.7.2] and [23, Theorem 2.7.2], it follows that if $r(t)$ has at least 2 spectral lines then $\omega_m(t) \in PE(2)$, that is, $\omega_m(t)$ is a persistently exciting function in a space of dimension 2 (see e.g., [23, Eq. (2.5.3)] for a definition of $PE(2)$).

Let us now focus in the vector $\omega(t)$. It can be seen that, according to their definitions,

$$\omega_m(t) - \omega(t) = \left[x_m(t) - x_p(t) \quad 0\right]^T = [-e_c(t) \quad 0]^T \tag{6.28}$$

Since $I^{\alpha}\left\{e_c^2\right\}(t) < \infty$, as it was proved in Sect. 6.4.1, by [11, Theorem 1(iv)] it can be concluded that $\omega(t) \in PE(2)$. From [12, Sect. 3], which allows to conclude that

$$\lim_{t\to\infty} \phi(t) = 0. \tag{6.29}$$

6.4.3 *Convergence of the Control Error in the DFOMRAC Scheme*

As it was stated in previous section, it can be concluded that, for the particular case when $k_p = \gamma_1 = \gamma_2 = 1$, if $r(t)$ has at least 2 spectral lines, then $\lim_{t\to\infty} \phi(t) = 0$. Since $\omega(t)$ is bounded, then it follows that $\lim_{t\to\infty} \phi^T(t)\,\omega(t) = 0$. Using [12, Theorem 1], it can be concluded that $\lim_{t\to\infty} e_c(t) = 0$. That is, it can be assured that in this case

$$\text{If}\quad r(t) \in PE(2) \quad \text{then} \quad \lim_{t\to\infty} e_c(t) = 0. \tag{6.30}$$

The statement in (6.30) does not mean that the control error will not converge to zero if the reference signal $r(t) \notin PE(2)$, but these are the only cases where convergence to zero of the control error has been analytically proved. In fact, extensive simulation studies have shown that $e_c(t)$ also converges to zero for many $r(t) \notin PE(2)$, the most usual reference signals such as steps, constant references, sinusoidal references, etc. It also converges to zero for those cases when k_p and $\gamma_1, \gamma_2 > 0$ are different from 1.

Although the convergence to zero of the control error $e_c(t)$ cannot be analytically assured for all bounded piecewise continuous reference signals at this moment, due to a lack of tools to perform the analysis, it can be proved that the mean value of the squared control error converges to zero as $t \to \infty$, under the same hypothesis of differentiability of $\phi_\theta(t), \phi_k(t)$ used for the proof of boundedness in Sect. 6.4.1. In what follows, the corresponding proof is detailed.

Proof Let us consider expression (6.19), which is derived based on the differentiability of $\phi_\theta^2(t)$ and $\phi_k^2(t)$. If the fractional integral of order α is applied to (6.19), then rearranging terms, the following inequality is obtained.

$$-a_m I_{t_0}^{\alpha}\left\{e_c^2\right\}(t) \le \frac{|k_p|}{2\gamma_1}\left[\phi_{\theta_0}^2 - \phi_\theta^2(t)\right] + \frac{|k_p|}{2\gamma_2}\left[\phi_{k_0}^2 - \phi_k^2(t)\right] - I_{t_0}^{\alpha}\left\{e_c\dot{e}_c\right\}(t) \tag{6.31}$$

Since $-a_m I_{t_0}^{\alpha}\left\{e_c^2\right\}(t) \ge 0$ and $\phi_\theta(t), \phi_k(t)$ are bounded, it follows from (6.31) that $I_{t_0}^{\alpha}\left\{e_c\dot{e}_c\right\}(t)$ is upper bounded. Also, given that $e_c(t)$ and $\dot{e}_c(t)$ are bounded, then it follows from [10, Proposition 1] that $I_{t_0}^{\alpha}\left\{e_c\dot{e}_c\right\}(t)$ is uniformly continuous.

Let us now assume that $I_{t_0}^{\alpha}\left\{e_c\dot{e}_c\right\}(t)$ is not bounded. Since it is upper bounded, the only possibility is that $I_{t_0}^{\alpha}\left\{e_c\dot{e}_c\right\}(t) \to -\infty$. Thus, since $I_{t_0}^{\alpha}\left\{e_c\dot{e}_c\right\}(t)$ is uniformly continuous, we can state that

$$I_{t_0}^{1-\alpha}\left\{I_{t_0}^{\alpha}\left\{e_c\dot{e}_c\right\}\right\}(t) \to -\infty,$$

that according to [15, Lemma 2.3] can be written as

$$\int_{t_0}^{t} e_c(\tau)\,\dot{e}_c(\tau)\,d\tau = \frac{1}{2}e_c^2(t) - \frac{1}{2}e_{c_0}^2 \to -\infty.$$

which contradicts the fact that $e_c(t)$ is bounded. Therefore, it can be concluded that $I_{t_0}^{\alpha}\{e_c\dot{e}_c\}(t)$ is lower bounded.

Thus, since $\phi_\theta(t)$, $\phi_k(t)$ and $I_{t_0}^{\alpha}\{e_c\dot{e}_c\}(t)$ are bounded, we can conclude from (6.31) that $-a_m I_{t_0}^{\alpha}\left\{e_c^2\right\}(t)$ is bounded as well.

Based on this result and using the fact that $e_c^2(t)$ is nonnegative and bounded, we can apply Lemma 6.2, which allows concluding that

$$\lim_{t\to\infty}\left[t^{\alpha-\varepsilon}\frac{\int_{t_0}^{t} e_c^2(\tau)\,d\tau}{t}\right] = 0, \quad \forall\varepsilon > 0, \tag{6.32}$$

that is, the mean value of the squared control error is $o\left(t^{\varepsilon-\alpha}\right)$, $\forall\varepsilon > 0$, which means that it converges asymptotically to zero, with a convergence speed higher than $t^{-\alpha}$. This concludes the proof.

6.5 The Combined Approach

In the direct approach presented in Sect. 6.4, the control of the plant was carried out directly adjusting the control parameters $\theta(t)$ and $k(t)$, without explicitly estimating the unknown plant parameters a_p, k_p.

However, as it was mentioned in Sect. 6.3, if the controller parameters are selected as

$$\theta(t) = \theta^* = \frac{a_m - a_p}{k_p} \quad \text{and} \quad k(t) = k^* = \frac{k_m}{k_p}, \tag{6.33}$$

then the plant transfer function would be equal to the reference model transfer function, fulfilling the control objective (6.8).

However, since the plant parameters a_p, k_p are unknown, the direct approach estimates the parameters $\theta(t), k(t)$ based on the control error $e_c(t)$ and available input signals $x_p(t), r(t)$, but no information regarding the unknown plant parameters is incorporated.

Nevertheless, the above relations (6.33) motivated an alternative method where the unknown plant parameters a_p, k_p are estimated through $\hat{a}_p(t)$ and $\hat{k}_p(t)$ and used inside the adaptive control scheme [8], namely the combined MRAC (CMRAC).

It can be seen from (6.33) that if

$$\lim_{t\to\infty}\left[\hat{a}_p(t)+\hat{k}_p(t)\theta(t)\right]=a_m$$
$$\lim_{t\to\infty}\left[\hat{k}_p(t)k(t)\right]=k_m \tag{6.34}$$

where $\hat{a}_p(t)\in\mathbb{R}$ and $\hat{k}_p(t)\in\mathbb{R}$ correspond to the estimates of the unknown plant parameters a_p and k_p, respectively. Then the control objective (6.8) is fulfilled.

With that information, the closed-loop estimation errors are defined as

$$\varepsilon_\theta(t)=\hat{a}_p(t)+\hat{k}_p(t)\theta(t)-a_m \tag{6.35}$$

$$\varepsilon_k(t)=\hat{k}_p(t)k(t)-k_m \tag{6.36}$$

The idea now is to use these closed-loop estimation errors to adaptively estimate the controller parameters $\theta(t), k(t)$ and the estimated plant parameters $\hat{a}_p(t), \hat{k}_p(t)$.

To estimate the unknown plant parameters, an estimator is used as in [18], of the form

$$\dot{\hat{x}}_p(t)=a_m\,e_i(t)+\hat{a}_p(t)x_p(t)+\hat{k}_p(t)u(t). \tag{6.37}$$

where $e_i(t)=\hat{x}_p(t)-x_p(t)$ corresponds to the identification error.

Adaptive laws for $\hat{a}_p$ and $\hat{k}_p$ are then proposed, which contain information about both, the identification process and the control process. These adaptive laws are proposed in the form of IODEs in [8], but in this work a generalization is used, where these estimated parameters can be obtained using FODEs, with the form

$$_C D_{t_0}^{\alpha}\hat{a}_p(t)=-\gamma_3 e_i(t)x_p(t)-\gamma_4\varepsilon_\theta(t) \tag{6.38}$$

$$_C D_{t_0}^{\alpha}\hat{k}_p(t)=-\gamma_3 e_i(t)u(t)-\gamma_4\theta(t)\varepsilon_\theta(t)-\gamma_4 k(t)\varepsilon_k(t) \tag{6.39}$$

where $\gamma_3,\gamma_4\in\mathbb{R}^+$ are adaptive gains used to handle the convergence speed of the estimated parameters.

In the case of the controller parameters, they are estimated using the following fractional adaptive laws

$$_C D_{t_0}^{\alpha}\theta(t)=-\gamma_1\mathrm{sgn}\left(k_p\right)e_c(t)x_p(t)-\gamma_2\mathrm{sgn}\left(k_p\right)\varepsilon_\theta(t) \tag{6.40}$$

$$_C D_{t_0}^{\alpha}k(t)=-\gamma_1\mathrm{sgn}\left(k_p\right)e_c(t)r(t)-\gamma_2\mathrm{sgn}\left(k_p\right)\varepsilon_k(t) \tag{6.41}$$

Table 6.2 CFOMRAC implementation details

Plant	$\dot{x}_p(t) = a_p x_p(t) + k_p u(t)$
Reference model	$\dot{x}_m(t) = a_m x_m(t) + k_m r(t), \quad a_m < 0$
Control law	$u(t) = \theta(t) x_p(t) + k(t) r(t)$
Control error	$e_c(t) = x_p(t) - x_m(t)$
Estimator	$\dot{\hat{x}}_p(t) =$ $a_m\, e_i(t) + \hat{a}_p(t) x_p(t) + \hat{k}_p(t) u(t)$.
Identification error	$e_i(t) = \hat{x}_p(t) - x_p(t)$
Closed-loop estimation errors	$\varepsilon_\theta(t) = \hat{a}_p(t) + \hat{k}_p(t)\theta(t) - a_m$
	$\varepsilon_k(t) = \hat{k}_p(t) k(t) - k_m$
Adaptive laws	${}_C D_{t0}^{\alpha}\theta(t) =$ $-\gamma_1 \mathrm{sgn}(k_p) e_c(t) x_p(t) - \gamma_2 \mathrm{sgn}(k_p) \varepsilon_\theta(t)$
	${}_C D_{t0}^{\alpha} k(t) =$ $-\gamma_1 \mathrm{sgn}(k_p) e_c(t) r(t) - \gamma_2 \mathrm{sgn}(k_p) \varepsilon_k(t)$
	${}_C D_{t0}^{\alpha}\hat{a}_p(t) = -\gamma_3 e_i(t) x_p(t) - \gamma_4 \varepsilon_\theta(t)$
	${}_C D_{t0}^{\alpha}\hat{k}_p(t) = -\gamma_3 e_i(t) u(t) - \gamma_4 \theta(t) \varepsilon_\theta(t) -$ $\gamma_4 k(t) \varepsilon_k(t), \quad \alpha \in (0, 1]$

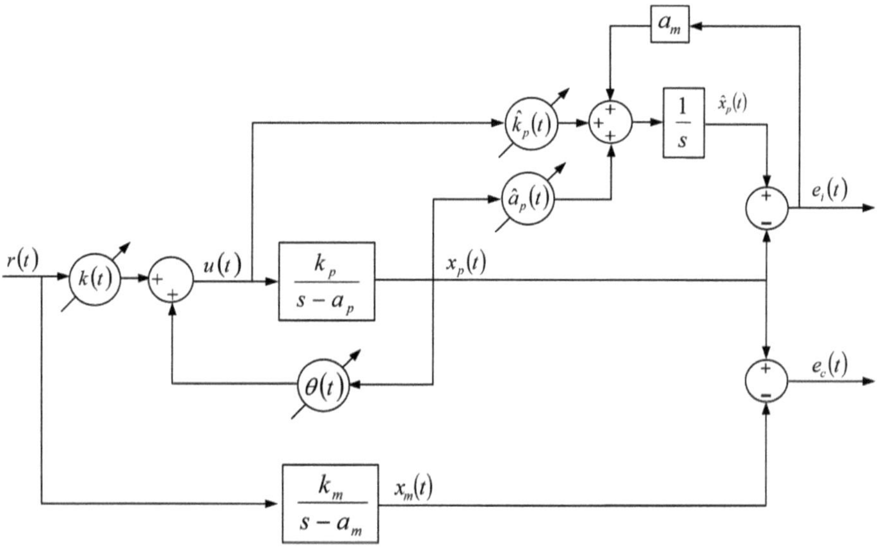

Fig. 6.2 Block diagram for the CFOMRAC scheme

where $e_c(t) = x_p(t) - x_m(t)$ is the control error and $\gamma_1, \gamma_2 \in \mathbb{R}^+$ are the adaptive gains used to handle the convergence speed of the controller parameters.

To summarize, Table 6.2 contains the details of the adaptive scheme detailed in this section, which will be referred as Combined FOMRAC scheme (CFOMRAC). Also, Fig. 6.2 shows the corresponding block diagram.

6.5.1 Boundedness of the Signals in the CFOMRAC Scheme

For this CFOMRAC, the evolution of the control error $e_c(t)$ is described by the same IODE than in the DFOMRAC (6.15).

In the case of the identification error $e_i(t)$, if we subtract (6.6) from (6.37), it can be obtained that

$$\dot{e}_i(t) = a_m e_i(t) + \phi_a(t)\,x_p(t) + \phi_p(t)\,u(t), \quad a_m < 0 \tag{6.42}$$

where

$$\phi_a(t) \triangleq \hat{a}_p(t) - a_p \qquad \text{and} \qquad \phi_p(t) \triangleq \hat{k}_p(t) - k_p \tag{6.43}$$

correspond to the plant parameter errors.

Regarding the parameter errors $\phi_a(t)$, $\phi_p(t)$, $\phi_\theta(t)$ and $\phi_k(t)$, according to their definitions, their evolution can be described by the same FODE than the estimated parameters $\hat{a}_p(t)$, $\hat{k}_p(t)$ and the controller parameters $\theta(t)$ and $k(t)$, (6.38)–(6.41), respectively.

Thus, the behavior of the adaptive scheme in this CFOMRAC is represented by the following set of differential equations

$$\begin{aligned}
\dot{e}_c(t) &= a_m e_c(t) + k_p\phi_\theta(t)\,x_p(t) + k_p\phi_k(t)\,r(t), \quad a_m < 0, \quad e_c(t_0) = e_{c0}\\
\dot{e}_i(t) &= a_m e_i(t) + \phi_a(t)\,x_p(t) + \phi_p(t)\,u(t), \quad e_i(t_0) = e_{i_0}\\
\varepsilon_\theta(t) &= \hat{a}_p(t) + \hat{k}_p(t)\,\theta(t) - a_m\\
\varepsilon_k(t) &= \hat{k}_p(t)\,k(t) - k_m\\
{}_C D_{t_0}^{\alpha}\phi_\theta(t) &= -\gamma_1 \operatorname{sgn}(k_p)\,e_c(t)\,x_p(t) - \gamma_2 \operatorname{sgn}(k_p)\,\varepsilon_\theta(t), \quad \phi_\theta(t_0) = \phi_{\theta_0}\\
{}_C D_{t_0}^{\alpha}\phi_k(t) &= -\gamma_1 \operatorname{sgn}(k_p)\,e_c(t)\,r(t) - \gamma_2 \operatorname{sgn}(k_p)\,\varepsilon_k(t), \quad \phi_k(t_0) = \phi_{k_0}\\
{}_C D_{t_0}^{\alpha}\phi_a(t) &= -\gamma_3 e_i(t)\,x_p(t) - \gamma_4 \varepsilon_\theta(t), \quad \phi_a(t_0) = \phi_{a0}\\
{}_C D_{t_0}^{\alpha}\phi_p(t) &= -\gamma_3 e_i(t)\,u(t) - \gamma_4 \theta(t)\,\varepsilon_\theta(t) - \gamma_4 k(t)\,\varepsilon_k(t), \quad \phi_p(t_0) = \phi_{p0}, \quad \alpha \in (0,1]
\end{aligned} \tag{6.44}$$

To prove boundedness of signals in the adaptive scheme (6.44), the following hypotheses are needed

1. The parameter errors ϕ_θ, ϕ_k, ϕ_a and ϕ_p are differentiable and uniformly continuous functions.
2. The control error e_c and the identification error e_i are uniformly continuous functions.

These hypotheses are needed for the same reasons already exposed in the DFOMRAC. We should mention that also in this CFOMRAC, although these hypotheses have not been analytically proved yet, extensive simulation studies using bounded piecewise continuous reference signals have shown that these hypotheses hold.

Proof Since $\phi_\theta(t)$, $\phi_k(t)$, $\phi_a(t)$ and $\phi_p(t)$ are differentiable functions, Lemma 6.1 can be applied to obtain the following expression

$$\begin{bmatrix} \frac{\gamma_1}{2\gamma_2}\frac{\mathrm{d}}{\mathrm{dt}}e_c^2(t) + \frac{\gamma_3}{2\gamma_4}\frac{\mathrm{d}}{\mathrm{dt}}e_i^2(t) + \\ +\frac{|k_p|}{2\gamma_2}{}_CD_{t_0}^{\alpha}\{\phi_\theta^2\}(t) + \frac{|k_p|}{2\gamma_2}{}_CD_{t_0}^{\alpha}\{\phi_k^2\}(t) + \\ +\frac{1}{2\gamma_4}{}_CD_{t_0}^{\alpha}\{\phi_a^2\}(t) + \frac{1}{2\gamma_4}{}_CD_{t_0}^{\alpha}\left\{\phi_p^2\right\}(t) \end{bmatrix} \le \begin{bmatrix} \frac{\gamma_1}{\gamma_2}e_c(t)\dot{e}_c(t) + \frac{\gamma_3}{\gamma_4}e_i(t)\dot{e}_i(t) + \\ +\frac{|k_p|}{\gamma_2}\phi_\theta(t)\,{}_CD_{t_0}^{\alpha}\phi_\theta(t) + \\ \frac{|k_p|}{\gamma_2}\phi_k(t)\,{}_CD_{t_0}^{\alpha}\phi_k(t) + \\ +\frac{1}{\gamma_4}\phi_a(t)\,{}_CD_{t_0}^{\alpha}\phi_a(t) + \\ \frac{1}{\gamma_4}\phi_p(t)\,{}_CD_{t_0}^{\alpha}\phi_p(t) \end{bmatrix}. \tag{6.45}$$

Using (6.44) in the right-hand side of (6.45), it follows that

$$\begin{bmatrix} \frac{\gamma_1}{2\gamma_2}\frac{\mathrm{d}}{\mathrm{dt}}e_c^2(t) + \frac{\gamma_3}{2\gamma_4}\frac{\mathrm{d}}{\mathrm{dt}}e_i^2(t) + \\ +\frac{|k_p|}{2\gamma_2}{}_CD_{t_0}^{\alpha}\{\phi_\theta^2\}(t) + \frac{|k_p|}{2\gamma_2}{}_CD_{t_0}^{\alpha}\{\phi_k^2\}(t) + \\ +\frac{1}{2\gamma_4}{}_CD_{t_0}^{\alpha}\{\phi_a^2\}(t) + \frac{1}{2\gamma_4}{}_CD_{t_0}^{\alpha}\left\{\phi_p^2\right\}(t) \end{bmatrix} \le \begin{bmatrix} \frac{\gamma_1}{\gamma_2}a_m e_c^2(t) + \frac{\gamma_3}{\gamma_4}a_m e_i^2(t) - \\ -\varepsilon_\theta(t)\left[k_p\phi_\theta(t) + \phi_a(t) + \theta(t)\phi_p(t)\right] - \\ -\varepsilon_k(t)\left[k_p\phi_k(t) + k(t)\phi_p(t)\right] \end{bmatrix}. \tag{6.46}$$

Definitions of the closed-loop estimation errors (6.35) and (6.36) can be algebraically manipulated to obtain

$$\begin{aligned} \varepsilon_\theta(t) &= \hat{a}_p(t) + \hat{k}_p(t)\theta(t) - a_m \\ &= \hat{a}_p(t) + a_p - a_p + \hat{k}_p(t)\theta(t) - k_p\theta(t) + k_p\theta(t) - a_m \\ &= \left(\hat{a}_p(t) - a_p\right) + \theta(t)\left(\hat{k}_p(t) - k_p\right) + k_p\left(\theta(t) - \frac{a_m - a_p}{k_p}\right) \\ &= \phi_a(t) + \theta(t)\phi_p(t) + k_p\phi_\theta(t) \end{aligned} \tag{6.47}$$

and

$$\begin{aligned} \varepsilon_k(t) &= \hat{k}_p(t)k(t) - k_m \\ &= \hat{k}_p(t)k(t) - k_pk(t) + k_pk(t) - k_m \\ &= \left(\hat{k}_p(t) - k_p\right)k(t) + k_p\left(k(t) - \frac{k_m}{k_p}\right). \\ &= \phi_p(t)k(t) + k_p\phi_k(t) \end{aligned} \tag{6.48}$$

Thus, using (6.47), (6.48) in the right-hand side of (6.46) results as follows:

$$\begin{bmatrix} \frac{\gamma_1}{2\gamma_2}\frac{d}{dt}e_c^2(t) + \frac{\gamma_3}{2\gamma_4}\frac{d}{dt}e_i^2(t) + \\ +\frac{|k_p|}{2\gamma_2}{}_C D_{t_0}^{\alpha}\{\phi_\theta^2\}(t) + \frac{|k_p|}{2\gamma_2}{}_C D_{t_0}^{\alpha}\{\phi_k^2\}(t) + \\ +\frac{1}{2\gamma_4}{}_C D_{t_0}^{\alpha}\{\phi_a^2\}(t) + \frac{1}{2\gamma_4}{}_C D_{t_0}^{\alpha}\left\{\phi_p^2\right\}(t) \end{bmatrix} \le \frac{\gamma_1}{\gamma_2}a_m e_c^2(t) + \frac{\gamma_3}{\gamma_4}a_m e_i^2(t) - \varepsilon_\theta^2(t) - \varepsilon_k^2(t) \tag{6.49}$$

If the integer-order integral is applied to expression (6.49), using the same procedure as in the DFOMRAC case, the following expression can be obtained

$$\begin{bmatrix} \frac{\gamma_1}{2\gamma_2}e_c^2(t) - \frac{\gamma_1}{2\gamma_2}e_{c0}^2 + \frac{\gamma_3}{2\gamma_4}e_i^2(t) - \frac{\gamma_3}{2\gamma_4}e_{i0}^2 - \frac{\gamma_1 a_m}{\gamma_2}\int_{t_0}^{t} e_c^2(\tau)\,d\tau - \\ -\frac{\gamma_3 a_m}{\gamma_4}\int_{t_0}^{t} e_i^2(\tau)\,d\tau + \int_{t_0}^{t}\varepsilon_\theta^2(\tau)\,d\tau + \int_{t_0}^{t}\varepsilon_k^2(\tau)\,d\tau + \frac{1}{2}I_{t_0}^{1-\alpha}\left\{\phi^T\phi - \phi_0^T\phi_0\right\}(t) \end{bmatrix} \le 0, \tag{6.50}$$

where

$$\phi(t) \triangleq \left[\sqrt{\frac{|k_p|}{\gamma_2}}\phi_\theta(t) \quad \sqrt{\frac{|k_p|}{\gamma_2}}\phi_k(t) \quad \frac{1}{\sqrt{\gamma_4}}\phi_a(t) \quad \frac{1}{\sqrt{\gamma_4}}\phi_p(t)\right]^T$$

and

$$\phi_0 \triangleq \left[\sqrt{\frac{|k_p|}{\gamma_2}}\phi_{\theta_0} \quad \sqrt{\frac{|k_p|}{\gamma_2}}\phi_{k_0} \quad \frac{1}{\sqrt{\gamma_4}}\phi_{a_0} \quad \frac{1}{\sqrt{\gamma_4}}\phi_{p_0}\right]^T.$$

Using the same proof by contradiction than in the DFOMRAC, it can be proved, based on expression (6.50), that $\phi(t)$, $e_c(t)$, $e_i(t)$ remain bounded $\forall t \ge t_0$.

Based on (6.47) and (6.48), this implies that $\varepsilon_\theta(t)$, $\varepsilon_k(t)$ remain bounded as well.

Since $e_c(t) = x_p(t) - x_m(t)$ is bounded and $x_m(t)$ is the output of an asymptotically stable system with a bounded input $r(t)$, then this implies that $x_p(t)$ remains bounded as well.

Given that $x_p(t)$, $e_i(t) = \hat{x}_p(t) - x_p(t)$ remain bounded, then it follows that $\hat{x}_p(t)$ is also bounded.

Since the estimated parameters $\theta(t)$, $k(t)$ remain bounded, then the control signal $u(t)$ (6.11) will also remain bounded.

Finally, using all these conclusions in (6.44), it follows that $\dot{e}_c(t)$, $\dot{e}_i(t)$, ${}_C D_{t_0}^{\alpha}\phi_\theta(t)$, ${}_C D_{t_0}^{\alpha}\phi_k(t)$, ${}_C D_{t_0}^{\alpha}\phi_a(t)$ and ${}_C D_{t_0}^{\alpha}\phi_p(t)$ remain bounded as well.

Thus, all the signals in the CFOMRAC remain bounded, and this concludes the proof. □

6.5.2 Convergence of the Control Error in the CFOMRAC Scheme

So far, conclusions about the convergence to zero of parameter errors $\phi_\theta(t)$, $\phi_k(t)$, $\phi_a(t)$ and $\phi_p(t)$ cannot be established for the CFOMRAC scheme, since the results used in Sect. 6.4.2 cannot be directly applied in this case. Currently, research is being carried out in this direction.

Also, convergence of the control error $e_c(t)$ to zero has not been analytically established yet for this CFOMRAC scheme, neither for the identification error $e_i(t)$ and the closed-loop estimation errors $\varepsilon_\theta(t)$ and $\varepsilon_k(t)$. However, simulation studies show that these errors converge to zero for the most usual reference signals such as steps, constant references, sinusoidal references, etc. These are topics currently under investigation.

Nevertheless, we can analytically prove that the mean value of the control error, the identification error and the closed-loop estimation errors converge to zero as $t \to \infty$, under the hypotheses of differentiability for the parameter errors $\phi_\theta(t)$, $\phi_k(t)$, $\phi_a(t)$ and $\phi_p(t)$. The proof is stated in what follows.

Proof Let us consider expression (6.49), which can be obtained based on the differentiability of the parameter errors and the application of Lemma 6.1. If the fractional integral of order α is applied to (6.49), by rearranging terms the following inequality can be obtained.

$$\begin{bmatrix} -\dfrac{\gamma_1}{\gamma_2} a_m I_{t_0}^{\alpha}\left\{e_c^2\right\}(t) - \\ -\dfrac{\gamma_3}{\gamma_4} a_m I_{t_0}^{\alpha}\left\{e_i^2\right\}(t) + \\ +I_{t_0}^{\alpha}\left\{\varepsilon_\theta^2\right\}(t) + I_{t_0}^{\alpha}\left\{\varepsilon_k^2\right\}(t) \end{bmatrix} \leq \begin{bmatrix} \dfrac{\gamma_1}{\gamma_2} I_{t_0}^{\alpha}\left\{e_c \dot{e}_c\right\}(t) + \dfrac{\gamma_3}{\gamma_4} I_{t_0}^{\alpha}\left\{e_i \dot{e}_i\right\}(t) + \\ +\dfrac{|k_p|}{2\gamma_2}\left[\phi_\theta^2(t) + \phi_k^2(t) - \phi_{\theta_0}^2 - \phi_{k_0}^2\right] + \\ +\dfrac{1}{2\gamma_4}\left[\phi_a^2(t) + \phi_p^2(t) - \phi_{a_0}^2 - \phi_{p_0}^2\right] \end{bmatrix} \tag{6.51}$$

It can be seen that, since $a_m < 0$, the left-hand side of (6.51) will always be greater or equal than zero. Thus, since $\phi_\theta(t)$, $\phi_k(t)$, $\phi_a(t)$ and $\phi_p(t)$ are bounded, it follows from (6.51) that $I_{t_0}^{\alpha}\{e_c\dot{e}_c\}(t) + I_{t_0}^{\alpha}\{e_i\dot{e}_i\}(t)$ is upper bounded. Also, given that $e_c(t)$, $\dot{e}_c(t)$, $e_i(t)$, $\dot{e}_i(t)$ are bounded, then it follows from [10, Proposition 1] that $I_{t_0}^{\alpha}\{e_c\dot{e}_c\}(t)$ and $I_{t_0}^{\alpha}\{e_i\dot{e}_i\}(t)$ are uniformly continuous.

Let us now assume that $I_{t_0}^{\alpha}\{e_c\dot{e}_c\}(t) + I_{t_0}^{\alpha}\{e_i\dot{e}_i\}(t)$ is not bounded. Since it is upper bounded, the only possibility is that $I_{t_0}^{\alpha}\{e_c\dot{e}_c\}(t) + I_{t_0}^{\alpha}\{e_i\dot{e}_i\}(t) \to -\infty$. Thus, since $I_{t_0}^{\alpha}\{e_c\dot{e}_c\}(t)$ and $I_{t_0}^{\alpha}\{e_i\dot{e}_i\}(t)$ are uniformly continuous, we can state that

$$I_{t_0}^{1-\alpha}\left\{I_{t_0}^{\alpha}\{e_c\dot{e}_c\} + I_{t_0}^{\alpha}\{e_i\dot{e}_i\}\right\}(t) \to -\infty,$$

and according to [15], Lemma 2.3 can be written as

$$\int_{t_0}^{t} e_c(\tau)\,\dot{e}_c(\tau)\,d\tau + \int_{t_0}^{t} e_i(\tau)\,\dot{e}_i(\tau)\,d\tau = \frac{1}{2}e_c^2(t) + \frac{1}{2}e_i^2(t) - \frac{1}{2}e_{c0}^2 - \frac{1}{2}e_{i0}^2 \to -\infty.$$

which contradicts the fact that $e_c(t)$ and $e_i(t)$ are bounded. Thus, it can be concluded that $I_{t_0}^{\alpha}\{e_c\dot{e}_c\}(t) + I_{t_0}^{\alpha}\{e_i\dot{e}_i\}(t)$ is lower bounded.

Since $\phi_\theta(t)$, $\phi_k(t)$, $\phi_a(t)$, $\phi_p(t)$ and $I_{t_0}^{\alpha}\{e_c\dot{e}_c\}(t) + I_{t_0}^{\alpha}\{e_i\dot{e}_i\}(t)$ are bounded, we can conclude from (6.51) that all the fractional integrals in the left-hand side of (6.51) are bounded as well.

Based on this result and using the fact that $e_c^2(t)$, $e_i^2(t)$, $\varepsilon_\theta^2(t)$ and $\varepsilon_k^2(t)$ are nonnegative and bounded, we can apply Lemma 6.2, to conclude that

$$\lim_{t\to\infty}\left[t^{\alpha-\epsilon}\frac{\int_{t_0}^{t} e_c^2(\tau)\,d\tau}{t}\right] = 0, \quad \forall\epsilon > 0, \tag{6.52}$$

$$\lim_{t\to\infty}\left[t^{\alpha-\epsilon}\frac{\int_{t_0}^{t} e_i^2(\tau)\,d\tau}{t}\right] = 0, \quad \forall\epsilon > 0, \tag{6.53}$$

$$\lim_{t\to\infty}\left[t^{\alpha-\epsilon}\frac{\int_{t_0}^{t} \varepsilon_\theta^2(\tau)\,d\tau}{t}\right] = 0, \quad \forall\epsilon > 0, \tag{6.54}$$

$$\lim_{t\to\infty}\left[t^{\alpha-\epsilon}\frac{\int_{t_0}^{t} \varepsilon_k^2(\tau)\,d\tau}{t}\right] = 0, \quad \forall\epsilon > 0, \tag{6.55}$$

The four expressions above can be interpreted as that the mean values of e_c^2, e_i^2, ε_θ^2 and ε_k^2 are $o\left(t^{\epsilon-\alpha}\right)$, $\forall\epsilon > 0$, which means that they converge asymptotically to zero, with a convergence speed higher than $t^{-\alpha}$. This concludes the proof. □

6.6 A Comparative Simulation Study

This section presents a simulation study where the proposed DFOMRAC and CFOMRAC are used to control an integer-order scalar and unstable plant. Simulation studies include different fractional orders for the adaptive laws, different reference signals and also different adaptive gains.

Let the plant to be controlled be given by the following differential equation

$$\dot{x}_p(t) = x_p(t) + u(t), \quad x_p(0) = 0 \tag{6.56}$$

and the reference model be given by

$$\dot{x}_m(t) = -x_m(t) + r(t), \quad x_m(0) = 0 \tag{6.57}$$

For this plant and reference model, the controllers detailed in Tables 6.1 and 6.2 were implemented using Matlab Simulink. In the case of the CFROMAC, the estimator used corresponds to

$$\dot{\hat{x}}_p(t) = -e_i(t) + \hat{a}_p(t)\,x_p(t) + \hat{k}_p(t)\,u(t), \quad \hat{x}_p(0) = 0 \tag{6.58}$$

The fractional adaptive laws were implemented using the NID block included in the Ninteger toolbox for Matlab [25]. The Oustaloup numerical approximation [22] was used for the fractional operator in the NID block, where it is included as the Crone approximation. The initial conditions for the estimated parameters were set to zero in all cases.

6.6.1 *Simulation Environment 1: Influence of the Fractional Orders*

In this first simulation environment, a unit step is used as reference signal and all the adaptive gains were set to 1, that is $\gamma_1 = \gamma_2 = \gamma_3 = \gamma_4 = 1$. The idea here is to show how the fractional order of the adaptive laws influence the signals behavior in the scheme. To that extent, some representative values of the fractional order are used, which correspond to $\alpha = 0.3,\ 0.5,\ 0.7,\ 0.9,\ 1$.

Figure 6.3 shows the evolution of the control error $e_c(t)$ for these values of α in the DFOMRAC, while Fig. 6.4 shows the results for the CFOMRAC.

It can be seen from Figs. 6.3 and 6.4 that the control error remains bounded for every α used, as it was analytically proved in Sects. 6.4.1 and 6.5.1. Moreover, it can be observed that it converges to zero, and the convergence speed is higher as α gets closer to 1, which results advantageous for the classic scheme ($\alpha = 1$). However, the amplitude of the initial oscillations of the control error and consequently of the output of the plant $x_p(t)$, is also higher as α gets closer to 1. Thus, the fractional-order α can be used by the designer to achieve smoother transients or rapid convergence speed, or a midpoint between both characteristics.

Figures corresponding to the evolution of the identification error $e_i(t)$ and the closed-loop estimation errors $\varepsilon_\theta(t)$, $\varepsilon_k(t)$ for the CFOMRAC are not shown here for the sake of conciseness. Nevertheless, we should mention that they all converge

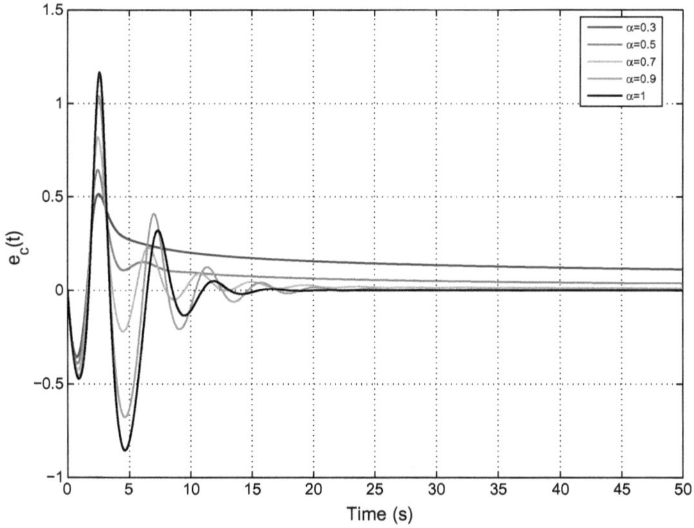

Fig. 6.3 Evolution of the control error $e_c\,(t)$ for the DFOMRAC scheme, using different values of the fractional-order α

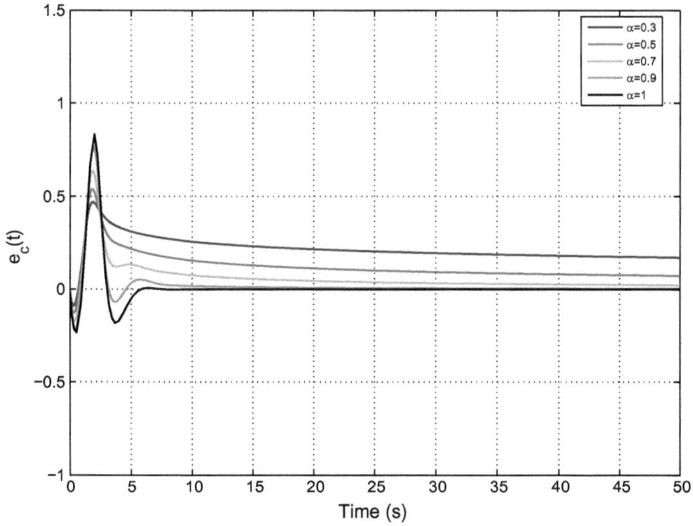

Fig. 6.4 Evolution of the control error $e_c\,(t)$ for the CFOMRAC scheme, using different values of the fractional-order α

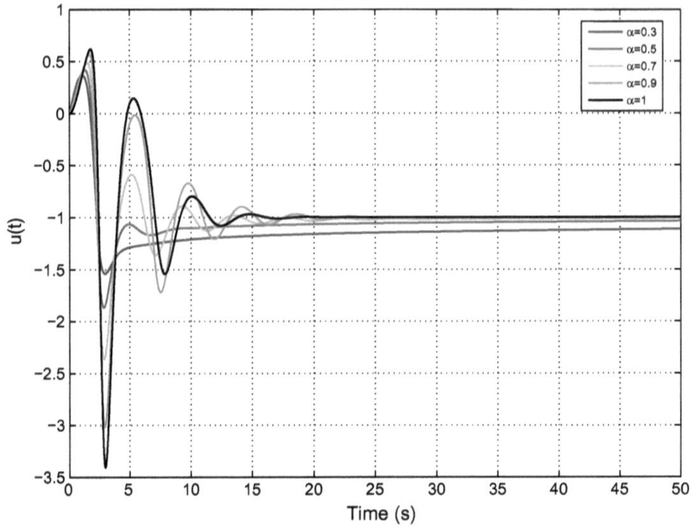

Fig. 6.5 Evolution of the control signal $u(t)$ for the DFOMRAC scheme, using different values of the fractional-order α

to zero. Also, the conclusions about convergence speed and transients of these errors, depending on the α that is used, are the same as for the control error $e_c(t)$.

The smoothness of the transient for the control error $e_c(t)$ for lower α is directly related with the control signal $u(t)$, whose evolution can be observed in Figs. 6.5 and 6.6, for the DFOMRAC and CFOMRAC, respectively.

It can be seen from Figs. 6.5 and 6.6 that, in fact, the control signal is smoother for the cases when α gets closer to 0, characteristic that may be appreciated in many applications. When the control signal is smooth, not only the transients of the control error are also smooth, but the integral of the squared control signal can result in lower values, which can be related to the energy used to achieve the control goal. Since the energy efficiency is becoming more important every day, the use of fractional orders different from 1 in these schemes could be advantageous.

Comparing the DFOMRAC and the CFOMRAC, it can be seen from Figs. 6.3, 6.4, 6.5, and 6.6 that for a fixed α, the convergence times are smaller for the CFOMRAC. Also, CFOMRAC presents less initial oscillations with lower amplitudes, showing that the CFOMRAC behaves better than the DFOMRAC. Of course, it must be noted that the CFOMRAC uses more information about the system, since it estimates not only the controller parameters but also the plant unknown parameters, thus it would be expected that it behaves better.

Finally, Figs. 6.7 and 6.8 show the corresponding parameter errors for this simulation environment, for the DFOMRAC and CFOMRAC, respectively.

It can be seen from Figs. 6.7 and 6.8 that the parameter errors remain bounded, as it was analytically proved for both DFOMRAC and CFOMRAC. It is interesting to observe that, although for all α the same initial conditions were set for the estimated

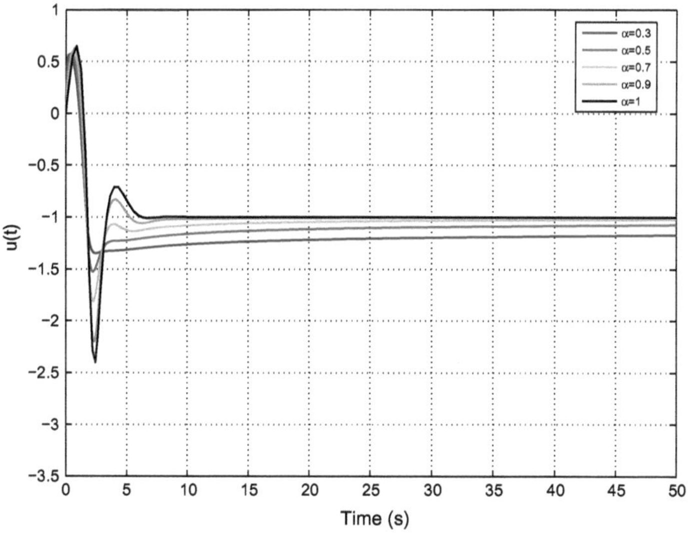

Fig. 6.6 Evolution of the control signal $u\ (t)$ for the CFOMRAC scheme, using different values of the fractional-order α

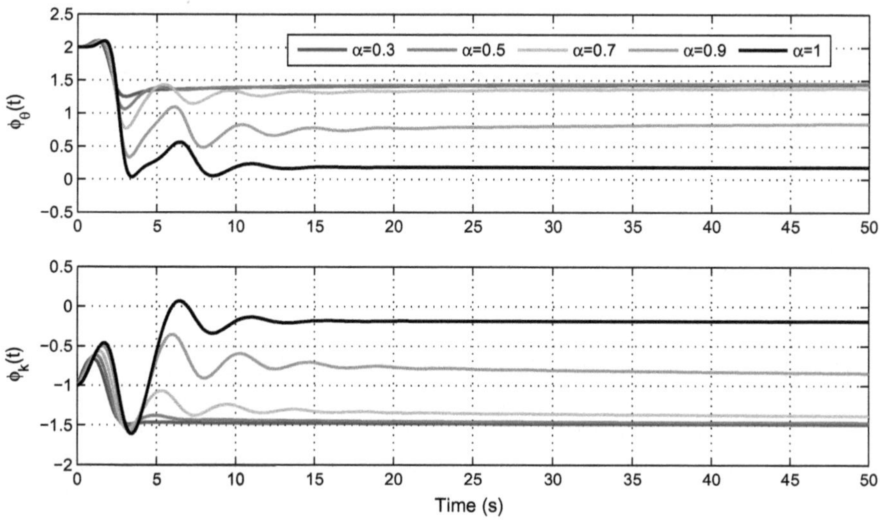

Fig. 6.7 Evolution of the parameter errors $\phi_\theta\ (t)$ and $\phi_k\ (t)$ for the DFOMRAC scheme, using different values of the fractional-order α

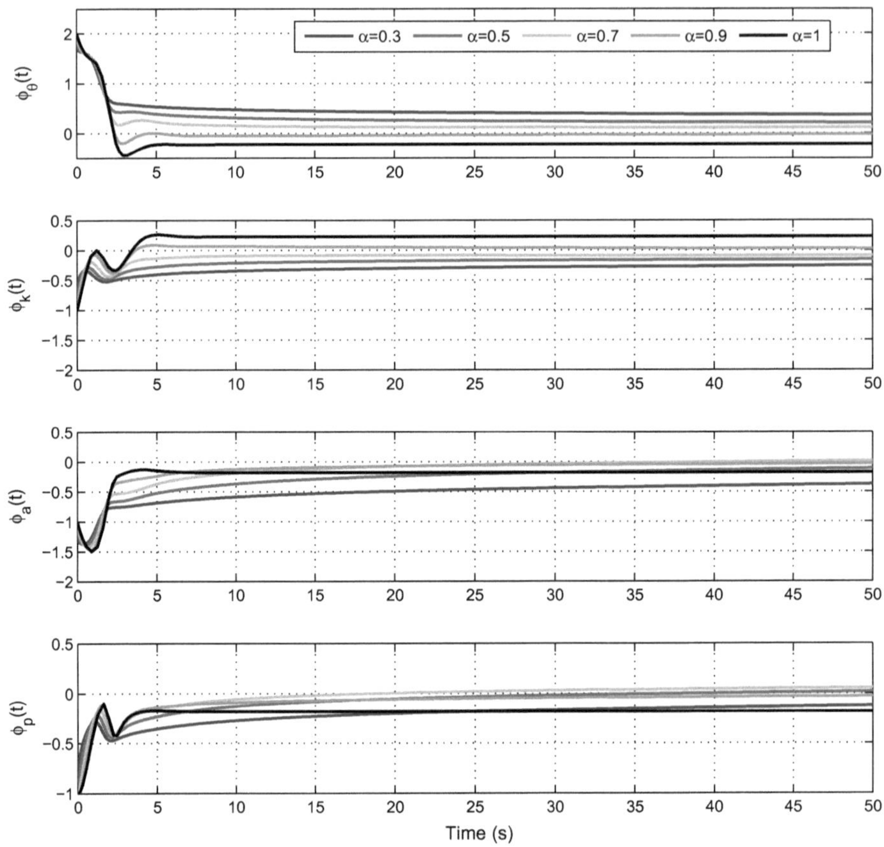

Fig. 6.8 Evolution of the parameter errors $\phi_\theta(t)$, $\phi_k(t)$, $\phi_a(t)$ and $\phi_p(t)$ for the CFOMRAC scheme, using different values of the fractional-order α

parameters, they can converge to different values for different values of α. This, however, does not change the fact that the control error $e_c(t)$ converges to zero for all α used, of course with different convergence speeds. The parameter errors do not converge to zero because the reference signal used in this simulation environment is a unit step, which does not have enough spectral lines to guarantee the convergence to zero of the parameter errors, as it was explained in Sect. 6.4.2 for the DFOMRAC. For the CFOMRAC, no analytical results were provided for the convergence of the parameter errors, but we can see here that it is also not achieved for this reference signal.

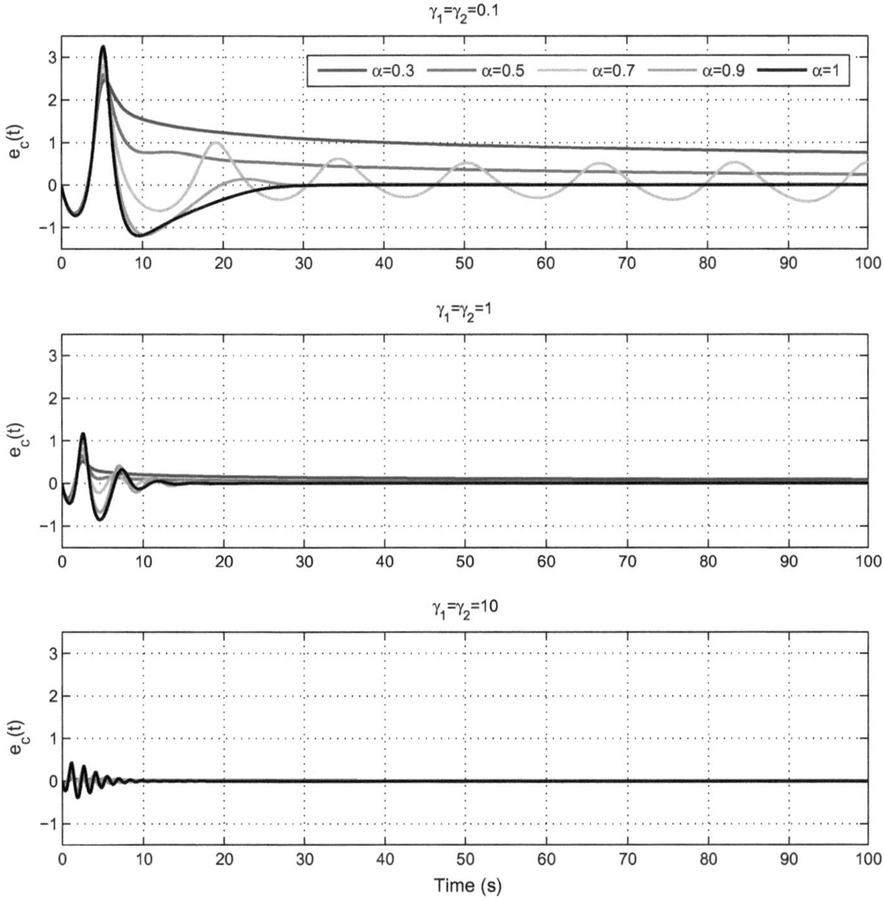

Fig. 6.9 Evolution of the control error $e_c(t)$ for the DFOMRAC scheme, using different values of the fractional-order α and the adaptive gains γ_1, γ_2

6.6.2 *Simulation Environment 2: Influence of the Adaptive Gains*

In the previous simulation environment, the adaptive gains were set to 1 in all cases, and only the influence of the fractional-order α was analyzed. In this simulation environment, the reference signal and the fractional orders used are the same as in Sect. 6.6.1, however different values of adaptive gains are employed, to show how this parameter influences the evolution of the signals behavior in the schemes. To that extent, three values of γ are used, which correspond to $\gamma_1 = \gamma_2 = \gamma_3 = \gamma_4 = \gamma = 0.1,\ 1,\ 10$.

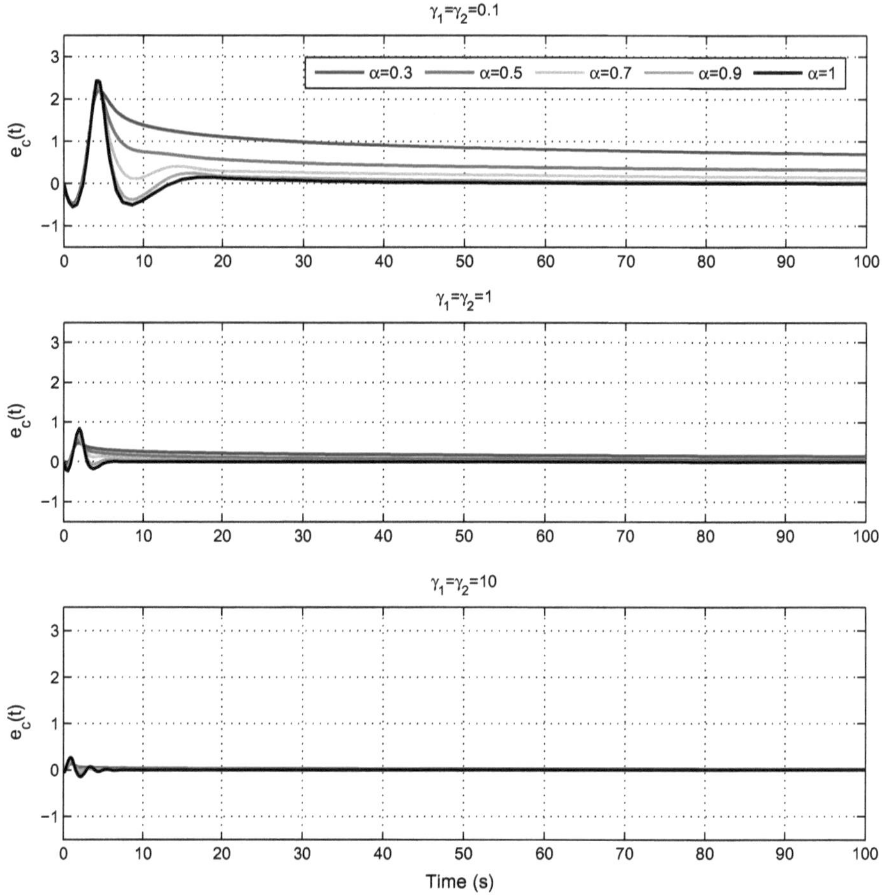

Fig. 6.10 Evolution of the control error $e_c(t)$ for the CFOMRAC scheme, using different values of the fractional-order α and the adaptive gains γ_1, γ_2, γ_3, γ_4

Figure 6.9 shows the evolution of the control error $e_c(t)$ for these values of γ in the DFOMRAC, while Fig. 6.10 shows the result for the CFOMRAC.

It can be seen from Figs. 6.9 and 6.10 that the value of the adaptive gain does not affect the boundedness of the control error, no matter what fractional-order α is used. However, it can be observed that for every α used, the convergence speed of the control error increases as γ increases. Also, it can be seen that the amplitude of the initial oscillations decreases as γ increases, but those cases with lower values of α remain with lower amplitudes. This is a very interesting behavior, since it tells us that the convergence speed can be increased using the adaptive gain γ, and it can be done without deteriorating the transient, just using fractional orders in the adaptive laws.

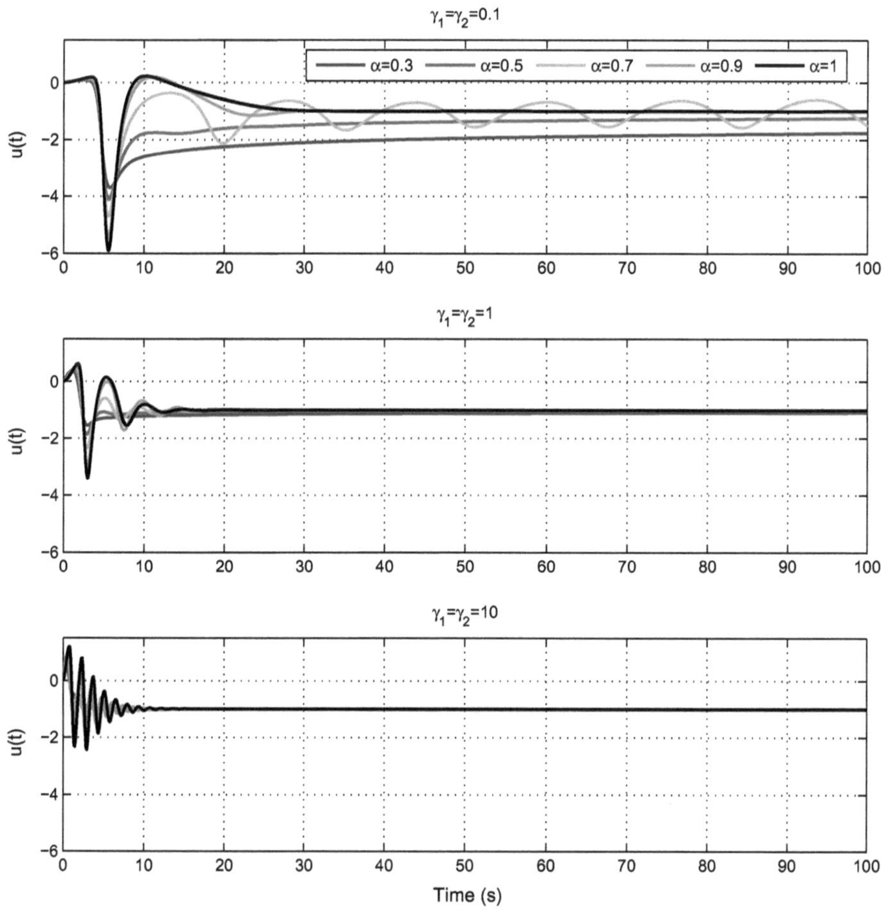

Fig. 6.11 Evolution of the control signal $u(t)$ for the DFOMRAC scheme, using different values of the fractional-order α and the adaptive gains γ_1, γ_2

Of course, some other facts need to be taken into account, like the evolution of the control signal $u(t)$, which can be observed in Figs. 6.11 and 6.12, for the DFOMRAC and CFOMRAC respectively.

In this case, it can be seen from Figs. 6.11 and 6.12 that the increase of the adaptive gain γ has a similar effect on the control signal than that for the control error, with respect to convergence speed and amplitude of the initial oscillations. Thus, the advantage of using higher adaptive gains together with fractional orders in the adaptive laws is evident in this problem.

Also, it can be seen from Figs. 6.9, 6.10 and 6.12 that the CFOMRAC behaves better than the DFOMRAC for fixed α and adaptive gains, since the convergence speed of the control error and control signal is higher for CFOMRAC, and also their transients show less oscillations with smaller amplitudes.

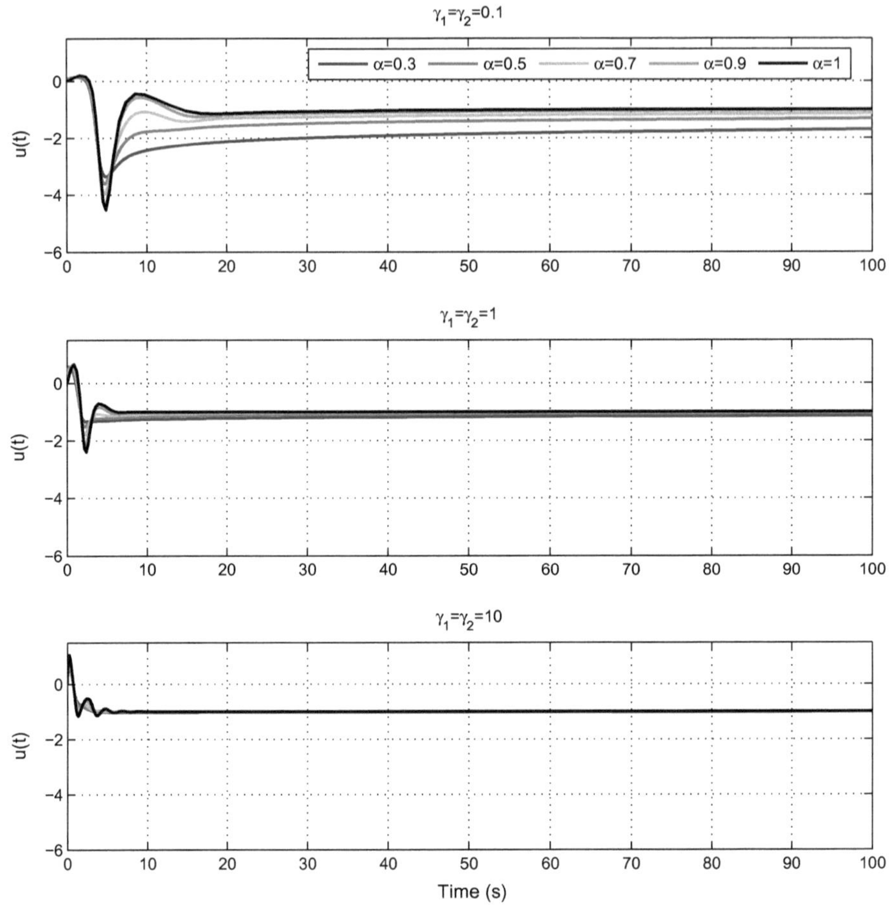

Fig. 6.12 Evolution of the control signal $u(t)$ for the CFOMRAC scheme, using different values of the fractional-order α and the adaptive gains $\gamma_1, \gamma_2, \gamma_3, \gamma_4$

Regarding the parameter errors, they are not shown in this case for the sake of conciseness. But in summary, they do not converge to zero, since the reference signal is a unit step, which is not persistently exciting, and the increase of the adaptive gain does not modify this fact. However, the convergence speed to their final values increases as γ increases, as it happened for the control error and the control signal.

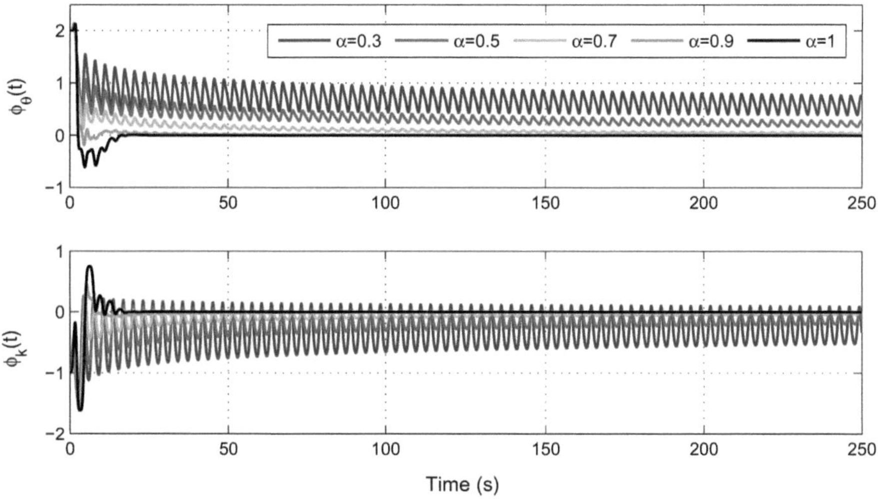

Fig. 6.13 Evolution of the parameter errors $\phi_\theta(t)$ and $\phi_k(t)$ for the DFOMRAC scheme, using different values of the fractional-order α and a persistently exciting reference signal

6.6.3 *Simulation Environment 3: Influence of the Reference Signal in the Parameters Convergence*

Although the main goal of the control schemes proposed in this chapter is not the parameter convergence but the control error convergence, it is interesting to show how the reference signal can influence the parameter convergence in these schemes. To that extent, this simulation environment uses a reference signal given by $r(t) = 2\sin(t)$, which has two spectral lines, and consequently, the parameter convergence should be achieved according to the analytical results, at least for the DFOMRAC. The fractional orders used here are the same as in the previous simulations, and the adaptive gains are $\gamma_1 = \gamma_2 = \gamma_3 = \gamma_4 = 1$.

Figures 6.13 and 6.14 show the corresponding parameter errors for this simulation environment, for the DFOMRAC and CFOMRAC respectively.

It can be seen from Figs. 6.13 and 6.14 that in this case, all the parameter errors converge to zero. For the lower values of α, the convergence cannot be observed in the time window plotted, but they can be seen to converge if the simulation time is increased enough. The convergence to zero of the parameter error in the case of DFOMRAC was analytically proved in Sect. 6.4.2 for the particular case where $k_p = \gamma_1 = \gamma_2 = 1$. Thus, this simulation only shows what we already knew.

However, in the case of the CFOMRAC, no analytical results are available at this moment regarding the parameter convergence to zero. Nevertheless, simulations show that even though more parameters are adjusted in the CFOMRAC, the spectral content of the reference input need not be increased to achieve parameter errors convergence to zero, as can be observed in Fig. 6.14.

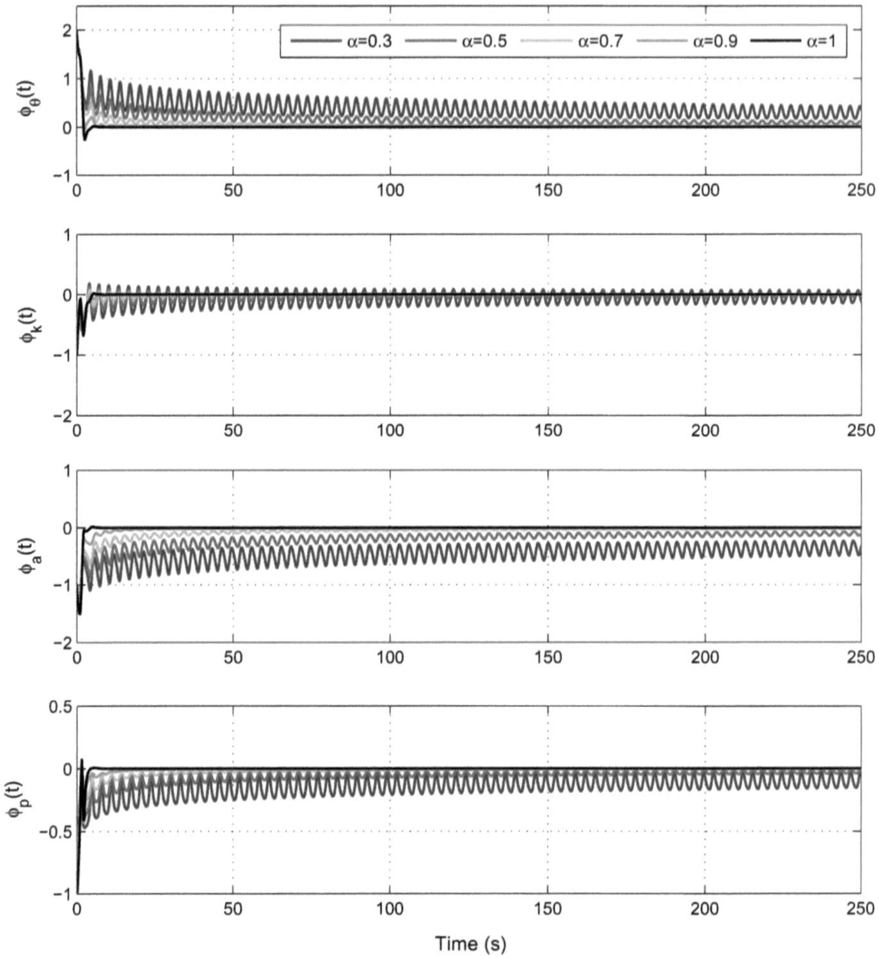

Fig. 6.14 Evolution of the parameter errors $\phi_\theta(t)$, $\phi_k(t)$, $\phi_a(t)$ and $\phi_p(t)$ for the CFOMRAC scheme, using different values of the fractional-order α and a persistently exciting reference signal

Acknowledgements The results reported in this chapter have been financed by CONICYT- Chile, under the Basal Financing Program FB0809 "Advanced Mining Technology Center", FONDECYT Project 1150488, "Fractional Error Models in Adaptive Control and Applications", and FONDE-CYT 3150007, "Postdoctoral Program 2015".

References

1. Aguila-Camacho, N., Duarte-Mermoud, M.A.: Fractional adaptive control for an automatic voltage regulator. ISA Trans. **52**, 807–815 (2013)
2. Aguila-Camacho, N., Duarte-Mermoud, M.A.: Boundedness of the solutions for certain classes of fractional differential equations with applications to adaptive systems. ISA Trans. **60**, 82–88 (2016)

3. Aguila-Camacho, N., Duarte-Mermoud, M.A., Gallegos, J.A.: Lyapunov functions for fractional order systems. Commun. Nonlinear Sci. Numer. Simul. **19**, 2951–2957 (2014)
4. Alikhanov, A.: A priori estimates for solutions of boundary value problems for fractional-order equations. Differ. Equ. **46**, 660–666 (2010)
5. Astrom, K.J., Bohlin, T.: Numerical Identification of Linear Dynamic Systems From Normal Operating Records, pp. 96–111. Springer, Boston (1966)
6. Diethelm, K.: The Analysis of Fractional Differential Equations. Springer, Heidelberg (2004)
7. Duarte-Mermoud, M.A., Aguila-Camacho, N., Gallegos, J.A., Castro-Linares, R.: Using general quadratic lyapunov functions to prove lyapunov uniform stability for fractional order systems. Commun. Nonlinear Sci. Numer. Simul. **22**, 650–659 (2015)
8. Duarte-Mermoud, M.A., Narendra, K.S.: Combined direct and indirect approach to adaptive control. IEEE Trans. Autom. Control **34**, 1071–1075 (1989)
9. Duarte-Mermoud, N.A., Narendra, K.S.: A new approach to model reference adaptive control. Int. J. Adapt. Control Signal Process. **3**(1), 53–73 (1989)
10. Gallegos, J.A., Duarte-Mermoud, M.A.: Boundedness and convergence on fractional order systems. J. Comput. Appl. Math. **296**, 815–826 (2016)
11. Gallegos, J.A., Duarte-Mermoud, M.A.: Mixed order robust adaptive control. Submitt. J. Franklin Inst. (2017)
12. Gallegos, J.A., Duarte-Mermoud, M.A.: Robustness and convergence of fractional systems and their applications to adaptive systems. Submitted to Fractional Calculus and Applied Analysis (2017)
13. Gallegos, J.A., Duarte-Mermoud, M.A., Aguila-Camacho, N.: Smoothness and boundedness on system of mixed-order fractional differential equations. Technical Report 2016-1, Santiago, Chile: Department of Electrical Engineering, University of Chile (2016)
14. Goodwin, G.C., Ramadge, P.J., Caines, P.E.: Discrete time multivariable adaptive control. IEEE Trans. Autom. Control **AC-25**(3), 449–456 (1980)
15. Kilbas, A.A., Srivastava, H.M., Trujillo, J.J.: Theory and Applications of Fractional Differential Equations. Elsevier, Amsterdam (2006)
16. Morse, A.S.: Global stability of parameter-adaptive control systems. IEEE Trans. Autom. Control **AC-25**(3), 433–439 (1980)
17. Narendra, K.S., Annaswamy, A.M.: Persistent excitation of adaptive systems. Int. J. Control **45**, 127–160 (1987)
18. Narendra, K.S., Annaswamy, A.M.: Stable Adaptive Systems. Dover Publications Inc., Mineola (2005)
19. Narendra, K.S., Duarte-Mermoud, M.A.: Robust adaptive control using direct and indirect methods. In: A.A.C. Council (ed.) Proceedings of the 7th American Control Conference, vol. 3, pp. 2429–2433. Atlanta, Georgia, USA (1988)
20. Narendra, K.S., Lin, Y.H.: Stable discrete adaptive control. IEEE Trans. Autom. Control **AC-25**(3), 456–461 (1980)
21. Narendra, K.S., Lin, Y.H., Valavani, L.S.: Stable adaptive controller design part ii: Proof of stability. IEEE Trans. Autom. Control **AC-25**(3), 440–480 (1980)
22. Oustaloup, A.: La commande CRONE: commande robuste d'ordre non entier. Hermes, Paris (1991)
23. Sastry, S., Bodson, M.: Adaptive Control: Stability Convergence and Robustness. Prentice Hall, Upper Saddle River (1994)
24. Suárez, J.I., Vinagre, B.M., Chen, Y.Q.: A fractional adaptation scheme for lateral control of an AGV. J. Vibr. Control **14**, 1499–1511 (2008)
25. Valério, D., Da Costa, J.S.: Ninteger: a non-integer control toolbox for matlab. In: IFAC (ed.) Fractional Derivatives and Applications. Bordeaux, France (2004)
26. Vinagre, B.M., Petrás, I., Podlubny, I., Chen, Y.Q.: Using fractional order adjustment rules and fractional order reference models in model-reference adaptive control. Nonlinear Dyn. **29**, 269–279 (2002)
27. Whitaker, H.P., Yamron, J., Kezer, A.: Design of model reference adaptive control systems for aircraft. Technical Report R-164, Instrumentation Laboratory, MIT, Cambridge, MA (1958)

Chapter 7
Sliding Modes for Switched Uncertain Linear Time Invariant Systems: An Output Integral Sliding Mode Approach

Leonid Fridman, Rosalba Galván-Guerra, Juan-Eduardo Velázquez-Velázquez and Rafael Iriarte

Abstract A robustifying methodology for switched uncertain linear time invariant systems with matched uncertainties/perturbations and state-dependent location transitions using only output information is presented. An output integral sliding mode control technique, based on an algebraic hierarchical observer is proposed. This approach allows the theoretically exact compensation of the matched uncertainties/perturbations right after the initial time; but it requires the use of filters to reconstruct the state vector and produces a high level of chattering. To eliminate the necessity of filtering and to diminish the chattering, a continuous output integral sliding mode controller is designed. This controller is based on the super-twisting algorithm and it compensates the matched uncertainties/perturbations after a finite transient. For this case, sufficient conditions to ensure the convergence of the controller and the observer before every switching are given. The proposed approach is illustrated via numerical simulations.

L. Fridman · R. Iriarte
Facultad de Ingeniería, Universidad Nacional Autonoma de México,
Av Universidad 3000, Ciudad Universitaria, 04510 Mexico City, Mexico
e-mail: lfridman@unam.mx

R. Iriarte
e-mail: ririarte@unam.mx

R. Galván-Guerra (✉) · J.-E. Velázquez-Velázquez
Unidad Profesional Interdisciplinaria de Ingeniería Campus Hidalgo, Instituto Politécnico Nacional, Carretera Pachuca-Actopan Km 1+500, San Agustin Tlaxiaca, Hidalgo, Mexico
e-mail: rgalvang@ipn.mx

J.-E. Velázquez-Velázquez
e-mail: jvelazquezv@ipn.mx

J.B. Clempner and W. Yu (eds.), *New Perspectives and Applications of Modern Control Theory*, https://doi.org/10.1007/978-3-319-62464-8_7

7.1 Introduction

7.1.1 State of the Art

The switched system modeling framework is used to represent a system composed of continuous dynamical subsystems, which interact using discrete switching rules [7, 25]. Each subsystem is active in a specific state space subset, called *location*, which is defined by a set of *switching manifolds*. When a switching manifold is activated, the system passes between locations and it is said that a *location transition* occurs at a *switching moment*. These transitions conform a *location trajectory*. Normally, the location transitions are considered as autonomous or controlled. If the switching manifolds depend on the state vector it is said that the system is a *switched system with state-dependent location transitions*.

In the context of switched systems, several control architectures are reported (see e.g., [16, 20]) that assume the state vector is available, and others that use only output information [8, 11, 18]. These strategies provide a nominal controller that makes the closed-loop system behaves in a nominal way. In the presence of matched uncertainties/perturbations, the trajectory of the system could deviate from the nominal behavior. In this work, using only output information, we will design a robustifying methodology that makes the system insensitive to the matched uncertainties/perturbations, i.e., given a nominal controller we design a new controller that, together with the nominal one, ensure the location trajectory of the system and the robustification of the nominal behavior.

7.1.2 Methodology

It is well known that the sliding modes methodology compensates the matched uncertainties/perturbations. But, if it is required for this to occur just after the initial time, the Integral Sliding Mode (ISM) framework [27] is the right option. Moreover, ISM allows to robustify a given nominal controller. However, it has two main disadvantages:

- The need of all the states and the initial conditions.
- And the discontinuous control of ISM produces high-level chattering.

In the switched framework in [19] an ISM is proposed with the same disadvantages than in the non-switched case.

The presence of uncertainties/perturbations in a switched system does not only affect the trajectory of the system, but also the location trajectory and the switching moments. This problem becomes more difficult when only output information is available. Works like [5, 10, 12] based on sliding modes, address the problem of state vector reconstruction in the presence of matched uncertainties/perturbations. However, most of the sliding mode observers are not capable to reconstruct the state vector right after the initial time. The Output Integral Sliding Mode (OISM) methodology ([4] and [13, p. 31]) is based on a hierarchical observer that allows theoretically exact reconstruction of the states just after the initial time when the uncertainties/perturbations are completely compensated.

The OISM method eliminates the need of the states and the initial conditions. This method robustifies the nominal behavior of the system using only output information and allows to reconstruct the system states theoretically exactly just after the initial time by reconstructing step-by-step the output and its derivatives. Moreover, the controller and the observer work simultaneously and the sliding motions start just after the initial time. Since the OISM still uses a discontinuous control law, it has two weak points:

- Presence of high-level chattering.
- To reconstruct the output and its derivatives a first-order filter is used. This affects its ability to reconstruct the state vector and to compensate the uncertainties/perturbations right after the initial time.

For switched systems in [3, 6, 24], different observers capable to reconstruct theoretically exactly both the states and the location trajectory are proposed, using the Super-Twisting Algorithm (STA) [17]. The continuous state is reconstructed using classical observation techniques and the location trajectory is identified using residual-based strategies. However, these results are based on the existence of STA gains that assure convergence before the switching and they do not provide any methodology for the design of such gains.

When the ISM methodology is combined with the STA a Continuous ISM (CISM) is obtained. This strategy allows to robustify in a continuous way the nominal behavior of the system. But, to assure exact compensation of the Lipschitz uncertainties-perturbations right after the initial time it is necessary to know the initial conditions of the uncertainties/perturbations, otherwise the methodology has reaching phase.

Continuous OISM (COISM) methodology is used in [22] for fault tolerant control of uncertain linear systems. There, it is assumed that the faults are Lipschitz and at the initial time there are not faults, eliminating the reaching phase of the controller. The states are reconstructed using a high-order differentiator, that must converge before the fault occurs. The design of the gains to assure the convergence time is not provided.

Hence, to assure the location trajectory and the robustification of the nominal behavior of switched systems despite the presence of matched uncertainties/perturbations using only output information. It is necessary to extend the existent robustifying output-based ISM approaches (OISM/COISM) to the switched case and to give conditions for the design of controller and observer gains guaranteeing the convergence properties.

7.1.3 Contribution

In this chapter, the OISM/COISM robustifying methodology is extended for Switched Uncertain Linear Time Invariant Systems (SULTIS) with matched uncertainties/perturbations and state-dependent location transitions. Two strategies are designed:

- The OISM concept is extended ensuring theoretically exact compensation of matched uncertainties/perturbations right after the initial time and preserving the switching moments. The observer has a cascade structure, composed by a family of reduced order observers, turned on at the same time, that reconstruct the outputs not affected by the uncertainties/perturbations and their derivatives theoretically exactly just after the initial time. To reconstruct the states, first-order filters are needed. As the states are reconstructed theoretically exactly just after the initial time, the observer and the controller can work simultaneously. The switching moments are detected online by using the observed state vector and the given switching manifolds.
- The COISM concept is extended ensuring theoretically exact compensation of matched uncertainties/perturbations right after a finite transient in every location and preserving the switching moments. The observer has the same structure as in the OISM case but the observers are turned on sequentially, i.e., the observers are turned on once the lower derivatives have been theoretically exactly reconstructed. Sufficient conditions are developed to assure the convergence of the full observer before half of the dwell time. After the observer has converged, a *COISM controller* that uses an OISM sliding variable together with a STA control law is turned on and the gains are designed to assure its convergence before the dwell time. This ensures exact compensation of the matched uncertainties/perturbations before the dwell time. Sufficient restrictions to the state vector at the switching times are developed to guarantee the absence of switching during the convergence. Under these conditions, the robustification methodology assures that the location trajectory is preserved. Here, the STA gains are adjusted as a function of the convergence time, using an estimation of the reaching time. For the implementation, the STA controller is modified to assure the non-Lipschitz terms of the observer do not affect the STA behavior of the controller. The integral structure of the sliding dynamics preserves the continuity of the control law, allowing to adjust the chattering.

7.1.4 Chapter Organization

Section 7.2 presents the problem formulation where the SULTIS and the nominal system are defined. Section 7.3 is dedicated to the design of the OISM controller for SULTIS in the presence of matched uncertainties/perturbations highlighting the switched hierarchical observer and the OISM controller design. Section 7.4 is devoted to the design of the COISM controller, giving sufficient conditions for the controller and the observer such that the matched uncertainties/perturbations are compensated theoretically exactly after the dwell-time condition. In this section, also the restrictions to the initial conditions to guarantee the robustification of the SULTIS are discussed. Finally, Sect. 7.5 concludes the chapter.

7.2 Problem Formulation

Consider a SULTIS with state-dependent location transitions, composed of r different locations[1]

$$\begin{aligned}\dot{\bar{x}}(t) &= \bar{A}_i\bar{x}(t) + \bar{B}_i\left(u(t) + \phi(t)\right),\\ \bar{y}(t) &= \bar{C}_i\bar{x}(t), \quad \bar{x}(0) = \bar{x}_0,\end{aligned} \tag{7.1}$$

where $i = \sigma(\bar{x}, t) : \mathbb{R}^n \times \mathbb{R}^+ \to \mathcal{I} = \{1, 2, \ldots, r\}$ denotes the active location,

$$\bar{A}_i \in \mathbb{R}^{n\times n},\ \bar{B}_i \in \mathbb{R}^{n\times m} \text{ and } \bar{C}_i \in \mathbb{R}^{p\times n}$$

are known matrices, $\phi : \mathbb{R}_+ \to \mathbb{R}^m$ are uncertainties/perturbations and $u(t)$ is the control.

Using a state transformation $x = \left[x_1^T(t), x_2^T(t)\right]^T = T_{1,i}\bar{x}$ and an output transformation $y = T_{y,i}\bar{y}$, where

$$T_{1,i} = \begin{bmatrix} \bar{B}_i^{\perp^+} \\ C_{t_i}\bar{B}_i^{\perp^+} + \bar{B}_i^+ \end{bmatrix}$$

with $C_{t_i} = (\bar{C}_i\bar{B}_i)^+\bar{C}_i(I_n - \bar{B}_i\bar{B}_i^+)\bar{B}_i^\perp$, and

$$T_{y,i} = \begin{bmatrix} (\bar{C}_i\bar{B}_i)^{\perp^+} \\ (\bar{C}_i\bar{B}_i)^+ \end{bmatrix};$$

Equation (7.1) is transformed into

$$\begin{aligned}\dot{x}(t) &= \underbrace{\begin{bmatrix} A_{11,i} & A_{12,i} \\ A_{21,i} & A_{22,i} \end{bmatrix}}_{A_i} x(t) + \underbrace{\begin{bmatrix} 0 \\ I_m \end{bmatrix}}_{B_i}(u(t) + \phi(t)),\\ \begin{bmatrix} y_1(t) \\ y_2(t) \end{bmatrix} &= \begin{bmatrix} C_{1,i} \\ C_{2,i} \end{bmatrix} x(t) = \underbrace{\begin{bmatrix} C_{11,i} & 0 \\ 0 & I_m \end{bmatrix}}_{C_i} x(t), \qquad x(0) = x_0,\end{aligned} \tag{7.2}$$

with $i = \sigma(T_{1,i}^{-1}x, t)$.

[1] See [26] for a general formal definition.

A transition between each location is ruled by a set of given switching manifolds $m_{i,i'} : \mathbb{R}^n \to \mathbb{R}$, with $i' \in \mathcal{I}$ and $i \neq i'$, such that

$$\bigcap_{\substack{i' \in \mathcal{I} \\ i' \neq i}} \{x : m_{i,i'}(x) = 0\} = \emptyset.$$

When the state x is in location i and $m_{i,i'}(x) = 0$ it is said that a transition to location i' occurs. It is considered that these switching manifolds are known.

Under the assumption that $\phi(t) = 0$ a nominal system is defined as

$$\begin{aligned} \dot{x}_{nom}(t) &= A_i x_{nom}(t) + B_i u_{nom}(t), \\ y_{nom}(t) &= C_i x_{nom}(t), \; x_{nom}(0) = x_0, \end{aligned} \tag{7.3}$$

where $u_{nom}(t) = -\bar{K}_i x_{nom}(t)$, with $\bar{K}_i \in \mathbb{R}^{m \times n}$, is a given stabilizing nominal control.

The switching moments t_j, $j = 1, 2, \ldots$ are not defined a priori but, by using the given switching manifolds and the nominal trajectory x_{nom}, it is possible to detect these transitions such that

$$t_j = \min_{\substack{i' \in \mathcal{I} \\ i' \neq i}} \{t > t_j : m_{i,i'}(x_{nom}(t)) = 0\}.$$

Suppose that the switching moments for (7.3) conform an ordered sequence

$$0 = t_0 < t_1 < t_2 < \cdots < t_{j-1} < t_j < \ldots.$$

such that $t_j - t_{j-1} > \delta$, i.e., the nominal switched system (7.3) satisfies the dwell time condition. To assure the nominal trajectory does not present Zeno behavior, we assume

$$\lim_{t \to t_j^-} \frac{\mathrm{d}}{\mathrm{d}t} m_{i,i'}(x_{nom}(t)) > \rho \text{ and } \lim_{t \to t_j^+} \frac{\mathrm{d}}{\mathrm{d}t} m_{i,i'}(x_{nom}(t)) > \rho$$

or

$$\lim_{t \to t_j^-} \frac{\mathrm{d}}{\mathrm{d}t} m_{i,i'}(x_{nom}(t)) < -\rho \text{ and } \lim_{t \to t_j^+} \frac{\mathrm{d}}{\mathrm{d}t} m_{i,i'}(x_{nom}(t)) < -\rho$$

with $\rho > 0$. Let $\sigma(T_{1,i_0}^{-1} x_{nom}(0), 0) = i_0 \in \mathcal{I}$. Then, the location trajectory $\sigma(T_{1,i}^{-1} x_{nom}(t), t)$ can be restated as

$$\sigma(T_{1,i}^{-1} x_{nom}(t), t) := \begin{cases} i \;\text{ if } \sigma(T_{1,i}^{-1} x_{nom}(t^-), t^-) = i \text{ and } m_{i,i'}(T_{1,i}^{-1} x_{nom}(t)) \neq 0, \\ i' \;\text{ if } \sigma(T_{1,i}^{-1} x_{nom}(t^-), t^-) = i \text{ and } m_{i,i'}(T_{1,i}^{-1} x_{nom}(t)) = 0. \end{cases}$$

By using this information the location trajectory of (7.3) can be characterized by the switching sequence

$$\Sigma = \{x_0; (i_0, t_0), (i_1, t_1), \dots, (i_j, t_j), \dots | i_j \in \mathcal{I}, j = 0, 1, \dots\}.$$

For (7.1), in general, the switching sequence and the state trajectory are affected by the uncertainties/perturbations $\phi(t)$.

7.2.1 General Assumptions

In order to robustify the SULTIS by using only output information, we need to make the following assumptions.

1. The initial location is known and the initial condition is unknown but bounded, i.e., there exists $\mu \in \mathbb{R}_+$, such that $\|\bar{x}(0)\| \leq \mu$.
2. The SULTIS without any controller satisfies the dwell time condition.
3. $\text{rank}\bar{B}_i = \text{rank}\bar{C}_i\bar{B}_i = m$ for all $i \in \mathcal{I}$.
4. The uncertainties/disturbances ϕ are bounded

$$\|\phi(t)\| \leq \Psi_{i_{max}}, \quad \forall t \in [t_{j-1}, t_j],$$

   where $\Psi_{i_{max}}$ is given for all $i \in \mathcal{I}$.
5. The switched system (7.1) has more outputs than inputs, i.e., $p > m$.
6. The switched system is controllable in every location [23], i.e., the pair $(\bar{A}_i, \bar{B}_i)$ is controllable for $i \in \mathcal{I}$.
7. The switched system is strongly observable in every location [15], i.e., the triple $(\bar{A}_i, \bar{B}_i, \bar{C}_i)$ is strongly observable for $i \in \mathcal{I}$.

7.3 Output Integral Sliding Mode Control for SULTIS

Let us start this section by designing an output-based robustifying methodology of the nominal trajectory of the SULTIS (7.2) by using an OISM strategy. This methodology uses a controller that compensates the matched uncertainties/perturbations and an observer that reconstructs the state vector of the system. The controller and the observer work at the same time, guaranteeing the convergence to the nominal trajectory since the initial time. The overall structure of the proposed methodology is depicted in Fig. 7.1.

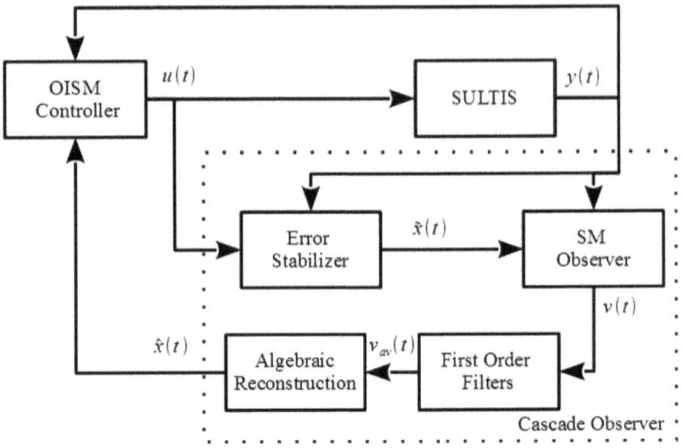

Fig. 7.1 Block diagram of the OISM robustifying strategy for SULTIS

7.3.1 Switched Hierarchical Observer

7.3.1.1 General Form of the Switched Observer

Let us start this section designing a cascade structure observer capable to reconstruct the state vector of (7.2) theoretically exactly since the initial time. Note that we are measuring all the states that are directly affected by the uncertainties/perturbations (y_2). Hence, we only need to reconstruct the remaining ones. Recall assumptions A3, A5–A7, since the pair (A_i, B_i, C_i) is strongly observable we can assure that the pair $(A_{11,i}, C_{11,i})$ is observable (see [4] for a proof of this claim) and assume it has observability index l_i. Note that since the discrete state is reconstructed using the observed states, the switched system is observable with observability index $l = \max_{i \in \mathcal{I}} l_i$.

The proposed reduced order observer is composed by:

1. An auxiliary *Luenberger error stabilizer*

$$\dot{\tilde{x}}_1(t) = A_{11,i}\tilde{x}_1(t) + A_{12,i}y_2(t) + K_i\left(y_1(t) - C_{11,i}\tilde{x}_1(t)\right) \tag{7.4}$$

   where K_i is the observer gain that is designed to assure the observation error $e_1(t) = x_1(t) - \tilde{x}_1(t)$ is bounded.
2. A *switched version of the hierarchical observer* [4, 13] composed by

   a. A family of sliding mode observers

$$\dot{x}_{ak} = A_{11,i}\tilde{x}_1(t) + A_{12,i}y_2(t) + L_{k,i}(t)\left(C_{11,i}A_{11,i}^{k-1}L_{k,i}\right)^{-1}v_k(t), \tag{7.5}$$

   $k = 1, \ldots, l-1$. These observers reconstruct the output error $y_1(t) - C_{11,i}\tilde{x}_1(t)$ and its $l-1$ derivatives. The matrices $L_{k,i}$ are design parameters and v_k is a first-order sliding mode output injection signal that contains since the initial time the output error k-th derivative.

b. An algebraic part

$$\hat{x}_1(t) = \tilde{x}_1(t) + \mathcal{O}^{+}_{1,i,l} v_{eq}(t). \tag{7.6}$$

where $\mathcal{O}_{1,i,l}$ is the observability matrix of the pair $(A_{11,i}, C_{11,i})$ and

$$v_{eq}(t) = \begin{bmatrix} y_1(t) - C_{11,i}\tilde{x}_1(t) \\ v_{1_{eq}}(t) \\ v_{2_{eq}}(t) \\ \vdots \\ v_{l-1_{eq}}(t) \end{bmatrix} \in \mathbb{R}^{pl}.$$

The full reconstructed state vector is given by

$$\hat{x}(t) = \begin{bmatrix} \hat{x}_1(t) \\ y_2(t) \end{bmatrix};$$

This part reconstructs the state vector of (7.1) theoretically exactly just after the initial time by using the information provided by (7.4)–(7.5), i.e., $\hat{x}(t) = x(t)$.

Let us design each part of the observer.

7.3.1.2 Luenberger Error Stabilizer

Consider the Luenberger-type error stabilizer (7.4) with initial condition $C_{11,i}\tilde{x}_1(0) = y_1(0)$, the matrix K_i must be designed such that the error dynamics

$$\dot{e}_1(t) = \left(A_{11,i} - K_i C_{11,i}\right) e_1(t) = \hat{A}_i e_1(t), \quad e_1(0) = x_1(0) - \tilde{x}_1(0),$$

is exponentially stable. The trajectory of the observer is assumed continuous, then at the switching moments

$$\tilde{x}_1(t_j^-) = \tilde{x}_1(t_j^+).$$

Since the dynamical error $e_1(t)$ is stable there exist positive constants γ_i, η_i such that

$$\|e_1(t)\| \le \gamma_i e^{-\eta_i(t-t_j)} \|e_1(0)\| \le \gamma_i e^{-\eta_i(t-t_j)} \left(\mu + \|\tilde{x}(0)\|\right).$$

Note that we are not assuming that x neither $\tilde{x}$ are bounded. The Luenberger observer only assures e_1 is bounded.

7.3.1.3 Switched Hierarchical Observer in Algebraic Form

In this subsection a sliding mode-based step-by-step observer is proposed to reconstruct the output error $y_1(t) - C_{11,i}\tilde{x}_1(t)$ and its $l-1$ time-derivatives just after the initial time. This observer recovers the vectors $C_{11,i}A_{11,i}^{k-1}x_1(t)$, $k = 1, \ldots, l-1$ using a family of first-order sliding mode observers. The observer design is given by the following proposition.

Proposition 7.1 *Assume*

1. *The auxiliary variable x_{ak} is designed as* (7.5)*, where $L_{k,i}(t) \in \mathbb{R}^{n\times p}$ is a design matrix such that* $\det\left(C_{11,i}A_{11,i}^{k-1}L_{k,i}\right) \neq 0$.
2. *The auxiliary input $v_k(t)$ is defined as*

$$v_k(t) = M_{k,\sigma}(t)\frac{\bar{s}_k(t)}{\|\bar{s}_k(t)\|},$$

where the scalar gain $M_{k,i}$ should satisfy the condition

$$\|C_{11,i}A_{11,i}^{k}\|\|e_1(t)\| < M_{k,i}(t).$$

3. *At the switching moments, t_j all the variables x_{ak} satisfy the conditions*

$$C_{11,i}x_{a1}(t_j^+) = y_1(t_j^+), \tag{7.7}$$

and

$$C_{11,i}A_{11,i}^{k-1}\tilde{x}_1(t_j^+) + v_{k-1_{eq}}(t_j^+) = C_{11,i}A_{11,i}^{k-1}x_{ak}(t_j^+). \tag{7.8}$$

4. *The sliding variables $\bar{s}_k$ are designed as*

$$\bar{s}_1(t) = y_1(t) - C_{11,i}x_{a1}(t), \tag{7.9}$$

and

$$\bar{s}_k(t) = C_{11,i}A_{11,i}^{k-1}\tilde{x}_1(t) + v_{k-1_{eq}}(t) - C_{11,i}A_{11,i}^{k-1}x_{ak}(t). \tag{7.10}$$

Then, for all $t > 0$

$$v_{k_{eq}}(t) = C_{11,i}A_{11,i}^{k}e_1(t),$$

and it is possible to reconstruct completely all the vector functions $C_{11,i}A_{11,i}^{k-1}x(t)$.

Proof Since the auxiliary variables x_{ak} fulfill the conditions (7.7)–(7.8) the sliding variables (7.9)–(7.10) are zero at each switching time. The establishment of the sliding mode $\forall t \in [t_j, t_{j+1}]$ follows from the results presented in [4] and [13, p. 34]. Then it is possible to assure $\bar{s}_k(t) = \dot{\bar{s}}_k(t) = 0, \;\; \forall t \geq 0$.

By using the equivalent control of the sliding mode observers we can construct the vector

$$v_{eq}(t) = \mathcal{O}_{1,i,l}e_1(t), \quad \forall t > 0.$$

The equivalent control v_{eq} is not available, but it can be reconstructed by filtering the high-frequency signals v_k. Note that we have reconstructed the output error in every location and its $l-1$ time-derivatives. Since the pair $(A_{11,i}, C_{11,i})$ is observable, the pseudo-inverse of $\mathcal{O}_{1,i,l}$ is well defined and the states can be recovered by means of the equation

$$x_1(t) = \tilde{x}_1(t) + \mathcal{O}^+_{1,i,l}v_{eq}(t). \tag{7.11}$$

The hierarchical observer is defined as

$$\hat{x}_1(t) = \tilde{x}_1(t) + \mathcal{O}^+_{1,i,l}v_{eq}(t). \tag{7.12}$$

Assuming an ideal output integral sliding mode, $\hat{x}(t) \equiv x(t)$ for all $t > 0$ and the states have been reconstructed theoretically exactly just after the initial time.

7.3.2 OISM Control Design

Now that the state has been reconstructed theoretically exactly, we can design an OISM controller that compensates the uncertainties/perturbations, right after the initial time. The proposed sliding variable only depends on output information,

$$s(y,t) = G_i(y(t) - y(t_j)) - \int_{t_j}^{t} G_iC_i(A_i\hat{x}(\rho) + B_iu_{nom}(\rho))\mathrm{d}\rho. \tag{7.13}$$

G_i is a projection matrix and without loss of generality, it is designed such that $G_iC_iB_i = I$.

Taking the derivative of the sliding variable (7.13) along the trajectories of (7.2)

$$\begin{aligned} \dot{s}(y,t) &= G_iC_iA_i\left(x(t) - \hat{x}(t)\right) + u_{int}(t) + \phi(x,t), \\ s(y(t_j),t_j) &= 0; \end{aligned} \tag{7.14}$$

the equivalent control is

$$u_{int_{eq}} = -\phi(x,t) - G_iC_iA_i\left(x(t) - \hat{x}(t)\right);$$

and the sliding dynamics of the SULTIS (7.2) takes the form

$$\begin{aligned} \dot{x}(t) &= \tilde{A}_ix(t) + B_iu_{nom}(t) + B_iG_iC_iA_i\hat{x}(t), \\ y(t) &= C_ix(t), \quad x(0) = x_0, \end{aligned} \tag{7.15}$$

where

$$\tilde{A}_i = [I_n - B_i G_i C_i] A_i.$$

Note that if the state vector is exactly reconstructed, (7.15) is equivalent to (7.3).

The next theorem gives the design conditions of the OISM controller.

Theorem 7.1 *Assume A1–A7 are fulfilled, the observation error $\|x(t) - \hat{x}(t)\|$ is bounded and the scalar β_i satisfies the inequality*

$$\beta_i - \|G_i C_i A_i\| \, \|x(t) - \hat{x}(t)\| - \Psi_{i_{Max}} \geq \lambda_i > 0, \tag{7.16}$$

where λ_i is a constant. Then the sliding mode is established from the initial moment and the theoretically exact compensation of the matched uncertainties/perturbations just after the initial time is assured.

Proof Let us select

$$V(s(y,t)) = \frac{1}{2}\|s(y,t)\|^2,$$

as the candidate Lyapunov function. The derivative of V along the trajectories of the sliding variable (7.14) is:

$$\dot{V}(s(y,t)) = s^T(y,t)\dot{s}(y,t) = s^T(y,t)\left(G_i C_i A_i\left(x(t) - \hat{x}(t)\right) + u_{int}(t) + \phi(x,t)\right).$$

Consider a first-order sliding mode control

$$u_{int}(t) = -\beta_i \frac{s(y,t)}{\|s(y,t)\|}.$$

Then, the derivative of the candidate Lyapunov function takes the form

$$\begin{aligned}\dot{V}(s(y,t)) &\leq -\|s(y,t)\|\left(\beta_i - \|G_i C_i A_i\| \, \|x(t) - \hat{x}(t)\| - \Psi_{\sigma Max}\right)\\ &= -2\left(\beta_i - \|G_i C_i A_i\| \, \|x(t) - \hat{x}(t)\| - \Psi_{\sigma Max}\right) V(s(y,t))^{1/2}, \quad \forall t \in (t_j, t_{j+1}]\end{aligned}$$

If (7.16) is satisfied, the proposed sliding mode control assures finite-time stability of the sliding variable (7.14) for all $t \in (t_j, t_{j+1}]$. Note that V is not increasing along the trajectories of the sliding variable and since $s(y(t_j), t_j) = 0$, then

$$\frac{1}{2}\|s(y(t),t)\|^2 = V(s(y(t),t)) \leq V(s(y(t_j),t_j)) = \frac{1}{2}\|s(y(t_j),t_j)\|^2 = 0, \quad t \in [t_j, t_{j+1}]$$

Now, it is clear that $s(y(t),t) = 0$ for all $t \geq 0$ and the sliding mode is established from the initial time.

The next algorithm summarizes the proposed robustifying output control theoretical approach.

OISM Algorithm

1. In order to assure a bounded observation error, design the matrix function K_i, assuring that the error dynamics of the Luenberger-based stabilizer is exponentially stable.
2. Compute the scalar gain β_i satisfying (7.16) and assuring the existence of the sliding mode in every location.
3. Design the auxiliary systems x_{ak} (7.5) with the sliding variable $\bar{s}_k$ (7.10) and compute the constants $M_{k,i}$. This ensures that the proposed observer reconstructs the output and its derivatives right after the initial time.
4. Reconstruct the control input $v_{k_{eq}}$ from the high-frequency signal v_k.
5. In the nominal control substitute x by $\hat{x}$, i.e., use only output information. The switching sequences are identified with the approximated state vector $\hat{x}$.
6. Run simultaneously the observer $\hat{x}$ according to (7.17), and the controllers u_{nom} and u_{int}.

7.3.3 *Hierarchical Observer Realization*

The proposed sliding mode observer needs a low-pass filter to reconstruct the equivalent output injection v_{eq} from the high-frequency signal $v_k(t)$. We consider a first-order low-pass filter [27, p. 23] where the filtered signal $v_{k_{av}}(t)$ would approximate $v_{k_{eq}}(t)$, i.e.,

$$F_k \dot{v}_{k_{av}}(t) = -v_{k_{av}}(t) + v_k(t), \quad v_{k_{av}}(t_j^+) = C_{11,i} A_{11,i}^k (\hat{x}_1(t_j) - \tilde{x}_1(t_j)),$$

where $F_k = \sqrt{\Delta t^k}$ is the time constant of the filter and Δt is the sample step.

We define a filter for every location and we assume that the filtered output $v_{k_{av}}$ is a Lipschitz function, considering that at the switching moments the observation error is very small and can be depreciated.

The realization of the observer (7.12) takes the form

$$\hat{x}_1(t) = \tilde{x}_1(t) + \mathcal{O}_{1,i,l}^{+} v_{av}(t), \quad \forall t \geq 0,$$
$$v_{av} = \begin{bmatrix} y_1(t) - C_{11,i}\tilde{x}_1(t) \\ v_{1_{av}}(t) \\ v_{2_{av}}(t) \\ \vdots \\ v_{l-1_{av}}(t) \end{bmatrix} \in \mathbb{R}^{p(l-1)}. \tag{7.17}$$

Due to the effects of the first-order filters there is an error $\epsilon(t) = v_{eq}(t) - v_{av}(t)$ and (7.17) takes the form $\mathcal{O}_{1,i,l}\hat{x}_1(t) = \mathcal{O}_{1,i,l}\tilde{x}_1(t) + v_{av}(t) + \epsilon(t)$. The observed states $\hat{x}_1(t)$ satisfy the identity $\hat{x}_1(t) = \arg\min_{\hat{x}_1 \in \mathbb{R}^n} \|\mathcal{O}_{1,i,l}\hat{x}_1(t) - \mathcal{O}_{1,i,l}\tilde{x}_1(t) - v_{av}(t)\|$.

In [13, p. 88] it is shown that the value of the error constant ϵ depends on the step size Δt. Hence, with a small value of Δt, ϵ can be depreciated as well as the term $\|x(t) - \hat{x}(t)\|$ in (7.16).

7.3.4 Simulation Results

Consider a system of the form (7.1) with two locations, where

$$\bar{A}_1 = \begin{bmatrix} 0 & 1 & 0 & 0 \\ 0 & 0 & 1 & 0 \\ 0 & 0 & 0 & 1 \\ 1 & 2 & 3 & 4 \end{bmatrix}, \quad \bar{A}_2 = \begin{bmatrix} 0 & 0 & -1 & 0 \\ -50 & -200 & -365 & -250 \\ 0 & 0 & 0 & -1 \\ 0 & -1 & 0 & 0 \end{bmatrix}$$

$$\bar{B}_1 = \begin{bmatrix} 0 \\ 0 \\ 0 \\ 1 \end{bmatrix}, \quad \bar{B}_2 = \begin{bmatrix} 0 \\ 2 \\ 0 \\ 0 \end{bmatrix}, \quad \bar{C}_1 = \begin{bmatrix} 0 & 0 & 0 & 1 \\ 1 & 0 & 0 & 0 \end{bmatrix}, \quad \bar{C}_2 = \begin{bmatrix} 1 & 0 & 0 & 0 \\ 0 & 1 & 0 & 0 \end{bmatrix},$$

and

$$\sigma(\bar{x}, t) = \begin{cases} 2 & \sigma(\bar{x}(t^-), t^-) = 1 \text{ and } |\bar{x}_4| \leq 0.5, \\ 1 & \sigma(\bar{x}(t^-), t^-) = 2 \text{ and } |\bar{x}_4| > 1, \\ \sigma(\bar{x}(t^-), t^-) & \text{otherwise.} \end{cases}$$

It is considered that the system starts in location $i_0 = 1$. The initial conditions $x_0 = [0.1, -0.1, 0.2, -1.5]^T$ are unknown with a bound $\mu = 2$ and the uncertainties/perturbations used in the simulations are

$$\phi(t) = 5\sin(\pi\cos(3\pi t)) + 10.$$

All the simulations are done in Simulink, using the Euler method with a sample step $\Delta t = 1e - 5$. Note that the system is unstable.

The system is transformed into the form (7.2), with

$$T_{1,1} = \begin{bmatrix} 0 & 1 & 0 & 0 \\ 0 & 0 & 1 & 0 \\ -1 & 0 & 0 & 0 \\ 0 & 0 & 0 & 1 \end{bmatrix}, \quad T_{1,2} = \begin{bmatrix} -1 & 0 & 0 & 0 \\ 0 & 0 & 1 & 0 \\ 0 & 0 & 0 & 1 \\ 0 & 0.5 & 0 & 0 \end{bmatrix},$$

$$T_{y,1} = \begin{bmatrix} 0 & 1 \\ 1 & 0 \end{bmatrix}, \quad \text{and} \quad T_{y,2} = \begin{bmatrix} -1 & 0 \\ 0 & 0.5 \end{bmatrix},$$

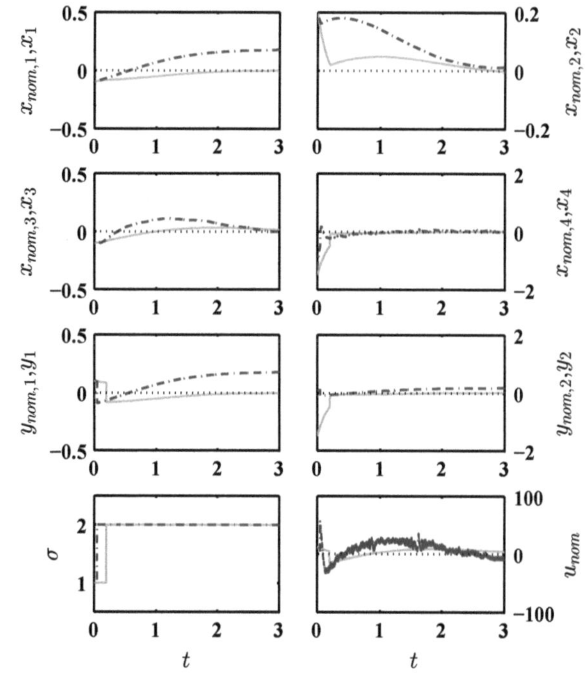

Fig. 7.2 Nominal behavior of (7.3) versus SULTIS (7.2) with the nominal controller only: Nominal behavior (*continuous line*), SULTIS (*dashed line*)

For the transformed nominal system (7.3), an LQR nominal controller is proposed in each location:

$$\bar{K}_1 = \begin{bmatrix} 2.4142\ 6.0210\ 9.0699\ 9.9279 \end{bmatrix}$$

and

$$\bar{K}_2 = \begin{bmatrix} -50.0200\ 1.2563\ 0.2688\ -251.8423 \end{bmatrix}.$$

This nominal controller guarantees the stability of the nominal system (see Fig. 7.2) and a dwell time $\delta = 0.1$. When the nominal controller is applied and there are uncertainties/perturbations in the system, the nominal controller is incapable to take the states to zero, see Fig. 7.2. This shows the necessity of the robustifying methodology.

Let's apply the OISM controller to the system. The Luenberger stabilizer gains were designed as in [1]. This stabilizer is shown in Fig. 7.3 assuring the existence of a bound for the observation error $e_1(t)$. In Fig. 7.4 the reconstructed state is presented and in Fig. 7.5 the sliding variables $\bar{s}_k$ are given. Note that the observer is in the sliding mode since the initial time. Hence the observer is capable to reconstruct the state vector theoretically exactly right after the initial time. In Fig. 7.6 a comparison of the observation error norm for different sample steps is given. Note that, the filters affect the state reconstruction and the lower the sample step is the lower the observation error will be.

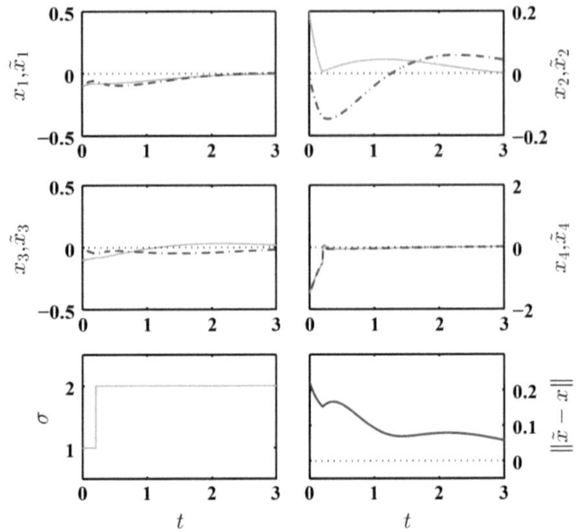

Fig. 7.3 Luenberger stabilizer behavior: Real state (*continuous line*), observed state $\tilde{x}$ (*dashed line*)

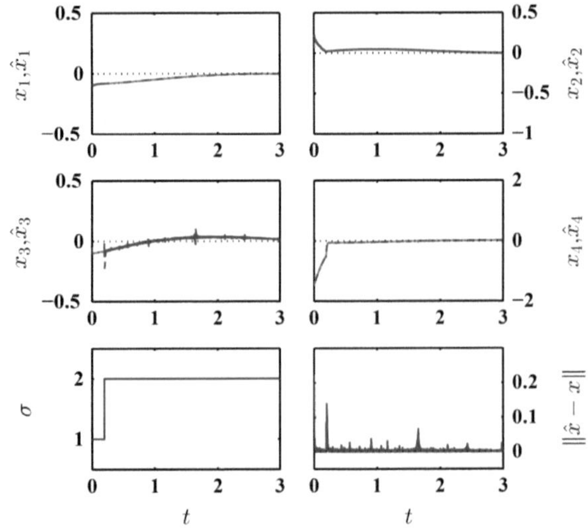

Fig. 7.4 OISM cascade observer: Real state (*continuous line*), observed state $\hat{x}$ (*dashed line*)

The norm of the error between the nominal system and the SULTIS is illustrated for different sample steps in Fig. 7.7. Notice that the lower the sample step is the lower this error will be. Hence, the OISM controller compensates theoretically exactly the matched uncertainties/perturbations since the initial time. But the control signal is discontinuous, generating high-level chattering (see Fig. 7.8).

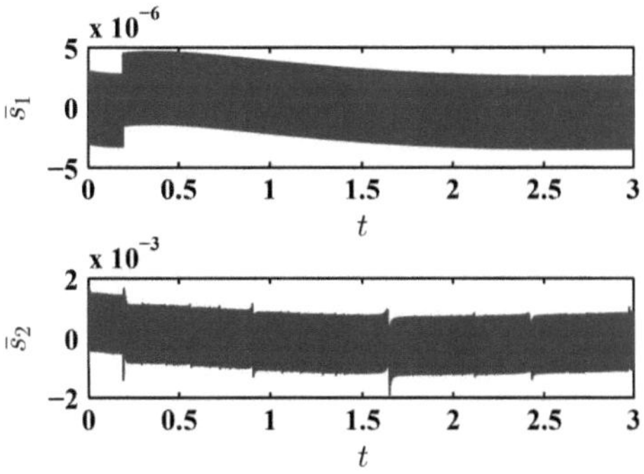

Fig. 7.5 OISM cascade observer: Virtual sliding mode variables

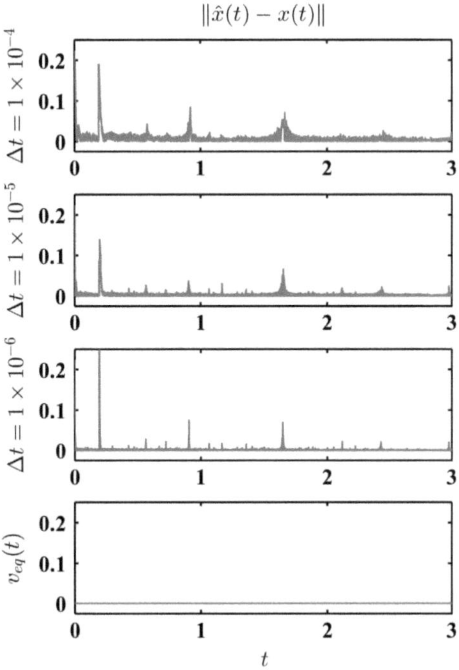

Fig. 7.6 Comparison of the observation error norm different sample steps

7.4 Continuous Output Integral Sliding Mode for SULTIS

To eliminate the necessity of filters and to diminish the chattering a COISM-based robustifying methodology for SULTIS is proposed. This methodology uses a STA-based cascade observer that reconstructs the state vector theoretically exactly before half of the dwell time. After the observer has converged the COISM controller is turn on, assuring theoretically exact compensation of the matched uncertainties/perturbations before the dwell time. For this methodology, it is necessary to add the following assumptions.

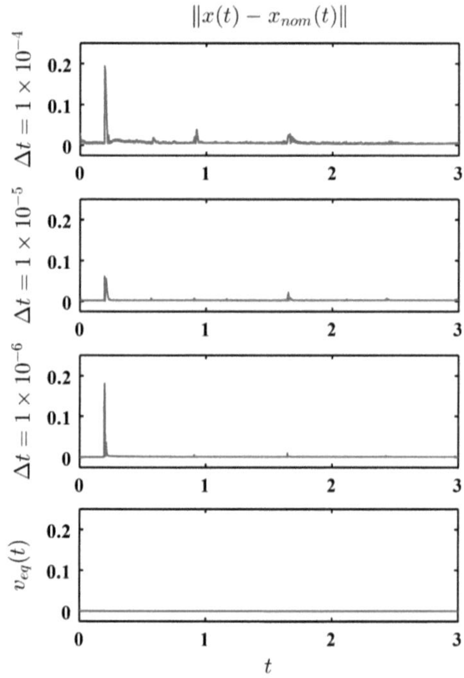

Fig. 7.7 Comparison of the control error norm different sample steps

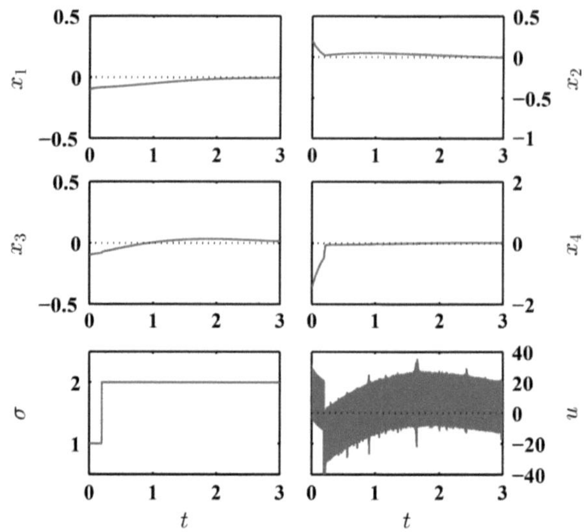

Fig. 7.8 SULTIS with OISM controller

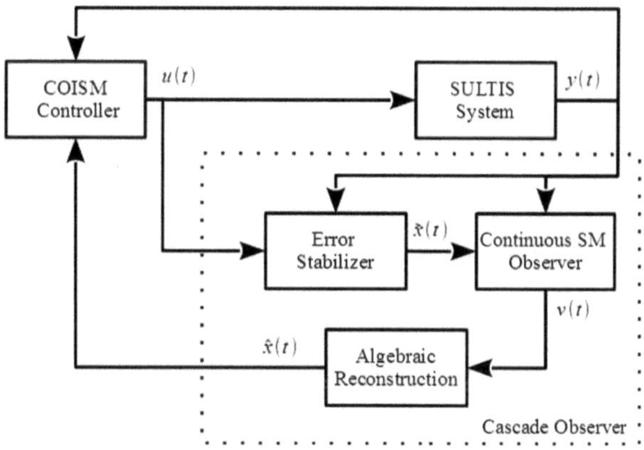

Fig. 7.9 Robustifying COISM block diagram

A8. The uncertainties/disturbances ϕ conform a bounded Lipschitz function:

$$\|\phi(t)\| \leq \Psi_{i_{max}}, \quad \forall t \in [t_{j-1}, t_j];$$
$$\left\|\dot{\phi}(t)\right\| \leq \Phi_{i_{max}}, \quad \forall t \in [t_{j-1}, t_j];$$

where $\Phi_{i_{max}}, \Psi_{i_{max}} \in \mathbb{R}_+$ are given for all $i \in \mathcal{I}$.

A9. The SULTIS without any controller satisfies the dwell time condition.

The main structure of the COISM robustifying methodology is very similar as the one presented in the OISM section, but the filter block is no longer necessary (see Fig. 7.9). Assumption A9 ensures the absence of a switching before the dwell time even when the controller is turn off. Before design each part of the robustifying methodology we introduce a result that is used in the sequel (for the proof see the appendix)

Lemma 7.1 *Consider the scalar system*

$$\dot{\chi}(t) = u(t) + \zeta(t), \tag{7.18}$$

where $\zeta(t)$ is a bounded lipschitz unknown signal such that $\left\|\dot{\zeta}(t)\right\| \leq \vartheta_1$ and $\|\zeta(t)\| \leq \vartheta_2$. $u(t)$ is an STA control law

$$u(t) = -\kappa_1 \lceil \chi(t) \rfloor^{\frac{1}{2}} + w(t),$$
$$\dot{w}(t) = -\kappa_2 \lceil \chi(t) \rfloor^{0}, \quad w(0) = 0. \tag{7.19}$$

If the STA gains are designed such that

$$(\kappa_1, \kappa_2) \in K = \left\{ (\kappa_1, \kappa_2) \left| \begin{array}{c} \frac{2\sqrt{2\bar{k}}\vartheta_2}{k_{min}\kappa_1} - t_d < 0, \\ \kappa_2 > 5\vartheta_1, \\ 32\vartheta_1 < \kappa_1^2 < 8(\kappa_2 - \vartheta_1) \end{array} \right. \right\},$$

where t_d is a given time and $\chi(0) = 0$, then χ has a reaching time $t_R < t_d$.

Remark 7.1 Any given reaching time can be achieved if the STA gains satisfies Lemma 7.1. But the lower the reaching time, the bigger the gains should be, causing an increment in the chattering. By using a switched strategy, as the one proposed in [14], the chattering can be attenuated.

7.4.1 Cascade Structure Observer

Recall assumptions A3, A5–A7, A8. Once again, since the pair (A_i, B_i, C_i) is strongly observable we can assure the pair $(A_{11,i}, C_{11,i})$ is observable (see [4]) and assume that the SULTIS (7.2) has observability index l.

To reconstruct theoretically exactly the states without filtration an STA-based hierarchical observer is used. This cascade observer has the same structure as in the OISM strategy, but using the STA. Here, the structure of the cascade observer is presented emphasizing its peculiarities:

- A *Luenberger stabilizer* of the form (7.4), designed as in Sect. 3.1.2. Under these conditions, the first derivative of the observation error remains bounded, i.e.,

$$\|\dot{e}_1(t)\| < \|A_{11,i} - K_i C_{11,i}\|\gamma.$$

- A *hierarchical observer* composed by:
  - A family of $l-1$ super-twisting based observers (7.5), that reconstruct theoretically exactly before half of the dwell time, step-by-step, the output uncertainties/perturbations free part and its derivatives. For this observer, $v_k(t)$ is an output injection signal based on the STA.
  - And an algebraic part

$$\hat{x}(t) = \begin{bmatrix} \hat{x}_1(t) \\ y_2(t) \end{bmatrix}; \tag{7.20}$$

  where $\hat{x}_1(t) = \tilde{x}_1(t) - \mathcal{O}^+_{1,i,l} v(t)$ with $\mathcal{O}_{1,i,l}$ the observability matrix of the pair (C_{11}, A_{11}) and

$$v(t) = \begin{bmatrix} C_{11}\tilde{x}_1(t) - y_1(t) \\ v_1(t) \\ v_2(t) \\ \vdots \\ v_{l-1}(t) \end{bmatrix}.$$

  The algebraic part (7.20) reconstructs theoretically exactly the states of the system before half of the dwell time.

The next subsections are devoted to the design of each part of the observer.

7.4.1.1 Hierarchical Observer Design

To reconstruct the states $x(t)$ theoretically exactly before half of the dwell time without the use of filters, it is necessary to recover the vectors $C_{11}A_{11,i}^{k-1}x_1(t)$ by using a family of super-twisting based observers with a convergence time $t_R < \frac{\delta}{2(l-1)}$. Each observer is turn on once the lower derivative observers have converged. Assuring theoretically exact reconstruction of the output error and its $l-1$-derivatives for $t_{j-1} + \frac{\delta}{2} \leq t \leq t_j$.

The observer design is given by the following theorem.

Theorem 7.2 *Assume*

(a) The auxiliary state vectors x_{ak}, for all $k = 1, \ldots, l-1$, $i \in \mathcal{I}$ and $\tau_k = t_{j-1} + \frac{\delta(k-1)}{2(l-1)} \leq t \leq t_j$ is designed as in (7.5), where $L_{k,i}(t) \in \mathbb{R}^{n-m \times p-m}$ is a design matrix such that $\det\left(C_{11,i}A_{11,i}^{k-1}L_{k,i}\right) \neq 0$.

(b) At $t = \tau_k$ the k-th variable x_{ak} satisfies

$$C_{11,i}x_{a1}(t_{j-1}^{+}) = y_1(t_{j-1}^{+}),$$
$$C_{11,i}A_{11,i}^{k-1}x_{ak}(\tau_k) = C_{11,i}A_{11,i}^{k-1}\tilde{x}_1(\tau_k) - v_{k-1}(\tau_k);$$

$i = \mathcal{I}$, $j = 1, 2, \ldots,$ and $k = 2, \ldots, l-1$.

(c) The sliding variables s_k are designed as

$$s_k(y_1(t), x_{ak}(t)) = \begin{cases} y_1(t) - C_{11,i}x_{a1}(t), & k = 1, \\ C_{11,i}A_{11,i}^{k-1}\tilde{x}_1(t) - v_{k-1}(t) - C_{11,i}A_{11,i}^{k-1}x_{ak}(t), & k = 2, \ldots, l-1. \end{cases} \tag{7.21}$$

(d) The output injection v_k is designed as an STA of the form

$$\begin{aligned} v_k(t) &= -\kappa_{k,i_1}\lceil s_k(y_1(t), t)\rfloor^{\frac{1}{2}} + \varpi_k(t), \\ \dot{\varpi}_k &= -\kappa_{k,i_2}\lceil s_k(y_1(t), t)\rfloor^{0}, \\ \varpi_k(t_0) &= 0, \quad \varpi_k(\tau_k^{+}) = \varpi_k(\tau_k^{-}) + \bar{\varpi}_{i,k}; \end{aligned} \tag{7.22}$$

where $\bar{\varpi}_{i,k} \in \mathbb{R}^p$.

(e) And $(\kappa_{k,i_1}, \kappa_{k,i_2}) \in K_{O,k}$, where

$$K_{O,k} = \left\{ (\kappa_1, \kappa_2) \left| \begin{array}{c} \frac{2\sqrt{2\bar{k}}M_{2,i,k}}{k_{min}\kappa_1} - \frac{\delta}{2(l-1)} < 0, \\ \kappa_2 > 5M_{1,i,k}, \\ 32M_{1,i,k} < \kappa_1^2 < 8(\kappa_2 - M_{1,i,k}) \end{array} \right. \right\},$$

with

$$M_{1,i,k} \geq \left\| C_{11,i} A_{11,i}^{k} \right\| \left(\| A_i - K_i C_i \| \gamma_i + \| B_i \| \Psi_{i_{max}} \right);$$
$$M_{2,i,k} \geq \left\| C_{11,i} A_{11,i}^{k} \right\| \gamma_i .$$

Then,

$$v_k(t) = -C_{11,i} A_{11,i}^{k} \left(x_1(t) - \tilde{x}_1(t) \right),$$

for $t_{j-1} + \frac{\delta}{2} \leq t \leq t_j$ *and it is possible to reconstruct theoretically exactly all the vector functions* $C_{11,i} A_{11,i}^{k-1} x_1(t)$ *in the same time interval.*

Proof Remember that the observers are turn on sequentially whenever the observers that reconstruct the lowers derivatives have converged. The proof is constructive and can be obtained iteratively.

Let's start by recovering the first vector $C_{11,i} A_{11,i} x_1(t)$. Let k=1, since the conditions (a)–(c) are fulfilled; it is clear that $s_1(y_1(t_{j-1}), x_{a1}(t_{j-1})) = 0$. Moreover, the time derivative of the sliding variable along the trajectories of (7.2) and (7.5) has the form

$$\dot{s}_1(y_1(t), x_{a1}(t)) = C_{11,i} A_{11,i} \left(x_1(t) - \tilde{x}_1(t) \right) + v_1(t), \tag{7.23}$$

and once it has converged, the equivalent control that maintains the trajectory on the surface has the form

$$v_{1_{eq}}(t) = -C_{11,i} A_{11,i} \left(x_1(t) - \tilde{x}_1(t) \right).$$

Now, we design the output injection (d) assuring the sliding variable s_1 and its derivatives converge to the origin before each switching. Substituting the STA output injection (7.22) on the auxiliary state vector x_{a1}, then (7.23) can be restated as

$$\begin{aligned} \dot{s}_1(y_1(t), x_{a1}(t)) &= -\kappa_{1,i_1} \lceil s_1(y_1(t), t) \rfloor^{\frac{1}{2}} + \Lambda_1(t), \\ \dot{\Lambda}_1 &= -\kappa_{1,i_2} \lceil s_1(y_1(t), t) \rfloor^{0} + C_{11,i} A_{11,i} \left(\dot{x}_1(t) - \dot{\tilde{x}}_1(t) \right), \\ \Lambda_1(t_{jR}) &= C_{11,i} A_{11,i} \left(x_1(t_{jR}) - \tilde{x}_1(t_{jR}) \right); \end{aligned}$$

where $\Lambda_1(t) = C_{11,i} A_{11,i} \left(x_1(t) - \tilde{x}_1(t) \right) + \varpi_1(t)$. Observe that this dynamical system has the form (7.18).

Since $(\kappa_{1,i_1}, \kappa_{1,i_2})$ satisfies Lemma 7.1 (condition (e)) and $s_1(y_1(t_{j-1}), x_{a1}(t_{j-1}) = 0$, s_1 converges to the origin with a reaching time $\bar{t}_{1,jR} < \tau_2$. Then, the exact convergence of s_1 and its derivatives to the origin for $\tau_2 \leq t \leq t_j$ is assured.

Assume the result is true for $k = \mathbf{k}$, and let us prove the result for $k = \mathbf{k} + 1$. Once again our aim is to recover the k-vector $C_{11,i} A_{11,i}^{k} x(t)$. Conditions (a)–(c) are satisfied. Taking the time derivative of the sliding variable along the trajectories of (7.2) and (7.5) for $\tau_k \leq t < t_j$.

$$\dot{s}_k(y_1(t), x_{a_k}(t)) = C_{11,i} A_{11,i}^{k} \left(x_1(t) - \tilde{x}_1(t) \right) + v_k(t). \tag{7.24}$$

Moreover, once the k-th sliding variable and its derivatives have converged,

$$C_{11,i}A_{11,i}^{\mathbf{k}}x_{a_k}(\tau_k) = C_{11,i}A_{11,i}^{\mathbf{k}}x_1(\tau_k).$$

and the equivalent control is

$$v_{\mathbf{k}_{eq}}(t) = -C_{11,i}A_{11,i}^{k}\left(\tilde{x}_1(t) - x_1(t)\right).$$

Designing the output injection (d) the sliding variable dynamics (7.24) can be rewritten as

$$\begin{aligned}
\dot{s}_k(y_1(t), x_{a_k}(t)) &= -\kappa_{k_1}\lceil s_k(y_1(t), t)\rfloor^{\frac{1}{2}} + \Lambda_k(t),\\
\dot{\Lambda}_k &= -\kappa_{k,i_2}\lceil s_k(y_1(t), t)\rfloor^{0} + C_{11,i}A_{11,i}^{k}\left(\dot{x}_1(t) - \dot{\tilde{x}}_1(t)\right),\\
\Lambda_k(\tau_k) &= C_{11,i}A_{11,i}^{k}x_1(\tau_k) - \tilde{x}_1(\tau_k)),
\end{aligned}$$

where $\Lambda_k(t) = C_{11,i}A_{11,i}^{k}\left(x_1(t) - \tilde{x}_1(t)\right) + \varpi_k(t)$.

Since condition (e) is satisfied and $s_k(y_1(\tau_{\mathbf{k}}), x_{ak}(\tau_{\mathbf{k}})) = 0$, once again from Lemma 7.1, s_k converges to the origin with a reaching time $\bar{t}_{k,jR} < \tau_{k+1}$. And the exact convergence of s_k and its derivatives to the origin for $\tau_{k+1} \leq t \leq t_j$ is guaranteed.

Note that we proved exact reconstruction of the output error and its $(l-1)$ time-derivatives for $t_{j-1} + \frac{\delta}{2} \leq t \leq t_j$.

7.4.1.2 State Reconstruction

Using the family of STA observers the output y_1 and its $l-1$ time-derivatives at every location have been reconstructed theoretically exactly in the time interval $[t_{j-1} + \frac{\delta}{2}, t_j]$. Using this information we can construct the vector

$$\mathcal{O}_{1,i,l}x_1(t) = \mathcal{O}_{1,i,l}\tilde{x}_1(t) - v(t).$$

Since the pair $(A_{11,i}, C_{11,i})$ is observable, the pseudo-inverse of $\mathcal{O}_{1,i,l}$ is well defined and the states can be recovered by means of the equation

$$x_1(t) = \tilde{x}_1(t) - \mathcal{O}_{1,i,l}^{+}v(t). \tag{7.25}$$

Then, the algebraic observer is suggested as

$$\hat{x}_1(t) = \tilde{x}_1(t) - \mathcal{O}_{1,i,l}^{+}v(t); \tag{7.26}$$

and we are able to reconstruct theoretically exactly the states for $t_{j-1} + \frac{\delta}{2} \leq t \leq t_j$ by using (7.20).

7.4.2 COISM Control Design

In the previous section, we designed an observer capable to reconstruct theoretically exactly the states of the SULTIS before half of the dwell time, i.e., $\hat{x}(t) = x(t)$ for all $t_{j-1} + \frac{\delta}{2} \leq t < t_j$. Once the states have been reconstructed, at $t = t_{j-1} + \frac{\delta}{2}$ the controllers are turn on. In this section, the integral part of the control law is designed.

Consider the SULTIS (7.2) and define the output-based integral sliding dynamics

$$s(y,t) = G_i\left(y(t) - y\left(t_{j-1} + \frac{\delta}{2}\right)\right) - \int_{t_{j-1}+\frac{\delta}{2}}^{t} G_i C_i (A_i \hat{x}(\tau) + B_i u_{nom}(\tau))\mathrm{d}\tau, \tag{7.27}$$

$i \in \mathcal{I}$; where $G_i \in \mathcal{G}_i$ is a design projection matrix, chosen, without loss of generality, such that $G_i C_i B_i = I$. Observe that $s\left(y\left(t_{j-1} + \frac{\delta}{2}\right), t_{j-1} + \frac{\delta}{2}\right) = 0$. Taking the first derivative of the sliding variable along the trajectory of (7.2) we obtain

$$\dot{s}(y,t) = G_i C_i A_i \left(x(t) - \hat{x}(t)\right) + u_{int}(t) + \phi(t). \tag{7.28}$$

Since $x(t) = \hat{x}(t)$, then

$$\dot{s}(y,t) = u_{int}(t) + \phi(t); \tag{7.29}$$

the equivalent control [27] that maintains the trajectory on the sliding mode is

$$u_{Ieq} = -\phi(t), \tag{7.30}$$

and the sliding mode dynamics of the SULTIS takes the form

$$\begin{aligned} \dot{x}(t) &= A_i x(t) + B_i u_{nom}(t) \\ y(t) &= C_i x(t), \quad x(0) = x_0; \end{aligned} \tag{7.31}$$

7.4.2.1 Implementation of STA Controller Based on STA Observer

The control law u_{int} depends on the observed state $\hat{x}$, so it is necessary to design this controller in such a way that the observer STA dynamics does not affect the properties of the controller [9], i.e., the uncertain/perturbations should be Lipschitz without affecting the continuity of the control law. Let us design the COISM term using the STA [17]

$$\begin{aligned} u_{int} &= -\kappa_{i_1}\lceil s(y(t),t)\rfloor^{\frac{1}{2}} + \omega(t), \\ \dot{\omega} &= -\kappa_{i_2}\lceil s(y(t),t)\rfloor^{0}, \quad \omega\left(t_{j-1} + \frac{\delta}{2}\right) = 0; \end{aligned} \tag{7.32}$$

where $\kappa_{i_\iota} \in \mathbb{R}_+$ with $i \in \mathcal{I}$ and $\iota = 1, 2$. Substituting the STA control (7.32) in (7.28) and with a simple change of variable.

$$\dot{s}(y,t) = -\kappa_{i_1}\lceil s(y(t),t)\rfloor^{\frac{1}{2}} + \Omega(t) + G_iC_iA_i\begin{bmatrix}\mathcal{O}^+_{1,i,l}v(t)\\0\end{bmatrix},$$
$$\dot{\Omega}(t) = -\kappa_{i_2}\lceil s(y(t),t)\rfloor^{0} + G_iC_iA_i\begin{bmatrix}\left(\dot{x}_1(t)-\dot{\tilde{x}}_1(t)\right)\\0\end{bmatrix} + \dot{\phi}(t),$$
$$s\left(y, t_{j-1}+\frac{\delta}{2}\right) = 0; \tag{7.33}$$

where

$$\Omega(t) = G_iC_iA_i\begin{bmatrix}(x_1(t)-\tilde{x}_1(t))\\0\end{bmatrix} + \phi(t) + \omega(t).$$

The term $\mathcal{O}^+_{1,i,l}v(t)$ is non-Lipschitz, so it is not possible to consider it as uncertainty/perturbation and to assure the convergence of the STA. If the proposed Super-Twisting controller is changed to

$$u_{int} = -G_iC_iA_i\begin{bmatrix}\mathcal{O}^+_{1,i,l}v(t)\\0\end{bmatrix} - \kappa_{i_1}\lceil s(y(t),t)\rfloor^{\frac{1}{2}} + \omega(t),$$
$$\dot{\omega} = -\kappa_{i_2}\lceil s(y(t),t)\rfloor^{0},\ \ \omega(0)=0,\ \ \omega(t_j^+) = \omega(t_j^-) + \bar{\omega}_i; \tag{7.34}$$

we compensate the term $G_iC_iA_i\mathcal{O}^+_{i,l}v(t)$ and the sliding dynamics have the typical STA form

$$\dot{s}(y,t) = -\kappa_{i_1}\lceil s(y(t),t)\rfloor^{\frac{1}{2}} + \Omega(t),$$
$$\dot{\Omega}(t) = -\kappa_{i_2}\lceil s(y(t),t)\rfloor^{0} + G_iC_iA_i\begin{bmatrix}\left(\dot{x}_1(t)-\dot{\tilde{x}}_1(t)\right)\\0\end{bmatrix} + \dot{\phi}(t), \tag{7.35}$$
$$s(y, t_{j-1}) = 0;$$

Remark 7.2 Thanks to the integral structure of $s(y,t)$ the non-Lipschitz dynamics of the observer that affect the controller, only depends on v, the continuous part of the observer dynamics. Hence, the continuity of u_{int} is preserved.

The following Lemma gives the design of the STA gains $(\kappa_{i_1}, \kappa_{i_2})$ for the controller.

Lemma 7.2 *Suppose assumptions A1–A6 are satisfied and* $(\kappa_{i_1}, \kappa_{i_2}) \in K_C$

$$K_C = \left\{ (\kappa_1,\kappa_2) \;\middle|\; \begin{array}{c} \frac{2\sqrt{2}\bar{k}L_{2,i}}{k_{min}\kappa_1} - \frac{\delta}{2} < 0, \\ \kappa_2 > 5L_{1,i}, \\ 32L_{1,i} < \kappa_1^2 < 8(\kappa_2 - L_{1,i}) \end{array} \right\},$$

with

$$L_{1,i} \geq \Phi_{i_{max}} + \|G_i C_i A_i\| \left(\|A_i - K_i C_i\|\gamma_i + \|B_i\|\Psi_{i_{max}}\right);$$
$$L_{2,i} \geq \|G_i C_i A_i\|\gamma_i + \Phi_{i_{max}}.$$

Then, the sliding mode dynamics converges to the origin with a reaching time $t_{jR} < t_{j-1} + \delta < t_j$.

Proof This result comes directly from Lemma 7.1.

With this, we assure the robustification of the nominal trajectory before the dwell time.

Remark 7.3 In comparison with the OISM strategy, we are not able to reconstruct the states right after the initial time. Due to the presence of the reaching phase, the controller and the observer will reconverge after every switching. But, the proposed methodology does not need the use of filters.

7.4.3 Robustifying Conditions: Restrictions to the Initial Conditions

Along this chapter, we assume the existence of the dwell time. However, we are considering state-dependent location transitions and any deviation of the state trajectory may provoke an undesirable switching. In this section, we analyze under which condition we can guarantee the existence of the dwell time, and hence the robustification of the nominal trajectory.

Recall that the controller has been designed such that it converges before $t = t_{j-1} + \delta$. But, during the reaching phase, the COISM controller causes a transient in the SULTIS that could produce an unexpected switching. To avoid this issue, it is necessary to restrict the value of the states at the switching times.

Lemma 7.3 *If*

$$\|x(t_{j-1})\| \leq \mu_{j-1} < \frac{\xi_{j-1,max} - \left\| \int\limits_{t_{j-1}}^{t_{j-1}+\delta/2} e^{A_i(t_{j-1}+\frac{\delta}{2}-\tau)} \mathrm{d}\tau \right\| \|B_i\| \Phi_{max}}{\left\| e^{A_i \delta/2} \right\|}. \tag{7.36}$$

where

$$\xi_{j-1,max} = \frac{\|A_i - B_i K_i\|}{e^{\|A_i - B_i K_i\| \frac{\delta}{2}}} M_{i,min} - \|B_i\| \frac{\delta}{2} u_{max};$$
$$M_{i,min} = \min_{\substack{i'\mathcal{I},\\ i \neq i'}} \left\| arg\ min_{x \in \mathbb{R}^n} |M_{i,i'}(x)| \right\|;$$

$$u_{max} = \|u_{int}(t) + \phi(t)\| \leq \kappa_1 m^{\frac{1}{4}} s_{max}^{\frac{1}{2}} + \Omega_{max}; \tag{7.37}$$
$$\Omega_{max} = \Phi_{max} + (\kappa_2 + \Psi_{max})\frac{\delta}{2};$$
$$s_{max} = \left(\Omega_{max} + \alpha_0 \kappa_1 m^{\frac{1}{4}}\right)\frac{\delta}{2};$$

with α_0 the unique positive root of the equation

$$\alpha - a_0 - b_0 \alpha^{\frac{1}{2}} = 0$$

with

$$a_0 = \frac{\delta^2 \Omega_{max}}{8}, \text{ and } b_0 = \frac{\sqrt{2}\delta^{\frac{3}{2}} \kappa_1 m^{\frac{1}{4}}}{6}.$$

Then the SULTIS does not switch before the dwell time.

Proof The proof of this lemma comes directly from the bound of the state vector and the sliding variable for all $t_{j-1} < t \leq t_{j-1} + \delta$. Recall that $s(y, t) \in \mathbb{R}^p$. Majoring Ω, we can show that $\|\Omega(t)\| \leq \Omega_{max}$ and by applying the Martynyuk-Kosolapov's integral inequality [Theorem 4.1] [2] to the integral equation of $\dot{s}$ we get that $\|s(y, t)\| \leq s_{max}$. Note that at $t = t_{j-1} + \frac{\delta}{2}$

$$\|x(t_{j-1} + \delta/2)\| \leq \left\|e^{A_i \delta/2}\right\| \mu_{j-1} + \left\| \int_{t_{j-1}}^{t_{j-1}+\delta/2} e^{A_i(t_{j-1}+\delta/2-\tau)} d\tau \right\| \|B_i\| \Phi_{max} = \xi_{j-1}.$$

Hence,

$$\|x(t)\| \leq \frac{\xi_{j-1} + \|B_i\| \frac{\delta}{2} u_{max}}{\|A_i - B_i K_i\|} e^{\|A_i - B_i K_i\| \frac{\delta}{2}}$$

by direct application of the Gronwall-Bellman inequality to the integral equation of the SULTIS for $t_{j-1} \leq t < \delta$. Furthermore, if (7.36) is satisfied then $\xi_{j-1} < \xi_{j-1,max}$ and the state trajectory does not hit any switching manifold for all $t_{j-1} \leq t < \delta$.

With this lemma, we can analyze if for a given set of switching manifolds and a bound of the initial condition, the SULTIS with state-dependent location transitions can be robustified.

7.4.4 Simulation Results

For comparison purposes, consider the system proposed in Sect. 7.3.4. This system has two locations, initial conditions $x_0 = [0.1; -0.1; 0.2; -1.5]^T$ are unknown with a known bound $\mu = 2$ and the uncertainties/perturbations used in the simulations are

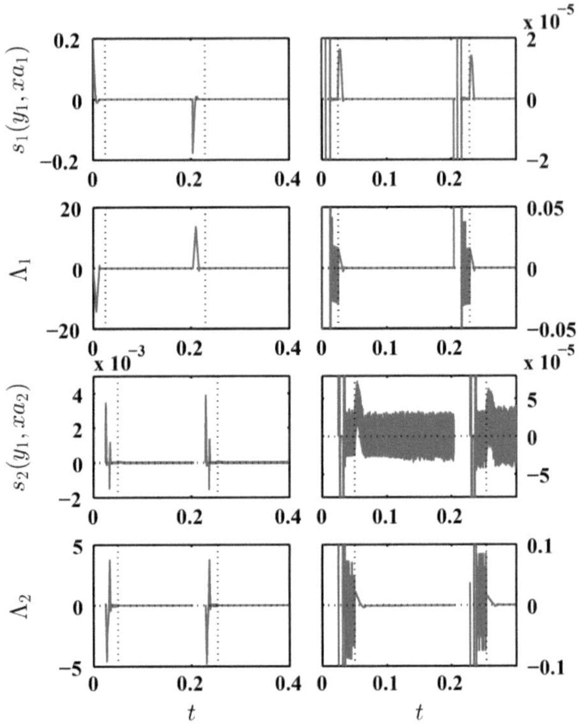

Fig. 7.10 Sliding variables of the hierarchical observer

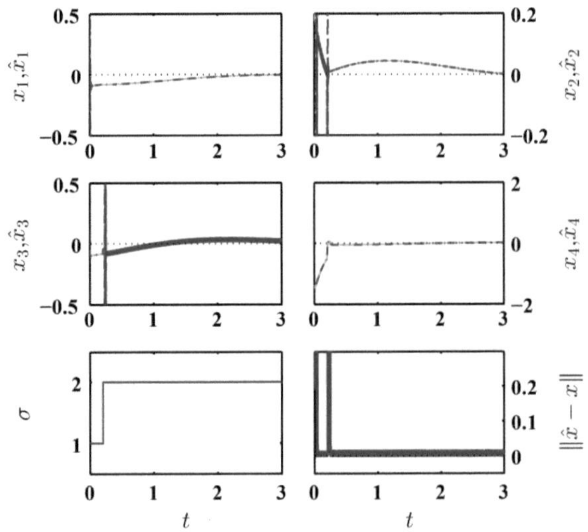

Fig. 7.11 Cascade observer: Real states (*continuous line*) versus observed states $\hat{x}$ (*dashed line*)

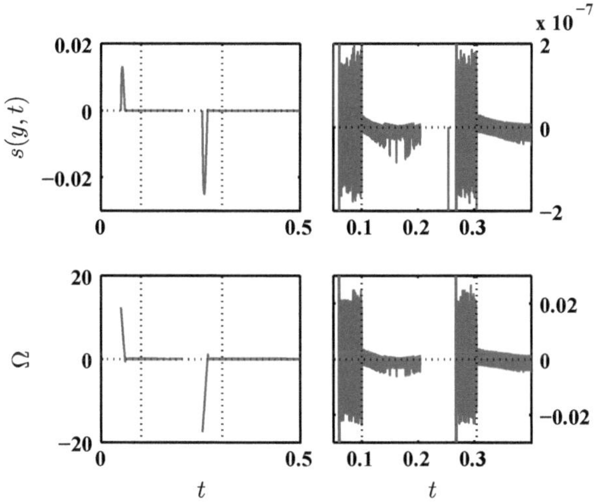

Fig. 7.12 Controller sliding variables

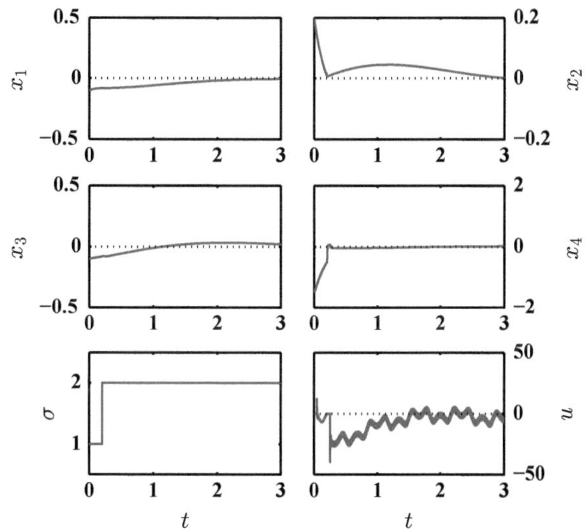

Fig. 7.13 SULTIS with the COISM controller

$$\phi(t) = 5\sin(\pi\cos(3\pi t)) + 10.$$

All the simulations are done in Simulink, using the Euler method with a sample step $\Delta t = 1e - 5$. Remember that the system is unstable and it can be shown that satisfy assumption A9 with dwell time $\delta = 0.1$. The system is transformed into the form (7.2), with the same transformation and for the transformed nominal system (7.3) an LQR nominal controller is proposed in each location. Now, let's apply the COISM methodology. First, we design the cascade observer. Recall that this observer is composed by two parts: The Luenberger-type error stabilizer and the hierarchical observer. The Luenberger stabilizer was designed using the methodology proposed in [1]. This stabilizer is shown in Fig. 7.3 and it is designed for the uncertainties/perturbations free part of the SULTIS assuring the existence of a bound for the observation error e_1 and its first derivative.

Now, let's design the second part of the cascade observer. Since the considered system has observability index $l = 2$, it is necessary to design a hierarchical observer composed by two sliding modes observers. The observers are designed such that they guarantee a convergence time lower than 0.025.

At every switching time, the controllers are turned off and the first observer starts to evolve. At $t = t_j + 0.025$, the second observer is turn on. The reaching phase of both observers is illustrated in Fig. 7.10. Notice that the first observer converges before $t = t_j + 0.025$. Moreover, the full observer converges before $t = t_j + 0.05$. To attenuate the chattering a switched STA gains strategy [14] is applied to both observers, the attenuation of the chattering can be seen in the second column of the figure. The complete behavior of the cascade observer is presented in Fig. 7.11. Hence, we have reconstruct the states of the system without the use of filters theoretically exactly before half of the dwell time. Moreover, note that the observation error is lower than the OISM one.

Once the observer has converged the COISM controller is turn on. This is designed to accomplish a convergence time $t_d = 0.05$ and after the dwell time the gains are reduced to attenuate the chattering. The sliding variables of the controller are given in Fig. 7.12. Observe that the sliding variables converge before the dwell time, assuring exact compensation of the uncertainties-perturbations before the dwell time and the robustification of the nominal controller. Moreover, the generated control signal is continuous, attenuating the chattering (see Fig. 7.13).

7.5 Concluding Remarks

A robustifying methodology based on the OISM/COISM theory for SULTIS with state-dependent location transition is developed. Two strategies are considered.

- An output feedback controller for SULTIS with matched uncertainties/perturbations based on an OISM methodology is presented. The uncertainties are compensated and the state vector is reconstructed theoretically exactly just after the initial time. The switching sequences of the SULTIS and the nominal system coincide. But, in the practice to reconstruct the state it is necessary to use filters that affect the performance of the observer and the controller. Moreover, the used first-order sliding mode generates high level of chattering.
- To eliminate the disadvantages of the OISM controller, a continuous output feedback controller for SULTIS with matched uncertainties/perturbations based on a COISM methodology is presented. The chattering is attenuated and the necessity of filters is eliminated. But, in general, the observer and the controller reconverge after each switching. Hence, the STA is adjusted to the switched case satisfying

a specific reaching time in each location. The COISM controller implementation is modified to preserve the STA behavior of the controller by eliminating the non-Lipschitz terms of the STA observer. The continuity of the controller is kept, thanks to the integral structure of the sliding variable. Moreover, sufficient conditions are given for the states at the switching times to guarantee the absence of switchings before the dwell time. The location trajectory remains beside the effects of the reaching phase and the robustification of the nominal trajectory is assured after the dwell time.

Acknowledgements My co-authors and myself are happy to prepare this chapter for this book reflecting some new results developing the concept of output integral sliding mode that we proposed with you, Professor Alexander Poznyak, 10 years ago. Thank you for your creativity, kindness, and constant support. Happy birthday and long life.

Leonid Fridman would like to acknowledge the financial support of the project 113216 of PAPIIT-UNAM.

Leonid Fridman

Appendix: Proof Lemma 4

To design the STA gains with a desired convergence time an estimation of the reaching time t_R is needed. This estimation can be obtained by using the continuous candidate Lyapunov function [21]

$$V(\chi,\Pi)=\begin{cases}\frac{k^2}{4}\left(\frac{\Pi(t)\lfloor\chi(t)\rceil^0}{\bar{\gamma}}+k_0e^{\bar{m}(\chi,\Pi)}\sqrt{\bar{\chi}(\chi,\Pi)}\right)^2, & \chi(t)\Pi(t)\neq 0,\\ \frac{2\bar{k}^2\Pi^2(t)}{\kappa_1}, & \chi(t)=0,\\ \frac{|\chi(t)|}{2}, & \Pi(t)=0.\end{cases}$$

where

$$\begin{aligned}\bar{\gamma}&=\kappa_2-\vartheta_1\lfloor\chi(t)\Pi(t)\rceil^0,\\ \bar{m}(\chi,\Pi)&=\frac{1}{\sqrt{g-1}}\arctan\left(\frac{\kappa_1 g\lfloor\chi(t)\rceil^{1/2}-2\Pi(t)}{2\sqrt{g-1}\Pi(t)}\right),\\ \bar{\chi}(\chi,\Pi)&=2\bar{\gamma}|\chi(t)|-\kappa_1\lfloor\chi(t)\rceil^{1/2}\Pi(t)+\Pi(t)^2,\\ k&=\sqrt{\bar{\gamma}}\left|\bar{k}\sqrt{g}-e^{\frac{-(\arctan((g-1)^{-1/2})+0.5\pi\lfloor\chi(t)\Pi(t)\rceil^0)}{\sqrt{g-1}}}\right|,\\ k_0&=\frac{\frac{\lfloor\chi(t)\Pi(t)\rceil^0}{\bar{\gamma}}}{\bar{k}\sqrt{g}e^{\frac{\pi\lfloor\chi(t)\Pi(t)\rceil^0}{\sqrt{g-1}}}-e^{\arctan(\frac{-1}{g-1})}},\\ g&=\frac{8\bar{\gamma}}{\kappa_1^2},\ \ \bar{k}\in I(g^-)\cap I(g^+),\ \ g^{\pm}=\frac{8(\kappa_2\pm\vartheta_1)}{\kappa_1^2},\end{aligned}$$

$$I(g) = \left(\frac{2}{g} + \frac{e^{\left(-\frac{\pi}{2} - \arctan\left(\frac{1}{\sqrt{g-1}}\right)\right)\frac{1}{g-1}}}{\sqrt{g}}, \frac{e^{\left(\frac{\pi}{2} - \arctan\left(\frac{1}{\sqrt{g-1}}\right)\right)\frac{1}{g-1}}}{\sqrt{g}} \right).$$

The next statement provides sufficient conditions for the STA gains to assure finite-time stability and an estimation of the reaching time.

Theorem 7.3 [21] *If* $\kappa_2 > 5\vartheta_1$ *and* $32\vartheta_1 < \kappa_1^2 < 8(\kappa_2 - \vartheta_1)$ *then:*

- $I(g^-) \cap I(g^+) \neq \emptyset$
- $\chi(t)$ *converge in finite time to the origin*
- $V(\chi, \Pi)$ *is a positive definite absolutely continuous function in* $\mathbb{R}^2$ *and continuously differentiable when* $\chi \Pi \neq 0$
- *The time derivative of* $V(\chi, \Pi)$ *along the trajectory of Eq. 7.18 satisfies*

$$\dot{V}_{\mathrm{i}}(\chi, \Pi) \leq -k\sqrt{V(\chi, \Pi)} \leq -k_{min}\sqrt{V(\chi, \Pi)},$$

almost everywhere with

$$k_{min} = \frac{\kappa_1}{\sqrt{8}} \min_{g \in (g^-, g^+)} \left\{ \left| g\bar{k} - \sqrt{g} e^{\frac{\arctan\left(\frac{-1}{\sqrt{g-1}}\right) + \frac{\pi(\kappa_1^2 g - 8\kappa_2)}{16\vartheta_1}}{\sqrt{g-1}}} \right| \right\}$$

- *The corresponding guaranteed reaching time is*

$$t_R \leq 2k_{min}^{-1}\sqrt{V(\chi(0), \Pi(0))}. \tag{7.38}$$

In order to assure that the system has converged before a given time t_d we need to design the gains κ_ι, $\iota = 1, 2$ such that the reaching time t_R fulfills the condition $0 < t_R < t_d$. Since $\kappa_2 > 5\vartheta_1$ and $32\vartheta_1 < \kappa_1^2 < 8(\kappa_2 - \vartheta_1)$, $\chi(t)$ converge in finite time with a reaching time t_R. Now, since $\chi(0) = 0$,

$$V(\chi(0), \Pi(0)) \leq \frac{2\bar{k}^2\vartheta_2^2}{\kappa_1^2}$$

and

$$t_R \leq \frac{2\sqrt{2}\bar{k}\vartheta_2}{k_{min}\kappa_1}.$$

Hence since $(\kappa_1, \kappa_2) \in K$, we can assure $t_R < t_d$.

References

1. Alessandri, A., Coletta, P.: Switching observers for continuous-time and discrete-time linear systems. In: American Control Conference, 2001. Proceedings of the 2001, vol. 3, pp. 2516–2521. IEEE, New York (2001)

2. Bainov, D., Simeonov, P.: Integral inequalities and applications. In: Mathematics and its Applications, vol. 57, 1 edn. Springer, Netherlands (1992)
3. Bejarano, F.J., Fridman, L.: State exact reconstruction for switched linear systems via a super-twisting algorithm. Int. J. Syst. Sci. **42**(5), 717–724 (2011)
4. Bejarano, F.J., Fridman, L., Poznyak, A.S.: Output integral sliding mode control based on algebraic hierarchical observer. Int. J. Control **80**, 443–453 (2007)
5. Bejarano, F.J., Pisano, A.: Switched observers for switched linear systems with unknown inputs. IEEE Trans. Autom. Control **56**(3), 681–686 (2011)
6. Bejarano, F.J., Pisano, A., Usai, E.: Finite-time converging jump observer for switched linear systems with unknown inputs. Nonlinear Anal. Hybrid Syst. **5**(2), 174–188 (2011)
7. Branicky, M.S.: Introduction to hybrid systems. In: Handbook of Networked and Embedded Control Systems, pp. 91–116. Birkhäuser Boston (2005)
8. Caravani, P., De Santis, E.: Observer-based stabilization of linear switching systems. Int. J. Robust Nonlinear Control **19**(14), 1541–1563 (2009)
9. Chalanga, A., Kamal, S., Fridman, L., Bandyopadhyay, B., Moreno, J.A.: How to implement super-twisting controller based on sliding mode observer? In: 2014 13th International Workshop on Variable Structure Systems (VSS), pp. 1–6 (2014)
10. Dávila, J., Ríos, H., Fridman, L.: State observation for nonlinear switched systems using non-homogeneous high-order sliding mode observers. Asian J. Control **14**(4), 911–923 (2012)
11. El-Farra, N.H., Mhaskar, P., Christofides, P.D.: Output feedback control of switched nonlinear systems using multiple lyapunov functions. Syst. Control Lett. **54**(12), 1163–1182 (2005)
12. Floquet, T., Barbot, J.P.: Super twisting algorithm-based step-by-step sliding mode observers for nonlinear systems with unknown inputs. Int. J. Syst. Sci. **38**(10), 803–815 (2007)
13. Fridman, L., Poznyak, A., Bejarano, F.J.: Robust Output LQ Optimal Control via Integral Sliding Modes. Birkhäuser (2014)
14. Galván-Guerra, R., Fridman, L., Velázquez-Velázquez, J., Kamal, S., Bandyopadhyay, B.: Continuous output integral sliding mode control for switched linear systems. Nonlinear Anal. Hybrid Syst. **22**, 284–305 (2016)
15. Hautus, M.: Strong detectability and observers. Linear Algebra Appl. **50**, 353–368 (1983)
16. Hespanha, J.P., Morse, A.: Switching between stabilizing controllers. Automatica **38**(11), 1905–1917 (2002)
17. Levant, A.: Sliding order and sliding accuracy in sliding mode control. Int. J. Control **58**(6), 1247–1263 (1993)
18. Li, Z., Wen, C., Soh, Y.: Observer-based stabilization of switching linear systems. Automatica **39**(3), 517–524 (2003)
19. Lian, J., Zhao, J., Dimirovski, G.M.: Integral sliding mode control for a class of uncertain switched nonlinear systems. Eur. J. Control **16**(1), 16–22 (2010)
20. Mohrenschildt, M.V.: Hybrid systems: solutions, stability, control. In: American Control Conference (ACC), vol. 1, pp. 692 –698 (2000)
21. Polyakov, A., Poznyak, A.: Reaching time estimation for super-twisting second order sliding mode controller via lyapunov function designing. IEEE Trans. Autom. Control **54**(8), 1951–1955 (2009)
22. Ríos, H., Kamal, S., Fridman, L., Zolghadri, A.: Fault tolerant control allocation via continuous integral sliding-modes: a hosm-observer approach. Automatica **51**, 318–325 (2015)
23. Rugh, W.J.: Linear System Theory. Prentice Hall, Upper Saddle River (1993)
24. Saadaoui, H., Manamanni, N., Djemaï, M., Barbot, J., Floquet, T.: Exact differentiation and sliding mode observers for switched lagrangian systems. Nonlinear Anal. Theory, Methods Appl. **65**(5), 1050–1069 (2006)
25. Van der Schaft, A., Schumacher, H.: An Introduction to Hybrid Dynamical Systems. Springer, Berlin (2000)
26. Shaikh, M.S., Caines, P.E.: On the hybrid optimal control problem: theory and algorithms. IEEE Trans. Autom. Control **52**(9), 1587–1603 (2007)
27. Utkin, V.I.: Sliding Modes in Control and Optimization. Springer, Berlin (1992)

Chapter 8
Discontinuous Integral Control for Systems with Relative Degree Two

Jaime A. Moreno

Abstract For systems with relative degree two we propose a homogeneous controller capable of tracking a smooth but unknown reference signal, despite of a Lipschitz continuous perturbation, and by means of a continuous control signal. The proposed control scheme consists of two terms: (i) a continuous and homogeneous state feedback, and (ii) a discontinuous integral term. The state feedback term aims at stabilizing (in finite time) the closed loop while the (discontinuous) integral term estimates the perturbation and the unknown reference signal in finite time and provides for perfect compensation in closed loop. By adding a continuous and homogeneous observer we complete an output feedback scheme, when not all states are available for measurement. The global finite time stability of the closed loop and its insensitivity with respect to matched Lipschitz continuous perturbations is proved in detail using a smooth and homogeneous Lyapunov function.

8.1 Introduction

Second Order Sliding Modes (SOSM) [13, 23, 25, 27, 28] can ensure perfect output tracking for uncertain systems with relative degree two despite of bounded perturbations. Combined with the first order differentiator [24, 26] this completes an output feedback control strategy (see also [30] for a Lyapunov based approach). In the presence of arbitrary but bounded perturbations/uncertainties it is well known that this objective can only be achieved by a discontinuous control law. The price to pay is the high frequency switching of the control signal, a phenomenon usually called *chattering*, and which is undesirable since it has a negative effect in the actuator, and excites unmodeled dynamics of the plant. To reduce the effect of chattering, a natural alternative consists in adding an integrator to the plant and designing a third order Higher Order Sliding Modes (HOSM) controller for the new control variable. This leads to perfect tracking in finite time, and the system will be insensitive to Lipschitz

J.A. Moreno (✉)
Instituto de Ingeniería, Universidad Nacional Autónoma de Mexico (UNAM), 04510 Coyoacán, Ciudad de Mexico, Mexico
e-mail: JMorenoP@ii.unam.mx

J.B. Clempner and W. Yu (eds.), *New Perspectives and Applications of Modern Control Theory*, https://doi.org/10.1007/978-3-319-62464-8_8

continuous (but not necessarily bounded) perturbations. In this form a continuous control signal is obtained, reducing the chattering effect. However, this requires to feedback not only the two plant states but also an extra state, which is unknown due to the unknown perturbation. Moreover, to implement an output feedback controller it is necessary to differentiate two times the variable to be tracked, with the consequent noise amplification effect.

For systems with relative degree one, first order Sliding Mode (SM) controllers [40, 41] are able to solve the perfect tracking problem for non vanishing (or persistently acting) bounded perturbations, again at the expense of the presence of chattering. To attenuate the chattering effect, besides the addition of an extra integrator and the application of a SOSM controller, there is an alternative approach using the Super-Twisting (ST) algorithm [13, 23, 28]: it allows to obtain a continuous control signal achieving the objective in finite time despite Lipschitz continuous perturbation/uncertainty signals, and without the necessity of taking the derivative of the sliding variable. This solution has many advantages and is one of the main virtues of the ST controller.

An analogous solution for systems with relative degree two has been presented for the first time in [42], and then a more restricted version was given in [14]. A different proof of the algorithm in [14] was proposed in [21] and a Twisting-Like algorithm (more in the spirit of [42]) was proposed in [39], where again a different proof technique was developed. All these solutions require the full state. In [31] the algorithm developed in [39] is analyzed using a novel and more general Lyapunov-based approach. Moreover, an output feedback version is developed, which uses a *continuous* rather than a discontinuous observer. In the present work we follow the ideas of [31] to analyze and design the algorithm presented in [42], which includes the results of [14, 39] as particular cases. We also develop an output feedback version with a continuous observer for that algorithm. Moreover, we provide complete and detailed proofs, in contrast to [31], where only some of the proofs were given due to space restrictions.

In the case of (almost) constant perturbations $\rho(t)$ a classical solution to the robust regulation problem is the use of integral action, as for example in the PID control [22]. The linear solution would consist of a state feedback plus an integral action, $u = -k_1 x_1 - k_2 x_2 + z$, $\dot{z} = -k_3 x_1$. This controller requires only to feedback the position and the velocity. For an output feedback it would be only necessary to estimate the velocity (with the D action for example). In contrast to the HOSM controller this PID control is only able to reject constant perturbations, instead of Lipschitz ones, and it will reach the target only exponentially fast, and not in finite time. By the Internal Model Principle it would be possible to reject exactly any kind of time varying perturbations $\rho(t)$, for which a dynamical model (so called an exosystem) is available. However, this would increase the complexity (order) of the controller, since this exosystem has to be included in the control law.

Here we provide a solution to the problem, that is somehow an intermediate solution between HOSM and PID control. Similar to the HOSM control our solution uses a discontinuous integral action, it can compensate perturbations with bounded derivative ($\rho(t)$ is Lipschitz) and the origin is reached in finite time. So it can solve not only regulation problems (where ρ is constant) but also tracking problems (with ρ time varying) in finite time and with the same complexity of the controller. Similar to the PID control the proposed controller provides a continuous control signal (avoiding chattering) and it requires only to feedback position and velocity. We also provide for a (non classical) D term, i.e. a finite time converging observer, to estimate the velocity. This basic idea has been already presented in our previous work [42]. In the present one we give a much simpler Lyapunov-based proof, and we also include an observer in the closed loop together with its Lyapunov-based proof. Our solution can be seen as a generalization of the Super Twisting control for systems of relative degree one [13, 23, 26, 28, 32] to systems with relative degree two.

The rest of the paper is organized as follows. In next Sect. 8.2 we recall some facts about differential inclusions, homogeneity and Lyapunov functions. Section 8.3 presents the problem to be solved in the paper while in Sect. 8.4 we present the main result: The Discontinuous Integral Controller with and without observer and give some discussion on the algorithms. Section 8.5 is dedicated to present a detailed Lyapunov-based proof of the convergence of the closed loop for the proposed control algorithms. In Sect. 8.6 we give an illustrative example with some simulations and in Sect. 8.7 some conclusions are drawn.

8.2 Preliminaries: Differential Inclusions and Homogeneity

We recall some important concepts about Differential Inclusions (DI's), homogeneity and homogeneous DI's [1–6, 10–12, 18, 27, 29], which are used in the chapter.

Uncertain or discontinuous systems are more appropriately described by DI $\dot{x} \in F(t, x)$ than by Differential Equations (DE). A solution of this DI is any function $x(t)$, defined in some interval $I \subseteq [0, \infty)$, which is absolutely continuous on each compact subinterval of I and such that $\dot{x}(t) \in F(t, x(t))$ almost everywhere on I. Thus, for a discontinuous DE $\dot{x} = f(t, x)$ the function $x(t)$ is said to be a generalized solution of the DE if and only if it is a solution of the associated DI $\dot{x} \in F(t, x)$. We will consider the DI $\dot{x} \in F(t, x)$ associated to $\dot{x} = f(t, x)$, as the one given by the approach of A.F. Filippov [2, 12, Sect. 1.2]. So, we refer to such DI as Filippov DI and to its solutions as Filippov solutions. The multivalued map $F(t, x)$ satisfies the *standard assumptions* if: (H1) $F(t, x)$ is a nonempty, compact, convex subset of $\mathbb{R}^n$, for each $t \geq 0$ and each $x \in \mathbb{R}^n$; (H2) $F(t, x)$ as a set valued map of x, is upper semi-continuous for each $t \geq 0$; (H3) $F(t, x)$ as a set valued map of t, is Lebesgue measurable for each $x \in \mathbb{R}^n$. (H4) $F(t, x)$ is locally bounded. Recall that a set valued map $G : \mathbb{R}^{n_1} \rightrightarrows \mathbb{R}^{n_2}$ with compact values is *upper-semicontinuous* if for each x_0 and for each $\epsilon > 0$ there exists $\delta > 0$ such that $G(x) \subseteq G(x_0) + B_\epsilon$, provided that $x \in B_\delta(x_0)$. It is well-known that, see [12] or [2, Theorem 1.4], if the multivalued map $F(t, x)$ satisfies the standard assumptions then for each pair $(t_0, x_0) \in [0, \infty) \times \mathbb{R}^n$ there is an interval I and at least a solution $x(t) : I \to \mathbb{R}^n$ such that $t_0 \in I$ and $x(t_0) = x_0$. A DI $\dot{x} \in F(x)$ (a DE $\dot{x} = f(x)$) is called *globally uniformly finite-time stable* (GUFTS) at 0, if $x(t) = 0$ is a Lyapunov-stable solution and for any $R > 0$ there exists $T > 0$ such that the trajectory starting within the ball $\|x\| < R$ reaches zero in the time T.

Assume that the origin $x = 0$ is an equilibrium position of the the DI $\dot{x} \in F(t, x)$, this means that $0 \in F(t, 0)$ for almost every (a.e.) $t \geq 0$. It is possible to characterize the asymptotic stability of the origin by means of a (strict) Lyapunov function, in the same spirit as the classical second Lyapunov theorem for smooth systems (see e.g. [2, Theorem 4.1]). In particular, if the Lyapunov function $V(t, x)$ is of class $\mathcal{C}^1$ (once continuously differentiable), and it satisfies

$$a(\|x\|) \leq V(t, x) \leq b(\|x\|)$$
$$\frac{\partial V(t, x)}{\partial t} + \frac{\partial V(t, x)}{\partial x}\nu \leq -c(\|x\|)$$

for a.e. $t \geq 0$, for all $x \in \mathbb{R}^n$ and all $\nu \in F(t, x)$, and for some functions $a, b, c \in \mathcal{K}_\infty$, then the origin is Uniformly Globally Asymptotically Stable (UGAS)[1] for the DI.

Continuous and discontinuous homogeneous functions and systems have a long history [1–4, 6, 11, 15, 17, 27, 29, 33–36, 43]. We recall briefly this important property. For a given vector $x = [x_1, ..., x_n]^\top \in \mathbb{R}^n$ and for every $\epsilon > 0$, the dilation operator is defined as $\Delta_\epsilon^r x := [\epsilon^{r_1} x_1, ..., \epsilon^{r_n} x_n]^\top$, where $r_i > 0$ are the weights of the coordinates, and let $\mathbf{r} = [r_1, ..., r_n]^\top$ be the vector of weights. A function $V : \mathbb{R}^n \to \mathbb{R}$ (respectively, a vector field $f : \mathbb{R}^n \to \mathbb{R}^n$, or a vector-set field $F(x) \subset \mathbb{R}^n$) is called $\mathbf{r}$-homogeneous of degree $l \in \mathbb{R}$ if the identity $V(\Delta_\epsilon^r x) = \epsilon^l V(x)$ holds for every $\epsilon > 0$ (resp., $f(\Delta_\epsilon^r x) = \epsilon^l \Delta_\epsilon^r f(x)$, or $F(\Delta_\epsilon^r x) = \epsilon^l \Delta_\epsilon^r F(x)$). Along this paper we refer to this property as $\mathbf{r}$-*homogeneity* or simply *homogeneity*. A system is called homogeneous if its vector field (or vector-set field) is $\mathbf{r}$-homogeneous of some degree.

Given a vector $\mathbf{r}$ and a dilation $\Delta_\epsilon^r x$, the homogeneous norm is defined by $\|x\|_{\mathbf{r},p} := \left(\sum_{i=1}^n |x_i|^{\frac{p}{r_i}}\right)^{\frac{1}{p}}$, $\forall x \in \mathbb{R}^n$, for any $p \geq 1$, and it is an $\mathbf{r}$-homogeneous function of degree 1. The set $S = \{x \in \mathbb{R}^n : \|x\|_{\mathbf{r},p} = 1\}$ is the corresponding homogeneous unit sphere. The following Lemma provides some important properties of homogeneous functions and vector fields (some others are recalled in the Appendix).

Lemma 8.1 ([2, 6, 18]) *For a given family of dilations $\Delta_\epsilon^r x$, and continuous real-valued functions V_1, V_2 on $\mathbb{R}^n$ (resp., a vector field f) which are* $\mathbf{r}$*-homogeneous of degrees $m_1 > 0$ and $m_2 > 0$ (resp., $l \in \mathbb{R}$), we have: (i) $V_1 V_2$ is homogeneous of degree $m_1 + m_2$. (ii) For every $x \in \mathbb{R}^n$ and each positive-definite function V_1, we have $c_1 V_1^{\frac{m_2}{m_1}}(x) \leq V_2(x) \leq c_2 V_1^{\frac{m_2}{m_1}}(x)$, where $c_1 \triangleq \min_{\{z: V_1(z)=1\}} V_2(z)$ and $c_2 \triangleq \max_{\{z: V_1(z)=1\}} V_2(z)$. Moreover, if V_2 is positive definite, there exists $c_1 > 0$. (iii)*

[1] Since uniqueness of solutions is in general not assumed, this means that *all* trajectories starting at any initial point (t_0, x_0) is uniformly stable and uniformly attractive. This concept is sometimes refered to as "strong stability", in contrast to the "weak stability" which is valid only for *some* trajectory for every initial point.

$\partial V_1(x)/\partial x_i$ is homogeneous of degree $m_1 - r_i$, with r_i being the weight of x_i. (iv) The Lie's derivative of $V_1(x)$ along $f(x)$, $L_f V_1(x) := \frac{\partial V_1(x)}{\partial x} \cdot f(x)$, is homogeneous of degree $m_1 + l$.

It is worth to recall that for homogeneous systems the local stability implies global stability and if the homogeneous degree is negative, asymptotic stability implies finite-time stability [2, 4, 27, 29]. (Asymptotic) stability of homogeneous systems and homogeneous DI's can be studied by means of homogeneous LFs (HLFs), see for example [2–4, 6, 15, 17, 27, 29, 33, 37, 43]: Assume that the origin of a homogeneous Filippov DI $\dot{x} \in F(x)$ is strongly globally AS. Then, there exists a $\mathcal{C}^\infty$ homogeneous strong LF.

The following robustness result of Asymptotically stable homogeneous Filippov Differential Inclusions is of paramount importance for the assertion of the accuracy properties of HOSM algorithms in presence of measurement or discretization noise or also delay and external perturbations. They have been established by Levant [4, 26, 27, 29].

Proposition 8.1 *Let $\dot{x} \in F(x)$ be a globally uniformly finite-time stable homogeneous Filippov inclusion with homogeneity degrees $\mathbf{r} = (r_1, \cdots, r_n)$ and degree $l < 0$, and let $\tau > 0$. Suppose that a continuous function $x(t)$ is defined for any $t \geq -\tau^l$ and satisfy some initial conditions $x(t) = \xi(t)$ for $t \in \left[-\tau^l, 0\right]$. Then if $x(t)$ is a solution of the perturbed differential inclusion*

$$\dot{x}(t) \in F_\tau\left(x\left(t + \left[-\tau^l, 0\right]\right)\right), \; 0 < t < \infty,$$

then the inequalities $|x_i| < \gamma_i \tau^{r_i}$ are established in finite time with some positive constants γ_i independent of τ and ξ.

Along this paper we use the following notation. For a real variable $z \in \mathbb{R}$ and a real number $p \in \mathbb{R}$ the symbol $\lceil z \rfloor^p = |z|^p sign(z)$ is the sign preserving power p of z. According to this $\lceil z \rfloor^0 = \text{sign}(z)$, $\frac{d}{dz}\lceil z \rfloor^p = p\,|z|^{p-1}$ and $\frac{d}{dz}|z|^p = p\lceil z \rfloor^{p-1}$ almost everywhere for z. Note that $\lceil z \rfloor^2 = |z|^2 \text{sign}(z) \neq z^2$, and if p is an odd number then $\lceil z \rfloor^p = z^p$ and $|z|^p = z^p$ for any even integer p. Moreover, $\lceil z \rfloor^p \lceil z \rfloor^q = |z|^{p+q}$, $\lceil z \rfloor^p \lceil z \rfloor^0 = |z|^p$, and $\lceil z \rfloor^0 |z|^p = \lceil z \rfloor^p$.

8.3 SISO Regulation and Tracking Problem

Consider a SISO dynamical system affine in the control

$$\dot{z} = f(t, z) + g(t, z)\,u\,, \; y = h(t, z)\,, \tag{8.1}$$

where $z \in \mathbb{R}^n$ defines the state vector, $u \in \mathbb{R}$ is the control input, $y \in \mathbb{R}$ is the output and $h(t, z) : \mathbb{R} \times \mathbb{R}^n \to \mathbb{R}$ is a smooth output function. A standard problem of control is the output tracking problem [20], consisting in forcing the output y to track a (time-varying) signal $y_R(t)$. Usually this problem has associated a (robust) disturbance decoupling or attenuation property [19, 20]. For our purposes we assume that the functions $f(t, z)$ and $g(t, z)$ are uncertain smooth vector fields on $\mathbb{R}^n$ and the dimension n can also be unknown. The control objective, i.e. the standard HOSM control problem [25, 27, 38], consists in making the output $\sigma = y - y_R$ vanish in finite time and to keep $\sigma \equiv 0$ exactly by a bounded (discontinuous) feedback control. All differential equations are understood in the Filippov's sense [12].

When the relative degree δ with respect to σ is known, well defined and constant of value $\delta = 2$ and the zero dynamics is e.g. Input-to-State Stable, this is equivalent to designing a controller for the uncertain system

$$\Sigma : \begin{cases} \dot{x}_1 = x_2, \\ \dot{x}_2 = \phi(t, z) + b(t, z)u, \end{cases} \tag{8.2}$$

or equivalently for the (Filippov) DI [12]

$$\Sigma_{ID} : \begin{cases} \dot{x}_1 = x_2, \\ \dot{x}_2 \in [-C, C] + [K_m, K_M]u, \end{cases} \tag{8.3}$$

where $x = (x_1, x_2)^T = (\sigma, \dot{\sigma})^T$ and $\dot{\sigma} = \frac{d}{dt}h(z, t)$ and $\ddot{\sigma} = \frac{d^2}{dt^2}h(z, t) = \phi(t, z) + b(t, z)u$, and we assume that $|\phi(t, z)| \le C$, $b(t, z) \in [K_m, K_M]$. Recall that $[a, b]$ represents the closed interval in the real line with limits a and b. Note that Σ_{ID} does not depend on the particular properties of the original systems' dynamics and the DI only retains the constants $\rho = 2$, C, K_m and K_M. Due to the persisting uncertainty/perturbation causing the constant $C > 0$ the stabilization of $x = 0$ requires a control discontinuous at $x = 0$, and therefore the classical nonlinear control techniques, that aim at designing a continuous controller as e.g. [19, 20, 22], cannot be applied.

For homogenous HOSM [27], and in particular for Second Order Sliding Modes (SOSM) [23, 28], the problem is solved by designing a bounded memoryless feedback **r**-homogeneous control law of degree 0 (called also 2-sliding homogeneous)

$$u = \varphi(x_1, x_2) = \varphi\left(\epsilon^{r_1}x_1, \epsilon^{r_2}x_2\right), \ \forall \epsilon > 0, \tag{8.4}$$

with $\mathbf{r} = (2, 1)$, that renders the origin $x = 0$ finite-time stable for Σ_{DI}. The motion on the set $x = 0$, which consists of Filippov trajectories [12], is called a *2nd-order sliding mode*. The function φ is discontinuous at the 2-sliding set ($x = 0$). The closed-loop inclusion (8.3)–(8.4) is an **r**-homogeneous of degree -1 Filippov DI satisfying *standard assumptions*. Some well-known SOSM controllers are the twisting (8.5) and the terminal (8.6) controllers [23, 28] (see also [8, 9] for a Lyapunov-based approach)

$$u = -k_1\lceil x_1 \rfloor^0 - k_2\lceil x_2 \rfloor^0 \tag{8.5}$$

$$u = -k_1\left\lceil x_1 + k_2\lceil x_2 \rfloor^2 \right\rfloor^0 . \tag{8.6}$$

The previous controllers are discontinuous and cause the high frequency switching of the control signal $u(t)$ (the so called chattering), that has a negative effect on the actuator and excites the unmodelled dynamics of the plant. In order to avoid the switching of the control signal with the aim of attenuating the chattering effect, it is natural to look for a controller producing a continuous control signal. In this case we have to change the assumptions on the class of uncertainties, since it is impossible to fully compensate a possibly discontinuous perturbation $\phi(t, z)$ with a continuous control signal.

If we assume for example, that the perturbation/uncertainty $\rho(t) \triangleq \phi(t, z)$ is a Lipschitz function of time, i.e. $\left|\frac{d\phi(t,z)}{dt}\right| \leq L$ and (for simplicity) that $K_M = K_m = 1$, then (8.2) becomes

$$\begin{aligned} \dot{x}_1 &= x_2 \\ \dot{x}_2 &= u + \rho(t) \ , \end{aligned} \tag{8.7}$$

and a natural alternative consists in extending the system adding an integrator to (8.7)

$$\Sigma_e : \begin{cases} \dot{x}_1 = x_2, \\ \dot{x}_2 = z \ , \\ \dot{z} \in v + [-L, \ L] \ , \end{cases} \tag{8.8}$$

that is, defining a new state $z = u + \rho(t)$, and designing a third order HOSM controller for the new control variable v. A particular example of such an algorithm is given by the "relay polynomial" controller [8, 9, 11]

$$v = -k_1 \left\lceil x_1 + k_2 \lceil x_2 \rfloor^{\frac{3}{2}} + k_3 \lceil z \rfloor^3 \right\rfloor^0 . \tag{8.9}$$

This allows to reach the origin in finite time, and it will be insensitive to Lipschitz perturbations, i.e. with $\dot{\rho}(t)$ bounded. In this form a continuous control signal u will be obtained, so that the chattering effect is reduced. However, this requires to feedback not only the two states x_1 and x_2 but also the state z, which is unknown due to the unknown perturbation. z has to be obtained (even if x_1 and x_2 are measured) by using Levant's robust and exact differentiator [26, 27] (see also [7] for a Lyapunov-based approach), i.e.

$$\begin{aligned} \dot{\hat{x}}_1 &= -l_1 \left\lceil \hat{x}_1 - x_1 \right\rfloor^{\frac{2}{3}} + \hat{x}_2 \\ \dot{\hat{x}}_2 &= -l_2 \left\lceil \hat{x}_1 - x_1 \right\rfloor^{\frac{1}{3}} + \hat{z} \\ \dot{\hat{z}} &= -l_3 \left\lceil \hat{x}_1 - x_1 \right\rfloor^{0} . \end{aligned} \tag{8.10}$$

This also allows to implement an output feedback controller, assuming that only the sliding variable x_1 is measured. Since it is necessary to differentiate two times x_1 there is a noise amplification effect.

8.4 Main Result: Discontinuous Integral Controller

For the robust finite time stabilization of the origin of system (8.7) we propose a nonlinear, homogeneous state feedback control law, which is able to stabilize the origin in finite time in the absence of non vanishing perturbations, and a discontinuous integral controller is added to compensate for the persistently acting perturbations. In contrast to the continuous integral controller, that can only compensate for (almost) constant perturbations, the discontinuous one can deal with time varying perturbations which are Lipschitz continuous, that is, their derivatives exist almost everywhere and it is uniformly bounded. The control signal of the controller is continuous, so that the chattering effect of the SM and HOSM controllers is avoided.

Theorem 8.1 *Consider the plant (8.7) with Lipschitz continuous perturbation signal* $\rho(t)$ *with Lipschitz constant L, i.e.* $\forall t \geq 0$, $\left|\frac{d\rho(t)}{dt}\right| \leq L$. *Then the control law*

$$\begin{aligned} u &= -k_1 \left\lceil x_1 + k_2 \left\lceil x_2 \right\rfloor^{\frac{3}{2}} \right\rfloor^{\frac{1}{3}} + z \\ \dot{z} &= -k_3 \left\lceil x_1 + k_4 \left\lceil x_2 \right\rfloor^{\frac{3}{2}} \right\rfloor^{0} \end{aligned} \tag{8.11}$$

can stabilize the origin in finite time for any k_4 *and appropriate designed gains* k_1, k_2, k_3. *The control signal* $u(t)$ *is continuous.* ■

A procedure to design the gains is described in Sect. 8.5.1.1. Theorem 8.1 shows that with the addition of the discontinuous integral term it is possible to eliminate completely the effect of the Lipschitz perturbation $\rho(t)$, that cannot be (fully) compensated by the state feedback $u = -k_1 \left\lceil x_1 + k_2 \left\lceil x_2 \right\rfloor^{\frac{3}{2}} \right\rfloor^{\frac{1}{3}}$ alone. In fact, the integral controller can be interpreted as a perturbation estimator, since $z(t) = -\rho(t)$ after a finite time. Note that if both x_1 and x_2 are available for measurement, it is possible to implement the controller (8.11), and this is in contrast with the HOSM control alternative (8.9), that cannot be implemented without the differentiator (8.10).

Note also that the input to the discontinuous integrator $y = x_1 + k_4 \left\lceil x_2 \right\rfloor^{\frac{3}{2}}$ can be a combination of the "position" x_1 (with relative degree two) and the "velocity" x_2 (with a relative degree one). The value of k_4 can be arbitrary (including zero), so that the velocity is not necessary for the integral action. For $k_4 > 0$ this output can be seen as a passive output of the system (x_1, x_2). However, it is necessary to have the position in this signal, otherwise the closed loop will be unstable.

The implementation of controller (8.11) requires the measurement of both states x_1, x_2. If only the position is measured a finite time convergent observer for x_2 can be implemented, so that an output feedback control is obtained. There are different alternatives to doing this.

A first option consists in designing an observer for x_2 as a modification of Levant's differentiator (8.10) given by

$$
\begin{aligned}
\dot{\hat{x}}_1 &= -l_1 \lceil \hat{x}_1 - x_1 \rfloor^{\frac{2}{3}} + \hat{x}_2 \\
\dot{\hat{x}}_2 &= -l_2 \lceil \hat{x}_1 - x_1 \rfloor^{\frac{1}{3}} - k_1 \left\lceil x_1 + k_2 \lceil \hat{x}_2 \rfloor^{\frac{3}{2}} \right\rfloor^{\frac{1}{3}} + z + \hat{\varsigma} \\
\dot{\hat{\varsigma}} &= -l_3 \lceil \hat{x}_1 - x_1 \rfloor^{0} .
\end{aligned}
\tag{8.12}
$$

It is possible to show that the closed loop with this observer (8.11)–(8.12) is homogeneous and achieves the desired output feedback controller. However, this structure is somehow redundant in the sense that both the integral controller and the observer estimate the perturbation, which is therefore estimated twice.

In order to avoid this double estimation of the perturbation a further alternative is to use a first order differentiator [24] to estimate x_2 based on x_1. This indeed avoids to estimate twice the perturbation but requires from $\rho(t)$ to be not only Lipschitz continuous but also uniformly bounded. Moreover, the closed loop is not longer homogeneous and the precision and the global stability has to be studied in particular.

We present in the next Theorem a further alternative, avoiding the double estimation of the perturbation and keeping the homogeneity of the closed loop system.

Theorem 8.2 *Consider the plant (8.7) with Lipschitz continuous perturbation signal $\rho(t)$ with Lipschitz constant L. Assume that only x_1 is measured. Then the output feedback control law*

$$
\begin{aligned}
\dot{\hat{x}}_1 &= -l_1 \lceil \hat{x}_1 - x_1 \rfloor^{\frac{2}{3}} + \hat{x}_2 \\
\dot{\hat{x}}_2 &= -l_2 \lceil \hat{x}_1 - x_1 \rfloor^{\frac{1}{3}} - k_1 \left\lceil x_1 + k_2 \lceil \hat{x}_2 \rfloor^{\frac{3}{2}} \right\rfloor^{\frac{1}{3}} \\
u &= -k_1 \left\lceil x_1 + k_2 \lceil \hat{x}_2 \rfloor^{\frac{3}{2}} \right\rfloor^{\frac{1}{3}} + z \\
\dot{z} &= -k_3 \left\lceil x_1 + k_4 \lceil \hat{x}_2 \rfloor^{\frac{3}{2}} \right\rfloor^{0} ,
\end{aligned}
\tag{8.13}
$$

can stabilize the origin in finite time for appropriate designed gains k_1, k_2, k_3, k_4, l_1 and l_2. In particular, the gains have to fulfill the following inequality

$$
k_4 \neq k_2 + \left(\frac{l_2}{l_1^{\frac{1}{2}} k_1} \right)^3 . \tag{8.14}
$$

■

A procedure to design the gains is described in Sect. 8.5.3.5. We notice that the gains $k_1 > 0$, $k_2 > 0$, $l_1 > 0$, $l_2 > 0$ while k_4 can be selected freely, except for the value given in (8.14). Moreover, if $k_4 < k_2 + \left(\frac{l_2}{l_1^{\frac{1}{2}} k_1} \right)^3$ then $k_3 > 0$ and if $k_4 > k_2 + \left(\frac{l_2}{l_1^{\frac{1}{2}} k_1} \right)^3$ then $k_3 < 0$. Condition (8.14) corresponds to the existence of a transmission zero of the system consisting of the plant (8.7) and the observer (see Sect. 8.5.3.4 below for this interpretation). It is well-known, that integral control is not feasible if there exists a transmission zero at $s = 0$, and this is exactly what happens if condition (8.14) is violated.

Note that the (continuous) observer in (8.13) is not able to reconstruct x_2 in presence of the (possibly) unboundedly growing perturbation $\rho(t)$ in open loop. But the interaction of the observer and the discontinuous integral controller allows the estimation of both, the variable x_2 by the observer and the perturbation $\rho(t)$ by the integral controller.

The next Corollary shows how to scale the gains of the controller and observer to maintain the properties of the closed loop system, but with the objective of e.g. accelerating the convergence without the necessity of calculating a new set of feasible gains.

Corollary 8.1 *Consider controller* (8.13) *(or in particular controller* (8.11) *with* $l_1 = l_2 = 0$*) with the set of parameters* $(k_1, k_2, l_1, l_2, k_3, k_4, L)$*. Assume that with this parameters the closed loop system has a finite-time stable equilibrium point. Then, the equilibrium point of the closed loop system will be finite time stable with the set of parameters given by* $\left(\lambda^{\frac{2}{3}}k_1, \lambda^{-\frac{1}{2}}k_2, \lambda^{\frac{1}{3}}l_1, \lambda^{\frac{2}{3}}l_2, \lambda k_3, \lambda^{-\frac{1}{2}}k_4, \lambda L\right)$*, for every* $\lambda > 0$.

The closed loop system consisting of plant (8.7) and integral controller (8.11), excluding the perturbation, is $\mathbf{r}$–homogeneous with weights $\mathbf{r} = (3, 2, 1)$ for the variables (x_1, x_2, z) and negative homogeneous degree $l = -1$. Also the closed loop system consisting of plant (8.7) and the observer and integral controller (8.13) without perturbation is $\mathbf{r}$–homogeneous with weights $\mathbf{r} = (3, 2, 3, 2, 1)$ for the variables (x_1, x_2, e_1, e_2, z) and negative homogeneous degree $l = -1$. From homogeneity arguments [13, 26, 27] (see Proposition 8.1) one expects that the controllers have precision of order $|x_1| \leq \nu_1 \tau^3$ and $|x_2| \leq \nu_2 \tau^2$, where τ is the discretization step and ν_1 and ν_2 are constants depending only on the gains of the algorithm. Moreover, it is easy to show that for the Lyapunov functions used in the proofs of the previous results, the following inequality is satisfied

$$\dot{V}(x) \leq -\kappa V^{\frac{4}{5}}(x),$$

from which finite time convergence can be deduced. With the value of κ it is possible to estimate the convergence time as

$$T(x_0) \leq \frac{5}{\kappa} V^{\frac{1}{5}}(x_0).$$

8.5 Proof of the Main Results Using the Lyapunov Method

We prove Theorems 8.1 and 8.2 using smooth and homogeneous Lyapunov functions.

We consider first the properties of some functions that will appear in the following proofs.

Lemma 8.2 *Consider for every real value $p > 0$ the continuous function of two real variables $x,\ y \in \mathbb{R}$*

$$\psi(x,\ y) = \lceil x + \lceil y \rfloor^{p} \rfloor^{\frac{1}{p}} - y\,.$$

ψ has the following properties:

(i) $\psi(0,\ y) = 0,\ \forall y \in \mathbb{R}$.

(ii) $\left\{(x,\ y) \in \mathbb{R}^2 \mid \psi(x,\ y) = 0\right\} = \left\{(x,\ y) \in \mathbb{R}^2 \mid x = 0\right\}$, i.e. $\psi(x,\ y)$ vanishes only when $x = 0$.

(iii) $\psi(x,\ y)$ is monotonically increasing in x for every y. Moreover, if $p \geq 1$

$$0 < \psi(x,\ y)\, x < 2^{\frac{p-1}{p}} \left|x\right|^{\frac{p+1}{p}},\ \forall y \in \mathbb{R},\ \forall x \in \mathbb{R} \setminus 0\,.$$

Proof Item (i) is immediate. Item (ii) follows from the equivalences

$$\psi(x,\ y) = 0 \Leftrightarrow \lceil x + \lceil y \rfloor^{p} \rfloor^{\frac{1}{p}} = y \Leftrightarrow x + \lceil y \rfloor^{p} = \lceil y \rfloor^{p} \Leftrightarrow x = 0\,.$$

The first part of item (iii) is implied by the fact that

$$\frac{\partial \psi(x,\ y)}{\partial x} = \frac{1}{p\left|x + \lceil y \rfloor^{p}\right|^{\frac{p-1}{p}}} > 0\,,$$

where the derivative exists. The left part of the inequality follows from the monotonicity of $\psi(x,\ y)$. Since the function $\lceil \cdot \rfloor^{\frac{1}{p}}$ is Hölder continuous for $p \geq 1$ we have that

$$\left|\psi(x,\ y)\right| = \left|\lceil x + \lceil y \rfloor^{p} \rfloor^{\frac{1}{p}} - y\right| \leq 2^{\frac{p-1}{p}} \left|x\right|^{\frac{1}{p}}\,.$$

And this implies the right-hand side of the inequality. □

8.5.1 *Proof of Theorem 8.1*

By introducing as state variable $x_3 = z + \rho(t)$ the dynamics of the closed loop system formed by the plant (8.7) and the controller (8.11) is given by

$$
\begin{aligned}
\dot{x}_1 &= x_2 \\
\dot{x}_2 &= -k_1 \left\lceil x_1 + k_2 \lceil x_2 \rfloor^{\frac{3}{2}} \right\rfloor^{\frac{1}{3}} + x_3 \\
\dot{x}_3 &= -k_3 \left\lceil x_1 + k_4 \lceil x_2 \rfloor^{\frac{3}{2}} \right\rfloor^{0} + \dot{\rho}(t) \, .
\end{aligned}
$$

Since $|\dot{\rho}(t)| \leq L$ we can write the discontinuous differential equation as the Filippov differential inclusion (DI)

$$
\begin{aligned}
\dot{x}_1 &= x_2 \\
\dot{x}_2 &= -k_1 \left\lceil x_1 + k_2 \lceil x_2 \rfloor^{\frac{3}{2}} \right\rfloor^{\frac{1}{3}} + x_3 \\
\dot{x}_3 &\in -k_3 \left\lceil x_1 + k_4 \lceil x_2 \rfloor^{\frac{3}{2}} \right\rfloor^{0} + [-1,\, 1]\, L ,
\end{aligned}
\tag{8.15}
$$

where the discontinuous function $\lceil x \rfloor^0$ is now interpreted as the multivalued map

$$
\lceil x \rfloor^0 = \begin{cases} 1 & \text{if } x > 0 \\ [-1,\, 1] & \text{if } x = 0 \\ -1 & \text{if } x < 0 \end{cases} \, .
\tag{8.16}
$$

(8.15) is $\mathbf{r}$−homogenous of degree -1, with weights $\mathbf{r} = (3,\, 2,\, 1)$ for the state variables $x = (x_1,\, x_2,\, x_3)$. To proof the Theorem we show that the origin $x = 0$ is the only equilibrium point for (8.15) and that it is finite time stable. Note that $x = 0$ is an equilibrium point for (8.15) only if $0 \in F(0)$, where $F(x)$ is the multivalued map on the right-hand side of the DI (8.15). $x = 0$ is the unique equilibrium point of the DI if $k_3 > L$. To show that it is asymptotically stable we use Lyapunov's theorem for DI [2] and due to homogeneity finite time convergence follows [3, 11, 27, 29, 35].

For convenience, we introduce the diffeomorphic change of state variables

$$
\xi_1 = x_1 - \left\lceil \frac{x_3}{k_1} \right\rfloor^3 , \; \xi_2 = x_2 , \; \xi_3 = \frac{x_3}{k_1} ,
$$

and the change of parameters (assuming that $k_1 > 0$),

$$
\kappa_3 = \frac{k_3}{k_1} , \; \ell = \frac{L}{k_3} .
$$

Thus the dynamics of (8.15) can be written as

$$
\begin{aligned}
\dot{\xi}_1 &= \xi_2 - 3\, |\xi_3|^2\, \dot{\xi}_3 \\
\dot{\xi}_2 &= -k_1 \psi \left(\xi_1 + k_2 \lceil \xi_2 \rfloor^{\frac{3}{2}} ,\, \xi_3 \right) \\
\dot{\xi}_3 &\in -\kappa_3 \left(\left\lceil \xi_1 + \lceil \xi_3 \rfloor^3 + k_4 \lceil \xi_2 \rfloor^{\frac{3}{2}} \right\rfloor^{0} - [-1,\, 1]\, \ell \right) ,
\end{aligned}
\tag{8.17}
$$

where

$$\psi(x, y) = \left\lceil x + \lceil y \rfloor^3 \right\rfloor^{\frac{1}{3}} - y\,.$$

Consider the homogeneous and smooth Lyapunov Function

$$V(\xi) = \frac{3}{5}\gamma_1 |\xi_1|^{\frac{5}{3}} + \xi_1\xi_2 + \frac{2}{5}k_2 |\xi_2|^{\frac{5}{2}} + \frac{1}{5}|\xi_3|^5\,.$$

From Lemma 8.4 it follows easily that given $k_2 > 0$ it is always possible to render V positive definite selecting γ_1 sufficiently large.

Its derivative along the trajectories of (8.15) is given by

$$\dot{V} \in W_1(\xi_1, \xi_2) - k_1 W_2(\xi) - \kappa_3 W_3(\xi)\,,$$

where

$$\begin{aligned} W_1(\xi) &= \left(\gamma_1 \lceil \xi_1 \rfloor^{\frac{2}{3}} + \xi_2\right)\xi_2 \\ W_2(\xi_1, \xi_2) &= \left(\xi_1 + k_2 \lceil \xi_2 \rfloor^{\frac{3}{2}}\right)\psi\left(\xi_1 + k_2 \lceil \xi_2 \rfloor^{\frac{3}{2}}, \xi_3\right) \\ W_3(\xi) &= \left(-3|\xi_3|^2\left(\gamma_1 \lceil \xi_1 \rfloor^{\frac{2}{3}} + \xi_2\right) + \lceil \xi_3 \rfloor^4\right)\left(\left\lceil \xi_1 + \lceil \xi_3 \rfloor^3 + k_4 \lceil \xi_2 \rfloor^{\frac{3}{2}} \right\rfloor^0 - [-1, 1]\,\ell\right). \end{aligned}$$

Note that W_1 and W_2 are single-valued functions, while W_3 is a multivalued (or set valued) one. The latter function can be rewritten as

$$\begin{aligned} W_3(\xi) = {} & |\xi_3|\left(\xi_1 + \lceil \xi_3 \rfloor^3 + k_4 \lceil \xi_2 \rfloor^{\frac{3}{2}}\right)\left(\left\lceil \xi_1 + \lceil \xi_3 \rfloor^3 + k_4 \lceil \xi_2 \rfloor^{\frac{3}{2}} \right\rfloor^0 - [-1, 1]\,\ell\right) - \\ & \left(3|\xi_3|\left(\gamma_1 \lceil \xi_1 \rfloor^{\frac{2}{3}} + \xi_2\right) + \xi_1 + k_4 \lceil \xi_2 \rfloor^{\frac{3}{2}}\right)|\xi_3|\left(\left\lceil \xi_1 + \lceil \xi_3 \rfloor^3 + k_4 \lceil \xi_2 \rfloor^{\frac{3}{2}} \right\rfloor^0 - [-1, 1]\,\ell\right). \end{aligned}$$

Note that, since $0 \leq \ell = \frac{L}{k_3} < 1$, we have that

$$\left\lceil \xi_1 + \lceil \xi_3 \rfloor^3 + k_4 \lceil \xi_2 \rfloor^{\frac{3}{2}} \right\rfloor^0 - [-1, 1]\,\ell = \begin{cases} > 0 & \text{if } \xi_1 + \lceil \xi_3 \rfloor^3 + k_4 \lceil \xi_2 \rfloor^{\frac{3}{2}} > 0 \\ < 0 & \text{if } \xi_1 + \lceil \xi_3 \rfloor^3 + k_4 \lceil \xi_2 \rfloor^{\frac{3}{2}} < 0 \end{cases}.$$

This implies that the first term in W_3 is non negative, and it follows that W_3 can be bounded from below by a continuous and homogeneous function

$$W_3(\xi) \geq \tilde{W}_3(\xi) = W_{31}(\xi) - W_{32}(\xi)\,,$$

where

$$\begin{aligned} W_{31}(\xi) &= (1-\ell)\,|\xi_3|\left|\xi_1 + \lceil \xi_3 \rfloor^3 + k_4 \lceil \xi_2 \rfloor^{\frac{3}{2}}\right| \\ W_{32}(\xi) &= (1-\ell)\left|3|\xi_3|\left(\gamma_1 \lceil \xi_1 \rfloor^{\frac{2}{3}} + \xi_2\right) + \xi_1 + k_4 \lceil \xi_2 \rfloor^{\frac{3}{2}}\right||\xi_3|\,. \end{aligned}$$

Thus, for asymptotic stability it suffices to show that

$$\dot{V} \leq W_1(\xi_1, \xi_2) - k_1 W_2(\xi) - \kappa_3 \tilde{W}_3(\xi) < 0\,,$$

where the right-hand side is a continuous homogeneous function.

In the latter expression we conclude, thanks to Lemma 8.2, that the second term is non negative, i.e. $W_2(\xi) \geq 0$, and it vanishes on the set

$$\mathcal{S}_1 = \left\{ \psi\left(\xi_1 + k_2 \lceil \xi_2 \rfloor^{\frac{3}{2}}, \xi_3\right) = 0 \Leftrightarrow \xi_1 + k_2 \lceil \xi_2 \rfloor^{\frac{3}{2}} = 0 \Leftrightarrow \xi_2 = -\left\lceil \frac{\xi_1}{k_2} \right\rfloor^{\frac{2}{3}} \right\}.$$

On this set we have that

$$\dot{V}\,|_{\mathcal{S}_1} \leq -\left(\gamma_1 k_2^{\frac{2}{3}} - 1\right) \left|\frac{\xi_1}{k_2}\right|^{\frac{4}{3}} - \kappa_3 (1-\ell) |\xi_3| \times \\ \left(\left| \left(1 - \frac{k_4}{k_2}\right) \xi_1 + \lceil \xi_3 \rfloor^3 \right| - \left| 3 |\xi_3| \left(\gamma_1 k_2^{\frac{2}{3}} - 1\right) \left\lceil \frac{\xi_1}{k_2} \right\rfloor^{\frac{2}{3}} + \left(1 - \frac{k_4}{k_2}\right) \xi_1 \right| \right).$$

In this latter expression, the first term on the right-hand side is negative if we select

$$\gamma_1 > \frac{1}{k_2^{\frac{2}{3}}}$$

and it only vanishes on the set $\mathcal{S}_2 = \{\xi_1 = 0\}$. On this set we obtain

$$\dot{V}\,|_{\mathcal{S}_1 \cap \mathcal{S}_2} \leq -\kappa_3 (1-\ell) |\xi_3|^4 < 0\,,$$

which is negative, since $0 \leq \ell < 1$. According to Lemma 8.5 we conclude that selecting κ_3 small we can render $\dot{V}\,|_{\mathcal{S}_1} < 0$ and then selecting k_1 sufficiently large we can render $\dot{V} < 0$.

Moreover, from Lemmas 8.1 and 8.5 we obtain that

$$\dot{V}(x) \leq -\kappa V^{\frac{4}{5}}(x)\,,$$

for some $\kappa > 0$, and from this inequality we conclude finite time convergence. With the value of κ it is possible to estimate the convergence time as

$$T(x_0) \leq \frac{5}{\kappa} V^{\frac{1}{5}}(x_0)\,.$$

8.5.1.1 Gain Selection for Controller (8.11)

From the proof of Theorem 8.1 we derive the following procedure to design the gains k_1, k_2, k_3, k_4 given a Lipschitz constant $L \geq 0$, to assure the convergence of the controller. Note that this procedure does not guarantee a good performance.

1. Select $k_2 > 0$ and $k_4 \in \mathbb{R}$ arbitrary. Select also a value for $0 < \ell < 1$ (or $\ell = 0$ if $L = 0$).
2. Choose γ_1 large enough such that $\gamma_1 > k_2^{-\frac{2}{3}}$ and $V > 0$.
3. Calculate κ_3 small enough such that the inequality $\dot{V}\ |_{\mathcal{S}_1} < 0$ is fulfilled, i.e.

$$\frac{1}{\kappa_3} > \frac{1}{\kappa_3^*} \triangleq \frac{(1-\ell)}{\left(\gamma_1 k_2^{\frac{2}{3}} - 1\right)} \times$$
$$\max\left\{ \frac{|\xi_3| \left(\left| 3\,|\xi_3| \left(\gamma_1 k_2^{\frac{2}{3}} - 1\right) \left\lceil \frac{\xi_1}{k_2} \right\rfloor^{\frac{2}{3}} + \left(1 - \frac{k_4}{k_2}\right) \xi_1 \right| - \left| \left(1 - \frac{k_4}{k_2}\right) \xi_1 + \lceil \xi_3 \rfloor^3 \right| \right)}{\left| \frac{\xi_1}{k_2} \right|^{\frac{4}{3}}} \right\}.$$

4. Find k_1 sufficiently large such that the inequality $W_1(\xi_1, \xi_2) - k_1 W_2(\xi) - \kappa_3 \tilde{W}_3(\xi) < 0$ is fulfilled, i.e.

$$k_1 > k_1^* \triangleq \max\left\{ \frac{W_1(\xi_1, \xi_2) - \kappa_3 \tilde{W}_3(\xi)}{W_2(\xi)} \right\}.$$

5. Finally, the actual gains are calculated as follows: (i) $k_2 > 0$ and $k_4 \in \mathbb{R}$ are freely chosen. (ii) $k_3 = \frac{L}{\ell}$, with $\ell \neq 0$ selected. (iii) $k_1 > \max\left\{\frac{k_3}{\kappa_3^*}, k_1^*\right\}$.

Note that the maximizations above are feasible, since the functions to be maximized are continuous everywhere, except when the denominator vanishes. In a neighborhood of these points the function is negative and tends to $-\infty$ as the root of the denominator is approached. Moreover, the function is $\mathbf{r}-$homogeneous of degree 0 and is bounded above, so that a maximum exists.

8.5.2 *Proof of Theorem 8.2*

System Description

System (8.7) with controller (8.13) is given by the dynamics

$$
\begin{aligned}
\dot{x}_1 &= x_2 \\
\dot{x}_2 &= -k_1 \left\lceil x_1 + k_2 \lceil \hat{x}_2 \rfloor^{\frac{3}{2}} \right\rfloor^{\frac{1}{3}} + z + \rho(t) \,, \\
\dot{\hat{x}}_1 &= -l_1 \lceil \hat{x}_1 - x_1 \rfloor^{\frac{2}{3}} + \hat{x}_2 \\
\dot{\hat{x}}_2 &= -l_2 \lceil \hat{x}_1 - x_1 \rfloor^{\frac{1}{3}} - k_1 \left\lceil x_1 + k_2 \lceil \hat{x}_2 \rfloor^{\frac{3}{2}} \right\rfloor^{\frac{1}{3}} \\
\dot{z} &= -k_3 \left\lceil x_1 + k_4 \lceil \hat{x}_2 \rfloor^{\frac{3}{2}} \right\rfloor^{0}
\end{aligned}
\tag{8.18}
$$

which is a discontinuous system, i.e. the right-hand side is discontinuous as a function of the state. If we introduce the estimation errors $e_1 = \hat{x}_1 - x_1$, $e_2 = \hat{x}_2 - x_2$ and the state variable combining the output of the integrator and the perturbation $x_3 = z + \rho(t)$, recalling that $|\dot{\rho}(t)| \leq L$, and applying Filippov's regularization procedure [3, 12, 27] the dynamics of the system can be described by the Filippov DI

$$
\begin{aligned}
\dot{x}_1 &= x_2 \\
\dot{x}_2 &= -k_1 \left\lceil x_1 + k_2 \lceil x_2 + e_2 \rfloor^{\frac{3}{2}} \right\rfloor^{\frac{1}{3}} + x_3 \,, \\
\dot{e}_1 &= -l_1 \lceil e_1 \rfloor^{\frac{2}{3}} + e_2 \\
\dot{e}_2 &= -l_2 \lceil e_1 \rfloor^{\frac{1}{3}} - x_3(t) \\
\dot{x}_3 &\in -k_3 \left\lceil x_1 + k_4 \lceil x_2 + e_2 \rfloor^{\frac{3}{2}} \right\rfloor^{0} + [-1,\ 1] L \,,
\end{aligned}
\tag{8.19}
$$

where the discontinuous function $\lceil x \rfloor^0$ is now interpreted as a multivalued map (8.16), as in the previous proof. The trajectories of (8.19) are defined in the sense of Filippov [12]. To prove Theorem 8.2 we show (i) that $(x_1,\ x_2,\ e_1,\ e_2,\ x_3) = 0$ is the only equilibrium point of the DI, and (ii) that it is Globally Finite Time Stable, for what we use Lyapunov's theorem for DIs [2], with a smooth and homogeneous Lyapunov function. Uniqueness of the equilibrium point at $x = 0$ is guaranteed under the condition that $k_3 > L$.

Again, for convenience we introduce the following diffeomorphic change of variables

$$
\xi_1 = x_1 - \left(1 + k_2 \left(\frac{k_1 l_1^{\frac{1}{2}}}{l_2}\right)^3\right) \left\lceil \frac{x_3}{k_1} \right\rfloor^3 = x_1 - \lceil \alpha z_3 \rfloor^3 \,,\ \xi_2 = x_2,\ z_3 = \frac{x_3}{l_2},
$$

$$
\epsilon_1 = e_1 + \left\lceil \frac{x_3}{l_2} \right\rfloor^3 ,\ \epsilon_2 = \frac{e_2}{l_1} + \left\lceil \frac{x_3}{l_2} \right\rfloor^2 ,
$$

with

$$
\alpha = \left(\left(\frac{l_2}{k_1}\right)^3 + k_2 l_1^{\frac{3}{2}}\right)^{\frac{1}{3}} ,\ \tilde{l}_2 = \frac{l_2}{l_1},\ \kappa_3 = \frac{k_3}{l_2},\ \ell = \frac{L}{k_3} .
$$

For the dynamics of ξ we will use

$$\zeta_3 = l_1^{\frac{1}{2}} z_3,\ \eta = \frac{l_1^{\frac{1}{2}}}{k_1},\ \frac{\alpha}{l_1^{\frac{1}{2}}} = \left(\left(\tilde{l}_2\eta\right)^3 + k_2\right)^{\frac{1}{3}}.$$

The dynamics become

$$\begin{aligned}
\dot{\xi}_1 &= \xi_2 - 3\alpha^3 |z_3|^2 \dot{z}_3 \\
\dot{\xi}_2 &= -k_1\psi_1(\xi,\ z_3) - k_1\omega(\epsilon_2,\ \xi) \\
\dot{\epsilon}_1 &= -l_1(\varphi_1(\epsilon_1,\ z_3) - \epsilon_2) + 3|z_3|^2 \dot{z}_3 \\
\dot{\epsilon}_2 &= -\tilde{l}_2\varphi_2(\epsilon_1,\ z_3) + 2|z_3|\dot{z}_3 \\
\dot{z}_3 &\in -\kappa_3\left(\lceil\psi_2(\xi,\ \epsilon_2,\ z_3)\rfloor^0 - [-1,\ 1]\,\ell\right)
\end{aligned}$$

where

$$\begin{aligned}
\varphi_1(\epsilon_1,\ z_3) &= \left\lceil \epsilon_1 - \lceil z_3\rfloor^3 \right\rfloor^{\frac{2}{3}} + \lceil z_3\rfloor^2 \\
\varphi_2(\epsilon_1, z_3) &= \left\lceil \epsilon_1 - \lceil z_3\rfloor^3 \right\rfloor^{\frac{1}{3}} + z_3 \\
\psi_1(\xi,\ z_3) &= \left\lceil \xi_1 + k_2\left\lceil \xi_2 - \lceil\zeta_3\rfloor^2\right\rfloor^{\frac{3}{2}} + \left(\left(\tilde{l}_2\eta\right)^3 + k_2\right)\lceil\zeta_3\rfloor^3 \right\rfloor^{\frac{1}{3}} - \tilde{l}_2\eta\zeta_3 \\
\psi_2(\xi,\ \epsilon_2,\ z_3) &= \xi_1 + \lceil\alpha z_3\rfloor^3 + k_4\left\lceil \xi_2 + l_1\epsilon_2 - l_1\lceil z_3\rfloor^2\right\rfloor^{\frac{3}{2}}, \\
\omega(\epsilon_2,\ \xi) &= \left\lceil \xi_1 + k_2\left\lceil \xi_2 + l_1\epsilon_2 - l_1\lceil z_3\rfloor^2\right\rfloor^{\frac{3}{2}} + \lceil\alpha z_3\rfloor^3 \right\rfloor^{\frac{1}{3}} - \\
&\quad \left\lceil \xi_1 + k_2\left\lceil \xi_2 - l_1\lceil z_3\rfloor^2\right\rfloor^{\frac{3}{2}} + \lceil\alpha z_3\rfloor^3 \right\rfloor^{\frac{1}{3}}.
\end{aligned}$$

The Lyapunov Function Candidate

We use the following Lyapunov function candidate

$$V(\xi,\ \epsilon,\ z_3) = V_1(\xi,\ z_3) + \mu V_2(\epsilon,\ z_3),$$

where $\mu > 0$,

$$V_1(\xi,\ z_3) = \frac{3}{5}\gamma_1|\xi_1|^{\frac{5}{3}} + \xi_1\xi_2 + \frac{2}{5}k_2\left|\xi_2 - \lceil\zeta_3\rfloor^2\right|^{\frac{5}{2}} + k_2\xi_2\lceil\zeta_3\rfloor^3 + \frac{1}{5}\gamma_3|z_3|^5$$

$$V_2(e,\ z_3) = |\varsigma_1|^{\frac{5}{3}} + \gamma_2|\epsilon_2|^{\frac{5}{2}},$$

and

$$\varsigma_1 = \epsilon_1 - \phi(\epsilon_2, z_3),$$
$$\phi(\epsilon_2, z_3) = \left\lceil \epsilon_2 - \lceil z_3 \rfloor^2 \right\rfloor^{\frac{3}{2}} + \lceil z_3 \rfloor^3,$$

with $\gamma_i > 0$ for $i = 1, 2, 3$. Given values of l_1 and k_2 function V is positive definite for any $\gamma_2 > 0$, any $\mu > 0$, and for large values of γ_1 and γ_3.

The derivative of V along the trajectories of the system can be written in the following form

$$\dot{V}(\xi, \epsilon, z_3) \in -W_1^{z_3}(\xi) + W_2(\xi, \epsilon_2) - \mu W_3^{z_3}(\epsilon) - \kappa_3 W_4(\xi, \epsilon_2, z_3)$$

where

$$W_1^{z_3}(\xi_1, \xi_2) = k_1 \chi_1(\xi, z_3)\psi_1(\xi, z_3) - \left(\gamma_1 \lceil \xi_1 \rfloor^{\frac{2}{3}} + \xi_2\right)\xi_2$$
$$W_2(\xi, \epsilon_2) = -k_1 \chi_1(\xi, z_3)\omega(\epsilon_2, \xi)$$
$$W_3^{z_3}(\epsilon) = \frac{5}{3} l_1 \lceil \epsilon_1 - \phi(\epsilon_2, z_3) \rfloor^{\frac{2}{3}} (\varphi_1(\epsilon_1, z_3) - \epsilon_2) - \frac{5}{2}\tilde{l}_2 \chi_2(\epsilon, z_3)\varphi_2(\epsilon_1, z_3)$$
$$W_4(\xi, \epsilon_2, z_3) = \chi_3(\xi, \epsilon, z_3)|z_3|\left(\lceil \psi_2(\xi, \epsilon_2, z_3) \rfloor^0 - [-1, 1]\ell\right),$$

and

$$\chi_1(\xi, z_3) = \xi_1 + k_2\left(\left\lceil \xi_2 - \lceil \zeta_3 \rfloor^2 \right\rfloor^{\frac{3}{2}} + \lceil \zeta_3 \rfloor^3\right),$$
$$\chi_2(\epsilon, z_3) = \lceil \epsilon_1 - \phi(\epsilon_2, z_3) \rfloor^{\frac{2}{3}} \left|\epsilon_2 - \lceil z_3 \rfloor^2\right|^{\frac{1}{2}} - \gamma_2 \lceil \epsilon_2 \rfloor^{\frac{3}{2}}$$
$$\chi_3(\xi, \epsilon, z_3) = -3\alpha^3 |z_3| \left(\gamma_1 \lceil \xi_1 \rfloor^{\frac{2}{3}} + \xi_2\right) - 2k_2 l_1 \left\lceil \xi_2 - l_1 \lceil z_3 \rfloor^2 \right\rfloor^{\frac{3}{2}} +$$
$$3k_2 l_1^{\frac{3}{2}} \xi_2 |z_3| + \gamma_3 \lceil z_3 \rfloor^3 + 5\mu\gamma_2 \lceil \epsilon_2 \rfloor^{\frac{3}{2}}.$$

The proof of the theorem is based on the following properties for the W_i functions appearing in $\dot{V}$.

Lemma 8.3 *Functions W_i have the following properties:*

(i) All are $\mathbf{r}$*−homogeneous functions of degree 4, with weights* $\mathbf{r} = (3, 2, 3, 2, 1)$ *for the states* $(\xi_1, \xi_2, \epsilon_1, \epsilon_2, z_3)$.

(ii) $W_1^{z_3}(\xi_1, \xi_2)$, $W_2(\xi, \epsilon_2)$ *and* $W_3^{z_3}(\epsilon)$ *are single-valued and continuous while* $W_4(\xi, \epsilon_2)$ *is a multivalued function.*

(iii) $W_3^{z_3}(\epsilon)$*: Given any* $\tilde{l}_2 > 0$ *it is possible to render* $W_3^{z_3}(\epsilon) > 0$ *for every* $(\epsilon_1, \epsilon_2) \neq 0$ *and for every* $z_3 \in \mathbb{R}$, *and* $W_3^{z_3}(\epsilon) = 0$ *if and only if* $\epsilon = (\epsilon_1, \epsilon_2) = 0$, *by selecting* l_1 *sufficiently large.*

(iv) $W_1^{z_3}(\xi_1, \xi_2)$*: Given any* $\eta > 0$ *and any* $k_2 > 0$, *it is possible to render* $W_1^{z_3}(\xi_1, \xi_2) > 0$ *for every* $(\xi_1, \xi_2) \neq 0$ *and for every* $z_3 \in \mathbb{R}$, *and* $W_1^{z_3}(\xi_1, \xi_2) = 0$ *if and only if* $(\xi_1, \xi_2) = 0$, *by selecting* k_1 *sufficiently large.*

(v) $W_2(\xi, \epsilon_2)$ *has no defined sign, but* $W_2(\xi, 0) = 0$, *since* $\omega(0, \xi) = 0$.

(vi) $W_4(\xi, \epsilon_2)$ *is a multivalued function. It can be bounded from below by a continuous and homogeneous function* $\tilde{W}_4(\xi, \epsilon_2, z_3)$

$$W_4(\xi, \epsilon_2, z_3) \geq \tilde{W}_4(\xi, \epsilon_2, z_3)$$

where

$$\tilde{W}_4(\xi, \epsilon_2, z_3) = (1-\ell)\,\beta\,|z_3|\left|\tilde{\psi}_2(\xi, \epsilon_2, z_3)\right| - (1-\ell)\,|z_3|\left|\tilde{\chi}_3(\xi, \epsilon, z_3) - \beta\tilde{\psi}_2(\xi, \epsilon_2, z_3)\right|,$$

$$\tilde{\psi}_2(\xi, \epsilon_2, z_3) = \left\lceil \left\lceil \xi_1 + \lceil \alpha z_3 \rfloor^3 \right\rfloor^{\frac{2}{3}} + k_4^{\frac{2}{3}}\left(\xi_2 - l_1 \lceil z_3 \rfloor^2\right) \right\rfloor^{\frac{3}{2}} + k_4 \lceil \epsilon_2 \rfloor^{\frac{3}{2}},$$

$$\beta = \frac{2k_2 l_1^{\frac{5}{2}} + \gamma_3}{\left(\alpha^2 - k_4^{\frac{2}{3}} l_1\right)^{\frac{3}{2}}}.$$

Moreover, if $\alpha^2 - k_4^{\frac{2}{3}} l_1 > 0$

$$\begin{aligned}\tilde{W}_4(0, 0, z_3) &= (1-\ell)\,\beta\,|z_3|\left|\tilde{\psi}_2(0, 0, z_3)\right| \\ &= (1-\ell)\,\beta\left|\left(\left(\left(\frac{l_2}{k_1}\right)^3 + k_2 l_1^{\frac{3}{2}}\right)^{\frac{2}{3}} - k_4^{\frac{2}{3}} l_1\right)^{\frac{3}{2}}\right| |z_3|^4 > 0,\end{aligned}$$

which is positive.[2] □

With these properties we can prove that $\dot{V} < 0$. First note that one can select the gains k_1, k_2, l_1 and l_2 such that $W_1^{z_3}(\xi) > 0$ and $W_3^{z_3}(\epsilon) > 0$. Since $W_3^{z_3}(\epsilon) = 0$ only on the set $\mathcal{S}_1 = \{(\xi, \epsilon, z_3) \in \mathbb{R}^5 \mid \epsilon = 0\}$, then the value of the function

$$-W_1^{z_3}(\xi) + W_2(\xi, \epsilon_2) - \mu W_3^{z_3}(\epsilon)\big|_{\mathcal{S}_1} = -W_1^{z_3}(\xi) < 0,$$

since $W_2(\xi, 0) = 0$. Thus, according to Lemma 8.5 one can render the function $-W_1^{z_3}(\xi) + W_2(\xi, \epsilon_2) - \mu W_3^{z_3}(\epsilon) < 0$ selecting μ large enough. Moreover, $-W_1^{z_3}(\xi) + W_2(\xi, \epsilon_2) - \mu W_3^{z_3}(\epsilon) = 0$ only on the set $\mathcal{S}_2 = \{(\xi, \epsilon, z_3) \in \mathbb{R}^5 \mid \xi = \epsilon = 0\}$.

Finally, since

$$\dot{V}(\xi, \epsilon, z_3) \leq -W_1^{z_3}(\xi) + W_2(\xi, \epsilon_2) - \mu W_3^{z_3}(\epsilon) - \kappa_3 \tilde{W}_4(\xi, \epsilon_2, z_3)$$

---

[2] We assume here that $\left(\frac{l_2}{k_1}\right)^3 + k_2 l_1^{\frac{3}{2}} > k_4 l_1^{\frac{3}{2}}$ and also that $k_3 > 0$. The case $\left(\frac{l_2}{k_1}\right)^3 + k_2 l_1^{\frac{3}{2}} < k_4 l_1^{\frac{3}{2}}$ and $k_3 < 0$ can be treated in the same manner. These conditions are derived from (8.14), which corresponds to the absence of a transmission zero located at the zero frequency.

and the value of $\dot{V}$ on the set $\mathcal{S}_2$ satisfies

$$\dot{V}\mid_{\mathcal{S}_2}\leq -\kappa_3 \tilde{W}_4\left(0,\ 0,\ z_3\right) < 0\,,$$

then it follows from Lemma 8.5 that selecting κ_3 small enough we obtain $\dot{V} < 0$.

Moreover, from Lemmas 8.1 and 8.5 we obtain that

$$\dot{V}\left(\xi,\ \epsilon,\ z_3\right) \leq -\kappa V^{\frac{4}{5}}\left(\xi,\ \epsilon,\ z_3\right),$$

for some $\kappa > 0$, and from this inequality we conclude finite time convergence. With the value of κ it is possible to estimate the convergence time as

$$T\left(x_0\right) \leq \frac{5}{\kappa} V^{\frac{1}{5}}\left(\xi_0,\ \epsilon_0,\ z_{30}\right)\,.$$

8.5.3 *Proof of Lemma 8.3*

Items (i) and (ii) are simple and can be obtained by checking the conditions on the functions.

8.5.3.1 Proof of Item (iii)

We first show that $\lceil \epsilon_1 - \phi\left(\epsilon_2, z_3\right)\rfloor^{\frac{2}{3}}\left(\varphi_1\left(\epsilon_1, z_3\right) - \epsilon_2\right) > 0$, except on the set $\mathcal{S}_3 = \{\epsilon_1 = \phi\left(\epsilon_2, z_3\right)\}$. The claim follows by simply checking (a) that the sets where each of the factors vanish are the same, i.e. $\{\epsilon_1 - \phi\left(\epsilon_2, z_3\right) = 0\} = \{\varphi_1\left(\epsilon_1, z_3\right) - \epsilon_2 = 0\}$, and then (b) evaluating the expression at some particular value. (a) follows from the following equivalences

$$\epsilon_1 = \phi\left(\epsilon_2, z_3\right) \Leftrightarrow \left\lceil \lceil \epsilon_1 \rfloor - \lceil z_3 \rfloor^3 \right\rfloor^{\frac{2}{3}} = \epsilon_2 - \lceil z_3 \rfloor^2 \Leftrightarrow \varphi_1\left(\epsilon_1, z_3\right) - \epsilon_2 = 0\,.$$

We check (b) at the point $z_3 = 0$

$$\lceil \epsilon_1 - \phi\left(\epsilon_2,\ 0\right)\rfloor^{\frac{2}{3}}\left(\varphi_1\left(\epsilon_1,\ 0\right) - \epsilon_2\right) = \left\lceil \epsilon_1 - \lceil \epsilon_2 \rfloor^{\frac{3}{2}} \right\rfloor^{\frac{2}{3}}\left(\lceil \epsilon_1 \rfloor^{\frac{2}{3}} - \epsilon_2\right) > 0\,.$$

To conclude the proof we observe that the first term in $W_3^{z_3}\left(\epsilon\right)$ is positive, and evaluating $W_3^{z_3}\left(\epsilon\right)$ on $\mathcal{S}_3$ (where it vanishes) we obtain

$$\begin{aligned} W_3^{z_3}(\epsilon)\,|_{\mathcal{S}_3} &= \frac{5}{2}\gamma_2\tilde{l}_2 \lceil \epsilon_2 \rfloor^{\frac{3}{2}} \varphi_2\left(\phi\left(\epsilon_2, z_3\right), z_3\right) \\ &= \frac{5}{2}\gamma_2\tilde{l}_2 \lceil \epsilon_2 \rfloor^{\frac{3}{2}} \left(\left\lceil \phi\left(\epsilon_2, z_3\right) - \lceil z_3 \rfloor^3 \right\rfloor^{\frac{1}{3}} + z_3\right) \\ &= \frac{5}{2}\gamma_2\tilde{l}_2 \lceil \epsilon_2 \rfloor^{\frac{3}{2}} \left(\left\lceil \epsilon_2 - \lceil z_3 \rfloor^2 \right\rfloor^{\frac{1}{2}} + z_3\right) > 0\,. \end{aligned}$$

The last inequality is a consequence of Lemma 8.2 with $p = 2$. Lemma 8.5 implies item (iii).

8.5.3.2 Proof of Item (iv)

This proof is similar to that of item (iii). We first show that the first term in $W_1^{z_3}(\xi)$ is non negative, i.e. $\chi_1(\xi, z_3)\,\psi_1(\xi, z_3) > 0$ except on the set

$$\mathcal{S}_4 = \left\{\chi_1(\xi, z_3) = 0 \Leftrightarrow \xi_1 = -k_2\left(\left\lceil \xi_2 - \lceil \zeta_3 \rfloor^2 \right\rfloor^{\frac{3}{2}} + \lceil \zeta_3 \rfloor^3\right)\right\}.$$

Note that we can write $\psi_1(\xi, z_3)$ as

$$\psi_1(\xi, z_3) = \left\lceil \chi_1(\xi, z_3) + \left\lceil \eta\tilde{l}_2\zeta_3 \right\rfloor^3 \right\rfloor^{\frac{1}{3}} - \eta\tilde{l}_2\zeta_3\,.$$

The claim is therefore a consequence of Lemma 8.2 with $p = 3$. In particular, we obtain that

$$0 \leq \chi_1(\xi, z_3)\,\psi_1(\xi, z_3) \leq 2^{\frac{2}{3}}\,|\chi_1(\xi, z_3)|^{\frac{4}{3}}\,.$$

Evaluating $W_1^{z_3}(\xi)$ on the set $\mathcal{S}_4$ (where its first term vanishes) we obtain

$$W_1^{z_3}(\xi)\,|_{\mathcal{S}_4} = \left(\gamma_1 k_2^{\frac{2}{3}} \left\lceil \left\lceil \xi_2 - \lceil \zeta_3 \rfloor^2 \right\rfloor^{\frac{3}{2}} + \lceil \zeta_3 \rfloor^3 \right\rfloor^{\frac{2}{3}} - \xi_2\right)\xi_2\,. \tag{8.20}$$

If $W_1^{z_3}(\xi)\,|_{\mathcal{S}_4} > 0$ we can conclude item (iv) by means of Lemma 8.5.

We show that for γ_1 large enough, then $W_1^{z_3}(\xi)\,|_{\mathcal{S}_4} > 0$. Note first that for $z_3 = 0$ this is true. Moreover, the term $\left\lceil \left\lceil \xi_2 - \lceil \zeta_3 \rfloor^2 \right\rfloor^{\frac{3}{2}} + \lceil \zeta_3 \rfloor^3 \right\rfloor^{\frac{2}{3}}$ is monotonicaly increasing as a function of ξ_2 and vanish only at $\xi_2 = 0$. This can be concluded since its derivative is

$$\frac{\partial \left\lceil \left\lceil \xi_2 - \lceil \zeta_3 \rfloor^2 \right\rfloor^{\frac{3}{2}} + \lceil \zeta_3 \rfloor^3 \right\rfloor^{\frac{2}{3}}}{\partial \xi_2} = \frac{\left|\xi_2 - \lceil \zeta_3 \rfloor^2\right|^{\frac{1}{2}}}{\left|\left\lceil \xi_2 - \lceil \zeta_3 \rfloor^2 \right\rfloor^{\frac{3}{2}} + \lceil \zeta_3 \rfloor^3\right|^{\frac{1}{3}}}$$

which is non negative everywhere, where it exists. When $z_3 \neq 0$ this derivative is very large in a neighborhood of $\xi_2 = 0$ (at $\xi_2 = 0$ it becomes infinite, since the function is not differentiable there). This implies that for every value of z_3 function $W_1^{z_3}(\xi)|_{\mathcal{S}_4} > 0$ in a neighborhood of $\xi_2 = 0$, and therefore if $W_1^{z_3}(\xi)|_{\mathcal{S}_4} \not> 0$ there is a value $\xi_2 \neq 0$ such that $W_1^{z_3}(\xi)|_{\mathcal{S}_4} = 0$. This is only possible if at that point

$$\gamma_1 k_2^{\frac{2}{3}} \left\lceil \left\lceil \xi_2 - \lceil \zeta_3 \rfloor^2 \right\rfloor^{\frac{3}{2}} + \lceil \zeta_3 \rfloor^3 \right\rfloor^{\frac{2}{3}} = \xi_2$$

or equivalently

$$\xi_2 = \left\lceil \left\lceil \frac{\xi_2}{\gamma_1 k_2^{\frac{2}{3}}} \right\rfloor^{\frac{3}{2}} - \lceil \zeta_3 \rfloor^3 \right\rfloor^{\frac{2}{3}} + \lceil \zeta_3 \rfloor^2 . \tag{8.21}$$

Using Lemma 8.2 (with $p = \frac{3}{2}$) we can conclude that the right-hand side satisfies

$$\left| \left\lceil \left\lceil \frac{\xi_2}{\gamma_1 k_2^{\frac{2}{3}}} \right\rfloor^{\frac{3}{2}} - \lceil \zeta_3 \rfloor^3 \right\rfloor^{\frac{2}{3}} + \lceil \zeta_3 \rfloor^2 \right| \leq \frac{2^{\frac{1}{3}}}{\gamma_1 k_2^{\frac{2}{3}}} |\xi_2| .$$

This implies, that if $\frac{2^{\frac{1}{3}}}{\gamma_1 k_2^{\frac{2}{3}}} < 1$, i.e. $\frac{2^{\frac{1}{3}}}{k_2^{\frac{2}{3}}} < \gamma_1$, the right-hand side of Eq. (8.21) grows slower than the left-hand side (i.e. ξ_2), and therefore the equation cannot have a solution different from $\xi_2 = 0$. This finishes the proof of item (iv).

8.5.3.3 Proof of Items (v) and (vi)

The proof of item (v) is direct and simple. To prove item (vi) we observe that $W_4(\xi, \epsilon_2, z_3)$ can be written as for any constant $\beta > 0$

$$\begin{aligned} W_4(\xi, \epsilon_2, z_3) = {} & |z_3| \beta \tilde{\psi}_2(\xi, \epsilon_2, z_3) \left(\lceil \psi_2(\xi, \epsilon_2, z_3) \rfloor^0 - [-1, 1] \ell \right) + \\ & |z_3| \left(\tilde{\chi}_3(\xi, \epsilon, z_3) - \beta \tilde{\psi}_2(\xi, \epsilon_2, z_3) \right) \left(\lceil \psi_2(\xi, \epsilon_2, z_3) \rfloor^0 - [-1, 1] \ell \right) \end{aligned}$$

Since $0 \leq \ell < 1$ function $\left(\lceil \psi_2(\xi, \epsilon_2, z_3) \rfloor^0 - [-1, 1] \ell \right)$ has the same sign of $\psi_2(\xi, \epsilon_2, z_3)$ when it does not vanish. Note further that $\tilde{\psi}_2(\xi, \epsilon_2, z_3) \psi_2(\xi, \epsilon_2, z_3) > 0$. This follows from the following equivalences

$$\tilde{\psi}_2 = 0 \Leftrightarrow \left\lceil \xi_1 + \lceil \alpha z_3 \rfloor^3 \right\rfloor^{\frac{2}{3}} = -k_4^{\frac{2}{3}} \left(\xi_2 + \epsilon_2 - l_1 \lceil z_3 \rfloor^2 \right) \Leftrightarrow \psi_2 = 0 ,$$

and for $z_3 = 0$ it becomes

$$\tilde{\psi}_2(\xi, \epsilon_2, 0)\,\psi_2(\xi, \epsilon_2, 0) = \left(\left\lceil \lceil \xi_1 \rfloor^{\frac{2}{3}} + k_4^{\frac{2}{3}} \xi_2 \right\rfloor^{\frac{3}{2}} + k_4 \lceil \epsilon_2 \rfloor^{\frac{3}{2}}\right)\left(\xi_1 + k_4 \lceil \xi_2 + \epsilon_2 \rfloor^{\frac{3}{2}}\right) > 0\,.$$

This implies that

$$W_4(\xi, \epsilon_2, z_3) \geq (1-\ell)\,\beta\,|z_3|\,\left|\tilde{\psi}_2(\xi, \epsilon_2, z_3)\right| - (1-\ell)\,|z_3|\,\left|\tilde{\chi}_3(\xi, \epsilon, z_3) - \beta\tilde{\psi}_2(\xi, \epsilon_2, z_3)\right|\,.$$

Since

$$\tilde{\chi}_3(0, 0, z_3) - \beta\tilde{\psi}_2(0, 0, z_3) = \left(2k_2 l_1^{\frac{5}{2}} + \gamma_3 - \beta\left(\alpha^2 - k_4^{\frac{2}{3}} l_1\right)^{\frac{3}{2}}\right)\lceil z_3 \rfloor^3$$

the selection of β as in the statement of item (vi) provides that $\tilde{\chi}_3(0, 0, z_3) - \beta\tilde{\psi}_2(0, 0, z_3) = 0$ and that $\tilde{W}_4(0, 0, z_3) > 0$, since

$$\begin{aligned}\psi_2(0, 0, z_3) &= \left(\alpha^3 - k_4 l_1^{\frac{3}{2}}\right)\lceil z_3 \rfloor^3 \\ &= \left(\left(\frac{l_2}{k_1}\right)^3 + k_2 l_1^{\frac{3}{2}} - k_4 l_1^{\frac{3}{2}}\right)\lceil z_3 \rfloor^3\,,\end{aligned}$$

and we have assumed that $\alpha^3 - k_4 l_1^{\frac{3}{2}} \neq 0$. In fact, we assume that $\alpha^3 - k_4 l_1^{\frac{3}{2}} > 0$ but the case $\alpha^3 - k_4 l_1^{\frac{3}{2}} < 0$ can be treated in the same manner by just assuming that $k_3 < 0$.

8.5.3.4 Interpretation of the Condition (8.14): Transmission Zero

We provide an interpretation of condition (8.14) as the absence of a transmission zero of the open loop system (see (8.19))

$$\begin{aligned}\dot{x}_1 &= x_2 \\ \dot{x}_2 &= -k_1\left\lceil x_1 + k_2 \lceil x_2 + e_2 \rfloor^{\frac{3}{2}} \right\rfloor^{\frac{1}{3}} + x_3\,, \\ \dot{e}_1 &= -l_1 \lceil e_1 \rfloor^{\frac{2}{3}} + e_2 \\ \dot{e}_2 &= -l_2 \lceil e_1 \rfloor^{\frac{1}{3}} - x_3 \\ y &= x_1 + k_4 \lceil x_2 + e_2 \rfloor^{\frac{3}{2}}\,,\end{aligned}$$

from the input x_3 to the output y. Since the right-hand side of this system is not smooth, then we cannot use the standard tools [20, 22] to find the relative degree and the zero dynamics. However, we know that a transmission zero at frequency $s = 0$

corresponds to the fact that for any constant input x_3 the output will be zero $y = 0$ for all the time (if the initial conditions are selected appropriately). We show for the previous system that if condition (8.14) is violated, then this is exactly what happens.

For this we calculate the (unique) equilibrium point $(\bar{x}_1, \bar{x}_2, \bar{e}_1, \bar{e}_2)$ corresponding to a constant input $\bar{x}_3$, that is given by

$$(\bar{x}_1, \bar{x}_2, \bar{e}_1, \bar{e}_2) = \left(\left(\frac{1}{k_1^3} + \frac{k_2 l_1^{\frac{3}{2}}}{l_2^{\frac{3}{2}}} \right) \lceil \bar{x}_3 \rfloor^3, 0, -\left\lceil \frac{\bar{x}_3}{l_2} \right\rfloor^3, -l_1 \left\lceil \frac{\bar{x}_3}{l_2} \right\rfloor^2 \right)$$

and therefore the output $y = \bar{y}$ at this point is given by

$$\bar{y} = \left(\frac{1}{k_1^3} + \frac{k_2 l_1^{\frac{3}{2}}}{l_2^{\frac{3}{2}}} - k_4 \left(\frac{l_1}{l_2^2} \right)^{\frac{3}{2}} \right) \lceil \bar{x}_3 \rfloor^3 .$$

We see immediately that if condition (8.14) is violated, then the output $\bar{y} = 0$ independently of the constant value $\bar{x}_3$, and so the system has a transmission zero at zero frequency.

8.5.3.5 Gain Selection for Controller (8.13)

From the proof of Theorem 8.2 we derive the following procedure to design the gains k_1, k_2, k_3, k_4, l_1, l_2 given a Lipschitz constant $L \geq 0$, to assure the convergence of the controled system. Note that this procedure does not guarantee a good performance.

1. Select arbitrary values for $\gamma_2 > 0$ and $\tilde{l}_2 > 0$ and calculate $l_{1\,\min}$ sufficiently large such that $W_3^{z_3}(\epsilon) > 0$, i.e.

$$l_{1\,\min} = \frac{3}{2} \tilde{l}_2 \max \left\{ \frac{\chi_2(\epsilon, z_3)\, \varphi_2(\epsilon_1, z_3)}{\lceil \epsilon_1 - \phi(\epsilon_2, z_3) \rfloor^{\frac{2}{3}} (\varphi_1(\epsilon_1, z_3) - \epsilon_2)} \right\} .$$

2. Select and arbitrary $k_2 > 0$ and choose γ_1 and $\tilde{\gamma}_3 = \gamma_3 / l_1^{\frac{5}{2}}$ large enough so that $W_1^{z_3}(\xi)\,|_{S_4} > 0$ (see Eq. (8.20)) and $V_1(\xi, z_3) > 0$. Calculate $k_{1\,\min}$ sufficiently large such that $W_1^{z_3}(\xi_1, \xi_2) > 0$, i.e.

$$k_{1\,\min} = \max \left\{ \frac{\left(\gamma_1 \lceil \xi_1 \rfloor^{\frac{2}{3}} + \xi_2 \right) \xi_2}{\chi_1(\xi, z_3)\, \psi_1(\xi, z_3)} \right\} .$$

From the expressions of $\chi_1(\xi, z_3)$ and $\psi_1(\xi, z_3)$ we observe that $k_{1\min}$ depends only on the selection of $\gamma_1, k_2, \tilde{l}_2, \eta$. We can fix the value of η arbitrarily in order to find $k_{1\min}$. Finally, we select $k_1 > k_{1\min}$ and $l_1 > l_{1\min}$ such that $\eta = \sqrt{l_1}/k_1$.

3. Select $\mu > 0$ sufficiently large so that $-W_1^{z_3}(\xi) + W_2(\xi, \epsilon_2) - \mu W_3^{z_3}(\epsilon) < 0$.
4. Choose a value for $0 < \ell < 1$ (or $\ell = 0$ if $L = 0$). Calculate κ_3 small enough such that the inequality $\dot{V}(\xi, \epsilon, z_3) < 0$ is fulfilled, i.e.

$$\frac{1}{\kappa_3} \geq \max\left\{\frac{-\tilde{W}_4(\xi, \epsilon_2, z_3)}{W_1^{z_3}(\xi) - W_2(\xi, \epsilon_2) + \mu W_3^{z_3}(\epsilon)}\right\}.$$

Note that the maximizations above are feasible, since the functions to be maximized are continuous everywhere, except when the denominator vanishes. In a neighborhood of these points the function is negative and tends to $-\infty$ as the root of the denominator is approached. Moreover, the functions are $\mathbf{r}$−homogeneous of degree 0 and are bounded above, so that a maximum exists.

8.5.3.6 Proof of Corollary 8.1

Consider system (8.19) and perform on it the linear change of states $(X_1, X_2, E_1, E_2, X_3) = \lambda(x_1, x_2, e_1, e_2, x_3)$ for some $\lambda > 0$. In the new variables the dynamics become

$$\begin{aligned}
\dot{X}_1 &= X_2 \\
\dot{X}_2 &= -k_1\lambda^{\frac{2}{3}}\left\lceil X_1 + k_2\lambda^{-\frac{1}{2}}\lceil X_2 + E_2\rfloor^{\frac{3}{2}}\right\rfloor^{\frac{1}{3}} + X_3, \\
\dot{E}_1 &= -l_1\lambda^{\frac{1}{3}}\lceil E_1\rfloor^{\frac{2}{3}} + E_2 \\
\dot{E}_2 &= -l_2\lambda^{\frac{2}{3}}\lceil E_1\rfloor^{\frac{1}{3}} - X_3(t) \\
\dot{X}_3 &\in -k_3\lambda\left\lceil X_1 + k_4\lambda^{-\frac{1}{2}}\lceil X_2 + E_2\rfloor^{\frac{3}{2}}\right\rfloor^{0} + [-1, 1]\lambda L.
\end{aligned}$$

Transforming the gains as

$$(k_1, k_2, l_1, l_2, k_3, k_4, L) \to \left(\lambda^{\frac{2}{3}}k_1, \lambda^{-\frac{1}{2}}k_2, \lambda^{\frac{1}{3}}l_1, \lambda^{\frac{2}{3}}l_2, \lambda k_3, \lambda^{-\frac{1}{2}}k_4, \lambda L\right)$$

we obtain an equation identical to (8.19), and therefore asymptotic stability with the set of parameters $(k_1, k_2, l_1, l_2, k_3, k_4, L)$ implies the asymptotic stability with the transformed set of parameters $\left(\lambda^{\frac{2}{3}}k_1, \lambda^{-\frac{1}{2}}k_2, \lambda^{\frac{1}{3}}l_1, \lambda^{\frac{2}{3}}l_2, \lambda k_3, \lambda^{-\frac{1}{2}}k_4, \lambda L\right)$.

8.6 Simulation Example

We illustrate the behavior of the proposed integral controllers by some simulations. Consider the dynamics of a simple pendulum without friction

$$\begin{aligned}\dot{x}_1 &= x_2\\ \dot{x}_2 &= -\tfrac{g}{l}\sin(x_1) + \tfrac{1}{ml^2}u + \rho(t)\ ,\end{aligned}$$

where $x_1 = \theta$ is the position angle, $x_2 = \dot{\theta}$ is the angular velocity, m is the mass of the bob, g is the gravity acceleration, l is the length of the bob, the control u is the torque applied to the pendulum, and the perturbation $\rho(t) = 0.4\sin(t)$, that can be interpreted also as the second derivative of a reference signal (in this case the state x corresponds to the tracking error). For the simulations we have used the following parameter values $l = 1\,[\mathrm{m}]$, $m = 1.1\,[\mathrm{Kg}]$, $g = 9.815\,[\mathrm{m/s^2}]$, and the initial conditions $x_1(0) = 2$, $x_2(0) = 2$.

We have implemented three controllers:

- A State Feedback (SF) controller with discontinuous integral term, as given by (8.11), with gains $k_1 = 2$, $k_2 = 5$, $k_3 = 0.5$, $k_4 = 0$, and initial value of the integrator $z(0) = 0$.
- An Output Feedback (OF) controller with discontinuous integral term, as given by (8.13), with controller gains $k_1 = 2\lambda^{\frac{2}{3}}$, $k_2 = 5\lambda^{\frac{1}{2}}$, $k_3 = 0.5\lambda$, $k_4 = 0$, $\lambda = 3$, observer gains $l_1 = 2L, l_2 = 1.1L^2, L = 4$, observer initial conditions $\hat{x}_1(0) = 0$, $\hat{x}_2(0) = 0$, and initial value of the integrator $z(0) = 0$.
- A Terminal controller [13, 23, 26], given by $u = -k_1\left\lceil x_1 + k_2\lceil x_2\rfloor^2\right\rfloor^0$, with gains $k_1 = 1.2$, $k_2 = 0.6$.

The simulations for the three controllers are presented in Figs. 8.1, 8.2, 8.3 and 8.4. In Fig. 8.1 the evolution of the position is presented and also the evolution of the estimated position given by the observer for the OF, which converges very fast. All controllers are able to bring the position to zero in finite time.

Figure 8.2 presents the time evolution of the velocity and its estimation by the observer for the OF, which converges in finite time around the time 15. We see also the typical zig-zag behavior for the Terminal controller. All controllers are able to bring the velocity to rest in finite time.

In Fig. 8.3 the integrator state is presented for both controllers OF and SF, and the (negative) value of the perturbation $(-\rho(t))$. We note the zig-zag behavior of the integral controller, due precisely to its discontinuous character. We appreciate also that the integrator signal reconstructs after a finite time the (negative value of the) perturbation, and this is the reason for it to be able to fully compensate its action on the plant.

Figure 8.4 presents the control signal u for the three controllers. We see that, while the OF and the SF controllers with discontinuous Integral action provide a continuous control signal, the Terminal controller provides a switching (discontinuous) control signal, with an extremely high frequency when the equilibrium has been reached, which corresponds to the (undesirable) chattering phenomenon. Finally, in Fig. 8.5 the trajectories for the three controllers are presented on the phase plane for $(x_1,\ x_2)$.

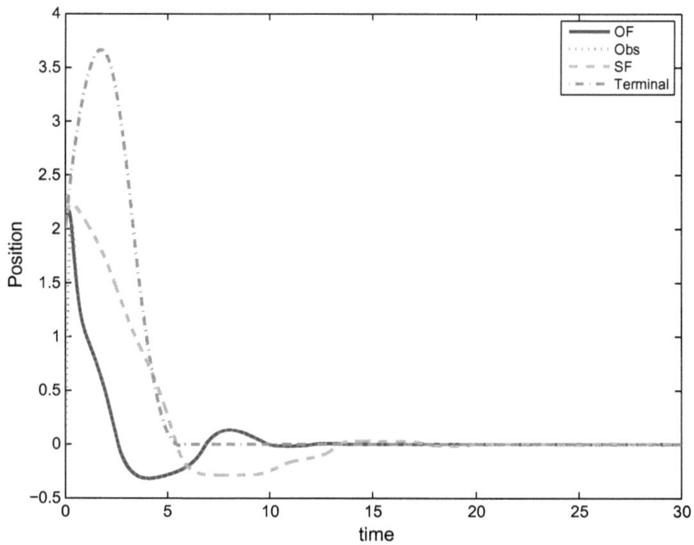

Fig. 8.1 Behavior of x_1 with terminal and the discontinuous integral controller

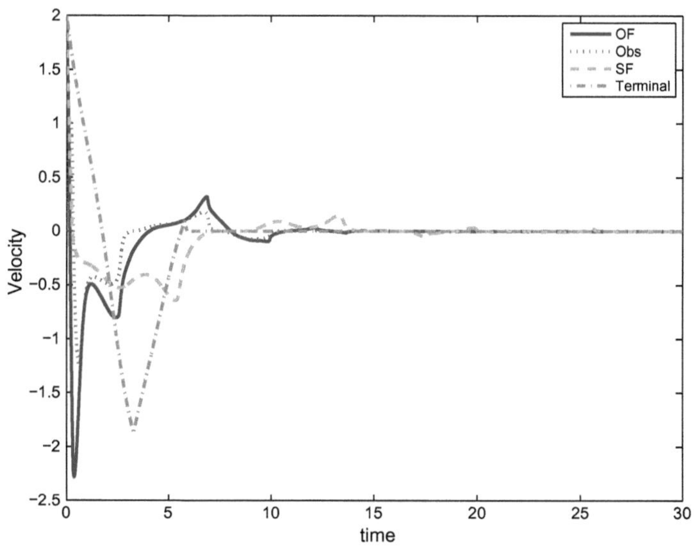

Fig. 8.2 Behavior of x_2 with terminal and the discontinuous integral controller

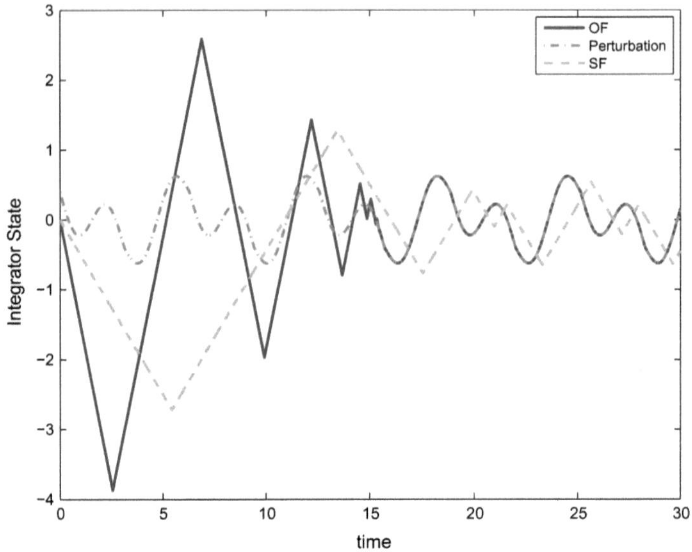

Fig. 8.3 Behavior of the state of the discontinuous integrator and the negative value of the perturbation $-\rho(t)$

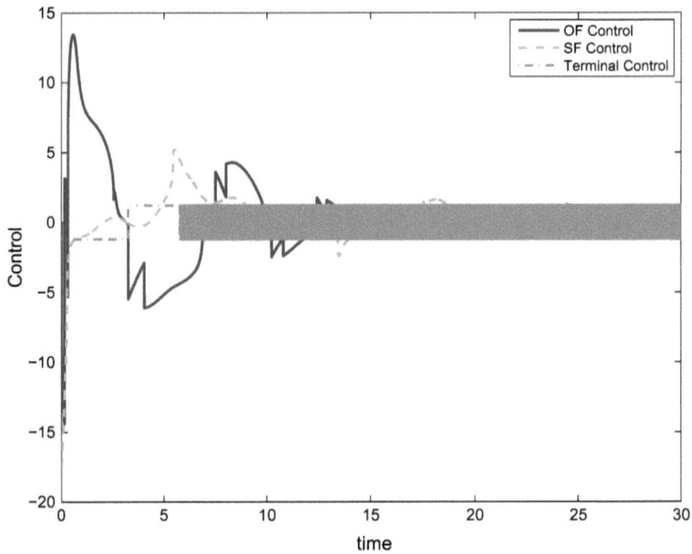

Fig. 8.4 Control signal for OF and SF controllers with discontinuous integral action and for the terminal controller

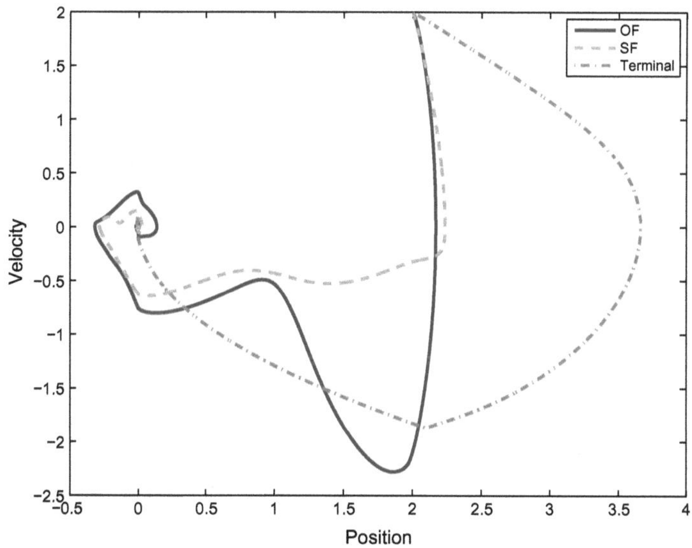

Fig. 8.5 Behavior of x_1 and x_2 with terminal and the discontinuous integral controller in the phase plane

8.7 Conclusion

We present in this paper a Discontinuous Integral Controller, which shares the properties of the classical PID control and the HOSM controllers: Similar to HOSM it is able to fully compensate a Lispchitz perturbation or to track an (unknown) time varying reference with bounded second derivative, it has high precision due to the homogeneity properties, and it stabilizes the origin globally and in finite time. Similar to the PID control it has a continuous control signal. In order to achieve an Output Feedback scheme we introduce a finite time converging observer. The stability proofs are performed with a novel Lyapunov method. It is possible to extend this idea to systems with higher relative degree, and this will be done in future work.

Acknowledgements The author would like to thank the financial support from PAPIIT-UNAM (Programa de Apoyo a Proyectos de Investigación e Innovación Tecnológica), projects IN113614 and IN113617; Fondo de Colaboración II-FI UNAM, Project IISGBAS-100-2015; CONACyT (Consejo Nacional de Ciencia y Tecnología), project 241171.

Appendix: Some Technical Lemmas on Homogeneous Functions

We recall or prove some useful Lemmas needed for the development of our main results. Lemmas 8.5 and 8.6 are extensions of classical results for homogeneous continuous functions to semicontinuous ones [18, Theorems 4.4 and 4.1].

Lemma 8.4 [16] (Young's inequality) *For any positive real numbers $a > 0$, $b > 0$, $c > 0$, $p > 1$ and $q > 1$, with $\frac{1}{p} + \frac{1}{q} = 1$, the following inequality is always satisfied*

$$ab \leq \frac{c^p}{p} a^p + \frac{c^{-q}}{q} b^q,$$

and equality holds if and only if $a^p = b^q$.

And the Lemma:

Lemma 8.5 *Let $\eta : \mathbb{R}^n \to \mathbb{R}$ and $\gamma : \mathbb{R}^n \to \mathbb{R}_+$ be two lower (upper) semicontinuous single-valued* **r***-homogeneous functions of degree $m > 0$. Suppose that $\gamma(x) \geq 0$ $(\gamma(x) \leq 0)$ on $\mathbb{R}^n$. If $\eta(x) > 0$ $(\eta(x) < 0)$ for all $x \neq 0$ such that $\gamma(x) = 0$, then there is a constant $\lambda^* \in \mathbb{R}$ and a constant $c > 0$ such that for all $\lambda \geq \lambda^*$ and for all $x \in \mathbb{R}^n \setminus \{0\}$,*

$$\eta(x) + \lambda\gamma(x) \geq c \|x\|_{\mathbf{r},p}^m$$

$$\left(\eta(x) + \lambda\gamma(x) \leq -c \|x\|_{\mathbf{r},p}^m\right).$$

Proof By virtue of the homogeneity of η and γ, it is sufficient to establish the result on the unit sphere $S = \{x \in \mathbb{R}^n : \|x\|_{\mathbf{r},p} = 1\}$. Suppose that this relation is not valid. Then for every integer q there is a point x_q in S such that

$$\eta(x_q) + q\gamma(x_q) < \frac{1}{q}. \tag{8.22}$$

The sequence $\{x_q\}$, being bounded, has a subsequence converging to a point x_0, and we can accordingly suppose that $\{x_q\}$ converges to x_0. Since $\gamma(x) \geq 0$ on S, it follows from (8.22) and the lower semicontinuity of η and γ, i.e. $\liminf_{x=x_0} \eta(x) \geq \eta(x_0)$, $\liminf_{x=x_0} \gamma(x) \geq \gamma(x_0)$, that $\eta(x_0) \leq 0$, $\gamma(x_0) = 0$. This contradicts our hypothesis, and the first part of the Lemma is established. The second part is proved in the same manner since if $\eta(x)$ is lower semicontinuous then $-\eta(x)$ is upper semicontinuous. □

Lemma 8.6 *Let $\eta : \mathbb{R}^n \to \mathbb{R}$ be an upper semicontinuous, single-valued* **r***-homogeneous function, with weights $\mathbf{r} = [r_1, ..., r_n]^\top$ and degree $m > 0$. Then there is a point x_2 in the unit homogeneous sphere $S = \{x \in \mathbb{R}^n : \|x\|_{\mathbf{r},p} = 1\}$ such that the following inequality holds for all $x \in \mathbb{R}^n$*

$$\eta(x) \leq \eta(x_2) \|x\|_{\mathbf{r},p}^{m} . \quad (8.23)$$

Under the same conditions, if η is lower semicontinuous, there is a point x_1 in the unit homogeneous sphere S such that the following inequality holds for all $x \in \mathbb{R}^n$

$$\eta(x_1) \|x\|_{\mathbf{r},p}^{m} \leq \eta(x) . \quad (8.24)$$

Proof By virtue of the homogeneity of η, it is sufficient to establish the inequality (8.23) on the unit homogeneous sphere $S = \{x \in \mathbb{R}^n : \|x\|_{\mathbf{r},p} = 1\}$, i.e. $\eta(x) \leq \eta(x_2)$. Since S is compact and non empty, the latter inequality is a consequence of the fact that an upper semicontinuous function has a finite maximum value on a compact set and it achieves it at some point x_2 [18, Theorem 3.2]. The second part of the Lemma, i.e. inequality (8.24), is obtained by applying the same arguments to $-\eta(x)$, which is upper semicontinuous. If η is continuous, then we obtain item (ii) in Lemma 8.1.

References

1. Andrieu, V., Praly, L., Astolfi, A.: Homogeneous approximation, recursive observer design, and output feedback. SIAM J. Control Optim. **47**(4), 1814–1850 (2008)
2. Bacciotti, A., Rosier, L.: Liapunov functions and stability in control theory. Springer Science & Business Media (2006)
3. Bernuau, E., Efimov, D., Perruquetti, W., Polyakov, A.: On an extension of homogeneity notion for differential inclusions. In: Control conference (ECC), 2013 European, pp. 2204–2209. IEEE (2013)
4. Bernuau, E., Efimov, D., Perruquetti, W., Polyakov, A.: On homogeneity and its application in sliding mode control. J. Frankl. Inst. **351**(4), 1866–1901 (2014)
5. Bernuau, E., Polyakov, A., Efimov, D., Perruquetti, W.: Verification of iss, iiss and ioss properties applying weighted homogeneity. Syst. Control Lett. **62**(12), 1159–1167 (2013)
6. Bhat, S.P., Bernstein, D.S.: Geometric homogeneity with applications to finite-time stability. Math. Control Signal Syst. (MCSS) **17**(2), 101–127 (2005)
7. Cruz-Zavala, E., Moreno, J.A.: Lyapunov functions for continuous and discontinuous differentiators. IFAC-Pap. OnLine **49**(18), 660–665 (2016)
8. Cruz-Zavala, E., Moreno, J.A.: Homogeneous high order sliding mode design: A lyapunov approach. Automatica **80**, 232–238 (2017)
9. Cruz-Zavala E, Moreno J.A..: Lyapunov approach to higher-order sliding mode design. Control Robot. Sens. 3–28 (2016)
10. Deimling, K.: Multivalued Differential Equations, vol. 1. Walter de Gruyter, Berlin (1992)
11. Ding, S., Levant, A., Li, S.: Simple homogeneous sliding-mode controller. Automatica **67**, 22–32 (2016)
12. Filippov, A.F.: Differential Equations With Discontinuous Right-hand Side. American Mathematical Society, 191–231 (1988)
13. Fridman, L., Levant, A.: Sliding mode control in engineering. In: High-order sliding modes, Marcel Dekker, USA (2002)
14. Fridman, L., Moreno, J.A., Bandyopadhyay, B., Kamal, S., Chalanga, A.: Continuous nested algorithms: The fifth generation of sliding mode controllers. In: Recent Advances in Sliding Modes: From Control to Intelligent Mechatronics, pp. 5–35. Springer, Berlin (2015)
15. Hahn, W.: Stability Of Motion, vol. 138. Springer, Berlin (1967)

16. Hardy, G.H., Littlewood, J.E., Pólya, G.: Inequalities. Cambridge University Press, Cambridge (1952)
17. Hermes, H.: Homogeneous coordinates and continuous asymptotically stabilizing feedback controls. Differ. Equ. Stab. Control **109** 249–260 (1991)
18. Hestenes, M.R.: Calculus Of Variations And Optimal Control Theory. John Wiley and Sons Inc, Gauthier-Villars (1966)
19. Isidori, A.: Nonlinear control systems II. Springer, London (1999)
20. Isidori, A.: Nonlinear Control Systems. Springer Science and Business Media (2013)
21. Kamal, S., Moreno, J.A., Chalanga, A., Bandyopadhyay, B., Fridman, L.M.: Continuous terminal sliding-mode controller. Automatica **69**, 308–314 (2016)
22. Khalil, H.K.: Noninear Systems. Prentice-Hall, New Jersey (1996)
23. Levant, A.: Sliding order and sliding accuracy in sliding mode control. Int. J. Control **58**(6), 1247–1263 (1993)
24. Levant, A.: Robust exact differentiation via sliding mode technique. Automatica **34**(3), 379–384 (1998)
25. Levant, A.: Universal single-input-single-output (siso) sliding-mode controllers with finite-time convergence. IEEE Trans. Autom. Control **46**(9), 1447–1451 (2001)
26. Levant, A.: Higher-order sliding modes, differentiation and output-feedback control. Int. J. Control **76**(9–10), 924–941 (2003)
27. Levant, A.: Homogeneity approach to high-order sliding mode design. Automatica **41**(5), 823–830 (2005)
28. Levant, A.: Principles of 2-sliding mode design. Automatica **43**(4), 576–586 (2007)
29. Levant, A., Livne, M.: Weighted homogeneity and robustness of sliding mode control. Automatica **72**, 186–193 (2016)
30. Moreno, J.A.: A lyapunov approach to output feedback control using second-order sliding modes. IMA J. Math. Control Inf. 291–308 (2012)
31. Moreno, J.A.: Discontinuous integral control for mechanical systems. In: Variable structure systems (VSS), 2016 14th International workshop, 142–147. IEEE, Newyork (2016)
32. Moreno, J.A., Osorio, M.: Strict lyapunov functions for the super-twisting algorithm. IEEE Trans. Autom. Control **57**(4), 1035–1040 (2012)
33. Nakamura, H., Yamashita, Y., Nishitani, H.: Smooth lyapunov functions for homogeneous differential inclusions. In: SICE 2002. Proceedings of the 41st SICE Annual Conference, vol. 3, pp. 1974–1979. IEEE, Newyork (2002)
34. Nakamura, N., Nakamura, H., Yamashita, Y., Nishitani, H.: Homogeneous stabilization for input affine homogeneous systems. IEEE Trans. Autom. Control **54**(9), 2271–2275 (2009)
35. Orlov, Y.: Finite time stability of homogeneous switched systems. In: Decision and Control. Proceedings of the 42nd IEEE Conference, vol. 4, pp. 4271–4276. IEEE, Newyork (2003)
36. Orlov, Y.V.: Discontinuous Systems: Lyapunov Analysis And Robust Synthesis Under Uncertainty Conditions. Springer Science and Business Media (2008)
37. Rosier, L.: Homogeneous lyapunov function for homogeneous continuous vector field. Syst. Control Lett. **19**(6), 467–473 (1992)
38. Shtessel, Y., Edwards, C., Fridman, L., Levant, A.: Sliding Mode Control And Observation. Springer, Berlin (2014)
39. Torres-González, V., Fridman, L.M., Moreno, J.A.: Continuous twisting algorithm. In: Decision and control (CDC), IEEE 54th annual conference on 2015, pp. 5397–5401. IEEE (2015)
40. Utkin, V., Guldner, J., Shi, J.: Sliding Mode Control In Electro-mechanical Systems, vol. 34. CRC press (2009)
41. Utkin, V.I.: Sliding Modes In Control And Optimization. Springer Science and Business Media (2013)
42. Zamora, C.A., Moreno, J.A., Kamal, S.: Control integral dis-continuo para sistemas mecanicos. In: Congreso Anual de Asociación Mexicana de Control Automatico (2013)
43. Zubov, V.I., Boron, L.F.: Methods of AM Lyapunov And Their Application. Noordhoff Groningen (1964)

Chapter 9
Theory of Differential Inclusions and Its Application in Mechanics

Maria Kiseleva, Nikolay Kuznetsov and Gennady Leonov

Abstract The following chapter deals with systems of differential equations with discontinuous right-hand sides. The key question is how to define the solutions of such systems. The most adequate approach is to treat discontinuous systems as systems with multivalued right-hand sides (differential inclusions). In this work, three well-known definitions of solution of discontinuous system are considered. We will demonstrate the difference between these definitions and their application to different mechanical problems. Mathematical models of drilling systems with discontinuous friction torque characteristics are considered. Here, opposite to classical Coulomb symmetric friction law, the friction torque characteristic is asymmetrical. Problem of sudden load chande is studied. Analytical methods of investigation of systems with such asymmetrical friction, based on the use of Lyapunov functions, are demonstrated. The Watt governor and Chua system are considered to show different aspects of computer modeling of discontinuous systems.

9.1 Introduction

Two hundred and thirty years ago, after numerous experiments, Coulomb has formulated a law of dry fiction (Coulomb friction, see Fig. 9.1). Since then, various problems stimulated the development of theory of mechanical systems with dry friction.

First of all, it is important to mention the well-known Penleve paradoxes [65], which provoked interesting discussions and showed contradiction of Coulomb's law with Newtons laws of classical mechanics. Nowadays, an independent research

M. Kiseleva · N. Kuznetsov (✉) · G. Leonov
St. Petersburg State University, Saint Peterburg, Russia
e-mail: nkuznetsov239@gmail.com

N. Kuznetsov
University of Jyväskylä, Jyväskylä, Finland

G. Leonov
Institute of Problems of Mechanical Engineering, Russian Academy of Sciences, Saint Peterburg, Russia

J.B. Clempner and W. Yu (eds.), *New Perspectives and Applications of Modern Control Theory*, https://doi.org/10.1007/978-3-319-62464-8_9

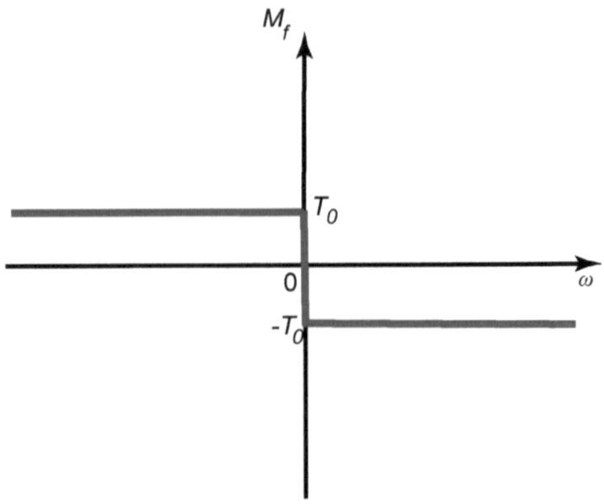

Fig. 9.1 Coulomb friction torque M_f

branch named "tribology" has grown out of these classical problems. Many researchers contributed to this branch of science, among them there are such famous scientists as F.P. Bowden and D. Tabor [8], E. Rabinovicz [73], P.J. Blau [6], K.C. Ludema [62], I.G. Goryacheva [23, 24], V.I. Kolesnikov [33].

From a mathematical point of view the problem of investigation of dynamics in models with dry friction is closely connected with the theory of differential inclusions and dynamical systems with discontinuous right-hand sides. Nowadays this theory is being actively developed and applied to investigation of different applications by such famous scientists as S.V. Emelyanov, A.S. Poznyak, V.I. Utkin and others (see e.g. [1, 5, 10, 16, 17, 30, 66, 71, 72, 75, 78]).

The following work is motivated by the problem of investigation of a drilling system. This problem was studied by the research group from the Eindhoven University of Technology [14, 64]. In these papers, the interaction of the drill with the bedrock is described by symmetric discontinuous characteristics. In the paper [45] a more precise model of friction is considered for simplified mathematical model of drilling system actuated by induction motor. Here the following assumption is made: the moment of resistance force with asymmetric characteristics (see Fig. 9.2, M is assumed to be large enough) is used instead of classical Coulomb friction with symmetric discontinuous characteristics. Such an asymmetric characteristic has a "locking" property—it allows rotation of the drill in one direction only. The considered simplified model corresponds to an, ordinary hand electric drill. In this case it is naturally to assume that the drilling takes place in one direction only.

The study of discontinuous systems with dry friction is a challenging task due to the need for a special theory for discontinuous systems to be developed. In particular, a proper definition of the solution on discontinuity surfaces is required. Now there are many definitions of solutions of discontinuous system, here three of them are considered following the works [19, 20, 22, 28, 54]. Analytical investigation of stability of simplified drilling systems will be performed. Additional examples of numerical modeling theory of differential inclusions and its application in discontinuous

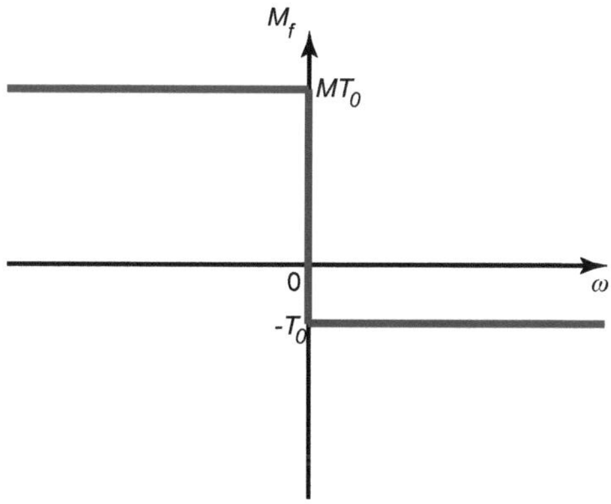

Fig. 9.2 Non-symmetric friction torque M_f

mechanical systems will be considered. It will be explained why it is necessary to use special methods of investigation for discontinuous systems.

9.2 Differential Equation with Discontinuous Right-Hand Sides and Differential Inclusions: Definitions of Solutions

The starting point of studies in theory of differential inclusions is usually connected with the works of French mathematician A. Marchaud and Polish mathematician S.K. Zaremba published in 1934–1936. They were studying equations of the form

$$Dx \subset f(t, x), \tag{9.1}$$

where $t \in \mathcal{D}_t \subset \mathbb{R}$, $x \in \mathcal{D}_x \subset \mathbb{R}^n$ and $f(t, x)$ is a multivalued vector function that maps each point (t, x) of some region $\mathcal{D} = \mathcal{D}_t \times \mathcal{D}_x$ to the set $f(t, x)$ of points from $\mathbb{R}^n$. For operator D the notions of contingent and paratingent were introduced by Marchaud and Zaremba respectively.

Definition 9.1 Contingent of vector function $x(t)$ at the point t_0 is a set Cont $x(t_0)$ of all limit points of sequences $\dfrac{x(t_i) - x(t_0)}{t_i - t_0}$, $t_i \to t_0$, $i = 1, 2, \ldots$

Definition 9.2 Paratingent of vector function $x(t)$ at the point t_0 is a set Parat $x(t_0)$ of all limit points of sequences $\dfrac{x(t_i) - x(t_j)}{t_i - t_j}$, $t_i \to t_0$, $t_j \to t_0$, $i = 1, 2, \ldots$

Wazhewski continued investigations of Marchaud and Zaremba and proved [80] that if $x(t)$ is a solution of differential inclusion (9.1) in the sense of Marchaud then vector function $x(t)$ is *absolutely continuous*.

Definition 9.3 Let $I \subset \mathcal{D}_t \subset \mathbb{R}$ be an interval of time. Function $x(t) : I \to \mathbb{R}^n$ is *absolutely continuous* on I if for every positive number ε there is a positive number δ such that whenever a finite sequence of pairwise disjoint sub-intervals (t_{1k}, t_{2k}) of I with $t_{1k}, t_{2k} \in I$ satisfies

$$\sum_k (t_{2k} - t_{1k}) < \delta$$

then

$$\sum_k ||x(t_{2k}) - x(t_{1k})|| < \varepsilon.$$

Important property of absolutely continuous function $x(t)$ is that $x(t)$ has derivative $\dot{x}(t)$ almost everywhere on I (see, e.g. [74]). This property played a key role in the development of theory of differential inclusions and equations with discontinuous right-hand side since it allowed to avoid artificial constructions in Definitions 9.1 and 9.2 and to consider usual derivative almost everywhere.

In 1960 paper [21] was published by A.F. Filippov, where he considered solutions of differential equations with discontinuous right-hand side as absolutely continuous functions. Filippov approach is one of the most popular among other notions of solutions of *systems with discontinuous right-hand sides*. Following [21], consider a system

$$\dot{x} = f(t, x), \qquad t \in \mathbb{R}, x \in \mathbb{R}^n, \tag{9.2}$$

where $f : \mathbb{R} \times \mathbb{R}^n \to \mathbb{R}^n$ is a piecewise continuous function such that measure of the set of discontinuity points is assumed to be zero.

Definition 9.4 Vector function $x(t)$, defined on an interval (t_1, t_2), is called a solution of (9.2) if it is absolutely continuous and for almost all $t \in (t_1, t_2)$ vector $\dot{x}(t)$ is within minimal closed convex set, which contains all $f(t, x')$ when x' is within almost all δ-neighbourhoods of the point $x(t)$ in $\mathbb{R}^n$ (for fixed t), i.e.

$$\dot{x} \in \prod_{\delta>0} \prod_{\mu N=0} \text{conv}\, f(t, U(x(t), \delta) - N). \tag{9.3}$$

Consider the case when system (9.2) is autonomous and vector function $f(x)$ is discontinuous on some smooth surface S in $\mathbb{R}^n$ and continuous in the neighbourhood of this surface. Let there exist limits $f_+(x)$ and $f_-(x)$ of vector function $f(x)$ when a point x approaches S from one or another side. Suppose that the vectors $f_+(x)$ and $f_-(x)$ are both pointing towards the discontinuity surface S. Then the so-called sliding mode appears. According to Definition 9.4, the vector field of sliding mode on the discontinuity surface can be defined as follows. The plane tangent to the surface S at the point x and the segment l, which connects the terminal points of vectors $f_+(x)$ and $f_-(x)$, are constructed. Then the vector with initial point at x and terminal point at the point of intersection of the segment and tangent plane is constructed: $f_0 = f_0(x)$. According to Definition 9.4, vector $f_0(x)$ defines vector field at the point x.

The obtained solution of (9.2) satisfies Definition 9.4, but nevertheless there are important applied problems for which Definition 9.4 is unsuitable. As an example of such a problem we consider a problem of synthesis of controls u_1 and u_2, which are limited, $|u_1| \leq 1$, $|u_2| \leq 1$, and which transform optimally fast each point $(x_1(0), x_2(0))$ of the system

$$\dot{x}_1 = x_2 u_1, \qquad \dot{x}_2 = u_2 \tag{9.4}$$

to the origin of coordinates. It is well-known [7] that synthesis of such a control is possible for the whole plane (x_1, x_2). For example, for the first quadrant of the plane the optimal control is as follows

$$u_1 = \begin{cases} 1, & x_1 < 0.5{x_2}^2, \\ -1, & x_1 \geq 0.5{x_2}^2, \end{cases} \qquad u_2 = \begin{cases} -1, & x_1 \leq 0.5{x_2}^2, \\ 1, & x_1 > 0.5{x_2}^2. \end{cases} \tag{9.5}$$

In particular, the trajectory $x_1 = 0.5{x_2}^2$ is optimal and for this trajectory system (9.4) takes the form $\dot{x}_1 = -x_2$, $\dot{x}_2 = -1$. Let us take the point $x = (x_1, x_2)$ on this trajectory and approach to this trajectory from the side $x_1 < 0.5{x_2}^2$. The limit value of the right-hand sides of system (9.4) is $f_+(x) = (x_2, -1)$. If we approach the trajectory from the side $x_1 > 0.5{x_2}^2$, then the limit is $f_-(x) = (-x_2, 1)$. Since $f_+(x) = -f_-(x)$, in this particular case the segment l passes through the point x, i.e. $f_0(x) = 0$ and according to Definition 9.4 the solution on sliding mode is equilibrium state. At the same time $(-x_2, -1)$ is a velocity vector on optimal trajectory. Thus, optimal trajectory is not a solution in the sense of Definition 9.4 by Filippov.

M.A. Aizerman and E.S. Pyatnitskiy [61] offered other definition of solution of equations with discontinuous right-hand sides which allows one to deal with usual derivative. We consider their approach in the particular case when $f(t, x)$ is discontinuous on the surface Σ. Consider a sequence of continuous vector functions $f_\varepsilon(t, x)$, which coincide with $f(t, x)$ outside of ε-neighbourhood of surface Σ, and tend to $f(t, x)$ for $\varepsilon \to 0$ at each point, which does not belong to Σ. Let $x_\varepsilon(t)$ be a solution of the system

$$\dot{x} = f_\varepsilon(t, x). \tag{9.6}$$

Then the solution of system (9.2) in the sense of Aizerman and Pyatnitskiy is a limit of any uniformly converging subsequence of solutions $x_{\varepsilon_k}(t)$:

$$x_{\varepsilon_k}(t) \rightrightarrows x(t).$$

In general, there may exist more than one such limit. Nevertheless this notion of solution, introduced in [61], does not always suitable for applications.

For example, consider a system

$$\dot{x} = Ax + b\phi(\sigma), \quad \sigma = c^* x, \tag{9.7}$$

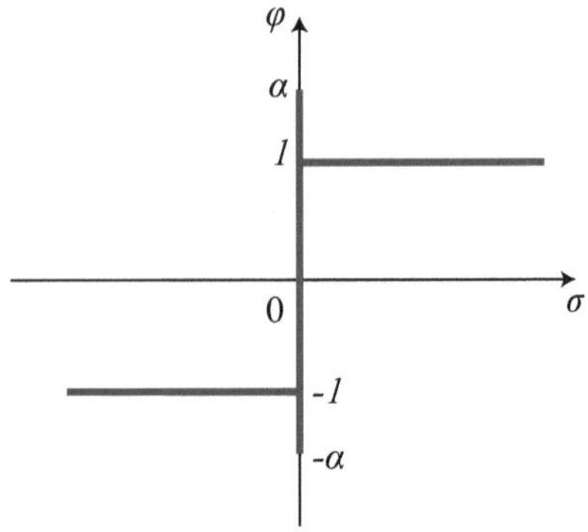

Fig. 9.3 Friction model, where static friction exceeds dynamic friction

where $\phi(\sigma)$ is a dry friction characteristic, shown in Fig. 9.1 or in Fig. 9.3, i.e.

$$\phi(\sigma) = \begin{cases} \operatorname{sign} \sigma, & \sigma \neq 0, \\ [-1, 1], & \sigma = 0, \end{cases} \quad \text{or} \quad \phi(\sigma) = \begin{cases} \operatorname{sign} \sigma, & \sigma \neq 0, \\ [-\alpha, \alpha], & \sigma = 0. \end{cases} \tag{9.8}$$

Since the definitions suggested by Filippov and by Aizerman and Pyatnitskiy deal only with those values of a nonlinearity for which $\sigma \neq 0$, the solutions of system (9.7) with dry friction characteristics, shown in Fig. 9.1 and in Fig. 9.3 coincide. This result does not match the physics of this phenomena.

To take into account dynamics on the discontinuity surface, the most adequate approach is to consider system with discontinuous right-hand side (9.2) as system with multivalued right-hand side, called *differential inclusion* [18, 20]:

$$\dot{x} \in f(t, x), \tag{9.9}$$

where $t \in \mathcal{D}_t \subset \mathbb{R}$, $x \in \mathcal{D}_x \subset \mathbb{R}^n$ and $f(t, x)$ is a multivalued vector function that maps each point (t, x) of some region $\mathcal{D} = \mathcal{D}_t \times \mathcal{D}_x$ to the set $f(t, x)$ of points from $\mathbb{R}^n$.

Definition 9.5 The vector function $x(t)$ is called a solution of differential inclusion (9.9), if it is absolutely continuous and for those t for which derivative $\dot{x}(t)$ exists, the following inclusion holds:

$$\dot{x}(t) \in f(t, x(t)). \tag{9.10}$$

To build a substantive theory, it is assumed that multivalued function $f(t, x)$ is semicontinuous. *Filippov approach* [19] requires additionally that $\forall (t, x) \in \mathcal{D}$ the set $f(t, x)$ is a minimal closed bounded set. This conditions coincide with Definition 9.4.

Definition 9.6 Function $f(t, x)$ is called *semicontinuous* (upper semicontinuous, β-continuous) at the point (t_0, x_0) if for any $\varepsilon > 0$ there exists $\delta(\varepsilon, t, x)$ such that the set $f(t, x)$ is contained in the ε-neighbourhood of set $f(x_0, x_0)$, provided that the point (t, x) belongs to δ-neighbourhood of the point (t_0, x_0).

As was shown above for some physical problems Filippov definition may gives wrong results, thus a more general class of multivalued functions $f(t, x)$ was considered by A. Kh. Gelig, G.A. Leonov and V.A. Yakubovich [20] (*Gelig-Leonov-Yakubovich approach*): $\forall (t, x) \in \mathcal{D}$ the set $f(t, x)$ is a bounded, closed, and convex set.

The following local theorem on the existence of solutions of differential inclusion holds true [20].

Theorem 9.1 *Suppose that multivalued the function $f(t, x)$ is semicontinuous at every point (t_1, x_1) of a region*

$$\mathcal{D}_1 \subset \mathcal{D}: \qquad |t_1 - t_0| \leq \alpha, \qquad |x_1 - a| \leq \rho,$$

and set $f(t_1, x_1)$ is bounded, closed, and convex. In addition, suppose

$$\sup |y| = c \quad \text{for} \quad y \in f(t_1, x_1), \qquad (t_1, x_1) \in \mathcal{D}_1.$$

Then for $|t - t_0| \leq \tau = \min(\alpha, \rho/c)$ there exists at least one solution $x(t)$ with initial condition $x(t_0) = a$, which satisfies (9.9) in the sense of Definition 9.5.

For differential inclusion (9.9) theorem on the continuation of solution remaining in a bounded region holds true. Also the theorem, which states that for every ω-limiting point of trajectory $x(t)$ there exists at least one trajectory that entirely consists of ω-limiting points, and some other theorems of qualitative theory, are valid [19, 20, 77].

For generalizations of classic results of stability theory on solutions of differential inclusion (9.9) in Gelig-Leonov-Yakubovich approach the existence of procedure of determination of discontinuous right-hand side according to a chosen solution (i.e. existence of extended nonlinearity, which allows one to replace differential inclusion with differential equation) was proven by B.M. Makarov specially for the monograph [20].

Let us demonstrate now the methods of theory of differential equations with discontinuous right-hand sides, described above, in concrete problems.

9.3 Analytical Methods of Investigation of Discontinuous Systems: An Example of Mathematical Model of Drilling System with "Locking Friction"

Consider the simplified mathematical model of drilling system actuated by induction motor (here we follow the works [26, 45]). Assume that the drill is absolutely rigid body stiffly connected to the rotor, which rotates by means of the magnetic field created by the stator of the induction motor. The value of interaction of the drill with the bedrock is defined as a value of resistance torque, which appears during the drilling process. Such a system experiences rapidly changes loads during the drilling, thus it is necessary to investigate the behavior of induction motor during load jumps, i.e. when resistance torque acting on the drill suddenly changes.

The following problem of stability is urgent since decrease of drilling systems failures plays important role in the oil and gas industry [76, 79].

As the equations of electromechanical model of the drilling system we consider the equations of induction motor, proposed in [34, 52], supplemented with the resistance torque M_f of drilling:

$$
\begin{aligned}
&L\frac{di_1}{dt} + Ri_1 = SB(sin\theta)\dot{\theta},\\
&L\frac{di_2}{dt} + Ri_2 = SB(cos\theta)\dot{\theta},\\
&I\ddot{\theta} \in -\beta SB(i_1 sin\theta + i_2 cos\theta) + M_f\left(\frac{R}{L} + \dot{\theta}\right).
\end{aligned}
\tag{9.11}
$$

Here θ is a rotation angle of the drill about the magnetic field created by the stator, which rotates with a constant angular speed $\frac{R}{L}$, $i_1(t)$, $i_2(t)$ are currents in rotor windings, R is resistance of windings, L is inductance of windings, B is the induction of magnetic field, S is an area of one wind, I is an inertia torque of drill, β is a proportionality factor, $\omega = \dot{\theta} + \frac{R}{L}$ is an angular velocity of the drill rotation with respect to a fixed coordinate system. The resistance torque M_f is assumed to be of the Coulomb type [20, 65]. Unlike the classic Coulomb friction law with symmetrical discontinuous characteristic the friction torque M_f has non-symmetrical discontinuous characteristics shown in Fig. 9.2.:

$$
M_f(\omega) = \begin{cases} -T_0 & \text{for } \omega > 0 \\ [-T_0, MT_0] & \text{for } \omega = 0 \\ MT_0 & \text{for } \omega < 0. \end{cases}
$$

For $T_0 \geq 0$ the number $M >$ is assumed to be large enough. That means that the drilling process only takes place when $\omega > 0$. Such characteristics does not allow for ω to switch from positive to negative values during the transient process in real drilling systems. In this case the system only gets stuck for $\omega = 0$ for a long enough period of time. These effects happen frequently during drilling operation and are studied by the analysis of system (9.11).

Performing the nonsingular change of variables

$$
\begin{aligned}
s &= -\dot{\theta}, \\
x &= \frac{L}{SB}(i_1 cos\theta - i_2 sin\theta), \\
y &= \frac{L}{SB}(i_1 sin\theta + i_2 cos\theta)
\end{aligned}
$$

we reduce system (9.11) to the following one:

$$
\begin{aligned}
\dot{s} &\in ay + \widetilde{M}_f(s), \\
\dot{y} &= -cy - s - xs, \\
\dot{x} &= -cx + ys,
\end{aligned} \tag{9.12}
$$

where $a = \dfrac{\beta(SB)^2}{IL}$, $c = \dfrac{R}{L}$. Here variables x, y define electric values in rotor windings and the variable s defines the sliding of the rotor. $\widetilde{M}_f$ has the following form

$$
\widetilde{M}_f(s) = \begin{cases} \gamma, & s < c; \\ [-\gamma M, \gamma], & s = c; \\ -\gamma M, & s > c; \end{cases}
$$

where $\gamma = \dfrac{T_0}{I}$.

According to Makarov's theorem for any solution of (9.12) in the sense of the Gelig-Leonov-Yakubovich approach there exists extended nonlinearity $\widetilde{M}_{f0}$ such that the following system is valid

$$
\begin{aligned}
\dot{s}(t) &= ay(t) + \widetilde{M}_{f0}(t), \\
\dot{y}(t) &= -cy(t) - s(t) - x(t)s(t), \\
\dot{x}(t) &= -cx(t) + y(t)s(t)
\end{aligned} \tag{9.13}
$$

for almost all t.

Let us conduct local analysis of equilibrium states of system (9.13).

Proposition 9.1 *For $0 \le \gamma < \dfrac{a}{2}$ system (9.13) has a unique asymptotically stable equilibrium state.*

Indeed, for $\gamma = 0$ the system (9.13) has one asymptotically stable equilibrium state $s = 0$, $y = 0$, $x = 0$, which occurs when the rotation of drill with constant angular speed is congruent to the rotation speed of the magnetic field (idle speed operation).

For $\gamma \in (0, \dfrac{a}{2})$ system (9.13) has one equilibrium state

$$
s_0 = \frac{c(a - \sqrt{a^2 - 4\gamma^2})}{2\gamma}, \qquad y_0 = -\frac{\gamma}{a}, \qquad x_0 = -\frac{\gamma s_0}{ac},
$$

where s_0 is the smallest root of the equation

$$\frac{acs}{c^2 + s^2} = \gamma.$$

In this case the direction of rotation of the drill and the magnetic field are the same, but the drill rotates with a lower angular speed $s_0 < c$.

Assume that there is a sudden change in load at the moment $t = \tau$ from value γ_0 to value γ_1, where $0 < \gamma_0 < \gamma_1$. This occurs at the moment when the drill comes in contact with the bedrock. For $\gamma = \gamma_0$ the system experiences a unique state of stable equilibrium

$$s_0 = \frac{c(a - \sqrt{a^2 - 4{\gamma_0}^2})}{2\gamma_0}, \quad y_0 = -\frac{\gamma_0}{a}, \quad x_0 = -\frac{\gamma_0 s_0}{ac}.$$

It is essential that after the transient process the solution $s(t)$, $x(t)$, $y(t)$ of the system (9.12) with $\gamma = \gamma_1$ and the initial data $s(\tau) = \dfrac{c(a - \sqrt{a^2 - 4{\gamma_0}^2})}{2\gamma_0}$, $y(\tau) = -\dfrac{\gamma_0}{a}$, $x(\tau) = -\dfrac{\gamma_0 s_0}{ac}$ tends to the equilibrium state

$$s_1 = \frac{c(a - \sqrt{a^2 - 4{\gamma_1}^2})}{2\gamma_1}, \quad y_1 = -\frac{\gamma_1}{a}, \quad x_1 = -\frac{\gamma_1 s_1}{ac}$$

as $t \to +\infty$.

The following theorem holds.

Theorem 9.2 *Let the following conditions be fulfilled*

$$\gamma_0 < \frac{a}{2}, \tag{9.14}$$

$$\gamma_1 < min\left\{\frac{a}{2},\ 2c^2\right\}, \tag{9.15}$$

$$3(M^2 + 2M)\gamma_1^2 - 8c^2\gamma_1 + 3ac^2 \geq 0. \tag{9.16}$$

Then the solution of system (9.13) with $\gamma = \gamma_1$ *and the initial data* $s(\tau) = \dfrac{c(a - \sqrt{a^2 - 4{\gamma_0}^2})}{2\gamma_0}$, $y(\tau) = -\dfrac{\gamma_0}{a}$, $x(\tau) = -\dfrac{\gamma_0 s_0}{ac}$ *tends to an equilibrium state of this system as* $t \to +\infty$.

Let us give the scheme of the proof of this theorem. We consider the region $\{s(t) < c\}$ of the phase space of system (9.13).

Performing the change of variables

$$\eta(t) = ay(t) + \gamma_1, \qquad z(t) = -x(t) - \frac{\gamma_1}{ac}s(t),$$

we reduce system (9.13) to the following system

$$\begin{aligned} \dot{s}(t) &= \eta(t), \\ \dot{\eta}(t) &= -c\eta(t) + az(t)s(t) - \psi(s(t)), \qquad \text{for almost all } t \\ \dot{z}(t) &= -cz(t) - \frac{1}{a}s(t)\eta(t) - \frac{\gamma_1}{ac}\eta(t). \end{aligned} \tag{9.17}$$

Here $\psi(s) = -\frac{\gamma_1}{c}s^2 + as - c\gamma_1$.

Introduce a function

$$V(s, \eta, z) = \frac{a^2}{2}z^2 + \frac{1}{2}\eta^2 + \int_{s_1}^{s} \psi(s)ds.$$

For every solution of system (9.17) from region $s(t) < c$ the following condition

$$\dot{V}(s(t), \eta(t), z(t)) = -a^2cz(t)^2 - \frac{a\gamma_1}{c}\eta(t)z(t) - c\eta(t)^2 \leq 0 \quad \text{for almost all } t \tag{9.18}$$

is fulfilled.

Quadratic form in the right-hand side of system (9.18) is definitely negative under condition (9.15).

We introduce a set

$$\Omega = \left\{ V(s, \eta, z) \leq \int_{s_1}^{c} \psi(s)ds + \frac{(1+M)^2}{2}\gamma_1^2, s \in [s_2, c] \right\},$$

where the point $s_2 < c$ is such that

$$\int_{s_2}^{c} \psi(s)ds + \frac{(1+M)^2}{2}\gamma_1^2 = 0.$$

The set Ω is limited and for $s = c$ it becomes

$$\frac{a^2}{2}z^2 + \frac{1}{2}\eta^2 \leq \frac{(1+M)^2}{2}\gamma_1^2.$$

Returning to the initial coordinates (x, y, s), we obtain

$$(x + \frac{\gamma_1}{a})^2 + (y + \frac{\gamma_1}{a})^2 \leq \frac{(1+M)^2}{a^2}\gamma_1^2.$$

Note that this circle is below the upper boundary $y = \frac{M\gamma_1}{a}$ of the sliding region $\Delta = \left\{s = c, -\frac{\gamma_1}{a} \leq y \leq \frac{M\gamma_1}{a}\right\}$ of system (9.13).

In the sliding region Δ system (9.13) can be reduced to the system of ordinary differential equations

$$\begin{aligned}\dot{y}(t) &= -cy(t) - c - cx(t),\\ \dot{x}(t) &= -cx(t) + cy(t),\end{aligned}$$

which is reduced by replacement of time $t = \frac{t_1}{c}$ to

$$\begin{aligned}\dot{y}(t) &= -y(t) - x(t) - 1,\\ \dot{x}(t) &= -x(t) + y(t).\end{aligned} \tag{9.19}$$

We introduce a function

$$W(x, y) = \left(x + \frac{\gamma_1}{a}\right)^2 + \left(y + \frac{\gamma_1}{a}\right)^2.$$

Semicircles $\left\{W(x, y) = R^2, y > -\frac{\gamma_1}{a}\right\}$, where $R \leq \frac{M+1}{a}\gamma_1$, are noncontact for system (9.19). Indeed, for the solutions of the system (9.19) under condition (9.15) the following relation

$$\begin{aligned}&\frac{1}{2}\dot{W}(x(t), y(t)) = -y^2(t) - y(t) - \frac{\gamma_1}{a} + \frac{\gamma_1^2}{a^2} - \left(x(t) + \frac{\gamma_1}{a}\right)^2\\ &= \left(\frac{2\gamma_1}{a} - 1\right)y(t) + \frac{\gamma_1}{a}\left(\frac{2\gamma_1}{a} - 1\right) - R^2 < 0\end{aligned}$$

is valid.

The solution, which falls into the sliding region, necessarily comes out through the lower boundary $y = -\frac{\gamma_1}{a}$ into the region $s < c$ due to the fact that $\dot{s} < 0$ for $s = c, y < -\frac{\gamma_1}{a}$. From condition (9.18) it follows that this solution proves to be inside the region $\left\{V(s, \eta, z) \leq \int_{s_1}^{c} \psi(s)ds\right\}$. Then it does not fall further into the sliding region, and tends to a unique equilibrium state (s_1, y_1, x_1) of the system due to the limitation of Ω. It is obvious that the trajectories, which fall into Ω, but not existing in the sliding region, also tend to the equilibrium state.

For details about the classical results of Lyapunov (see [20]) allow one to prove that the system is dichotomic[1] if condition (9.15) is fulfilled.

---

[1]System is called dichotomic if every solution bounded for $t > 0$ tends to stationary set for $t \to +\infty$ [20, 46, 47, 52].

The set Ω contains the point $s = s_0$, $\eta = \gamma_1 - \gamma_0$, $z = \dfrac{\gamma_0 - \gamma_1}{ac} s_0$ if

$$\frac{(\gamma_1 - \gamma_0)^2}{2c^2} {s_0}^2 + \frac{(\gamma_1 - \gamma_0)^2}{2} \le \int_{s_0}^{c} \psi(s)ds + \frac{(1+M)^2}{2} {\gamma_1}^2. \tag{9.20}$$

For $\gamma_0 < \gamma_1$ and condition (9.16) we have

$$\frac{(\gamma_1 - \gamma_0)^2}{2} \le \int_{0}^{c} \psi(s)ds + \frac{(1+M)^2}{2} {\gamma_1}^2. \tag{9.21}$$

Let us show that

$$\frac{(\gamma_1 - \gamma_0)^2}{2c^2} {s_0}^2 \le \int_{s_0}^{0} \psi(s)ds. \tag{9.22}$$

Indeed, taking into account that $\gamma_0 \le 2\gamma_1$, we obtain:
$\frac{\gamma_1}{3c} {s_0}^2 - \frac{a}{2} s_0 - \frac{(\gamma_1 - \gamma_0)^2}{2c^2} s_0 + c\gamma_1 = \frac{1}{12c^2{\gamma_0}^2}(c^2(a - \sqrt{a^2 - 4{\gamma_0}^2})^2\gamma_1 - 3a^2$
$c^2\gamma_0 + 3a^2c\sqrt{a^2 - 4{\gamma_0}^2}\gamma_0 - 3(\gamma_1 - \gamma_0)^2(a - \sqrt{a^2 - 4{\gamma_0}^2})\gamma_0 + 12c^2\gamma_1{\gamma_0}^2) \ge$
$\frac{1}{12c^2{\gamma_0}^2}(2a^2c^2 - 2ac^2\sqrt{a^2 - 4{\gamma_0}^2}\gamma_1 + 3ac^2\sqrt{a^2 - 4{\gamma_0}^2} - 3a^2c^2\gamma_0 +$
$3\sqrt{a^2 - 4{\gamma_0}^2}\gamma_0{\gamma_1}^2 - 3a{\gamma_1}^2\gamma_0 + 8c^2{\gamma_0}^2\gamma_1) \ge 0.$

Hence, from inequalities (9.21) and (9.22) we obtain condition (9.20).

Thus solution $s(t)$, $\eta(t)$, $z(t)$ with the initial data $s(\tau) = s_0$, $\eta(\tau) = \gamma_1 - \gamma_0$, $z(\tau) = \dfrac{\gamma_0 - \gamma_1}{ac} s_0$ tends to equilibrium state of the system.

Let M be a reasonably large number such that condition (9.16) of the theorem is fulfilled. In this case the following statement is valid.

Corollary 9.1 *Let $\gamma_0 = 0$ and*

$$\gamma_1 < min\left\{\frac{a}{2},\ 2c^2\right\}. \tag{9.23}$$

Then the solution of system (9.12) with $\gamma = \gamma_1$ and the initial data $s(\tau) = 0$, $y(\tau) = 0$, $x(\tau) = 0$ tends to equilibrium state of this system as $t \to +\infty$.

For the values $\gamma_1 \in \left\{2c^2, \frac{a}{2}\right\}$ (i.e., condition (9.23) is not fulfilled) the computer modeling of system (9.12) (region 2 in Fig. 9.4), which shows that the statement of consequence is retained, is carried out. Further we will discuss the aspects of modeling of systems with multivalued right-hand sides.

It can be checked that the extended nonlinearity can be written down in explicit form.

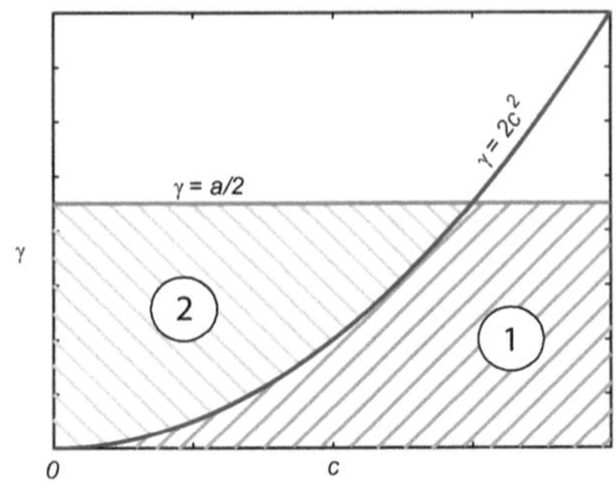

Fig. 9.4 Safe load region: 1—due to the theorem, 2—obtained by numerical modeling of the system

Corollary 9.2 *For system (9.12) the extended nonlinearity is of the following form:*

$$\widetilde{M}_{f0} = \begin{cases} \gamma, & if\, s = c,\, y < -\dfrac{\gamma}{a}\, or\, s < c; \\ -\gamma M, & if\, s = c,\, y > \dfrac{M\gamma}{a}\, or\, s > c; \\ -ay, & if\, s = c,\, -\dfrac{\gamma}{a} \leq y \leq \dfrac{M\gamma}{a}. \end{cases}$$

In the works [27, 29, 31, 32, 53, 55] more complex models of drilling systems were studied. Analytical investigation of such models is a challenging task, so it is necessary to use numerical methods. Let us further describe some aspects of numerical modeling in two other applied systems—Watt governor and Chua circuit.

9.4 Numerical Methods of Investigation of Discontinuous Systems

9.4.1 Difficulties of Numerical Modeling of Discontinuous Systems

Numerical modeling is one of the tools of investigation of differential equations with discontinuous equations with right-hand sides. Let us first show why it is important to use special methods developed for discontinuous systems. Consider the I.A. Vyshnegradsky problem. The following system of differential equations describes dynamics of Watt governor with dry friction

$$\begin{aligned} \dot{y}_1 &= -Ay_1 + y_2 - \mathrm{sign}(y_1), \\ \dot{y}_2 &= -By_1 + y_3, \\ \dot{y}_3 &= -y_1. \end{aligned} \tag{9.24}$$

Let sign be understood here in ordinary sense:

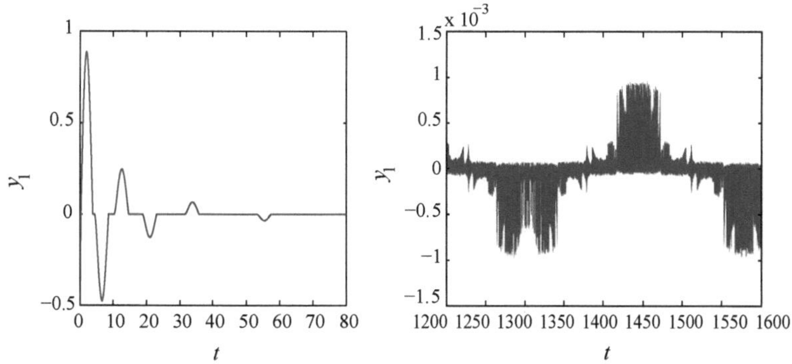

Fig. 9.5 Numerical modeling of Watt governor using built-in Matlab functions

$$\mathrm{sign}(y_1) = \begin{cases} 1, & \text{if } y_1 > 0; \\ 0, & \text{if } y_1 = 0; \\ -1, & \text{if } y_1 < 0. \end{cases}$$

Let us consider the values of parameters $A = 1.5$, $B = 1.1$ and conduct numerical modeling of trajectory of system (9.24) with initial data $y_1(0) = -0.5$, $y_2 = 1$, $y_3(0) = 1.2$, using standard Matlab build-in function *ode45* for solving ordinary differential equations. As can be seen in Fig. 9.5 numerical modeling shows that there are oscillations in system (9.24).

This nonlinear system was studied by A.A. Andronov and A.G. Mayer [4]. In particular, they proved that sliding segment of this system is globally stable if the following inequalities

$$A > 0, \qquad B > 0, \qquad AB > 1. \tag{9.25}$$

are satisfied.

Thus, the result of modeling with standard build-in Matlab functions may lead to wrong results. Moreover, the notation (9.24) is wrong and right notation is as follows

$$\begin{aligned} \dot{y}_1 &\in -Ay_1 + y_2 - \mathrm{Sign}(y_1), \\ \dot{y}_2 &= -By_1 + y_3, \\ \dot{y}_3 &= -y_1, \end{aligned} \tag{9.26}$$

and the model of dry friction is described in the following way

$$\mathrm{Sign}(y_1) = \begin{cases} 1, & \text{if } y_1 > 0; \\ [-1, 1], & \text{if } y_1 = 0; \\ -1, & \text{if } y_1 < 0. \end{cases}$$

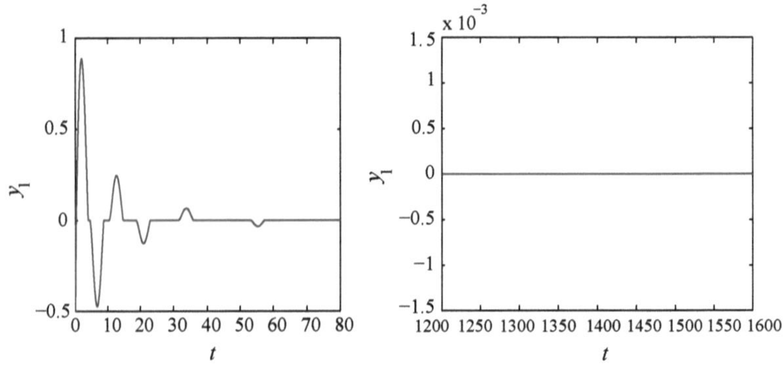

Fig. 9.6 Numerical modeling of Watt governor using Filippov definition

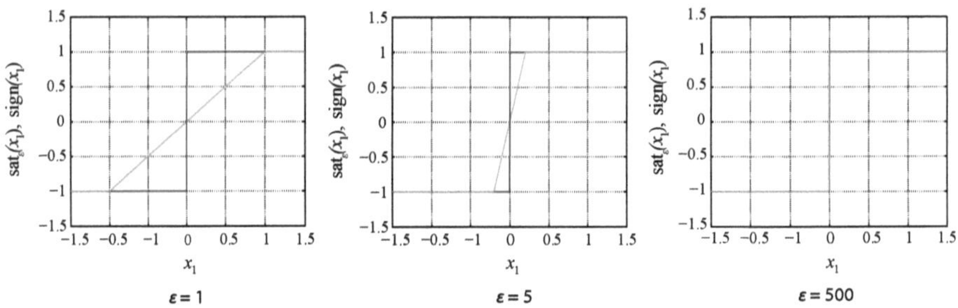

Fig. 9.7 Plots of $\mathrm{sat}_\varepsilon(x_1)$ and $\mathrm{sign}(x_1)$

Let us conduct the numerical modeling using Filippov definition [70]. The results of the modeling of system (9.26) correspond to theoretical results and are shown in Fig. 9.6.

9.4.2 *Numerical Modeling of Chua System*

We showed an example of numerical modeling of a discontinuous system based on Filippov's approach. Let us compare this method with modeling based on the Aizerman–Pyatnitskiy approach.

Consider the following example of discontinuous system—modified Chua system with discontinues characteristic [36, 37, 54]

$$\begin{aligned} \dot{x}_1 &\in -\alpha(m_1+1)x_1 + \alpha x_2 - \alpha(m_0 - m_1)\mathrm{Sign}(x_1), \\ \dot{x}_2 &= x_1 - x_2 + x_3, \\ \dot{x}_3 &= -\beta x_2 - \gamma x_3, \end{aligned} \tag{9.27}$$

where $\alpha, \beta, \gamma, m_0, m_1$ are parameters of the system.

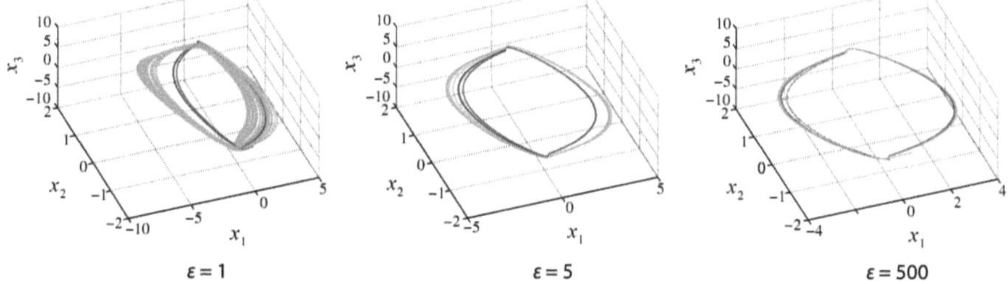

Fig. 9.8 Hidden attractor in Chua system: comparison of solutions

For parameters $\alpha = 8.4562$, $\beta = 12.0732$, $\gamma = 0.0052$, $m_0 = -0.1768$, $m_1 = -1.1468$ system (9.27) has a so-called *hidden attractor*[2] [35, 41, 49–51, 56].

So as to model system (9.27) with the help of both Filippov and Gelig-Leonov-Yakubovich definitions, the special event-driven numerical method, described in [70], was used. To model the system, using the Aizerman-Pyatnitskiy approach, one needs to replace $\text{sign}(x_1)$ by $\text{sat}_\varepsilon(x_1) = \frac{1}{2}\left(|\frac{x_1}{\varepsilon} + 1| - |\frac{x_1}{\varepsilon} - 1|\right)$, where $\varepsilon > 0$. Decrease of parameter ε allows one to obtain Aizerman-Pyatnitskiy solution ($\text{sat}_\varepsilon(x_1) \rightrightarrows \text{sign}(x_1)$ for $\varepsilon \neq 0$, see Fig. 9.7). In Fig. 9.8 hidden attractor modeled using Filippov's definition method is drawn in red colour and hidden attractor modeled using the Aizerman and Pyatnitskiy definition method is drawn in green. As one can see, the more ε is decreased, the more solutions (attractors) coincide with each other. This fact meets theorem proved in [28, 54].

Conclusion

We have discussed Filippov, Aizerman-Pyatnitskiy, and Gelig-Leonov-Yakubovich approaches to the study of differential equations with discontinuous right-hand sides and differential inclusions. While for a wide range of dynamical models with these three approaches give the same result (see, e.g. [28, 54]), there are models, where the

[2]From a computational point of view, it is natural to suggest the following classification of attractors, based on the simplicity of finding the basin of attraction in the phase space: [35, 41, 49–51, 56]: *An attractor is called a* self-excited attractor *if its basin of attraction intersects with any open neighbourhood of an equilibrium, otherwise it is called a* hidden attractor. For a *self-excited attractor* its basin of attraction is connected with an unstable equilibrium and, therefore, self-excited attractors can be localized numerically by the *standard computational procedure* in which after a transient process a trajectory, started in a neighbourhood of an unstable equilibrium, is attracted to the state of oscillation and traces it. For a hidden attractor its basin of attraction is not connected with equilibria. The hidden attractors, for example, are the attractors in the systems with no equilibria or with only one stable equilibrium (a special case of multistability—multistable systems and coexistence of attractors) [15, 31, 39]. The classical examples of the hidden oscillations are nested limit cycles in the study of 16th Hilbert problem (see, e.g., [42]), counter examples to the Aizerman and Kalman conjectures on the absolute stability of nonlinear control systems [9, 44, 57], and oscillations in electromechanical models without equilibria and with Sommerfeld effect [31]. Recent examples of hidden attractors can be found, e.g., in [2, 3, 11–13, 25, 27, 38, 40, 43, 48, 49, 53, 58–60, 63, 67–69, 81–85].

difference between these definitions is essential. As examples, we have considered te Chua circuit with discontinues characteristic, and mechanical systems with classical Coulomb symmetric friction law and the asymmetrical friction torque characteristic.

Acknowledgements This work was supported by the Russian Science Foundation (project 14-21-00041).

References

1. Agrachev, A., Sachkov, Y.: Control theory from the geometric viewpoint. Encyclopaedia of Mathematical Sciences, vol. 87. Control Theory and Optimization, II. Springer, Berlin (2004)
2. Aizerman, M.A., Pyatnitskii, E.S.: Fundamentals of the theory of discontinuous systems. I. Avtom. Tele mekh **7**, 33–47 (1974)
3. Andrievsky, B.R., Kuznetsov, N.V., Leonov, G.A., Pogromsky, A.Yu.: Hidden oscillations in aircraft flight control system with input saturation. IFAC Proc. **46**(12), 75–79 (2013)
4. Andrievsky, B.R., Kuznetsov, N.V., Leonov, G.A., Seledzhi, S.M.: Hidden oscillations in stabilization system of flexible launcher with saturating actuators. IFAC Proc. **46**(19), 37–41 (2013)
5. Andronov, A., Mayer, M.: Vyshnegradsky problem in control theory i. Automatica i Telemekhanika **8**(5), 314–334 (1947)
6. Arkin, R.C.: Behavior-Based Robotics. MIT Press, Cambridge (1998)
7. Bennett, S.: A History of Control Engineering (unknown) (1930)
8. Blau, P.J.: Friction Science and Technology: From Concepts to Applications. CRC Press, Boca Raton (2008)
9. Boltianskii, V.G.: Mathematical Methods of Optimal Control. Holt, Rinehart and Winston, New York (1971)
10. Bowden, F.P., Tabor, D.: The Friction and Lubrication of Solids, vol. 1. Oxford University Press, Oxford (2001)
11. Bragin, V.O., Vagaitsev, V.I., Kuznetsov, N.V., Leonov, G.A.: Algorithms for finding hidden oscillations in nonlinear systems. The Aizerman and Kalman conjectures and Chuas circuits. J. Comput. Syst. Sci. Int. **50**(4), 511–543 (2011)
12. Brogliato, B., Brogliato, B.: Nonsmooth Mechanics. Springer, Berlin (1999)
13. Burkin, I., Khien, N.N.: Analytical-numerical methods of finding hidden oscillations in multidimensional dynamical systems. Differ. Equs. **50**(13), 1695–1717 (2014)
14. Chaudhuri, U., Prasad, A.: Complicated basins and the phenomenon of amplitude death in coupled hidden attractors. Phys. Lett. A **378**(9), 713–718 (2014)
15. Danca, M.F.: Hidden transient chaotic attractors of Rabinovich-Fabrikant system. Nonlinear Dyn. **86**(2), 1263–1270 (2016)
16. De Bruin, J., Doris, A., van de Wouw, N., Heemels, W., Nijmeijer, H.: Control of mechanical motion systems with non-collocation of actuation and friction: a Popov criterion approach for input-to-state stability and set-valued nonlinearities. Automatica **45**(2), 405–415 (2009)
17. Dudkowski, D., Jafari, S., Kapitaniak, T., Kuznetsov, N.V., Leonov, G.A., Prasad, A.: Hidden attractors in dynamical systems. Phys. Rep. **637**, 1–50 (2016)
18. Edwards, C., Spurgeon, S.: Sliding Mode Control: Theory and Applications. CRC Press, Boca Raton (1998)
19. Emelyanov, S.: Variable Structure Control Systems. Nouka, Moscow (1967)
20. Filippov, A.: On existence of solutions of multivalued differential equations. Math. Notes **10**(3), 307–313 (1971)
21. Filippov, A.: Differential equations with discontinuous right-hand side. Math. Notes (1985)
22. Filippov, A.F.: Differential equations with discontinuous right-hand side. Matematicheskii sbornik **93**(1), 99–128 (1960)

23. Gelig, A.: Non-classical Differential Equations. Series 1. Mathematics. Mechanics. Astronomy, vol. 4 (2006)
24. Gelig, AKh, Leonov, G.A., Yakubovich, V.A.: Stability of Nonlinear Systems with Non-unique Equilibrium State. Nauka, Moscow (1978)
25. Goryacheva, I.: Contact Mechanics in Tribology. Series: Solid Mechanics and Its Applications, vol. 61 (1998)
26. Goryacheva, I., Rajeev, P., Farris, T.: Wear in partial slip contact. Trans.-Am. Soc. Mech. Eng. Tribol. **123**(4), 848–856 (2001)
27. Jafari, S., Pham, V.T., Golpayegani, S.M.R.H., Moghtadaei, M., Kingni, S.T.: The relationship between chaotic maps and some chaotic systems with hidden attractors. Int. J. Bifurc. Chaos **26**(13), 1650,211 (2016)
28. Kiseleva, M.: Oscillations of dynamical systems applied in drilling: analytical and numerical methods. Jyväskylä Stud. Comput. **181**, 1456–5390 (2013)
29. Kiseleva, M., Kondratyeva, N., Kuznetsov, N., Leonov, G., Solovyeva, E.: Hidden periodic oscillations in drilling system driven by induction motor. IFAC Proc. **47**(3), 5872–5877 (2014)
30. Kiseleva, M., Kondratyeva, N., Kuznetsov, N., Leonov, G.: Hidden oscillations in drilling systems with salient pole synchronous motor. IFAC-PapersOnLine **48**(11), 700–705 (2015)
31. Kiseleva, M.A., Kuznetsov, N.V.: Coinsidence of Gelig-Leonov-Yakubovich, Filippov, and Aizerman-Pyatnitskii definitions. Mathematics **48**(2), 66–71 (2015)
32. Kiseleva, M.A., Kuznetsov, N.V., Leonov, G.A., Neittaanmaki, P.: Drilling systems: stability and hidden oscillations. In: Discontinuity and Complexity in Nonlinear Physical Systems, vol. 6. Springer, Cham (2014)
33. Kiseleva, M.A., Kuznetsov, N.V., Leonov, G.A.: Hidden attractors in electromechanical systems with and without equilibria. IFAC-PapersOnLine **49**(14), 51–55 (2016)
34. Kloeden, P.E., Marín-Rubio, P.: Negatively invariant sets and entire trajectories of set-valued dynamical systems. Set-Valued Var. Anal. **19**(1), 43–57 (2011)
35. Kolesnikov, V.: Thermal-Physical Processes in Metal-polymeric Tribosystems. Nauka, Moscow (2003)
36. Kondrateva, N., Leonov, G., Rodyukov, F., Shepeljavyi, A.: Nonlocal analysis of differential equations of induction motors. Tech. Mech. **21**(1), 75–86 (2001)
37. Kuznetsov, A., Kuznetsov, S., Mosekilde, E., Stankevich, N.: Co-existing hidden attractors in a radio-physical oscillator system. J. Phys. A: Math. Theor. **48**(12), 125,101 (2015)
38. Kuznetsov, N., Leonov, G.: Hidden attractors in dynamical systems: systems with no equilibria, multistability and coexisting attractors. IFAC Proc. **47**(3), 5445–5454 (2014)
39. Kuznetsov, N., Kuznetsova, O., Leonov, G., Vagaitsev, V.: Analytical-numerical localization of hidden attractor in electrical Chuas circuit. Informatics in Control. Automation and Robotics, pp. 149–158. Springer, Berlin (2013)
40. Kuznetsov, N.V.: Hidden attractors in fundamental problems and engineering models: a short survey. In: AETA 2015: Recent Advances in Electrical Engineering and Related Sciences, pp. 13–25. Springer, Berlin (2016)
41. Kuznetsov, N.V., Leonov, G.A., Vagaitsev, V.I.: Analytical-numerical method for attractor localization of generalized chua's system. IFAC Proc. **43**(11), 29–33 (2010)
42. Kuznetsov, N.V., Kuznetsova, O.A., Leonov, G.A., Vagaytsev, V.: Hidden attractor in chua's circuits. ICINCO **1**, 279–283 (2011)
43. Kuznetsov, N.V., Kuznetsova, O.A., Leonov, G.A.: Visualization of four normal size limit cycles in two-dimensional polynomial quadratic system. Differ. Equs. Dyn. Syst. **21**(1–2), 29–34 (2013)
44. Kuznetsov, N.V., Leonov, G.A., Yuldashev, M., Yuldashev, R.: Nonlinear analysis of classical phase-locked loops in signal's phase space. IFAC Proc. **47**(3), 8253–8258 (2014)
45. Lao, S.K., Shekofteh, Y., Jafari, S., Sprott, J.C.: Cost function based on Gaussian mixture model for parameter estimation of a chaotic circuit with a hidden attractor. Int. J. Bifurc. Chaos **24**(01), 1450,010 (2014)
46. Leonov, G.A.: Mathematical Problems of Control Theory: An Introduction, vol. 6. World Scientific, Singapore (2001)

47. Leonov, G.A.: Phase synchronization: theory and applications. Autom. Remote Control **67**(10), 1573–1609 (2006)
48. Leonov, G.A., Kiseleva, M.A.: Stability of electromechanical models of drilling systems under discontinuous loads. In: Doklady Physics, vol. 57, pp. 206–209. Springer, Berlin (2012)
49. Leonov, G.A., Kondrateva, N.V.: Stability analysis of electric alternating current machines. SPb: Isd. St. Petersburg. univ 259 (2009)
50. Leonov, G.A., Kuznetsov, N.V.: Algorithms for searching for hidden oscillations in the aizerman and kalman problems. In: Doklady Mathematics, vol. 84, pp. 475–481. Springer, Berlin (2011)
51. Leonov, G.A., Kuznetsov, N.V.: Hidden attractors in dynamical systems: from hidden oscillations in Hilbert–Kolmogorov, Aizerman, and Kalman problems to hidden chaotic attractor in Chua circuits. Int. J. Bifurc Chaos **23**(01), 1330,002 (2013)
52. Leonov, G.A., Bragin, V.O., Kuznetsov, N.Y.: Algorithm for constructing counterexamples to the Kalman problem. In: Doklady Mathematics, vol. 82, pp. 540–542. Springer, Berlin (2010)
53. Leonov, G.A., Kuznetsov, N.V., Vagaitsev, V.I.: Localization of hidden chua's attractors. Phys. Lett. A **375**(23), 2230–2233 (2011)
54. Leonov, G.A., Kuznetsov, N.V., Vagaitsev, V.I.: Hidden attractor in smooth chua systems. Phys. D: Nonlinear Phenom. **241**(18), 1482–1486 (2012)
55. Leonov, G.A., Kiseleva, M.A., Kuznetsov, N.V., Neittaanmäki, P.: Hidden oscillations in drilling systems: torsional vibrations. J. Appl. Nonlinear Dyn. **2**(1), 83–94 (2013)
56. Leonov, G.A., Kuznetsov, N.V., Kiseleva, M.A., Solovyeva, E.P., et al.: Hidden oscillations in mathematical model of drilling system actuated by induction motor with a wound rotor. Nonlinear Dyn. **77**(1–2), 277–288 (2014)
57. Leonov, G.A., Kuznetsov, N.V., Mokaev, T.N.: Hidden attractor and homoclinic orbit in lorenz-like system describing convective fluid motion in rotating cavity. Commun. Nonlinear Sci. Numer. Simul. **28**(1), 166–174 (2015)
58. Leonov, G.A., Kuznetsov, N.V., Mokaev, T.N.: Homoclinic orbits, and self-excited and hidden attractors in a Lorenz-like system describing convective fluid motion. Eur. Phys. J. Spec. Top. **224**(8), 1421–1458 (2015)
59. Leonov, G.A., Kiseleva, M.A., Kuznetsov, N.V., Kuznetsova, O.A.: Discontinuous differential equations: comparison of solution definitions and localization of hidden chua attractors. IFAC-PapersOnLine **48**(11), 408–413 (2015)
60. Li, C., Sprott, J.C.: Coexisting hidden attractors in a 4-D simplified lorenz system. Int. J. Bifurc. Chaos **24**(03), 12 (2014)
61. Li, P., Zheng, T., Li, C., Wang, X., Hu, W.: A unique jerk system with hidden chaotic oscillation. Nonlinear Dyn. **86**(1), 197–203 (2016)
62. Li, Q., Zeng, H., Yang, X.S.: On hidden twin attractors and bifurcation in the Chuas circuit. Nonlinear Dyn. **77**(1–2), 255–266 (2014)
63. Ludema, K.C.: Friction, Wear, Lubrication: A Textbook in Tribology. CRC Press, Boca Raton (1996)
64. Menacer, T., Lozi, R., Chua, L.O.: Hidden bifurcations in the multispiral chua attractor. Int. J. Bifurc. Chaos **26**(14), 1630,039 (2016)
65. Mihajlovic, N., Van Veggel, A., Van de Wouw, N., Nijmeijer, H.: Analysis of friction-induced limit cycling in an experimental drill-string system. J. Dyn. Syst. Meas. Control **126**(4), 709–720 (2004)
66. Painlevé, P.: Lectures on Friction. Gostekhizdat, Moscow (1954)
67. Pfeiffer, F., Glocker, C.: Multibody Dynamics with Unilateral Contacts, vol. 9. Wiley, New York (1996)
68. Pham, V.T., Rahma, F., Frasca, M., Fortuna, L.: Dynamics and synchronization of a novel hyperchaotic system without equilibrium. Int. J. Bifurc. Chaos **24**(06), 1450,087 (2014)
69. Pham, V.T., Jafari, S., Volos, C., Wang, X., Golpayegani, S.M.R.H.: Is that really hidden? the presence of complex fixed-points in chaotic flows with no equilibria. Int. J. Bifurc. Chaos **24**(11), 1450,146 (2014)
70. Pham, V.T., Volos, C., Jafari, S., Vaidyanathan, S., Kapitaniak, T., Wang, X.: A chaotic system with different families of hidden attractors. Int. J. Bifurc. Chaos **26**(08), 1650,139 (2016)

71. Piiroinen, P.T., Kuznetsov, Y.A.: An event-driven method to simulate filippov systems with accurate computing of sliding motions. ACM Trans. Math. Softw. (TOMS) **34**(3), 13 (2008)
72. Plestan, F., Shtessel, Y., Bregeault, V., Poznyak, A.: New methodologies for adaptive sliding mode control. Int. J. Control **83**(9), 1907–1919 (2010)
73. Poznyak, A.S., Yu, W., Sanchez, E.N., Perez, J.P.: Nonlinear adaptive trajectory tracking using dynamic neural networks. IEEE Trans. Neural Netw. **10**(6), 1402–1411 (1999)
74. Rabinowicz, E.: Friction and Wear of Materials. Wiley, New York (1965)
75. Royden, H.L., Fitzpatrick, P.: Real Analysis, vol. 198. Macmillan, New York (1988)
76. Shokir, E.: A novel pc program for drill string failure detection and prevention before and while drilling specially in new areas. J. Oil Gas Bus **1**, 1–14 (2004)
77. Tolstonogov, A.: Differential inclusions in a Banach space, vol. 524. Springer Science & Business Media, Berlin (2012)
78. Utkin, V.I., Poznyak, A.S.: Adaptive sliding mode control with application to super-twist algorithm: equivalent control method. Automatica **49**(1), 39–47 (2013)
79. Vaisberg, O., Vincke, O., Perrin, G., Sarda, J., Fay, J.: Fatigue of drillstring: state of the art. Oil Gas Sci. Technol. **57**(1), 7–37 (2002)
80. Wazewski, T.: Sur une condition équivalente à l'équation au contingent. Bull. Acad. Polon. Sci. Sér. Sci. Math. Astronom. Phys. **9**, 865–867 (1961)
81. Wei, Z., Moroz, I., Liu, A.: Degenerate Hopf bifurcations, hidden attractors, and control in the extended Sprott E system with only one stable equilibrium. Turk. J. Math. **38**(4), 672–687 (2014)
82. Wei, Z., Wang, R., Liu, A.: A new finding of the existence of hidden hyperchaotic attractors with no equilibria. Math. Comput. Simul. **100**, 13–23 (2014)
83. Zelinka, I.: Evolutionary identification of hidden chaotic attractors. Eng. Appl. Artif. Intell. **50**, 159–167 (2016)
84. Zhao, H., Lin, Y., Dai, Y.: Hidden attractors and dynamics of a general autonomous van der pol–duffing oscillator. Int. J. Bifurc. Chaos **24**(06), 1450,080 (2014)
85. Zhusubaliyev, Z.T., Mosekilde, E.: Multistability and hidden attractors in a multilevel DC/DC converter. Math. Comput. Simul. **109**, 32–45 (2015)

Chapter 10
Stability Probabilities of Sliding Mode Control of Linear Continuous Markovian Jump Systems

Jiaming Zhu and Xinghuo Yu

Abstract In this chapter, an equivalent control-based sliding mode control is proposed for linear Markovian jump systems, which guarantees the asymptotical stability. The control objects are single-input single-output systems and multi-input multi-output systems. By using the stochastic system theory, a multi-step state transition conditional probability function is introduced for the continuous Markovian process, which is used to define the reaching and sliding probabilities. Furthermore, the formulas for calculating reaching and sliding probabilities are derived for situations where the control force may not be strong enough to ensure the fully asymptotical stability. In particular, for multi-input multi-output systems, by using the linear matrix inequality approach, sufficient conditions are proposed to guarantee the stochastically asymptotical stability of the systems on the sliding surfaces. Extensive simulations are conducted to validate the theoretical results and show the relationship between the control force and reaching and sliding probabilities.

10.1 Introduction

In recent years, continuous Markovian jump systems (MJSs) have gained particular research interest. Continuous MJSs are a class of continuous-time dynamical systems with stochastic jumps, in which jumping parameters are modeled as a continuous-time, discrete-state Markov chain with right continuous trajectories taking values in a finite set. A continuous-time Markov chain is a stochastic process that moves from state to state in accordance with a (discrete-time) Markov chain, but is such that the amount of time it spends in each state is exponentially distributed, before proceeding to the next state. These systems represent a class of stochastic systems that are popular in modeling practical systems which may experience random abrupt changes in

J. Zhu (✉)
Yangzhou University, Yangzhou, China
e-mail: qingzhu@yzu.edu.cn

X. Yu
RMIT University, Melbourne, VIC, Australia
e-mail: x.yu@rmit.edu.au

J.B. Clempner and W. Yu (eds.), *New Perspectives and Applications of Modern Control Theory*, https://doi.org/10.1007/978-3-319-62464-8_10

their structures and parameters. Such a system can be found in robotic manipulator systems, aircraft control, space stations, nuclear power plants, and wireless communication networks. Numerous results on MJSs have been reported in the literature [1–11]. These results cover a wide range of fields, including quadratic control [1], output feedback control [2], guarantee cost control [3], robust stabilization of MJS with uncertain switch probability [4] and partly unknown transition probability [5], $H\infty$ control and filter [6], robust Kalman filter [7], exponential filter [8], and state estimation and sliding mode control of singular MJSs [11].

Sliding mode control (SMC) has been regarded as an important robust control method for uncertain nonlinear systems. SMC is a nonlinear control method that alters the dynamics of a nonlinear system by application of a discontinuous control signal that forces the system to enter and then slide along a surface embedded with prescribed dynamical properties for all subsequent times. In the past decades, SMC has been successfully applied to a wide variety of practical engineering systems. Most recently, significant and substantial progress has been achieved in SMC for MJSs [12–25]. These include adaptive SMC for stochastic MJSs with actuator degradation [15], SMC for stochastic MJSs with incomplete transition rate [16], non-fragile observer-based $H\infty$ SMC for Ito stochastic systems with Markovian switching [18], asynchronous $H_2/H\infty$ filtering for discrete-time stochastic MJSs with randomly occurred sensor nonlinearities [19], observer-based $H\infty$ control on nonhomogeneous MJSs with nonlinear input [20], adaptive SMC with application to super-twist algorithm [21], robust $H\infty$ SMC for MJSs subject to intermittent observations and partially known transition probabilities [22], and finite-time stabilization for MJSs with Gaussian transition probabilities [23]. In [24], combining with LMI conditions of stochastic stability and integral sliding surface, a SMC law is synthesized for singular MJSs, which can achieve the bounded $L2$ gain performance. In [25], a bounded real lemma and an integral sliding surface are introduced, and then a SMC law is proposed for time-delay singular MJSs.

Although substantial progress has been made in MJSs under SMC, there are still some fundamental problems which have not been fully studied. The literatures only explored SMC of MJSs assuming that the required strong control force is available to overpower the stochastic uncertainty. However, in practice, controls are usually limited in power and sometimes insufficient, which may only make the controlled system reach the control target in a probabilistic sense. The outstanding question is what is the relationship between the control force and probabilistic jumping nature of MJSs. In [26], for SMC of second-order MJS;, the asymptotical stability probability was introduced to ensure the fully asymptotical stability.

In this chapter, we further explore the SMC of linear MJSs. The control objects include single-input single-output (SISO) systems and multi-input multi-output (MIMO) systems. We first derive conditions that guarantee the asymptotical stability of the MJSs under an equivalent control-based SMC paradigm. We then propose the reaching and sliding probabilities to deal with the situations where there is no sufficient control force. Particularly for MIMO systems, by using the linear matrix inequality(LMI) approach, sufficient conditions are proposed to guarantee the stochastically asymptotical stability of the systems on the sliding surfaces. Simulation studies are done to demonstrate the effectiveness of the results.

This paper is organized as follows. After the introduction in Sect. 10.1, the SMC of SISO linear MJSs is presented in Sect. 10.2. Next, the SMC of MIMO linear MJSs is presented in Sect. 10.3. Then, the numerical simulation result is given in Sect. 10.4.

10.2 Sliding Mode Control of SISO Linear Markovian Jump Systems

Before proceeding, we introduce some notations which will be used later for derivations and discussions. $P(\cdot)$ denotes the probability of an event. $P(A|B)$ denotes the conditional probability of event A given event B. $P(EA|B)$ denotes the conditional probability of events A and E given event B. $P(E|AB)$ denotes the conditional probability of event E given events A and B. $\|\cdot\|$ denotes the Euclidean norm of a vector or the Frobenius norm of a matrix. $M < 0$ denotes that matrix M is a negative definite matrix.

10.2.1 Problem Statement and Preliminaries

The SISO linear MJS under investigation is

$$\begin{cases} \dot{x}_1(t) &= A_{11}x_1(t) + A_{12}x_2(t), \\ dx_2(t) &= (A_{21}(\eta_t)x_1(t) + A_{22}(\eta_t)x_2(t) + B_2(\eta_t)u(t))dt + D(x)d\omega, \\ \eta_0 &= s_0, \ t \geq 0, \end{cases} \tag{10.1}$$

where

$$A_{11} = \begin{bmatrix} 0 & 1 & 0 & \cdots & 0 \\ 0 & 0 & 1 & \cdots & 0 \\ \vdots & \vdots & \vdots & \ddots & 1 \\ 0 & 0 & 0 & \cdots & 0 \end{bmatrix}, \quad A_{12} = \begin{bmatrix} 0 \\ \vdots \\ 0 \\ 1 \end{bmatrix}, \tag{10.2}$$

$$A_{21}(\eta_t) = \begin{bmatrix} -a_1(\eta_t), & -a_2(\eta_t) & \cdots, & -a_{n-1}(\eta_t) \end{bmatrix}, \tag{10.3}$$

$$A_{22}(\eta_t) = -a_n(\eta_t), \quad B_2(\eta_t) = b_2(\eta_t), \tag{10.4}$$

$$|D(x)| \leq H\|x\|, \tag{10.5}$$

$x(t) = [x_1^T(t),\ x_2(t)]^T$ is the system state, $x_1(t) \in R^{n-1}$, $x_2(t) \in R$; $u(t) \in R$ is the control input; $D(x)$ is a known continuous function; ω is a 1-dimensional Brownian motion; A_{11}, A_{12}, A_{21}, A_{22}, B_2 are coefficient matrices with compatible dimensions, H is a known constant and $\{\eta_t, t \in [0, T]\}$ is a finite-state Markovian process having a state-space $S = \{1, 2, \dots, \nu\}$, $B_2(j) > b_0 > 0,\ j \in S$, generator (α_{ij}) with transition probability from mode i at time t to mode j at time $t + \Delta$, $i, j \in S$,

$$P_{ij}(\Delta) = P(\eta_{t+\Delta} = j|\eta_t = i) = \begin{cases} \alpha_{ij}\Delta + o(\Delta), & \text{if } i \neq j, \\ 1 + \alpha_{ii}\Delta + o(\Delta), & \text{if } i = j, \end{cases} \tag{10.6}$$

where

$$\alpha_{ii} = -\sum_{m=1, m\neq i}^{\nu} \alpha_{im},\ \alpha_{ij} \geq 0, \forall i, j \in S, i \neq j, \tag{10.7}$$

$\Delta > 0$ and $\lim_{\Delta \to 0} o(\Delta)/\Delta = 0$.

Assumption 1. The Markovian process η_t is unmeasurable. Only the initial state η_{t_0} and the generators (α_{ij}) are available.

The sliding variable is defined as

$$s(x) = C_1 x_1 + x_2, \quad C_1 = [c_1, c_2, \dots, c_{n-1}] \tag{10.8}$$

where s is Hurwitz. Substituting (10.8) into system (10.1) yields

$$\begin{cases} \dot{x}_1 = (A_{11} - A_{12}C_1)x_1 + A_{12}s, \\ ds = (\bar{A}_{21}(\eta_t)x_1 + \bar{A}_{22}(\eta_t)s + B_2(\eta_t)u(t))dt + D(x)d\omega, \\ \eta_0 = s_0,\ t \geq 0. \end{cases} \tag{10.9}$$

where

$$\begin{aligned} \bar{A}_{21}(\eta_t) &= C_1 A_{11} - C_1 A_{12} C_1 + A_{21} - A_{22} C_1, \\ \bar{A}_{22}(\eta_t) &= C_1 A_{12} + A_{22}. \end{aligned} \tag{10.10}$$

Definition 10.1 The system (1) is called mean-square stable, if for each $\varepsilon > 0$, there is a $\delta > 0$, such that

$$\sup_{t_0 \leq t < \infty} E\|x(t)\|^2 < \varepsilon, \text{ for all } \|x_0\| < \delta. \tag{10.11}$$

In addition, the system (1) is called asymptotically mean-square stable, if it is mean-square stable and there is a $\delta_0 > 0$, such that

$$\lim_{t \to \infty} E\|x(t)\|^2 = 0, \text{ for all } \|x_0\| < \delta_0. \tag{10.12}$$

For conventional (non-Markovian jumping) linear systems, $D(x) = 0$, a well-known SMC is the equivalent control-based SMC [14] where

$$u(t) = u_{eq} - k\text{sgn}(s), \tag{10.13}$$

with $k > 0$ and

$$u_{eq} = -B_2^{-1}(\bar{A}_{21}x_1 + \bar{A}_{22}s). \tag{10.14}$$

which enables $\dot{s} = 0$. This SMC scheme can drive the system trajectory to reach the switching surface in a finite time, and then maintain it on the surface at all times. When in the sliding mode, the system state converges to zero exponentially.

Noting Assumption 1, this SMC scheme cannot be applied directly, for the stochastic coefficients $\bar{A}_{21}$, $\bar{A}_{22}$ are not available. The equivalent control u_{eq} has to be redesigned.

An important step to study MJSs under SMC is to derive multi-step state transition probability which will be very useful to study reaching and sliding probabilities. Equation (10.6) shows the probability of state transition from i to j at time interval from t to $t + \Delta$ whatever it jumps how many times and along which transition route. Also, it is not suitable for a large time interval, considering the existence of item $o(\Delta)$ in the equation. However, in order to achieve the asymptotical stability of the systems with different stochastic process parameters, the valid ranges of control parameters could be different. Therefore, we must derive the probability of stochastic parameter jumps along a known parameter transition route such as $(s_0, s_1, \ldots, s_n)$ from time t to $t + \Delta$, for any Δ. Note that it is different from the transition probability in (10.6). Without loss of generality, we consider the multi-step transition probability from time 0 to t_r where t_r is a given time by which the switching surface $s = 0$ will be reached.

First, we derive the one-step jump stochastic state transition probability density function. From [27, 28], we have the transition probability formula

$$P(\tau_1 \leq t, \eta_{\tau_1} = j|\eta_0 = i) = \frac{q_{ij}}{q_i}(1 - e^{-q_i t}), \tag{10.15}$$

$$P(\eta_{\tau_1} = j|\eta_0 = i) = \frac{q_{ij}}{q_i}, \tag{10.16}$$

where

$$q_i = -\alpha_{ii} = \lim_{t \to 0} \frac{1 - P_{ii}(t)}{t}, \tag{10.17}$$

$$q_{ij} = \alpha_{ij} = \lim_{t \to 0} \frac{P_{ij}(t)}{t}, \tag{10.18}$$

and τ_1 represents the time at which the Markovian process parameter first jumps.

Thus, we have the one-step jump state transition probability density function as

$$\begin{aligned} &f_{\mathrm{pdf}}(\tau_1 = t, \eta_{\tau_1} = j|\eta_0 = i) \\ &= \frac{d}{dt} P(\tau_1 \leq t, \eta_{\tau_1} = j|\eta_0 = i) \\ &= q_{ij} e^{-q_i t}. \end{aligned} \tag{10.19}$$

Events are defined as follows.

$$B: \text{ The initial condition is } \eta_0 = s_0. \tag{10.20}$$

$$A_j: \ \eta_t = j, 0 \leq t \leq t_r. \tag{10.21}$$

$$A^0(t): \text{ The stochastic process parameter does not jump, } \eta_\tau = s_0, 0 \leq \tau \leq t, \tag{10.22}$$

$$A_j^k(t): \text{ The stochastic process parameter jumps } k \text{ times } \eta_\tau = \begin{cases} s_0 \ 0 \leq \tau < t_1, \\ s_1 \ t_1 \leq \tau < t_2, \\ \quad \vdots \\ s_k \ t_k \leq \tau \leq t. \end{cases} \text{ where the route } r_j^k = (s_0, \ldots, s_k) \in S^k \tag{10.23}$$

$$S^k = \{(s_0, s_1, \ldots, s_k)|s_i \in S, s_{i-1} \neq s_i, 1 \leq i \leq k\}. \tag{10.24}$$

$$A^k(t): \ A^k(t) = \bigcup_{r_j^k \in S^k} A_j^k(t), k = 1, 2, \ldots, n. \tag{10.25}$$

The sample space is defined as

$$A(t): \ A(t) = \bigcup_{n=0}^{\infty} A^n(t), \tag{10.26}$$

which is the set of all possible stochastic jump events.

The total conditional probability is given as

$$P(A|B) = P\left(\bigcup_{n=0}^{\infty} A^n(t)|B\right) = \sum_{n=0}^{\infty} P(A^n|B) = 1, \tag{10.27}$$

where

$$P(A^n|B) = \sum_{r_j^n \in S^n} P(A_j^n|B). \tag{10.28}$$

The following Lemma is given.

Lemma 10.1 ([29]) *For the continuous Markovian stochastic process, n−step state transition conditional probability is*

$$P(A^0(t)|B) = e^{-q_{s_0}t}, \tag{10.29}$$

$$P(A^n_j(t)|B) = \int_0^t P(A^{n-1}_j(t-s)|B) q_{s_0 s_1} e^{-q_{s_0}s} ds, \tag{10.30}$$

and

$$\begin{aligned} P(A_j|B) &= P(\eta_t = j|\eta_0) \\ &= \sum_{n=0, s_n=j}^{n=\infty} \sum_{r^n_i \in S^n} P(A^n_i(t)|B). \end{aligned} \tag{10.31}$$

where events are defined by (10.20)–(10.25).

Proof First, we consider that the Markovian process parameter does not jump. The stochastic state transition conditional probability is given as

$$P(A^0(t)|B) = e^{-q_{s_0}t}. \tag{10.32}$$

Second, we consider that the Markovian process stochastic parameter jumps n times ($n \in [1, \infty)$).

Considering the Markov property, the stochastic state transition conditional probability is given as

$$P(A^n_j(t)|B) = \int_0^t P(A^{n-1}_j(t-s)|B) q_{s_0 s_1} e^{-q_{s_0}s} ds \tag{10.33}$$

where $P(A^{n-1}_j(t-s)|B)$ is the modification of $P(A^{n-1}_j(t)|B)$, with all t being replaced by $t-s$ and all s_i being replaced by s_{i+1}. Therefore, all $q_{s_{i-1}s_i}$ are replaced by $q_{s_i s_{i+1}}$, $0 \le i \le n-1$.

By (10.20)–(10.25), we have

$$\begin{aligned} P(A_j|B) = P(\eta_t = j|\eta_0) &= \sum_{n=0, s_n=j}^{n=\infty} P(A^n|B) \\ &= \sum_{n=0, s_n=j}^{n=\infty} \sum_{r^n_i \in S^n} P(A^n_i(t)|B). \end{aligned} \tag{10.34}$$

This completes the proof. □

10.2.2 Control Scheme and Stability Probability

There are two important phases in SMC, the reaching phase and sliding phase. If the control force is large enough, it can suppress any matched uncertainties to realize the sliding motion. However, if it is not sufficiently large, it will affect the reaching and sliding abilities of the SMC, that is, the reachability and sliding ability would be in a probabilistic sense. In this section, we will derive conditions to guarantee the globally asymptotical stability of the MJSs under the equivalent control-based SMC. We will also derive the reaching probability function and the sliding probability function. For convenience, we introduce the *reaching probability*, denoted by $P(E_r|B_r)$, which represents the condition probability of the system reaching the sliding mode in a finite time t_r, and the *sliding probability*, denoted by $P(E_s|B_s)$, which represents the condition probability of the system maintaining on the sliding mode.

Events are defined as

$$B_r:\ \text{The condition is } \eta_0 = s_0, x(0) = x_0,\quad u(t) = f_u(t), 0 \le t \le t_r. \tag{10.35}$$

$$B_s:\ \text{The condition is } \eta_{t_r} = \eta_{t_r}, x(t_r) = x_{t_r},\quad s(t_r) = 0, u(t) = f_u(t), t > t_r. \tag{10.36}$$

$$E_r:\ \text{The system first reaches the sliding mode at time } t_r,\ s(t_r) = 0. \tag{10.37}$$

$$E_s:\ \text{The system maintains on the sliding mode, } s(t) = 0,\ t > t_r. \tag{10.38}$$

where t_r is a given constant, $f_u(t)$ represents the designed control input.

From the above discussions, we know that we can easily design an equivalent control-based SMC for LTI non-MJS systems that is globally and asymptotically stable. Since MJSs can be regarded as a system stochastically jumping between a set of LTI systems, the first question arises that how to design an equivalent control-based SMC to make the SISO MJSs asymptotically stable.

The following SMC is proposed for the SISO MJS as

$$\begin{aligned} u(t) &= \hat{u}_{eq}(t) - k_1\|x_1\|\text{sgn}(s) - k_2 s - k_r\text{sgn}(s) - \frac{3}{2s}\frac{H^2}{b_0}\|x\|^2, \quad \text{if } s \ne 0,\\ u(t) &= \hat{u}_{eq}(t) - k_1\|x_1\|\text{sgn}(s) - k_2 s - k_r\text{sgn}(s), \quad \text{if } s = 0, \end{aligned} \tag{10.39}$$

where the variable s is defined by (10.8), k_1, k_2 are control parameters, and $\hat{u}_{eq}(t)$ is the approximation of the equivalent control defined as

$$\hat{u}_{eq}(t) = -\sum_{j\in S} \hat{P}_j B_2^{-1}(j)(\bar{A}_{21}(j)x_1 + \bar{A}_{22}(j)s). \tag{10.40}$$

where $\hat{P}_j$ is the approximation of $P(\eta_t = j|\eta_0)$. Some auxiliary parameters are defined as

$$k_{1i} = B_2^{-1}(i) \left\| \bar{A}_{21}(i) - B_2(i) \sum_{j \in S} \hat{P}_j B_2^{-1}(j) \bar{A}_{21}(j) \right\|, \tag{10.41}$$

$$k_{2i} = B_2^{-1}(i) \left\| \bar{A}_{22}(i) - B_2(i) \sum_{j \in S} \hat{P}_j B_2^{-1}(j) \bar{A}_{22}(j) \right\|, \tag{10.42}$$

for $i = 1, \ldots, \nu$.

The following theorem is given for the globally asymptotically stable MJS under SMC.

Theorem 10.1 ([29]) *For the SISO linear MJS (10.1) under SMC (10.39), if k_1, k_2 are selected as*

$$k_1 = \max_{i \in S}\{k_{1i}\},\ k_2 = \max_{i \in S}\{k_{2i}\}, \tag{10.43}$$

where k_{1i}, k_{2i} are defined by (10.41) and (10.42), then the close-loop system is asymptotically stable.

Proof The Lyapunov function candidate is given as

$$V(s) = \frac{1}{4}s^4. \tag{10.44}$$

For any time $t \geq 0, \eta_t = i,\ i \in S$. Thus system (10.9) can be written as

$$\begin{cases} \dot{x}_1 = (A_{11} - A_{12}C_1)x_1 + A_{12}s, \\ ds = (\bar{A}_{21}(i)x_1 + \bar{A}_{22}(i)s + B_2(i)u(t))dt + D(x)d\omega, \end{cases} \tag{10.45}$$

where $\eta_0 = s_0,\ t \geq 0$.

First, we study the reaching stage$(0 \leq t \leq t_r)$.

Substituting SMC (10.39) and (10.43) into the derivative of s yields

$$\begin{aligned} LV(s) &= s^3[\bar{A}_{21}(i)x_1 + \bar{A}_{22}(i)s + B_2(i)u(t)] + \frac{1}{2}Tr\left\{D^T(x)\frac{\partial^2 V}{\partial s^2}D(x)\right\} \\ &= s^3\left[\bar{A}_{21}(i)x_1 + \bar{A}_{22}(i)s - B_2(i)\sum_{j \in S}\hat{P}_j B_2^{-1}(j)\right. \\ &\quad \left. \cdot (\bar{A}_{21}(j)x_1 + \bar{A}_{22}(j)s)\right] - B_2(i)k_1\|x_1\||s|^3 \end{aligned}$$

$$
\begin{aligned}
&-B_2(i)k_2s^4 - B_2(i)k_r|s|^3 - \frac{3}{2}H^2\|x\|^2s^2 + \frac{3}{2}D^2(x)s^2 \\
\le& \left\| \bar{A}_{21}(i) - B_2(i)\sum_{j\in S}\hat{P}_j B_2^{-1}(j)\bar{A}_{21}(j) \right\| \|x_1\||s|^3 \\
&+ \left\| \bar{A}_{22}(i) - B_2(i)\sum_{j\in S}\hat{P}_j B_2^{-1}(j)\bar{A}_{22}(j) \right\| s^4 \\
&-B_2(i)k_1\|x_1\||s|^3 - B_2(i)k_2s^4 - B_2(i)k_r|s|^3 \\
\le& -B_2(i)k_r|s|^3 \le -\min_{i\in S}\{|B_2(i)|\}k_r|s|^3 \le 0,
\end{aligned} \tag{10.46}
$$

which means the system reaching the switching surface in finite time t_r where $t_r = \frac{|s(0)|}{\min_{i\in S}\{|B_2(i)|\}k_r}$. The equality holds if and only if $s = 0$. It means under SMC (10.39), the system can maintain it on the switching surface after reaching it.

Once the switching surface is reached, by (10.45), we have

$$
\dot{x}_1 = (A_{11} - A_{12}C_1)x_1. \tag{10.47}
$$

Noting $x_2 = -C_1x_1$, the entire system state x exponentially converges to the equilibrium. This completes the proof. □

Here a new question arises. If the control is insufficient, what will the system's asymptotical stability probability be? This question is answered in the following.

In the reaching stage, the control target is to drive the system state to reach the sliding mode in finite time t_r. With the following definitions of auxiliary parameters

$$
k_{1j}^n = \max_{i\in r_j^n}\{k_{1i}\},\ k_{2j}^n = \max_{i\in r_j^n}\{k_{2i}\}, \tag{10.48}
$$

where k_{1i}, k_{2i} are defined by (10.41) and (10.42), and $r_j^n = (s_{j0}, \ldots, s_{jn}) \in S^{n+1}$ represents the Markovian process state transition route, we derive the reaching probability formulae.

Theorem 10.2 ([29]) *For the SISO linear MJS (10.1) under SMC (10.39), the reaching probability is given as follows*

$$
P(E_r|B_r) = \sum_{n=0}^{m} P(E_rA^n|B_r), \tag{10.49}
$$

where

$$
P(E_rA^n|B) = \sum_{r_j^n\subset S_n} P(E_r|A_j^nB)P(A_j^n|B), \tag{10.50}
$$

$$
P(E_r|A_j^nB) = \begin{cases} 1 & k_1 \ge k_{1j}^n \text{ and } k_2 \ge k_{2j}^n, \\ 0 & k_1 < k_{1j}^n \text{or } k_2 < k_{2j}^n, \end{cases} \tag{10.51}
$$

where m is a positive integer which satisfies

$$\sum_{i=0}^{m} P(E_r A^i | B_r) > 1 - \varepsilon,$$

on condition $k_1 = \max_{j \in S}\{k_{1j}\}$, $k_2 = \max_{j \in S}\{k_{2j}\}$ *and* $\varepsilon > 0$ *is a small constant.*

Proof We consider the situations where the control parameters k_1 and k_2 are not sufficient to guarantee the asymptotical stability of the MJS under SMC. Therefore, there exist probabilities that the reaching phase does not exist.

First, consider that the stochastic parameter η_t does not jump. $\eta_t = s_0$. The reaching probability for no jump is

$$P(E_r A^0 | B_r) = P(E_r | A^0 B_r) P(A^0 | B) \tag{10.52}$$

$$= \begin{cases} e^{-q_{s_0} t} & k_1 \geq k_{1s_0} \text{ and } k_2 \geq k_{2s_0}, \\ 0 & k_1 < k_{1s_0} \text{or } k_2 < k_{2s_0}, \end{cases} \tag{10.53}$$

where $P(A^0 | B)$ is defined by (10.29) and

$$P(E_r | A^0 B_r) = \begin{cases} 1 & k_1 \geq k_{1s_0} \text{ and } k_2 \geq k_{2s_0}, \\ 0 & k_1 < k_{1s_0} \text{or } k_2 < k_{2s_0}. \end{cases} \tag{10.54}$$

Now consider that the stochastic parameter η_t jumps n times ($n \in [1, \infty)$). Denote k_{1j}^n, k_{2j}^n as the sufficient control parameters which enable the system to reach the control target while the stochastic parameter η_t jumps along the transition route r_j^n. It holds that

$$k_{1j}^n = \max_{i \in r_j^n} \{k_{1i}\}, \; k_{2j}^n = \max_{i \in r_j^n} \{k_{2i}\}. \tag{10.55}$$

The reaching probability for n jumps is

$$P(E_r A^n | B_r) = \sum_{r_j^n \subset S_n} P(E_r | A_j^n B_r) P(A_j^n | B). \tag{10.56}$$

where $P(A_j^n | B)$ is defined by (10.30) and

$$P(E_r | A_j^n B_r) = \begin{cases} 1 & k_1 \geq k_{1j}^n \text{ and } k_2 \geq k_{2j}^n, \\ 0 & k_1 < k_{1j}^n \text{or } k_2 < k_{2j}^n. \end{cases} \tag{10.57}$$

The total reaching probability is

$$P(E_r | B_r) = \sum_{i=0}^{\infty} P(E_r A^i | B_r). \tag{10.58}$$

Considering that when t_r is a constant, the stochastic parameter jump number j is limited, we have an approximation that

$$P(E_r|B_r) = \sum_{j=0}^{m} P(E_r A^j|B_r), \tag{10.59}$$

where m is a positive integer. In practice, we can set a rule that if $\sum_{i=0}^{m} P(E_r A^i|B_r) > 1 - \varepsilon$ on condition $k_1 = \max_{j\in S}\{k_{1j}\}$, $k_2 = \max_{j\in S}\{k_{2j}\}$, where ε is a small constant, then we can omit the residual items $P(E_r A^j|B_r)$, $j > m$. This completes the proof. □

In the sliding stage, the control target is to keep the system state sliding on the switching surface. Insufficient control force may result in violation of the existence condition of the sliding mode $\dot{s}s < 0$ in the neighborhood of $s = 0$.

The following theorem is given for the sliding probability.

Theorem 10.3 *For the SISO linear MJS (10.1) under SMC (10.39), the sliding probability is given as follows*

$$P(E_s|B_s) = \sum_{j\in S} P(E_s|A_j B_s)P(A_j|B_s) \tag{10.60}$$

where

$$P(E_s|A_j B_s) = \begin{cases} 1 & k_1 \geq k_{1j} \textit{ and } k_2 \geq k_{2j}, \\ 0 & k_1 < k_{1j} \textit{or } k_2 < k_{2j}. \end{cases} \tag{10.61}$$

Proof In the sliding stage, since the system state is already on the switching surface, the control target is to keep the system state to stay on it at any time.

For each time t, $\eta_t = j$, $j \in S$, as seen from (10.46), the MJS under SMC (10.39) can ensure that the system state stays in the switching surface. However, if the control parameters k_1 and k_2 are somehow insufficient, we get

$$P(E_s|A_j B_s) = \begin{cases} 1 & k_1 \geq k_{1j} \text{ and } k_2 \geq k_{2j}, \\ 0 & k_1 < k_{1j} \text{or } k_2 < k_{2j}. \end{cases} \tag{10.62}$$

Synthesizing all possibilities, we then have the sliding probability calculated as

$$P(E_s|B_s) = \sum_{j\in S} P(E_s|A_j B_s)P(A_j|B_s). \tag{10.63}$$

This completes the proof. □

10.3 Sliding Mode Control of MIMO Linear Markovian Jump Systems

10.3.1 Problem Statement and Preliminaries

The MIMO linear MJS under investigation is

$$\begin{cases} \dot{X}_1(t) = (A_{11}(\eta_t) + \Delta_{11}(\eta_t))X_1(t) \\ \qquad\quad +(A_{12}(\eta_t) + \Delta_{12}(\eta_t))X_2(t), \\ \dot{X}_2(t) = (A_{21}(\eta_t) + \Delta_{21}(\eta_t))X_1(t) \\ \qquad\quad +(A_{22}(\eta_t) + \Delta_{22}(\eta_t))X_2(t) + B_2(\eta_t)U(t), \\ Y(t) \;\; = X_1(t), \\ \eta_0 \quad\;\; = s_0, \;\; t \geq 0, \end{cases} \tag{10.64}$$

where $X(t) = [X_1^T(t),\ X_2^T(t)]^T$ is the system state, $X_1 \in R^{(n-m)}$, $X_2 \in R^m$; $U(t) \in R^m$ is the control input; $Y(t) \in R^{(n-m)}$ is the system output; $A_{ij}(\eta_t), i, j = 1, 2,\ B_2(\eta_t)$ are stochastic coefficient matrices with compatible dimensions, $\Delta_{ij}(\eta_t), i, j = 1, 2$ are stochastic uncertain matrices and $\{\eta_t, t \in [0, T]\}$ is a finite-state Markovian process having a state-space $S = \{1, 2, \ldots, \nu\}$, $\det(B_2(j)) \neq 0,\ j \in S$, generator (q_{ij}) with transition probability from mode i at time t to mode j at time $t + \delta$, $i, j \in S$,

$$\begin{aligned} P_{ij}(\delta) &= P(\eta_{t+\delta} = j | \eta_t = i) \\ &= \begin{cases} q_{ij}\delta + o(\delta), & \text{if } i \neq j, \\ 1 + q_{ii}\delta + o(\delta), & \text{if } i = j, \end{cases} \end{aligned} \tag{10.65}$$

where

$$q_{ii} = -\sum_{m=1, m\neq i}^{\nu} q_{im},\ q_{ij} \geq 0, \forall i, j \in S, i \neq j, \tag{10.66}$$

$\delta > 0$ and $\lim\limits_{\delta \to 0} o(\delta)/\delta = 0$.
Assumption 1. The uncertain matrices satisfy

$$\|\Delta_{ij}(k)\| \leq \delta_{ij}(k),\ i, j = 1, \ldots, 2,\ k = 1, \ldots, \nu, \tag{10.67}$$

where $\delta_{ij}(k)$ are known constants.

Definition 10.2 The system (10.64) is called mean-square stable, if for each $\varepsilon > 0$, there exists $\delta > 0$, such that

$$\sup_{t_0 \leq t < \infty} \mathbf{E}\|X(t)\|^2 < \varepsilon, \text{ for all } \|X(t_0)\| < \delta. \tag{10.68}$$

In addition, the system (10.64) is called asymptotically mean-square stable, if it is mean-square stable and

$$\lim_{t\to\infty} \mathbf{E}\|X(t)\|^2 = 0, \text{ for all } \|X(t_0)\| < \delta. \tag{10.69}$$

Furthermore, if Eq. (10.12) holds for arbitrary positive constant δ, then the system (1) is called globally asymptotically mean-square stable.

The sliding surfaces are defined as

$$S(X, \eta_t) = C_1(\eta_t)X_1 + X_2 = \mathbf{0}. \tag{10.70}$$

Denotes $\mathscr{L}$ as the weak infinitesimal operator of the process $\{X_1(t), \eta_t, t \geq 0\}$ at the point $\{t, X_1, j\}$. Substituting (10.70) and $\eta_t = j$ into system (10.64) yields

$$\begin{cases} \dot{X}_1 = \bar{A}_{11}(j)X_1 + \bar{A}_{12}(j)S, \\ \mathscr{L}S = \bar{A}_{21}(j)X_1 + \bar{A}_{22}(j)S + B_2(j)U(t), \\ \eta_0 = s_0, \ t \geq 0, \end{cases} \tag{10.71}$$

where $\eta_t = j, j = 1, \ldots v$, and

$$\begin{aligned} \bar{A}_{11}(j) &= A_{11}(j) + \Delta_{11}(j) - (A_{12}(j) + \Delta_{12}(j))C_1(j), \\ \bar{A}_{12}(j) &= A_{12}(j) + \Delta_{12}(j), \\ \bar{A}_{21}(j) &= C_1(j)(A_{11}(j) + \Delta_{11}(j)) \\ &\quad -C_1(j)(A_{12}(j) + \Delta_{12}(j))C_1(j) \\ &\quad +(A_{21}(j) + \Delta_{21}(j)) \\ &\quad -(A_{22}(j) + \Delta_{22}(j))C_1(j) + \sum_{i=1}^{v} \alpha_{ji}C_1(i), \\ \bar{A}_{22}(j) &= C_1(j)(A_{12}(j) + \Delta_{12}(j)) + (A_{22}(j) + \Delta_{22}(j)). \end{aligned} \tag{10.72}$$

An important step for analyzing the probability problems of MJSs under SMC is to derive multi-step state transition probability. Some events are defined as follows.

$$B: \text{ The initial condition is } \eta_0 = s_0. \tag{10.73}$$

$$A_j: \ \eta_t = j, 0 \leq t \leq t_r. \tag{10.74}$$

$$A^0(t): \text{ The stochastic process parameter does not jump, } \eta_\tau = s_0, 0 \leq \tau \leq t, \tag{10.75}$$

$A_j^k(t)$: The stochastic process parameter jumps k times

$$\eta_\tau = \begin{cases} s_0 \ 0 \le \tau < t_1, \\ s_1 \ t_1 \le \tau < t_2, \\ \vdots \\ s_k \ t_k \le \tau \le t. \end{cases}$$

where the route $r_j^k = (s_0, \ldots, s_k) \in S^k$ (10.76)

$$S^k = \{(s_0, s_1, \ldots, s_k) | s_i \in S, s_{i-1} \neq s_i, \\ 1 \le i \le k\}. \tag{10.77}$$

$A^k(t)$: The stochastic process parameter jumps k times

$$A^k(t) = \bigcup_{r_j^k \in S^k} A_j^k(t), k = 1, 2, \ldots, n. \tag{10.78}$$

10.3.2 Control Scheme and Stability Probability

There are two key steps in designing an SMC scheme. The first step is to design sliding surfaces which have desired system dynamics performance (e.g., asymptotical stability). The second step is to design a discontinuous controller which can drive the system to reach the sliding surface in a finite time.

At the beginning, we design sliding surfaces (10.70) and give sufficient conditions in terms of LMIs which can guarantee that the sliding motion is asymptotically stable. The result is given in the following theorem.

Theorem 10.4 ([30]) *The system (10.64) is asymptotically stable on the sliding surface (10.70), if there exist symmetric positive-definite matrices $P(j)$, general matrices $Q(j)$ and positive scalars $b_1(j)$, $b_2(j)$ such that the following inequalities hold for all $j \in S$.*

$$\begin{bmatrix} N_{11}(j) & N_{12}(j) \\ N_{12}^T(j) & N_{22}(j) \end{bmatrix} < 0, \tag{10.79}$$

where

$$\begin{aligned} N_{11}(j) &= A_{11}(j)P(j) + P(j)A_{11}^T(j) - A_{12}(j)Q(j) \\ &\quad - Q^T(j)A_{12}^T(j) + \alpha_{jj}P(j) + b_1(j)\delta_{11}^2(j)I \\ &\quad + b_2(j)\delta_{12}^2(j)I, &(10.80)\\ N_{12}(j) &= [\alpha_{j1}^{1/2}P(j), \ \ldots, \ \alpha_{jj-1}^{1/2}P(j), \ \alpha_{jj+1}^{1/2}P(j), \\ &\quad \ldots, \ \alpha_{jv}^{1/2}P(j), \ P(j), \ Q^T(j)], &(10.81)\\ N_{22}(j) &= diag(-P(1), \ \ldots, -P(j-1), \ -P(j+1), \\ &\quad -P(v), \ -b_1(j)I, \ -b_2(j)I), &(10.82)\\ Q(j) &= C_1(j)P(j). &(10.83) \end{aligned}$$

Proof By Schur complement lemma, we get that (10.79) is equivalent to

$$\begin{aligned}&A_{11}(j)P(j)+P(j)A_{11}^T(j)-A_{12}(j)Q(j)-Q^T(j)A_{12}^T(j)\\&+\alpha_{jj}P(j)+b_1\delta_{11}^2I+b_2\delta_{12}^2I+\sum_{i=1,i\neq j}^{v}\alpha_{ji}P(j)P^{-1}(i)P(j)\\&+b_1^{-1}P^2(j)+b_2^{-1}Q^T(j)Q(j)<0.\end{aligned}\tag{10.84}$$

It leads to

$$\begin{aligned}&A_{11}(j)P(j)+P(j)A_{11}^T(j)-A_{12}(j)Q(j)-Q^T(j)A_{12}^T(j)\\&+\alpha_{jj}P(j)+b_1\Delta_{11}\Delta_{11}^T+b_2\Delta_{12}\Delta_{12}^T\\&+\sum_{i=1,i\neq j}^{v}\alpha_{ji}P(j)P^{-1}(i)P(j)+b_1^{-1}P^2(j)\\&+b_2^{-1}Q^T(j)Q(j)<0.\end{aligned}\tag{10.85}$$

Let $M(j)=P^{-1}(j)$. Pre- and post multiplied by $M(j)$, it gives

$$\begin{aligned}&M(j)A_{11}(j)+A_{11}^T(j)M(j)-M(j)A_{12}(j)C_1(j)\\&-C_1^T(j)A_{12}^T(j)M(j)+\sum_{i=1}^{v}\alpha_{ji}M(i)\\&+b_1M_j\Delta_{11}\Delta_{11}^TM_j+b_2M_j\Delta_{12}\Delta_{12}^TM_j\\&+b_1^{-1}I+b_2^{-1}C_1^T(j)C_1(j)<0.\end{aligned}\tag{10.86}$$

The Lyapunov function candidate is given as

$$V_1=X_1^TM(j)X_1.\tag{10.87}$$

The weak infinitesimal operator $\mathscr{L}$ of the process $\{X_1(t),\eta_t,t\geq 0\}$ at the point $\{t,X_1,j\}$ is given by

$$\begin{aligned}\mathscr{L}V_1&=2X_1^TM(j)\dot{X}_1+\sum_{i=1}^{v}X_1^T\alpha_{ji}M(i)X_1\\&=2X_1^TM(j)(A_{11}(j)+\Delta_{11}(j)-A_{12}(j)C_1(j)\\&\quad-\Delta_{12}(j)C_1(j))X_1+\sum_{i=1}^{v}X_1^T\alpha_{ji}M(i)X_1.\end{aligned}\tag{10.88}$$

By (10.86), we get $\mathscr{L}V_1<0$, if $X_1\neq\mathbf{0}$. Referring (10.70), it can be drawn that the system state $[X_1^T,X_2^T]^T$ is stochastically asymptotically stable. This completes the proof. □

There are two phases in sliding mode dynamics, the reaching phase and sliding phase. If the control force is large enough, it can suppress any matched uncertainties to realize the sliding motion. However, if it is not sufficiently large, it will affect the reaching and sliding abilities of the SMC, that is, the reaching ability and sliding ability would be in a probabilistic sense. In this section, we will derive the conditions to guarantee the globally asymptotical stability of the MJSs under SMC. We will also derive the reaching probability function and the sliding probability function. For convenience, we introduce the *reaching probability* [29], denoted by $P(E_r|B_r)$, which represents the condition probability of the system reaching the sliding mode in a finite time t_r, and the *sliding probability* [29], denoted by $P(E_s|B_s)$, which represents the condition probability of the system maintaining on the sliding mode.

Some events are defined as

$$\begin{aligned} B_r : &\text{ The condition is } \eta_0 = s_0, X(0) = X_0, \\ &U(t) = F_u(t), 0 \le t \le t_r. \qquad (10.89)\\ B_s : &\text{ The condition is } \eta_{t_r} = \eta_{t_r}, X(t_r) = X_{t_r}, \\ &S(t_r) = \mathbf{0}, U(t) = F_u(t), t > t_r. \qquad (10.90)\\ E_r : &\text{ The system first reaches the sliding mode} \\ &\text{at time } t_r,\ S(t_r) = \mathbf{0}. \qquad (10.91)\\ E_s : &\text{ The system maintains on the sliding mode,} \\ &S(t) = \mathbf{0},\ t > t_r. \qquad (10.92) \end{aligned}$$

where t_r is a given constant, $F_u(t)$ represents the designed control input.

The following SMC is proposed for the MIMO MJS as

$$U(t) = \begin{cases} U_{eq}(t) - B_2^{-1}(j)\left(k_1\|X_1\|\frac{S}{\|S\|} + k_2 S + k_r\frac{S}{\|S\|}\right) \\ \qquad\qquad\qquad\qquad S \neq \mathbf{0}, \\ \mathbf{0} \qquad\qquad\qquad\quad S = \mathbf{0}, \end{cases}$$

where variable S is defined by (10.70), k_1, k_2 are control parameters, and $U_{eq}(t)$ is the equivalent control defined as

$$\begin{aligned} U_{eq}(t) = &-B_2^{-1}(j)[(C_1A_{11}(j) - C_1A_{12}(j)C_1 + A_{21}(j) \\ &-A_{22}(j)C_1 + \sum_{i=1}^{v}\alpha_{ji}C_1(i))X_1 \\ &+(C_1A_{12}(j) + A_{22}(j))S]. \end{aligned} \qquad (10.93)$$

The following theorem is given.

Theorem 10.5 *Assume the condition in Theorem 10.4 holds, and SMC is designed as (10.93), the close-loop system is asymptotically stable, if the control parameters in (10.93) are given as follows.*

$$k_1 = \max_{j\in S}\{k_{1j}\},\ k_2 = \max_{j\in S}\{k_{2j}\}, \tag{10.94}$$

$$k_{1j} = \|C_1(j)\|\delta_{11}(j) + \|C_1(j)\|^2\delta_{12}(j) + \delta_{21}(j) + \|C_1(j)\|\delta_{22}(j), \tag{10.95}$$

$$k_{2j} = \|C_1(j)\|\delta_{12}(j) + \delta_{22}(j),\ j = 1, \ldots, \nu. \tag{10.96}$$

Proof The Lyapunov function candidate is given as

$$V(S) = \frac{1}{2}S^T(X)S(X). \tag{10.97}$$

Substituting SMC (10.93) and (10.94) into it, the weak infinitesimal operator $\mathscr{L}$ of the process $\{X_1(t), \eta_t, t \geq 0\}$ at the point $\{t, X_1, j\}$ is given by

$$\begin{aligned}\mathscr{L}V(S) &= S^T[\bar{A}_{21}(j)X_1 + \bar{A}_{22}(j)S + B_2(j)U(t)]\\ &= S^T[(C_1\Delta_{11}(j) - C_1\Delta_{12}(j)C_1 + \Delta_{21}(j)\\ &\quad -\Delta_{22}(j)C_1)X_1 + (C_1\Delta_{12}(j) + \Delta_{22}(j))S]\\ &\quad -S^T\left(k_1\|X_1\|\frac{S}{\|S\|} + k_2 S + k_r\frac{S}{\|S\|}\right)\\ &\leq -k_r\|S\| \leq 0.\end{aligned} \tag{10.98}$$

The equality holds if and only if $S = \mathbf{0}$. It means under SMC (10.93), the system can reach the switching surface in a finite time $t_r = \frac{\|S(0)\|}{k_r}$ and then maintain its state on the switching surface afterwards.

Here a question arises. If the control is insufficient, what will the reaching probability and sliding probability be? This question is answered in the following.

In the reaching phase, the control objective is to drive the system state to reach the sliding mode in finite time t_r. With the following definitions of auxiliary parameters

$$k_{1j}^n = \max_{i\in r_j^n}\{k_{1i}\},\ k_{2j}^n = \max_{i\in r_j^n}\{k_{2i}\}, \tag{10.99}$$

where k_{1i}, k_{2i} are defined by (10.95) and (10.96), and $r_j^n = (s_{j0}, \ldots, s_{jn}) \in S^{n+1}$ represents the Markovian process state transition route, we derive the reaching probability formulae.

Theorem 10.6 ([30]) *For the MIMO linear MJS (10.64) under SMC (10.93), the reaching probability is given as follows*

$$P(E_r|B_r) = \sum_{n=0}^{m} P(E_rA^n|B_r), \tag{10.100}$$

where

$$m = arg \min_{n} \left\{ \sum_{i=0}^{n} P(E_r A^i | B_r) > 1 - \varepsilon,\ 0 < \varepsilon \ll 1, \right.$$

$$\left. if\, k_1 = \max_{j \in S} \{k_{1j}\},\, k_2 = \max_{j \in S} \{k_{2j}\} \right\}, \tag{10.101}$$

$$P(E_r A^n | B) = \sum_{r_j^n \subset S_n} P(E_r | A_j^n B) P(A_j^n | B), \tag{10.102}$$

$$P(E_r | A_j^n B) = \begin{cases} 1 & k_1 \geq k_{1j}^n \text{ and } k_2 \geq k_{2j}^n, \\ 0 & k_1 < k_{1j}^n \text{or } k_2 < k_{2j}^n. \end{cases} \tag{10.103}$$

Proof The proof is similar to the proof of Theorem 10.2, so it is omitted here. □

In the sliding phase, the control objective is to keep the system state sliding on the switching surface. Insufficient control force may result in violation of the existence condition of the sliding mode $S^T \dot{S} < 0$ in the neighborhood of $S = \mathbf{0}$.

The following theorem is given for the sliding probability.

Theorem 10.7 *For the MIMO linear MJS (10.64) under SMC (10.93), the sliding probability is given as follows*

$$P(E_s | B_s) = \sum_{j \in S} P(E_s | A_j B_s) P(A_j | B_s) \tag{10.104}$$

where

$$P(E_s | A_j B_s) = \begin{cases} 1 & k_1 \geq k_{1j} \text{ and } k_2 \geq k_{2j}, \\ 0 & k_1 < k_{1j} \text{or } k_2 < k_{2j}. \end{cases} \tag{10.105}$$

Proof The proof is similar to the proof of Theorem 10.3, so it is omitted here. □

10.4 Numerical Simulations

In this section, we present numerical simulations to show the effectiveness of the theoretical results.

We consider a second-order MJS (10.1), where $\eta_t \in S$ is a continuous Markovian process and

$$S = \{1, 2, 3, 4\},\ A(i) = \begin{bmatrix} A_{11} & A_{12} \\ A_{21}(i) & A_{22}(i) \end{bmatrix},$$

$$A(1) = \begin{bmatrix} 0 & 1 \\ 3 & 5 \end{bmatrix}, \quad B_2(1) = 1,$$
$$A(2) = \begin{bmatrix} 0 & 1 \\ 2 & 4 \end{bmatrix}, \quad B_2(2) = 2,$$
$$A(3) = \begin{bmatrix} 0 & 1 \\ 5 & 3 \end{bmatrix}, \quad B_2(3) = 1,$$
$$A(4) = \begin{bmatrix} 0 & 1 \\ 4 & 2 \end{bmatrix}, \quad B_2(4) = 2,$$

generator (α_{ij}) with transition probability from mode i at time t to mode j at time $t + \Delta, i, j \in S$ as

$$(\alpha_{ij}) = \begin{bmatrix} -8 & 3 & 2 & 3 \\ 1 & -1 & 0 & 0 \\ 2 & 1 & -5 & 2 \\ 0 & 2 & 2 & -4 \end{bmatrix}$$

The sliding variable is defined as

$$s = C_1 x_1 + x_2, \ C_1 = 2.$$

It is obvious that

$$A_{11} - A_{12} C_1 < 0, \ i = 1, \ldots, 4.$$

The initial parameters are set that $x(0) = [0.5 \ \ -0.2]^T$, $\eta_0 = 1$, and $t_r = 0.1$.
By (10.17) and (10.18), we get

$$q_{ij} = \alpha_{ij}, \ q_i = -\alpha_{ii}, \ 1 \leq i, j \leq 4,$$
$$q_1 = 8, \ q_2 = 1, \ q_3 = 4, \ q_4 = 4.$$

Using $\hat{P}_j = P(\eta_{\tau_1} = j | \eta_0 = i)$ to estimate $P(\eta_t = j | \eta_0 = i)$, we get

$$\hat{P}_1 = 0.5, \ \hat{P}_2 = 0.3, \ \hat{P}_2 = 0.125, \ \hat{P}_4 = 0.125.$$

The SMC law is given as (10.39).
According to Theorem 10.1, we have

$$k_{11} = 3.3750, \ k_{12} = 2.6250, \ k_{13} = 2.6250, \ k_{14} = 5.6250,$$
$$k_{21} = 1.8750, \ k_{22} = 2.1250, \ k_{23} = 0.1250, \ k_{24} = 3.1250,$$

If control parameters satisfy $k_1 \geq 5.6250$, $k_2 \geq 3.1250$, then the system is asymptotically stable. The results are shown in Figs. 10.4 and 10.5.

Table 10.1 xxx

j	r_j^1	k_{1j}^1	k_{2j}^1	$P(A_j^1\|B)$
1	(1,2)	3.3750	2.1250	0.2644
2	(1,3)	3.3750	1.8750	0.1107
3	(1,4)	5.6250	3.1250	0.1856

Next, we study the stability probability when the control is insufficient. It is investigated in two stages.

For the reaching state ($0 \le t \le t_r$), first, consider that the stochastic parameter does not jump. From Theorem 10.2, we get

$$P(EA^0|B) = P(E|A^0B)P(A^0|B) = \begin{cases} 0.2019 & k_1 \ge 3.3750 \text{ and } k_2 \ge 1.8750, \\ 0 & k_1 < 3.3750 \text{ or } k_2 < 1.8750, \end{cases}$$

where

$$P(E|A^0B) = \begin{cases} 1 & k_1 \ge 3.3750 \text{ and } k_2 \ge 1.8750, \\ 0 & k_1 < 3.3750 \text{ or } k_2 < 1.8750, \end{cases}$$

and $P(A^0|B)$ is defined by (10.29).

However, if the stochastic parameter jumps n times, we have the following scenarios.

For case $n = 1$, there are three parameter transition routes which are

$$r_1^1 = (1,2),\ r_2^1 = (1,3),\ r_3^1 = (1,4).$$

From Theorem 10.2, we have

$$\begin{aligned} k_{11}^1 &= \max\{k_{11}, k_{12}\} = 3.3750, \\ k_{21}^1 &= \max\{k_{21}, k_{22}\} = 2.1250, \\ k_{12}^1 &= \max\{k_{11}, k_{13}\} = 3.3750, \\ k_{22}^1 &= \max\{k_{21}, k_{23}\} = 1.8750, \\ k_{13}^1 &= \max\{k_{11}, k_{14}\} = 5.6250, \\ k_{23}^1 &= \max\{k_{21}, k_{24}\} = 3.1250. \end{aligned}$$

From Lemma 10.1, we have the solution of $P(A_j^1|B)$, which leads to (Tables 10.1 and 10.2)

Table 10.2 xxx

j	r_j^2	k_{1j}^2	k_{2j}^2	$P(A_j^2\|B)$
1	(1,2,1)	3.3750	2.1250	0.0205
2	(1,2,3)	3.3750	2.1250	0
3	(1,2,4)	5.6250	3.1250	0
4	(1,3,1)	3.3750	1.8750	0.0199
5	(1,3,2)	3.3750	2.1250	0.0164
6	(1,3,4)	5.6250	3.1250	0.0261
7	(1,4,1)	5.6250	3.1250	0
8	(1,4,2)	5.6250	3.1250	0.0525
9	(1,4,3)	5.6250	3.1250	0.0392

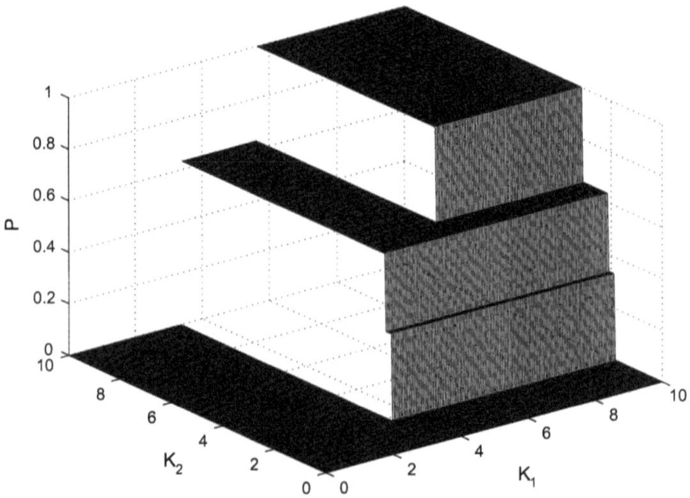

Fig. 10.1 Reaching probability

Thus, it gives that

$$P(EA^1|B) = \sum_{r_j^1 \in S_1} P(E|A_j^1 B)P(A_j^1|B)$$

$$= \begin{cases} 0 & k_1 < 3.3750 \text{ or } k_2 < 1.8750, \\ 0.1105 & 3.3750 \le k_1 < 5.6250 \text{ and } 1.8750 \le k_2 < 2.1250, \\ 0.3751 & \begin{array}{l} 3.3750 \le k_1 < 5.6250 \text{ and } k_2 \ge 2.1250, \\ \text{or } 2.1250 \le k_2 < 3.1250 \text{ and } k_1 \ge 3.3750, \end{array} \\ 0.5607 & k_1 \ge 5.6250 \text{ and } k_2 \ge 3.1250. \end{cases}$$

For case $n = 2$, similar to the above, we get

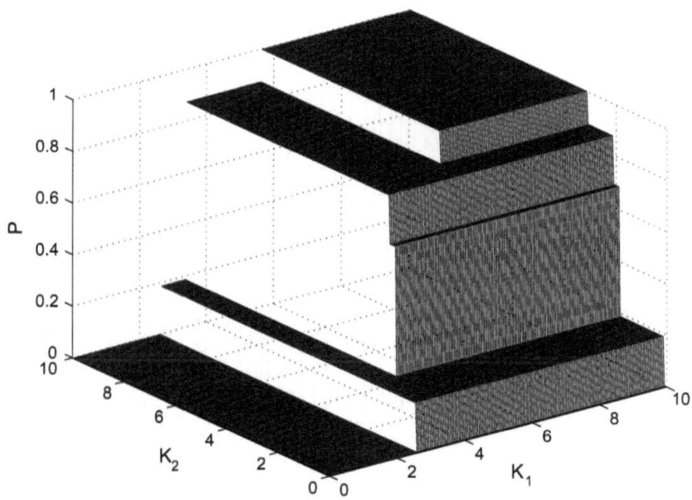

Fig. 10.2 Sliding probability

$$P(EA^2|B) = \sum_{r_j^2 \in S_2} P(E|A_j^2 B)P(A_j^2|B)$$

$$= \begin{cases} 0 & k_1 < 3.3750 \text{or } k_2 < 1.8750, \\ 0.0199 & 3.3750 \leq k_1 < 5.6250 \text{ and } 1.8750 \leq k_2 \leq 2.1250, \\ 0.0568 & \begin{array}{l} 3.3750 \leq k_1 < 5.6250 \text{ and } k_2 \geq 2.1250, \\ \text{or } k_1 \geq 3.3750 \text{ and } 2.1250 \leq k_2 \leq 3.1250, \end{array} \\ 0.1746 & k_1 \geq 5.6250 \text{ and } k_2 \geq 3.1250, \end{cases}$$

Finally, for case $n = 3$, the same derivations can be done and we have the total reaching probability approximated as

$$P(E_r|B_r) = \sum_{i=0}^{3} P(EA^i|B)$$

$$= \begin{cases} 0 & k_1 < 3.3750 \text{ or } k_2 < 1.8750, \\ 0.3323 & 3.3750 \leq k_1 \text{ and } 1.8750 \leq k_2 < 2.1250, \\ 0.6338 & \begin{array}{l} 3.3750 \leq k_1 < 5.6250 \text{ and } k_2 \geq 2.1250, \\ \text{or } 2.1250 \leq k_2 < 3.1250 \text{ and } k_1 \geq 3.3750, \end{array} \\ 1.0 & k_1 \geq 5.6250 \text{ and } k_2 \geq 3.1250, \end{cases}$$

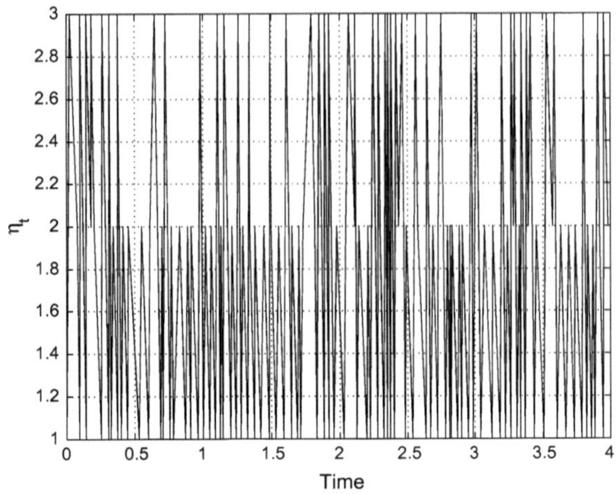

Fig. 10.3 Continuous Markovian process

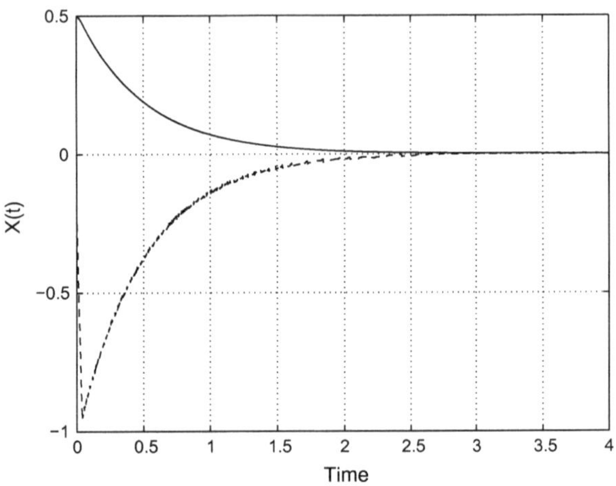

Fig. 10.4 System state

For the sliding stage($t > t_r$), let $t_s = 0.2$. By Theorem 10.1 and $\hat{P}_j = P_{ij}(t_s)$, we get

$$\hat{P}_1 = 0.5000,\ \ \hat{P}_2 = 0.1875,\ \ \hat{P}_3 = 0.1875,\ \ \hat{P}_4 = 0.1250.$$

Referring to Theorem 10.3, we get the sliding probability

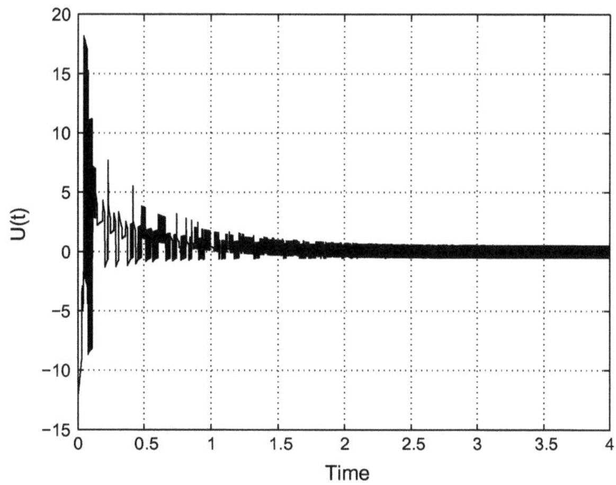

Fig. 10.5 Control input

$$
P(E_s|B_s) = \sum_{j\in S} P(E_s|A_jB_s)P(A_j|B_s)
$$

$$
= \begin{cases} 0 & k_1 < 2.6250 \text{ or } k_2 < 0.1250, \\ 0.1875 & \begin{array}{l} k_1 \geq 2.6250 \text{ and } 0.1250 \leq k_2 < 1.8750, \\ \text{or } 2.6250 \leq k_1 < 3.3750 \text{ and } k_2 \geq 0.1250, \end{array} \\ 0.6875 & k_1 \geq 3.3750 \text{ and } 1.8750 \leq k_2 < 2.1250, \\ 0.8750 & \begin{array}{l} k_1 \geq 3.3750 \text{ and } 2.1250 \leq k_2 < 3.1250, \\ \text{or } 3.3750 \leq k_1 < 5.6250 \text{ and } k_2 \geq 3.1250, \end{array} \\ 1.0 & k_1 \geq 5.6250 \text{ and } k_2 \geq 3.1250, \end{cases}
$$

The results are shown in Figs. 10.1, 10.2, 10.3, 10.4 and 10.5.

Acknowledgements The authors would like to thank the editor and the reviewers for their helpful and valuable comments and suggestions, which have improved the presentation. This work is supported by National Natural Science Foundation of China under Grant no's. 61573307, 61273352, 61473249, 61473250, 61175111, 61174046, and the Australian Research Council (No. DP140100544.)

References

1. Ji, Y., Chizeck, H.J.: Controllability, stabilizability, and continuous-time Markovian jump linear quadratic control. IEEE Trans. Autom. Control **35**(7), 777–788 (1990)
2. De Farias, D.P., Geromel, J.C., Do Val, J.B.R., Costa, O.L.V.: Output feedback control of Markov jump linear systems in continuous-time. IEEE Trans. Autom. Control **45**(5), 944–949 (2000)
3. Chen, W.H., Xu, J.X., Guan, Z.H.: Guaranteed cost control for uncertain Markovian jump systems with mode-dependent time-delays. IEEE Trans. Autom. Control **48**(12), 2270–2277 (2003)

4. Xiong, J., Lam, J., Gao, H., Ho, D.W.: On robust stabilization of Markovian jump systems with uncertain switching probabilities. Automatica **41**(5), 897–903 (2005)
5. Zhang, L., Boukas, E.K.: Stability and stabilization of Markovian jump linear systems with partly unknown transition probabilities. Automatica **45**(2), 463–468 (2009)
6. Xu, S., Lam, J., Mao, X.: Delay-dependent $H\infty$ control and filtering for uncertain Markovian jump systems with time-varying delays. IEEE Trans. Circuits Syst. I: Regular Papers **54**(9), 2070–2077 (2007)
7. Mahmoud, M.S., Shi, P.: Robust Kalman filtering for continuous time-lag systems with Markovian jump parameters. IEEE Trans. Circuits Syst. I: Fundam. Theory Appl. **50**(1), 98–105 (2003)
8. Wang, Z., Lam, J., Liu, X.: Exponential filtering for uncertain Markovian jump time-delay systems with nonlinear disturbances. IEEE Trans. Circuits Syst. II: Express Briefs **51**(5), 262–268 (2004)
9. Xu, S., Chen, T., Lam, J.: Robust $H\infty$ filtering for uncertain Markovian jump systems with mode-dependent time delays. IEEE Trans. Autom. Control **48**(5), 900–907 (2003)
10. Shi, P., Boukas, E., Agarwal, R.K.: Kalman filtering for continuous-time uncertain systems with Markovian jumping parameters. IEEE Trans. Autom. Control **44**(8), 1592–1597 (1999)
11. Wu, L., Shi, P., Gao, H.: State estimation and sliding-mode control of Markovian jump singular systems. IEEE Trans. Autom. Control **55**(5), 1213–1219 (2010)
12. Shi, P., Xia, Y., Liu, G.P., Rees, D.: On designing of sliding-mode control for stochastic jump systems. IEEE Trans. Autom. Control **51**(1), 97–103 (2006)
13. Niu, Y., Ho, D.W.: Design of sliding mode control subject to packet losses. IEEE Trans. Autom. Control **55**(11), 2623–2628 (2010)
14. Yu, X.: Sliding-mode control with soft computing: a survey. IEEE Trans. Ind. Electron. **56**(9), 3275–3285 (2009)
15. Chen, B., Niu, Y., Zou, Y.: Adaptive sliding mode control for stochastic Markovian jumping systems with actuator degradation. Automatica **49**(6), 1748–1754 (2013)
16. Chen, B., Niu, Y., Zou, Y.: Sliding mode control for stochastic Markovian jumping systems with incomplete transition rate. IET Control Theory Appl. **7**(10), 1330–1338 (2013)
17. Kao, Y., Li, W., Wang, C.: Nonfragile observer-based $H\infty$ sliding mode control for Ito stochastic systems with Markovian switching. Int. J. Robust Nonlinear Control (2013). https://doi.org/10.1002/rnc.2970
18. Niu, Y., Ho, D.W., Wang, X.: Sliding mode control for Ito stochastic systems with Markovian switching. Automatica **43**(10), 1784–1790 (2007)
19. Wu, Z.G., Shi, P., Su, H., Chu, J.: Asynchronous $l_2 - l\infty$ filtering for discrete-time stochastic Markov jump systems with randomly occurred sensor nonlinearities. Automatica **50**(1), 180–186 (2014)
20. Yin, Y., Shi, P., Liu, F., Teo, K.L.: Observer-based $H\infty$ control on nonhomogeneous Markov jump systems with nonlinear input. Int. J. Robust Nonlinear Control (2013). https://doi.org/10.1002/rnc.2974
21. Utkin, V.I., Poznyak, A.S.: Adaptive sliding mode control with application to super-twist algorithm: Equivalent control method. Automatica **49**(1), 39–47 (2013)
22. Zhang, H., Wang, J., Shi, Y.: Robust $H\infty$ sliding-mode control for Markovian jump systems subject to intermittent observations and partially known transition probabilities. Syst. Control Lett. **62**(12), 1114–1124 (2013)
23. Luan, X., Shi, P., Liu, F.: Finite-time stabilisation for Markov jump systems with Gaussian transition probabilities. IET Control Theory Appl. **7**(2), 298–304 (2013)
24. Wu, L., Ho, D.W.C.: Sliding mode control of singular stochastic hybrid systems. Automatica **46**(4), 779–783 (2010)
25. Wu, L., Su, X., Shi, P.: Sliding mode control with bounded L2 gain performance of Markovian jump singular time-delay systems. Automatica **48**(8), 1929–1933 (2012)
26. Zhu, Q., Yu, X., Song, A. et al.: Stability Probability in Sliding Mode Control of Second Order Markovian Jump Systems. CCDC China (2014)
27. Lin, Y.L.: Applied Stochastic Processes. Tsinghua University Press, Beijing (2001). ISBN 978-7-30205-958-5

28. Ross, S.M.: Introduction to Probability Models. Academic Press, Burlington (2010). ISBN 978-0-12-598062-3
29. Zhu, Q., Yu, X., Song, A., et al.: On sliding mode control of single input Markovian jump systems. Automatica **50**(11), 2897–2904 (2014)
30. Zhu, J., Yu, X., Zhang, T., et al.: Sliding mode control of MIMO Markovian jump systems. Automatica **68**, 286–293 (2016)
31. Karan, M., Shi, P., Kaya, C.Y.: Transition probability bounds for the stochastic stability robustness of continuous-and discrete-time Markovian jump linear systems. Automatica **42**(12), 2159–2168 (2006)
32. Wang, Z., Qiao, H., Burnham, K.J.: On stabilization of bilinear uncertain time-delay stochastic systems with Markovian jumping parameters. IEEE Trans. Autom. Control **47**(4), 640–646 (2002)
33. Zhang, L., Huang, B., Lam, J.: $H\infty$ model reduction of Markovian jump linear systems. Syst. Control Lett. **50**(2), 103–118 (2003)
34. Shi, P., Boukas, E.K.: $H\infty$-control for Markovian jumping linear systems with parametric uncertainty. J. Optim. Theory Appl. **95**(1), 75–99 (1997)
35. Wang, Z., Liu, Y., Yu, L., Liu, X.: Exponential stability of delayed recurrent neural networks with Markovian jumping parameters. Phys. Lett. A **356**(4), 346–352 (2006)
36. Mao, X.: Exponential stability of stochastic delay interval systems with Markovian switching. IEEE Trans. Autom. Control **47**(10), 1604–1612 (2002)
37. Yin, S., Li, X., Gao, H., Kaynak, O.: Data-based techniques focused on modern industry: an overview. IEEE Trans. Ind. Electron. (2014). https://doi.org/10.1109/TIE.2014.2308133
38. Yin, S., Ding, S., Xie, X., Luo, H.: A review on basic data-driven approaches for industrial process monitoring. IEEE Trans. Ind. Electron. (2014). https://doi.org/10.1109/TIE.2014.2301773

Chapter 11
Adaptive Sliding Mode Control Using Monitoring Functions

Liu Hsu, Tiago Roux Oliveira, Gabriel Tavares Melo and José Paulo V. S. Cunha

Abstract In this chapter, we propose an adaptive sliding mode control approach based on monitoring functions, to deal with disturbances of unknown bounds. An uncertain linear system is considered as well as a quite general class of non-smooth disturbances. Global tracking is demonstrated using only output feedback. The proposed adaptation method is able to make the control gain less conservative, but large enough when the disturbance grows and allows it to decrease if the latter vanishes, leading to reduced chattering effects. Simulations are presented to show the potential of the new adaptation scheme in this adverse scenario of possibly growing, temporarily large, or vanishing disturbances.

11.1 Introduction

Sliding mode control (SMC) is a very appealing control strategy to deal with disturbances and uncertainties. Many adaptive methodologies have been explored to remove the obstacle of considering disturbances of *known* bounds in the control design. Examples are the use of monotonically increasing gains [13, 15, 20], that, as the name suggests, increases the switching gains until a sliding mode is achieved. The price to be paid in these schemes is that, when the disturbances reduce to lower magnitudes, the gains keep unnecessarily high, thus increasing the effects of the *chattering* phenomenon.

L. Hsu
COPPE/Federal University of Rio de Janeiro, Rio de Janeiro, Brazil
e-mail: liu@coep.ufrj.br

T.R. Oliveira (✉) · G.T. Melo · J.P.V.S. Cunha
State University of Rio de Janeiro, Rio de Janeiro, Brazil
e-mail: tiagoroux@uerj.br

G.T. Melo
e-mail: gabrieltmelo@yahoo.com.br

J.P.V.S. Cunha
e-mail: jpaulo@ieee.org

J.B. Clempner and W. Yu (eds.), *New Perspectives and Applications of Modern Control Theory*, https://doi.org/10.1007/978-3-319-62464-8_11

Another common strategy found in literature to deal with disturbances of unknown bounds is the use of dynamic switching gains [1, 5, 17], such that these gains are increased until a sliding mode is obtained, then decreased until the same is lost following some predetermined rule. This strategy may have the disadvantage of loosing the sliding mode an excessive number of times, generating undesired oscillations in the sliding variable. Other adaptive control strategies based on the concepts of equivalent [4, 19] and average control [16] were also proposed to avoid excessive switching gains, thus reducing the chattering effects.

On the other hand, monitoring functions were proposed in earlier works to deal with a variety of problems, including unknown control direction [14, 21] and extremum seeking control [10]. In this chapter, we propose a novel application of monitoring functions to adapt the switching gains of sliding mode controllers so that the disturbances can be majored and, simultaneously, reduced chattering effects can be achieved, depending on the design parameters. In addition, this strategy seems to lead to better performance or to less restrictive assumptions for a wider class of disturbances in comparison with the existing literature. For instance, no upper bound for the disturbance derivatives [4, 19] is necessary for the implementation of our novel adaptation law.

Unlike most research done in the area of adaptive sliding mode control, we have considered objectives of trajectory tracking and output feedback in the presence not only of disturbances with unknown bounds but also of parametric uncertainties of the system as well.

In general, state-feedback framework is assumed to facilitate the control problem in earlier publications. Moreover, a stabilization scenario free of parametric uncertainties is usually assumed in the existing literature.

11.1.1 Notation and Terminology

In this chapter, we adopt the following notation:

- The element-wise absolute value of a vector $x = [x_1, x_2, \ldots, x_n]^T$ is defined as $|x| := [|x_1|, |x_2|, \ldots, |x_n|]^T$.
- The symbol "s" represents either the Laplace variable or the differential operator "d/dt", according to the context.
- As in [8, 11] the output y of a linear time-invariant (LTI) system with transfer function $G(s)$ and input u is given by $y = G(s)u$. Convolution operations $g(t) * u(t)$, with $g(t)$ being the impulse response from $G(s)$, will be eventually written, for simplicity, as $G(s) * u$.
- Filippov's definition [6] is assumed for the time-domain solution of discontinuous systems. Note that the control signal u is not necessarily a function of t in the usual sense when sliding modes take place. In order to avoid clutter, $u(t)$ denotes the locally integrable functions which are equivalent to u, in the sense of *equivalent control* [18], along any given Filippov solution $z(t)$ of the closed-loop system. It

should be stressed that $z(t)$ is, *by definition*, absolutely continuous. Then, along any solution, u can be replaced by $u(t)$ in the right-hand side of the governing differential equations. Although the equivalent control $u(t) = u_{\text{eq}}(t)$ is not directly available, the output signal of linear time-invariant systems with strictly proper transfer function $G(s)$ is given by $y(t) = G(s)u = G(s)u(t) = G(s)u_{\text{eq}}(t)$.
- Class $\mathcal{K}$ and class $\mathcal{KL}$ functions are defined as in [12]:

Definition 11.1 A continuous function $\alpha : [0, a) \to [0, +\infty)$ is said to belong to class $\mathcal{K}$ if it is strictly increasing and $\alpha(0) = 0$.

Definition 11.2 A continuous function $\beta : [0, a) \times [0, +\infty) \to [0, +\infty)$ is said to belong to class $\mathcal{KL}$ if, for each fixed s, the mapping $\beta(r, s)$ belongs to class $\mathcal{K}$ with respect to r and, for each fixed r, the mapping $\beta(r, s)$ is decreasing with respect to s and $\beta(r, s) \to 0$ as $s \to +\infty$.

11.2 Problem Statement

Consider an uncertain single-input-single-output (SISO) LTI system

$$y = G_p(s)[u + d(t)] \,, \tag{11.1}$$

where u is the control input, y is the output, $d(t)$ is a matched input disturbance and

$$G_p(s) = k_p \frac{N_p(s)}{D_p(s)} \,, \tag{11.2}$$

with $N_p(s)$ and $D_p(s)$ being monic polynomials of degree m and n, respectively.

The following assumptions are made:

(A1) $G_p(s)$ is minimum phase and its parameters are unknown but belong to a known compact set.

(A2) The degrees m and n of $N_p(s)$ and $D_p(s)$, respectively, are known constants. Moreover, $G_p(s)$ has relative degree one ($n^* = n - m = 1$).

The above assumptions **(A1)**–**(A2)** are slight modifications of those commonly found in model-reference adaptive control literature [11]. Consider the additional assumptions:

(A3) The sign of the high frequency gain $k_p \neq 0$ is known.

(A4) The disturbance $d(t)$ is piecewise continuous in t, and satisfies

$$|d(t)| \leq \bar{d} < +\infty \,, \quad \forall t \geq 0 \,, \tag{11.3}$$

where $\bar{d} > 0$ is an unknown constant.

11.2.1 Reference Model

The reference model is given by

$$y_m = M(s)r = \frac{k_m}{s + a_m} r \,, \tag{11.4}$$

where $a_m > 0$ and $k_m > 0$. The reference signal $r(t)$ is assumed piecewise continuous and uniformly bounded.

11.2.2 Control Objective

The control objective is to achieve global stability and convergence of the error state with respect to a small neighborhood of the origin of the error space. In particular, the tracking error

$$e_0(t) = y(t) - y_m(t) \tag{11.5}$$

must converge at least asymptotically to a small neighborhood of zero, i.e., *practical tracking* is desired.

11.3 Control Parameterization and Disturbance Upper Bound

Considering the usual model reference adaptive control (MRAC) approach [11], the output error e_0 satisfies [7]:

$$e_0 = k^* M(s)[u - u^*] \,, \tag{11.6}$$

where $k^* = k_p / k_m$,

$$u^* := \theta^{*T} \omega - W_d(s) * d \,, \tag{11.7}$$

is the model matching control in the presence of the disturbance d [2]. The *regressor vector* ω is composed by the states of the input/output filters, by the system output y and the reference signal r:

$$\omega := \left[\omega_1^T,\, y,\, \omega_2^T,\, r\right]^T \in \mathbb{R}^{2n} \,. \tag{11.8}$$

According to [11], the input and output filters are given by:

$$\dot{\omega}_1 = F\omega_1 + gu \,, \qquad \dot{\omega}_2 = F\omega_2 + gy \,, \tag{11.9}$$

where $F \in \mathbb{R}^{(n-1)\times(n-1)}$ and $g \in \mathbb{R}^{(n-1)}$.

The *ideal parameter vector* $\theta^* = [\theta_1^{*T}, \theta_0^*, \theta_2^{*T}, 1/k^*]^T$ is unknown but it is assumed to be element-wise bounded by a known constant vector $\bar{\theta}$ [7]:

$$\bar{\theta}_i \geq \left|\theta_i^*\right| , \quad \forall i \in \{1, \ldots, 2n\} . \tag{11.10}$$

The transfer function $W_d(s)$ is proper, stable, and given by

$$W_d(s) = [k^* M(s)]^{-1} \bar{W}_d(s) , \tag{11.11}$$

where $\bar{W}_d(s)$ is the closed-loop transfer function from the input disturbance d to e_0 [8].

The signal u^* will be regarded as the total input disturbance to be canceled by the sliding mode control law, thus an upper bound will be required for control design. If the upper bound $\bar{d}$ in assumption **(A4)** is known, since $W_d(s)$ is a proper and bounded-input–bounded-output (BIBO) stable transfer function and $d(t)$ satisfies **(A4)**, then applying [9, Lemma 2] to the convolution $W_d(s) * d(t)$, such that $|W_d(s) * d(t)| \leq \hat{d}_1(t)$, where $\hat{d}_1$ is defined by

$$\hat{d}_1(t) := \bar{d} + c_d e^{-\gamma_d t} * \bar{d} . \tag{11.12}$$

The constants $c_d > 0$ and $\gamma_d > 0$ are the coefficients of a first-order approximation filter (FOAF) with transfer function $c_d/(s + \gamma_d)$. The FOAF can be designed as described by [3].

However, the disturbance upper bound $\bar{d}$ is unknown, thus, the following adaptive law is proposed

$$\hat{d}(t) = \beta(k, t) + c_d e^{-\gamma_d t} * \beta(k, t) . \tag{11.13}$$

This is based on (11.12), where the unknown upper bound $\bar{d}$ is replaced by an appropriate class $\mathcal{KL}$ function $\beta(k, t)$, and $k \in \mathbb{N}$ is the switching number of a monitoring function to be defined later. The function $\beta(k, t)$ grows monotonically with the number of switchings k and decreases with time after each switching in a prescribed manner. After each new switching, the function $\beta(k, t)$ working together with the FOAF will give a norm bound for the disturbance $d(t)$ at least during a sufficiently large interval. The union of all these time intervals is denoted here by T^+. Thus, from (11.7), u^* satisfies

$$|u^*(t)| \leq \bar{\theta}^T \, |\omega(t)| + \hat{d}(t) , \quad \forall t \in T^+ . \tag{11.14}$$

The set denoted by T^- corresponds to the complementary set of T^+, i.e., $T^+ \cup T^- = [0, +\infty)$ and $T^+ \cap T^- = \emptyset$.

11.4 Output-Feedback Control Law

The adopted control law is given by

$$u = -f(t)\,\mathrm{sign}(k_p)\,\mathrm{sign}(e_0)\,. \tag{11.15}$$

According to (11.6), the *modulation function* $f(t)$ is designed to overcome the ideal matching control u^*, which is regarded as an input disturbance in (11.6). From (11.14), note that one possible choice for $f(t)$ that verifies the inequality $f(t) \geq |u^*(t)|$ is

$$f(t) = \bar{\theta}^T\,|\omega(t)| + \hat{d}(t) + \delta\,, \tag{11.16}$$

where $\delta > 0$ is an arbitrary constant. By invoking [8, Lemma 1], this modulation function guarantees that the output error $e_0(t)$ tends to zero, $\forall t \in T^+$. From (11.4)–(11.6), it can be shown that the output error signal is the solution of the differential equation

$$\dot{e}_0(t) = -a_m e_0(t) + k_p[u(t) - u^*(t)] + \pi(t)\,, \tag{11.17}$$

where $\pi(t)$ denotes an exponentially decaying transient term due to initial conditions of the observable but not controllable subsystem of the nonminimal realization (A_c, b_c, h_c^T) of $M(s)$ in (11.6), used in MRAC theory [11, p. 346]. Now, from (11.15), (11.6) and (11.7), noting that $\mathrm{sgn}(u - u^*) = -\mathrm{sgn}(e_0)$, if $f(t) > |u^*(t)|$, $\forall t \geq \bar{t}_0$, then by using the *Comparison Theorem* [6], $|e_0|$ would be bounded by the solution of the following differential equation

$$\dot{\xi}(t) = -a_m\xi(t) + \pi(t)\,, \quad \forall t \in [\bar{t}_0, +\infty)\,, \quad \xi(\bar{t}_0) = e_0(\bar{t}_0)\,, \tag{11.18}$$

i.e., one has

$$|e_0(t)| \leq |\xi(t)| \leq e^{-a_m(t-\bar{t}_0)}|e_0(\bar{t}_0)| + c_0 e^{-\lambda_0 t}\,, \quad \forall t \geq [\bar{t}_0, +\infty)\,, \tag{11.19}$$

where $\bar{t}_0$ denotes some initial time, and $|e^{-a_m t} * \pi(t)| \leq c_0 e^{-\lambda_0 t}$, for c_0 being an unknown positive constant depending on the initial conditions of the state variables and λ_0 being a known constant satisfying $0 < \lambda_0 < \min\{-\mathrm{Re}(\lambda_i[A_c]), a_m\}$, with $\lambda_i[A_c]$ being the spectrum of A_c referred above [11, p. 346].

11.5 Monitoring Function

Based on (11.19), consider the auxiliary function $\varphi_k(t)$ defined as follows:

$$\varphi_k(t) = e^{-a_m(t-t_k)}|e_0(t_k)| + a(k)e^{-\lambda_0 t}\,, \qquad t \in [t_k, +\infty)\,, \tag{11.20}$$

where $a(k)$ is any positive monotonically increasing unbounded sequence, $k \in \mathbb{N}$, and $t_0 = 0$.

The *monitoring function* φ_m can be defined as

$$\varphi_m(t) := \varphi_k(t)\,, \quad \forall t \in [t_k, t_{k+1}) \quad (\subset [0, +\infty))\ . \tag{11.21}$$

The motivation behind the introduction of $\varphi_m(t)$ is that $\pi(t)$ in (11.17) is not available for measurement. Reminding that the inequality (11.19) holds if the inequality $f(t) > |u^*(t)|$ is satisfied, it would seem natural to use the signal $\xi(t)$ in (11.18) as a benchmark to detect if the sliding mode is being lost and the error is increasing so that k in $\beta(k, t)$ must be increased, see Eq. (11.13). However, since $\pi(t)$ is not available, one has to use $\varphi_m(t)$ to replace $\xi(t)$ and invoke the switching of $\varphi_m(t)$. Note that from (11.21), one always has $|e_0(t_k)| < \varphi_k(t_k)$ at $t = t_k$. Hence, the switching time t_k is well defined (for $k \geq 0$):

$$t_{k+1} = \begin{cases} \min\{t > t_k : |e_0(t)| = \varphi_k(t)\}, & \text{if it exists}\,, \\ +\infty, & \text{otherwise}\,. \end{cases} \tag{11.22}$$

The whole scheme is depicted in Fig. 11.1.

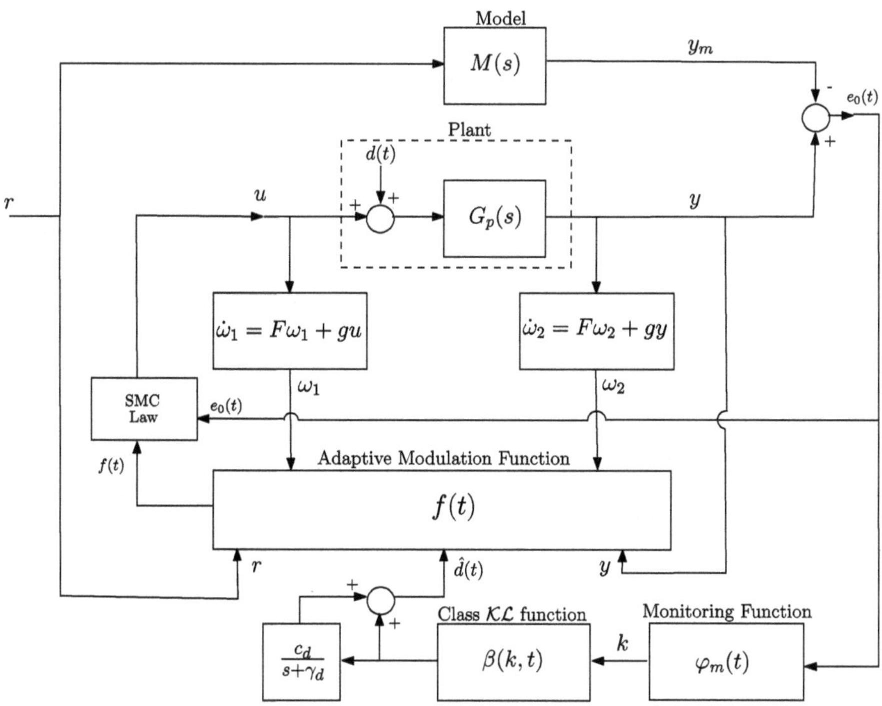

Fig. 11.1 Block diagram of the proposed output feedback adaptive sliding mode control strategy

11.6 Stability Analysis

The theorem below presents preliminary results of practical tracking of the proposed approach in Fig. 11.1. Although the ultimate residual set has not yet been fully characterized, it is possible to state that the tracking error becomes sufficiently small for an appropriate choice of the class $\mathcal{KL}$ function $\beta(k, t)$ in (11.13) applied to the modulation function (11.16) [20]. This is a current investigation topic.

Theorem 11.1 *Consider the plant (11.1), the reference model (11.4), the control law (11.15) with modulation function (11.16), and adaptive law (11.13) as well as the monitoring function (11.21). Assume that* **(A1)–(A4)** *hold. Then, practical tracking is achieved, and the tracking error $e_0(t)$ defined in (11.5) converges ultimately close to the origin.*

Proof If a class $\mathcal{K}$ function $\beta(k)$ was used in (11.13) instead of the class $\mathcal{KL}$ function assumed before, it is possible to verify from assumption **(A4)** and (11.19) that after a finite number k_i of switchings, the signal $\hat{d}(t)$ becomes ultimately greater than $\bar{d}$. Hence, the condition in (11.14) is verified for $t \geq t_{k_i}$. In addition, for k_i sufficiently large, $a(k_i)e^{-\lambda_0 t}$ of (11.20) will allow $\varphi_k(t)$ to be an upper bound valid for $\xi(t)$ in (11.19) since

$$|e^{-a_m t} * \pi(t)| \leq c_0 e^{-\lambda_0 t} < a(k_i)e^{-\lambda_0 t}\,, \tag{11.23}$$

consequently no switching will occur after that. Thus, the tracking error $e_0(t)$ converges to zero at least exponentially since the monitoring function (11.20) and (11.21) converges exponentially when the switching process stops. In addition, the sliding mode at $e_0 = 0$ is reached in finite time since the condition $\dot{e_0} e_0 < -\delta|e_0|$ can be obtained, with $\delta > 0$ included in the modulation function (11.16).

The application of a class $\mathcal{KL}$ function $\beta(k, t)$ instead of a class $\mathcal{K}$ just introduces a forgetting factor in the described scheme. According to [20], it is always possible to find an increasing sequence of switchings for $\beta(k, t)$ until the tracking error $e_0(t)$ enters inside any prescribed residual set. Since $\beta(k, t)$ always decreases with time, the monitoring function switches from time to time when necessary, in order to make $\varphi_k(t) > |e_0(t)|$. If the FOAF defined by the coefficients c_d, γ_d in (11.13) is appropriately designed to generate a valid upper bound for $d(t)$ from $\beta(k, t)$, then the sliding mode $e_0 = 0$ is lost only during small intervals of time ($t \in T^-$), and is immediately recovered after a finite number of successive switchings of the monitoring function. Thus, as $t \to +\infty$, the monitoring function never stops switching from time to time (unless $d(t)$ ultimately vanishes going to zero), and the tracking error $e_0(t)$ is never allowed to increase more than some prescribed small residual value since the time intervals belonging to T^- can be made arbitrarily short so that the monitoring function (11.20) and (11.21) assumes very small values for long intervals T^+.

Since the disturbance and the reference signals are uniformly bounded (see assumption **(A4)** and Sect. 11.2.1), then the error cannot ultimately diverge significantly from zero. □

Remark 11.1 (*Modulation Function Reset*) The term $\beta(k,t)$ in (11.16) plays a key role in the domination of the unknown disturbance $d(t)$ in (11.7). It allows that inequality $f(t) > |u^*(t)|$ is satisfied as the number of switchings increases. However, since $\beta(k,t) \to +\infty$ as $k \to +\infty$, within every fixed time interval, the modulation function may need a reset mechanism to reinitialize k, from time to time, in order to avoid that the controller amplitudes increase to very high values as $t \to +\infty$. An alternative to avoid a large number of switchings could be the inclusion of a small piecewise constant term in the modulation function, which would be properly increased every time the switching index achieves some prescribed threshold value.

In some practical applications, despite the fact the disturbance $d(t)$ has unknown norm bound, it may tend to some specific value or has a minimal upper bound after some finite time (for example, when the disturbance has a large transient and then goes to a small steady state). For these situations, the following corollary can be stated.

Corollary 11.1 *In Theorem 11.1, if the disturbance $d(t)$ has the additional property:*

$$|d(t)| < d_l, \quad \forall t \in [t_l, +\infty), \tag{11.24}$$

where $d_l \geq 0$ is a known constant, then replacing (11.16) by the new modulation function given by

$$f(t) = \bar{\theta}^T\,|\omega(t)| + \hat{d}(t) + \left(1 + \frac{c_d}{\gamma_d}\right) d_l + \delta\,, \quad \forall t \geq 0\,, \tag{11.25}$$

exact tracking is achieved and $e_0(t)$ is kept in the origin after some finite time. Moreover, $\hat{d}(t) \to 0$, $\forall t > t_l$, which decreases the amplitude of the control signal $u(t)$ needed to keep the sliding mode.

Proof For $t < t_l$, Theorem 11.1 can be applied such that at least practical tracking is achieved during this time interval. For all $t \geq t_l$, the redesigned modulation function (11.25) implies that $f(t) > |u^*(t)|$ is ultimately satisfied, the sliding manifold $e_0 = 0$ is necessarily reached in some finite time, and the monitoring function switching finally ceases. Hence, the tracking error becomes null and global exact tracking is obtained.

Since the term d_l is large enough to guarantee that $f(t) > |u^*(t)|$, $\forall \geq t_l$, and the function $\beta(k,t)$ vanishes when the switching process stops (k is fixed and finite), then $\hat{d}(t)$ in (11.13) tends to zero leading to the reduction of the amplitude of the control signal $u(t)$ given by (11.25). □

11.7 Numerical Examples

In this section, a numerical example is presented to illustrate the properties of the proposed adaptive sliding mode controller. The system being considered is an integrator, thus $G_p(s) = 1/s$. The chosen reference model is given by $M(s) = 1/(s+1)$.

In this example, no input/output filters ω_1 and ω_2 in the regressor vector ω are needed, since the order of the system is $n = 1$. The matching parameters are $\theta_0^* = -1$ and $1/k^* = 1$ in the vector θ^*, thus $\theta^{*T}\omega = -y + r$. Therefore, the upper bound $\bar{\theta}^T\,|\omega(t)| = |y| + |r|$ and $\delta = 0.01$ were chosen in the modulation function (11.16).

In the simulations, the initial condition of the system is $y(0) = 2$, the remaining initial conditions are null. The reference signal is chosen as $r(t) = \sin(0.5t)$, and the disturbance is $d(t) = 5\,\sin(t) + \mathcal{U}(t-10) - \mathcal{U}(t-20)$, where $\mathcal{U}(t)$ is the unit step function.

To avoid spurious switchings of the monitoring function due to numerical residues of the error (caused by numeric integration), a modification in (11.20) and (11.21) is introduced for simulation purposes:

$$\varphi_k(t) = e^{-a_m(t-t_k)}|e_0(t_k)| + (k+1)e^{-\lambda_0 t} + \epsilon\,, \tag{11.26}$$

where ϵ is an arbitrarily small positive constant. In the simulations, ϵ was set to 0.01 and the remaining design parameters in (11.13) were chosen as $a(k) = k + 1$, $a_m = \lambda_0 = 1$, $c_d = 2$, $\gamma_d = 2$, and $\beta(k,t) = (10k+1)e^{-3t/(10k+1)}$.

In Fig. 11.2, it can be verified that the exact tracking starts around 0.2 seconds, and stays in a very small neighborhood of the desired trajectory y_m, despite of the unknown disturbance d. Figure 11.3 shows the switchings of the monitoring function φ_m. As discussed before, every time the value of the absolute error $|e_0|$ tries to increase, the monitoring function switches.

As can be seen in Fig. 11.4, in this simulation example, the chosen function $\beta(k,t)$ fails to upper bound the disturbance most of the time, but the FOAF defined by c_d, γ_d

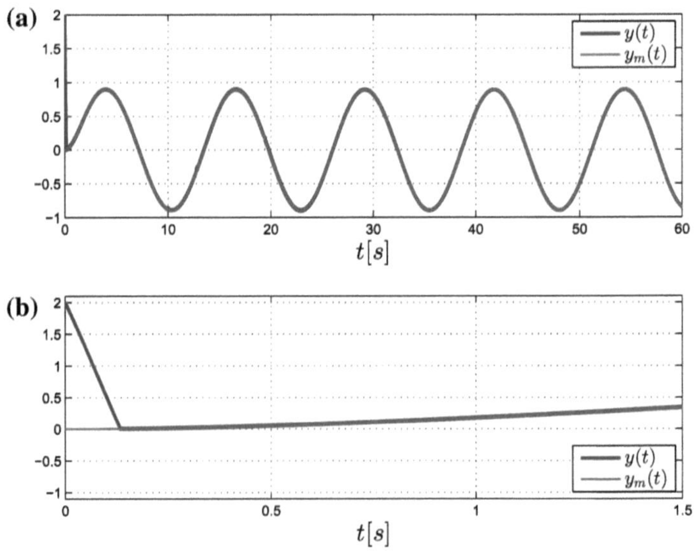

Fig. 11.2 **a** System output $y(t)$ and model output $y_m(t)$. **b** Convergence of $y(t)$ to $y_m(t)$ (zoom)

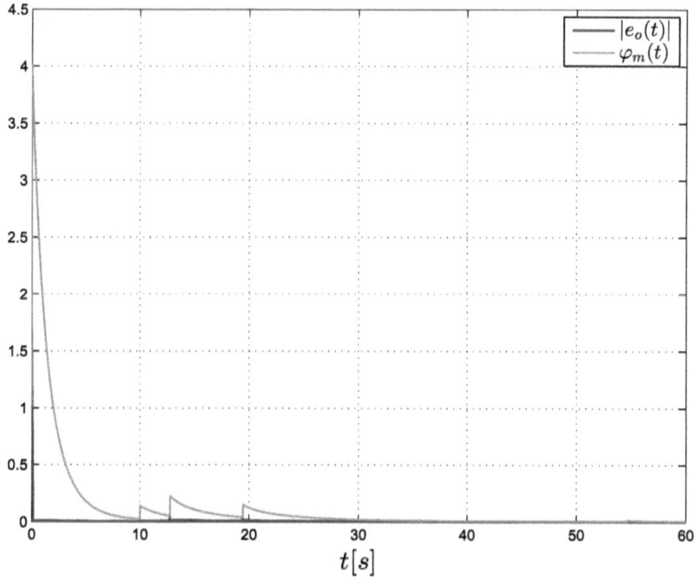

Fig. 11.3 Absolute tracking error $|e_0(t)|$ and monitoring function $\varphi_m(t)$

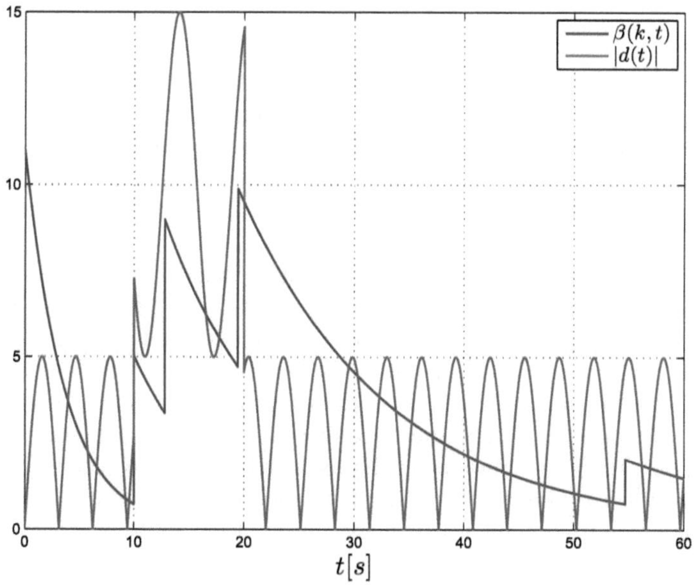

Fig. 11.4 Time evolution of the function $\beta(k, t)$ as well as the absolute value of the unknown disturbance $|d(t)|$

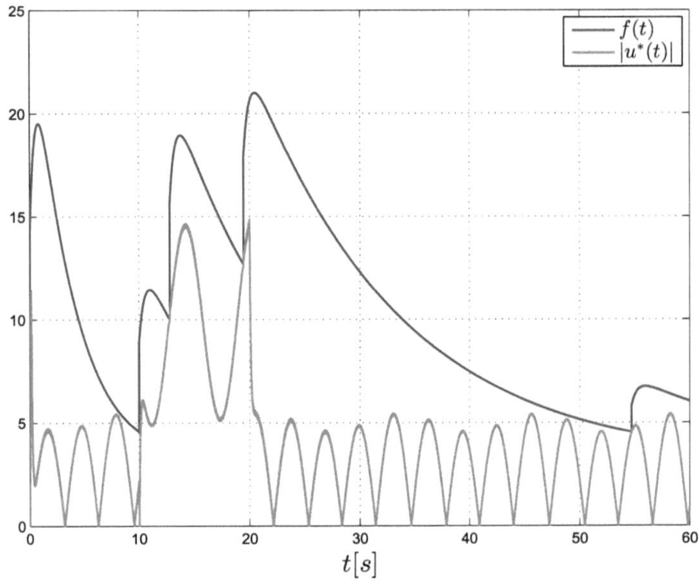

Fig. 11.5 Modulation function $f(t)$ and the absolute value of the ideal matching control $|u^*(t)|$ faced as an input disturbance

in (11.13) is conservative enough to generate a valid upper bound for the equivalent (total) input disturbance u^*, i.e., $f(t) > |u^*|$ (see Fig. 11.5).

Figure 11.6 shows by means of the signal u that the sliding mode is lost only in very short time intervals, and the amplitude of the switching (modulation function) changes according to the need imposed by the matched disturbance u^*, thus mitigating chattering effects.

For the sake of comparison, another simulation result is presented in the spirit of the approach proposed by [15, 20], using $\beta(k, t) = 3k + 1$ (a class $\mathcal{K}$ function instead of a class $\mathcal{KL}$) with the system, reference model, and parameters as defined before. In Fig. 11.7, it is possible to note when the disturbance returns to lower levels, the modulation function stays at unnecessarily high control amplitudes. This kind of control signal would increase chattering effects in practical situations.

At last, the proposed control algorithm is tested taking into account a piecewise constant disturbance with initial growing behavior, followed by a vanishing phase and ultimate large-persistent profile. This shows the capability of the proposed adaptation scheme in changing the modulation function for different amplitude conditions so that the tracking of the desired trajectory can be always preserved (Figs. 11.8, 11.9 and 11.10). According to Corollary 11.1, if additional information on the upper bound for the final amplitude of the disturbance ($d_l = 10$) is known, then the monitoring function switching stops so that the exact tracking ($e_0 = 0$) is perfectly obtained as shown in Fig. 11.11.

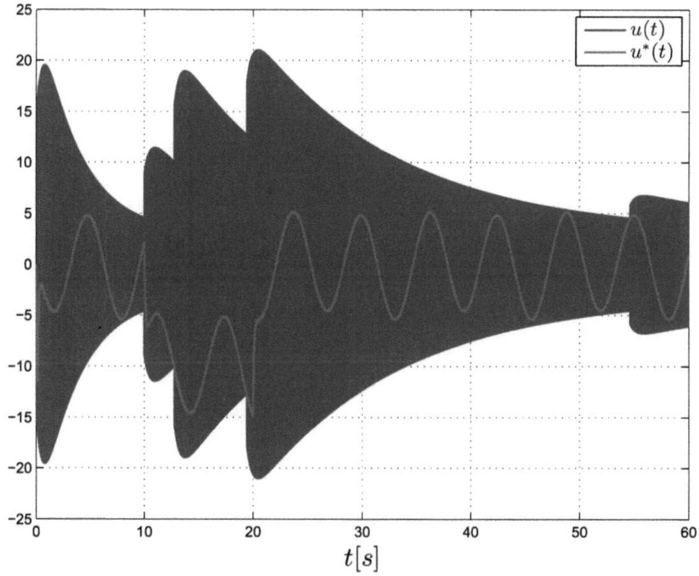

Fig. 11.6 Control signal $u(t)$ and the ideal matching control $u^*(t)$

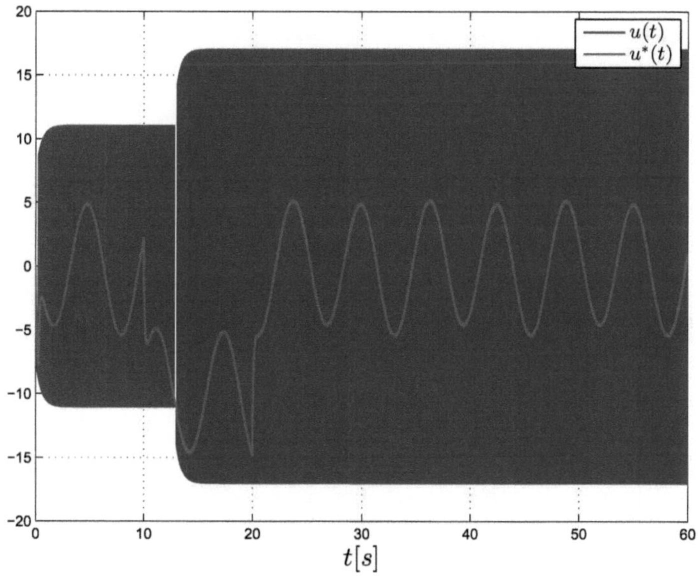

Fig. 11.7 Control signal $u(t)$ and the ideal matching control $u^*(t)$ when a class $\mathcal{K}$ function $\beta(k)$ is used rather than a class $\mathcal{KL}$ function $\beta(k, t)$. The control objectives are achieved at the expense of an unnecessarily large control signal amplitude

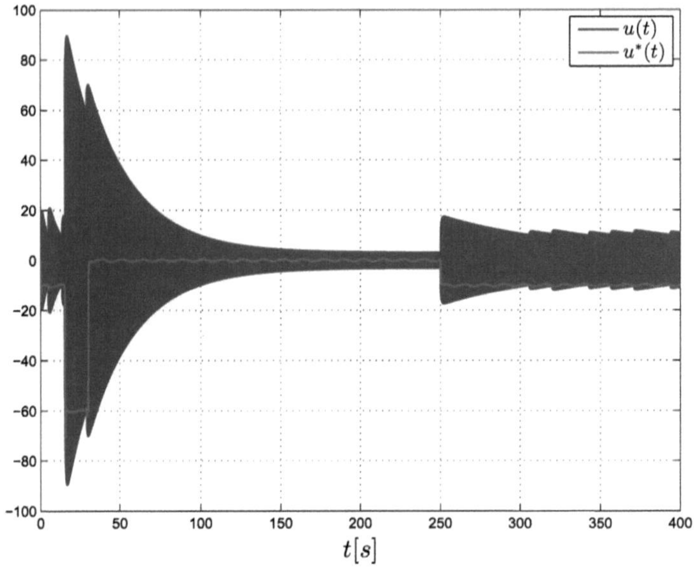

Fig. 11.8 Control signal $u(t)$ and the ideal matching control $u^*(t)$ when the disturbance is $d(t) = 10\mathcal{U}(t) + 50\mathcal{U}(t-10) - 60\mathcal{U}(t-30) + 10\mathcal{U}(t-250)$

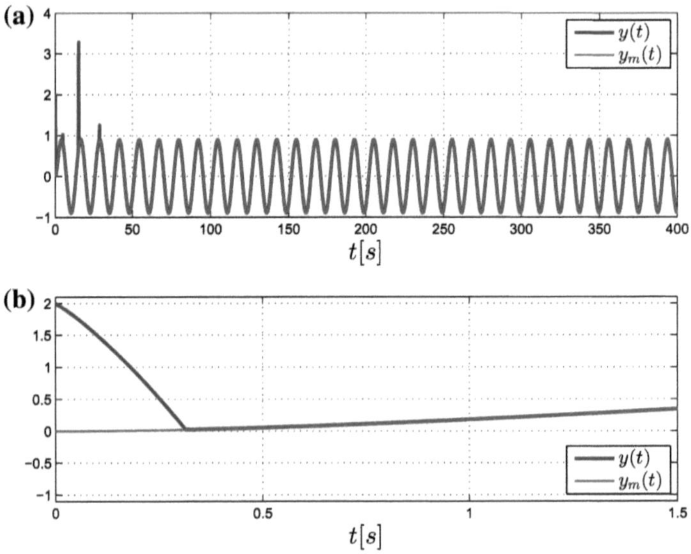

Fig. 11.9 **a** System output $y(t)$ and model output $y_m(t)$ when $d(t) = 10\mathcal{U}(t) + 50\mathcal{U}(t-10) - 60\mathcal{U}(t-30) + 10\mathcal{U}(t-250)$. **b** Convergence of $y(t)$ to $y_m(t)$ (zoom)

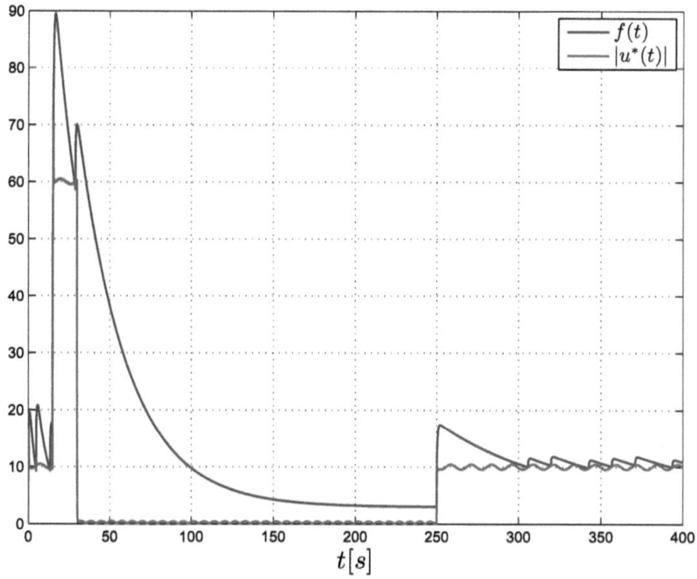

Fig. 11.10 Modulation function $f(t)$ and the absolute value of the ideal matching control $|u^*(t)|$ when $d(t) = 10\mathcal{U}(t) + 50\mathcal{U}(t-10) - 60\mathcal{U}(t-30) + 10\mathcal{U}(t-250)$

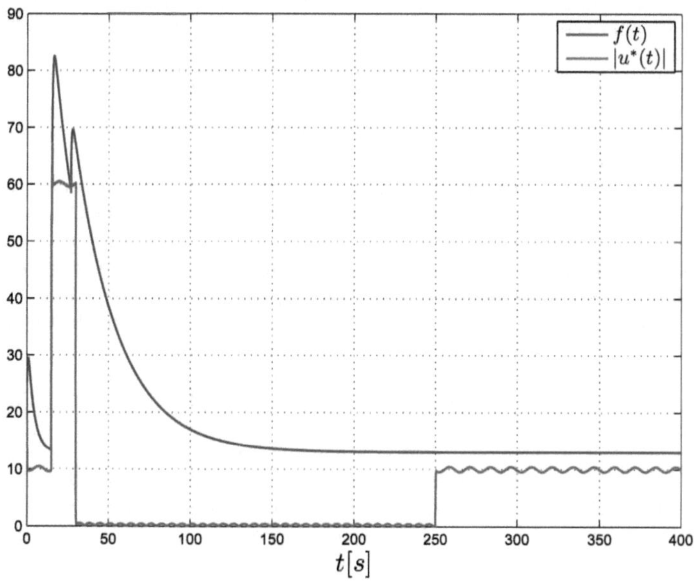

Fig. 11.11 Modulation function $f(t)$ and the absolute value of $|u^*(t)|$ when $d(t) = 10\mathcal{U}(t) + 50\mathcal{U}(t-10) - 60\mathcal{U}(t-30) + 10\mathcal{U}(t-250)$ and an ultimate upper bound $d_l = 10$ is assumed known for $d(t)$

11.8 Conclusion

A novel adaptive sliding mode controller to cope with parameter uncertainties and non-smooth disturbances with unknown constant upper bound has been developed. The proposed controller is based on a monitoring function as a tool for switching the control gain (modulation function adaptation). Disturbance domination as well as global convergence and practical trajectory tracking are guaranteed by using only output feedback. The ultimately decreasing characteristic of the adaptive modulation function leads to less conservative and smaller switching control gains, thus reducing undesirable chattering effects of the sliding mode controller, as illustrated by numerical simulations.

Acknowledgements The authors thank the Brazilian funding agencies CAPES, CNPq and FAPERJ for the financial support.

References

1. Bartolini, G., Levant, A., Plestan, F., Taleb, M., Punta, E.: Adaptation of sliding modes. IMA J Math. Control Inf. **30**(3), 285–300 (2013)
2. Cunha, J.P.V.S., Hsu, L., Costa, R.R., Lizarralde, F.: Output-feedback model-reference sliding mode control of uncertain multivariable systems. IEEE Trans. Autom. Control **48**(12), 2245–2250 (2003)
3. Cunha, J.P.V.S., Costa, R.R., Hsu, L.: Design of first-order approximation filters for sliding-mode control of uncertain systems. IEEE Trans. Ind. Electron. **55**(11), 4037–4046 (2008)
4. Edwards, C., Shtessel, Y.B.: Adaptive continuous higher order sliding mode control. Automatica **65**, 183–190 (2016)
5. Estrada, A., Plestan, F., Allouche, B.: An adaptive version of a second order sliding mode output feedback controller. In: 2013 European Control Conference (ECC), pp. 3228–3233. IEEE, Zurich, Switzerland (2013)
6. Filippov, A.F.: Differential equations with discontinuous right-hand side. Am. Math. Soc. Transl. **42**(2), 199–231 (1964)
7. Hsu, L., Araújo, A.D., Costa, R.R.: Analysis and design of I/O based variable structure adaptive control. IEEE Trans. Autom. Control **39**(1), 4–21 (1994)
8. Hsu, L., Lizarralde, F., Araújo, A.D.: New results on output-feedback variable structure model-reference adaptive control: design and stability analysis. IEEE Trans. Autom. Control **42**(3), 386–393 (1997)
9. Hsu, L., Costa, R.R., Cunha, J.P.V.S.: Model-reference output-feedback sliding mode controller for a class of multivariable nonlinear systems. Asian J. Control **5**(4), 543–556 (2003)
10. Hsu, L., Oliveira, T.R., Cunha, J.P.V.S.: Extremum seeking control via monitoring function and time-scaling for plants of arbitrary relative degree. In: Proceedings 13th International Workshop on Variable Structure System, Nantes, pp. 1–6 (2014)
11. Ioannou, P.A., Sun, J.: Robust Adaptive Control. Prentice-Hall, Upper Saddle River (1996)
12. Khalil, H.K.: Nonlinear Systems, 2nd edn. Prentice-Hall, Upper Saddle River (1996)
13. Moreno, J.A., Negrete, D.Y., Torres-González, V., Fridman, L.: Adaptive continuous twisting algorithm. Int. J. Control. **89**(9), 1798–1806 (2016)
14. Oliveira, T.R., Peixoto, A.J., Nunes, E.V.L., Hsu, L.: Control of uncertain nonlinear systems with arbitrary relative degree and unknown control direction using sliding modes. Int. J. Adapt. Control. Signal Process **21**(8/9), 692–707 (2007)

15. Oliveira, T.R., Leite, A.C., Peixoto, A.J., Hsu, L.: Overcoming limitations of uncalibrated robotics visual servoing by means of sliding mode control and switching monitoring scheme. Asian J. Control **16**(3), 752–764 (2014)
16. Oliveira, T.R., Cunha J.P.V.S., Hsu, L.: Adaptive sliding mode control for disturbances with unknown bounds. In: Proceedings 14th International Workshop on Variable Structure Systems, Nanjing, Jiangsu, China, pp. 59–64 (2016)
17. Plestan, F., Shtessel, Y., Brégeault, V., Poznyak, A.: New methodologies for adaptive sliding mode control. Int. J. Control **83**(9), 1907–1919 (2010)
18. Utkin, V.I.: Sliding Modes and Their Application in Variable Structure Systems. MIR Publishers, Moscow (1978)
19. Utkin, V.I., Poznyak, A.S.: Adaptive sliding mode control with application to super-twist algorithm: equivalent control method. Automatica **49**(1), 39–47 (2013)
20. Yan, L., Hsu, L., Xiuxia, S.: A variable structure MRAC with expected transient and steady-state performance. Automatica **42**(5), 805–813 (2006)
21. Yan, L., Hsu, L., Costa, R.R., Lizarralde, F.: A variable structure model reference robust control without a prior knowledge of high frequency gain sign. Automatica **44**(4), 1036–1044 (2008)

Chapter 12
Fast Control Systems: Nonlinear Approach

Andrey Polyakov

Abstract This chapter treats the problem of fast control design for nonlinear systems. First, we discuss the question: which nonlinear system can be called fast? Next, we develop some tools for analysis and design of such control systems. The method generalized homogeneity is mainly utilized for these purposes. Finally, we survey possible research directions of the fast control systems.

12.1 Introduction

The Olympic motto "Citius, Altius, Fortius" ("Faster, Higher, Stronger") precisely reflects the evolution trend of any known engineering invention. Each next generation of aircrafts (automobiles, trains, robots, and so on) has to be faster than the previous one. They also must demonstrate higher performance and stronger robustness. Any related renovation usually requires an update of automatic control system.

The control theory is an interdisciplinary branch of both mathematics and engineering sciences. Consequently, on the one hand, some engineering invention precedes some renovation of control design methodology. Indeed, digital controllers (computes in general) replaced analog devices implying the appearance of new research fields such as sampled, hybrid, event-driven control systems. On the other hand, new ideas to control theory may come from pure mathematics. For example, sliding mode control principles [47] are essentially based on ideas of the theory of differential equations with discontinuous right-hand sides [17] introduced in 1960, but the calculus of variations (see, e.g., [39]) initiated in seventeenth century underlies the optimal control design [39].

The fundamental background of the whole modern mathematical control theory has been presented in the seminal work [26] of A.M. Lyapunov "The general problem of the stability of motion" published in 1892. Today quantitative characteristics of stability (Lyapunov exponents, Lyapunov functions, etc.) specify performance of control systems (such as convergence rate, input-to-state stability, etc.). Exponen-

A. Polyakov (✉)
Inria Lille, 40 avenue Halley, 59650 Villenueve d'Ascq, France
e-mail: andrey.polyakov@inria.fr

J.B. Clempner and W. Yu (eds.), *New Perspectives and Applications of Modern Control Theory*, https://doi.org/10.1007/978-3-319-62464-8_12

tial stability determines the convergence rate of stable linear Ordinary Differential Equations (ODEs), which are still the most popular models of control systems.

Any mathematical model is just an approximation of a real (physical) plant process under some assumptions on its behavior. Frequently, linear approximation (model) is not appropriate if we deal with a "fast" control system, since linearizion neglects some nonlinear dynamics, which imply fast transitions. To define "fast" control, this paper uses linear systems as the reference point for comparison of convergence rate. Namely, *a nonlinear system is said to be fast if it demonstrates transients motions faster than any linear one*, i.e., convergence rate of a nonlinear system is faster than any exponential. In [37] such systems were called **hyper-exponential**.

12.1.1 Motivating Example

Here we present an example of a nonlinear model of the physical system in order to pick out nonlinearities which may invoke fast transitions.

Let us consider a mechanical system consisting of a rigid body moving laterally on a contact surface and in some viscous environment (fluid). The simplest real-life example of such a mechanical system is a car moving on a flat road with a sufficiently high velocity (more than 50 km/h).

Let z be position of the center of mass of the body in an inertial frame. The equation describing the motion of this system has the form

$$\dot{z}(t) = v(t), \quad m\,\dot{v}(t) = F(t), \quad t > 0, \quad z(t) \in \mathbb{R},$$

where $v(t)$ is the velocity, m is a mass of the body, and F is the sum of external forces.

We study only deceleration motion of this system assuming that at the initial instant of time the system has some nonzero velocity $\dot{z}(0) = v(0) \neq 0$. Dissipation of the energy may be caused by several external forces. To discover hyper-exponential behavior it is sufficient to consider only two of them:

- *drag force* (fluid resistance) is proportional to the square of the velocity [16]

$$F_{drag}(t) = -k_{drag}\; v^2(t)\; \mathrm{sign}[\mathrm{v(t)}],$$

  where $k_{drag} > 0$ is the coefficient of fluid resistance and the sign function is

$$\mathrm{sign}[\rho] = \begin{cases} 1 & \text{if } \rho > 0, \\ 0 & \text{if } \rho = 0, \\ -1 & \text{if } \rho < 0; \end{cases}$$

- *dry friction force* is nearly velocity independent and given by the next model [3]

$$F_{dry}(t) = -k_{dry}\; \mathrm{sign}[\mathrm{v(t)}],$$

  where $k_{dry} > 0$ is the coefficient of dry friction.

Usually, the friction models also contain some linear terms (proportional to velocity). We skip them for simplicity of analysis, since they will not effect the final conclusions about convergence rate.

The sum of external forces $F(t)$ can be represented as follows

$$F(t) = F_{drag}(t) + F_{dry}(t) = -\left(k_{dry} + k_{drag}\, v^2(t)\right) \text{sign}[\text{v(t)}].$$

and the differential equation describing the evolution of the velocity of the body has the form:

$$m\dot{v}(t) = -\left(k_{dry} + k_{drag}\, v^2(t)\right) \text{sign}[\text{v(t)}].$$

It is easy to see that $v = 0$ is the equilibrium point of the last equation, which is globally stable, $v(t) \to 0$ as $t \to +\infty$. The equation can be solved analytically:

$$v(t) = \tan\left(\arctan(|v(0)|) - \frac{\sqrt{k_{dry}\, k_{drag}}}{m} t\right) \text{sign}[\text{v(0)}].$$

This immediately implies $v(t) = 0$ for $t \geq \frac{m}{\sqrt{k_{dry}\, k_{drag}}} \arctan(|v(0)|)$. Since arctan is bounded function we conclude that *independently of initial velocity the motion of the body terminates no later than*

$$T_{\max} = \frac{m\pi}{2\sqrt{k_{dry}\, k_{drag}}}.$$

The deceleration rate of this mechanical system is **hyper-exponential** since it is faster than any exponential in the following sense:

$$\forall C > 0,\ \forall \alpha > 0 \ :\ \exists v(0) \in \mathbb{R},\ \exists t' > 0 \ \text{ such that } \ |v(t)| < C|v(0)|e^{-\alpha t},\ \forall t > t'.$$

Below we use this property for the rigorous definition of fast stability.

12.1.2 State of the Art

It seems that the first separation of fast and slow motions of dynamical systems have been systematically studied in the context of the so-called singularly perturbed ODEs [46], which contain a small parameter multiplied by the highest derivative. Tending this parameter to zero implies boosting system transitions in a certain subspace. In this paper, we follow another philosophy.

According to the motivating example given above fast motions are caused by the nonlinearity of the plant. Depending on of the type nonlinearity, the fast transitions can be guaranteed locally or globally. Description of the behavior of such a "fast nonlinear systems" can be efficiently embedded into Lyapunov's Theory of stability.

Some results in this context can be discovered in the literature. In particular, fast stability of ODEs is represented by the notions of finite-time and fixed-time stabilities [2, 6, 11, 13, 19, 25, 28, 32, 38, 43, 50], but hyper-exponential transitions are studied in [37] as fast behavior of time-delay systems. Fast models described by partial differential equations may demonstrate the so-called finite-time extinction property [10, 18, 30, 45] also known as super stability [5, 12]. This chapter studies systematically all the mentioned concepts and presents some tools for analysis and design of fast (in particular fixed-time) control systems.

12.2 Stability and Fast Convergence

The concept of stability introduced by A.M. Lyapunov [26] considers a *nominal motion* $x^*_{t_0,x_0}(t), t \geq t_0$ of a dynamic system with initial state $x^*_{t_0,x_0}(t_0) = x_0$ and *perturbed motions* obtained for initial conditions $x_0 + \delta$, where δ is a perturbation. If small perturbations of initial conditions imply small deviations of perturbed motions from $x^*_{t_0,x_0}(t)$ then the nominal motion is called stable. In this chapter, we deal only with *stability analysis of the zero solution* (i.e., the origin), since making the change of variables $y = x - x^*$ we transform stability analysis problem to the latter case.

Let us consider the nonlinear system

$$\dot{x}(t) = f(t, x(t)), t > t_0 \in \mathbb{R}, \tag{12.1}$$

$$x(t_0) = x_0 \in \mathbb{R}^n, \tag{12.2}$$

where $f : \mathbb{R} \times \mathbb{R}^n$ may be a non-Lipschitz or even discontinuous function. In the latter case, we assume that f satisfies conditions of existence of Filippov solutions [17], which almost everywhere satisfy the differential inclusion $\dot{x}(t) \in F(t, x(t)), t > 0$, where $F : \mathbb{R} \times \mathbb{R}^n \rightrightarrows \mathbb{R}^n$ is a set-valued function contracted from f using a proper regularization procedure. The three most popular regularization procedures are surveyed in [17, 38]. Note that in general the Cauchy problem (12.1), (12.2) may have nonunique solutions implying two types of stability: *weak stability* (a stability property holds for some solution) and *strong stability* (a stability property holds for all solutions) (see, for example, [17, 42, 43]). Weak stability usually is not enough for robust control purposes. *All conditions presented in definitions below are assumed to hold for all solutions of* (12.1), (12.2).

12.2.1 Nonrated Stability

Assume that the origin is the equilibrium point of (12.1), i.e., $f(t, 0) = 0$ (or $0 \in F(t, 0)$) for all $t \in \mathbb{R}$. This means that $x^*(t) \equiv 0$ is the solution to (12.1), (12.2) with $x_0 = 0$.

Definition 12.1 (*Lyapunov stability*) The origin of the system (12.1) is said to be *Lyapunov stable* if for $\forall \varepsilon \in \mathbb{R}_+$ and $\forall t_0 \in \mathbb{R}$ there exists $\delta = \delta(\varepsilon, t_0) \in \mathbb{R}_+$ such that for $\forall x_0 \in \mathbb{R}^n : \|x_0\| < \delta$ any solution $x_{t_0,x_0}(t)$ of the Cauchy problem (12.1), (12.2) exists for $t > t_0$ and $\|x_{t_0,x_0}(t)\| < \varepsilon$ for $t > t_0$.

If the function δ does not depend on t_0 then the origin is called *uniformly Lyapunov stable*. If $f(t, x)$ is independent of t (time-invariant) and the zero solution of (12.1) is Lyapunov stable, then it is uniformly Lyapunov stable.

Proposition 12.1 *If the origin of the system* (12.1) *is Lyapunov stable then* $x(t) = 0$ *is the unique solution of the Cauchy problem* (12.1), (12.2) *with* $x_0 = 0$ *and* $t_0 \in \mathbb{R}$.

The origin, which does not satisfy any condition from Definition 12.1, is *unstable*.

Definition 12.2 (*Asymptotic stability*) The origin of the system (12.1) is said to be asymptotically stable if it is Lyapunov stable and if for any $t_0 \in \mathbb{R}$ there exists an open set $U(t_0) \subseteq \mathbb{R}^n : 0 \in \text{int}(\text{U}(\text{t}_0))$ such that $\forall x_0 \in U(t_0),\ \lim\limits_{t\to+\infty} \|x_{t_0,x_0}(t)\| = 0$ holds.

The set $U(t_0)$ is called *domain of attraction*. It is always a neighborhood of the origin. If $U(t_0) = \mathbb{R}^n$ then the asymptotically stable origin of the system (12.1) is called *globally asymptotically stable*.

The uniform asymptotic stability asks for a more strong attractivity property.

Definition 12.3 (*Uniform asymptotic stability*) The origin of the system (12.1) is said to be *uniformly* asymptotically stable if it is asymptotically stable with a time-invariant attraction domain $U \subseteq \mathbb{R}^n$ and $\forall R > 0, \forall \varepsilon > 0$ there exists $t^* = t^*(R, \varepsilon) \in \mathbb{R}_+$ such that $\forall x_0 \in U : \|x_0\| < R$ and $\forall t_0 \in \mathbb{R}$, $\|x_{t_0,x_0}(t)\| < \varepsilon$ for $t > t_0 + t^*$ holds.

If $U = \mathbb{R}^n$ then a uniformly asymptotically stable origin of the system (12.1) is called *globally uniformly asymptotically stable*. Uniform asymptotic stability always implies asymptotic stability. The converse holds for time-invariant systems.

Proposition 12.2 ([9], Proposition 2.2, p. 78) *Let a set-valued function* $F : R^n \to \mathbb{R}^n$ *be defined and upper-semicontinuous in* $\mathbb{R}^n$. *Let* $F(x)$ *be nonempty, compact and convex for any* $x \in \mathbb{R}^n$. *If the origin of the system* $\dot{x} \in F(x)$ *is asymptotically stable then it is uniformly asymptotically stable.*

12.2.2 Rated Stability

In order to provide good performance to a control system, the rate of transition processes has to be adjusted. An asymptotic stability does not characterize **convergence rate**, which should be somehow specified. The exponential stability is the classical example of a "rated" stability.

Definition 12.4 (*Exponential stability*) The origin of the system (12.1) is said to be exponentially stable if it is asymptotically stable and for any $t_0 \in \mathbb{R}$, there exists an attraction domain $U(t_0)$ and $C = C(t_0) > 0$, $r = r(t_0) > 0$ such that

$$\|x_{t_0,x_0}(t)\| \leq C\|x_0\|e^{-r(t-t_0)}, \quad t > t_0, \quad x_0 \in U(t_0). \tag{12.3}$$

The exponential stability is *uniform* if $U(t_0)$, $C(t_0)$ and $r(t_0)$ are time-invariant. The parameter r defines the rate of exponential convergence. Obviously, exponential stability implies both Lyapunov stability and asymptotic stability, and it is usually exploited by linear control theory.

Definition 12.5 (*Hyper-Exponential Stability*) The origin of the system (12.1) is said to be hyper-exponentially stable if it is exponentially stable with $U(t_0) \subseteq \mathbb{R}^n$, $t_0 \in \mathbb{R}$ and

$$\forall C > 0, \ \forall r > 0, \quad \exists x_0 \in U(t_0), \ \exists t' = t'(t_0) \ : \ \|x_{t_0,x_0}(t)\| < C\|x_0\|e^{-r(t-t_0)}, \quad t > t_0 + t'. \tag{12.4}$$

The origin is *uniformly hyper-exponentially stable* if $U(t_0)$, $t'(t_0)$ are time-invariant. Definition 12.5 also introduces kind of "nonrated" stability since it does not provide any quantitative index to characterize (compare) the hyper-exponential convergence rates.

Given a vector $\alpha = (\alpha_0, \alpha_1, \ldots, \alpha_r)^\top \in \mathbb{R}_+^{r+1}$, $\alpha_i > 0$ with $r \geq 0$ let us define recursively the following family of functions

$$\rho_{0,\alpha}(s) = \alpha_0 s, \quad \rho_{i,\alpha}(s) = \alpha_i(e^{\rho_{i-1,\alpha}(s)} - e^{\rho_{i-1,\alpha}(0)}), \quad i = 1, 2, \ldots, r \ .$$

Obviously $\rho_{i,\alpha}(0) = 0$. The Fig. 12.1 depicts $e^{-\rho_{i,\alpha}(t)}$, $t > 0$ for $i = 1, 2$ and $\alpha_i = 1$ in a logarithmic scale in order to show the decay rate.

Definition 12.6 (*Rated Hyper-Exponential Stability*, [37]) The origin of the system (12.1) is said to be *hyper-exponentially stable of degree* $r \geq 0$, if it is hyper-exponentially stable with $U(t_0) \subseteq \mathbb{R}^n$, $t_0 \in \mathbb{R}$ and $\exists C = C(t_0) > 0$, $\exists \alpha = \alpha(t_0) \in \mathbb{R}_+^{r+1}$ such that

$$\|x_{t_0,x_0}(t)\| \leq C\|x_0\|e^{-\rho_{r,\alpha}(t-t_0))}, \quad t > t_0, \quad x_0 \in U(t_0). \tag{12.5}$$

The rated hyper-exponential stability becomes *uniform* if $U(t_0)$, $C(t_0)$ and $r(t_0)$ are time-invariant. By analogy with the exponential case let us call the vector α by the rate of hyper-exponential convergence. *If the degree r is equal to zero then rate of convergence becomes exponential.*

Example 12.1 The right-hand side of the system

$$\dot{x}(t) = -2x(t)|\ln|x(t)||, \quad t > t_0, \quad x(t_0) = x_0 \in U := (-0.5, 0.5),$$

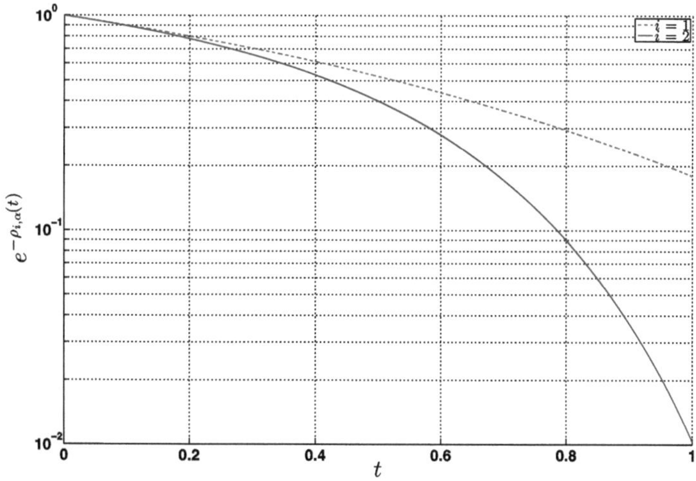

Fig. 12.1 Hyper-exponential rate of convergence

is continuous at the origin (and, in fact, in $\mathbb{R}$). The considered system has the following an explicit solution

$$x(t) = x_0 e^{\ln(|x_0|)(e^{2(t-t_0)}-1)} \quad \text{for} \quad x_0 \in U.$$

Since, obviously, we have $|x(t)| \leq |x_0| e^{-\ln(2)(e^{2(t-t_0)}-e^{2\cdot 0})}$ for all $x_0 \in U$ then the origin of the considered system is uniformly hyper-exponentially stable of degree 1 with the rate of hyper-exponential convergence given by the vector $\alpha = (2, \ln 2)^\top$.

Remark 12.1 Similarly to the exponential case, the index α (rate of hyper-exponential convergence) can be utilized for comparison of decay rates. Indeed, it is easy to show that for $\alpha, \beta \in \mathbb{R}_+^{r+1}$ the inequality $\alpha \geq \beta$ (understood in a component-wise sense) implies $e^{-\rho_{r,\alpha}(s)} \leq e^{-\rho_{r,\beta}(s)}$ for all $s \geq 0$.

12.2.3 Non-asymptotic Convergence

The motivating example considered in Sect. 12.1.1 presents the mechanical system, which has non-asymptotic transitions, i.e., any trajectory reach the equilibrium after a finite instant of time.

Definition 12.7 (*Finite-time stability* [6, 43]) The origin of the system (12.1) is said to be finite-time stable if it is Lyapunov stable in $U(t_0) \in \mathbb{R}^n$, $t_0 \in \mathbb{R}$ and *finite-time attractive*: $\forall x_0 \in U(t_0)$, $\exists T = T(t_0, x_0) \geq 0$ such that $x_{t_0,x_0}(t) = 0$, $\forall t \geq t_0 + T$.

The finite-time transitions are required for many control applications. For example, anti-missile control has to be designed only on a finite interval of time, since there is nothing to control after missile explosion.

Obviously, that finite-time stability always implies asymptotic stability. The **settling-time function** T of time-invariant finite-time stable system (12.1) is independent of t_0, i.e., $T = T(x_0)$. However, in contrast to asymptotic and Lyapunov stability, finite-time stability of a time-invariant system, in general, does not imply uniform finite-time stability, which asks at least for boundednees of the settling-time function in a neighborhood of the origin.

Example 12.2 (*[6], p. 756*) Let a vector field $f : \mathbb{R}^2 \to \mathbb{R}^2$ of a time-invariant system be defined on the quadrants

$$Q_I = \left\{x \in \mathbb{R}^2 \backslash \{0\} : x_1 \geq 0,\ x_2 \geq 0\right\}\ Q_{II} = \left\{x \in \mathbb{R}^2 : x_1 < 0,\ x_2 \geq 0\right\}$$
$$Q_{III} = \left\{x \in \mathbb{R}^2 : x_1 \leq 0,\ x_2 < 0\right\}\ Q_{IV} = \left\{x \in \mathbb{R}^2 : x_1 > 0,\ x_2 < 0\right\}$$

as shown in Fig. 12.2. The vector field f is continuous, $f(0) = 0$ and $x = (x_1, x_2)^T = (r\cos(\theta), r\sin(\theta))^T$, $r > 0$, $\theta \in [0, 2\pi)$. In [6] it was shown that this system is finite-time stable and, moreover, it is *uniformly asymptotically stable*. However, for the sequence of the initial conditions $x_0^i = (0, -1/i)^T, i = 1, 2, \ldots$ we have (see [6] for the details) $x_0^i \to 0$ and $T(x_0^i) \to +\infty$ as $i \to +\infty$. So, considering an open ball $B(r)$ of the radius r with the center at the origin we have for any $r > 0$ that

$$\sup_{x_0 \in B(r)} T(x_0) = +\infty,$$

i.e., the trajectories of the considered system converge to zero in finite-time, but non-uniformly with respect to the initial conditions.

Definition 12.8 (*Uniform finite-time stability*, [28, 38]) The origin of the system (12.1) is said to be uniformly finite-time attractive if it is finite-time stable in a time-invariant attraction domain $U \subseteq \mathbb{R}^n$ and the settling-time function $T : \mathbb{R} \times U \to \mathbb{R}$ is locally bounded on $\mathbb{R} \times U$ uniformly on the first argument, i.e.,

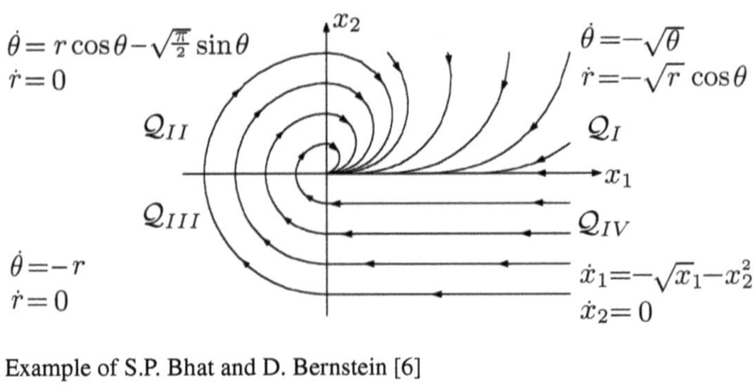

Fig. 12.2 Example of S.P. Bhat and D. Bernstein [6]

$$\forall y \in U, \quad \exists \varepsilon > 0 \quad \text{such that} \quad \sup_{t_0 \in \mathbb{R},\ \|x_0 - y\| < \varepsilon} T(t_0, x_0) < +\infty.$$

The mechanical system considered in the motivating example presented in Sect. 12.1.1 is obviously uniformly finite-time stable even if the drag force is assumed to be equal to zero. The quadratic term provided by drag force implies faster non-asymptotic transitions characterized by the next definition.

Definition 12.9 (*Fixed-time stability*, [32]) The origin of the system (12.1) is said to be fixed-time stable if it is uniformly finite-time stable in $U \subset \mathbb{R}^n$ and the settling-time function $T(t_0, x_0)$ is *bounded* on $\mathbb{R} \times U$, i.e.,

$$\exists T_{\max} > 0 : x_{t_0,x_0}(t) = 0, \quad t > t_0 + T_{\max}, \quad \forall t_0 \in \mathbb{R}, \quad \forall x_0 \in U.$$

If $U = \mathbb{R}^n$ then the origin of the system (12.1) is *globally stable* in the sense of the definitions given above.

Obviously, all finite-time and fixed-time stable systems are also hyper-exponentially stable (locally or globally).

12.3 Mathematical Tools for Analysis of Fast Nonlinear Systems

12.3.1 Generalized Homogeneity

Homogeneity, widely studied in control theory [14, 23, 31, 36, 40], is a sort of symmetry of an object (e.g., function or vector field) with respect to some group of transformations called *dilation*.

12.3.1.1 Dilation Group

Let $\|\cdot\|$ be a norm in $\mathbb{R}^n$ and $\|\cdot\|_{\mathbb{A}}$ be the matrix norm induced by $\|\cdot\|$, i.e., $\|A\|_{\mathbb{A}} = \sup_{u \in \mathbb{R}^n} \frac{\|Au\|}{\|u\|}$ if $A \in \mathbb{R}^{n \times n}$.

Definition 12.10 A map $\mathfrak{d} : \mathbb{R} \to \mathbb{R}^{n \times n}$ is called **dilation** in $\mathbb{R}^n$ if it satisfies

- **the group property:** $\mathfrak{d}(0) = I \in \mathbb{R}^{n \times n}$ and $\mathfrak{d}(t + s) = \mathfrak{d}(t)\mathfrak{d}(s)$ for $t, s \in \mathbb{R}$;
- **the continuity property:** the map $\mathfrak{d}$ is continuous in the norm $\|\cdot\|_{\mathbb{A}}$, i.e.,

$$\forall t > 0, \ \forall \varepsilon > 0, \quad \exists \delta = \delta(t, \varepsilon) > 0 \ : \ |s - t| < \delta \ \Rightarrow \ \|\mathfrak{d}(s) - \mathfrak{d}(t)\|_{\mathbb{A}} \leq \varepsilon.$$

- **the limit property:** $\lim_{s \to -\infty} \|\mathfrak{d}(s)u\| = 0$ and $\lim_{s \to +\infty} \|\mathfrak{d}(s)u\| = +\infty$ uniformly on $u \in S$, where $S = \{u \in \mathbb{R}^n : \|u\| = 1\}$ is the unit sphere in $\mathbb{R}^n$.

The dilation given by this definition was originally introduced in [34] for abstract Banach spaces **B** in the form a strongly continuous group of linear bounded operators [29], i.e., $\mathfrak{d}(s) \in \mathcal{L}(\mathbf{B}, \mathbf{B})$ and $\mathfrak{d}(\cdot)u : \mathbb{R} \to \mathbf{B}$ is continuous for any $u \in \mathbf{B}$. In a finite dimensional space any strongly continuous group is uniformly continuous, i.e., continuous in the norm of the Banach space $\mathcal{L}(\mathbf{B}, \mathbf{B})$ - the space of linear bounded operators $\mathbf{B} \to \mathbf{B}$. For $\mathbf{B} = \mathbb{R}^n$ we derive $\mathcal{L}(\mathbf{B}, \mathbf{B}) = \mathbb{R}^{n\times n}$ and the Definition 12.10 is equivalent Definition 1 from [34].

Example 12.3 The well-known dilations in $\mathbb{R}^n$

- *uniform dilation* (L. Euler 17th century): $\mathfrak{d}(s) = e^s$, $s \in \mathbb{R}$;
- *weighted dilation* (Zubov 1958, [49]): $\mathfrak{d}(s) = \begin{pmatrix} e^{r_1 s} & 0 & \dots & 0 \\ 0 & e^{r_2 s} & \dots & 0 \\ \dots & \dots & \dots & \dots \\ 0 & 0 & \dots & e^{r_n s} \end{pmatrix}$, $s \in \mathbb{R}$, $r_i > 0$;

obviously, satisfy Definition 12.10. The geometric dilation [23, 41, 48] is more general since it allows the map $\mathfrak{d}(s) : \mathbb{R}^n \to \mathbb{R}^n (s \in \mathbb{R})$ to be nonlinear.

The matrix $G_{\mathfrak{d}} \in \mathbb{R}^{n\times n}$ defined as $G_{\mathfrak{d}} = \lim_{s\to 0} \frac{\mathfrak{d}(s)-I}{s}$ is known (see, e.g., [29, Chapter 1]) as the **generator** of the group $\mathfrak{d}(s)$. It satisfies the following properties

$$\frac{d}{ds}\mathfrak{d}(s) = G_{\mathfrak{d}}\mathfrak{d}(s) = \mathfrak{d}(s)G_{\mathfrak{d}}, \; s \in \mathbb{R} \quad \text{and} \quad \mathfrak{d}(s) = e^{G_{\mathfrak{d}} s} = \sum_{i=0}^{+\infty} \frac{s^i G_{\mathfrak{d}}^i}{i!}.$$

If we denote $\lfloor A \rfloor_{\mathbb{A}} = \inf_{u\in\mathbb{R}^n} \frac{\|Au\|}{\|u\|}$, $A \in \mathbb{R}^{n\times n}$ then the limit property implies that

- $\mathfrak{d}(s) \neq I$ if $s \neq 0$;
- $\lfloor \mathfrak{d}(s) \rfloor_{\mathbb{A}} \to +\infty$ as $s \to +\infty$;
- $\|\mathfrak{d}(s)\|_{\mathbb{A}} \to 0$ as $s \to -\infty$;
- $\lfloor G_{\mathfrak{d}} \rfloor > 0$ (i.e. $\ker G_{\mathfrak{d}} = \{\mathbf{0}\}$).

Definition 12.11 The dilation $\mathfrak{d}$ is **monotone** on $\mathbb{R}^n$ if $\|\mathfrak{d}(s)\|_{\mathbb{A}} < 1$ for $s < 0$.

Monotonicity of the dilation depends on the norm $\|\cdot\|$.

Example 12.4 The dilation $\mathfrak{d}(s) = e^s \begin{pmatrix} \cos(s) & \sin(s) \\ -\sin(s) & \cos(s) \end{pmatrix}$ with $G_{\mathfrak{d}} = \begin{pmatrix} 1 & 1 \\ -1 & 1 \end{pmatrix}$ is monotone on $\mathbb{R}^2$ equipped with the weighted norm $\|u\|_P = \sqrt{u^\top P u}$ if $P = \begin{pmatrix} 1 & 1/\sqrt{2} \\ 1/\sqrt{2} & 1 \end{pmatrix} > 0$ and it is non-monotone if, for example, $P = \begin{pmatrix} 1 & 3/4 \\ 3/4 & 1 \end{pmatrix} > 0$. In the latter case, the curve $\{\mathfrak{d}(s)u : s \in \mathbb{R}\}$ may cross the unit sphere in two different points.

Theorem 12.1 *A dilation $\mathfrak{d}$ is monotone on $\mathbb{R}^n$ if and only if one of the following conditions holds:*

(1) $\lfloor \mathfrak{d}(s) \rfloor_{\mathbb{A}} > 1$ *for* $s > 0$;
(2) the continuous function $\|\mathfrak{d}(\cdot)u\| : \mathbb{R} \to \mathbb{R}_+$ *is strictly increasing for any* $u \in S$, *where* $S := \{u \in \mathbb{R}^n : \|u\| = 1\}$ *is the* ***unit sphere****;*
(3) for any $u \in \mathbb{R}^n$ *there exists a unique pair* $(s_0, u_0) \in \mathbb{R} \times S$ *such that* $u = \mathfrak{d}(s_0)u_0$.

Proof (**1**) For any $u \in S$ we have $1 = \|u\| = \|\mathfrak{d}(s)\mathfrak{d}(-s)u\| \le \|\mathfrak{d}(s)\|_{\mathbb{A}}\|\mathfrak{d}(-s)u\|$. Hence, $1 \le \|\mathfrak{d}(s)\|_{\mathbb{A}}\lfloor \mathfrak{d}(-s)\rfloor_{\mathbb{A}}$ for any $s \in \mathbb{R}$.
(**2**) On the one hand, since $\sup_{u\in S}\|\mathfrak{d}(s)u\| = \|\mathfrak{d}(s)u_s\|$ for some $u_s \in S$, then strict monotonicity of $\|\mathfrak{d}(\cdot)u\|$ for any $u \in S$ implies $\|\mathfrak{d}(s)\| < 1$ for $s < 0$. On the other hand, if $\mathfrak{d}$ is monotone then for $u \ne 0$ and $s_1 < s_2$ one has $\|\mathfrak{d}(s_1)u\| - \|\mathfrak{d}(s_2)u\| = \|\mathfrak{d}(s_1)u\| - \|\mathfrak{d}(s_2 - s_1)\mathfrak{d}(s_1)u\| \le (1 - \lfloor \mathfrak{d}(s_2 - s_1)\rfloor_{\mathbb{A}})\|\mathfrak{d}(s_1)u\| < 0$. This implies the function $\|\mathfrak{d}(\cdot)u\|$ is strictly increasing for any $u \in S$.
(**3**) *Necessity.* Existence and uniqueness of the pair (s_0, u_0) such that $u_0 = \mathfrak{d}(s_0)u \in S$ for $u \in \mathbb{R}^n$ immediately implies continuity of the dilation and the condition 2). *Sufficiency.* If $u \in S$ is an arbitrary vector from the unit sphere then $\mathfrak{d}(s)u \notin S$ for all $s \ne 0$. Indeed, otherwise the pair $(s_0, u_0) \in \mathbb{R} \times S : u_0 = \mathfrak{d}(s)u \in S$ is not unique. Hence, the limit property of the dilation (see, Definition 12.10) implies $\|\mathfrak{d}(s)u\| < 1$ for all $s < 0$ and all $u \in S$, i.e., $\|\mathfrak{d}(s)\|_{\mathbb{A}} < 1$ for $s < 0$.

Theorem 12.1 guarantees the functions $\|\mathfrak{d}(\cdot)\|_{\mathbb{A}} : \mathbb{R} \to \mathbb{R}_+$ and $\lfloor \mathfrak{d}(\cdot)\rfloor_{\mathbb{A}} : \mathbb{R} \to \mathbb{R}_+$ are also continuous and strictly increasing. Moreover, if $\|\cdot\|$ is C^k outside the origin then the identity $\frac{d}{ds}\mathfrak{d}(s) = G_{\mathfrak{d}}\mathfrak{d}(s)$ guarantees that these functions are also C^k.

Definition 12.12 The dilation $\mathfrak{d}$ is said to be **strictly monotone** on $\mathbb{R}^n$ if there exists $\beta > 0$ such that $\|\mathfrak{d}(s)\|_{\mathbb{A}} \le e^{\beta s}$ for $s \le 0$.

The dilation $\mathfrak{d}$ considered in Example 12.4 is strictly monotone on $\mathbb{R}^2$ equipped with the conventional Euclidian norm.

Theorem 12.2 *Let $\mathfrak{d}$ be a dilation in $\mathbb{R}^n$ then*

- *the matrix $-G_{\mathfrak{d}}$ is Hurwitz, i.e., all eigenvalues λ_i of $G_{\mathfrak{d}}$ are placed in the right complex half-plane;*
- *for any $\beta \in (0, \beta^*]$ there exists a symmetric matrix $P \in \mathbb{R}^{n\times n}$, $P = P^\top$ such that*

$$PG_{\mathfrak{d}} + G_{\mathfrak{d}}^\top P \ge 2\beta P, \quad P > 0; \tag{12.6}$$

where $-\beta^* < 0$ is the spectral abscissa of $-G_{\mathfrak{d}}$, i.e., $\beta^* = \min\Re(\lambda_i)$
- *the dilation $\mathfrak{d}$ is strictly monotone with respect to the weighted Euclidean norm $\|\cdot\| = \sqrt{\langle\cdot,\cdot\rangle}$ induced by the inner product $\langle u, v\rangle = u^\top P v$ with P satisfying* (12.6) *and*

$$e^{\alpha s} \le \lfloor \mathfrak{d}(s)\rfloor_{\mathbb{A}} \le \|\mathfrak{d}(s)\|_{\mathbb{A}} \le e^{\beta s} \textit{ if } s \le 0,\ e^{\beta s} \le \lfloor \mathfrak{d}(s)\rfloor_{\mathbb{A}} \le \|\mathfrak{d}(s)\|_{\mathbb{A}} \le e^{\alpha s} \textit{ if } s \ge 0,$$

$$\begin{aligned} \alpha &:= \sup_{z\in S}\langle G_{\mathfrak{d}}z, z\rangle = \frac{1}{2}\lambda_{\max}(P^{1/2}G_{\mathfrak{d}}P^{-1/2} + P^{-1/2}G_{\mathfrak{d}}^\top P^{1/2}) > 0, \\ \beta &:= \inf_{z\in S}\langle G_{\mathfrak{d}}z, z\rangle = \frac{1}{2}\lambda_{\min}(P^{1/2}G_{\mathfrak{d}}P^{-1/2} + P^{-1/2}G_{\mathfrak{d}}^\top P^{1/2}) > 0. \end{aligned} \tag{12.7}$$

Proof Since $\frac{d}{ds}\mathfrak{d}(s) = G_{\mathfrak{d}}\mathfrak{d}(s)$, $\mathfrak{d}(0) = I$ then $\mathfrak{d}(s)$ is the fundamental matrix of the linear system of ODEs with the matrix $G_{\mathfrak{d}}$. The limit property of the dilation implies that this system of ODEs is globally asymptotically stable in the inverse time, i.e., the matrix $-G_{\mathfrak{d}}$ is Hurwitz. Hence, there exists a symmetric positive definite matrix such that (12.6) holds and for any $u \in S$ one has $\frac{d}{ds}\|\mathfrak{d}(s)u\|^2 = u^\top \mathfrak{d}(s)^\top (G_{\mathfrak{d}}^\top P + PG_{\mathfrak{d}})\mathfrak{d}(s)u \geq 2\beta\|\mathfrak{d}(s)u\|^2$. Similarly we derive $\frac{d}{ds}\|\mathfrak{d}(s)u\|^2 \leq \alpha\|\mathfrak{d}(s)u\|^2$, i.e., the inequalities (12.7) hold.

Therefore, any dilation $\mathfrak{d}$ is strictly monotone on $\mathbb{R}^n$ equipped with the weighted Euclidian norm $\|u\| = \sqrt{u^\top Pu}$ if the matrix $P > 0$ satisfies (12.6).

12.3.1.2 Homogeneous Norm

The "homogeneous norm" is not a norm in the classical sense, since, in particular, the triangle inequality may not hold. However, it introduces a topology in $\mathbb{R}^n$:

$$S_{\mathfrak{d}}(r) = \left\{u \in \mathbb{R}^n : \|u\|_{\mathfrak{d}} = r\right\} \quad \text{and} \quad B_{\mathfrak{d}}(r) = \{u \in \mathrm{d} : \|u\|_{\mathfrak{d}} < r\}, \quad r > 0,$$

where $S_{\mathfrak{d}}(r)$ is the **homogeneous sphere** of the radius r, $B_{\mathfrak{d}}(r)$ is the **homogeneous ball** of the radius r and $\|\cdot\|_{\mathfrak{d}} : \mathbb{R}^n \to \mathbb{R}$ is a function introduced by the next definition.

Definition 12.13 A continuous function $\|\cdot\|_{\mathfrak{d}} : \mathbb{R}^n \to \mathbb{R}_+$ is said to be $\mathfrak{d}$-homogeneous norm if $\|u\|_{\mathfrak{d}} \to 0$ as $u \to \mathbf{0}$ and $\|\mathfrak{d}(s)u\|_{\mathfrak{d}} = e^s\|u\|_{\mathfrak{d}} > 0$ for $u \in \mathbb{R}^n\backslash\{\mathbf{0}\}$ and $s \in \mathbb{R}$.

There are many ways to construct a homogeneous norm in $\mathbb{R}^n$ (see, e.g., [21, 23, 34, 40]). For monotone dilations we can introduce the **canonical homogeneous norm** as follows:

$$\|u\|_{\mathfrak{d}} = e^{s_u} \quad : \quad \|\mathfrak{d}(-s_u)u\| = 1. \tag{12.8}$$

In [33] this homogeneous norm was called canonical since it is induced by the canonical norm $\|\cdot\|$ in $\mathbb{R}^n$ and $\|x\|_{\mathfrak{d}} = \|x\| = 1$ on the unit sphere S. Obviously

$$\lfloor \mathfrak{d}(\ln \|u\|_{\mathfrak{d}})\rfloor_{\mathbb{A}} \leq \|u\| \leq \|\mathfrak{d}(\ln \|u\|_{\mathfrak{d}})\|_{\mathbb{A}},$$

where $\lfloor \mathfrak{d}(\cdot)\rfloor_{\mathbb{A}}$ and $\|\mathfrak{d}(\cdot)\|_{\mathbb{A}}$ are continuous and strictly increasing functions (see, Theorem 12.1).

As a consequence of Theorem 12.2, the symbol $\|\cdot\|_{\mathfrak{d}}$ denotes the canonical homogeneous norm by default.

Proposition 12.3 *If $\mathfrak{d}$ is strictly monotone on $\mathbb{R}^n$ with dilation rate $\beta > 0$: $\|\mathfrak{d}(s)\|_{\mathbb{A}} \leq e^{\beta s}$ for $s < 0$ (see Definition 12.12) then*

- $\left| \|u_1\|_{\mathfrak{d}}^{\beta} - \|u_2\|_{\mathfrak{d}}^{\beta} \right| \leq \|u_1 - u_2\|$ *for* $u_1, u_2 \in \mathbb{R}^n \backslash B_{\mathfrak{d}}(1)$,
- *the homogeneous norm* $\|\cdot\|_{\mathfrak{d}}$ *is Lipschitz continuous outside the origin;*
- *if the norm* $\|\cdot\|$ *is smooth outside the origin then the homogeneous norm* $\|\cdot\|_{\mathfrak{d}}$ *is also smooth outside the origin,* $\frac{d}{ds}\|\mathfrak{d}(-s)u\| < 0$ *for* $s \in \mathbb{R}$, $u \in \mathbb{R}^n \backslash \{\mathbf{0}\}$ *and*

$$\frac{\partial \|u\|_{\mathfrak{d}}}{\partial u} = \|u\|_{\mathfrak{d}} \left. \frac{\left. \frac{\partial \|z\|}{\partial z} \right|_{z=\mathfrak{d}(-s)u}}{\left. \frac{\partial \|z\|}{\partial z} \right|_{z=\mathfrak{d}(-s)u} G_{\mathfrak{d}} \mathfrak{d}(-s)u} \right|_{s=\ln \|u\|_{\mathfrak{d}}} \quad \textit{for} \quad u \in \mathbb{R}^n \backslash \{\mathbf{0}\} \tag{12.9}$$

Proof

- Since for $u_i \in \mathbb{R}^n$ we have $\|u_i\|_{\mathfrak{d}} = e^{s_i} : \|\mathfrak{d}(-s_i)u\| = 1$ then $1 = \|\mathfrak{d}(-s_1)u_1\| = \|\mathfrak{d}(-s_1)(u_1 - u_2) + \mathfrak{d}(s_2 - s_1)\mathfrak{d}(-s_2)u_2\| \leq \|\mathfrak{d}(-s_1)\|_{\mathbb{A}} \|(u_1 - u_2)\| + \|\mathfrak{d}(s_2 - s_1)\|_{\mathbb{A}}$. For $1 < \|u_2\|_{\mathfrak{d}} < \|u_1\|_{\mathfrak{d}}$ we have $0 < s_2 < s_1$ and $1 \leq e^{-\beta s_1}\|u_1 - u_2\| + e^{\beta s_2 - \beta s_1}$ or equivalently, $\|u_1\|_{\mathfrak{d}} - \|u_2\|_{\mathfrak{d}} \leq \|u_1 - u_2\|$.
- Lipschitz continuity follows from the proven inequality, the identity $\|\mathfrak{d}(s)u\|_{\mathfrak{d}} = e^s\|u\|_{\mathfrak{d}}$ and monotonicity of the dilation.
- The existence of the unique function $s : \mathbb{R}^n \to \mathbb{R}$ such that $\|\mathfrak{d}(-s(u))u\| = 1$ has been proven in Theorem 12.1. Since the dilation is strictly monotone then $\frac{d}{ds}\|\mathfrak{d}(-s)u\| < 0$ on S (and, on $\mathbb{R}^n \backslash \{0\}$) for all $s \in \mathbb{R}$ (see, Theorem 12.1). Since the norm $\|\cdot\|$ is smooth them $\frac{d}{ds}\|\mathfrak{d}(-s)u\| = -\left.\frac{\partial \|z\|}{\partial z}\right|_{z=\mathfrak{d}(-s)u} G_{\mathfrak{d}}\mathfrak{d}(-s)u$. Taking into account $\frac{\partial}{\partial u}\|u\|_{\mathfrak{d}} = e^s \left.\frac{\partial s}{\partial u}\right|_{s=\ln \|u\|_{\mathfrak{d}}}$ the formula (12.9) can be derived using the Implicit Function Theorem [22] applied to the equality $\|\mathfrak{d}(-s)u\| = 1$.

12.3.1.3 Homogeneous Functions and Homogeneous Vectors Fields

Vector fields, which are symmetric in certain sense with respect to a dilation $\mathfrak{d}$, have a lot of properties useful for control design and state estimation of both linear and nonlinear plants as well as for analysis of the convergence rate.

Definition 12.14 (*Homogeneous vector field (function)* [34]) A vector field $f : \mathbb{R}^n \to \mathbb{R}^n$ (a function $h : \mathbb{R}^n \to \mathbb{R}$) is said to be $\mathfrak{d}$-homogeneous of degree $\nu \in \mathbb{R}$ if

$$f(\mathfrak{d}(s)u) = e^{\nu s}\mathfrak{d}(s)f(u), \quad \forall u \in \mathbb{R}^n \backslash \{\mathbf{0}\}, \quad \forall s \in \mathbb{R}. \tag{12.10}$$

(resp. $h(\mathfrak{d}(s)u) = e^{\nu s}h(u), \quad \forall u \in \mathbb{R}^n \backslash \{\mathbf{0}\}, \quad \forall s \in \mathbb{R}$.)

Example 12.5 Let us consider the dilation $\mathfrak{d}(s) = e^s \begin{pmatrix} 1 & 0 & 0 \\ 0 & \cos(s) & \sin(s) \\ 0 & -\cos(s) & \sin(s) \end{pmatrix}$ that is strictly monotone with respect to the Euclidean norm $\|x\| = \sqrt{x^T x}$ and $G_{\mathfrak{d}} = \begin{pmatrix} 1 & 0 & 0 \\ 0 & 1 & 1 \\ 0 & -1 & 1 \end{pmatrix}$. The vector field $f : \mathbb{R}^3 \to \mathbb{R}^3$ defined as $f(x) = \begin{pmatrix} x_2^2 + x_3^2 \\ x_1^2(\cos(\ln |x_1|) + \sin(\ln |x_1|)) \\ x_1^2(\cos(\ln |x_1|) - \sin(\ln |x_1|)) \end{pmatrix}$ and the function $h : \mathbb{R}^3 \to \mathbb{R}$ given by $h = x_1^3 + (x_2^2 + x_3^2)^{\frac{3}{2}}$ are $\mathfrak{d}$-homogeneous of degree 1 and 3, respectively.

Example 12.6 (Vector Field with Both Negative and Positive Homogeneity Degrees) Note that the vector field may have different degrees of homogeneity dependently of dilation group. Indeed, the linear vector field $f = Ax : \mathbb{R}^n \to \mathbb{R}^n$ defined by a chain of integrators $A = \begin{pmatrix} 0 & I_{n-1} \\ 0 & 0 \end{pmatrix} \in \mathbb{R}^{n\times n}$ is $\mathfrak{d}_1$-homogeneous of degree 1 with $\mathfrak{d}_1(s) = \mathrm{diag}\{e^{is}\}_{i=1}^n$ and $\mathfrak{d}_2$-homogeneous of degree -1 with $\mathfrak{d}_2(s) = \mathrm{diag}\{e^{(n-i+1)s}\}_{i=1}^n$. A similar conclusion can be drawn for the chain of power integrators.

The homogeneity allows local properties (e.g., smoothness) of vector fields (functions) to be extended globally.

Corollary 12.1 *Let the vector field $f : \mathbb{R}^n \to \mathbb{R}^n$ (a function $h : \mathbb{R}^n \to \mathbb{R}$) be $\mathfrak{d}$-homogeneous of degree $\nu \in \mathbb{R}$ and the norm $\|\cdot\|$ in $\mathbb{R}^n$ be defined according to Theorem 12.2 with $\beta = \beta^*$ and $\alpha \geq \beta$ given by* (12.7).

(i) If the function h is bounded on the unit sphere S then

(a) for $\nu > 0$ it is continuous at the origin, $h(\mathbf{0}) = 0$ and radially unbounded[1] *provided that $h(x) \neq 0$ on S;*
(b) for $\nu = 0$ it is globally bounded in $\mathbb{R}^n$ and continuity of h at the origin implies that $h = const$;
(c) for $\nu < 0$ it is discontinuous at the origin, unbounded in any neighborhood of the origin and $|h(x)| \to 0$ as $x \to \infty$;

(ii) If the vector field f is bounded on the unit sphere S then

(a) for $\nu+\beta > 0$ it is continuous at the origin, $f(\mathbf{0}) = 0$ and radially unbounded if $f(x) \neq 0$ on S;

– for $\nu + \beta = 0$ (resp. $\nu + \alpha = 0$) it is bounded on $B_{\mathfrak{d}}(r)$ (resp. on $\mathbb{R}^n \backslash B_{\mathfrak{d}}(r)$) for any fixed $r > 0$;
– for $\nu + \beta = \nu + \alpha = 0$ it is globally bounded on $\mathbb{R}^n$;
– for $\nu+\beta < 0$ it is discontinuous at the origin, unbounded in any neighborhood of the origin and $\|f(x)\| \to 0$ as $x \to \infty$;

(iii) The vector field f (resp. function h) is Lipschitz continuous on $\mathbb{R}^n\backslash\{\mathbf{0}\}$ if and only if it satisfies the Lipschitz condition on S.

(iv) If $h : \mathbb{R}^n \to \mathbb{R}$ is a $\mathfrak{d}$-homogeneous function of degree ν and differentiable on S then it differentiable on $\mathbb{R}^n\backslash\{\mathbf{0}\}$ and

$$\frac{\partial h(u)}{\partial u} G_{\mathfrak{d}} u = \nu h(u) \quad for \quad u \in \mathbb{R}^n\backslash\{\mathbf{0}\}. \tag{12.11}$$

[1]The vector field f (resp. function f) is radially unbounded if $x \to \infty$ implies $\|f(x)\| \to +\infty$ (resp. $|h(x)| \to +\infty$).

Proof

(i) Since $h(u) = h(\mathbf{d}(\ln \|u\|_{\mathbf{d}})z) = \|u\|_{\mathbf{d}}^{\nu} h(z)$ where $z = \mathbf{d}(-\ln \|u\|)u \in S$ and the homogeneous norm is continuous then h is continuous at the origin if $\nu > 0$, discontinuous at the origin if $\nu < 0$ and globally bounded if $\nu = 0$.

(ii) Similarly, for the vector field f we derive $f(u) = \|u\|_{\mathbf{d}}^{\nu}\mathbf{d}(\ln \|u\|_{\mathbf{d}})f(z)$ and $\|u\|_{\mathbf{d}}^{\nu}\lfloor \mathbf{d}(\ln \|u\|_{\mathbf{d}})\rfloor_{\mathbb{A}}\|f(z)\| \leq \|f(u)\| \leq \|u\|_{\mathbf{d}}^{\nu}\|\mathbf{d}(\ln \|u\|_{\mathbf{d}})\|_{\mathbb{A}}\|f(z)\|$. For $\|u\| < 1$ we have $\|u\|_{\mathbf{d}}^{\nu+\alpha}\|f(z)\| \leq \|f(u)\| \leq \|u\|_{\mathbf{d}}^{\nu+\beta}\|f(z)\|$ and $\|u\|_{\mathbf{d}}^{\nu+\beta}\|f(z)\| \leq \|f(u)\| \leq \|u\|_{\mathbf{d}}^{\nu+\alpha}\|f(z)\|$ if $\|u\| > 1$.

(iii) *Sufficiency.* Let $u_i \in \mathbb{R}^n \backslash \{\mathbf{0}\}$, $i = 1, 2$ then $u_i = \mathbf{d}(\ln \|u_i\|_{\mathbf{d}})z_i$ for some $z_i \in S$, $f(u_1) - f(u_2) = f(\mathbf{d}(\ln \|u_1\|_{\mathbf{d}})z_1) - f(\mathbf{d}(\ln \|u_2\|_{\mathbf{d}})z_2) = \|u_1\|_{\mathbf{d}}^{\nu}\mathbf{d}(\ln \|u_1\|_{\mathbf{d}}) f(z_1) - \|u_2\|_{\mathbf{d}}^{\nu}\mathbf{d}(\ln \|u_2\|_{\mathbf{d}})f(z_2) = \|u_1\|_{\mathbf{d}}^{\nu}\mathbf{d}(\ln \|u_1\|_{\mathbf{d}})(f(z_1) - f(z_2)) + (\|u_1\|_{\mathbf{d}}^{\nu}\mathbf{d} (\ln \|u_1\|_{\mathbf{d}}) - \|u_2\|^{\nu}\mathbf{d}(\ln \|u_1\|_{\mathbf{d}})f(z_2) + (\|u_2\|^{\nu}\mathbf{d}(\ln \|u_1\|_{\mathbf{d}}) - \|u_2\|_{\mathbf{d}}^{\nu}\mathbf{d}(\ln \|u_2\|_{\mathbf{d}})) f(z_2)$. If $L > 0$ is a Lipschitz constant on S then $\|f(u_1) - f(u_2)\| \leq L\|u_1\|_{\mathbf{d}}^{\nu}\mathbf{d}(\ln \|u_1\|_{\mathbf{d}})\|z_1 - z_2\| + \|\mathbf{d}(\ln \|u_1\|_{\mathbf{d}})f(z_2)\|(\|u_1\|_{\mathbf{d}}^{\nu} - \|u_2\|_{\mathbf{d}}^{\nu}) + \|f(z_2)\| \|u_2\|^{\nu}\|\mathbf{d}(\ln \|u_1\|_{\mathbf{d}}) - \mathbf{d}(\ln \|u_2\|_{\mathbf{d}})\|_{\mathbb{A}}$. Since $\mathbf{d}(s_1) - \mathbf{d}(s_2) = G_{\mathbf{d}} \int_{s_2}^{s_1} \mathbf{d}(s)ds$ and the function $\|\mathbf{d}(\cdot)\|_{\mathbb{A}}$ is strictly increasing then $\|\mathbf{d}(\ln \|u_1\|_{\mathbf{d}}) - \mathbf{d}(\ln \|u_2\|_{\mathbf{d}}))\|_{\mathbb{A}} \leq M\|G_{\mathbf{d}}\||\ln \|u_1\|_{\mathbf{d}} - \ln \|u_2\|_{\mathbf{d}}|$, where $M = \max\{\|\mathbf{d}(\ln \|u_1\|_{\mathbf{d}})\|_{\mathbb{A}}, \|\mathbf{d}(\ln \|u_2\|_{\mathbf{d}}) \|_{\mathbb{A}}\}$. Since the homogeneous norm is Lipschitz continuous (see Proposition 12.3) on $\mathbb{R}^n \backslash \{\mathbf{0}\}$, and power and logarithm functions are Lipschitz continuous outside zero, then f is Lipschitz continuous outside the origin.

Necessity. Suppose the contrary, i.e., f is Lipschitz continuous on $\mathbb{R}^n \backslash \{\mathbf{0}\}$, but it does not satisfy the Lipschitz condition of S. This means for any $L_n > 0$ there exists $u_n, v_n \in S$ such that $\|f(u_n) - f(v_n)\| > L_n\|u_n - v_n\|$, $n = 1, 2, ..,$ i.e., $L_n \to +\infty$ as $n \to +\infty$. Since S is compact then f is bounded on S. This means that $\|u_n - v_n\| \to 0$ as $n \to 0$. The latter contradicts Lipschitz continuity, since for any $u \in S$ there exists $\varepsilon_u > 0$ and $L_u > 0$ such that $\|u - v\| < \varepsilon$ implies $\|f(u) - f(v)\| \leq L_u\|u - v\|$.

(iv) Since $h(u) = h(\mathbf{d}(\ln \|u\|_{\mathbf{d}})z) = \|u\|_{\mathbf{d}}^{\nu} h(z)$ where $z = \mathbf{d}(-\ln \|u\|)u \in S$ then $\frac{\partial h}{\partial u} = \frac{\partial}{\partial u}\left(\|u\|_{\mathbf{d}}^{\nu} h(z)\right)$, $z = z(u) = \mathbf{d}(-\ln \|u\|_{\mathbf{d}})u$. Since $\|\cdot\|_{\mathbf{d}}$ is smooth outside the origin, then the differentiability of the function h on the sphere S implies its differentiability on $\mathbb{R}^n \backslash \{\mathbf{0}\}$. From $h(u) = \|u\|_{\mathbf{d}}^{\nu} h(\mathbf{d}(-\ln \|u\|_{\mathbf{d}})u)$ we derive $\frac{\partial h(u)}{\partial u} = \nu h\ (\mathbf{d}(-\ln \|u\|_{\mathbf{d}})u)\|u\|_{\mathbf{d}}^{\nu-1}\frac{\partial\|u\|_{\mathbf{d}}}{\partial u} + \|u\|_{\mathbf{d}}^{\nu}\frac{\partial h(\mathbf{d}(-\ln \|u\|_{\mathbf{d}} u))}{\partial u} =$
$$\nu h(\mathbf{d}(-\ln \|u\|_{\mathbf{d}})u)\|u\|_{\mathbf{d}}^{\nu-1}\frac{\partial\|u\|_{\mathbf{d}}}{\partial u} + \|u\|_{\mathbf{d}}^{\nu}\left.\frac{\partial h(z)}{\partial z}\right|_{z=\mathbf{d}(-\ln \|u\|_{\mathbf{d}})u} \frac{\partial}{\partial u}(\mathbf{d}(-\ln \|u\|_{\mathbf{d}})u)$$
$$= \nu h(\mathbf{d}(-\ln \|u\|_{\mathbf{d}} u)\|u\|_{\mathbf{d}}^{\nu-1}\frac{\partial\|u\|_{\mathbf{d}}}{\partial u} + \|u\|_{\mathbf{d}}^{\nu}\left.\frac{\partial h(z)}{\partial z}\right|_{z=\mathbf{d}(-\ln \|u\|_{\mathbf{d}})u} \left(\mathbf{d}(-\ln \|u\|_{\mathbf{d}}) - \frac{G_{\mathbf{d}}\mathbf{d}(-\ln \|u\|_{\mathbf{d}})u}{\|u\|_{\mathbf{d}}}\frac{\partial\|u\|_{\mathbf{d}}}{\partial u}\right).$$
For $\|u\| = 1$ we derive $\frac{\partial h(u)}{\partial u} G_{\mathbf{d}} u \frac{\partial\|u\|_{\mathbf{d}}}{\partial u} = \nu h(u)\frac{\partial\|u\|_{\mathbf{d}}}{\partial u}$. Hence, multiplying by u we derive that (12.11) holds for $\|u\| = 1$. Since $\|\cdot\|_{new} = \gamma\|\cdot\|$ with $\gamma > 0$ is again the norm satisfying Theorem 12.2, then the obtained identity holds on $\mathbb{R}^n \backslash \{0\}$. Finally, note that this identity immediately implies homogeneity of the gradient field of h.

Let $\deg_{\mathfrak{d}}(h)$ (resp. $\deg_{\mathfrak{d}}(f)$) denote the homogeneity degree of $\mathfrak{d}$-homogeneous function h (resp. $\mathfrak{d}$-homogeneous vector field f).

Theorem 12.3 ("Homogeneous arithmetics") *If h, w are $\mathfrak{d}$-homogeneous functions and f, g are vector fields then*

1. *if* $\deg_{\mathfrak{d}}(h) = \deg(w)$ *then* $\deg_{\mathfrak{d}}(w+h) = \deg_{\mathfrak{d}}(w) = \deg_{\mathfrak{d}}(h)$*;*
2. $\deg_{\mathfrak{d}}(wh) = \deg_{\mathfrak{d}}(w) + \deg_{\mathfrak{d}}(h)$*;*
3. *if $h \in C^1$ at least on $\mathbb{R}^n\backslash\{\mathbf{0}\}$ then*

$$e^{\deg(h)s}\frac{\partial h(u)}{\partial u} = \left.\frac{\partial h(z)}{\partial z}\right|_{z=\mathfrak{d}(s)u}\mathfrak{d}(s), \quad \forall u \in \mathbb{R}^n\backslash\{\mathbf{0}\}, \quad \forall s \in \mathbb{R}; \tag{12.12}$$

4. *if* $\deg_{\mathfrak{d}}(f) = \deg_{\mathfrak{d}}(g)$ *then* $\deg_{\mathfrak{d}}(f+g) = \deg_{\mathfrak{d}}(f) = \deg_{\mathfrak{d}}(g)$*;*
5. $\frac{\partial h}{\partial u}f \in \mathbb{H}_{\mathfrak{d}}$ *and* $\deg_{\mathfrak{d}}\left(\frac{\partial h}{\partial u}f\right) = \deg_{\mathfrak{d}}(h) + \deg_{\mathfrak{d}}(f)$ *if* $h \in C^1(\mathbb{R}^n\backslash\{\mathbf{0}\}, \mathbb{R})$.

Proof The Properties 1,2, and 4 are obvious. The Property 3 follows from the definition of the Frechét derivative, which coincides with $\frac{\partial h}{\partial u}$ if h is smooth. Namely,

$$\lim_{\|\Delta\|\to 0}\frac{\left\|h(u+\Delta)-h(u)-\frac{\partial h(u)}{\partial u}\Delta\right\|}{\|\Delta\|} = 0 \text{ and } \lim_{\|\Delta\|\to 0}\frac{\left\|h(\mathfrak{d}(s)u+\Delta)-h(\mathfrak{d}(s)u)-\left.\frac{\partial h(z)}{\partial z}\right|_{z=\mathfrak{d}(s)u}\Delta\right\|}{\|\Delta\|} = 0$$

with $\Delta \in \mathbb{R}^n$. Since h is $\mathfrak{d}$-homogeneous then $\frac{\left\|h(\mathfrak{d}(s)u+\Delta)-h(\mathfrak{d}(s)u)-\left.\frac{\partial h(z)}{\partial z}\right|_{z=\mathfrak{d}(s)u}\Delta\right\|}{\|\Delta\|} =$

$$e^{\nu s}\frac{\left\|h(u+\tilde{\Delta})-h(u)-e^{-\nu s}\left.\frac{\partial h(z)}{\partial z}\right|_{z=\mathfrak{d}(s)u}\mathfrak{d}(s)\tilde{\Delta}\right\|}{\|\mathfrak{d}(s)\tilde{\Delta}\|} \leq \frac{e^{\nu s}}{\lfloor\mathfrak{d}(s)\rfloor_{\mathbb{A}}}\frac{\left\|h(u+\tilde{\Delta})-h(u)-e^{-\nu s}\left.\frac{\partial h(z)}{\partial z}\right|_{z=\mathfrak{d}(s)u}\mathfrak{d}(s)\tilde{\Delta}\right\|}{\|\tilde{\Delta}\|},$$

where $\tilde{\Delta} = \mathfrak{d}(-s)\Delta$ such that $\|\tilde{\Delta}\| \to 0$ implies $\|\Delta\| \to 0$. Therefore, the identity (12.12) holds. The Property 5 is a straightforward corollary of the Property 3.

Let us denote by $\mathcal{L}_f^{(k)}h$ the Lie derivative of the function along the vector field f, i.e., $\mathcal{L}_f^{(0)}h = h$ and $\mathcal{L}_f^{(k)}h = \frac{\partial \mathcal{L}_f^{(k-1)}h}{\partial u}f$ for $k = 1, 2, \ldots$.

Corollary 12.2 *If the $\mathfrak{d}$-homogeneous vector field $f : \mathbb{R}^n \to \mathbb{R}^n$ and the $\mathfrak{d}$-homogeneous function $h : \mathbb{R}^n \to \mathbb{R}$ are sufficiently smooth to guarantee existence and continuity the Lie derivatives up to the order k at least on $\mathbb{R}^n\backslash\{\mathbf{0}\}$, then $\mathcal{L}_f^{(i)}h$ is $\mathfrak{d}$-homogeneous of degree* $\deg_{\mathfrak{d}}(f)\,i + \deg_{\mathfrak{d}}(h)$, $i = 0, 1, 2, \ldots$.

12.3.2 Homogeneous Differential Equations

12.3.2.1 Stability of Homogeneous Systems

The next theorem gives the most important result about scalability solutions to $\mathfrak{d}$-homogeneous evolution equations [7, 23, 27, 34, 40, 49].

Theorem 12.4 *Let* $f : \mathbb{R}^n \to \mathbb{R}^n$ *be continuous* $\mathfrak{d}$ *- homogeneous vector field of degree* $\nu \in \mathbb{R}$. *If* $\varphi_{x_0} : \mathbb{R}_+ \to \mathbb{R}^n$ *is a solution to*

$$\dot{x} = f(x), \tag{12.13}$$

with the initial condition $x(0) = x_0 \in \mathbb{R}^n$ *then* $\varphi_{\mathfrak{d}(s)x_0}(t) := \mathfrak{d}(s)\varphi(t_0 + e^{\nu s}t)$ *with* $t_0, s \in \mathbb{R}$ *is a solution to* (12.13) *with the initial condition* $x(0) = \mathfrak{d}(s)\varphi_{x_0}(t_0)$.

Proof Since $\frac{d}{dt}\varphi(t) = f(\varphi(t))$ then $\mathfrak{d}(s)\frac{d}{dt}\varphi(t) = \frac{d}{dt}\mathfrak{d}(s)\varphi(t) = \mathfrak{d}(s)f(\varphi(t)) = e^{-\nu s}f(\mathfrak{d}(s)\varphi(t))$. Making the change of time $t = t_0 + e^{\nu s}t^{new}$ we complete the proof.

This theorem has a lot of corollaries, which are very useful for qualitative analysis of homogeneous systems.

Corollary 12.3 *Let a continuous vector field* $f : \mathbb{R}^n \to \mathbb{R}^n$ *be* $\mathfrak{d}$ *- homogeneous of degree* $\nu \in \mathbb{R}$. *The origin of the system* (12.13) *is globally uniformly asymptotically stable if and only if one of the following three conditions holds*

(*1*) *it is locally attractive;*
(*2*) *there exists a strictly positively invariant compact set*[2] *to the system* (12.13)*;*
(*3*) *there exists a* $\mathfrak{d}$*-homogeneous Lyapunov function* $V : \mathbb{R}^n \to \mathbb{R}_+$ *such that* $V \in C^\infty$.

Proof (1) **Necessity** is obvious. Let us prove **Sufficiency**. Local attractivity implies that there exists a closed ball $\overline{B}$ with the center at the origin such that $\varphi_{x_0}(t) \to \mathbf{0}$ as $t \to +\infty$ for any $x_0 \in B$. Theorem 12.4 implies that the same property holds for any $\mathfrak{d}$-homogeneous ball $\mathfrak{d}(s)\overline{B}$, i.e., the origin is globally attractive. All trajectories of the system with initial conditions $x_0 \in \overline{B}$ and $t \in [0, t^*]$ form again a compact set $\overline{\Omega}_{t^*}$. Since the origin is globally attractive then the set $\overline{\Omega} = \bigcup_{t_* \geq 0} \overline{\Omega}_{t^*}$ is compact invariant set to the system (12.13). Using Theorem 12.4 we derive the same result for $\mathfrak{d}(s)\overline{\Omega}$, i.e., the origin of the system (12.13) is Lyapunov stable and, consequently, globally uniformly asymptotically stable.
(2) **Necessity** follows from the Converse Lyapunov theorem (see [4]) about existence of the Lyapunov function for any uniformly asymptotically stable continuous system. The level set of the corresponding Lyapunov function is strictly positively invariant compact set. **Sufficiency**. Since f is continuous and $\mathfrak{d}$-homogeneous then $f(\mathbf{0}) = \mathbf{0}$, i.e., the origin is the equilibrium point of the system (12.13). If Ω is strictly positively invariant then by Theorem 12.4 the set $\mathfrak{d}(s)\Omega$ as well as is also strictly positively invariant compact set for the system (12.13). This means that $0 \in$ int. Indeed, otherwise there exists s^* such that $\text{int}(\mathfrak{d}(s)\Omega) \cap \text{int}(\Omega) = \emptyset$ and $\partial(\mathfrak{d}(s)\Omega) \cap \partial(\Omega) \neq \emptyset$, but the latter contradicts the strict positive invariance of these sets. On the other hand, the origin is the unique equilibrium point of the system (12.13). Indeed, otherwise, $f(\mathfrak{d}(s)x^*) = 0$ for all $s \in \mathbb{R}$ if $f(x^*) = 0$ and the continuous curve $\{\mathfrak{d}(s)x^* : s \in \mathbb{R}\}$ crosses Ω, but this again contradicts the strict positive invariance of Ω. Taking into account continuity of the dilation $\mathfrak{d}$ and $\mathfrak{d}(s)\Omega \to \mathbf{0}$ as $s \to 0$ we

[2] A compact set Ω is strictly positively invariant for (12.13) if $x_0 \in \partial\Omega \Rightarrow \varphi_{x_0}(t) \in \text{int}()$, t > 0.

conclude that the origin is locally attractive and globally asymptotically stable.
(**3**) **Sufficiency** is obvious. Let us prove **necessity** using the idea proposed in [40]. The Converse Lyapunov Theorem implies that there exists a smooth Lyapunov function $V : \mathbb{R}^n \to \mathbb{R}_+$. Let the smooth function $a : \mathbb{R} \to \mathbb{R}_+$ be defined as $a(\rho) = e^{\frac{1}{1-\rho}}$ if $\rho > 0$ and $a(\rho) = 0$ if $\rho \le 1$. Obviously, $a'(\rho) > 0$ if $\rho > 1$. Then the function $V_{\text{hom}} : \mathbb{R}^n \to \mathbb{R}_+$ defined as $V_{hom}(x) = \int_{-\infty}^{+\infty} e^{-s} a(V(\mathfrak{d}(s)x))ds$ is $\mathfrak{d}$-homogeneous Lyapunov function to the system (12.13). Indeed, it is well-defined due to the cut-off function a), smooth, positive definite and radially unbounded. Finally, V is $\mathfrak{d}$-homogeneous $V(\mathfrak{d}(q)x) = \int_{-\infty}^{+\infty} e^{-s} a(V(\mathfrak{d}(-s+q)x))ds = e^q V(x)$ and $\dot{V}(x) = \int_{-\infty}^{+\infty} e^{-s} a'(V(\mathfrak{d}(-s)x)) \left.\frac{\partial V(z)}{\partial z}\right|_{z=\mathfrak{d}(s)x} \mathfrak{d}(s) f(x) ds < 0$ since f is $\mathfrak{d}$-homogeneous and $\left.\frac{\partial V(z)}{\partial z}\right|_{z=\mathfrak{d}(s)x} \mathfrak{d}(s) f(x) = e^{-\nu s} \left.\frac{\partial V(z)}{\partial z} f(z)\right|_{z=\mathfrak{d}(s)x} < 0.$

Corollary 12.4 *Let a continuous vector field* $f : \mathbb{R}^n \to \mathbb{R}^n$ *be* $\mathfrak{d}$ *- homogeneous of degree* $\nu \in \mathbb{R}$ *and the origin of* (12.13) *be globally uniformly asymptotically stable.*

- *If* $\nu < 0$ *then it is globally finite-time stable.*
- *If* $\nu > 0$ *then any ball* $B_{\mathfrak{d}}(r)$ *is fixed-time attractive, i.e.,*

$$\forall r > 0 \;\; \exists T = T(r) \ge 0 \;\; : \;\; \varphi_{x_0}(t) \in B_{\mathfrak{d}}(r), \;\; \forall t > T, \;\; \forall x_0 \in \mathbb{R}^n.$$

Proof Let us denote $T_0 = \sup_{\|x_0\|_{\mathfrak{d}} \le 2} \; \text{in}_{\tau \ge 0 : \|\varphi_{x_0}(t)\|_{\mathfrak{d}} \le 1, \forall t > \tau} \tau$. Since the origin is uniformly asymptotically stable then $T_0 < +\infty$ and $\varphi_{x_1}(t) := \mathfrak{d}(\ln 2)\varphi_{x_0}(T_0 + 2^\nu t)$ is a solution to (12.13) with $x(0) = x_1 := \mathfrak{d}(\ln 2)\varphi_{x_0}(T_0)$ such that $\|\varphi_{x_1}(t)\|_{\mathfrak{d}} \le 2$ for all $t \ge 0$. Hence, $\varphi_{x_2}(t) := \mathfrak{d}(\ln 2)\varphi_{x_1}(T_0 + 2^\nu t) = \mathfrak{d}(2\ln 2)\varphi_{x_0}(T_0 + 2^\nu(T_0 + 2^\nu t))$ is again a solution to (12.13) $x(0) = x_2 := \mathfrak{d}(\ln 2)\varphi_{x_1}(T_0)$ such that $\|\varphi_{x_2}(t)\|_{\mathfrak{d}} \le 2$. We derive $\varphi_{x_i}(t) = \mathfrak{d}(i \ln 2)\varphi_{x_0}\left(2^{i\nu} t + T_0 \sum_{j=0}^{i-1} 2^{j\nu}\right)$, $\|\varphi_{x_i}(T_0)\|_{\mathfrak{d}} = \left\|\mathfrak{d}(i \ln 2)\varphi_{x_0}\left(T_0 \sum_{j=0}^{i} 2^{j\nu}\right)\right\|_{\mathfrak{d}} = 2^i \|\varphi_{x_0}(T_i)\|_{\mathfrak{d}} \le 1$, where $T_i = T_0 \sum\limits_{j=0}^{i} 2^{j\nu}$. Hence, $\|\varphi_{x_0}(T_i)\|_{\mathfrak{d}} \to 0$ as $i \to +\infty$, but $T_i \to T := T_0 \sum_{j=0}^{+\infty} 2^{j\nu}$ as $i \to +\infty$. For $\nu < 0$ we have $T = \frac{T_0}{1-2^\nu} < +\infty$.

Since $\varphi_{\tilde{x}_1}(t) = \mathfrak{d}(\ln 2))\varphi_{x_0}(2^\nu t)$ is a solution to (12.13) with $x(0) = \tilde{x}_1 := \mathfrak{d}(\ln 2)x_0$ then we have $\|\varphi_{\tilde{x}_1}(t)\|_{\mathfrak{d}} = 2\|\varphi_{x_0}(2^\nu t)\|_{\mathfrak{d}} \le 2$ for $t \ge 2^{-\nu}T_0$ and $\|\varphi_{\tilde{x}_1}(t)\|_{\mathfrak{d}} \le 1$ for $t \ge 2^{-\nu}T_0 + T_0$ due to the definition of T_0. Similarly, $\varphi_{\tilde{x}_i}(t) = \mathfrak{d}(\ln 2))\varphi_{\tilde{x}_{i-1}}(2^\nu t)$ is a solution to (12.13) with $x(0) = \tilde{x}_2 := \mathfrak{d}(\ln 2)\tilde{x}_1$ and $\|\varphi_{\tilde{x}_i}(t)\|_{\mathfrak{d}} \le 1$ for all $t \ge T_0 \sum_{j=0}^{i} 2^{-j\nu}$ as $i \to +\infty$. We complete the proof with the remark: $\|\tilde{x}_i\|_{\mathfrak{d}} \to +\infty$ as $i \to +\infty$ but $T^* = T_0 \sum_{j=0}^{+\infty} 2^{-j\nu} < +\infty$ if $\nu > 0$.

12.3.2.2 Observability of Homogeneous Systems

Let us consider the nonlinear system

$$\dot{x} = f(x), \quad x \in \mathbb{R}^n, \qquad y = h(x), \quad y \in \mathbb{R}, \tag{12.14}$$

The points $x_1, x_2 \in \mathbb{R}^n$, $x_1 \neq x_2$ are ***indistinguishable*** if $h(\varphi_{x_1}(t)) = h(\varphi_{x_2}(t))$ for all $t \geq 0$. Otherwise, these points are said to be ***distinguishable***.

Definition 12.15 (*Observability*) The system (12.14) is locally observable at the point $x_0 \in \mathbb{R}^n$ if there exists a neighborhood $U(x_0)$ such that for any $y \in U(x_0)\backslash\{x_0\}$ the points x_0 and y are distinguishable. The system (12.14) is globally observable in $\mathbb{R}^n$ if it is locally observable at any $x_0 \in \mathbb{R}^n$.

Corollary 12.5 *Let $f : \mathbb{R}^n \to \mathbb{R}^n$ be $\mathfrak{d}$ - homogeneous continuous vector field and $h : \mathbb{R}^n \to \mathbb{R}$ be $\mathfrak{d}$-homogeneous continuous function. The nonlinear system* (12.14) *is observable on $\mathbb{R}^n\backslash\{\mathbf{0}\}$ if and only if it is locally observable on the sphere S.*

This corollary is the straightforward consequence of Definition 12.15 and Theorem 12.4.

Theorem 12.5 *Let the norm $\|\cdot\|$ be smooth in $\mathbb{R}^n\backslash\{\mathbf{0}\}$ and the dilation $\mathfrak{d}$ be strictly monotone on $\mathbb{R}^n$. Let $h : \mathbb{R}^n \to \mathbb{R}$ be $\mathfrak{d}$-homogeneous of degree $\mu > 0$ and $f : \mathbb{R}^n \to \mathbb{R}^n$ be $\mathfrak{d}$ - homogeneous of degree $\nu > -\frac{\mu}{n-1}$. Let the vector field $H = \begin{pmatrix} h \\ L_f h \\ \cdots \\ L_f^{(n-1)} h \end{pmatrix} : \mathbb{R}^n \to \mathbb{R}^n$ be C^1 on S. The nonlinear system* (12.14) *is globally observable if one of the following conditions holds*

(i) $n \geq 3$, $\inf_{u \in S}\|H(u)\| > 0$ *and* $\det\left(\frac{\partial H}{\partial u}\big|_{u \in S}\right) \neq 0$;
(ii) H *is C^1 at zero,* $\det\left(\frac{\partial H}{\partial u}\big|_{u=0}\right) \neq 0$ *and* $n\mu + \frac{n(n-1)}{2}\nu = \mathrm{trace}(G_{\mathfrak{d}})$.

Proof To guarantee observability it is sufficient to show that the map H is homeomorphism (or diffeomorphism) on $\mathbb{R}^n$.

We have that $L_f^{(i)} h$ is homogeneous of degree $\mu + i\nu$ (see, Corollary 12.2). Hence, $H(u) = \begin{pmatrix} \|u\|_{\mathfrak{d}}^{\mu} & 0 & \dots & 0 \\ 0 & \|u\|_{\mathfrak{d}}^{\mu+\nu} & \dots & 0 \\ \dots & \dots & \dots & \dots \\ 0 & 0 & \dots & \|u\|_{\mathfrak{d}}^{\mu+(n-1)\nu} \end{pmatrix} H(z)$, where $z = \mathfrak{d}(-\ln\|u\|_{\mathfrak{d}})u \in S$. Since the norm $\|\cdot\|$ is selected to be smooth on $\mathbb{R}^n$ and H is C^1 on the sphere S then H is C^1 on $\mathbb{R}^n\backslash\{\mathbf{0}\}$ (see, Corollary 12.1). Due to (12.12) we derive $\begin{pmatrix} e^{\mu s} & 0 & \dots & 0 \\ 0 & e^{(\mu+\nu)s} & \dots & 0 \\ \dots & \dots & \dots & \dots\dots \\ 0 & 0 & \dots & e^{(\mu+(n-1)\nu)s} \end{pmatrix} \frac{\partial H(u)}{\partial u} = \frac{\partial H(\mathfrak{d}(s)u)}{\partial u}\mathfrak{d}(s)$ and $\frac{e^{(n\mu+0.5n(n-1)\nu)s}\det\left(\frac{\partial H(u)}{\partial u}\right)}{\det(\mathfrak{d}(s))} = \det\left(\frac{\partial H(\mathfrak{d}(s)u)}{\partial u}\right)$. Note that $\det(\mathfrak{d}(s)) = \det(e^{G_{\mathfrak{d}}s}) = e^{s\,\mathrm{trace}(G_{\mathfrak{d}})}$ for all $s \in \mathbb{R}$.

(i) Therefore, $\det\left(\frac{\partial H(u)}{\partial u}\right) \neq 0)$ on $\mathbb{R}^n\backslash\{\mathbf{0}\}$. On the other hand, since $H(z) \neq \mathbf{0}$ for $z \in S$ then $\|H(u)\| \to +\infty$ as $\|u\| \to +\infty$ (i.e., H radially unbounded and, consequently, proper) and $\|H(u)\| \to 0$ as $\|u\| \to 0$ (i.e., H is continuous at zero). According to Theorem of Hadamard (see, e.g., Theorem 2.1, [44]) we derive that H is the bijection on $\mathbb{R}^n\backslash\{\mathbf{0}\}$ provided that $\mathbb{R}^n\backslash\{\mathbf{0}\}$ is simply connected C^1-manifold (that is the case for $n \geq 3$). Continuity of H at the origin proven above as well as $H(\mathbf{0}) = \mathbf{0}$ together with $H(u) \neq \mathbf{0}$ on $\mathbb{R}^n\backslash\{\mathbf{0}\}$ implies that H is the global homeomorphism.

(ii) If $n\mu + \frac{n(n-1)}{2}\nu = \text{trace}(G_{\mathfrak{d}})$ then $\det\left(\frac{\partial H(u)}{\partial u}\right) = \det\left(\frac{\partial H(\mathfrak{d}(s)u)}{\partial u}\right)$ for all $s \in \mathbb{R}$. Since H is C^1 at zero then for all $u \in \mathbb{R}^n$ we have $\det\left(\frac{\partial H(u)}{\partial u}\right) = \det\left(\frac{\partial H}{\partial u}\big|_{u=0}\right) \neq 0$, i.e., H is C^1 diffeomorphism on $\mathbb{R}^n$.

The condition (**ii**) of Theorem 12.5 covers linear systems if $\nu = 0$, $\mu = 1$ and $G_{\mathfrak{d}} = I$.

12.3.3 *Implicit Lyapunov Function Method*

Lyapunov function method [26] is the main tool for analysis of nonlinear dynamical systems. Frequently, it is very difficult to find an appropriate Lyapunov function. However, stability analysis of homogeneous differential equation can be reduced to convergence analysis from a sphere S. Indeed, this sphere can be assigned to be the unit level set of a Lyapunov function, which can be constructed by means of the homogeneous dilation of S yielding $\mathfrak{d}(s)S$ to be its e^s-level set. The latter implies that the Lyapunov function of homogeneous system can always be designed implicitly

$$V : \mathbb{R}^n \to \mathbb{R}_+ \quad \text{such that} \quad \mathfrak{d}(-\ln V(x))x \in S, \quad x \in \mathbb{R}\backslash\{\mathbf{0}\}. \tag{12.15}$$

using the so-called Implicit Lyapunov Function method [1, 24, 35].

Theorem 12.6 [35] *Let continuous function $Q : \mathbb{R}_+ \times \mathbb{R}^n \to \mathbb{R}$ satisfy the conditions*

(C1) Q is continuously differentiable outside the origin of $\mathbb{R}_+ \times \mathbb{R}^n$;
(C2) for any $x \in \mathbb{R}^n\backslash\{0\}$ there exists $V \in \mathbb{R}_+$ such that $Q(V, x) = 0$;
(C3) $\lim\limits_{\substack{x\to 0\\(V,x)\in\Omega}} V = 0$, $\lim\limits_{\substack{V\to 0+\\(V,x)\in\Omega}} \|x\| = 0$, $\lim\limits_{\substack{\|x\|\to\infty\\(V,x)\in\Omega}} V = +\infty$, where $\Omega = \{(V, x) \in \mathbb{R}_+ \times \mathbb{R}^n : Q(V, x) = 0\}$;
(C4) $\frac{\partial Q(V,x)}{\partial V} < 0$ for all $V \in \mathbb{R}_+$ and $x \in \mathbb{R}^n\backslash\{0\}$.
If $\frac{\partial Q(V,x)}{\partial x} f(x) < 0$ for all $(V, x) \in \Omega$ then the origin of (12.13) is globally uniformly asymptotically stable.

Proof The conditions (*C1*), (*C2*), (*C4*) and the implicit function theorem [22] imply that the equation $Q(V, x) = 0$ implicitly defines a unique positive definite function $V : \mathbb{R}^n\backslash\{0\} \to \mathbb{R}_+$ such that $Q(V(x), x) = 0$ for all $x \in \mathbb{R}^n\backslash\{0\}$. The function V is continuously differentiable outside the origin and $\frac{\partial V}{\partial x} = -\left[\frac{\partial Q(V,x)}{\partial V}\right]^{-1} \frac{\partial Q(V,x)}{\partial x}$ for $Q(V, x) = 0$, $x \neq 0$. Hence, since $\frac{\partial Q(V,x)}{\partial V} < 0$ then the condition $\frac{\partial Q(V,x)}{\partial x} f(x) < 0$ implies $\dot{V}(x) = \frac{\partial V}{\partial x} f(x) < 0$. Note that due to the condition (*C3*) the function V is radially unbounded and it can be continuously prolonged at the origin by $V(0) = 0$.

Evidently, the conditions of Theorem 12.6 mainly repeat (in the implicit form) the requirements of the classical theorem on global asymptotic stability (see, for example,

[4]). Indeed, Condition *(C1)* asks for smoothness of the Lyapunov function. Condition *(C2)* and the first two limits from Condition *(C3)* provide its positive definiteness. The last limit from Condition *(C3)* implies radial unboundedness of the Lyapunov function. Condition *(C5)* guarantees the negative definiteness of the total derivative of the Lyapunov function calculated along trajectories of the system (12.13). The only specific condition is *(C4)*, which is imposed by implicit function theorem (see, for example, [22]). This condition is required in order to guarantee that the Lyapunov function is (uniquely) well-defined by the equation $Q(V, x) = 0$.

Corollary 12.6 [35] *Let a continuous function* $Q : \mathbb{R}_+ \times \mathbb{R}^n \to \mathbb{R}$ *satisfy the conditions (C1)–(C4) of Theorem 12.6. If there exist* $c > 0$ *and* $0 < \mu \le 1$ *such that* $\frac{\partial Q(V,x)}{\partial x} f \le cV^{1-\mu} \frac{\partial Q(V,x)}{\partial V}$ *for* $(V, x) \in \Omega$ *then the origin of the system* (12.13) *is* ***globally uniformly finite-time stable*** *and* $T(x_0) \le \frac{V_0^{\mu}}{c\mu}$, *where* $Q(V_0, x_0) = 0$.

Proof Theorem 12.6 implies global uniform asymptotic stability of the origin of (12.13). The uniform finite-time stability of the origin follows from the differential inequality $\dot{V}(x)) \le -cV^{1-\mu}(x)$, which, due to the condition of this corollary, holds.

Corollary 12.7 [35] *Let there exist two functions* Q_1 *and* Q_2 *satisfying the conditions (C1)–(C4) of Theorem 12.6 and*

(C5) $Q_1(1, x) = Q_2(1, x)$ *for all* $x \in \mathbb{R}^n \backslash \{0\}$;
(C6) there exist $c_1 > 0$ *and* $0 < \mu < 1$ *such that the inequality* $\frac{\partial Q_1}{\partial x} f(x) \le c_1 V^{1-\mu} \frac{\partial Q_1}{\partial V}$, *holds for all* $V \in (0, 1]$ *and* $x \in \mathbb{R}^n \backslash \{0\}$ *satisfying the equation* $Q_1(V, x) = 0$;
(C7) there exist $c_2 > 0$ *and* $\nu > 0$ *such that the inequality* $\frac{\partial Q_2}{\partial x} f(x) \le c_2 V^{1+\nu} \frac{\partial Q_2}{\partial V}$, *holds for all* $V \ge 1$ *and* $x \in \mathbb{R}^n \backslash \{0\}$ *satisfying the equation* $Q_2(V, x) = 0$, *then the system* (12.13) *is* ***globally fixed-time stable*** *with the settling-time estimate* $T(x_0) \le \frac{1}{c_1\mu} + \frac{1}{c_2\nu}$.

Proof Let the two functions V_1 and V_2 be defined by the equations $Q_1(V, x) = 0$ and $Q_2(V, x) = 0$ (see, the proof of Theorem 12.6). Consider the sets $\Sigma_1 = \{x \in \mathbb{R}^n : V_1(x) > 1\}$, $\Sigma_2 = \{x \in \mathbb{R}^n : V_2(x) > 1\}$ and prove that $\Sigma_1 = \Sigma_2$. Suppose the contrary, i.e., $\exists z \in \mathbb{R}^n$ such that $z \in \Sigma_1$ and $z \notin \Sigma_2$.

On the one hand, $Q_1(V_1, z) = 0$ implies $V_1 > 1$ and $Q_1(1, z) > Q_1(V_1, z) = 0$ due to Condition C4). On the other hand, $Q_2(V_2, z) = 0$ implies $V_2 \le 1$ and $Q_2(1, z) \le Q_2(V_2, z) = 0$. The contradiction follows from *(C5)*.

Therefore, due to (*C5*) and (*C4*) the function $V : \mathbb{R}^n \to \mathbb{R}$ defined by the equality

$$V(x) = \begin{cases} V_1(x) & \text{for} \quad V_1(x) < 1, \\ V_2(x) & \text{for} \quad V_2(x) > 1, \\ 1 & \text{for} \quad V_1(x) = V_2(x) = 1, \end{cases}$$

is positive definite, continuous in $\mathbb{R}^n$ and continuously differentiable for $x \notin \{0\} \cup \{x \in \mathbb{R}^n : V(x) = 1\}$. The function V is Lipschitz continuous outside the origin and has the following Clarke's gradient [8]:

$$\nabla_C V(x) = \xi \nabla V_1(x) + (1 - \xi) \nabla V_2(x), \quad x \in \mathbb{R}^n,$$

where $\xi = 1$ for $0 < V_1(x) < 1$, $\xi = 0$ for $V_2(x) > 1$, $\xi = [0, 1]$ for $V_1(x) = V_2(x) = 1$ and ∇V_i is the gradient of the function V_i, $i = 1, 2$. Hence, due to conditions (*C6*) and (*C7*), the inequality

$$\frac{dV(\varphi_{x_0}(t))}{dt} \leq \begin{cases} -c_1 V^{1-\mu}(\varphi_{x_0}(t)) & \text{for} \quad V(\varphi_{x_0}(t)) < 1, \\ -c_2 V^{1+\nu}(\varphi_{x_0}(t)) & \text{for} \quad V(\varphi_{x_0}(t)) > 1, \\ -\min\{c_1, c_2\} & \text{for} \quad V(\varphi_{x_0}(t)) = 1, \end{cases}$$

holds for almost all t such that $\varphi_{x_0}(t) \neq 0$, where $\varphi_{x_0}(t)$ is a solution of the system (12.13) with the initial condition $x(0) = x_0$. This implies the fixed-time stability of the origin of the system (12.13) with the estimate of settling-time function given above. Please see [32] or [38] for more details.

Corollary 12.8 *Let* Q_1, Q_2 *satisfy Conditions (C1-C5) of Corollary 12.7 and (C6*) for* $(V, x_h(t)) \in \Omega$ *such that* $x(t)$ *satisfies* (12.16) *one has*

$$\tfrac{\partial Q_i}{\partial x} \leq 2\alpha_0 V \ln\left(eV^{(-1)^i}\right) \tfrac{\partial Q_i(V, x_h(t))}{\partial V}, \qquad \forall t \in \mathbb{R}_+,$$

where $\alpha_0 > 0$ *and* $i = 1, 2$*;*
then the origin of system (12.16) *is globally hyper-exponentially stable with degree* $r = 1$ *and convergence rate* $\alpha = (\alpha_0, 1)$.

The proof this corollary is similar to Corollary 12.7.

The next theorem provides the important topological characterization of homogeneous systems. In particular, it says that any asymptotically stable homogeneous system is isomorphic to some "quadratically stable" system.

Theorem 12.7 *Let* $\mathfrak{d}$ *be dilation on* $\mathbb{R}^n$ *and* $f : \mathbb{R}^n \to \mathbb{R}^n$ *be a continuous* $\mathfrak{d}$*-homogeneous vector field. The origin of the system* (12.13) *is globally asymptotically stable* ***if and only if*** *there exist a positive definite symmetric matrix* $P \in \mathbb{R}^{n\times n}$ *satisfying* (12.6) *and a* $\mathfrak{d}$*-homogeneous diffeomorphism* $\Psi : \mathbb{R}^n \backslash \{\mathbf{0}\} \to \mathbb{R}^n \backslash \{\mathbf{0}\}$ *of degree zero such that*

$$\frac{\partial \Psi^\top(x) P\, \Psi(x)}{\partial x} f(x) < 0 \quad \textit{for} \quad x \in \mathbb{R}^n : \Psi^\top(x) P\, \Psi(x) = 1.$$

Moreover, $\|\Psi(\cdot)\|_{\mathfrak{d}}:\mathbb{R}^n\to\mathbb{R}_+$ *is the* $\mathfrak{d}$*-homogeneous Lyapunov function of degree 1, where* $\|\cdot\|_{\mathfrak{d}}$ *is the* $\mathfrak{d}$*-homogeneous norm induced by the norm* $\|u\|=\sqrt{u^\top Pu}$.

Proof ***Sufficiency*** Since P satisfies (12.6) then the dilation $\mathfrak{d}$ is strictly monotone on $\mathbb{R}^n$ equipped with the norm $\|x\|=\sqrt{x^TPx}$. Since $\Psi(\mathfrak{d}(s)u)=\mathfrak{d}(s)\Psi(u)$ then Ψ can be prolonged to the origin by continuity $\Psi(\mathbf{0})=\mathbf{0}$. Note also that $\Psi(u)\neq\mathbf{0}$ for all $u\neq 0$, otherwise (i.e., $\exists u^*\neq 0:\Psi(u^*)$), due to homogeneity we derive that $\Psi(u)=\mathbf{0}$ on a smooth curve $\{\mathfrak{d}(s)u^*, s\in\mathbb{R}\}$, which starts at the origin goes to ∞. The latter contradicts the assumption that Ψ is diffeomorphism (continuously differentiable invertible map with continuously differentiable inverse) on $\mathbb{R}^n\backslash\{\mathbf{0}\}$. Since $\|\Psi(\mathfrak{d}(s)u)\|_{\mathfrak{d}}=\|\mathfrak{d}(s)\Psi(u)\|_{\mathfrak{d}}=e^s\|\Psi(u)\|_{\mathfrak{d}}$ then the function $\|\Psi(\cdot)\|_{\mathfrak{d}}$ is $\mathfrak{d}$-homogeneous of degree 0, radially unbounded, continuous at the origin and continuously differentiable outside the origin. Due to (12.9) the inequality $\left.\frac{\partial\|\Psi(x)\|}{\partial x}f(x)\right|_{\|\Psi(x)\|=1}<0$ implies $\left.\frac{\partial\|\Psi(x)\|_{\mathfrak{d}}}{\partial x}f(x)\right|_{\|\Psi(x)\|=1}<0$. Applying homogeneity we derive $\frac{\partial\|\Psi(x)\|_{\mathfrak{d}}}{\partial x}f(x)<0$ for $x\in\mathbb{R}^n\backslash\{\mathbf{0}\}$.
Necessity Since $\mathfrak{d}$ is a dilation on $\mathbb{R}^n$ then due to Theorem 12.2 it is strictly monotone on $\mathbb{R}^n$ equipped with the smooth norm $\|x\|=\sqrt{x^\top Px}$, where $P>0$ satisfies the inequality (12.6). Since the origin of the system (12.13) is asymptotically stable then according to Corollary 12.3 there exists a smooth $\mathfrak{d}$-homogeneous Lyapunov function $\tilde{V}:\mathbb{R}^n\to\mathbb{R}_+$ of degree $\mu>0$. The function $V=\tilde{V}^{1/\mu}$ is also Lyapunov function to (12.13) that is homogeneous of degree 0, continuous at the origin and smooth outside the origin. Let us consider the map $\Psi:\mathbb{R}^n\backslash\{\mathbf{0}\}\to\mathbb{R}^n\backslash\{\mathbf{0}\}$ defined as $\Psi(x)=\mathfrak{d}\left(\ln\left(\frac{V(x)}{\|x\|_{\mathfrak{d}}}\right)\right)x$ for $x\in\mathbb{R}^n\backslash\{\mathbf{0}\}$. Obviously, the inverse map $\Psi^{-1}:\mathbb{R}^n\backslash\{\mathbf{0}\}\to\mathbb{R}^n\backslash\{\mathbf{0}\}$ defined as $\Psi^{-1}(z)=\mathfrak{d}\left(-\ln\left(\frac{V(z)}{\|z\|_{\mathfrak{d}}}\right)\right)z$. Finally, $\Psi(\mathfrak{d}(s)x)=\mathfrak{d}(s)\Psi(x)$ and $\|\Psi(x)\|_{\mathfrak{d}}=V(x)$, Using (12.9) we complete the proof.

Corollary 12.9 *Let* $\mathfrak{d}$ *be a strictly monotone dilation on* $\mathbb{R}^n$ *equipped with a smooth norm* $\|\cdot\|$. *Let a map* $H:\mathbb{R}^n\backslash\{\mathbf{0}\}\to\mathbb{R}^n\backslash\{\mathbf{0}\}$ *be a diffeomorphism and* $H(\mathfrak{d}(s)u)=\mathfrak{d}(s)H(u)$, $s\in\mathbb{R}$, $u\in\mathbb{R}^n$. *If* $f:\mathbb{R}^n\to\mathbb{R}^n$ *is* $\mathfrak{d}$*-homogeneous vector field and*

$$\left.\frac{\partial\|H(x)\|}{\partial x}f(x)\right|_{\|H(x)\|=1}<0$$

then $\|H(\cdot)\|_{\mathfrak{d}}:\mathbb{R}^n\to\mathbb{R}_+$ *is the homogeneous Lyapunov function to the system* (12.13).

Proof The homogeneous norm is defined implicitly by (12.8). It satisfies the conditions (C1)–(C3). The condition (C4) is also satisfied due to strict monotonicity of $\mathfrak{d}$. Since H is the diffeomorphism on $\mathbb{R}^n\backslash\{\mathbf{0}\}$ then $\|H\cdot\|_{\mathfrak{d}}$ also satisfy conditions (C1)–(C4). Due to (12.9) the inequality $\left.\frac{\partial\|H(x)\|}{\partial x}f(x)\right|_{\|H(x)\|=1}<0$ implies $\left.\frac{\partial\|H(x)\|_{\mathfrak{d}}}{\partial x}f(x)\right|_{\|x\|=1}<0$. Finally, applying homogeneity we derive $\frac{\partial\|H(x)\|_{\mathfrak{d}}}{\partial x}f(x)<0$ for all $x\in\mathbb{R}^n\backslash\{\mathbf{0}\}$.

12.4 Fast Control for Nonlinear Plants

Let us consider the nonlinear control system

$$\dot{x} = f(x,u), \tag{12.16}$$

$$y = h(x), \tag{12.17}$$

where $x \in \mathbb{R}^n$ is the state vector of a plant, $u \in \mathbb{R}^m$ is the vector of control inputs, $f : \mathbb{R}^n \times \mathbb{R}^m \to \mathbb{R}^n$ is a continuous vector field of the plant, $y \in \mathbb{R}^k$ is a measured output given by means of measurement operator $h : \mathbb{R}^n \to \mathbb{R}^k$.

The control aim is to stabilize the origin of the system (12.16) in a fixed-time independently of the initial condition $x(0) = x_0$.

Theorem 12.8 (Dynamical State Feedback) *Let $\overline{\mathfrak{d}}_1$ and $\overline{\mathfrak{d}}_2$ be dilations in $\mathbb{R}^n$ and $\mathfrak{d}_1^*$ and $\mathfrak{d}_2^*$ be dilations in $\mathbb{R}^m$. Let the joint dilations $\mathfrak{d}_1 = \begin{pmatrix} \overline{\mathfrak{d}}_1(s) \\ \mathfrak{d}_1^*(s) \end{pmatrix}$ and $\mathfrak{d}_2 = \begin{pmatrix} \overline{\mathfrak{d}}_2(s) \\ \mathfrak{d}_2^*(s) \end{pmatrix}$ be strictly monotone on $\mathbb{R}^{n+m}$ equipped with a smooth norm $\|\cdot\|$. Let the vector field $\tilde{f} = \begin{pmatrix} f \\ 0_m \end{pmatrix} : \mathbb{R}^{n+m} \to \mathbb{R}^{n+m}$ be continuous, $\mathfrak{d}_1$-homogeneous of negative degree $\mu_1 < 0$ and $\mathfrak{d}_2$-homogeneous of positive degree $\mu_2 > 0$. If there exist $\gamma > 0$ and $c > 0$:*

$$\tfrac{\partial \|\xi\|}{\partial x} f(\xi) + c \leq \gamma \tfrac{\partial \|\xi\|}{\partial u} \left[\tfrac{\partial \|\xi\|}{\partial u}\right]^{\top} \quad \text{for} \quad \|\xi\|=1, \quad \xi = \left(\begin{smallmatrix} x \\ u \end{smallmatrix}\right) \in \mathbb{R}^{n+m} \tag{12.18}$$

then the dynamical state feedback

$$\dot{u} = \begin{cases} g_1(\xi) & \text{if } \|\xi\| \leq 1, \\ g_2(\xi) & \text{if } \|\xi\| > 1, \end{cases} \quad u(0) = 0, \quad \xi = \begin{pmatrix} x \\ u \end{pmatrix} \tag{12.19}$$

$$g_i(\xi) = -\gamma \, \|\xi\|_{\mathfrak{d}_i}^{\mu_i} \, \mathfrak{d}_i^*(\ln \|\xi\|_{\mathfrak{d}_i}) \left[\left.\tfrac{\partial \|\xi\|}{\partial u}\right|_{\xi=\xi_i} \right]^{\top}, \quad \xi_i = \mathfrak{d}_i(-\ln \|\xi\|_{\mathfrak{d}_i})\xi, \quad i=1,2$$

stabilizes the origin of the system (12.16) in a fixed-time $T(x_0) \leq T_{\max} := \frac{c}{-\mu_1\alpha_1} + \frac{c}{\mu_2\alpha_2}$, $\alpha_1 = \inf_{\|\xi\|=1} \frac{\partial \|\xi\|}{\partial \xi} G_{\mathfrak{d}_1}\xi$, $\alpha_2 = \inf_{\|\xi\|=1} \frac{\partial \|\xi\|}{\partial \xi} G_{\mathfrak{d}_2}\xi$, and $\|u(t)\| < +\infty$ for $t \geq 0$, $\|u(t)\| \to 0$ as $t \to T(x_0)$.

Proof Let us consider the system $\dot{\xi} = f_i(\xi)$, $f_i = \left(\begin{smallmatrix} f \\ g_i \end{smallmatrix}\right)$, $i = 1, 2$. By construction f_i is $\mathfrak{d}_i$-homogeneous of degree μ_i. From (12.18) we derive $\frac{\partial \|\xi\|}{\partial \xi} f_i(\xi) \leq -c$ for $\|\xi\| = 1$. Corollary 12.7 implies that in this case the homogeneous norm $\|\cdot\|_{\mathfrak{d}_i}$ is the Lyapunov function to the corresponding system. The closed-loop system (12.16), (12.19) has the right-hand side $f^* = \begin{cases} f_1 & \text{if } 0 < \|\xi\| \leq 1, \\ f_2 & \text{if } \|\xi\| \geq 1, \end{cases}$ Since $f_1(\xi) = f_2(\xi)$

for $\|\xi\| = 1$ then f^* is continuous at least outside the origin. The function $V(\xi) = \begin{cases} \|\xi\|_{\mathbf{d}_1} & \text{if } \|\xi\| \le 1, \\ \|\xi\|_{\mathbf{d}_2} & \text{if } \|\xi\| > 1 \end{cases}$ is the Lyapunov function of the closed-loop system. Indeed, it is continuous at zero, Lipschitz continuous on $\mathbb{R}^{n+m}\backslash\{\mathbf{0}\}$ and continuously differentiable on $\mathbb{R}^{n+m}\backslash S\backslash\{\mathbf{0}\}$, where S is a unit sphere in $\mathbb{R}^{n+m}$. Using formula (12.9) we derive $\dot{V}(z) \le -\frac{c}{\alpha_1}V^{1+\mu_1}$ if $0 < V < 1$, $\dot{V}(z) \le -\frac{c}{\alpha_2}V^{1+\mu_2}$ if $V > 1$ and $\dot{V}(z) = \limsup_{h\to 0^+}\frac{V(z+hf^*)-V(z)}{h} \le -\frac{c}{\max\{\alpha_1,\alpha_2\}}$ if $V = 1$. Taking into account $\mu_1 < 0$ and $\mu_2 > 0$ we immediately derive fixed-time stability of the closed-loop system.

The similar result can be extended to the output-based control design for affine single input single output system with relative degree n.

Theorem 12.9 (Dynamical Output-Feedback for SISO System ($m = k = 1$))

- *Let $\overline{\mathbf{d}}_1$ and $\overline{\mathbf{d}}_2$ be dilations in $\mathbb{R}^n$.*
- *Let the vector field $\tilde{f} = \binom{f}{0} : \mathbb{R}^{n+1} \to \mathbb{R}^n$ be continuous, $\begin{pmatrix} \overline{\mathbf{d}}_1(s) \\ e^{q_1 s}\end{pmatrix}$-homogeneous of negative degree $\mu_1 < 0$ and $\begin{pmatrix} \overline{\mathbf{d}}_2(s) \\ e^{q_2 s}\end{pmatrix}$-homogeneous of positive degree $\mu_2 > 0$, where $q_1 > 0, q_2 > 0$.*
- *Let the measurement function $h : \mathbb{R}^n \to \mathbb{R}$ be $\overline{\mathbf{d}}_1$-homogeneous of degree $\kappa_1 > -(n-1)\mu_1$ and $\overline{\mathbf{d}}_2$-homogeneous of degree $\kappa_2 > 0$.*
- *Let $H := \begin{pmatrix} h \\ L_f h \\ \dots \\ L_f^{(n-1)} h \end{pmatrix}$ be independent of u and the map $H : \mathbb{R}^n \to \mathbb{R}^n$ be global diffeomorphism in $\mathbb{R}^n$.*
- *Let the dilations $\mathbf{d}_1(s) = \begin{pmatrix} e^{\kappa_1 s} \\ e^{(\kappa_1+\mu_1)s} \\ \dots \\ e^{(\kappa_1+(n-1)\mu_1)s} \\ e^{q_1 s}\end{pmatrix}$ and $\mathbf{d}_2(s) = \begin{pmatrix} e^{\kappa_2 s} \\ e^{(\kappa_2+\mu_2)s} \\ \dots \\ e^{(\kappa_2+(n-1)\mu_2)s} \\ e^{q_2 s}\end{pmatrix}$ be strictly monotone on $\mathbb{R}^{n+1}$ equipped with a smooth norm $\|\cdot\|$.*

If there exist $\gamma > 0$ and $c > 0$ such that

$$\frac{\partial\left\|\binom{H(x)}{u}\right\|}{\partial x} f(x,u) + c \le \gamma \frac{\partial\left\|\binom{H(x)}{u}\right\|}{\partial u}\left[\frac{\partial\left\|\binom{H(x)}{u}\right\|}{\partial u}\right]^\top \quad \text{for} \quad \left\|\binom{H(x)}{u}\right\| = 1 \tag{12.20}$$

then the dynamical output-based control

$$\dot{u} = \begin{cases} g_1(z) & \text{if } \|z\| \le 1, \\ g_2(z) & \text{if } \|z\| > 1, \end{cases} \quad u(0) = 0, \quad z = \left(y, \dot{y}, \dots, y^{(n-1)}, u\right)^\top, \tag{12.21}$$

$$g_i(z) = -\gamma\, \|z\|_{\mathbf{d}_i}^{q_i+\mu_i}\left[\left.\frac{\partial\|z\|}{\partial u}\right|_{z=z_i}\right]^\top, \quad z_i = \mathbf{d}_i(-\ln\|z\|_{\mathbf{d}_i})z, \quad i = 1, 2$$

stabilizes the origin of the system (12.16) *in a fixed-time* $T(x_0) \leq T_{\max} := \frac{c}{-\mu_1\alpha_1} + \frac{c}{\mu_2\alpha_2}$, $\alpha_1 = \inf_{\|z\|=1} \frac{\partial\|z\|}{\partial\xi} G_{\mathfrak{d}_1} z$, $\alpha_2 = \inf_{\|z\|=1} \frac{\partial\|z\|}{\partial\xi} G_{\mathfrak{d}_2} z$, *and* $\|u(t)\| < +\infty$ *for* $t \geq 0$, $\|u(t)\| \to 0$ *as* $t \to T(x_0)$.

Proof Let us consider the system $\dot{\xi} = f_i(\xi) := \begin{pmatrix} f(x,u) \\ g_i(H(x),u) \end{pmatrix}$, $\xi = \binom{x}{u}$, $i = 1, 2$. Taking into account $y^{(j)} = \mathcal{L}_f^{(j)} h$ and $\deg_{\mathfrak{d}_i}(\mathcal{L}_f^{(j)} h) = \kappa_i + j\mu_i$ we derive by construction f_i is $\begin{pmatrix} \bar{\mathfrak{d}}_i(s) \\ e^{q_i s} \end{pmatrix}$-homogeneous of degree μ_i. Since the map $H : \mathbb{R}^n \to \mathbb{R}^n$ is the global diffeomorphism then the map $\tilde{H} : \mathbb{R}^{n+1} \to \mathbb{R}^{n+1}$ defined as $\tilde{H}(\xi) := \binom{H(x)}{u}$ is global diffeomorphism on $\mathbb{R}^{n+1}$. Note that $\kappa_i + (n-1)\mu_i > 0$ implies $\tilde{H}(\mathbf{0}) = \mathbf{0}$ and $\tilde{H}(\xi) \neq \mathbf{0}$ if $\xi \neq \mathbf{0}$. Since $\left\| \tilde{H}\left(\begin{pmatrix} \bar{\mathfrak{d}}_i(s) \\ e^{q_i s} \end{pmatrix} \xi \right) \right\|_{\mathfrak{d}_i} = \left\| \mathfrak{d}_i(s)\tilde{H}(\xi) \right\|_{\mathfrak{d}_i} = e^s \left\| \tilde{H}(\xi) \right\|_{\mathfrak{d}_i}$ then the function $V_i : \mathbb{R}^{n+1} \to \mathbb{R}_+$ defined as $V_i(\xi) = \left\| \tilde{H}(\xi) \right\|_{\mathfrak{d}_i}$ is $\begin{pmatrix} \bar{\mathfrak{d}}_i(s) \\ e^{q_i s} \end{pmatrix}$-homogeneous of degree 1, continuous on $\mathbb{R}^{n+1}$, smooth outside the origin (due to strict monotonicity of the dilation $\mathfrak{d}_i$ and differentiability of $\tilde{H}$) and radially unbounded. From (12.18) we derive $\frac{\partial \left\| \tilde{H}(\xi) \right\|}{\partial \xi} f_i(\xi) \leq -c$ for $\left\| \tilde{H}(\xi) \right\| = 1$. Note that $f_1(\xi) = f_2(\xi)$ for $\|\xi\| = 1$, so the closed-loop system (12.16), (12.19) has continuous right-hand side. The function $V(\xi) = \begin{cases} V_1(\xi) & \text{if } \|\tilde{H}(\xi)\| \leq 1, \\ V_2(\xi) & \text{if } \|\tilde{H}(\xi)\| > 1 \end{cases}$ is the Lyapunov function for the closed-loop system. This Lyapunov function is Lipschitz continuous on $\mathbb{R}^{n+m}$, continuously differentiable on $\mathbb{R}^{n+1} \backslash \tilde{S} \backslash \{\mathbf{0}\}$, where $\tilde{S} = \{\xi \in \mathbb{R}^{n+1} : \|\xi\| = 1\}$ is a unit level set of the function V. Using formula (12.9) we derive $\dot{V}(z) \leq -\frac{c}{\alpha_1} V^{1+\mu_1}$ if $0 < V < 1$, $\dot{V}(z) \leq -\frac{c}{\alpha_2} V^{1+\mu_2}$ if $V > 1$ and $\dot{V}(z) = \limsup_{h \to 0^+} \frac{V(z+hf^*) - V(z)}{h} \leq -\frac{c}{\max\{\alpha_1, \alpha_2\}}$ if $V = 1$. Taking into account $\mu_1 < 0$ and $\mu_2 > 0$ we immediately derive fixed-time stability of the closed-loop system.

12.5 Discussions and Conclusions

12.5.1 Summary of the Obtained Results

- The notions of fast stability are surveyed. Quantitative characteristics of fast stability (like hyper-exponential convergence rate) are introduced.
- The concept of generalized homogeneity is introduced for ODE as a main tool for fast control design.
- Two (state-based and output-based) schemes of dynamical fixed-time control design are proposed for plants described by nonlinear ODE.

12.5.2 *On Drawback Finite-Time Stability for Time-Delay Systems*

The examples given above present hyper-exponential systems with non-asymptotic transitions (solutions vanish in finite-time). However, finite-time convergence to zero is rather rare or sometimes impossible for time-delay models. For example, let us consider the scalar time-delay system

$$\dot{x}(t) = f_0(x(t)) + f_1(x(t-h)), \quad t > 0,$$

where $x(t) \in \mathbb{R}$, $f_1, f_2 : \mathbb{R} \to \mathbb{R}$ are continuous functions, and $h > 0$ is a constant delay. This system has continuously differentiable solutions [20] for any smooth initial conditions

$$x(\tau) = \psi(\tau), \quad \tau \in [-h, 0], \quad \phi \in C([-h, 0], \mathbb{R}).$$

Let us assume that $x = 0$ is the equilibrium of the system and f_1 is nontrivial in any neighbourhood of $x = 0$, i.e., for $\forall \varepsilon > 0$, $\exists x \in \mathbb{R} : |x| < \varepsilon$ such that $f_1(x) \neq 0$. If we omit the latter assumption then the considered time-delay system degenerates to delay-free one in some neighborhood of the origin.

We claim that the zero solution of the time-delay system is never finite-time stable. In other words, there is no continuous function f_0 such that the considered system is finite-time sable. Indeed, let us suppose the contrary, i.e.,

$$\exists T > 0,\ \exists \delta > 0 \quad \text{such that} \quad x(t) \neq 0,\ t \in (T-\delta, T) \quad \text{and} \quad x(t) = 0,\ t > T.$$

In this case, $f_0(x(t)) = 0$ and $\dot{x}(t) = f_1(x(t-h))$ for $t > T$. Since $x(t) \neq 0$ for $t \in (T-\delta, T)$ then $\dot{x}(t)$ is not identically zero for some $t > T$. This together with continuous differentiability of x implies the contradiction to $x(t) = 0$ for $t > T$.

Therefore, ideas of non-asymptotic (finite-time/fixed-time) stabilization are inconsistent with some time-delay systems (see also [15] for more details). However, they may demonstrate asymptotic hyper-exponential transitions [37].

12.5.3 *Fast Transition of PDEs*

12.5.3.1 Fixed-Time Extinction of Waves

Hyper-exponential transitions also appear in distributed-parameter systems. Indeed, let us consider the wave equation

$$u_{tt} = u_{xx}, \quad t > 0, \quad x \in [0, 1], \quad u : \mathbb{R}_+ \times [0, 1] \to \mathbb{R},$$

with the so-called transparent boundary condition

$$u_x(t,0) = u_t(t,0), \qquad u_x(t,1) = -u_t(t,1)$$

and the initial conditions

$$u(0,x) = \phi(x), \quad u_t(0,x) = \psi$$

from $\{(\phi,\psi) \in H^1((0,1),\mathbb{R}) \times L^2((0,1),\mathbb{R}) : \phi(0) + \phi(1) + \int_0^1 \psi(s)ds = 0\}$, where L^2 and H^1 are Lebesgue and Sobolev spaces, respectively. The boundary conditions are transparent in the sense that the wave $u(t,x) = f(x-t)$ traveling to the right leaves the domain at $x = 1$ without generating any reflected wave and the wave $u(t,x) = f(x+t)$ traveling to the left leaves the domain at $x = 0$ similarly.

It is well-known (see, e.g., [30]) that the solution of such wave equation vanishes for $t \geq 1$ independently of the initial condition. In the context of distributed parameters systems, this effect is known as finite-time extinction. Obviously, the norm of u tends to zero **hyper-exponentially** with respect to time variable t.

This example shows that in the case of distributed parameters systems the *hyper-exponential and finite-time/fixed-time transitions can be observed in linear models* (see also [5, 12]).

12.5.3.2 Generalized Homogeneity of Infinite-Dimensional Systems

The extension of the homogeneity concept to infinite-dimensional systems has been presented in [34], where a lot of important properties of homogeneous dilations, functionals and operators have been discovered in the case of Banach spaces. The developed homogeneous framework looks promising for the extension of the existing results to fast control and estimation to PDEs and Time-Daly Systems.

Acknowledgements This study is partially supported by The French National Research Agency, Grant ANR Finite4SoS (ANR 15 CE23 0007) and the Russian Federation Ministry of Education and Science, contract/grant numbers 02.G25.31.0111 and 14.Z50.31.0031.

References

1. Adamy, J., Flemming, A.: Soft variable-structure controls: a survey. Automatica **40**(11), 1821–1844 (2004)
2. Andrieu, V., Praly, L., Astolfi, A.: Homogeneous approximation, recursive observer design, and output feedback. SIAM J. Control Optim. **47**(4), 1814–1850 (2008)
3. Armstrong-Helouvry, B.: Control of Machines with Friction, vol. 128. Springer Science & Business Media (2012)
4. Bacciotti, A., Rosier, L.: Liapunov Functions and Stability in Control Theory. Springer Science & Business Media (2006)
5. Balakrishnan, A.: Superstability of systems. Appl. Math. Comput. **164**(2), 321–326 (2005)

6. Bhat, S.P., Bernstein, D.S.: Finite-time stability of continuous autonomous systems. SIAM J. Control Optim. **38**(3), 751–766 (2000)
7. Bhat, S.P., Bernstein, D.S.: Geometric homogeneity with applications to finite-time stability. Math. Control, Signals, Syst. (MCSS) **17**(2), 101–127 (2005)
8. Clarke, F.H.: Optimization and Nonsmooth Analysis. SIAM (1990)
9. Clarke, F.H., Ledyaev, Y.S., Stern, R.J.: Asymptotic stability and smooth lyapunov functions. J. Differ. Equ. **149**(1), 69–114 (1998)
10. Coron, J.M., Nguyen, H.M.: Null controllability and finite time stabilization for the heat equations with variable coefficients in space in one dimension via backstepping approach. Journal (2015)
11. Coron, J.M., Praly, L.: Adding an integrator for the stabilization problem. Syst. Control Lett. **17**(2), 89–104 (1991)
12. Creutz, D., Mazo Jr, M., Preda, C.: Superstability and finite time extinction for c_0-semigroups. arXiv:0907.4812 (2009)
13. Cruz-Zavala, E., Moreno, J.A., Fridman, L.M.: Uniform robust exact differentiator. IEEE Trans. Autom. Control **56**(11), 2727–2733 (2011)
14. Efimov, D., Perruquetti, W.: Oscillations conditions in homogenous systems. IFAC Proc. Vol. **43**(14), 1379–1384 (2010)
15. Efimov, D., Polyakov, A., Fridman, E., Perruquetti, W., Richard, J.P.: Comments on finite-time stability of time-delay systems. Automatica **50**(7), 1944–1947 (2014)
16. Falkovich, G.: Fluid Mechanics: A Short Course for Physicists. Cambridge University Press (2011)
17. Filippov, A.F.: Differential Equations with Discontinuous Righthand Sides: Control Systems, vol. 18. Springer Science & Business Media (2013)
18. Galaktionov, V.A., Vazquez, J.L.: Necessary and sufficient conditions for complete blow-up and extinction for one-dimensional quasilinear heat equations. Arch. Rational Mech. Anal. **129**(3), 225–244 (1995)
19. Haimo, V.T.: Finite time controllers. SIAM J. Control Optim. **24**(4), 760–770 (1986)
20. Hale, J.: Functional differential equations. In: Analytic Theory of Differential Equations, pp. 9–22 (1971)
21. Hermes, H.: Nilpotent approximations of control systems and distributions. SIAM J. Control Optim. **24**(4), 731–736 (1986)
22. John, R.C.: Introduction to Calculus and Analysis, vol. I, XXIII **661** (1965)
23. Kawski, M.: Geometric homogeneity and stabilization. Nonlinear Control Syst. Des. **1995**, 147 (2016)
24. Korobov, V.: A general approach to synthesis problem. Doklady Academii Nauk SSSR **248**, 1051–1063 (1979)
25. Levant, A.: Homogeneity approach to high-order sliding mode design. Automatica **41**(5), 823–830 (2005)
26. Lyapunov, A.M.: The general problem of the stability of motion. Int. J. Control **55**(3), 531–534 (1992)
27. Moulay, E., Perruquetti, W.: Finite time stability of differential inclusions. IMA J. Math. Control Inf. **22**(4), 465–475 (2005)
28. Orlov, Y.: Finite time stability and robust control synthesis of uncertain switched systems. SIAM J. Control Optim. **43**(4), 1253–1271 (2004)
29. Pazy, A.: Semigroups of Linear Operators and Applications to Partial Differential Equations, vol. 44. Springer Science & Business Media (2012)
30. Perrollaz, V., Rosier, L.: Finite-time stabilization of 2x2 hyperbolic systems on tree-shaped networks. SIAM J. Control Optim. **52**(1), 143–163 (2014)
31. Perruquetti, W., Floquet, T., Moulay, E.: Finite-time observers: application to secure communication. IEEE Trans. Autom. Control **53**(1), 356–360 (2008)
32. Polyakov, A.: Nonlinear feedback design for fixed-time stabilization of linear control systems. IEEE Trans. Autom. Control **57**(8), 2106–2110 (2012)

33. Polyakov, A., Coron, J.M., Rosier, L.: On finite-time stabilization of evolution equations: a homogeneous approach. In: 2016 IEEE 55th Conference on Decision and Control (CDC), pp. 3143–3148. IEEE (2016)
34. Polyakov, A., Efimov, D., Fridman, E., Perruquetti, W.: On homogeneous distributed parameters equations. IEEE Trans. Autom. Control **61**(11), 3657–3662 (2016)
35. Polyakov, A., Efimov, D., Perruquetti, W.: Finite-time and fixed-time stabilization: Implicit lyapunov function approach. Automatica **51**, 332–340 (2015)
36. Polyakov, A., Efimov, D., Perruquetti, W.: Robust stabilization of mimo systems in finite/fixed time. Int. J. Robust Nonlinear Control **26**(1), 69–90 (2016)
37. Polyakov, A., Efimov, D., Perruquetti, W., Richard, J.P.: Implicit lyapunov-krasovski functionals for stability analysis and control design of time-delay systems. IEEE Trans. Autom. Control **60**(12), 3344–3349 (2015)
38. Polyakov, A., Fridman, L.: Stability notions and lyapunov functions for sliding mode control systems. J. Frankl. Inst. **351**(4), 1831–1865 (2014)
39. Poznyak, A.S.: Advanced Mathematical Tools for Automatic Control Engineers: Deterministic Technique, vol. 1. Elsevier, Amsterdam (2008)
40. Rosier, L.: Homogeneous lyapunov function for homogeneous continuous vector field. Syst. Control Lett. **19**(6), 467–473 (1992)
41. Rosier, L.: Etude de quelques problemes de stabilisation. Ph.D. thesis, School (1993)
42. Roxin, E.: On stability in control systems. J. Soc. Industrial Appl. Math. Ser. A: Control **3**(3), 357–372 (1965)
43. Roxin, E.: On finite stability in control systems. Rendiconti del Circolo Matematico di Palermo **15**(3), 273–282 (1966)
44. Ruzhansky, M., Sugimoto, M.: On global inversion of homogeneous maps. Bull. Math. Sci. **5**(1), 13–18 (2015)
45. Sabinina, E.: A class of non-linear degenerating parabolic equations. Doklady Akademii Nauk SSSR **143**(4), 794 (1962)
46. Tikhonov, A.N.: Systems of differential equations containing small parameters in the derivatives. Matematicheskii sbornik **73**(3), 575–586 (1952)
47. Utkin, V.I.: Sliding Modes in Control and Optimization. Springer Science & Business Media (2013)
48. Zubov, V.: On systems of ordinary differential equations with generalized homogenous right-hand sides. Mathematica **1**, 80–88 (1958)
49. Zubov, V.: On systems of ordinary differential equations with generalized homogenous right-hand sides. izvestia vuzov. mathematica, J. **1**, 80–88 (1958)
50. Zubov, V.I., Boron, L.F.: Methods of AM Lyapunov and their application. Noordhoff Groningen (1964)

Chapter 13
Finite-Time Sliding Mode Controller with State-Dependent Gain Parameter

Cesar U. Solis and Julio B. Clempner

Abstract This brief text proposes a state-dependent gain parameter added to a Fast Terminal Sliding Mode Control in order to restrict the input signal amplitude. The suggested controller involves the following properties: convergence in finite time to the equilibrium point, robustness against bounded persistent state perturbations and uncertainties in the model. In order to exemplify the contributions we exhibit an application to an underactuated rotational inverted pendulum, to stabilize it upwards.

13.1 Introduction

13.1.1 Brief Review

The main idea of the Terminal Sliding Mode Control (TSMC) is guaranteeing the convergence to the equilibrium point in finite time reaching a structure of sequential sliding manifolds [17]. This control law works fine in a neighborhood of the attractor (equilibrium point), but when the initial condition is far away from equilibrium point the convergence is slow and there is a singularity, i.e. the control law diverges when the initial condition is outside of the convergence region [15, 20].

The solution to these problems is the non-singular Fast Terminal Sliding Mode Control (FTSMC), which adds a linear damping term and a specific selection of the coefficients accelerating the convergence from far away initial conditions and removing the singularity [3, 18, 20].

C.U. Solis (✉)
Departamento de Control Automático, CINVESTAV-IPN, Av. IPN 2508, Col. San Pedro Zacatenco, 07360 Ciudad de México, México
e-mail: csolis@ctrl.cinvestav.mx

J.B. Clempner
Centro de Investigaciones Económicas, Administrativas y Sociales, Instituto Politécnico Nacional, Lauro Aguirre 120, col. Agricultura, Del. Miguel Hidalgo, 11360 Ciudad de México, México
e-mail: julio@clempner.name

J.B. Clempner and W. Yu (eds.), *New Perspectives and Applications of Modern Control Theory*, https://doi.org/10.1007/978-3-319-62464-8_13

Several important research works on TSMC are already presented in the literature. Venkataraman et al. [17] developed the TSMC at the Jet Propulsion Laboratory (JPL). Zhihong et al. [21] proposed a TSMC for MIMO (multiple-input multiple-output) systems, by inserting a nonlinear term of the system in the linear sliding mode. Feng et al. [5] applied the non-singular TSMC to n-link rigid manipulators. Yang et al. [19] proposed new norms for FTSMC strategies, and a faster convergence rate is established. Mobayen et al. [9] developed a new LMI-based robust finite time sliding mode control strategies for a class of uncertain nonlinear systems. Cruz-Zavala et al. [4] designed a second-order uncertain system FTSMC with a single control input proposing a design based on a Lyapunov method. Solis et al. [14] designed a FTSMC including an integral filter applied to a chaotic van der Poll oscillator. Related works on adaptive laws applied to Sliding Mode Control are for instance the following: Plestan et al. [11] exposed new methodologies for adaptive gain robust control via Sliding Modes; Utkin et al. [16] presented an adaptive gain for Super-Twist control; Shtessel et al. [13] suggested a new adaptive method for Super-Twist. For Furuta-pendulum mathematical models and control we have that Jadlovsky et al. [6] designed and implemented a general procedure which yields the mathematical model for a classical or rotational inverted pendulum system with an arbitrary number of pendulum links. Khanesar et al. [7] applied a conventional sliding mode control to a rotational inverted pendulum. Azar et al. [1] developed an adaptive sliding mode technique for a Furuta pendulum.

13.1.2 Main Results

We propose a brief contribution to the FTSMC Theory involving a state-dependent gain parameter. We apply the control to an underactuated rotational inverted pendulum, well-known as Furuta pendulum. To our knowledge, this is the first time that a FTSMC is applied to a rotational inverted pendulum. This dynamical system consists of an actuated rotational horizontal arm with a concentrated mass linked to a non-actuated vertical rotational arm with a concentrated mass.

Summarized contributions:

- Presenting a Theorem that shows the stabilization of a controllable linear system using the FTSMC with recursive structure.
- Showing that the FTSMC converges in a bounded finite time.
- Showing the disturbance rejection with matching condition for controllable nonlinear systems with a state-dependent gain parameter.
- Applying the proposed controller to a linearized-approach dynamical model for a underactuated rotational inverted pendulum and the nonlinear terms are considered as uncertainties and perturbations.
- Presenting a numerical example to demonstrate the effectiveness of the controller.

13.1.3 Organization of the Text

The remainder of the contribution is organized as follows. Section 13.2 presents a motivation exposing an existent Terminal Sliding Mode control that solves the requirements, but considering that it is slow when the initial condition is far away from the equilibrium point. In Sect. 13.3, we suggest a Fast Terminal Sliding Mode Control technique and show the finite time convergence, the stability of the control law, and the disturbance rejection with a state-dependent gain parameter. In Sect. 13.4 we present the application of the algorithm to an underactuated rotational inverted pendulum and a numerical example that validate the effectiveness of the controller. In Sect. 13.5 we present a simplified guide to solve practical exercises. Finally, we close in Sect. 13.6 with conclusions.

13.2 Motivation: The Terminal Sliding Mode Control

Let us consider a controllable linear system given by

$$\dot{\mathbf{x}} = A\mathbf{x} + Bu \tag{13.1}$$

with $\mathbf{x} \in \mathbb{R}^n$, $A \in \mathbb{R}^{n\times n}$, $B \in \mathbb{R}^{n\times 1}$, $u \in \mathbb{R}$. Then, there exists a similarity transformation matrix T such that it transforms the dynamical system (13.1) into the form:

$$\begin{aligned} \dot{z}_1 &= z_2 \\ \vdots &\;\;\vdots \\ \dot{z}_n &= r(\mathbf{z}) + u \end{aligned} \tag{13.2}$$

with $\mathbf{z} = T\mathbf{x}$, $r(\mathbf{z})$ a linear scalar function in $\mathbf{z}$, and the transformation is given by

$$T := \begin{bmatrix} h \; hA \; hA^2 \cdots hA^{n-1} \end{bmatrix}^{\mathsf{T}} \tag{13.3}$$

where $h = \begin{bmatrix} 0 \; 0 \cdots 0 \; 1 \end{bmatrix} \mathcal{C}_{(A,B)}^{-1}$, and

$$\mathcal{C}_{(A,B)} := \begin{bmatrix} B \; AB \; A^2B \cdots A^{n-1}B \end{bmatrix}$$

denotes the controllability matrix (see [2, 12]).

Now, taking into account the dynamical system given in (13.2), let us design the control in the form $u = u_{eq} + u_c$.

To design u_{eq}, let us consider the following recursive structure of manifolds:

$$s_0 = z_1$$
$$\vdots$$
$$s_j = \dot{s}_{j-1} + \beta_{j-1} s_{j-1}^{q_{j-1}/p_{j-1}} \tag{13.4}$$
$$\vdots$$
$$s_{n-1} = \dot{s}_{n-2} + \beta_{n-2} s_{n-2}^{q_{n-2}/p_{n-2}}$$

where $\beta_j > 0$ and $p_j, q_j > 0$ are odd integers. When $s_j = 0$ is reached [18], the dynamics becomes:

$$\dot{s}_{j-1} + \beta_{j-1} s_{j-1}^{q_{j-1}/p_{j-1}} = 0 \tag{13.5}$$

then solving for an initial state $z_1(0) \neq 0$, the dynamics will reach $z_1 = 0$ in a finite time determined by:

$$t_r = \frac{p_0}{\beta_0 (p_0 - q_0)} |z_1(0)|^{(p_0 - q_0)/p_0}$$

The equilibrium point is a terminal attractor. Note that the terminal dynamics (13.5) is not Lipschitz, for any initial condition, but [10] proved that the forward-in-time solution exists and it is unique.

13.3 Fast Terminal Sliding Mode Control

When the states are far away from the equilibrium point, the term $z_1^{q_0/p_0}$ tends to reduce the magnitude of the convergence rate at a distance from equilibrium. To attenuate this problem we consider the following fast terminal sliding mode dynamics:

$$\dot{z} + \alpha z + \beta z^{q/p} = 0 \tag{13.6}$$

where $\alpha,\ \beta > 0$. If z is far away from zero, the approximate dynamics becomes $\dot{z} = -\alpha z$ that is faster than the dynamics given in (13.5). Now, when z nearest 0, the approximate dynamics becomes $\dot{z} = -\beta z^{q/p}$ which is a terminal attractor. We can solve the ODE (13.6) analytically, obtaining the zero-dynamics time is given by:

$$t_r = \frac{p}{\alpha (p - q)} \left(\ln \left(\alpha z_1 (0)^{(p-q)/p} + \beta \right) - \ln \beta \right) \tag{13.7}$$

The following recursive procedure for FTSMC of higher order systems accelerate the convergence:

$$
\begin{aligned}
s_0 &= z_1 \\
&\vdots \\
s_j &= \dot{s}_{j-1} + \alpha_{j-1} s_{j-1} + \beta_{j-1} s_{j-1}^{q_{j-1}/p_{j-1}} \\
&\vdots \\
s_{n-1} &= \dot{s}_{n-2} + \alpha_{n-2} s_{n-2} + \beta_{n-2} s_{n-2}^{q_{n-2}/p_{n-2}}
\end{aligned}
\tag{13.8}
$$

where $\alpha_j > 0, \ \beta_j > 0$.

13.3.1 Finite-Time Convergence

To compute the time to reach the equilibrium point let us solve and sum the time to reach every manifold given in (13.7) and (13.8), and we obtain:

$$
\begin{aligned}
T_r &= \sum_{j=1}^{n} t_j \\
&= t_n + \sum_{j=1}^{n-1} \frac{p_{j-1}}{\alpha_{j-1}\left(p_{j-1} - q_{j-1}\right)} \cdot \left(\ln\left(\alpha_{j-1} s_{j-1} t_j^{\frac{p_{j-1} - q_{j-1}}{p_{j-1} + p_{j-1}}} \right) - \ln \beta_{j-1} \right)
\end{aligned}
\tag{13.9}
$$

Below in Theorem 13.3 we implement a formula to obtain the coefficients in order to guarantee the convergence of the sum. Theorem 13.5 proofs $t_n < \infty$, whereby it is possible to guarantee convergence in finite time.

13.3.2 Control Law Design

The sliding mode control should be designed such that $s_{n-1}\dot{s}_{n-1} < -K\,|s_{n-1}|, K > 0$ therefore $s_{n-1} = 0$ can be reached in finite time. The dynamics (13.2) can be written as $\dot{\mathbf{z}} = A\mathbf{z} + Bu$.

Theorem 13.1 *For the dynamical system (13.2), if the control is designed as*

$$u = u_{eq} + u_c \tag{13.10}$$

where

$$u_{eq} = -\sum_{k=0}^{n-2}\left(\alpha_k \mathcal{L}^{n-k-1}_{\mathbf{Az}+\mathbf{B}u} s_k + \beta_k \mathcal{L}^{n-k-1}_{\mathbf{Az}+\mathbf{B}u} s_k^{q_k/p_k}\right)$$

$$u_c = -K sign\ s_{n-1}$$

where $\mathcal{L}$ is the Lie derivative and $K > 0$, then the system will reach the sliding mode $s_{n-1} = 0$ in finite time.

Proof Following [20], it is necessary to take the time-derivative of s_{n-1}:

$$\dot{s}_{n-1} = \ddot{s}_{n-2} + \alpha_{n-2}\dot{s}_{n-2} + \beta_{n-2}\mathcal{L}_{\mathbf{Az}+\mathbf{B}u} s_{n-2}^{q_{n-2}/p_{n-2}}$$

since $s_j = \dot{s}_{j-1} + \alpha_{j-1}s_{j-1} + \beta_{j-1}s_{j-1}^{q_{j-1}/p_{j-1}}$, for $i = n-1, n-2, \ldots, 1$ and the l-th-order Lie derivative of s_j is given by:

$$\mathcal{L}^{l}_{\mathbf{Az}+\mathbf{B}u} s_j = \mathcal{L}^{l+1}_{\mathbf{Az}+\mathbf{B}u}\dot{s}_{j-1} + \alpha_{j-1}\mathcal{L}^{l}_{\mathbf{Az}+\mathbf{B}u} s_{j-1} + \beta_{j-1}\mathcal{L}^{l}_{\mathbf{Az}+\mathbf{B}u} s_{j-1}^{q_{j-1}/p_{j-1}}$$

then applying to the time-derivative on s_{n-1} yields

$$\begin{aligned}\dot{s}_{n-1} &= \mathcal{L}^{n}_{\mathbf{Az}+\mathbf{B}u} s_0 + \sum_{k=0}^{n-2}\beta_k \mathcal{L}^{n-k-1}_{\mathbf{Az}+\mathbf{B}u} s_k + \sum_{k=0}^{n-2}\beta_k \mathcal{L}^{n-k-1}_{\mathbf{Az}+\mathbf{B}u} s_k^{q_k/p_k}\\ &= \dot{z}_n + \sum_{k=0}^{n-2}\alpha_k \mathcal{L}^{n-k-1}_{\mathbf{Az}+\mathbf{B}u} s_k + \sum_{k=0}^{n-2}\beta_k \mathcal{L}^{n-k-1}_{\mathbf{Az}+\mathbf{B}u} s_k^{q_k/p_k}\end{aligned}$$

substituting the control yields $s_{n-1}\dot{s}_{n-1} = -K\,|s_{n-1}|$, which means that the sliding mode $s_{n-1} = 0$ will be reached in finite time.

The parameters q_k, p_k must be chosen carefully in order to avoid the singularity, because there are terms in $d^{n-k-1}/dt^{n-k-1} s_k^{q_k/p_k}$ which may contain negative powers so that, when $s_{k-1} \to 0$, $u \to \infty$. This problem can be remedied by the following theorems.

Theorem 13.2 *For any $k \in 0, 1, \ldots, n-2$,*

$$\mathcal{L}^{n-k-1}_{\mathbf{Az}+\mathbf{B}u} s_k^{q_k/p_k} = f_k(z)$$

where f_k is a continuous nonlinear function.

Proof (Mathematical induction) Let $k = 0$, then $\mathcal{L}^{n-1}_{\mathbf{Az}+Bu} s_0^{q_0/p_0} = \mathcal{L}^{n-1}_{\mathbf{Az}+Bu} z_1^{q_0/p_0} = f_0(z)$. Let $k = 1$, then $\mathcal{L}^{n-2}_{\mathbf{Az}+Bu} s_1^{q_1/p_1} = \mathcal{L}^{n-2}_{\mathbf{Az}+Bu}(z_2 + \alpha_0 z_1 + \beta_0 z_1^{q_0/p_0})^{q_1/p_1}$ which can be apparently expressed as $f_1(z)$. Assume for $k = k_0$, $\mathcal{L}^{n-k_0-1}_{\mathbf{Az}+Bu} s_{k_0}^{q_{k_0}/p_{k_0}} = f_{k_0}(z)$. Let us consider $k = k_0 + 1$. Since

$$
\begin{aligned}
s_{k_0} &= f_{k_0}(\dot{s}_{k_0}, s_{k_0-1}) \\
&= f_{k_0}(\ddot{s}_{k_0-2}, \dot{s}_{k_0-2}, s_{k_0-2}) \\
&\vdots \\
&= f_{k_0}(s_0^{(k_0)}, s_0^{(k_0-1)}, \ldots, s_0)
\end{aligned}
$$

and $s_0 = z_1$, then $s_{k_0}^{(l)}$ is a function of $z_{k_0+l+1}, z_{k_0+l}, \ldots, z_1$. Therefore, $\mathcal{L}^{n-k_0-2}_{\mathbf{Az}+Bu} s_{k_0+1}^{q_{k_0+1}/p_{k_0+1}}$ is a function of z.

Theorem 13.3 *If*

$$
\frac{q_k}{p_k} > \frac{n-k-1}{n-k}
$$

when $s_k \to 0$ sequentially from $k = n-2$ to $k = 0$, then u is bounded.

Proof The n-derivative of a composite function is given by

$$
\frac{d^n}{dt^n} F(s) = \sum \frac{n!}{i_1! \ldots i_l!} \frac{d^m F}{ds^s} \left(\frac{\dot{s}}{1!}\right)^{i_1} \left(\frac{\ddot{s}}{2!}\right)^{i_2} \cdots \left(\frac{s^{(l)}}{l!}\right)^{i_l}
$$

over all solutions in non-negative integers of the equation $i_1 + 2i_2 + \cdots + li_l = n$ and $m = i_1 + i_2 + \cdots + i_l$. Let $r = q_k/p_k$, and dropping the index k. Since in the sliding mode $s_{k+1} = 0$,

$$
\begin{aligned}
s_{k+1} &= \dot{s}_k + \alpha_k s_k + \beta_k s_k^{q_k/p_k} \\
&= \dot{s} + \alpha s + \beta s^r \\
&= 0
\end{aligned}
$$

then $\dot{s} = \mathcal{O}(s^r)$ when $s \to 0$. We therefore have $d^m F/ds^m = \mathcal{O}\left(s^{r-m}\right)$ and $s^{(d)} = (\dot{s})^{d-1} = \mathcal{O}\left(s^{dr-(d-1)}\right)$. Hence when $s \to 0$, i.e., $s_k \to 0$, $(n+1)r - n = (n+1)q_k/p_k - n > 0$ will ensure that these terms are bounded. From the above analysis, we have $d^n/dt^n s = (\dot{s}^{(n-1)}) = \mathcal{O}(s^r)^{(n-1)} = \mathcal{O}(s^{nr-n+1})$. Hence when $s \to 0$, $nr - n + 1 = nq_k/p_k - n + 1 > 0$ will ensure that $d^n/dt^n s$ is bounded. With the above expressions in mind, the control (13.10) can be written as

$$u = a(z) + \sum \left(\mathcal{O}\left(s_k^{\frac{(n-k-1)q_k}{p_k-(n-k-1)+1}} \right) + \mathcal{O}\left(s_k^{\frac{(n-k)q_k}{p_k-(n-k-1)}} \right) \right) + K \operatorname{sign} s_{n-1}$$

For the second term to be bounded while $s_k \to 0$, it is sufficient that $(n-k)q_k/q_k - (n-k-1) > 0$ for $s_k \to 0$ sequentially from $k = n-2$ to $k = 0$ so that the control u is bounded.

13.3.3 Matching Disturbance Rejection with State-Dependent Gain Parameter

Theorem 13.4 *Let us consider the dynamical system:*

$$\dot{\mathbf{x}} = A\mathbf{x} + Bu + \zeta(\mathbf{x}, t) \tag{13.11}$$

where the pair (A, B) *is controllable,* $\zeta(\mathbf{x}, t) := B\eta(\mathbf{x}, t)$ *represents the uncertainties and perturbations satisfying the matching condition,* $\eta(\mathbf{x}, t)$ *is a Lipschitz locally function in* $\mathbf{x}$*,* $\|\mathbf{x}\|$ *is bounded and we have that:*

$$\|\eta(\mathbf{x}, t)\| \le \sum_{j=0}^{m} L_j \|\mathbf{x}\|^j$$

where $L_j > 0$ *for all* $j \in \{0, 1, \dots, m\}$*, for some* $m \ge 0$ *(note that if* $m = 0$ *then we have a constant bound). Given the control law (13.10) we may select the chattering parameter as follows:*

$$K(\mathbf{z}(t)) := \sum_{j=0}^{m} K_j \left\| T^{-1}\mathbf{z} \right\|^j \tag{13.12}$$

where $\mathbf{z} = T\mathbf{x}$ *a similarity transformation is given by (13.3),* $K_j > L_j$ *for all* $j \ne 1$ *(if* $L_0 = 0$*, then we take* $K_0 > 0$ *for ensuring the existence of the sliding manifold) and for* $K_1 > L_1 \left\| T^{-1} \right\| + \|H\|$ *where* H *is the bottom elements of* T*, then the control law stabilizes the system.*

Proof If (A, B) is controllable then there exists a similarity transformation T applied to the system (13.11) we obtain:

$$\begin{aligned}\dot{z}_1 &= z_2\\ &\vdots\\ \dot{z}_n &= r(\mathbf{z}) + u + \eta\left(T^{-1}\mathbf{z}, t\right)\end{aligned}$$

where $r(\mathbf{z}) = H\mathbf{z}$ is a scalar function in $\mathbf{z}$, i.e., H is a constant row vector, then $\|r(\mathbf{z})\| \leq \|H\| \, \|\mathbf{z}\|$.

Then, using the control law given in (13.10), and selecting the chattering adaptive parameter $K(\mathbf{z}(t))$ we have:

$$K(\mathbf{z}(t)) := \sum_{j=0}^{m} K_j \left\|T^{-1}\mathbf{z}\right\|^j \tag{13.13}$$

Then, we have (see Theorem 13.1):

$$\begin{aligned} s_{n-1}\dot{s}_{n-1} &= \left(-\sum_{j=0}^{m} K_j \left\|T^{-1}\mathbf{z}\right\|^j + \eta\left(T^{-1}\mathbf{z}, t\right) + r(\mathbf{z})\right) |s_{n-1}| \\ &\leq \left(\left(L_1\left\|T^{-1}\right\| + \|H\| - K_1\right)\|\mathbf{z}\| + \sum_{j=0, j\neq 1}^{m} \left(L_j - K_j\right)\left\|T^{-1}\mathbf{z}\right\|^j\right) \\ &\quad \times |s_{n-1}| \end{aligned}$$

Taking $K_j > L_j$ for all $j \neq 1$ (if $L_0 = 0$, take $K_0 > 0$ for the existence of the sliding manifold) and $K_1 > L_1\left\|T^{-1}\right\| + \|H\|$, we can guarantee the stability of the system.

Theorem 13.5 *Given $K(\mathbf{z}(t))$ by (13.12) applied to the control law (13.10), then $t_n < \infty$, i.e., convergence occurs in finite time.*

Proof Consider the Lyapunov function $V(t) := \frac{1}{2}s_{n-1}^2$, then time-derivative is as follows:

$$\begin{aligned}\dot{V}(t) &= s_{n-1}\dot{s}_{n-1}\\ &\leq -K(\mathbf{z}(t))\sqrt{2V(t)}\end{aligned}$$

solving with $V(0) = V_0$ (remember $V_0 > 0$ because a Lyapunov function is positive definite), we obtain:

$$\sqrt{V(t)} \leq \sqrt{V_0} - \frac{1}{\sqrt{2}} \int_0^t K(\mathbf{z}(\tau))\, d\tau$$

We have proven above that the manifold is reached, i.e., $V(t_n) = 0$ for some $t_n > 0$, we need to prove $t_n < \infty$, substituting we have:

$$\sqrt{2V_0} \geq \int_0^{t_n} K(\mathbf{z}(\tau))\, d\tau$$

but $K(\mathbf{z}(\tau)) = K_0 + \nu(\mathbf{z}(\tau))$ where $\nu(\mathbf{z}(\tau))$ is a non-negative function obtained from (13.13), then:

$$\sqrt{2V_0} \geq t_n K_0 + \int_0^{t_n} \nu(\mathbf{z}(\tau))\, d\tau \geq 0$$

The integral is non-negative for all $t_n > 0$, if $t_n \to \infty$ the right-hand side of the inequality is not finite, but $\sqrt{2V_0}$ is a finite upper bound for this, then there is a contradiction, concluding that t_n is finite.

13.4 The FTSMC Applied to an Underactuated Rotational Pendulum

13.4.1 Rotational Inverted Pendulum Description

The goal is to stabilize and hold in upward position an underactuated rotational inverted pendulum. The task must be accomplished in finite time with a continuous-time robust feedback control law based on terminal sliding manifolds.

The rotational inverted pendulum is described in Fig. 13.1. Let us consider the coordinate system as follows:

$$\begin{aligned} x_{m_1} &= l_1 \cos\alpha \\ y_{m_1} &= l_1 \sin\alpha \\ z_{m_1} &= 0 \end{aligned} \tag{13.14}$$

$$\begin{aligned} x_{m_2} &= x_{m_1} - l_2 \sin\beta \sin\alpha \\ y_{m_2} &= y_{m_1} + l_2 \sin\beta \cos\alpha \\ z_{m_2} &= l_2 \cos\beta \end{aligned} \tag{13.15}$$

where m_1 and m_2 are the mass of the links of the pendulum respectively, l_1 and l_2 its lengths. Note that the generalized coordinates are given by the angular coordinates α and β.

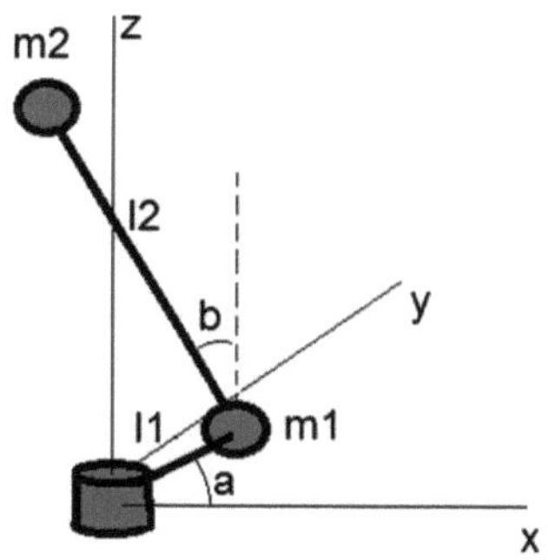

Fig. 13.1 Underactuated rotational inverted pendulum

The velocity of the links are given by the time-derivative of (13.14) and (13.15) as follows:

$$\begin{aligned} \dot{x}_{m_1} &= -l_1\dot{\alpha}\sin\alpha \\ \dot{y}_{m_1} &= l_1\dot{\alpha}\cos\alpha \\ \dot{z}_{m_1} &= 0 \end{aligned} \tag{13.16}$$

$$\begin{aligned} \dot{x}_{m_2} &= \dot{x}_{m_1} - l_2\left(\dot{\beta}\cos\beta\sin\alpha + \dot{\alpha}\sin\beta\cos\alpha\right) \\ \dot{y}_{m_2} &= \dot{y}_{m_1} + l_2\left(\dot{\beta}\cos\beta\cos\alpha - \dot{\alpha}\sin\beta\sin\alpha\right) \\ \dot{z}_{m_2} &= -l_2\dot{\beta}\sin\beta \end{aligned} \tag{13.17}$$

13.4.2 Dynamical Model for the Underactuated Rotational Pendulum

We proceed using Euler–Lagrange formulation to obtain the Newton equations. We can construct the lagrangian function in the generalized coordinates presented above, obtaining:

$$\begin{aligned} \mathcal{L}\left(\alpha,\beta,\dot{\alpha},\dot{\beta}\right) :=& \frac{1}{2}\left(m_1 l_1^2 + m_2 l_1^2 + m_2 l_2^2\right)\dot{\alpha}^2 + \frac{1}{2}m_2 l_2^2\dot{\beta}^2 \\ & - m_2 l_2 g\cos\beta - \frac{1}{2}m_2 l_2^2\dot{\alpha}^2\cos^2\beta \\ & + m_2 l_1 l_2\dot{\alpha}\dot{\beta}\cos\beta \end{aligned} \tag{13.18}$$

we can consider the viscous friction on the links with the coefficients k_1 and k_2, via the Rayleigh function:

$$\mathcal{R}\left(\dot{\alpha},\dot{\beta}\right) := \frac{1}{2}k_1\dot{\alpha}^2 + \frac{1}{2}k_2\dot{\beta}^2 \tag{13.19}$$

A different kind of friction might be represented by a perturbation. Let us solve the Euler–Lagrange formulation:

$$\frac{d}{dt}\left(\frac{\partial \mathcal{L}}{\partial \dot{q}}\right) - \frac{\partial \mathcal{L}}{\partial q} + \frac{\partial \mathcal{R}}{\partial \dot{q}} = \mathcal{F} \tag{13.20}$$

where $q := \begin{bmatrix} \alpha \ \beta \end{bmatrix}^{\mathsf{T}}$ and the input generalized force is $\mathcal{F} := \begin{bmatrix} u \ 0 \end{bmatrix}^{\mathsf{T}}$. The system is underactuated because there is one actuator represented by u and two degrees of freedom. Solving (13.20) for the acceleration $\ddot{q}$, we obtain the dynamical model of the rotational inverted pendulum:

$$\begin{aligned}\ddot{\alpha} =& \Big[-k_1 l_2 \dot{\alpha} + l_2 u + k_2 l_1 \dot{\beta} \cos\beta - \frac{1}{2} m_2 g l_1 l_2 \sin(2\beta) \\ & - m_2 l_2^3 \dot{\alpha}\dot{\beta} \sin(2\beta) - m_2 l_1 l_2^2 \sin\beta \left(\dot{\alpha}^2 \cos^2\beta - \dot{\beta}^2\right) \Big] \\ & / \left[l_2 \left(m_1 l_1^2 + m_2 \left(l_1^2 + l_2^2 \right) \sin^2\beta \right) \right]\end{aligned} \tag{13.21}$$

$$\begin{aligned}\ddot{\beta} =& \Big[m_2^2 g l_2^3 \sin^3\beta - \left(m_2 \left(l_1^2 + l_2^2 \right) + m_1 l_1^2 \right) k_2 \dot{\beta} + \\ & \frac{1}{2} m_2^2 l_2^4 \dot{\alpha}^2 \sin(2\beta) + m_2 l_2^2 k_2 \dot{\beta} \cos^2\beta + m_2^2 g l_2 l_1^2 \sin\beta \\ & - m_2^2 l_2^4 \dot{\alpha}^2 \cos^3\beta \sin\beta + \frac{1}{2} m_2^2 l_1^2 l_2^2 \left(\dot{\alpha}^2 - \dot{\beta}^2 \right) \sin(2\beta) \\ & - m_2 l_1 l_2 u \cos\beta + m_2 l_1 l_2 k_1 \dot{\alpha} \cos\beta \\ & - 2 m_2^2 l_1 l_2^3 \dot{\alpha}\dot{\beta} \sin^3\beta + m_1 m_2 g l_1^2 l_2 \sin\beta \\ & + 2 m_2 l_1 l_2^3 \dot{\alpha}\dot{\beta} \sin\beta + \frac{1}{2} m_1 m_2 l_1^2 l_2^2 \dot{\alpha}^2 \sin(2\beta) \Big] \\ & / \left[m_2 l_2^2 \left(m_1 l_1^2 + m_2 \left(l_1^2 + l_2^2 \right) \sin^2\beta \right) \right]\end{aligned} \tag{13.22}$$

It is possible to write the above dynamical system in the quasi-linear format:

$$\dot{\mathbf{x}} = f(\mathbf{x}) + g(\mathbf{x})\, u$$

with $\mathbf{x} = \begin{bmatrix} \alpha \ \dot{\alpha} \ \beta \ \dot{\beta} \end{bmatrix}^{\mathsf{T}} \in \mathbb{R}^{4\times 1}$ called the state vector, $f(\mathbf{x}),\ g(\mathbf{x}) \in \mathbb{R}^{4\times 1}$ functions that depend only on $\mathbf{x}$, $u \in \mathbb{R}$. The state $\mathbf{x}_0 = \begin{bmatrix} 0\ 0\ 0\ 0 \end{bmatrix}^{\mathsf{T}}$ is the equilibrium point that we want to reach and hold with a robust control law.

13.4.3 Approximated Linearization

The dynamical system given in (13.21) and (13.22) can be linearized around the equilibrium point $\mathbf{x}_0$ through the Non-hyperbolic Hartman-Grobman Theorem [8], then there exists a neighborhood N of $\mathbf{x}_0$ where the system has a similar behavior as the following linear system:

$$\dot{\mathbf{x}} = A\mathbf{x} + Bu \tag{13.23}$$

where $A = \frac{\partial f}{\partial \mathbf{x}}(\mathbf{x}_0)$ (the Jacobian matrix) and $B = g(\mathbf{x}_0)$.

In our case, we have:

$$A = \begin{bmatrix} 0 & 1 & 0 & 0 \\ 0 & -\frac{k_1}{m_1 l_1^2} & -\frac{m_2 g}{m_1 l_1} & \frac{k_2}{m_1 l_1 l_2} \\ 0 & 0 & 0 & 1 \\ 0 & \frac{k_1}{m_1 l_1 l_2} & \frac{g(m_2+m_1)}{m_1 l_2} & -\frac{k_2(m_1+m_2)}{m_1 m_2 l_2^2} \end{bmatrix} \tag{13.24}$$

$$B = \begin{bmatrix} 0 \\ \frac{1}{m_1 l_1^2} \\ 0 \\ -\frac{1}{m_1 l_1 l_2} \end{bmatrix} \tag{13.25}$$

Lemma 13.1 (A, B) *is controllable.*

Proof We can construct the controllability matrix $\mathcal{C}_{(A,B)} := \begin{bmatrix} B & AB & A^2B & A^3B \end{bmatrix}$. Then, we have that the rank $\mathcal{C}_{(A,B)} = 4\ (= n)$.

If (A, B) is controllable then there exists a non-singular linear transformation T (similarity transformation) that takes (13.23) into $\mathbf{z} = T\mathbf{x}$.

13.4.4 Numerical Example

Let us consider the dynamical system given in (13.21)–(13.22) with the following parameters: $m_1 = 0.05$ (kg), $l_1 = 0.05$ (m), $l_2 = 0.2$ (m), $g = 9.81$ (m/s^2), $k_1 = 0.001$ (kg $\cdot$ m^2/s^2) and $k_2 = 0.0005$ (kg $\cdot$ m^2/s^2). In order to exemplify the robustness of the controller let us add matching uncertainties and perturbations:

$$\zeta(\mathbf{x}, t) := B\left(\frac{1}{2}\sin(4\pi t) + \text{sign}(\sin(10\pi t))\,\|\mathbf{x}\|\right)$$

Substituting in the linearization we have:

$$A = \begin{bmatrix} 0.00 & 1.00 & 0.00 & 0.00 \\ 0.00 & -8.00 & -784.80 & 1.00 \\ 0.00 & 0.00 & 0.00 & 1.00 \\ 0.00 & 2.00 & 245.25 & -0.31 \end{bmatrix}$$

$$B = \begin{bmatrix} 0.00 \\ 8000.00 \\ 0.00 \\ -2000.00 \end{bmatrix}$$

It is possible to determine that this dynamical system is locally controllable, then we can transform it as follows:

$$T = 10^{-6} \begin{bmatrix} -2.50 & 0.00 & -10.20 & 0.00 \\ 0.00 & -2.50 & -0.60 & -10.20 \\ 0.00 & 0.00 & -500.00 & 0.00 \\ 0.00 & 0.00 & 0.00 & -500.00 \end{bmatrix}.$$

The new linear system is given by:

$$A_z = \begin{bmatrix} 0.00 & 1.00 & 0.00 & 0.00 \\ 0.00 & 0.00 & 1.00 & 0.00 \\ 0.00 & 0.00 & 0.00 & 1.00 \\ 0.00 & 392.40 & 244.75 & -8.31 \end{bmatrix}$$

For the control law we select the following parameters: $q_0, q_1, q_2 = 25$, $p_0, p_1, p_2 = 13, \alpha_0, \alpha_1, \alpha_2 = 5, \beta_0, \beta_1, \beta_2 = 3, K(\mathbf{x}) = 1 + 2\|\mathbf{x}\|$, and the initial state $\mathbf{x}_0 = [0\ 0\ 0\ 10]^{\top}$.

In Fig. 13.2 we can see the convergence of the angles of the pendulum to the equilibrium point and the disturbance and uncertainties rejection; this attractor is in finite time because the sliding surfaces reach it in finite time (see Fig. 13.4). Finally, in Fig. 13.3 it is possible to see the control law and the decreasing chattering effect because the $K(\mathbf{x})$ converges to 1 when the equilibrium point converges to $\mathbf{x} = 0$.

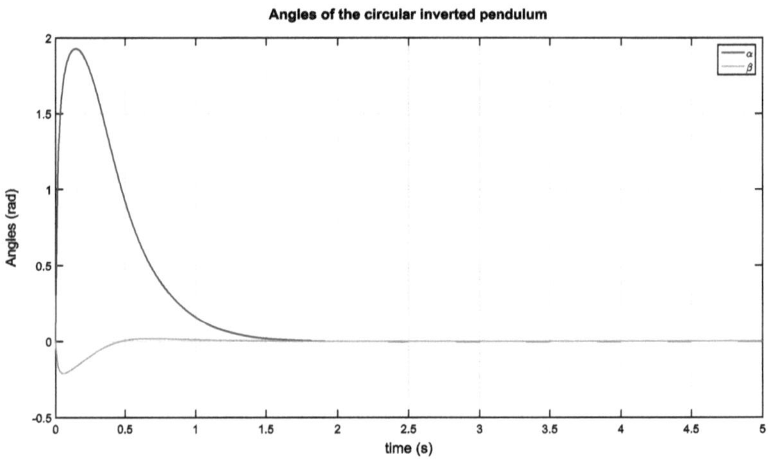

Fig. 13.2 Angles of the rotational inverted pendulum with FTSMC

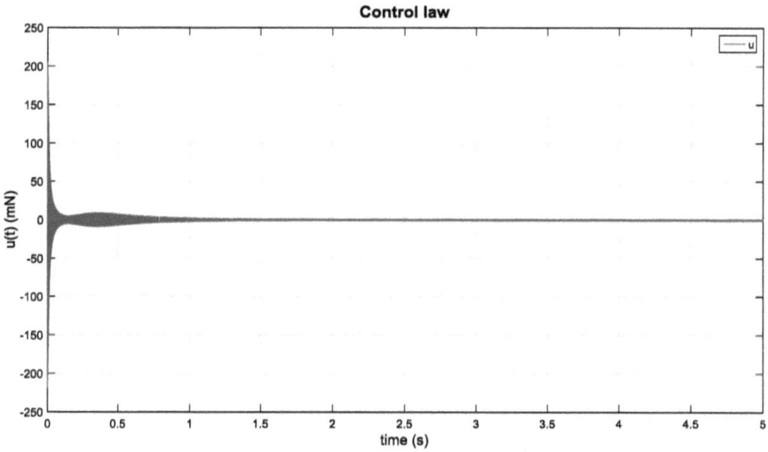

Fig. 13.3 Input FTSMC on the rotational inverted pendulum

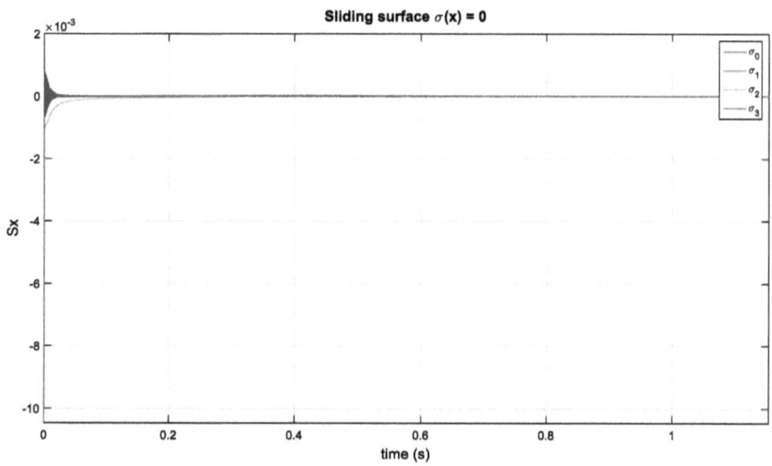

Fig. 13.4 Reaching the sliding surfaces in finite time with FTSMC

13.5 Practical Considerations

In order to put into practice the main ideas or solve some academic exercises, let us present to the reader the following simplified checkpoints:

- Determine if the dynamical system under study is controllable.
- Make sure that the perturbations and uncertainties satisfy the matching condition, otherwise it is not possible to guarantee convergence to a point.

- Obtain a linearization (or approximated linear version) in the form (13.11).
- Apply the transformation (13.3) to the dynamical system obtained and the parameters K that satisfy the Theorem 13.4.
- Compute the control law presented in Theorem 13.1 fulfilling the conditions in Theorem 13.3.
- Finally compute the remaining parameters to set a bound in the convergence time presented in Eq. (13.3), obviously set $T_r > 0$. If T_r is near to zero will obtain abrupt and larger control signals.

Clearly this is a brief guide to set over all parameters of the controller, maybe with experience and perseverance it will be possible to improve it in better way.

13.6 Conclusions

We have presented a FTSMC involving a state-dependent gain parameter which was proposed in order to limit the chattering effect. Showed the stability analysis of the controller, the uncertainty and disturbance rejection with matching condition. Next, we presented a proof of convergence of the method in finite time. The proposed solution decreases, in time, the chattering effect reaching a lower bound given by K. In order to validate the approach we presented a numerical example, where a rotational inverted pendulum reaches the equilibrium in upward position. Practical considerations were shown in order to apply the ideas to exercises.

In addition, we have made some comparisons with another adaptive method based on Sliding Mode Control. Shtessel [13] and Utkin [16] suggested two different algorithms employing an adaptive Super-Twist method with the benefit of presenting a chattering effect that decreases considerably better than our method but with the disadvantage of converging in exponential time. As well as, Plestan [11] presented a method where an adaptive gain is given by a piece-wise ordinary differential equation which convergence is in exponential time (with a correct selection of the parameters). In our case, the state-dependent gain is much more simple to obtain and the convergence time is finite.

References

1. Azar, A.T., Serrano, F.E.: Adaptive sliding mode control of the furuta pendulum. In: Advances and Applications in Sliding Mode Control systems, pp. 1–42. Springer, Berlin (2015)
2. Brunovskỳ, P.: A classification of linear controllable systems. Kybernetika **6**(3), 173–188 (1970)

3. Cheng, Y.: Nonsingular fast terminal sliding mode controller based on states in nonlinear system. In: 2012 4th International Conference on Intelligent Human-Machine Systems and Cybernetics (IHMSC), vol. 1, pp. 262–264. IEEE, New York (2012)
4. Cruz-Zavala, E., Moreno, J.A., Fridman, L.: Fast second-order sliding mode control design based on lyapunov function. In: 52nd IEEE Conference on Decision and Control, pp. 2858–2863. IEEE, New York (2013)
5. Feng, Y., Yu, X., Man, Z.: Non-singular terminal sliding mode control of rigid manipulators. Automatica **38**(12), 2159–2167 (2002)
6. Jadlovský, S., Sarnovský, J.: Modelling of classical and rotary inverted pendulum systems-a generalized approach. J. Electr. Eng. **64**(1), 12–19 (2013)
7. Khanesar, M.A., Teshnehlab, M., Shoorehdeli, M.A.: Sliding mode control of rotary inverted pendulm. In: Mediterranean Conference on Control & Automation, 2007. MED'07. pp. 1–6. IEEE, New York (2007)
8. Meiss, J.D.: Differential Dynamical Systems, vol. 14. SIAM (2007)
9. Mobayen, S., Tchier, F.: A new lmi-based robust finite-time sliding mode control strategy for a class of uncertain nonlinear systems. Kybernetika **51**(6), 1035–1048 (2015)
10. Nemytskii, V.V.: Qualitative Theory of Differential Equations. Princeton University Press, Princeton (2015)
11. Plestan, F., Shtessel, Y., Bregeault, V., Poznyak, A.: New methodologies for adaptive sliding mode control. Int. J. Control **83**(9), 1907–1919 (2010)
12. Poznyak, A.: Advanced Mathematical Tools for Control Engineers: Volume 1: Deterministic Systems, vol. 1. Elsevier, Amsterdam (2010)
13. Shtessel, Y.B., Moreno, J.A., Plestan, F., Fridman, L.M., Poznyak, A.S.: Super-twisting adaptive sliding mode control: A lyapunov design. In: CDC, pp. 5109–5113 (2010)
14. Solis, C.U., Clempner, J.B., Poznyak, A.: Fast terminal sliding mode control with integral filter applied to a van der pol oscillator. IEEE Trans. Indust. Electron. **PP**(99), 1 (2017). doi:10.1109/TIE.2017.2677299
15. Solis, C.U., Clempner, J.B., Poznyak, A.S.: Designing a terminal optimal control with an integral sliding mode component using a saddle point method approach: a cartesian 3d-crane application. Nonlinear Dyn. **86**(2), 911 (2016). doi:10.1007/s11071-016-2932-9
16. Utkin, V.I., Poznyak, A.S.: Adaptive sliding mode control with application to super-twist algorithm: equivalent control method. Automatica **49**(1), 39–47 (2013)
17. Venkataraman, S., Gulati, S.: Control of nonlinear systems using terminal sliding modes. J. Dyn. Syst. Meas. Control **115**(3), 554–560 (1993)
18. Wu, Y., Yu, X., Man, Z.: Terminal sliding mode control design for uncertain dynamic systems. Syst. Control Lett. **34**(5), 281–287 (1998)
19. Yang, L., Yang, J.: Nonsingular fast terminal sliding-mode control for nonlinear dynamical systems. Int. J. Robust Nonlinear Control **21**(16), 1865–1879 (2011)
20. Yu, X., Zhihong, M.: Fast terminal sliding-mode control design for nonlinear dynamical systems. IEEE Trans. Circuits Syst.^{2}I Fundem. Theory Appl. **49**(2) (2002)
21. Zhihong, M., Yu, X.H.: Terminal sliding mode control of mimo linear systems. In: Proceedings of the 35th IEEE Conference on Decision and Control, 1996, vol. 4, pp. 4619–4624. IEEE, New York (1996)

Chapter 14
Setting Nash Versus Kalai–Smorodinsky Bargaining Approach: Computing the Continuous-Time Controllable Markov Game

Kristal K. Trejo and Julio B. Clempner

Abstract The bargaining game refers to a situation in which players have the possibility of concluding a mutually beneficial agreement. Here there is a conflict of interests about which agreement to conclude or no-agreement may be imposed on any player without that player's approval. Remarkably, bargaining and its game-theoretic solutions has been applied in many important contexts like corporate deals, arbitration, duopoly market games, negotiation protocols, etc. Among all these research applications, equilibrium computation serves as a basis. This chapter examines bargaining games from a theoretical perspective and provides a solution method for the game-theoretic models of bargaining presented by Nash and Kalai–Smorodinsky which propose an elegant axiomatic approach to solve the problem depending on different principles of fairness. Our approach is restricted to a class of continuous-time, controllable and ergodic Markov games. We first introduce and axiomatize the Nash bargaining solution. Then, we present the Kalai–Smorodinsky approach that improves the Nash's model by introducing the monotonicity axiom. For the solution of the problem we suggest a bargaining solver implemented by an iterated procedure of a set of nonlinear equations described by the Lagrange principle and the Tikhonov regularization method to ensure convergence to a unique equilibrium point. Each equation in this solver is an optimization problem for which the necessary condition of a minimum is solved using the projection gradient method. An important result of this chapter shows the equilibrium computation in bargaining games. In particular, we present the analysis of the convergence as well as the rate of convergence of the proposed method. The usefulness of our approach is demonstrated by a numerical example contrasting the Nash and Kalai–Smorodinsky bargaining solution problem.

K.K. Trejo (✉)
Department of Automatic Control, Center for Research and Advanced Studies,
Av. IPN 2508, Col. San Pedro Zacatenco, Del. Gustavo A. Madero,
07360 Mexico, Mexico
e-mail: ktrejo@ctrl.cinvestav.mx

J.B. Clempner
Centro de Investigaciones Económicas, Administrativas y Sociales,
Instituto Politécnico Nacional, Lauro Aguirre 120, Col. Agricultura,
Del. Miguel Hidalgo, 11360 Mexico, Mexico
e-mail: julio@clempner.name

J.B. Clempner and W. Yu (eds.), *New Perspectives and Applications of Modern Control Theory*, https://doi.org/10.1007/978-3-319-62464-8_14

14.1 Introduction

14.1.1 Brief Review

The bargaining model has been applied in many important contexts including arbitration, supply chain contracts, duopoly market games, negotiation protocols, etc. It is related to negotiation and group decision processes, and introduces a solution concept for cooperative games. Cooperation concerns to coalitions of two or more players acting together with a specific common purpose taking into account the objective of maximizing their own individual payoffs. The bargaining game dynamics refers to a situation in which players have the possibility of concluding a mutually beneficial agreement. Here there is a conflict of interests about which agreement to conclude or no-agreement may be imposed on any player without that player's approval. There are two theoretical perspectives that provide a solution for the cooperative game-theoretic bargaining models that employ the axiomatic method to evaluate bargaining: Nash [28] and Kalai–Smorodinsky [20]. It is important to note that the two bargaining solution approaches have the same feasible payoff set and disagreement point are considered to be the same bargaining problem in Nash's model.

The bargaining model was first presented as a game in John Nash's seminal 1950 paper [28], using the framework of game theory proposed by von Neumann and Morgenstern [42]. The von Neumann and Morgenstern theory supposes that when players form a coalition, they expect that a complementary coalition responds by damaging them in the worst way. This statement finds disapprovements in the literature. In this sense, Nash improved von Neumann and Morgenstern's work extending the idea by proposing axioms that characterize a unique result and a solution to the problem called the Nash bargaining solution. The formal description consists of two main components: a feasible set of utility allocations reached via cooperation, and the disagreement point occurring when players do not cooperate. A solution is a function that selects a feasible utility allocation for every problem. It is interesting to note that bargaining is one of the first situations of conflict of interest presented in the literature of game theory [19, 30].

The Kalai–Smorodinsky [20] bargaining solution differs from the Nash approach [28]. The fundamental difference between the two approaches resides in the fact that the Nash solution complies with independence of irrelevant alternatives instead the Kalai–Smorodinsky solution fits monotonicity. Kalai and Smorodinsky argue that the entire set of alternatives must affect the agreement reached.

14.1.2 Related Work

The Nash bargaining problem and the Kalai–Smorodinsky approach has attracted the attention of researchers from different disciplines and it is still a relevant topic which is receiving a growing amount of attention for both practitioners and academics in game theory. Merlo and Wilson [24] studied an n-player sequential bargaining model for Markov processes. They investigated the uniqueness and efficiency of the equilibrium outcomes, the conditions under which agreement is delayed, and the advantage to proposing. The sequential bargaining model in which the size of the cake and the order in which players move is characterized by the sets of subgame perfect and stationary subgame perfect equilibria. Bolt and Houba [6] suggested a model for the bargaining process based on an offer model with an exogenous risk of breakdown for Markov games. They analyzed a modified version of the variable threat game without commitment within a dynamic context. Cai [7] presented an alternating offer bargaining game, showing that the proposed model has a finite number of Markov perfect equilibria, some of which exhibit wasteful delays. The maximum number of delay periods that can be supported in Markov perfect equilibria increases in the order of the square of the number of players and that these results are robust to a relaxing of the Markov requirements and to more general surplus functions. Rubinstein and Wolinsky [36] developed a bargaining problem treated with a strategic approach in which the market where pairs of agents who are interested in carrying out a transaction, are brought together by a stochastic process and, upon meeting, initiate a bargaining process over the terms of the transaction. They derived the steady-state equilibrium agreements and analyzed their dependence on market conditions. Kalandrakis [21] studied an infinitely repeated divide-the-dollar bargaining game with an endogenous reversion point characterizing a Markov equilibrium which is such that irrespective of the discount factor or the initial division of the dollar, the proposer eventually extracts the whole dollar in all periods. They showed that proposed strategies are weakly continuous in the status quo, and the correspondence of voters' acceptance set fails lower hemicontinuity. Cripps [15] analyzed an alternating offer bargaining game which is played by a risk neutral buyer and seller, where the value of the good to be traded follows a Markov process. He showed that if the buyer is less patient than the seller, then there will be delays in the players reaching an agreement, the buyer is forced into a suboptimal consumption policy and the equilibrium is ex-ante inefficient. In addition, he proved that if the buyer is more patient than the seller, then there is a unique and efficient equilibrium where agreement is immediate. Kenan [22] presented repeated contract negotiations involving the same buyer and seller where the size of the surplus being divided, is specified as a two-state Markov chain with transitions that are synchronized with contract negotiation dates. The contracts are linked because the buyer has persistent private information. Since there is persistence in the Markov chain generating the surplus, a successful demand induces the seller to make another aggressive demand in the next negotiation, since the buyer's acceptance reveals that the current surplus is large. Coles and Muthoo [14] introduced an alternating offers-bargaining model in which the set of possible utility pairs evolves through time in a non-stationary manner. They showed that in the limit, as the time interval between two consecutive offers becomes arbitrarily small, there exists a unique subgame perfect equilibrium and derived a characterization of the unique subgame perfect equilibrium payoffs. Naidu et al. [27] studied intentional idiosyncratic play in a standard stochastic evolutionary model of equilibrium selection in a class of bargaining games showing existence and uniqueness of a stochastically stable bargaining outcome under intentional idiosyncratic play in a class of games that nests contract games and the Nash demand game. Abreu and Manea [1] studied in the bargaining model the Markov perfect equilibria of an infinite horizon game in which pairs of players connected in a network are randomly matched. They established the existence of Markov perfect equilibria and show that Markov perfect equilibria payoffs are not necessarily unique. In addition, developed a method for constructing pure strategy Markov perfect equilibria for high discount factors. Agastya [2] studied the dual issues of allocation and coalition formation in a model of social learning showing that all self-perpetuating allocations realized from a simple bargaining game must be core allocations, although players make simultaneous demands for surplus, and only on their own behalf. Furthermore, they provided a sufficient condition under which the society eventually learns to divide the surplus according to some core allocation, regardless of the initial history.

Anant et al. [4] generalized the Kalai–Smorodinsky's result defining a property called Nash equilibrium regularity and showed that the result is true as long as the feasibility sets happen to be Nash equilibrium regular. Dubra [17] established a restricted version of Nash's Independence that overcomes its major criticisms and then show that a one parameter class of asymmetric Kalai–Smorodinsky solutions is characterized by restricted independence, scale invariance, Pareto optimality and Kalai–Smorodinsky's individual monotonicity axiom. Köbberling and Peters [23] demonstrated for the Kalai–Smorodinsky bargaining solution that for two forms of risk aversion (utility risk aversion and probabilistic risk aversion) can have surprisingly opposite consequences for bargaining solutions that exhibit a weak monotonicity property. Driesen et al. [16] analyzed bargaining problems under the assumption that players are loss-averse showing that n-player bargaining problems have a unique self-supporting outcome under the Kalai–Smorodinsky solution. They established the bargaining solutions that give exactly these outcomes, and characterized them by the standard axioms: scale invariance, individual monotonicity, and strong individual rationality, and proposed a new axiom called proportional concession invariance. Roth [35] concluded that Kalai–Smorodinsky game for two-person bargaining cannot be transformed into a general n-person bargaining games in a straightforward manner showing that the solution is not Pareto Optimal; no solution, whatever it is, can possess the other properties which characterize in the two-person case. In this sense, Peters and Tijs [31] proved that there exists a unique bargaining solution, defined on the whole class of two-person bargaining games, having the following properties: individual rationality, Pareto optimality independence of equivalent utility representation, individual monotonicity and symmetry. They introduced a rather large subclass of n-person bargaining games showing that having the four axioms of the Kalai–Smorodinsky solution, and exactly one of these solutions is symmetric. They also proved that all these solutions are risk-sensitive. Moulin [25] introduced a bargaining game where players bid fractions of dictatorship that are used by all nonwinners of the auction to threaten acceptance of the winner's proposal. He described the equilibrium behavior as a subgame perfect equilibrium. Alexander [3] characterized the Kalai–Smorodinsky bargaining solution when firms and unions negotiate over wages alone, and firms set the level of employment in order to maximize profits given the agreed wage. He analyzed the case that the wage elasticity of employment and the union's risk aversion are both constant showing that in this case there is a simple relationship between the Kalai–Smorodinsky and the Nash solutions.

14.1.3 Nash Versus Kalai–Smorodinsky

In this chapter, we analyze the bargaining solutions presented by Nash [28, 29] and Kalai–Smorodinsky [20], which depend on different principles of fairness. The bargaining solution allows players to solve the bargaining problem that result in a "fair" improved position.

Following Nash [28], a solution to the bargaining problems $\mathcal{B}$ is a function f that takes as input any bargaining problem and returns a vector of utilities that belongs to the set of possible agreements Φ. Several solutions can be proposed for solving the problem, but some of them can present inconsistencies. For example, one solution can go against symmetry by proposing a total improvement of the position of one player obtaining a point in the Pareto frontier of the utility and the other player receives no improvement [12]. A different solution to the problem could be a disagreement point. The first solution violates symmetry, so the solution is unfair, and the second solution is not Pareto-efficient, and does not take advantage of the cooperation related to an agreement situation. For solving the inconsistencies in the solution of the problem, Nash [28] proposed several axioms: (a) *Invariant to affine transformations* (or Invariant to equivalent utility representations): an affine transformation of the utility and disagreement point should not alter the outcome of the bargaining process; (b) *Pareto optimality*: the solution selects a point of the Pareto frontier such that the players can be made "better" off without making other players "worse off"; (c) *Symmetry*: if the players are indistinguishable, the solution should not discriminate between them; and (d) *Independence of irrelevant alternatives*: if the solution is chosen from a feasible set which is an element of a subset of the original set but containing the point selected earlier by the solution, then the solution must still assign the same point chosen from the subset. As a result, Nash [28] proposed the Nash bargaining solution: we say that there is a unique solution b to the bargaining problem that satisfies the four axioms (a to d) which is given by the point that maximizes the product of utilities of the players.

While three of Nash's axioms are quite uncontroversial, the fourth one (*Independence of irrelevant alternatives*) raised some criticism, which lead to a different line of research. Kalai and Smorodinsky [20] looked for characterizations of an alternative solution which do not use the controversial axiom. The solution idea can be represented geometrically in the following way. Let $a(\Phi)$ be the utopia point, typically not feasible, which gives the maximum payoff. Now, connect the point of disagreement d and that ideal point $a(\Phi)$ by a line segment. The Kalai–Smorodinsky solution is the maximal point in Φ on that line segment. They replaced Nash arguable fourth axiom by a monotonicity axiom: (e) If the set of possible agreements Φ is enlarged such that the maximum utilities of the players remain unchanged, then neither of the players must not suffer from it. Then, Kalai and Smorodinsky [20] proposed the following solution: we say that there is a unique solution b to the bargaining problem that satisfies the four axioms (a, b, c and e) which is given by the intersection point of the Pareto frontier and the straight line segment connecting d and the utopia point $a(\Phi)$.

Nash [28] showed that there exists a unique standard independent solution for the bargaining model, while Kalai and Smorodinsky [20] showed that a different solution is the unique standard monotonic one.

14.1.4 Organization of the Chapter

The rest of this chapter is organized in the following manner. In the next section, the motivation introduces the archetypal bargaining problem. In Sect. 14.3, the preliminaries present the Markov continuous-time model, the game theory Markov model and the c-variable method. The Nash bargaining model and the Kalai–Smorodinsky bargaining model are presented in Sects. 14.4 and 14.5, respectively. In Sect. 14.6, there are suggested bargaining solver for the Nash and the Kalai–Smorodinsky bargaining models. In addition, it discusses the equilibrium computation in bargaining games: we present the analysis of the convergence as well the rate of convergence of the proposed method. The model for the disagreement point is presented in Sect. 14.7. Section 14.8 shows the usefulness of our approach presenting a numerical example contrasting the Nash and Kalai–Smorodinsky bargaining solution. Section 14.9 concludes and discusses future work.

14.2 Motivation

The most basic definition of bargaining refers to a socio-economic class of problems involving several players who cooperate in terms of obtaining a mutually better position of a desirable surplus whose distribution is in conflict. The features of the cooperation of the players in terms of reaching an agreement and the initial situations of the players in the status quo before an agreement has effect will determine how the surplus will be distributed. Several social, political and economic problems are related to the bargaining problem.

For instance, let us consider the case of selling a used car. When it comes to selling the car, the seller naturally wants to obtain the most money possible. It is practical to trade the car at a dealer or make a quick sale to a used car dealership, but these options usually leave the seller with significantly less than what the car is actually worth. Selling a car by himself allows the seller to get its full value. Then, the seller values his car at 3,000 which is the minimum price at which he would sell it. On the other hand, there exists a buyer that values the car at 5,000 which is the maximum price at which he would buy it. If trade occurs, the price lies between 3,000 and

5,000, then both the seller and the buyer would become better-off and a conflict of interests arises. In any trade the seller and the buyer have the possibility of achieving a mutually beneficial agreement by having conflicting interests over the terms of the trade.

The formal theory of bargaining originated with John Nash's work in the early 1950s [28, 29]. The term bargaining is usually employed to refer to situations in which players have the possibility of achieving a mutually beneficial agreement, there is a conflict of interests about when an agreement should conclude, and no agreement may be imposed on any individual without their approval.

Let us consider two players $l = 1, 2$. A *bargaining* problem is a pair $\mathcal{B} = (\Phi, d)$ in the utility space where Φ is a set of possible agreements in terms of utilities u that player 1 and player 2 can yield. The player's utility function u^l is strictly increasing and concave. The set of possible agreements is Φ, which is a compact and convex set of $\mathbb{R}^2$. An element of Φ is a pair $u = (u^1, u^2) \in \Phi$ and $d = (d^1, d^2)$ is called the disagreement utility point. Compactness arises from the assumptions related to closed production sets and bounded factor endowments. Convexity is obtained from the fact that expected utility over outcomes. Also, the set Φ involves points that dominate the disagreement point, i.e., there is a positive surplus to be enjoyed if agreement is reached. The function f takes as input any bargaining problem and returns a pair of utilities $u = (u^1, u^2) \in \Phi$. When we need to refer to the components of f, we write $u^1 = f^1(\mathcal{B})$ and $u^2 = f^2(\mathcal{B})$. The interpretation is that given a bargaining problem $\mathcal{B} = (\Phi, d)$ there exists an agreement $u = f(\Phi, d) \in \Phi$ such that $u^1 \geq d^1$ and $u^2 \geq d^2$ which ensures that there exists a mutually beneficial agreement. Figure 14.1 shows the bargaining problem.

Fig. 14.1 Bargaining model

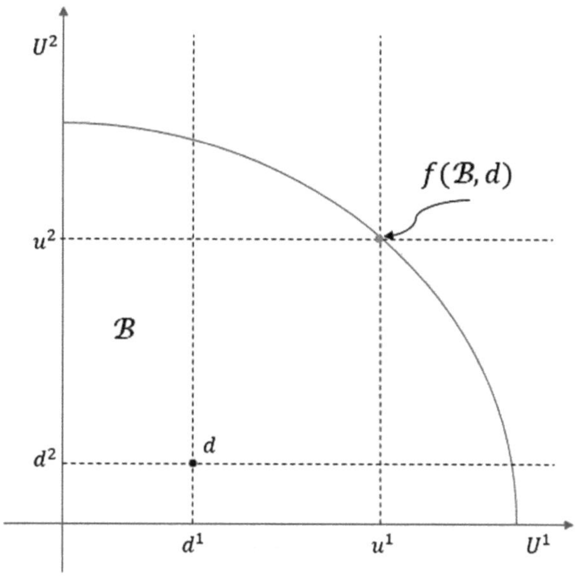

Fig. 14.2 Bargaining axioms

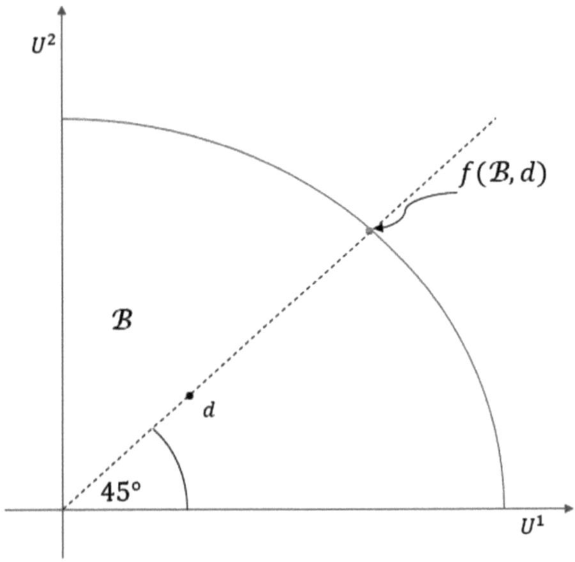

Fig. 14.3 Nash's bargaining solution

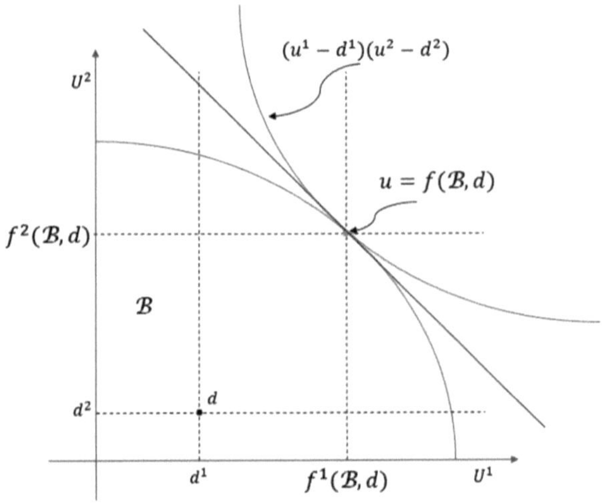

Two fundamental axioms impose the most important restrictions over the solution of the bargaining problem (see Fig. 14.2). Pareto optimality: the function $f(\Phi, d)$ has the property that there does not exist a point $u = (u^1, u^2) \in \Phi$ such that $u^1 \geq f^1(\Phi, d)$ and $u^2 \geq f^2(\Phi, d)$ such that $(u^1, u^2) \neq f(\Phi, d)$. Symmetry: suppose that $\mathcal{B}$ is such that U is symmetric around the 45° line and $d^1 = d^2$, then $f^1(\mathcal{B}) = f^2(\mathcal{B})$. The rest of the axioms will be presented in the formalization of the model.

Nash's bargaining solution of the bargaining situation described above is the unique pair of utilities, denoted by $(u^1, u^2) \in \Phi$, that solves the following maximization problem

$$\max_{u^1,u^2 \in \Theta} (u^1 - d^1)(u^2 - d^2)$$

where $\Theta \equiv \left\{ (u^1, u^2) \in \Phi \,\middle|\, u^1 \geq d^1 \text{ and } u^2 \geq d^2 \right\}$. The maximization problem stated above has a unique solution, because the result of $(u^1 - d^1)(u^2 - d^2)$, which is referred to as the Nash product, is continuous and strictly concave. Figure 14.3 illustrates Nash's bargaining solution.

14.3 Preliminaries

In this section, we introduce the (continuous-time, time-homogeneous) game model in which we are interested and the formulation of the problem [8]. As usual, $\mathbb{R}$ and $\mathbb{N}$ stand for the sets of real numbers and nonnegative integers, respectively.

Throughout the remainder

$$\mathcal{G} = (\mathcal{N}, S^l, A^l, \{A^l(s)\}_{s \in S}, V^l, Q^l)_{l=\overline{1,\mathcal{N}}} \tag{14.1}$$

stands for a *continuous-time Markov game* (CTMG), where $\mathcal{N}$ is the number of players and each player is indexed by $l = \overline{1, \mathcal{N}}$, the state space S^l is a *finite* set $\left\{ s^l_{(1)}, \ldots, s^l_{(N)} \right\}$, $N \in \mathbb{N}$, endowed with the discrete topology and the action set A^l is the action (or control) space, a *finite space* endowed with the corresponding Borel σ-algebra $\mathcal{B}(A^l)$.

For each $s^l \in S^l$, $A^l(s^l) \subset A^l$ is the nonempty set of admissible actions at s^l and we shall suppose that it is *compact*. Whereas, the set $\mathbb{K}^l := \left\{ (s^l, a^l) : s^l \in S^l, a^l \in A^l(s^l) \right\}$ is the class of admissible pairs, which is considered as a topological subspace of $S^l \times A^l$ and, similarly, the set $\mathbb{K} := \left\{ \mathbf{k} : \mathbf{k} \in \times_{l=1}^{\mathcal{N}} \mathbb{K}^l \right\}$. $V^l \in \mathcal{B}(\times_{l=1}^{\mathcal{N}} S^l \times \mathbb{K}^l)$ is the (measurable) one-stage cost function.

The function Q^l in (14.1) is the matrix $\left[q^l_{(j_l|i_l,k_l)} \right]$ of the game's transition rates, satisfying $q^l_{(j_l|i_l,k_l)} \geq 0$ for all $(s^l, a^l) \in \mathbb{K}^l$ and $i \neq j$ such that

$$\left[q^l_{(j_l|i_l,k_l)} \right] = \begin{cases} -\sum_{i \neq j}^{N} \lambda^l_{(i,j)}(a^l), & if\ i = j \\ \lambda^l_{(i,j)}(a^l), & if\ i \neq j \end{cases}$$

where $\lambda^l_{(i,j)}$ is a transition rate between state i and j, $\lambda^l_{(i)} = \sum_{j \neq i}^{N} \lambda_{(i,j)}$. This matrix is assumed to be conservative $\sum_{j_l=1}^{N_l} q^l_{(j_l|i_l,k_l)} = 0$ and stable, which means that

$$q^{l*}_{(i_l)} := \sup_{a^l \in A^l} q^l_{(i_l)}(a_l) < \infty \; \forall i_l \in S^l$$

where $q^l_{(i_l)}(a_l) := -q^l_{(i_l,i_l)}(a_l) \geq 0$ for all $a^l \in A^l$.

Now, we denote the probability *transition matrix* by

$$\Pi^l(t) = [\pi^l_{(s,i_l,\tau,j_l,k_l)}]_{i_l,j_l,k_l}, \ \tau \geq s$$

such that, $\pi^l_{(s,i_l,\tau,j_l,k_l)} = \pi^l_{(0,i_l,t,j_l,k_l)}$, $t = \tau - s \ \forall i_l, j_l \in S^l$ and where $\sum_{j_l=1}^{N_l} \pi^l_{(j_l|i_l,k_l)} = 1$. The Kolmogorov forward equations can be written as the matrix differential equation as follows:

$$\Pi'(t) = \Pi(t)Q; \quad \Pi(0) = I$$

$\Pi(t) \in \mathbb{R}^{N\times N}$, $I \in \mathbb{R}^{N\times N}$ is the identity matrix. This system can be solved by

$$\Pi(t) = \Pi(0)e^{Qt} = e^{Qt} := \sum_{t=0}^{\infty} \frac{t^n Q^n}{n!} \tag{14.2}$$

and at the stationary state, the probability transition matrix is defined as

$$\Pi^* = \lim_{t\to\infty} \Pi(t)$$

Definition 14.1 The vector $P^l \in \mathbb{R}^N$ is called *stationary distribution vector* if

$$\left(\Pi^{l\top}\right)^* P^l = P^l$$

where $\sum_{i_l=1}^{N_l} P^l_{(i_l)} = 1$.

This vector can be seen as the long-run proportion of time that the process is in state $i_l \in S^l$.

Theorem 14.1 *The following statements are equivalent:*

- $Q^{l\top} P^l = 0$
- $\Pi^{l\top}(t)P^l = P^l; \forall t \geq 0$.

The proof of this fact is easy in the case of a finite state space, recalling the Kolmogorov backward equation.

A *strategy* for player l is then defined as a sequence $d^l = \left\{d^l(t), t \geq 0\right\}$ of stochastic kernels $d^l(t)$ such that: (a) for each time $t \geq 0$, $d^l_{(k_l|i_l)}(t)$ is a probability measure on A^l such that $d^l_{(A^l(i_l)|i_l)}(t) = 1$ and, (b) for every $E^l \in \mathcal{B}(A^l)$ $d^l_{(E^l|i_l)}(t)$ is a Borel-measurable function in $t \geq 0$. We denoted by D^l the family of all strategies for player l. A multistrategy is a vector $\mathbf{d} = (d^1, \ldots, d^{\mathcal{N}}) \in D := \times_{l=1}^{\mathcal{N}} D^l$. From now on, we will consider only stationary strategies $d^l_{(k_l|i_l)}(t) = d^l_{(k_l|i_l)}$. For each strategy, $d^l_{(k_l|i_l)}$ the associated transition rate matrix is defined as:

$$Q^l(d^l) := [q^l_{(i_l,j_l)}(d^l)] = \sum_{k_l=1}^{M_l} q^l_{(j_l|i_l,k_l)} d^l_{(k_l|i_l)}$$

such that on a stationary state distribution for all $d^l_{(k_l|i_l)}$ and $t \geq 0$ we have that $\Pi^{l*}(d) = \lim_{t\to\infty} e^{Q^l(d^l)t}$, where $\Pi^{l*}(d^l)$ is a stationary transition controlled matrix.

The cost function of each player, depending on the states and actions of all the other players, is given by the values $W^l_{(i_1,k_1;\ldots;i_{\mathcal{N}},k_{\mathcal{N}})}$, so that the "average cost function" $\mathbf{V}^l$ in the stationary regime can be expressed as

$$\mathbf{V}^l(\mathbf{d}) := \sum_{i_1,k_1}^{N_1,M_1} \cdot\cdot \sum_{i_{\mathcal{N}},k_{\mathcal{N}}}^{N_{\mathcal{N}},M_{\mathcal{N}}} W^l_{(i_1,k_1,\ldots,i_{\mathcal{N}},k_{\mathcal{N}})} \prod_{l=1}^{\mathcal{N}} d^l_{(k_l|i_l)} P^l\left(s^l = s_{(i_l)}\right)$$

where

$$W^l_{(i_1,k_1,\ldots,i_{\mathcal{N}},k_{\mathcal{N}})} = \sum_{j_1}^{N_1} \cdot\cdot \sum_{j_{\mathcal{N}}}^{N_{\mathcal{N}}} V^l_{(i_1,j_1,k_1,\ldots,i_{\mathcal{N}},j_{\mathcal{N}},k_{\mathcal{N}})} \prod_{l=1}^{\mathcal{N}} \pi^l_{(j_l|i_l,k_l)}$$

Given that

$$c^l_{(i_l,k_l)} = d^l_{(k_l|i_l)} P^l\left(s^l = s_{(i_l)}\right)$$

we have

$$\mathbf{V}^l(\mathbf{c}) := \sum_{i_1,k_1}^{N_1,M_1} \cdot\cdot \sum_{i_{\mathcal{N}},k_{\mathcal{N}}}^{N_{\mathcal{N}},M_{\mathcal{N}}} W^l_{(i_1,k_1,\ldots,i_{\mathcal{N}},k_{\mathcal{N}})} \prod_{l=1}^{\mathcal{N}} c^l_{(i_l,k_l)} \tag{14.3}$$

$\mathbf{c} = (c^1, \ldots, c^{\mathcal{N}})$.

The variable $c^l_{(i_l,k_l)}$ satisfies the following restrictions [9, 33]:

1. Each vector from the matrix $c^l := \left[c^l_{(i_l,k_l)}\right]_{i_l=\overline{1,N_l};k_l=\overline{1,M_l}}$ that represents a stationary mixed-strategy that belongs to the simplex

$$\mathcal{S}^{N_l\times M_l} := \begin{cases} c^l_{(i_l,k_l)} \in \mathbb{R}^{N_l\times M_l} \text{ for } c^l_{(i_l,k_l)} \geq 0, \\ \text{where } \sum_{i_l,k_l}^{N_l,M_l} c^l_{(i_l,k_l)} = 1 \end{cases} \tag{14.4}$$

2. The variable $c^l_{(i_l,k_l)}$ satisfies the continuous time and the ergodicity constraints, and belongs to the convex, closed and bounded set defined as follows:

$$c^l \in C^l_{adm} = \begin{cases} h^l_{(j_l)}(c^l) = \sum_{i_l,k_l}^{N_l,M_l} \pi^l_{(j_l|i_l,k_l)} c^l_{(i_l,k_l)} - \sum_{k_l}^{M_l} c^l_{(j_l,k_l)} = 0 \\ \sum_{i_l,k_l}^{N_l,M_l} q^l_{(j_l|i_l,k_l)} c^l_{(i_l,k_l)} = 0 \end{cases} \tag{14.5}$$

Notice that by (14.5) it follows that

$$P^l\left(s^l{=}s_{(i_l)}\right) = \sum_{k_l}^{M_l} c^l_{(i_l,k_l)} \qquad d^l_{(k_l|i_l)} = \frac{c^l_{(i_l,k_l)}}{\sum_{k_l}^{M_l} c^l_{(i_l,k_l)}} \tag{14.6}$$

In the ergodic case $\sum_{k_l}^{M_l} c^l_{(i_l,k_l)} > 0$ for all $l = \overline{1,\mathcal{N}}$.

14.4 The Nash Bargaining Model

The Nash bargaining solution is based on a model in which the players are assumed to negotiate on which point of the set of feasible payoffs $\Phi \subset \mathbb{R}^{\mathcal{N}}$ will be agreed upon and realized by concerted actions of the members of the coalition $l = 1, \ldots, \mathcal{N}$. A pivotal element of the model is a fixed disagreement vector $d \in \mathbb{R}^{\mathcal{N}}$ which plays the role of a deterrent. If negotiations break down and no agreement is reached, then the disagreement point will take effect (the players are committed to the disagreement point in the case of failing to reach a consensus on which feasible payoff to realize). Thus the whole bargaining problem $\mathcal{B}$ will be concisely given by the pair $\mathcal{B} = (\Phi, d)$. We will call this form the condensed form of the bargaining problem (see [18, 28]).

A bargaining problem can be derived from the normal form of an $\mathcal{N}$-person game $\mathcal{G} = C^1, \ldots, C^{\mathcal{N}}; u^1, \ldots, u^{\mathcal{N}}$ in a natural way. The set of all feasible payoffs (outcomes) is defined as

$$\Theta = u : u = \left(u^1(c), \ldots, u^{\mathcal{N}}(c)\right), c \in C$$

where $C = C^1 \times \cdots \times C^{\mathcal{N}}$

Given a disagreement vector $d \in \mathbb{R}^{\mathcal{N}}$, $\mathcal{B} = (\Theta, d)$ it is a bargaining problem in condensed form. We can derive another bargaining problem $\mathcal{B} = (\Phi, d)$ from $\mathcal{G}$ by extending the set of feasible outcomes Θ to its convex hull Φ. Notice that any element $\varphi \in \Phi$ can be represented as

$$\varphi = \sum_{l=1}^{\mathcal{N}} \lambda^l u^l$$

where $u = (u^1(c), \ldots, u^{\mathcal{N}}(c))$, $(c \in C)$, $\lambda^l \geq 0$ for all l, and $\sum_{l=1}^{\mathcal{N}} \lambda^l = 1$.

The payoff vector φ can be realized by playing the strategies c with probability λ, and so φ is the expected payoff of the players. Thus, when the players face the bargaining problem $\mathcal{B}$ the question is, which point of Φ should be selected, taking into account the different position and strength of the players that is reflected in the set Φ of extended payoffs and the disagreement point d.

Nash approached this problem by assigning a one-point solution to $\mathcal{B}$ in an axiomatic manner. Let $\mathcal{B}$ denote the set of all pairs (Φ, d) such that

1. $\Phi \subset \mathbb{R}^{\mathcal{N}}$ is compact, convex;
2. there exists at least one $u \in \Phi$ such that $u > d$.

A Nash solution to the bargaining problem is a function $f : \mathcal{B} \to \mathbb{R}^{\mathcal{N}}$ such that $f(\Phi, d) \in \Phi$. We shall confine ourselves to functions satisfying the following axioms (see [18, 20, 26, 28]).

1. Feasibility: $f(\Phi, d) \in \Phi$.
2. Rationality: $f(\Phi, d) \geq d$.
3. Pareto Optimality: For every $(\Phi, d) \in \mathcal{B}$ there is $u \in \Phi$ such that $u \geq f(\Phi, d)$ and imply $u = f(\Phi, d)$.
4. Symmetry: If for a bargaining problem $(\Phi, d) \in \mathcal{B}$, there exist indices i, j such that $\varphi = (\varphi^1, \ldots, \varphi^{\mathcal{N}}) \in \Phi$ if and only if $\bar{\varphi} = (\bar{\varphi}^1, \ldots, \bar{\varphi}^{\mathcal{N}}) \in \Phi$, $(\bar{\varphi}^l = \varphi^l, l \neq i, l \neq j, \bar{\varphi}^i = \varphi^j, \bar{\varphi}^j = \varphi^i)$ and $d^i = d^j$ for $d = (d^1, \ldots, d^{\mathcal{N}})$, then $f^i = f^j$ for the solution vector $f(\Phi, d) = (f^1, \ldots, f^{\mathcal{N}})$.
5. Invariance with respect to affine transformations of utility: Let $\alpha^l > 0, \beta^l, (l = 1, \ldots, \mathcal{N})$ be arbitrary constants and let

$$d' = (\alpha^1 d^1 + \beta^1, \ldots, \alpha^{\mathcal{N}} d^{\mathcal{N}} + \beta^{\mathcal{N}}) \quad \text{with } d = (d^1, \ldots, d^{\mathcal{N}})$$

and

$$\Phi' = (\alpha^1 \varphi^1 + \beta^1, \ldots, \alpha^{\mathcal{N}} \varphi^{\mathcal{N}} + \beta^{\mathcal{N}}) : (\varphi^1, \ldots, \varphi^{\mathcal{N}}) \in \Phi.$$

Then $f(\Phi', d') = (\alpha^1 d^1 + \beta^1, \ldots, \alpha^{\mathcal{N}} f^{\mathcal{N}} + \beta^{\mathcal{N}})$, where $f(\Phi, d) = (f^1, \ldots, f^{\mathcal{N}})$.
6. Independence of irrelevant alternatives: If (Φ, d) and (T, d) are bargaining pairs such that $\Phi \subset T$ and $f(T, d) \in \Phi$, then $f(T, d) = f(\Phi, d)$.

Theorem 14.2 *There is a unique function f satisfying axioms 1–6, furthermore for all $(\Phi, d) \in \mathcal{B}$, the vector $f(\Phi, d) = (f^1, \ldots, f^{\mathcal{N}}) = (u^1, \ldots, u^{\mathcal{N}})$ is the unique solution of the optimization problem*

$$\begin{aligned} &\textit{maximize} \qquad g(u) = \prod_{l=1}^{\mathcal{N}} (u^l - d^l) \\ &\textit{subject to} \qquad u \in \Phi,\ u \geq d \end{aligned} \tag{14.7}$$

The objective function of problem (14.7) is usually called the Nash product.

Proof See [18].

Remark 14.1 There are exactly two solutions satisfying axioms 1, 2, 4, 5, and 6. One is the Nash's solution and the other is the disagreement solution.

For the next conjectures consider a bargaining problem as a pair (Φ, d) where $\Phi \subset \mathbb{R}^2$ and $d \in \mathbb{R}^2$.

Conjecture 14.1 *The Pareto frontier Ω^e of the set Φ is the graph of a concave function, denoted by h, whose domain is a closed interval $\mathcal{B} \subseteq R$. Furthermore, there exists $f^1 \in \mathcal{B}$ such that $u^1 > d^1$ and $h(u^1) > d^2$ [26].*

Conjecture 14.2 *The set Ω^w of weakly Pareto-efficient utility pairs is closed [26].*

14.5 The Kalai–Smorodinsky Bargaining Model

With the property of independence of irrelevant alternatives, Nash's solution is not sensitive to the range of outcomes contained in the feasible set, for instance, by the utopia point $a(\Phi) = (a^1(\Phi), \ldots, a^{\mathcal{N}}(\Phi))$ defined by

$$a^l(\Phi) = \max\{u^l | u^l \in \Phi \ , u \geq d\}$$

this point is the highest possible utility payoff player l can attain in the bargaining problem (Φ, d). Raiffa [34] proposed a solution for two-player games which is sensitive to changes in $a(\Phi)$, he proposed the solution u for two-player games such that $u = f(\Phi, d)$ is the Pareto-optimal point at which $(u^1 - d^1)/(a^1 - d^1) = (u^2 - d^2)/(a^2 - d^2)$. The solution u selects the maximal point on the line joining d to a, yielding each player the largest reward consistent with the constraint that the players' actual gains should be in proportion to their maximum gains, as measured by the ideal point $a(\Phi)$.

The Kalai–Smorodinsky solution for the bargaining problem amounts to normalizing the utility function of each agent in such a way that it is worth zero at the status quo and one at this agent's best outcome, given that all others get at least their status quo utility level; and to share equally the benefit from cooperation. This solution has been axiomatically characterized when society $\mathcal{N}$ contains only two agents, i.e., $l = 1, 2$. To every two-person game we associate a pair (Φ, d), where d is a point in the plane $d = (d^1, d^2)$ called the status quo and Φ is a subset of the plane, every point $u = (u^1, u^2) \in \Phi$ represents levels of utility for players 1 and 2 that can be reached by an outcome of the game which is feasible for the two players when they do cooperate.

Let $\mathcal{B}$ denote the set of all pairs (Φ, d) such that

1. $\Phi \subset \mathbb{R}^2$ is compact, convex;
2. there exists at least one point $u \in \Phi$ such that $u^l > d^l$, for $l = 1, 2$.

A solution to the bargaining problem is a function $f : \mathcal{B} \to \mathbb{R}^2$ such that $f(\Phi, d) \in \Phi$. We shall confine ourselves to functions satisfying the following axioms (see [20]).

1. Pareto Optimality: For every $(\Phi, d) \in \mathcal{B}$ there is no $u \in \Phi$ such that $u \geq f(\Phi, d)$ and imply $u \neq f(\Phi, d)$.

2. Symmetry: We let $f : \mathbb{R}^2 \to \mathbb{R}^2$ be defined by $T((u^1, u^2)) = (u^2, u^1)$ and we require that for every $(\Phi, d) \in \mathcal{B}$, $f(T(\Phi), T(d)) = T(f(\Phi, d))$
3. Invariance with respect to affine transformations of utility: A is an affine transformation of utility if $A = (A^1, A^2) : \mathbb{R}^2 \to \mathbb{R}^2$, $A((u^1, u^2)) = (A^1(u^1), A^2(u^2))$, and the maps $A^l(u)$ are of the form $c^l u + d^l$ for some positive constant c^l and some constant d^l. We require that for such a transformation A, $f(A(\Phi), A(d)) = A(f(\Phi, d))$.
4. Monotonicity: For a pair $(\Phi, d) \in \mathcal{B}$, let $a(\Phi) = (a^1(\Phi), a^2(\Phi))$ and $g_\Phi(u^1)$ be a function defined for $u^1 \leq a^1(\Phi)$ in the following way

$$\begin{aligned} g_\Phi(u^1) &= u^2 \text{ if } (u^1, u^2) \text{ is the Pareto of } (\Phi, d) \\ &= a^2(\Phi) \text{ if there is no such } u^2. \end{aligned}$$

If (Φ^2, d) and (Φ^1, d) are bargaining pairs such that $a^1(\Phi^1) = a^1(\phi^2)$ and $g_{\Phi^1} \leq g_{\Phi^2}$, then $f^2(\Phi^1, d) \leq f^2(\Phi^2, d)$, where $f(\Phi, d) = (f^1(\Phi, d), f^2(\Phi, d))$.

The axiom of monotonicity states that if, for every utility level that player 1 may demand, the maximum feasible utility level that player 2 can simultaneously reach is increased, then the utility level assigned to player 2 according to the solution should also be increased.

Theorem 14.3 *Let f be a bargaining solution. Then f satisfies Pareto optimality, symmetry, invariance with respect to affine transformations of utility and monotonicity if, and only if, f is Kalai–Smorodinsky solution. (See the proof in Roth [34]).*

14.5.1 The $\mathcal{N}$-Person Kalai–Smorodinsky Solution

Kalai and Smorodinsky [20] defined their solution only on two-player bargaining problems. We consider the set of all $\mathcal{N}$-player bargaining problems defined by Peters and Tijs [31], and on this set we define a class of asymmetric $\mathcal{N}$-person Kalai–Smorodinsky solutions. The set of players is denoted by $l = (1, \ldots, \mathcal{N})$, with $\mathcal{N} \geq 2$. A set $\Phi \subseteq \mathbb{R}^{\mathcal{N}}$ is comprehensive if $x \in \Phi$ and $x \geq y$ imply $y \in \Phi$, for all $x, y \in \mathbb{R}^{\mathcal{N}}$. A bargaining problem for $\mathcal{N}$ is a pair (Φ, d) where:

1. $\Phi \subseteq \mathbb{R}^{\mathcal{N}}$ is compact, convex, and comprehensive,
2. there exists a $u \in \Phi$ such that $u > d$ and $d \in \Phi$,

We talk about comprehensiveness in the sense that any player can choose a lower utility payoff without this leading to an infeasible outcome. Players seek agreement on an outcome $u \in \Phi$, yielding utility u^l to player l. In case no agreement is reached the disagreement outcome d results. For all bargaining problem $(\Phi, d) \in \mathcal{B}$ we define the Pareto set of Φ as

$$P(\Phi) = \{u \in \Phi |\text{ for all } x \in \mathbb{R}^{\mathcal{N}}, \text{ if } x \geq u \text{ and } x \neq u, \text{ then } x \notin \Phi\}$$

A bargaining solution is a map $f : \mathcal{B} \to \mathbb{R}^{\mathcal{N}}$ that assigns to each bargaining problem $(\Phi, d) \in \mathcal{B}$ a single point $f(\Phi, d) \in \Phi$.

Raiffa [34] and Kalai and Smorodinsky [20] defined and characterized the Kalai–Smorodinsky solution for two-person bargaining problems. Roth [35] observed that the $\mathcal{N}$-player extension of the solution is not Pareto-optimal on all bargaining problems in $\mathcal{B}$, i.e., does not assign an element of $P(\Phi)$ to each $(\Phi, d) \in \mathcal{B}$. Therefore, Peters and Tijs [31] introduced a subclass of bargaining problems in $\mathcal{B}$ for which this shortcoming does not occur.

Theorem 14.4 *For bargaining games with three or more players, no solution exists which possesses the properties of Pareto optimality, symmetry, and restricted monotonicity. (See the proof in Roth [35]).*

Condition 14.5 *For all* $u \in \Phi$, $u \geq d$, $l = (1, \ldots, \mathcal{N})$: $u \notin P(\Phi)$ *and* $u^l < a^l(\Phi) \Rightarrow \exists\, \varepsilon > 0$ *with* $u + \varepsilon e^l \in \Phi$, *where the vector* e^l *in* $\mathbb{R}^{\mathcal{N}}$ *has the l-th coordinate equal to* 1 *and all other coordinates equal to* 0.

If a feasible outcome u is not Pareto optimal, then for any player l who receives less than his utopia payoff it is possible to increase his utility while all other players still receive u. Let $\mathcal{I} \subseteq \mathcal{B}$ consists of all bargaining problems satisfying Condition 14.5. The class of bargaining problems $(\Phi, 0) \in \mathcal{I}$ is denoted by $\mathcal{I}_0$.

Peters and Tijs [31] defined the $\mathcal{N}$-player extension of the solution by making use of monotonic curves. A monotonic curve for $\mathcal{N}$ is a map

$$\psi : [1, \mathcal{N}] \to \left\{ u \in \mathbb{R}_+^{\mathcal{N}} | u^l \leq 1 \text{ for all player } l, \text{ and } 1 \leq \sum_{l=1}^{\mathcal{N}} u^l \right\}$$

such that for all $1 \leq s \leq t \leq \mathcal{N}$ we have $\psi(s) \leq \psi(t)$ and $\sum_{l=1}^{\mathcal{N}} \psi^l(s) = s$. The set of all monotonic curves for $\mathcal{N}$ is denoted by Ψ.

Lemma 14.1 *For each* $\psi \in \Psi$ *and* $(\Phi, 0) \in \mathcal{I}_0$ *with* $f(\Phi, 0) = e^{\mathcal{N}}$, *the set*

$$P(\Phi) \cap \{\psi(t) | t \in [1, \mathcal{N}]\}$$

contains exactly one point (see [31]).

Let ψ be some monotonic curve in Ψ. Following the Lemma 14.1 we can define $\rho^{\psi} : \mathcal{I} \to \mathbb{R}^{\mathcal{N}}$, the solution associated with ψ. Let $(\Phi, 0) \in \mathcal{I}_0$; if $a(\Phi, 0) = e^{\mathcal{N}}$, then

$$\left\{\rho^{\psi}(\Phi, 0)\right\} := P(\Phi) \cap \{\psi(t) | t \in [1, \mathcal{N}]\}$$

and if $a(\Phi, 0) = a$, then $\rho^{\psi}(\Phi, 0) := a\rho^{\psi}(a^{-1}\Phi)$. For $(\Phi, d) \in \mathcal{I}$, we define $\rho^{\psi}(\Phi, d) = d + \rho^{\psi}(\Phi - d)$. The class of all solutions associated with a monotonic curve in Ψ is referred to as the class of individually monotonic bargaining solutions, the Kalai–Smorodinsky solution is an element of this class. Observe that $\hat{\psi}$ (the monotonic curve of the Kalai-Smorodinsky solution) defines a straight line in $\mathbb{R}^{\mathcal{N}}$, which for bargaining games $(\Phi, 0) \in \mathcal{I}_0$ with $a(\Phi, 0) = e^{\mathcal{N}}$, coincides with the line connecting the disagreement point 0 and the utopia point $e^{\mathcal{N}}$. For general bargaining problems $(\Phi, d) \in \mathcal{I}$, the solution is the intersection of the Pareto set $P(\Phi)$ and the straight line that connects the disagreement point d and the utopia point $a(\Phi, d)$.

14.6 The Bargaining Solver

14.6.1 The Nash Bargaining Solver

Stated in general terms, a $\mathcal{N}$-person bargaining situation is a situation in which $\mathcal{N}$ players have a common interest to cooperate, but have conflicting interests over exactly how to cooperate. This process involves the players making offers and counteroffers to each other.

Consider a $\mathcal{N}$-person bargaining problem. Let us denote the disagreement utility that depends on the strategies $c^l_{(i_l,k_l)}$ as $d^l(c^1, \ldots, c^{\mathcal{N}})$ for each player $(l = 1, \ldots, \mathcal{N})$, and the solution for the Nash bargaining problem as the point $(u^1, \ldots, u^{\mathcal{N}})$. Following (14.3) the utilities u^l, in the same way that the disagreement utilities, are for Markov chains as follows

$$u^l = u^l\left(c^1, \ldots, c^{\mathcal{N}}\right) := \sum_{i_1,k_1}^{N_1,M_1} \cdots \sum_{i_{\mathcal{N}},k_{\mathcal{N}}}^{N_{\mathcal{N}},M_{\mathcal{N}}} W^l_{(i_1,k_1,\ldots,i_{\mathcal{N}},k_{\mathcal{N}})} \prod_{l=1}^{\mathcal{N}} c^l_{(i_l,k_l)} \tag{14.8}$$

where the matrices W^l represent the behavior of each player. This point is better than the disagreement point, therefore must satisfy that $u^l > d^l$.

The function for finding the solution to the Nash bargaining problem is

$$g(c^1, \ldots, c^{\mathcal{N}}) = \prod_{l=1}^{\mathcal{N}} (u^l - d^l)^{\alpha^l \chi(u^l > d^l)} \tag{14.9}$$

where $\alpha^l \geq 0$ and $\sum_l^{\mathcal{N}} \alpha^l = 1$, $(l = 1, .., \mathcal{N})$, which are weighting parameters for each player. We can rewrite (14.9) for purposes of implementation as follows

$$\tilde{g}(c^1, \ldots, c^{\mathcal{N}}) = \sum_{l=1}^{\mathcal{N}} \alpha^l \chi(u^l > d^l) \ln(u^l - d^l)$$

Thus, the strategy x^*, which is the vector $x^* = (c^1, \ldots, c^{\mathcal{N}}) \in X_{adm} := \bigotimes_{l=1}^{\mathcal{N}} C^l_{adm}$, is the solution for the Nash bargaining problem

$$x^* \in Arg \max_{x \in X_{adm}} \left\{\tilde{g}(c^1, .., c^{\mathcal{N}})\right\}$$

the strategies c^l satisfy the restrictions (14.4) and (14.5). Applying the Lagrange principle, (see, for example, [32, 33])

$$\mathcal{L}(x,\mu,\xi,\eta) = \tilde{g}(c^1,\ldots,c^{\mathcal{N}}) - \sum_{l=1}^{\mathcal{N}} \sum_{j_l=1}^{N_l} \mu^l_{(j_l)} h^l_{(j_l)}(c^l) -$$

$$\sum_{l=1}^{\mathcal{N}} \sum_{i_l=1}^{N_l} \sum_{j_l=1}^{N_l} \sum_{k_l=1}^{M_l} \xi^l_{(j_l)} q^l_{(j_l|i_l,k_l)} c^l_{(i_l,k_l)} - \sum_{l=1}^{\mathcal{N}} \sum_{i_l=1}^{N_l} \sum_{k_l=1}^{M_l} \eta^l \left(c^l_{(i_l,k_l)} - 1 \right)$$

The approximative solution obtained by the Tikhonov's regularization with $\delta > 0$ (see [33]) is given by

$$x^*, \mu^*, \xi^*, \eta^* = \arg \max_{x \in X_{adm}} \min_{\mu,\xi,\eta \geq 0} \mathcal{L}_\delta(x,\mu,\xi,\eta)$$

where

$$\mathcal{L}_\delta(x,\mu,\xi,\eta) = \tilde{g}(c^1,\ldots,c^{\mathcal{N}}) - \sum_{l=1}^{\mathcal{N}} \sum_{j_l=1}^{N_l} \mu^l_{(j_l)} h^l_{(j_l)}(c^l) -$$

$$\sum_{l=1}^{\mathcal{N}} \sum_{i_l=1}^{N_l} \sum_{j_l=1}^{N_l} \sum_{k_l=1}^{M_l} \xi^l_{(j_l)} q^l_{(j_l|i_l,k_l)} c^l_{(i_l,k_l)} - \sum_{l=1}^{\mathcal{N}} \sum_{i_l=1}^{N_l} \sum_{k_l=1}^{M_l} \eta^l \left(c^l_{(i_l,k_l)} - 1 \right) - \tag{14.10}$$

$$\frac{\delta}{2} \left(\|x\|^2 - \|\mu\|^2 - \|\xi\|^2 - \|\eta\|^2 \right)$$

Notice that the Lagrange function (14.10) satisfies the saddle-point [32] condition, namely, for all $x \in X_{adm}$ and $\mu, \xi, \eta \geq 0$ we have

$$\mathcal{L}_\delta(x_\delta, \mu^*_\delta, \xi^*_\delta, \eta^*_\delta) \leq \mathcal{L}_\delta(x^*_\delta, \mu^*_\delta, \xi^*_\delta, \eta^*_\delta) \leq \mathcal{L}_\delta(x^*_\delta, \mu_\delta, \xi_\delta, \eta_\delta)$$

14.6.2 Kalai–Smorodinsky Solver

Consider a $\mathcal{N}$-person bargaining problem. Let us denote the disagreement utility that depends on the strategies $c^l_{(i_l,k_l)}$ as $d^l(c^1,\ldots,c^{\mathcal{N}})$ for each player $(l = 1,\ldots,\mathcal{N})$, and the solution for the bargaining problem as the point $(u^1,\ldots,u^{\mathcal{N}})$. Following (14.3) the utilities u^l are for Markov chains as follows

$$u^l = u^l\left(c^1,\ldots,c^{\mathcal{N}}\right) := \sum_{i_1,k_1}^{N_1,M_1} \cdots \sum_{i_{\mathcal{N}},k_{\mathcal{N}}}^{N_{\mathcal{N}},M_{\mathcal{N}}} W^l_{(i_1,k_1,\ldots,i_{\mathcal{N}},k_{\mathcal{N}})} \prod_{l=1}^{\mathcal{N}} c^l_{(i_l,k_l)} \tag{14.11}$$

where the matrices W^l represent the behavior of each player.

The Kalai–Smorodinsky solution chooses the maximum individually rational payoff profile at which each player's payoff has the same proportion from disagreement point to the utopia point. For solving the bargaining problem we consider that there exists an optimal solution that is a strong Pareto optimal point and it is the closest solution to the utopia point. To find the Pareto optimal solution, we formulate the problem as the L_p-norm that reduces the distance to the utopia point in the Euclidean space. Following [41], the function for finding the solution to the bargaining problem is

$$g(c^1,\ldots,c^{\mathcal{N}}) = \left[\left|\sum_{l=1}^{\mathcal{N}} \lambda^l \frac{(u^l - d^l)^{\alpha^l \chi(u^l > d^l)}}{(a^l - d^l)^{\alpha^l \chi(a^l > d^l)}}\right|^p\right]^{1/p} \tag{14.12}$$

where a^l is the utopia point, $\alpha^l \geq 0$ are weighting parameters for each player, and $\lambda \in \Lambda^{\mathcal{N}}$ such that

$$\Lambda^{\mathcal{N}} := \left\{\lambda \in \mathbb{R}^{\mathcal{N}} : \lambda \in [0,1],\ \sum_{l=1}^{\mathcal{N}} \lambda^l = 1\right\}$$

We can rewrite (14.12) for purposes of implementation as follows

$$\tilde{g}(c^1,\ldots,c^{\mathcal{N}}) = \left[\left|\sum_{l=1}^{\mathcal{N}} \lambda^l \left(\alpha^l \chi(u^l > d^l)\ln(u^l - d^l) - \alpha^l \chi(a^l > d^l)\ln(a^l - d^l)\right)\right|^p\right]^{1/p}$$

Thus, the strategy x^*, which is the vector $x^* = (c^1,\ldots,c^{\mathcal{N}}) \in X_{adm} := \bigotimes_{l=1}^{\mathcal{N}} C^l_{adm}$, is the solution for the bargaining problem

$$x^* \in Arg \max_{x \in X_{adm}, \lambda \in \Lambda^{\mathcal{N}}} \{\tilde{g}(c^1,\ldots,c^{\mathcal{N}})\}$$

the strategies c^l satisfy the restrictions (14.4) and (14.5). Applying the Lagrange principle

$$\mathcal{L}(x,\lambda,\mu,\xi,\eta) = \tilde{g}(c^1,\ldots,c^{\mathcal{N}}) - \sum_{l=1}^{\mathcal{N}} \sum_{j_l=1}^{N_l} \mu^l_{(j_l)} h^l_{(j_l)}(c^l) -$$

$$\sum_{l=1}^{\mathcal{N}} \sum_{i_l=1}^{N_l} \sum_{j_l=1}^{N_l} \sum_{k_l=1}^{M_l} \xi^l_{(j_l)} q^l_{(j_l|i_l,k_l)} c^l_{(i_l,k_l)} - \sum_{l=1}^{\mathcal{N}} \sum_{i_l=1}^{N_l} \sum_{k_l=1}^{M_l} \eta^l \left(c^l_{(i_l,k_l)} - 1\right)$$

The approximative solution obtained by the Tikhonov's regularization with $\delta > 0$ is given by

$$x^*, \lambda^*, \mu^*, \xi^*, \eta^* = \arg \max_{x \in X_{adm}, \lambda \in \Lambda^{\mathcal{N}}} \min_{\mu,\xi,\eta \geq 0} \mathcal{L}_\delta(x,\lambda,\mu,\xi,\eta)$$

where

$$\mathcal{L}_\delta(x,\lambda,\mu,\xi,\eta) = \tilde{g}(c^1,\ldots,c^{\mathcal{N}}) - \sum_{l=1}^{\mathcal{N}}\sum_{j_l=1}^{N_l}\mu^l_{(j_l)}h^l_{(j_l)}(c^l) -$$

$$\sum_{l=1}^{\mathcal{N}}\sum_{i_l=1}^{N_l}\sum_{j_l=1}^{N_l}\sum_{k_l=1}^{M_l}\xi^l_{(j_l)}q^l_{(j_l|i_l,k_l)}c^l_{(i_l,k_l)} - \sum_{l=1}^{\mathcal{N}}\sum_{i_l=1}^{N_l}\sum_{k_l=1}^{M_l}\eta^l\left(c^l_{(i_l,k_l)}-1\right) - \tag{14.13}$$

$$\frac{\delta}{2}\left(\|x\|^2+\|\lambda\|^2-\|\mu\|^2-\|\xi\|^2-\|\eta\|^2\right)$$

Notice that the Lagrange function (14.13) satisfies the saddle-point condition, namely, for all $x \in X_{adm}$, $\lambda \in \Lambda^{\mathcal{N}}$ and $\mu, \xi, \eta \geq 0$ we have

$$\mathcal{L}_\delta(x_\delta,\lambda_\delta,\mu^*_\delta,\xi^*_\delta,\eta^*_\delta) \leq \mathcal{L}_\delta(x^*_\delta,\lambda^*_\delta,\mu^*_\delta,\xi^*_\delta,\eta^*_\delta) \leq \mathcal{L}_\delta(x^*_\delta,\lambda^*_\delta,\mu_\delta,\xi_\delta,\eta_\delta)$$

14.6.3 The Extraproximal Solver Method

In the proximal format (see, [5]) the relation (14.10) can be expressed as

$$\begin{aligned}
\mu^*_\delta &= \arg\min_{\mu\geq 0}\left\{\tfrac{1}{2}\|\mu-\mu^*_\delta\|^2+\gamma\mathcal{L}_\delta(x^*_\delta,\mu,\xi^*_\delta,\eta^*_\delta)\right\}\\
\xi^*_\delta &= \arg\min_{\xi\geq 0}\left\{\tfrac{1}{2}\|\xi-\xi^*_\delta\|^2+\gamma\mathcal{L}_\delta(x^*_\delta,\mu^*_\delta,\xi,\eta^*_\delta)\right\}\\
\eta^*_\delta &= \arg\min_{\eta\geq 0}\left\{\tfrac{1}{2}\|\eta-\eta^*_\delta\|^2+\gamma\mathcal{L}_\delta(x^*_\delta,\mu^*_\delta,\xi^*_\delta,\eta)\right\}\\
x^*_\delta &= \arg\max_{x\in X}\left\{-\tfrac{1}{2}\|x-x^*_\delta\|^2+\gamma\mathcal{L}_\delta(x,\mu^*_\delta,\xi^*_\delta,\eta^*_\delta)\right\}
\end{aligned} \tag{14.14}$$

for the relation (14.13) the proximal format will be extended with $\mathcal{L}_\delta(x,\lambda,\mu,\xi,\eta)$ and the following equation

$$\lambda^*_\delta = \arg\max_{\lambda\in\Lambda^{\mathcal{N}}}\left\{-\tfrac{1}{2}\|\lambda-\lambda^*_\delta\|^2+\gamma\mathcal{L}_\delta(x^*_\delta,\lambda,\mu^*_\delta,\xi^*_\delta,\eta^*_\delta)\right\}$$

where the solutions x^*_δ, λ^*_δ, μ^*_δ, ξ^*_δ and η^*_δ depend on the parameters $\delta > 0$ and $\gamma > 0$.

The Extraproximal Method for the conditional optimization problems was suggested in [5, 38]. We design the method for the static bargaining game in a general format with some fixed admissible initial values ($x_0 \in X$, $\lambda_0 \in \Lambda^{\mathcal{N}}$ and $\mu_0, \xi_0, \eta_0 \geq 0$), considering that we want to maximize the function as follows:

1. The *first half-step* (prediction):

$$\begin{aligned}
\bar{\mu}_n &= \arg\max_{\mu \geq 0} \left\{ -\tfrac{1}{2}\|\mu - \mu_n\|^2 - \gamma \mathcal{L}_\delta(x_n, \mu, \xi_n, \eta_n) \right\} \\
\bar{\xi}_n &= \arg\max_{\xi \geq 0} \left\{ -\tfrac{1}{2}\|\xi - \xi_n\|^2 - \gamma \mathcal{L}_\delta(x_n, \bar{\mu}_n, \xi, \eta_n) \right\} \\
\bar{\eta}_n &= \arg\max_{\eta \geq 0} \left\{ -\tfrac{1}{2}\|\eta - \eta_n\|^2 - \gamma \mathcal{L}_\delta(x_n, \bar{\mu}_n, \bar{\xi}_n, \eta) \right\} \\
\bar{x}_n &= \arg\max_{x \in X} \left\{ -\tfrac{1}{2}\|x - x_n\|^2 + \gamma \mathcal{L}_\delta(x, \bar{\mu}_n, \bar{\xi}_n, \bar{\eta}_n) \right\}
\end{aligned} \tag{14.15}$$

2. The *second half-step* (basic)

$$\begin{aligned}
\mu_{n+1} &= \arg\max_{\mu \geq 0} \left\{ -\tfrac{1}{2}\|\mu - \mu_n\|^2 - \gamma \mathcal{L}_\delta(\bar{x}_n, \mu, \bar{\xi}_n, \bar{\eta}_n) \right\} \\
\xi_{n+1} &= \arg\max_{\xi \geq 0} \left\{ -\tfrac{1}{2}\|\xi - \xi_n\|^2 - \gamma \mathcal{L}_\delta(\bar{x}_n, \bar{\mu}_n, \xi, \bar{\eta}_n) \right\} \\
\eta_{n+1} &= \arg\max_{\eta \geq 0} \left\{ -\tfrac{1}{2}\|\eta - \eta_n\|^2 - \gamma \mathcal{L}_\delta(\bar{x}_n, \bar{\mu}_n, \bar{\xi}_n, \eta) \right\} \\
x_{n+1} &= \arg\max_{x \in X} \left\{ -\tfrac{1}{2}\|x - x_n\|^2 + \gamma \mathcal{L}_\delta(x, \bar{\mu}_n, \bar{\xi}_n, \bar{\eta}_n) \right\}
\end{aligned} \tag{14.16}$$

For the Kalai–Smorodinsky solution the presented extraproximal method will be extended employing the relation (14.13) and the following equations:

1. The *first half-step* (prediction):

$$\bar{\lambda}_n = \arg\max_{\lambda \in \Lambda^{\mathcal{N}}} \left\{ -\tfrac{1}{2}\|\lambda - \lambda_n\|^2 + \gamma \mathcal{L}_\delta(x_n, \lambda, \bar{\mu}_n, \bar{\xi}_n, \bar{\eta}_n) \right\}$$

2. The *second half-step* (basic)

$$\lambda_{n+1} = \arg\max_{\lambda \in \Lambda^{\mathcal{N}}} \left\{ -\tfrac{1}{2}\|\lambda - \lambda_n\|^2 + \gamma \mathcal{L}_\delta(\bar{x}_n, \lambda, \bar{\mu}_n, \bar{\xi}_n, \bar{\eta}_n) \right\}$$

The following theorem presents the convergence conditions of (14.15) and (14.16) and gives the estimate of its rate of convergence for the bargaining equilibrium. As well, we prove that the extraproximal method converges to a unique equilibrium point. Let us define the following extended vectors

$$\tilde{x} = \begin{pmatrix} x \\ \lambda \end{pmatrix} \in \tilde{X} := X \times \mathbb{R}^+ \quad \tilde{\mu} = \begin{pmatrix} \mu \\ \xi \\ \eta \end{pmatrix} \in \mathbb{R}^+ \times \mathbb{R}^+ \times \mathbb{R}^+$$

Then, the regularized Lagrange function can be expressed as

$$\tilde{\mathcal{L}}_\delta(\tilde{x}, \tilde{\mu}) := \mathcal{L}_\delta(x, \lambda, \mu, \xi, \eta)$$

The equilibrium point that satisfies (14.14) can be expressed as

$$\tilde{\mu}_\delta^* = \arg\min_{\tilde{\mu}\geq 0} \left\{ \tfrac{1}{2}\|\tilde{\mu} - \tilde{\mu}_\delta^*\|^2 + \gamma \tilde{\mathcal{L}}_\delta(\tilde{x}_\delta^*, \tilde{\mu}) \right\}$$

$$\tilde{x}_\delta^* = \arg\max_{\tilde{x}\in\tilde{X}} \left\{ -\tfrac{1}{2}\|\tilde{x} - \tilde{x}_\delta^*\|^2 + \gamma \tilde{\mathcal{L}}_\delta(\tilde{x}, \tilde{\mu}_\delta^*) \right\}$$

Now, let us introduce the following variables

$$\tilde{y} = \begin{pmatrix} \tilde{y}_1 \\ \tilde{y}_2 \end{pmatrix} \in \tilde{X} \times \mathbb{R}^+, \quad \tilde{z} = \begin{pmatrix} \tilde{z}_1 \\ \tilde{z}_2 \end{pmatrix} \in \tilde{X} \times \mathbb{R}^+$$

and let define the Lagrange function in terms of $\tilde{y}$ and $\tilde{z}$

$$L_\delta(\tilde{y}, \tilde{z}) := \mathcal{L}_\delta(\tilde{y}_1, \tilde{z}_2) - \mathcal{L}_\delta(\tilde{z}_1, \tilde{y}_2)$$

For $\tilde{y}_1 = \tilde{x}$, $\tilde{y}_2 = \tilde{\mu}$, $\tilde{z}_1 = \tilde{z}_1^* = \tilde{x}_\delta^*$ and $\tilde{z}_2 = \tilde{z}_2^* = \tilde{\mu}_\delta^*$ we have

$$L_\delta(\tilde{y}, \tilde{z}^*) := \tilde{\mathcal{L}}_\delta(\tilde{x}, \tilde{\mu}_\delta^*) - \tilde{\mathcal{L}}_\delta(\tilde{x}_\delta^*, \tilde{\mu})$$

In these variables the relation (14.14) can be represented by

$$\tilde{z}^* = \arg\max_{\tilde{y}\in\tilde{X}\times\mathbb{R}^+} \left\{ -\tfrac{1}{2}\|\tilde{y} - \tilde{z}^*\|^2 + \gamma L_\delta(\tilde{y}, \tilde{z}^*) \right\} \tag{14.17}$$

Finally, we have that the extraproximal method can be expressed by

1. First step

$$\hat{z}_n = \arg\max_{\tilde{y}\in\tilde{X}\times\mathbb{R}^+} \left\{ -\tfrac{1}{2}\|\tilde{y} - \tilde{z}_n\|^2 + \gamma L_\delta(\tilde{y}, \tilde{z}_n) \right\} \tag{14.18}$$

2. Second step

$$\tilde{z}_{n+1} = \arg\max_{\tilde{y}\in\tilde{X}\times\mathbb{R}^+} \left\{ -\tfrac{1}{2}\|\tilde{y} - \tilde{z}_n\|^2 + \gamma L_\delta(\tilde{y}, \hat{z}_n) \right\} \tag{14.19}$$

Lemma 14.2 *Let $\tilde{\mathcal{L}}_\delta(\tilde{x}, \tilde{\mu})$ be differentiable in $\tilde{x}$ and $\tilde{\mu}$, whose partial derivative with respect to μ satisfies the Lipschitz condition with positive constant K_0. Then,*

$$\|\tilde{z}_{n+1} - \hat{z}_n\| \leq \gamma K_0 \|\tilde{z}_n - \hat{z}_n\|$$

Proof See [37, 41].

Lemma 14.3 *Let us consider the set of regularized solutions of a nonempty game. The behavior of the regularized function is described by the following inequality:*

$$L_\delta(\tilde{y}, \tilde{y}) - L_\delta(\tilde{z}_\delta^*, \tilde{y}) \geq \delta \|\tilde{y} - \tilde{z}_\delta^*\|$$

for all $\tilde{y} \in \{\tilde{y} \mid \tilde{y} \in X \times \mathbb{R}^+\}$ *and* $\delta > 0$.

Proof See [41].

Theorem 14.6 (Convergence and rate of convergence) *Let* $\tilde{\mathcal{L}}_\delta(\tilde{x}, \tilde{\mu})$ *be differentiable in* $\tilde{x}$ *and* $\tilde{\mu}$, *whose partial derivative with respect to* $\tilde{\mu}$ *satisfies the Lipschitz condition with positive constant* K. *Then, for any* $\delta > 0$ *there exists a small-enough*

$$\gamma_0 = \gamma_0(\delta) < K := \min\left\{\frac{1}{\sqrt{2}K_0}, \frac{1+\sqrt{1+2(K_0)^2}}{2(K_0)^2}\right\}$$

where such that, for any $0 < \gamma \leq \gamma_0$, *sequence* $\{\tilde{z}_n\}$, *which generated by the equivalent extraproximal procedure (14.18) and (14.19), monotonically converges with exponential rate* $r \in (0, 1)$ *to a unique equilibrium point* $\tilde{z}^*$, *i.e.,*

$$\|\tilde{z}_n - \tilde{z}^*\|^2 \leq e^{n \ln r}\|\tilde{z}_0 - \tilde{z}^*\|^2$$

where

$$r = 1 + \frac{4(\delta\gamma)^2}{1+2\delta\gamma-2\gamma^2K^2} - 2\delta\gamma < 1$$

and $r_{\min}$ *is given by*

$$r_{\min} = 1 - \frac{2\delta\gamma}{1+2\delta\gamma} = \frac{1}{1+2\delta\gamma}.$$

Proof Following Theorem 18 in [41] we obtain that

$$r = 1 - 2\gamma\delta + \frac{(2\gamma\delta)^2}{1+2\gamma\delta-2\gamma^2K^2} < 1.$$

Iterating over the previous inequality we have

$$\|\tilde{z}_\delta^* - \tilde{z}_{n+1}\|^2 \leq r\|\tilde{z}_\delta^* - \tilde{z}_n\|^2 \leq \cdots \leq e^{n+1 \ln r}\|\tilde{z}_\delta^* - \tilde{z}_0\|^2 \tag{14.20}$$

That implies that the series converge and also that the trajectories are bounded. Then, by (14.20) we have that

$$\|\tilde{z}_\delta^* - \tilde{z}_{n+1}\|^2 \underset{n\to\infty}{\to} 0.$$

Given that $\tilde{z}$ is a bounded sequence, by the Weierstrass Theorem there exists a point $\tilde{z}'$ such that any subsequence $\tilde{z}_{n_i}$ satisfies that $\tilde{z}_{n_i} \underset{n_i\to\infty}{\to} \tilde{z}'$. In addition, we have that $\left\|\tilde{z}_{n_i} - \tilde{z}_{n_i+1}\right\|^2 \to 0$. Fixing, $n = n_i$ in (14.17) and computing the limit when $n_i \to \infty$ we have

$$\tilde{z}' = \arg\min_{\tilde{y}\in\tilde{X}\times\mathbb{R}^+}\left\{\tfrac{1}{2}\|\tilde{y} - \tilde{z}'\|^2 + \gamma L_\delta(\tilde{y}, \tilde{z}')\right\}$$

Then, we have that $\tilde{z}' = \tilde{z}_\delta^*$, i.e., any limit point of the sequence $\tilde{z}_n$ is a solution of the problem. Given that $\left\|\tilde{z}_n - \tilde{z}_\delta^*\right\|^2$ is monotonically decreasing then, there exists a unique limit point (equilibrium point). As a consequence, we have that the sequence $\tilde{z}_n$ satisfies that $\tilde{z}_n \underset{n\to\infty}{\to} \tilde{z}^*_\delta$ with a convergence velocity of $e^{n \ln r}$.

See the complete proof in [41]

Remark 14.2 The exponential rate $r \in (0, 1)$ satisfies

$$r \simeq r_0 \left(1 + \tfrac{1}{N^2}\right).$$

14.7 The Model for the Disagreement Point

A pivotal element of the model is a fixed disagreement vector (also called status quo or threat point). If negotiations break down and no agreement is reached, then inevitably the disagreement point will take effect (the players are committed to the disagreement point in the case of failing to reach a consensus on which feasible payoff to realize).

Let us introduce the variables (see [38])

$$x := col\ c^l,\ \hat{x} := col\ c^{\hat{l}},\ \ (l = 1, \ldots, \mathcal{N})$$

The strategies of the players are denoted by the vector x, and $\hat{x}$ is a strategy of the rest of the players adjoint to x. For reaching the goal of the game, players try to find a join strategy $x^* = (c^1, \ldots, c^{\mathcal{N}}) \in X_{adm} := \bigotimes_{l=1}^{\mathcal{N}} C^l_{adm}$ satisfying

$$g(x, \hat{x}) := \sum_{l=1}^{\mathcal{N}} \left[d^l \left(c^l, c^{\hat{l}}\right) - \left(\max_{c^l \in C^l} d^l \left(c^l, c^{\hat{l}}\right) \right) \right] \tag{14.21}$$

Here $d^l \left(c^l, c^{\hat{l}}\right)$ (see 14.3) is the utility function of the player l which plays the strategy $c^l \in C^l$ and the other plays the strategy $c^{\hat{l}} \in C^{\hat{l}}$. If we consider the utopia point

$$\bar{c}^l := \arg\max_{c^l \in C^l} d^l \left(c^l, c^{\hat{l}}\right)$$

then, we can rewrite (14.21) as follows

$$g(x, \hat{x}) := \sum_{l=1}^{\mathcal{N}} \left[d^l \left(c^l, c^{\hat{l}}\right) - d^l \left(\bar{c}^l, c^{\hat{l}}\right) \right] \tag{14.22}$$

The functions $d^l \left(c^l, c^{\hat{l}}\right)$ $(l = 1, \ldots, \mathcal{N})$ are assumed to be concave in all their arguments.

Condition 14.7 *The function $g(x, \hat{x})$ satisfies the* ***Nash condition***

$$d^l\left(c^l, c^{\hat{l}}\right) - d^l\left(\bar{c}^l, c^{\hat{l}}\right) \leq 0$$

for any $c^l \in C^l$ *and all* $l = 1, \ldots, \mathcal{N}$

Definition 14.2 A strategy x^* is said to be a Nash equilibrium if

$$x^* \in Arg \max_{x \in X_{adm}} \{g(x, \hat{x})\}$$

Applying the regularized Lagrange principle we have the solution for the Nash equilibrium

$$x^*, \hat{x}^*, \mu^*, \xi^*, \eta^* = \arg \max_{x \in X, \hat{x} \in \hat{X}} \min_{\mu, \xi, \eta \geq 0} \mathcal{L}_{\theta,\delta}(x, \hat{x}, \mu, \xi, \eta)$$

where

$$\begin{gathered}\mathcal{L}_{\theta,\delta}(x, \hat{x}, \mu, \xi, \eta) := (1-\theta) g(x, \hat{x}) - \sum_{l=1}^{\mathcal{N}} \sum_{j_l=1}^{N_l} \mu^l_{(j_l)} h^l_{(j_l)}(c^l) - \\ \sum_{l=1}^{\mathcal{N}} \sum_{i_l=1}^{N_l} \sum_{j_l=1}^{N_l} \sum_{k_l=1}^{M_l} \xi^l_{(j_l)} q^l_{(j_l|i_l,k_l)} c^l_{(i_l,k_l)} - \sum_{l=1}^{\mathcal{N}} \sum_{i_l=1}^{N_l} \sum_{k_l=1}^{M_l} \eta^l \left(c^l_{(i_l,k_l)} - 1\right) - \\ \frac{\delta}{2}\left(\|x\|^2 + \left\|\hat{x}\right\|^2 - \|\mu\|^2 - \|\xi\|^2 - \|\eta\|^2\right)\end{gathered} \tag{14.23}$$

Notice also that the Lagrange function (14.23) satisfies the saddle-point condition, namely, for all $x \in X$, $\hat{x} \in \hat{X}$, and $\mu, \xi, \eta \geq 0$ we have

$$\mathcal{L}_{\theta,\delta}(x_\delta, \hat{x}_\delta, \mu^*_\delta, \xi^*_\delta, \eta^*_\delta) \leq \mathcal{L}_{\theta,\delta}(x^*_\delta, \hat{x}^*_\delta, \mu^*_\delta, \xi^*_\delta, \eta^*_\delta) \leq \mathcal{L}_{\theta,\delta}(x^*_\delta, \hat{x}^*_\delta, \mu_\delta, \xi_\delta, \eta_\delta)$$

14.7.1 The Extraproximal Solver Method

In the proximal format the relation (14.23) can be expressed as

$$\begin{aligned}
\mu^*_\delta &= \arg\min_{\mu \geq 0} \left\{\tfrac{1}{2}\|\mu - \mu^*_\delta\|^2 + \gamma \mathcal{L}_{\theta,\delta}(x^*_\delta, \hat{x}^*_\delta, \mu, \xi^*_\delta, \eta^*_\delta)\right\} \\
\xi^*_\delta &= \arg\min_{\xi \geq 0} \left\{\tfrac{1}{2}\|\xi - \xi^*_\delta\|^2 + \gamma \mathcal{L}_{\theta,\delta}(x^*_\delta, \hat{x}^*_\delta, \mu^*_\delta, \xi, \eta^*_\delta)\right\} \\
\eta^*_\delta &= \arg\min_{\eta \geq 0} \left\{\tfrac{1}{2}\|\eta - \eta^*_\delta\|^2 + \gamma \mathcal{L}_{\theta,\delta}(x^*_\delta, \hat{x}^*_\delta, \mu^*_\delta, \xi^*_\delta, \eta)\right\} \\
x^*_\delta &= \arg\max_{x \in X} \left\{-\tfrac{1}{2}\|x - x^*_\delta\|^2 + \gamma \mathcal{L}_{\theta,\delta}(x, \hat{x}^*_\delta, \mu^*_\delta, \xi^*_\delta, \eta^*_\delta)\right\} \\
\hat{x}^*_\delta &= \arg\max_{\hat{x} \in \hat{X}} \left\{-\tfrac{1}{2}\|\hat{x} - \hat{x}^*_\delta\|^2 + \gamma \mathcal{L}_{\theta,\delta}(x^*_\delta, \hat{x}, \mu^*_\delta, \xi^*_\delta, \eta^*_\delta)\right\}
\end{aligned}$$

where the solutions x_δ^*, $\hat{x}_\delta^*(u)$, μ_δ^*, ξ_δ^* and η_δ^* depend on the parameters $\delta > 0$ and $\gamma > 0$.

We design the method for the static Nash game in a general format with some fixed admissible initial values ($x_0 \in X, \hat{x}_0 \in \hat{X}$, and $\mu_0, \xi_0, \eta_0 \geq 0$), considering that we want to maximize the function, as follows:

1. The *first half-step*:

$$\begin{aligned}
\bar{\mu}_n &= \arg\max_{\mu \geq 0} \left\{-\tfrac{1}{2}\|\mu - \mu_n\|^2 - \gamma \mathcal{L}_{\theta,\delta}(x_n, \hat{x}_n, \mu, \xi_n, \eta_n)\right\} \\
\bar{\xi}_n &= \arg\max_{\xi \geq 0} \left\{-\tfrac{1}{2}\|\xi - \xi_n\|^2 - \gamma \mathcal{L}_{\theta,\delta}(x_n, \hat{x}_n, \bar{\mu}_n, \xi, \eta_n)\right\} \\
\bar{\eta}_n &= \arg\max_{\eta \geq 0} \left\{-\tfrac{1}{2}\|\eta - \eta_n\|^2 - \gamma \mathcal{L}_{\theta,\delta}(x_n, \hat{x}_n, \bar{\mu}_n, \bar{\xi}_n, \eta)\right\} \\
\bar{x}_n &= \arg\max_{x \in X} \left\{-\tfrac{1}{2}\|x - x_n\|^2 + \gamma \mathcal{L}_{\theta,\delta}(x, \hat{x}_n, \bar{\mu}_n, \bar{\xi}_n, \bar{\eta}_n)\right\} \\
\bar{\hat{x}}_n &= \arg\max_{\hat{x} \in \hat{X}} \left\{-\tfrac{1}{2}\|\hat{x} - \hat{x}_n\|^2 + \gamma \mathcal{L}_{\theta,\delta}(x_n, \hat{x}, \bar{\mu}_n, \bar{\xi}_n, \bar{\eta}_n)\right\}
\end{aligned} \tag{14.24}$$

2. The *second half-step*

$$\begin{aligned}
\mu_{n+1} &= \arg\max_{\mu \geq 0} \left\{-\tfrac{1}{2}\|\mu - \mu_n\|^2 - \gamma \mathcal{L}_{\theta,\delta}(\bar{x}_n, \bar{\hat{x}}_n, \mu, \bar{\xi}_n, \bar{\eta}_n)\right\} \\
\xi_{n+1} &= \arg\max_{\xi \geq 0} \left\{-\tfrac{1}{2}\|\xi - \xi_n\|^2 - \gamma \mathcal{L}_{\theta,\delta}(\bar{x}_n, \bar{\hat{x}}_n, \bar{\mu}_n, \xi, \bar{\eta}_n)\right\} \\
\eta_{n+1} &= \arg\max_{\eta \geq 0} \left\{-\tfrac{1}{2}\|\eta - \eta_n\|^2 - \gamma \mathcal{L}_{\theta,\delta}(\bar{x}_n, \bar{\hat{x}}_n, \bar{\mu}_n, \bar{\xi}_n, \eta)\right\} \\
x_{n+1} &= \arg\max_{x \in X} \left\{-\tfrac{1}{2}\|x - x_n\|^2 + \gamma \mathcal{L}_{\theta,\delta}(x, \bar{\hat{x}}_n, \bar{\mu}_n, \bar{\xi}_n, \bar{\eta}_n)\right\} \\
\hat{x}_{n+1} &= \arg\max_{\hat{x} \in \hat{X}} \left\{-\tfrac{1}{2}\|\hat{x} - \hat{x}_n\|^2 + \gamma \mathcal{L}_{\theta,\delta}(\bar{x}_n, \hat{x}, \bar{\mu}_n, \bar{\xi}_n, \bar{\eta}_n)\right\}
\end{aligned} \tag{14.25}$$

14.8 Numerical Illustration

Consider a two-person bargaining problem in a class of continuous-time controllable Markov chains. Let us denote the disagreement cost that depends on the strategies $c^l_{(i_l,k_l)}$ $(l = 1, 2)$ for players 1 and 2 as $d^1(c^1, c^2)$ and $d^2(c^1, c^2)$ respectively, and the solution for the bargaining problem as the point (u^1, u^2).

The process to solve the bargaining problem consists of two main steps, firstly to find the disagreement point we define it as the Nash equilibrium point of the problem [29]; while for the solution of the bargaining process we follow the models presented by Nash and Kalai–Smorodinsky.

Let the states $N_1 = N_2 = 6$, and the number of actions $M_1 = M_2 = 3$. The individual utility for each player is defined by

$$U^1_{(i,j|1)} = \begin{bmatrix} 34 & 45 & 1 & 28 & 7 & 23 \\ 27 & 43 & 25 & 47 & 26 & 24 \\ 15 & 45 & 14 & 15 & 43 & 48 \\ 36 & 47 & 12 & 17 & 20 & 5 \\ 20 & 41 & 22 & 43 & 35 & 14 \\ 29 & 29 & 18 & 18 & 32 & 23 \end{bmatrix} \quad U^2_{(i,j|1)} = \begin{bmatrix} 31 & 1 & 30 & 38 & 2 & 17 \\ 18 & 41 & 10 & 13 & 42 & 11 \\ 5 & 8 & 34 & 33 & 12 & 31 \\ 2 & 44 & 13 & 43 & 3 & 40 \\ 25 & 5 & 22 & 5 & 28 & 10 \\ 13 & 18 & 7 & 29 & 48 & 3 \end{bmatrix}$$

$$U^1_{(i,j|2)} = \begin{bmatrix} 30 & 44 & 14 & 47 & 25 & 31 \\ 44 & 24 & 45 & 37 & 11 & 30 \\ 24 & 25 & 12 & 20 & 32 & 22 \\ 22 & 25 & 44 & 50 & 12 & 33 \\ 38 & 12 & 36 & 33 & 27 & 22 \\ 24 & 5 & 44 & 45 & 37 & 1 \end{bmatrix} \quad U^2_{(i,j|2)} = \begin{bmatrix} 15 & 15 & 43 & 9 & 18 & 14 \\ 13 & 13 & 2 & 36 & 32 & 30 \\ 25 & 25 & 15 & 42 & 18 & 22 \\ 39 & 23 & 45 & 2 & 11 & 5 \\ 18 & 41 & 27 & 38 & 40 & 2 \\ 29 & 5 & 7 & 18 & 17 & 25 \end{bmatrix}$$

$$U^1_{(i,j|3)} = \begin{bmatrix} 28 & 27 & 48 & 8 & 16 & 27 \\ 43 & 47 & 33 & 24 & 22 & 28 \\ 21 & 37 & 19 & 28 & 15 & 42 \\ 24 & 29 & 24 & 3 & 50 & 42 \\ 42 & 49 & 46 & 33 & 31 & 42 \\ 50 & 42 & 51 & 45 & 13 & 11 \end{bmatrix} \quad U^2_{(i,j|3)} = \begin{bmatrix} 14 & 11 & 31 & 48 & 50 & 11 \\ 17 & 34 & 14 & 39 & 39 & 20 \\ 15 & 23 & 28 & 31 & 24 & 2 \\ 9 & 22 & 48 & 48 & 35 & 24 \\ 20 & 9 & 36 & 3 & 21 & 17 \\ 35 & 10 & 34 & 14 & 20 & 49 \end{bmatrix}$$

The transition rate matrices for each player are defined as follows

$$q^1_{(i,j|1)} = \begin{bmatrix} -0.5371 & 0.0444 & 0.2305 & 0.0946 & 0.0705 & 0.0970 \\ 0.0208 & -0.5381 & 0.0294 & 0.0665 & 0.0471 & 0.3743 \\ 0.1179 & 0.0965 & -0.6554 & 0.0939 & 0.1042 & 0.2429 \\ 0.1871 & 0.0965 & 0.1622 & -0.5826 & 0.0285 & 0.1083 \\ 0.0825 & 0.1871 & 0.0671 & 0.0431 & -0.4624 & 0.0827 \\ 0.0831 & 0.1685 & 0.1221 & 0.3425 & 0.0432 & -0.7593 \end{bmatrix}$$

$$q^1_{(i,j|2)} = \begin{bmatrix} -1.6112 & 0.1333 & 0.6916 & 0.2839 & 0.2114 & 0.2911 \\ 0.0624 & -1.6142 & 0.0881 & 0.1996 & 0.1412 & 1.1228 \\ 0.3538 & 0.2894 & -1.9662 & 0.2817 & 0.3127 & 0.7287 \\ 0.5614 & 0.2894 & 0.4867 & -1.7477 & 0.0855 & 0.3248 \\ 0.2474 & 0.5614 & 0.2012 & 0.1292 & -1.3873 & 0.2482 \\ 0.2492 & 0.5055 & 0.3662 & 1.0275 & 0.1295 & -2.2780 \end{bmatrix}$$

$$q^1_{(i,j|3)} = \begin{bmatrix} -0.5371 & 0.0444 & 0.2305 & 0.0946 & 0.0705 & 0.0970 \\ 0.0208 & -0.5381 & 0.0294 & 0.0665 & 0.0471 & 0.3743 \\ 0.1179 & 0.0965 & -0.6554 & 0.0939 & 0.1042 & 0.2429 \\ 0.1871 & 0.0965 & 0.1622 & -0.5826 & 0.0285 & 0.1083 \\ 0.0825 & 0.1871 & 0.0671 & 0.0431 & -0.4624 & 0.0827 \\ 0.0831 & 0.1685 & 0.1221 & 0.3425 & 0.0432 & -0.7593 \end{bmatrix}$$

$$q^2_{(i,j|1)}=\begin{bmatrix} -0.8499 & 0.2201 & 0.3707 & 0.1271 & 0.0374 & 0.0947 \\ 0.3467 & -0.6729 & 0.1271 & 0.0376 & 0.0970 & 0.0644 \\ 0.2831 & 0.0856 & -0.6306 & 0.0706 & 0.0376 & 0.1537 \\ 0.0703 & 0.1577 & 0.1369 & -0.8573 & 0.3673 & 0.1250 \\ 0.3727 & 0.0964 & 0.0944 & 0.1298 & -0.8026 & 0.1092 \\ 0.1627 & 0.1095 & 0.1237 & 0.0754 & 0.4537 & -0.9250 \end{bmatrix}$$

$$q^2_{(i,j|2)}=\begin{bmatrix} -0.8499 & 0.2201 & 0.3707 & 0.1271 & 0.0374 & 0.0947 \\ 0.3467 & -0.6729 & 0.1271 & 0.0376 & 0.0970 & 0.0644 \\ 0.2831 & 0.0856 & -0.6306 & 0.0706 & 0.0376 & 0.1537 \\ 0.0703 & 0.1577 & 0.1369 & -0.8573 & 0.3673 & 0.1250 \\ 0.3727 & 0.0964 & 0.0944 & 0.1298 & -0.8026 & 0.1092 \\ 0.1627 & 0.1095 & 0.1237 & 0.0754 & 0.4537 & -0.9250 \end{bmatrix}$$

$$q^2_{(i,j|3)}=\begin{bmatrix} -1.1332 & 0.2934 & 0.4942 & 0.1694 & 0.0498 & 0.1263 \\ 0.4623 & -0.8972 & 0.1694 & 0.0502 & 0.1294 & 0.0859 \\ 0.3774 & 0.1141 & -0.8408 & 0.0942 & 0.0501 & 0.2050 \\ 0.0938 & 0.2102 & 0.1825 & -1.1431 & 0.4898 & 0.1667 \\ 0.4970 & 0.1286 & 0.1258 & 0.1730 & -1.0701 & 0.1456 \\ 0.2169 & 0.1460 & 0.1650 & 0.1005 & 0.6049 & -1.2334 \end{bmatrix}$$

14.8.1 Computing the Disagreement Point

Given δ and γ and applying the extraproximal method we obtain the convergence of the strategies for the disagreement point in terms of the variable $c^1_{(i_1,k_1)}$ for the player 1 (see Fig. 14.4) and the convergence of the strategies $c^2_{(i_2,k_2)}$ for the player 2 (see Fig. 14.5).

$$c^1 = \begin{bmatrix} 0.0517 & 0.0540 & 0.0523 \\ 0.0560 & 0.0605 & 0.0607 \\ 0.0542 & 0.0548 & 0.0514 \\ 0.0660 & 0.0672 & 0.0635 \\ 0.0332 & 0.0372 & 0.0385 \\ 0.0582 & 0.0679 & 0.0727 \end{bmatrix} \quad c^2 = \begin{bmatrix} 0.0824 & 0.0766 & 0.0830 \\ 0.0669 & 0.0449 & 0.0584 \\ 0.0840 & 0.0736 & 0.0823 \\ 0.0407 & 0.0215 & 0.0325 \\ 0.0399 & 0.0564 & 0.0503 \\ 0.0371 & 0.0329 & 0.0366 \end{bmatrix}$$

Following (14.6) the mixed strategies obtained for the players are as follows

$$d^1 = \begin{bmatrix} 0.3273 & 0.3416 & 0.3311 \\ 0.3160 & 0.3416 & 0.3424 \\ 0.3378 & 0.3416 & 0.3205 \\ 0.3354 & 0.3416 & 0.3230 \\ 0.3051 & 0.3416 & 0.3533 \\ 0.2926 & 0.3416 & 0.3658 \end{bmatrix} \quad d^2 = \begin{bmatrix} 0.3405 & 0.3166 & 0.3429 \\ 0.3933 & 0.2637 & 0.3429 \\ 0.3503 & 0.3068 & 0.3429 \\ 0.4295 & 0.2275 & 0.3429 \\ 0.2723 & 0.3847 & 0.3429 \\ 0.3484 & 0.3087 & 0.3429 \end{bmatrix}$$

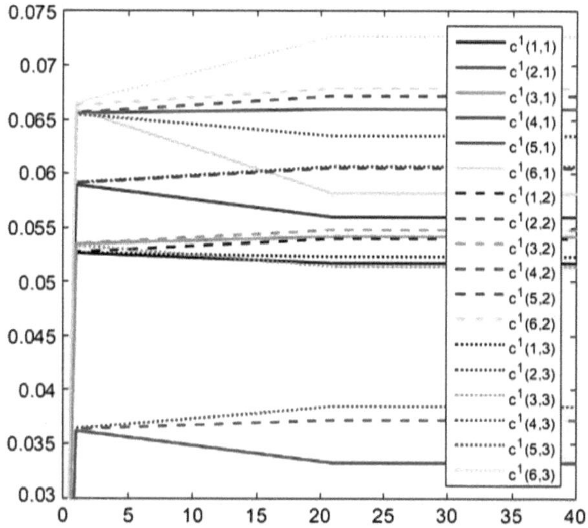

Fig. 14.4 Convergence of the strategies for player 1 to the disagreement point

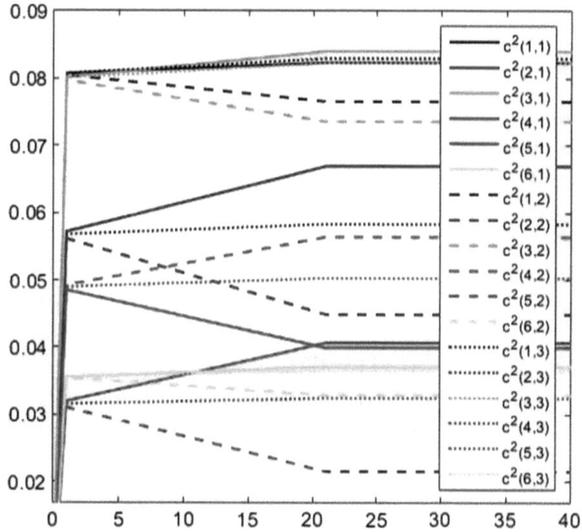

Fig. 14.5 Convergence of the strategies for player 2 to the disagreement point

With the strategies calculated, the resulting utilities following (14.3), at the disagreement point for each player $d^l(c^1, c^2)$, are as follows:

$$d^1(c^1, c^2) = 905.6447 \qquad d^2(c^1, c^2) = 704.2493$$

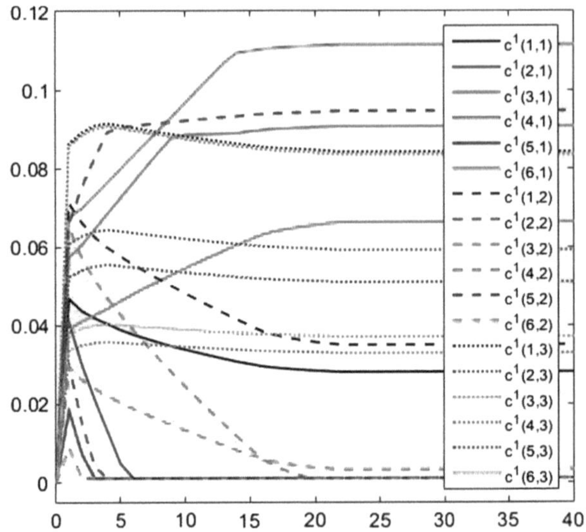

Fig. 14.6 Convergence of the strategies for player 1 to the Nash solution

14.8.2 Computing the Nash Bargaining Solution

Given δ, γ, α^l and applying the extraproximal method for the Nash bargaining solution, we obtain the convergence of the strategies in terms of the variable $c^1_{(i_1,k_1)}$ for the player 1 (see Fig. 14.6) and the convergence of the strategies $c^2_{(i_2,k_2)}$ for the player 2 (see Fig. 14.7).

$$c^1 = \begin{bmatrix} 0.0281 & 0.0677 & 0.0623 \\ 0.0010 & 0.0758 & 0.1003 \\ 0.0907 & 0.0686 & 0.0010 \\ 0.1115 & 0.0842 & 0.0010 \\ 0.0010 & 0.0466 & 0.0613 \\ 0.0010 & 0.0851 & 0.1127 \end{bmatrix} \quad c^2 = \begin{bmatrix} 0.1227 & 0.0350 & 0.0842 \\ 0.1100 & 0.0010 & 0.0592 \\ 0.1555 & 0.0010 & 0.0835 \\ 0.0607 & 0.0010 & 0.0329 \\ 0.0010 & 0.0946 & 0.0510 \\ 0.0663 & 0.0032 & 0.0371 \end{bmatrix}$$

The mixed strategies obtained for the players are as follows

$$d^1 = \begin{bmatrix} 0.1778 & 0.4280 & 0.3942 \\ 0.0056 & 0.4280 & 0.5663 \\ 0.5658 & 0.4280 & 0.0062 \\ 0.5669 & 0.4280 & 0.0051 \\ 0.0092 & 0.4280 & 0.5628 \\ 0.0050 & 0.4280 & 0.5670 \end{bmatrix} \quad d^2 = \begin{bmatrix} 0.5073 & 0.1447 & 0.3479 \\ 0.6462 & 0.0059 & 0.3479 \\ 0.6479 & 0.0042 & 0.3479 \\ 0.6415 & 0.0106 & 0.3479 \\ 0.0068 & 0.6453 & 0.3479 \\ 0.6221 & 0.0300 & 0.3479 \end{bmatrix}$$

With the strategies calculated, the resulting utilities at the Nash bargaining solution for each player, are as follows:

$$u^1(c^1, c^2) = 958.0281 \qquad u^2(c^1, c^2) = 813.2879$$

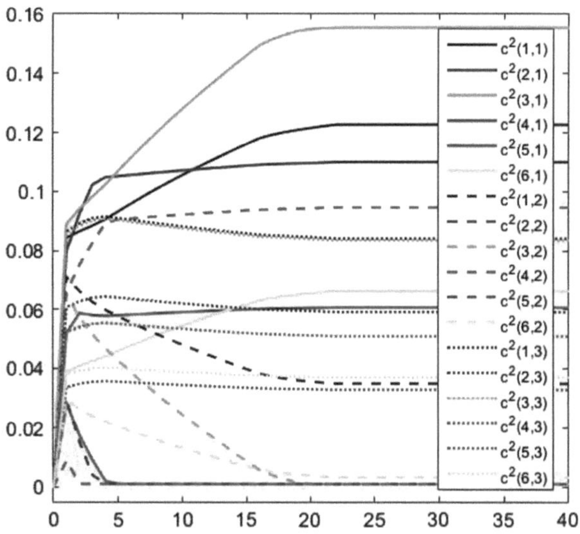

Fig. 14.7 Convergence of the strategies for player 2 to the Nash solution

14.8.3 Computing the Kalai–Smorodinsky Bargaining Solution

Given δ, γ, α^l and applying the extraproximal method for the Kalai–Smorodinsky bargaining solution with the L_1-norm, we obtain the convergence of the strategies in terms of the variable $c^1_{(i_1,k_1)}$ for the player 1 (see Fig. 14.8) and the convergence of the strategies $c^2_{(i_2,k_2)}$ for the player 2 (see Fig. 14.9).

$$c^1 = \begin{bmatrix} 0.0010 & 0.0432 & 0.1139 \\ 0.0010 & 0.0484 & 0.1278 \\ 0.1156 & 0.0438 & 0.0010 \\ 0.1420 & 0.0537 & 0.0010 \\ 0.0010 & 0.0297 & 0.0782 \\ 0.0010 & 0.0543 & 0.1435 \end{bmatrix} \quad c^2 = \begin{bmatrix} 0.2061 & 0.0010 & 0.0349 \\ 0.1447 & 0.0010 & 0.0245 \\ 0.2044 & 0.0010 & 0.0346 \\ 0.0800 & 0.0010 & 0.0136 \\ 0.0010 & 0.1245 & 0.0211 \\ 0.0903 & 0.0010 & 0.0154 \end{bmatrix}$$

The mixed strategies obtained for the players are as follows

$$d^1 = \begin{bmatrix} 0.0063 & 0.2730 & 0.7207 \\ 0.0056 & 0.2730 & 0.7213 \\ 0.7207 & 0.2730 & 0.0062 \\ 0.7219 & 0.2730 & 0.0051 \\ 0.0092 & 0.2730 & 0.7178 \\ 0.0050 & 0.2730 & 0.7219 \end{bmatrix} \quad d^2 = \begin{bmatrix} 0.8518 & 0.0041 & 0.1441 \\ 0.8500 & 0.0059 & 0.1441 \\ 0.8517 & 0.0042 & 0.1441 \\ 0.8454 & 0.0106 & 0.1441 \\ 0.0068 & 0.8491 & 0.1441 \\ 0.8465 & 0.0094 & 0.1441 \end{bmatrix}$$

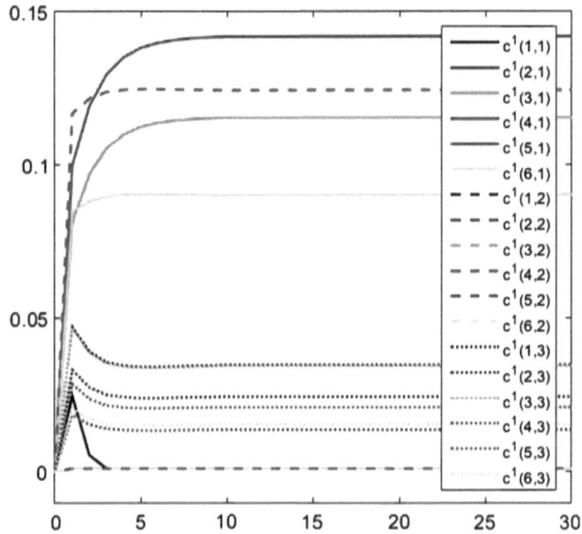

Fig. 14.8 Convergence of the strategies for player 1 to the KS solution

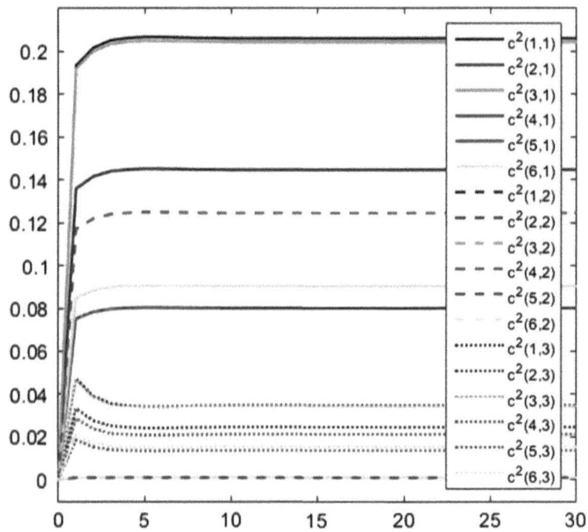

Fig. 14.9 Convergence of the strategies for player 2 to the KS solution

With the strategies calculated, the resulting utilities at the Kalai–Smorodinsky bargaining solution for each player are as follows:

$$u^1(c^1, c^2) = 960.5554 \qquad u^2(c^1, c^2) = 841.0831$$

Figure 14.10 shows the straight line linking the utilities obtained at the disagreement point and those obtained at the utopia point. We can also observe that the Nash solution approaches this line while the Kalai–Smorodinsky solution is exactly on this

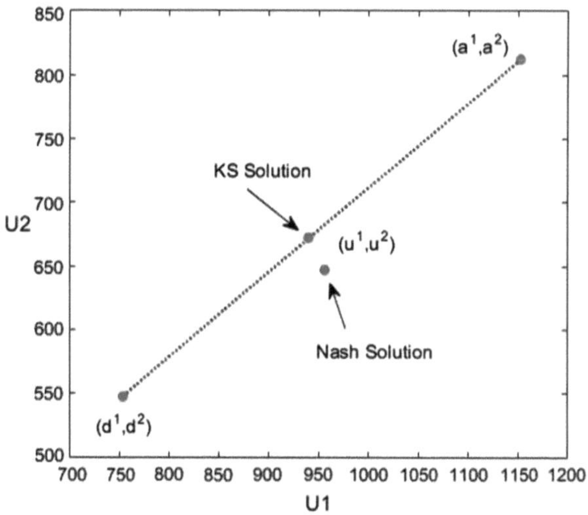

Fig. 14.10 The bargaining solution

line. The utilities at the utopia point for the bargaining problem are for each player as follows:

$$a^1(c^1, c^2) = 964.3472 \qquad a^2(c^1, c^2) = 849.8365$$

14.9 Conclusions and Future Work

This chapter focused on a general class of bargaining game models restricted to continuous-time, controllable and ergodic Markov games. We examined the bargaining approach from a theoretical perspective and provided a computational solution of the bargaining game for the Nash and Kalai–Smorodinsky models. We encapsulated both models, first focusing on some of the early results suggested in the literature, and then extending the Nash and Kalai–Smorodinsky analysis to continuous-time Markov games. For solving the problem we proposed a bargaining solver implemented by an iterated procedure of a set of nonlinear equations implemented by the Lagrange principle and including a regularization method to ensure convergence to a unique equilibrium point. In particular, we studied the computational complexity of equilibrium computation. We believe that our results form a theoretical basis for further algorithm design and complexity analysis in bargaining games.

Future research can take a number of directions. First, given the fact that bargaining theory is relevant for and applicable to real life situations (corporate deals, labor disputes and contracts, supply chain contracts, security games [10, 11, 13, 39, 40], etc.) it will be interesting to develop an application of these models. Since we conceptualize the bargaining games as a special class of poly-linear games, we will develop new methods for computing the equilibrium point and improving the regularization technique. Finally, we will propose several ways to generalize our model.

References

1. Abreu, D., Manea, M.: Markov equilibria in a model of bargaining in networks. Games Econ. Behav. **75**(1), 1–16 (2012)
2. Agastya, M.: Adaptive play in multiplayer bargaining situations. Rev. Econ. Stud. **64**(3), 411–426 (1997)
3. Alexander, C.: The Kalai-Smorodinsky bargaining solution in wage negotiations. J. Oper. Res. Soc. **43**(8), 779–786 (1992)
4. Anant, T.C.A., Mukherji, B., Basu, K.: Bargaining without convexity: generalizing the Kalai-Smorodinsky solution. Econ. Lett. **33**(2), 115–119 (1990)
5. Antipin, A.S.: An extraproximal method for solving equilibrium programming problems and games. Comput. Math. Math. Phys. **45**(11), 1893–1914 (2005)
6. Bolt, W., Houba, H.: Strategic bargaining in the variable threat game. Econ. Theor. **11**(1), 57–77 (1998)
7. Cai, H.: Inefficient Markov perfect equilibria in multilateral bargaining. Econ. Theor. **22**(3), 583–606 (2003)
8. Carrillo, L., Escobar, J., Clempner, J.B., Poznyak, A.S.: Solving optimization problems in chemical reactions using continuous-time Markov chains. J. Math. Chem. **54**, 1233–1254 (2016)
9. Clempner, J.B., Poznyak, A.S.: Simple computing of the customer lifetime value: a fixed local-optimal policy approach. J. Syst. Sci. Syst. Eng. **23**(4), 439–459 (2014)
10. Clempner, J.B., Poznyak, A.S.: Stackelberg security games: computing the shortest-path equilibrium. Expert Syst. Appl. **42**(8), 3967–3979 (2015)
11. Clempner, J.B., Poznyak, A.S.: Conforming coalitions in Stackelberg security games: setting max cooperative defenders vs. non-cooperative attackers. Appl. Soft Comput. **47**, 1–11 (2016)
12. Clempner, J.B., Poznyak, A.S.: Multiobjective Markov chains optimization problem with strong pareto frontier: principles of decision making. Expert Syst. Appl. **68**, 123–135 (2017)
13. Clempner, J.B., Poznyak, A.S.: Using the extraproximal method for computing the shortest-path mixed Lyapunov equilibrium in Stackelberg security games. Math. Comput. Simul. (2017). doi:10.1016/j.matcom.2016.12.010
14. Coles, M.G., Muthoo, A.: Bargaining in a non-stationary environment. J. Econ. Theory **109**(1), 70–89 (2003)
15. Cripps, M.W.: Markov bargaining games. J. Econ. Dyn. Control **22**(3), 341–355 (1998)
16. Driesen, B., Perea, A., Peters, H.: The Kalai-Smorodinsky bargaining solution with loss aversion. Math. Soc. Sci. **61**(1), 58–64 (2011)
17. Dubra, J.: An asymmetric Kalai-Smorodinsky solution. Econ. Lett. **73**(2), 131–136 (2001)
18. Forgó, F., Szép, J., Szidarovszky, F.: Introduction to the Theory of Games: Concepts, Methods, Applications. Kluwer Academic Publishers, Dordrecht (1999)
19. Kalai, E.: Solutions to the bargaining problem. Social Goals and Social Organization, pp. 75–105. Cambridge University Press, Cambridge (1985)
20. Kalai, E., Smorodinsky, M.: Other solutions to Nash's bargaining problem. Econometrica **43**(3), 513–518 (1975)
21. Kalandrakis, A.: A three-player dynamic majoritarian bargaining game. J. Econ. Theory **116**(2), 294–322 (2004)
22. Kennan, J.: Repeated bargaining with persistent private information. Rev. Econ. Stud. **68**, 719–755 (2001)

23. Köbberling, V., Peters, H.: The effect of decision weights in bargaining problems. J. Econ. Theory **110**(1), 154–175 (2003)
24. Merlo, A., Wilson, C.: A stochastic model of sequential bargaining with complete information. Econometrica **63**(2), 371–399 (1995)
25. Moulin, H.: Implementing the Kalai-Smorodinsky bargaining solution. J. Econ. Theory **33**(1), 32–45 (1984)
26. Muthoo, A.: Bargaining Theory with Applications. Cambridge University Press, Cambridge (2002)
27. Naidu, S., Hwang, S., Bowles, S.: Evolutiogame bargaining with intentional idiosyncratic play. Econ. Lett. **109**(1), 31–33 (2010)
28. Nash, J.F.: The bargaining problem. Econometrica **18**(2), 155–162 (1950)
29. Nash, J.F.: Two person cooperative games. Econometrica **21**, 128–140 (1953)
30. Osborne, M., Rubinstein, A.: Bargaining and Markets. Academic Press Inc., San Diego (1990)
31. Peters, H., Tijs, S.: Individually monotonic bargaining solutions for n-person bargaining games. Methods Oper. Res. **51**, 377–384 (1984)
32. Poznyak, A.S.: Advance Mathematical Tools for Automatic Control Engineers. Vol 2 Stochastic Techniques. Elsevier, Amsterdam (2009)
33. Poznyak, A.S., Najim, K., Gomez-Ramirez, E.: Self-learning Control of Finite Markov Chains. Marcel Dekker, New York (2000)
34. Raiffa, H.: Arbitration schemes for generalized two-person games. Ann. Math. Stud. **28**, 361–387 (1953)
35. Roth, A.E.: An impossibility result converning n-person bargaining games. Int. J. Game Theory **8**(3), 129–132 (1979)
36. Rubinstein, A., Wolinsky, A.: Equilibrium in a market with sequential bargaining. Econometrica **53**(5), 1133–1150 (1985)
37. Trejo, K.K., Clempner, J.B., Poznyak, A.S.: Computing the Lp-strong Nash equilibrium looking for cooperative stability in multiple agents Markov games. In: 12th International Conference on Electrical Engineering, Computing Science and Automatic Control, pp. 309–314. Mexico City. Mexico (2015)
38. Trejo, K.K., Clempner, J.B., Poznyak, A.S.: Computing the Stackelberg/Nash equilibria using the extraproximal method: convergence analysis and implementation details for Markov chains games. Int. J. Appl. Math. Comput. Sci. **25**(2), 337–351 (2015)
39. Trejo, K.K., Clempner, J.B., Poznyak, A.S.: A Stackelberg security game with random strategies based on the extraproximal theoretic approach. Eng. Appl. Artif. Intell. **37**, 145–153 (2015)
40. Trejo, K.K., Clempner, J.B., Poznyak, A.S.: Adapting strategies to dynamic environments in controllable Stackelberg security games. In: 55th IEEE Conference on Decision and Control, pp. 5484–5489. Las Vegas, USA (2016)
41. Trejo, K.K., Clempner, J.B., Poznyak, A.S.: Computing the strong L_p-Nash equilibrium for Markov chains games: convergence and uniqueness. Appl. Math. Model. **41**, 399–418 (2017)
42. von Neumann, J., Morgenstern, O.: Theory of Games and Economic Behavior. Princeton University Press, Princeton (1944)

Chapter 15
$\mathcal{H}_\infty$- Stabilization of a 3D Bipedal Locomotion Under a Unilateral Constraint

Oscar Montano, Yury Orlov, Yannick Aoustin and Christine Chevallereau

Abstract The applicability of the $\mathcal{H}_\infty$ control technique to a fully actuated 3D biped robot is addressed. In contrast to previous studies, this investigation contributes to the study of robustness of bipedal locomotion while assuming an imperfect knowledge of the restitution rule at the collision time instants in addition to external disturbance forces applied during the single support phases. Performance issues are illustrated in a numerical study performed with an emulator of the 32-DOF biped robot ROMEO, of Aldebaran Robotics.

15.1 Introduction

Bipedal robots form a subclass of legged robots. Their design is naturally inspired from the functional mobility of the human body. On the practical side, the study of mechanical legged locomotion has been motivated by its potential use as means of locomotion in rough terrains, but in particular, the interest arises from diverse sociological and commercial interests, ranging from the desire to replace humans in hazardous occupations (de-mining, nuclear power plant inspection, military interventions, etc.), to the restoration of motion in the disabled [36].

For practical implementation, good mechanical design and good modeling, play a very important role in achieving good performance. However, in real world appli-

O. Montano · Y. Orlov (✉)
CICESE, Carretera Ensenada-Tijuana No. 3918 Zona Playitas,
22860 Ensenada, Baja California, Mexico
e-mail: yorlov@cicese.mx

O. Montano
e-mail: montano.oscar@gmail.com

Y. Aoustin · C. Chevallereau
LS2N, UMR CNRS 6004, École Centrale, University of Nantes,
1 Rue de la Noe, 44300 Nantes Cedex 3, France
e-mail: yannick.aoustin@ls2n.fr

C. Chevallereau
e-mail: christine.chevallereau@ls2n.fr

J.B. Clempner and W. Yu (eds.), *New Perspectives and Applications of Modern Control Theory*, https://doi.org/10.1007/978-3-319-62464-8_15

cations bipedal robots are subject to many sources of uncertainty during walking; these could include a push from a human, an unexpected gust of wind, geometric perturbations of terrain height, or parametric uncertainties of non-modeled friction forces [5]. For these reasons, the design of feedback control systems, capable of attenuating the effect of these uncertainties, is critical to achieve the desired walking gait.

For simplicity, the complete model of the biped robot considered in this work consists of two parts: the differential equations describing the dynamics of the robot during the single support phase (one foot swinging in the air, the other staying as a pivot on the ground), and an impulse model of the contact event (the impact between the swing leg and the ground, which is modeled as a contact between two rigid bodies as in the work by [36]), thus considering the double support phase instantaneous. Therefore, the complete model of the biped robot considered in this work is simplified as a hybrid system consisting of free-motion phases separated by impacts. The study of hybrid dynamical systems has recently attracted a significant research interest, basically due to the wide variety of applications and the complexity that arises from the analysis of this type of systems (see, e.g. [7, 11], and references quoted therein). Particularly, the disturbance attenuation problem for hybrid dynamical systems has been addressed by [10, 24], where impulsive control inputs were admitted to counteract/compensate disturbances/uncertainties at time instants of instantaneous changes of the underlying state. However, the drawbacks of these approaches are (1) the complexity of finding a solution to the equations that allow the synthesis of the control law, and (2) the physical implementation of impulsive control inputs is impossible in many practical situations, e.g., while controlling walking biped robots.

In this regard, many authors have dealt with the problem of robustly controlling the walking gait of biped robots: [29] presents an iterative approach to find Control Lyapunov Functions, such that a mechanical system subject to impacts and friction remains stable. In [6, 30], PD control laws and a torque-optimization method are combined to increase the robustness against joint-torque disturbances. Other robust control techniques, such as sliding mode control, have been designed for this kind of systems (see e.g., the works by [25, 28]). While providing both finite-time convergence to a desired reference trajectory and disturbance rejection, these approaches also entail the well-known problem of chattering in the actuators. This further motivates the study of robust control techniques such as the one presented in this work, which attenuate the effect of disturbances while avoiding undesirable and harmful effects on both the actuators, and the joints.

This chapter further develops the results presented in [23], studying the applicability of the $\mathcal{H}_\infty$ control technique (extended in [20] towards mechanical systems operating under unilateral constraints) to the 32 degrees-of-freedom biped robot ROMEO, of Aldebaran Robotics [32]. Results dealing with the orbital stabilization of a simpler 2D, fully actuated biped can be found in [19]. In contrast to previous studies, this investigation contributes to the study of robustness of bipedal locomotion while assuming an imperfect knowledge of the restitution rule at the collision time instants, in addition to external disturbance forces applied during the single support phases.

15.2 Notation

The notation used throughout this work is rather standard. The argument t^+ is used to denote the right-hand side value $\mathbf{x}(t^+)$ of a trajectory $\mathbf{x}(t)$ at an impact time instant t whereas $\mathbf{x}(t^-)$ stands for the left-hand side value of the same; by default, $\mathbf{x}(t)$ is reserved for $\mathbf{x}(t^-)$, thus implying an underlying trajectory to be continuous on the left. Vectors are represented by bold, lowercase letters, whereas matrices are represented by bold, uppercase letters.

15.3 Background Materials

In this section, the $\mathcal{H}_\infty$ control problem under unilateral constraints is stated, and sufficient conditions for the existence of a solution are presented. Later on, the applicability of these results to the biped robot of interest will be studied.

15.3.1 Problem Statement

Given a scalar unilateral constraint $F(x_1, t) \geq 0$, consider a nonlinear system, evolving within the above constraint, which is governed by continuous dynamics of the form

$$\begin{aligned} \dot{\mathbf{x}}_1 &= \mathbf{x}_2 \\ \dot{\mathbf{x}}_2 &= \boldsymbol{\Phi}(\mathbf{x}_1, \mathbf{x}_2, t) + \boldsymbol{\Psi}_1(\mathbf{x}_1, \mathbf{x}_2, t)\mathbf{w} + \boldsymbol{\Psi}_2(\mathbf{x}_1, \mathbf{x}_2, t)\mathbf{u} \end{aligned} \tag{15.1}$$

$$\mathbf{z} = \mathbf{h_1}(\mathbf{x}_1, \mathbf{x}_2, t) + \mathbf{k_{12}}(\mathbf{x}_1, \mathbf{x}_2, t)\mathbf{u} \tag{15.2}$$

beyond the surface $F(x_1, t) = 0$ when the constraint is inactive, and by the algebraic relations

$$\begin{aligned} \mathbf{x}_1(t_i^+) &= \mathbf{x}_1(t_i^-) \\ \mathbf{x}_2(t_i^+) &= \boldsymbol{\mu}_0(\mathbf{x}_1(t_i),\mathbf{x}_2(t_i^-), t_i) + \boldsymbol{\omega}(\mathbf{x}_1(t_i), \mathbf{x}_2(t_i^-), t_i)\mathbf{w}_\mathbf{d}^i \end{aligned} \tag{15.3}$$

$$\mathbf{z}_i^\mathbf{d} = \mathbf{x}_2(t_i^+) \tag{15.4}$$

at a priori unknown collision time instants $t = t_i,\ i = 1, 2, \ldots,$ when the system trajectory hits the surface $F(x_1, t) = 0$. In the above relations, $\mathbf{x}^\top = [\mathbf{x}_1^\top, \mathbf{x}_2^\top] \in \mathbb{R}^{2n}$ represents the state vector with components $\mathbf{x}_1 \in \mathbb{R}^n$ and $\mathbf{x}_2 \in \mathbb{R}^n$; $\mathbf{u} \in \mathbb{R}^n$ is the control input of dimension n; $\mathbf{w} \in \mathbb{R}^l$ and $\mathbf{w}_\mathbf{d}^i \in \mathbb{R}^q$ collect exogenous signals affecting the motion of the system (external forces, including impulsive ones, as well as model imperfections). The variable $\mathbf{z} \in \mathbb{R}^s$ represents a continuous-time component of the system output to be controlled. The discrete component $\mathbf{z}_i^{\mathbf{d}}$ of this output is pre-specified by the post-impact value $\mathbf{x}_2(t)$. The overall system in the closed-loop should be dissipative with respect to the output thus specified. Throughout, the functions $\boldsymbol{\Phi}$, $\boldsymbol{\Psi}_1$, $\boldsymbol{\Psi}_2$, $\mathbf{h_1}$, $\mathbf{k_{12}}$, F, $\boldsymbol{\mu}_0$, and $\boldsymbol{\omega}$ are of appropriate dimensions, which are continuously differentiable in their arguments and uniformly bounded in t. The origin is assumed to be an equilibrium point of the unforced system (15.1)–(15.10), which is located beyond the unilateral constraint, i.e., $F(0,t) \neq 0$, $\boldsymbol{\Phi}(0,0,t) = 0$, $\mathbf{h_1}(0,0,t) = 0$, for all t and $\boldsymbol{\mu}_0(0,0,0) = 0$.

Clearly, the above system (15.1)–(15.10) is an affine control system of the vector relative degree $[2,\ldots,2]^\top$, and it governs a wide class of mechanical systems with impacts. Since the control input $\mathbf{u}$ has the same dimension as that of the generalized position $\mathbf{x}$, the present investigation is confined to the fully actuated case, though it could readily be extended to the over-actuated case with a correct choice of the control inputs [18]. The treatment in the underactuated case is also possible using the virtual constraint approach [1, 36] whenever it is applicable (e.g., for the undegraduation degree one similar to that of [21, 22]).

Admitting the above time-varying representation is particularly invoked to deal with tracking problems where the plant description is given in terms of the state deviation from the reference trajectory to track [4]. Therefore, if interpreted in terms of mechanical systems, Eq. (15.1) describes the continuous dynamics in terms of the position and velocity tracking errors ($\mathbf{x_1}$ and $\mathbf{x_2}$, respectively), before the underlying system hits the reset surface $F(\mathbf{x}_1,t) = 0$, depending on the position error $\mathbf{x_1}$ only, whilst the restitution law, given by Eq. (15.3), is a physical law for the instantaneous change of the velocity error when the resetting surface is hit.

Consider a causal feedback controller

$$\mathbf{u} = \kappa(\mathbf{x},t) \tag{15.5}$$

with the function $\kappa(\mathbf{x},t)$ of class C^1 such that $\kappa(0,t) = 0$. Such a controller is said to be a locally (globally) *admissible controller* if the undisturbed ($\mathbf{w}, \mathbf{w}_{\mathbf{d}}^i = \mathbf{0}$) closed-loop system (15.1)–(15.10) is uniformly (globally) asymptotically stable.

The $\mathcal{H}_\infty$-control problem of interest consists in finding an admissible global controller (if any exists) such that the $\mathcal{L}_2$-gain of the disturbed system (15.1)–(15.10) is less than a certain attenuation level $\gamma > 0$, that is the inequality

$$\int_{t_0}^{T} \|\mathbf{z}\|^2 \mathrm{d}t + \sum_{i=1}^{N_T} \|\mathbf{z}_i^{\mathbf{d}}\|^2 \leq$$
$$\gamma^2 \left[\int_{t_0}^{T} \|\mathbf{w}\|^2 \mathrm{d}t + \sum_{i=1}^{N_T} \|\mathbf{w}_{\mathbf{d}}^i\|^2 \right] + \sum_{j=0}^{N} \beta_j(\mathbf{x}(t_j^-), t_j) \tag{15.6}$$

locally holds for some positive definite functions $\beta_j(\mathbf{x},t)$, $j = 0,\ldots,N_T$, for all segments $[t_0, T]$ and a natural N_T such that $t_{N_T} \leq T < t_{N_T+1}$, and for all piecewise continuous disturbances $\mathbf{w}(t)$ and discrete ones $\mathbf{w}_{\mathbf{d}}^i$, $i = 1, 2, \ldots$. In turn, a locally admissible controller (15.5) is said to be a local solution of the $\mathcal{H}_\infty$-control problem if there exists a neighborhood $\mathcal{U} \in \mathbb{R}^{2n}$ of the origin, validating inequality (15.6) for some positive definite functions $\beta_j(\mathbf{x}, t)$, $j = 0, \ldots, N_T$, for all segments $[t_0, T]$ and a natural N_T such that $t_{N_T} \leq T < t_{N_T+1}$, for all piecewise continuous disturbances $\mathbf{w}(t)$ and discrete ones $\mathbf{w}_\mathbf{d}^i$, $i = 1, 2, \ldots$, for which the state trajectory of the closed-loop system starting from an initial point $(\mathbf{x}(t_0) = \mathbf{x_0}) \in \mathcal{U}$ remains in $\mathcal{U}$ for all $t \in [t_0, T]$.

It is worth noticing that the above $\mathcal{L}_2$-gain definition is consistent with the notion of dissipativity, introduced by [14, 37], and with the iISS notion [13]. It represents a natural extension to hybrid systems (see, e.g. the works by [2, 16, 24, 38]). In mechanical terms, for the disturbed case, even if the output $\mathbf{z}$ is not driven to zero, the $\mathcal{L}_2$-gain of the system is still locally less than the specified value γ, so the output will be bounded around zero and in consequence the state trajectories of the plant will evolve around the trajectory to track. To facilitate the expositions the underlying system, chosen for treatment, has been pre-specified with the post-impact velocity value $\mathbf{x}_2(t)$ in the discrete output (15.10) to be controlled. The general case of a certain function $\kappa_d(\mathbf{x}_2(t))$ of the post-impact velocity value in the discrete output (15.10) can be treated in a similar manner because the $\mathcal{L}_2$-gain inequality (15.6) is flexible in the choice of positive definite functions $\beta_k(\mathbf{x}, t)$, $k = 0, \ldots, N_T$.

15.3.2 Nonlinear $\mathcal{H}_\infty$-Control Synthesis Under Unilateral Constraints

For later use, the continuous dynamics (15.1) are rewritten in the form

$$\dot{\mathbf{x}} = \mathbf{f}(\mathbf{x}, t) + \mathbf{g}_1(\mathbf{x}, t)\mathbf{w} + \mathbf{g}_2(\mathbf{x}, t)\mathbf{u} \tag{15.7}$$

whereas the restitution rule is represented as follows

$$\mathbf{x}(t_i^+) = \boldsymbol{\mu}(\mathbf{x}(t_i^-), t_i) + \Omega(\mathbf{x}(t_i^-), t_i)\mathbf{w}_\mathbf{d}^i, \; i = 1, 2, \ldots \tag{15.8}$$

with $\mathbf{x}^\top = [\mathbf{x}_1^\top, \mathbf{x}_2^\top]$, $\mathbf{f}^\top(\mathbf{x}, t) = [\mathbf{x}_2^\top, \boldsymbol{\Phi}^\top(\mathbf{x}, t)]$, $\mathbf{g_1}^\top(\mathbf{x}, t) = [\mathbf{0}, \boldsymbol{\Psi}_1^\top(\mathbf{x}, t)]$, $\mathbf{g}_2^\top(\mathbf{x}, t) = [\mathbf{0}, \boldsymbol{\Psi}_2^\top(\mathbf{x}, t)]$, $\boldsymbol{\mu}^\top(\mathbf{x}, t) = [\mathbf{x}_1^\top, \boldsymbol{\mu}_0^\top(\mathbf{x}, t)]$, and $\Omega^\top(\mathbf{x}, t) = [\mathbf{0}, \omega(\mathbf{x}, \mathbf{t})]$. In combination with the output equations, which are repeated here for the sake of clarity

$$\mathbf{z} = \mathbf{h}_1(\mathbf{x}_1, \mathbf{x}_2, t) + \mathbf{k}_{12}(\mathbf{x}_1, \mathbf{x}_2, t)\mathbf{u} \tag{15.9}$$

$$\mathbf{z}_i^\mathbf{d} = \mathbf{x}_2(t_i^+), \tag{15.10}$$

the nonlinear system is completely described beyond and at the contact with the unilateral constraint $F(\mathbf{x}, t)$. In order to simplify the synthesis to be developed and to provide reasonable expressions for the controller design, the assumptions

$\mathbf{h_1}^\top \mathbf{k}_{12} = \mathbf{0}$, $\mathbf{k}_{12}{}^\top \mathbf{k}_{12} = \mathbf{I}$, which are standard in the literature (see, e.g., [26]) are made. Relaxing these assumptions is indeed possible, but it would substantially complicate the formulas to be worked out.

15.3.3 Non-local State-Space Solution

Below we list the hypotheses under which a solution to the problem in question is derived. Given $\gamma > 0$, in a domain $\mathbf{x} \in B_\delta^{2n}$, $t \in \mathbb{R}$, where $B_\delta^{2n} \in \mathbb{R}^{2n}$ is a ball of radius $\delta > 0$, centered around the origin,

(H1) The norm of the matrix function ω is upper bounded by $\frac{\sqrt{2}}{2}\gamma$, i.e.,

$$\|\omega(x,t)\| \leq \frac{\sqrt{2}}{2}\gamma. \tag{15.11}$$

(H2) there exist a smooth, positive definite, decrescent function $V(\mathbf{x}, t)$ and a positive definite function $R(x)$ such that the Hamilton–Jacobi–Isaacs inequality

$$\begin{aligned}\frac{\partial V}{\partial t} + \frac{\partial V}{\partial \mathbf{x}}(\mathbf{f}(\mathbf{x},t) + \mathbf{g}_1(\mathbf{x},t)\boldsymbol{\alpha}_1 + \mathbf{g}_2(\mathbf{x},t)\boldsymbol{\alpha}_2) +\\ \mathbf{h_1}^\top \mathbf{h_1} + \boldsymbol{\alpha}_2{}^\top \boldsymbol{\alpha}_2 - \gamma^2 \boldsymbol{\alpha}_1{}^\top \boldsymbol{\alpha}_1 \leq -R(\mathbf{x})\end{aligned} \tag{15.12}$$

holds with

$$\boldsymbol{\alpha}_1 = \frac{1}{2\gamma^2}\mathbf{g}_1^\top(\mathbf{x},t)\left(\frac{\partial V}{\partial \mathbf{x}}\right)^\top, \quad \boldsymbol{\alpha}_2 = -\frac{1}{2}\mathbf{g}_2^\top(\mathbf{x},t)\left(\frac{\partial V}{\partial \mathbf{x}}\right)^\top$$

(H3) Hypotheses H1 is satisfied with the function $V(\mathbf{x}, t)$ which decreases along the direction μ in the sense that the inequality

$$V(\mathbf{x},t) \geq V(\mu(\mathbf{x}),t) \tag{15.13}$$

holds in the domain of V.

The main result of the present work is as follows.

Theorem 15.1 *Consider system* (15.1)–(15.10) *subject to* (15.13)*. Given* $\gamma > 0$, *suppose Hypotheses (H1) and (H2) are satisfied in a domain* $\{\mathbf{x} \in B_\delta^{2n}, t \in \mathbb{R}\}$*. Then, the closed-loop system* (15.1)–(15.10)*, driven by the controller*

$$\mathbf{u} = \boldsymbol{\alpha}_2(\mathbf{x},t), \tag{15.14}$$

locally possesses a $\mathcal{L}_2$*-gain less than* γ*. Once Hypothesis (H3) is satisfied as well, the function* $V(\mathbf{x}, t)$ *constitutes a Lyapunov function of the disturbance-free closed-*

loop system (15.1)–(15.10), (15.14) *the uniform asymptotic stability of which is thus additionally guaranteed.*

Proof The proof of Theorem 15.1 is preceded by an instrumental lemma which extends the powerful Lyapunov approach to impact systems. The following result specifies [9, Theorem 2.4] to the present case with $x_1 = x$ and $x_2 = t$.

Lemma 15.1 *Consider the unforced* ($\mathbf{u} = \mathbf{0}$) *disturbance-free* ($\mathbf{w} = \mathbf{0},\ \mathbf{w}_\mathbf{i}^\mathbf{d} = \mathbf{0},\ i = 1, 2, \ldots$) *system* (15.7), (15.8) *with the assumptions above. Assume that there exists a positive definite decrescent function* $V(\mathbf{x}, t)$ *such that its time derivative, computed along* (15.7), *is negative definite whereas* $V(\mathbf{x}, t) \geq V(\boldsymbol{\mu}(\mathbf{x},t), t)$ *for all* $t \in \mathbb{R}$ *and all* $\mathbf{x} \in \mathbb{R}^{2n}$ *such that* $F(\mathbf{x}_1, t) = 0$. *Then the system is uniformly asymptotically stable.*

Since the proof follows the same line of reasoning as that in the book by [26] for the impact-free case here we provide only a sketch. Similar to the proof of [26, Theorem 7.1], let us consider the function $V(\mathbf{x}, t)$ whose time derivative, computed along the disturbed closed-loop system (15.1)–(15.10) between collision time instants $t \in (t_k, t_{k+1}),\ \ k = 0, 1, \ldots,$ is estimated as follows [26, p. 138]:

$$\frac{dV}{dt} \leq -\|\mathbf{z}\|^2 + \gamma^2\|\mathbf{w}\|^2 - R(\mathbf{x}). \tag{15.15}$$

Then integrating (15.15) from t_k to $t_{k+1},\ \ k = 0, 1, \ldots,$ yields

$$\int_{t_k}^{t_{k+1}} [\gamma^2\|\mathbf{w}\|^2 - \|\mathbf{z}\|^2]dt \geq$$
$$\int_{t_k}^{t_{k+1}} R(\mathbf{x}(t))dt + \int_{t_k}^{t_{k+1}} \frac{dV(\mathbf{x}(t), t)}{dt} dt > 0. \tag{15.16}$$

Skipping positive terms on the right-hand side of (15.16), it follows that

$$\int_{t_0}^{T} (\gamma^2\|\mathbf{w}\|^2 - \|\mathbf{z}\|^2)dt \geq V(\mathbf{x}(T), T)$$
$$+ \sum_{i=1}^{N_T}[V(\mathbf{x}(t_i^-), t_i) - V(\mathbf{x}(t_i^+), t_i)] - V(\mathbf{x}(t_0), t_0). \tag{15.17}$$

Since the function V is smooth by Hypothesis (H2), the following relation

$$|V(\mathbf{x}(t_i^-), t_i) - V(\mathbf{x}(t_i^+), t_i)| \leq L_i^V |\mathbf{x}(t_i^-) - \mathbf{x}(t_i^+)| \tag{15.18}$$

holds true with $L_i^V > 0$ being a local Lipschitz constant of V, in the ball of radius $\|\mathbf{x}(t_i^+)\|$, centered around $\mathbf{x}(t_i^-)$. Relations (15.17) and (15.18), coupled together, result in

$$\int_{t_0}^{T} (\gamma^2\|\mathbf{w}\|^2 - \|\mathbf{z}\|^2)\mathrm{d}t \geq -\sum_{i=1}^{N_T}[2(L_i^V)\|\mathbf{x}(t_i^-)\| - V(\mathbf{x}(t_0), t_0) \tag{15.19}$$

Apart from this, the inequality

$$\begin{aligned}\sum_{i=1}^{N_T} \|\mathbf{z}_i^{\mathbf{d}}\|^2 = \sum_{i=1}^{N_T} \|\mathbf{x_2}(t_i^+)\|^2 \leq \sum_{i=1}^{N_T}[2\|\boldsymbol{\mu}_\mathbf{0}\|^2] \\ +2\sum_{i=1}^{N_T}[\|\omega\mathbf{w}_\mathbf{d}^i\|^2] \leq \gamma^2\sum_{i=1}^{N_T} \|\mathbf{w}_\mathbf{d}^i\|^2 + \sum_{i=1}^{N_T}[2\|\boldsymbol{\mu}_\mathbf{0}\|^2]\end{aligned} \tag{15.20}$$

is ensured by H1. Thus, combining (15.19)–(15.20), one derives

$$\begin{aligned}\int_{t_0}^{T} \|\mathbf{z}\|^2\mathrm{d}t + \sum_{i=1}^{N_T} \|\mathbf{z}_i^{\mathbf{d}}\|^2 \leq V(\mathbf{x}(t_0), t_0) + \sum_{i=1}^{N_T}[2\|\mu_\mathbf{0}\|^2] \\ +\gamma^2\left[\int_{t_0}^{T} \|\mathbf{w}\|^2\mathrm{d}t + \sum_{i=1}^{N_T} \|\mathbf{w}_\mathbf{d}^i\|^2\right] + \sum_{i=1}^{N_T}[(2L_i^V)\|\mathbf{x}(t_i^-)\|\end{aligned} \tag{15.21}$$

i.e., the disturbance attenuation inequality (15.6) is established with

$$\begin{aligned}\beta_0(\mathbf{x}(t_0), t_0) = V(\mathbf{x}(t_0), t_0), \\ \beta_i(\mathbf{x}(t_i), t_i) = (2L_i^V)\|\mathbf{x}(t_i^-)\| + 2\|\boldsymbol{\mu}_\mathbf{0}(\mathbf{x}(t_i^-), t_i)\|^2\end{aligned} \tag{15.22}$$

with $i = 1, \ldots, N$.

To complete the proof it remains to establish the asymptotic stability of the undisturbed version of the closed-loop system (15.1)–(15.10), (15.14). Indeed, the negative definiteness (15.15) of the time derivative of the Lyapunov function $V(\boldsymbol{x}, t)$ between the collision time instants, coupled to Hypothesis (H3), ensures that Lemma 15.1 is applicable to the undisturbed version of the closed-loop system (15.1)–(15.10), (15.14). By applying Lemma 15.1, the required asymptotic stability is thus validated. □

15.3.4 Local State-Space Solution

To present a local solution to the problem in question the underlying system is linearized to

$$\dot{\mathbf{x}} = \mathbf{A}(t)\mathbf{x} + \mathbf{B_1}(t)\mathbf{w} + \mathbf{B_2}(t)\mathbf{u}, \tag{15.23}$$

$$\mathbf{z} = \mathbf{C_1}(t)\mathbf{x} + \mathbf{D_{12}}(t)\mathbf{u}, \tag{15.24}$$

within impact-free time intervals (t_{i-1}, t_i) where t_0 is the initial time instant and t_i, $i = 1, 2, \ldots$ are the collision time instants, whereas $\mathbf{A}(t) = \left.\frac{\partial \mathbf{f}}{\partial \mathbf{x}}\right|_{\mathbf{x}=\mathbf{0}}$, $\mathbf{B}_1(t) = \mathbf{g}_1(0,t)$, $\mathbf{B}_2(t) = \mathbf{g}_2(0,t)$, $\mathbf{C}(t) = \left.\frac{\partial \mathbf{h}}{\partial \mathbf{x}}\right|_{\mathbf{x}=\mathbf{0}}$, $\mathbf{D}_{12}(t) = \mathbf{k}_{12}(0,t)$.

By the time-varying strict bounded real lemma [27, p. 46], the following condition is necessary and sufficient for the linear $\mathcal{H}_\infty$ control problem (15.23), (15.24) to possess a solution: given $\gamma > 0$,

C) there exists a positive constant ε_0 such that the differential Riccati equation

$$
\begin{aligned}
-\dot{\mathbf{P}}_\varepsilon(t) &= \mathbf{P}_\varepsilon(t)\mathbf{A}(t) + \mathbf{A}^\top(t)\mathbf{P}_\varepsilon(t) + \mathbf{C}_1{}^\top(t)\mathbf{C}_1(t) \\
&+ \mathbf{P}_\varepsilon(t)[\frac{1}{\gamma^2}\mathbf{B}_1\mathbf{B}_1{}^\top - \mathbf{B}_2\mathbf{B}_2{}^\top](t)\mathbf{P}_\varepsilon(t) + \varepsilon\mathbf{I}
\end{aligned} \tag{15.25}
$$

has a uniformly bounded symmetric positive definite solution $\mathbf{P}_\varepsilon(t)$ for each $\varepsilon \in (0, \varepsilon_0)$;

As shown below, this condition, if coupled to Hypothesis H1 and a certain monotonicity condition, is also sufficient for a local solution to the nonlinear $\mathcal{H}_\infty$ control problem to exist under unilateral constraints.

Theorem 15.2 *Let condition C be satisfied with some $\gamma > 0$. Then Hypothesis H2 hold locally around the equilibrium* ($\mathbf{x} = 0$) *of the nonlinear system* (15.1)–(15.10) *with*

$$V(\mathbf{x}, t) = \mathbf{x}^\top \mathbf{P}_\varepsilon(t)\mathbf{x}, \quad R(\mathbf{x}) = \frac{\varepsilon}{2}\|\mathbf{x}\|^2 \tag{15.26}$$

and the closed-loop system driven by the state feedback

$$\mathbf{u} = -\mathbf{g}_2(\mathbf{x}, t)^\top \mathbf{P}_\varepsilon(t)\mathbf{x} \tag{15.27}$$

locally possesses a $\mathcal{L}_2$-gain less than γ provided that Hypothesis H1 holds as well. If in addition, Hypothesis H3 is satisfied with the quadratic function $V(\mathbf{x}, t)$, given in (15.26)*, then the disturbance-free closed-loop system* (15.1)–(15.10), (15.27) *is uniformly asymptotically stable.*

Proof Due to [27, Theorem 24], Hypothesis H2 locally holds with (15.26). Then by applying Theorem 15.1, the validity of Theorem 15.2 is established.

15.3.5 Remarks on the Synthesis of Periodic Systems

For the periodic tracking of period T with periodic impact instants $t_{i+1} = t_i + T$, $i = 1, 2, \ldots$, Theorem 15.2 admits a time-periodic synthesis (15.27) which is based on an appropriate periodic solution $\mathbf{P}_\varepsilon(t)$ of the periodic differential Riccati equation

(15.25). It should be noted that $P_\varepsilon(t_{i+1}^+) = P_\varepsilon(t_i^+)$, due to the periodicity, and inequality (15.13) of H3 is then specified to the boundary condition

$$\mathbf{x}^\top \mathbf{P}_\varepsilon(t_2^-)\mathbf{x} \geq \boldsymbol{\mu}^\top(\mathbf{x}, t_1^+))\mathbf{P}_\varepsilon(t_1^+)\boldsymbol{\mu}(\mathbf{x}, t_1^+)), \tag{15.28}$$

on the Riccati equation (15.25).

This result will be used in the following section to robustly track a reference trajectory for a fully actuated 3D biped robot.

15.4 Robust Trajectory Tracking of a 3D Biped Robot

In this section, the results on $\mathcal{H}_\infty$ control of mechanical systems under unilateral constraints are implemented on the 32-DOF biped robot ROMEO, from Aldebaran Robotics. In order to comply with all the conditions for the existence of the controller, an on-line trajectory adaptation method is introduced so as to ensure asymptotic tracking of the biped dynamics to the desired walking gait.

15.4.1 Model of a Biped with Feet

The bipedal robot considered in this section is walking on a rigid and horizontal surface. It consists of the 32-DOF robot Romeo, of Aldebaran Robotics, depicted in Fig. 15.1. Similar to the planar biped form the previous section, the walking gait takes place in the sagittal plane and is composed of single support phases and impacts.

15.4.2 Dynamic Model in Single Support

The configuration of the biped robot in single support can be described only by the vector $\mathbf{q} = (q_0, q_1, \ldots, q_{32})^\top$. We use the modified Denavit–Hartenberg notation [15] to define the frame position for each joint (see Fig. 15.2). To define the geometric structure of the biped we assume that the link 0 (stance foot) is the base of the bipedal robot while the link 12 (swing foot) is the terminal link. Considering the torso, head and arms, one obtains a tree structure. To take into account explicitly the contact with the ground, we have to add six more variables to describe the position and orientation of the frame 0 with respect to a fixed Galilean frame R_g. Thus, we can define the position, velocity, and acceleration vectors, $\mathbf{X} = (\mathbf{X_0}^\top, \boldsymbol{\alpha_0}^\top, \mathbf{q}^\top)^\top$, $\mathbf{V} = ({}^0\mathbf{V}_0^\top, {}^0\boldsymbol{\omega}_0^\top, \dot{\mathbf{q}}^\top)^\top$, and $\dot{\mathbf{V}} = ({}^0\dot{\mathbf{V}}_0^\top, {}^0\dot{\boldsymbol{\omega}}_0^\top, \ddot{\mathbf{q}}^\top)^\top$. $\mathbf{X_0}$ and $\boldsymbol{\alpha_0}$ are the position and orientation variables of frame R_0, while ${}^0\mathbf{V}_0$ and ${}^0\boldsymbol{\omega}_0$ are the linear and angular velocities of R_0, with respect to the Galilean frame. Therefore, the complete dynamic model of the biped can be written as follows:

Fig. 15.1 32-DOF Robot Romeo, of Aldebaran Robotics

$$\mathbf{D_e(X)\dot{V}} + \mathbf{C_e(X, V)} + \mathbf{G_e(X)} = \mathbf{D_{e\tau}}\boldsymbol{\tau} + \mathbf{J_1}^\top\mathbf{R_1} + \mathbf{J_2}^\top\mathbf{R_2} + \mathbf{D_w w_1}, \tag{15.29}$$

with the complementarity condition

$$0 \le F(\mathbf{X}) \perp \mathbf{R_2} \ge \mathbf{0} \tag{15.30}$$

and the constraint equation

$$\mathbf{J_1\dot{V}} = \mathbf{0}, \tag{15.31}$$

where $\mathbf{D_e}$ is the symmetric, positive definite 38×38 inertia matrix; $\mathbf{D_{e\tau}}$ and $\mathbf{D_w}$ are 38×32 constant matrices, composed of zeros and ones; $\boldsymbol{\tau} = (\tau_1, \ldots, \tau_{32})^\top$ is the 32×1 vector of joint torques; terms $\mathbf{C_e(X, V)}$, and $\mathbf{G_e(X)}$ are the 38×1 vector of the centrifugal and Coriolis forces, and the 38×1 vector of gravity forces, respectively; $\mathbf{w_1}$ is the 32×1 vector of external disturbances; $\mathbf{R_1}$ and $\mathbf{R_2}$ represent the wrench of reaction forces on foot 1 and foot 2, respectively, whereas $\mathbf{J_1}$ and $\mathbf{J_2}$ are 6×38 Jacobian matrices converting these efforts to the corresponding joint torques. Equations (15.29), (15.30), in addition to a restitution law to be defined later, form a Lagrangian Complementarity System.

Due to the difficulty of the analytic calculation of the dynamic model (15.29), it is numerically computed by means of the Newton–Euler algorithm [17], which is based on recursive calculations associated with the choice of the reference frames from Fig. 15.1b. Then, matrices $\mathbf{D_e}$, $\mathbf{C_e(X, V)}$ and $\mathbf{G_e(X)}$ can be easily and rapidly computed using the method of [35]. The same algorithm also allows to find the ground reaction forces.

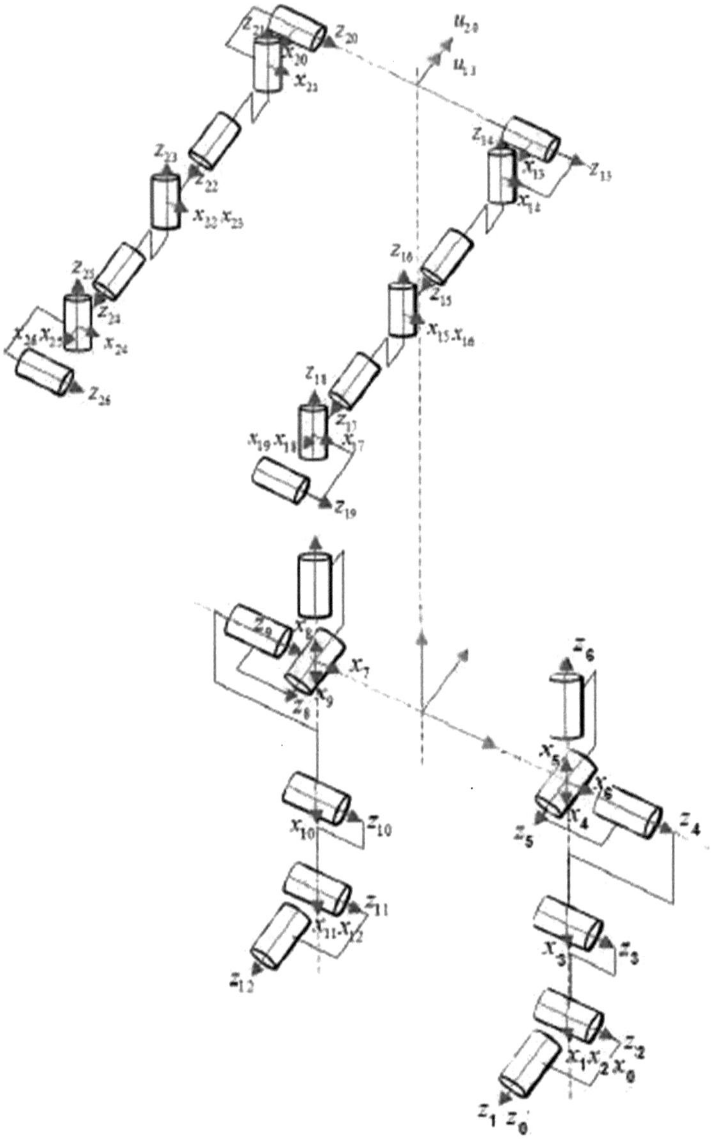

Fig. 15.2 Frames placement for the main limbs; the remaining 6 frames not appearing belong to the hands (2 frames) and the neck and head (4 frames), thus completing the 32 degrees of freedom. The zero frame R_0 is attached to the *left foot*

In the single support phase, considering a flat foot contact of the support foot with the ground, and assuming no take off, no sliding and no rotation of the support foot, the model (15.29)–(15.31) can be reduced to:

$$\mathbf{D}(\mathbf{q})\ddot{\mathbf{q}} + \mathbf{H}(\mathbf{q}, \dot{\mathbf{q}}) = \boldsymbol{\tau} + \mathbf{w_1} \tag{15.32}$$

where $\boldsymbol{\tau} = (\boldsymbol{\tau}_1, \ldots, \boldsymbol{\tau}_{32})^\top$ is the 32×1 vector of joint torques. The term $\mathbf{H}(\mathbf{q}, \dot{\mathbf{q}})$ is the 32×1 vector of the centrifugal, Coriolis and gravity forces.

The above-mentioned assumptions are verified during the numerical study using model (15.29). If at least one of these conditions are not satisfied, the conditions to construct (15.32) are not met and thus it is not valid. Thus, the reference trajectories are designed taking these conditions as restrictions.

15.4.3 Impact Model

The impact is assumed to be inelastic with complete surface of the foot sole touching the ground. This means that the velocity of the swing foot impacting the ground is zero after impact. The double support phase is instantaneous and it can be modeled through passive impact equations. An impact occurs at a time $t = T_I$ when the swing leg touches the ground.

Since the impact is assumed to be passive, absolutely inelastic, and that the legs do not slip [33], the following equations:

$$\mathbf{J_1}\mathbf{V}^- = \mathbf{0} \tag{15.33}$$

$$\mathbf{J_2}\mathbf{V}^+ = \mathbf{0} \tag{15.34}$$

hold. This means that the feet perfectly stick on the ground after the shock, thus avoiding multiple impacts. Given these conditions, the ground reactions can be viewed as impulsive forces. The algebraic equations, allowing one to compute the jumps of the velocities, can be obtained through integration of the dynamic equations of the motion, taking into account the ground reactions during an infinitesimal time interval from T_I^- to T_I^+ around an instantaneous impact.

The impact is assumed to be with complete surface of the foot sole touching the ground. This means that the velocity of the swing foot impacting the ground is zero after impact. After an impact, the right foot (previous stance foot) takes off the ground, so the vertical component of the velocity of the taking-off foot must be directed upwards right after an impact and the impulsive ground reaction in this foot equals zeros. Thus, the impact dynamic model can be represented as follows [33]:

$$\begin{aligned}\mathbf{V}^+ &= (\mathbf{I} - \mathbf{D_e^{-1}}\mathbf{J_2^\top}(\mathbf{J_2}\mathbf{D_e^{-1}}\mathbf{J_2^\top})^{-1}\mathbf{J_2})\mathbf{V}^- + \mathbf{w_{ei}^d}\\ \mathbf{V}^+ &= \phi_e(\mathbf{X})\mathbf{V}^- + \mathbf{w_{ei}^d}\end{aligned} \tag{15.35}$$

where $\mathbf{V}^-$ is the velocity of the robot before the impact, and $\mathbf{V}^+$ is the velocity after the impact; $\phi_e(\mathbf{X})$ represents a restitution law that determines the relations between the velocities before, and after the impacts; $\mathbf{X}$ is the configuration of the robot at the impact. The additive term $\mathbf{w}^{\mathbf{d}}_{\mathbf{ei}}$ is introduced to account for inadequacies in this restitution law. Thus, Eq. (15.35) renders the complementarity system (15.29)–(15.34) complete.

Considering that the support foot velocity is zero before the impact, and the swing foot velocity is zero after the impact, and combining (15.33)–(15.35), it is possible to obtain the following expression:

$$\dot{\mathbf{q}}^+ = \phi(\mathbf{q})\dot{\mathbf{q}}^- + \mathbf{w}_i^{\mathbf{d}} \tag{15.36}$$

Also, considering the reference frame x_0, y_0, z_0 shown in Fig. 15.2, and since we are assuming flat foot contact with the ground, the time-invariant unilateral constraint $F_0(\mathbf{q}) \geq 0$ is determined by the height of the swing foot's sole.

Remark 15.1 It is important to remark at this point, that if all the assumptions mentioned above are met, the so-called complementarity system [12] (15.29), (15.35), subject to the unilateral constraint (15.30), is simplified to the form equations (15.32), (15.36), which define a hybrid system that can be controlled using the methodology developed in this work. For the latter, the unilateral constraint can be defined as $F(\mathbf{q})$, which represents the height of swing foot, as a function of the generalized coordinates of the implicit-contact model (15.32).

15.4.4 Motion Planning

Since a walking biped gait is a periodical process, the objective is to design a cyclic gait. A complete walking cycle is composed of two phases: a single-support phase, and an instantaneous support phase, which is modeled through passive impact equations. The single support phase has a duration of 0.31 s, and it begins with one foot which stays on the ground while the other foot swings from the rear to the front. The double support phase is assumed instantaneous. This means that when the swing leg touches the ground the stance leg takes off. The reference trajectories, allowing a symmetric step, are obtained by an off-line optimization, minimizing a Sthenic criteria, as presented in the work of [33]. The restitution law during the impact phase is given by:

$$\dot{\mathbf{q}}^r(t_k^+) = \boldsymbol{\phi}(\mathbf{q}^{\mathbf{r}}(t_k))\dot{\mathbf{q}}^r(t_k^-), \quad k = 1, 2, \ldots. \tag{15.37}$$

15.4.5 Pre-feedback Design

Our objective is to design a pre-feedback controller of the form

$$\boldsymbol{\tau} = \mathbf{D}(\ddot{\mathbf{q}}^r - \mathbf{u}) + \mathbf{H} \tag{15.38}$$

that imposes on the undisturbed biped motion desired stability properties around $\mathbf{q}^{\mathrm{r}}$ while also locally attenuating the effect of the disturbances. Thus, the controller to be constructed consists of the feedback linearizing terms of (15.38) subject to $\mathbf{u} = \mathbf{0}$, which are responsible for the trajectory compensation, and a disturbance attenuator $\mathbf{u}$, internally stabilizing the closed-loop system around the desired trajectory. In what follows, we confine our research to the trajectory tracking control problem where the output to be controlled is given by

$$\mathbf{z} = \begin{bmatrix} \mathbf{0} \\ \rho_p(\mathbf{q}^{\mathrm{r}} - \mathbf{q}) \\ \rho_v(\dot{\mathbf{q}}^r - \dot{\mathbf{q}}) \end{bmatrix} + \begin{bmatrix} \mathbf{I} \\ \mathbf{0} \\ \mathbf{0} \end{bmatrix} \mathbf{u}, \; \mathbf{z}_{\mathrm{d}} = \mathbf{q}^{\mathrm{r}}(t_k^+) - \mathbf{q}(t_k^+) \tag{15.39}$$

with positive weight coefficients ρ_p, ρ_v.

Now, let us introduce the state deviation vector $\mathbf{x} = (\mathbf{x}_1, \mathbf{x}_2)^\top$, where $\mathbf{x}_1(t) = \mathbf{q}^{\mathrm{r}}(t) - \mathbf{q}(t)$ is the position deviation from the desired trajectory, and $\mathbf{x}_2(t) = \dot{\mathbf{q}}^r(t) - \dot{\mathbf{q}}(t)$ is the velocity deviation from the desired velocity.

Then, rewriting the state equations (15.32), (15.39) in terms of the errors $\mathbf{x}_1$ and $\mathbf{x}_2$, we obtain an error system in the form (15.1)–(15.10), being specified with

$$\mathbf{f}(\mathbf{x}, t) = \begin{bmatrix} \mathbf{x}_2 \\ \mathbf{0} \end{bmatrix}, \; \mathbf{g}_1(\mathbf{x}, t) = \begin{bmatrix} \mathbf{0} \\ \mathbf{D}^{-1}(\mathbf{q}^{\mathrm{r}} - \mathbf{x}_1) \end{bmatrix}, \tag{15.40}$$

$$\mathbf{g}_2(\mathbf{x}, t) = \begin{bmatrix} \mathbf{0} \\ \mathbf{I} \end{bmatrix}, \; \mathbf{h}(\mathbf{x}) = \begin{bmatrix} \mathbf{0} \\ \rho_p \mathbf{x}_1 \\ \rho_v \mathbf{x}_2 \end{bmatrix}, \; \mathbf{k}_{12}(\mathbf{x}) = \begin{bmatrix} \mathbf{I} \\ \mathbf{0} \\ \mathbf{0} \end{bmatrix}, \tag{15.41}$$

$$\boldsymbol{\mu}(\mathbf{x}, t) = \begin{bmatrix} \mathbf{x}_1 \\ \boldsymbol{\phi}(\mathbf{q}^{\mathrm{r}})\dot{\mathbf{q}}^r - \boldsymbol{\phi}(\mathbf{q}^{\mathrm{r}} - \mathbf{x}_1)(\dot{\mathbf{q}}^r - \mathbf{x}_2) \end{bmatrix}, \tag{15.42}$$

$$\boldsymbol{\omega}(\mathbf{x}, t) = -\mathbf{I} \tag{15.43}$$

where as a matter of fact, the zero symbols stand for zero matrices and $\mathbf{I}$ for identity matrices of appropriate dimensions.

15.4.6 Hybrid Error Dynamics

The transitions occur in the error dynamics according to the following scenarios:

(T1) The reference trajectory reaches its predefined impact time instant $t = t^k$, $k = 1, 2, \dots$ when it hits the unilateral constraint whereas the plant remains beyond this constraint, that is, $F_0(\mathbf{q^r}(t^k)) = 0$, $F_0(\mathbf{x_1}(t^k) + \mathbf{q^r}(t^k)) \neq 0$;
(T2) The plant hits the unilateral constraint at $t = t^j$, $j = 1, 2, \dots$ while the reference trajectory is beyond this constraint, that is, $F_0(\mathbf{q^r}(t^j)) \neq 0$, $F_0(\mathbf{x_1}(t^j) + \mathbf{q^r}(t^j)) = 0$;
(T3) Both the reference trajectory and the plant hits the unilateral constraint at the same time instant $t = t^l$, $l = 1, 2, \dots$ (what can deliberately be enforced by modifying the pre-specified reference trajectory on-line), that is, $F_0(\mathbf{q^r}(t^l)) = 0$, $F_0(\mathbf{x_1}(t^l) + \mathbf{q^r}(t^l)) = 0$.

These scenarios are illustrated in Fig. 15.3. Transition errors are then represented as follows.

Scenario T1:

$$\begin{aligned} \mathbf{x_1}(t^{k+}) &= \mathbf{x_1}(t^{k-}) \\ \mathbf{x_2}(t^{k+}) &= \boldsymbol{\mu}^1(\mathbf{x}(t^{k-}), t^k) + \mathbf{w}_k^{\mathbf{d}}, \end{aligned} \tag{15.44}$$

provided that $F_0(\mathbf{q^r}(t^k)) = 0$ and $F_0(\mathbf{x_1}(t^k) + \mathbf{q^r}(t^k)) \neq 0,\ k = 1, 2, \dots;$
Scenario T2:

$$\begin{aligned} \mathbf{x_1}(t^{j+}) &= \mathbf{x_1}(t^{j-}) \\ \mathbf{x_2}(t^{j+}) &= \boldsymbol{\mu}^2(\mathbf{x}(t^{j-}), t^j) + \mathbf{w}_j^{\mathbf{d}}, \end{aligned} \tag{15.45}$$

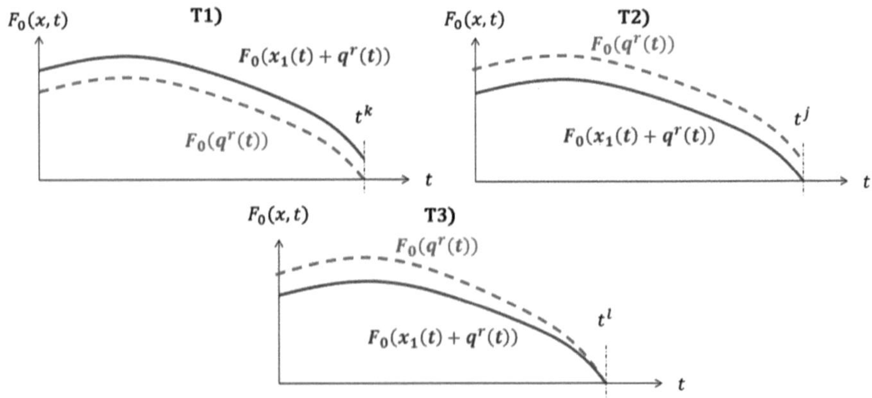

Fig. 15.3 The three different scenarios for the transitions in the error dynamics

provided that $F_0(\mathbf{q^r}(t^j)) \neq 0$ and $F_0(\mathbf{x_1}(t^j) + \mathbf{q^r}(t^j)) = 0,\ j = 1, 2, \ldots$;
Scenario T3:

$$\begin{aligned} \mathbf{x_1}(t^{l+}) &= \mathbf{x_1}(t^{l-}) \\ \mathbf{x_2}(t^{l+}) &= \boldsymbol{\mu}^3(\mathbf{x}(t^{l-}), t^l) + \mathbf{w}_l^{\mathbf{d}},\ l = 1, 2, \ldots \end{aligned} \tag{15.46}$$

provided that $F_0(\mathbf{q^r}(t^l)) = 0$ and $F_0(\mathbf{x_1}(t^l) + \mathbf{q^r}(t^l)) = 0,\ l = 1, 2, \ldots$

where $\mathbf{w}_k^{\mathbf{d}}$, $\mathbf{w}_j^{\mathbf{d}}$, $\mathbf{w}_l^{\mathbf{d}}$ are discrete perturbations, counting for restitution inadequacies, and functions $\boldsymbol{\mu}^1$, $\boldsymbol{\mu}^2$, and $\boldsymbol{\mu}^3$ are given by

$$\boldsymbol{\mu}^1(\mathbf{x}, t) = \mathbf{x_2} + [\mathbf{I} - \boldsymbol{\phi}(\mathbf{q^r}(t))]\dot{\mathbf{q}}^r(t^-) \tag{15.47}$$

$$\boldsymbol{\mu}^2(\mathbf{x}, t) = \boldsymbol{\phi}(\mathbf{x_1} + \mathbf{q^r}(t))[\mathbf{x_2} + \dot{\mathbf{q}}^r(t^-)] - \dot{\mathbf{q}}^r(t^-) \tag{15.48}$$

$$\boldsymbol{\mu}^3(\mathbf{x}, t) = \boldsymbol{\phi}(\mathbf{x_1} + \mathbf{q^r}(t)[\mathbf{x_2} + \dot{\mathbf{q}}^r(t^-)] - \boldsymbol{\phi}(\mathbf{q^r}(t))\dot{\mathbf{q}}^r(t^-). \tag{15.49}$$

To put the previous equations into the form (15.3), it suffices to set

$$F(\mathbf{x}, t) = F_0(\mathbf{x_1} + \mathbf{q^r}(t)), \quad \boldsymbol{\omega}(\mathbf{x}, t) = \mathbf{I}, \tag{15.50}$$

and specify the function $\boldsymbol{\mu}_0(\mathbf{x},t)$ by means of

$$\boldsymbol{\mu}_0(\mathbf{x},t) = \begin{cases} \boldsymbol{\mu}^1(\mathbf{x}, t) \text{ if } F_0(\mathbf{q^r}(t)) = 0, & F_0(\mathbf{x_1} + \mathbf{q^r}) \neq 0 \\ \boldsymbol{\mu}^2(\mathbf{x}, t) \text{ if } F_0(\mathbf{q^r}(t)) \neq 0, & F_0(\mathbf{x_1} + \mathbf{q^r}) = 0 \\ \boldsymbol{\mu}^3(\mathbf{x}, t) \text{ if } F_0(\mathbf{q^r}(t)) = 0, & F_0(\mathbf{x_1} + \mathbf{q^r}) = 0. \end{cases} \tag{15.51}$$

Clearly, the functions $\boldsymbol{\mu}_0(\mathbf{x},t)$, $\boldsymbol{\omega}(\mathbf{x}, t)$, $F(\mathbf{x}, t)$, thus specified, meet the assumptions, imposed on the generic system (15.1)–(15.10) to be twice continuously differentiable in the state domain for all t, and to be piece wise continuous, and uniformly bounded in t, for all state variables $\mathbf{x}$ in some neighborhood around the origin.

15.4.7 State Feedback $\mathcal{H}_\infty$ Synthesis Using Trajectory Adaptation

To respect Condition C of Theorem 15.2 for the error system (15.1)–(15.10), the controlled output (15.39) is specified with $\rho_p = 3500$ and $\rho_v = 500$, and then, following the standard $\mathcal{H}_\infty$ design procedure (see, e.g., [27, Sect. 6.2.1]), the disturbance attenuation level and the perturbation parameter are set to $\gamma = 200$ and $\epsilon = 0.01$ to ensure an appropriate solvability of the perturbed differential Riccati equation (15.25), subject to the boundary condition (15.28). Next, hypothesis H1 of

Theorem 15.2 is straightforwardly verified with γ, thus specified, and with $\boldsymbol{\omega}$, being an identity matrix. Finally, to comply with hypothesis H3 of Theorem 15.2 to be verified at the impact time instants the previously defined reference trajectory is adapted on-line in such a manner that the state error dynamics possess no jumps. Thus, inequality (15.13) becomes redundant for the adapted trajectory because only trivial transitions with $\boldsymbol{\mu}(\mathbf{x},t) = 0$ are feasible.

For hybrid systems with state-triggered jumps, the jump times of the plant and the reference trajectory are in general not coinciding. During the time interval caused by this jump-time mismatch, the tracking error is large, even in the undisturbed case. Since this behavior also occurs for arbitrarily small initial errors, the error dynamic displays unstable behavior in the sense of Lyapunov. This behavior is known in the literature as "peaking". It is expected to occur in all hybrid systems with state-triggered jumps when considering tracking or observer design problems [3], and imposes a difficulty in guaranteeing that the norm of the tracking error converges to zero. In order to achieve synchronization (Scenario T3 from the previous section) in our biped application, the reference trajectory is adapted, as illustrated in Fig. 15.4 for the first joint q_1. Provided that the impact is detectable (e.g., by using a force or touch sensor) it happens that either the reference trajectory hits the constraint before the plant does, or the plant hits the constraint before the reference trajectory does. In the former scenario, the reference trajectory is continuously extrapolated until the plant

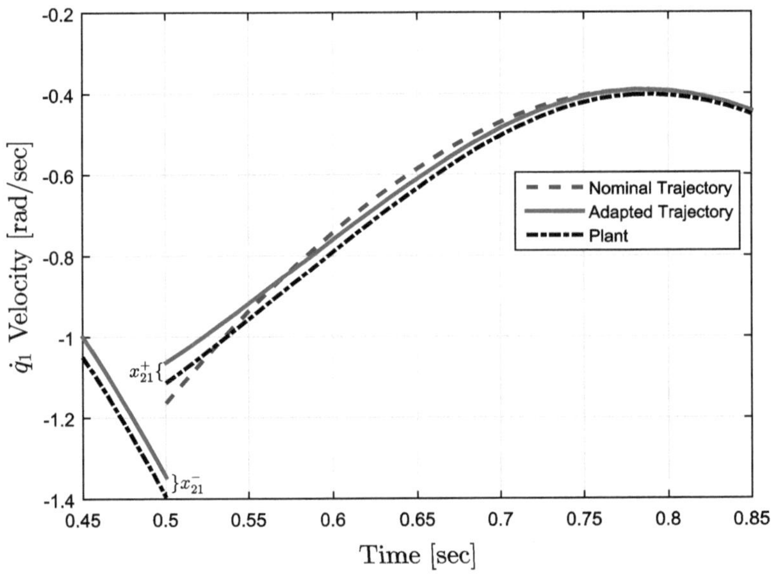

Fig. 15.4 Reference velocity adaptation for the first joint, with an impact at $t^I = 0.5\,\text{s}$

collision occurs, whereas in the latter scenario, the reference trajectory is restarted on-line once the plant collision is detected. Either way, both the plant trajectory and the adapted reference trajectory exhibit impacts at the same time instants. Before the collision, the nominal reference trajectory and the adapted one, are forced to be equivalent by the adaptation proposed. The position and velocity tracking errors are measured, and once the impact of the plant is detected, the adapted trajectory is updated on-line in such a manner that the new post-impact error, x_{21}^{+} in Fig. 15.4, coincides with the error measured before the impact ($x_{21}(t^{l-})$ in Fig. 15.4), thereby rendering the evolution of the error to exhibit no jump, so as to ensure a smooth control action. Following the idea of [8], a new polynomial in terms of the tracking error is defined for the adapted trajectory, that starts from this imposed condition, and will join the nominal reference trajectory in the middle of the step with the same velocity, and will continue to be the same until the end of the step. While the reference trajectory is recalculated after the impact, the perturbed differential Riccati equation (15.25) is also updated, and its corresponding solution is recomputed on-line.

15.4.8 Numerical Study

To illustrate the performance issues of the developed stable bipedal gait synthesis, numerical simulations were performed for the model of a laboratory prototype, whose parameters were drawn from the Aldebaran's ROMEO documentation. The contact constraints (no take off, no rotation, and no sliding during the single support phase) are verified on-line to confirm the validity of (15.32), (15.36). The well-known constraint complementarity-based approach [31] is used to simulate the biped contact with the ground, using a Matlab emulator. The latter approach belongs to the family of time-stepping approaches and it is often invoked for biped dynamics simulations (see, e.g., the works by [34, 39]).

First, the undisturbed-case results are presented,[1] where the plant initial conditions deviate 5% from the reference gait's initial conditions. The control law calculation time averaged at 1.5 ms. As predicted by the theory, Fig. 15.5 depicts the Lyapunov function $V(\mathbf{x}, t)$, decreasing smoothly and asymptotically towards zero, illustrating that no peaking phenomena is present, thanks to the trajectory adaptation (Figs. 15.6 and 15.7).

As a next step, a persistent disturbance of $10 \sin(t)$ Nm was applied to the hip, while the velocities after the impact are deviated 5% from their nominal values (given by (15.36)), thus considering disturbances on both the single support and impact phases. Six joints among the 32 were selected to clearly illustrate the effect

[1] Simulation video available at https://youtu.be/vzh01Vh-RcI.

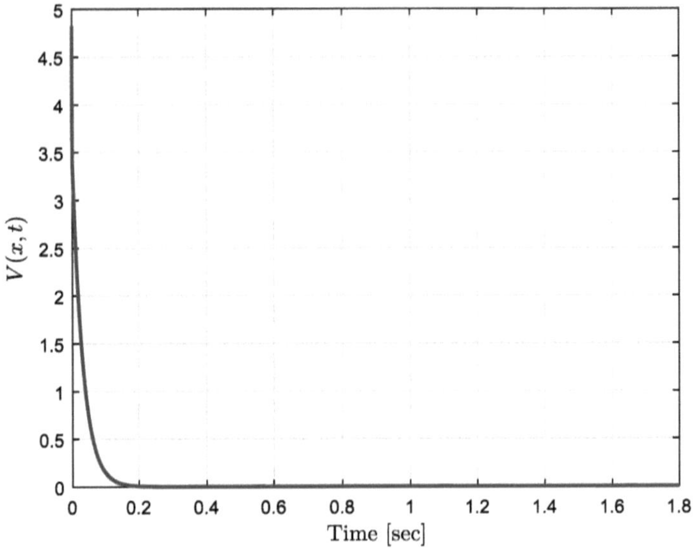

Fig. 15.5 Lyapunov function for the undisturbed system, with nonzero initial conditions

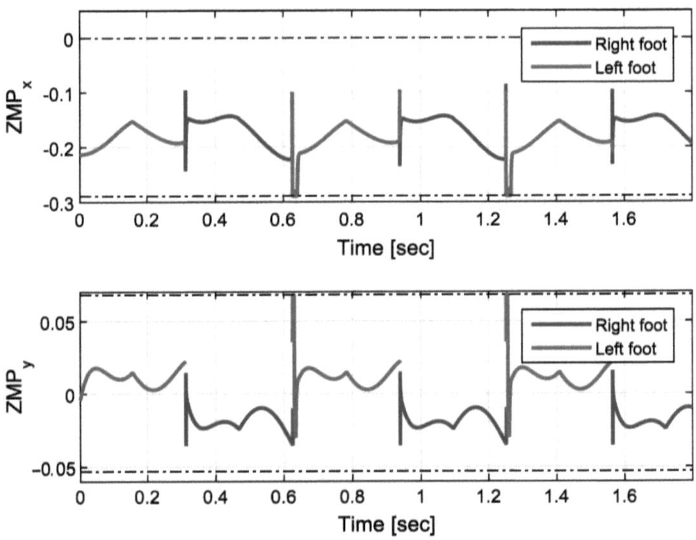

Fig. 15.6 Zero moment point locations for both feet during the disturbance-free walking gait. The *dashed lines* represent the feet geometrical limits and the ZMP always rests inside of them, thus illustrating no rotation of the support foot at each step

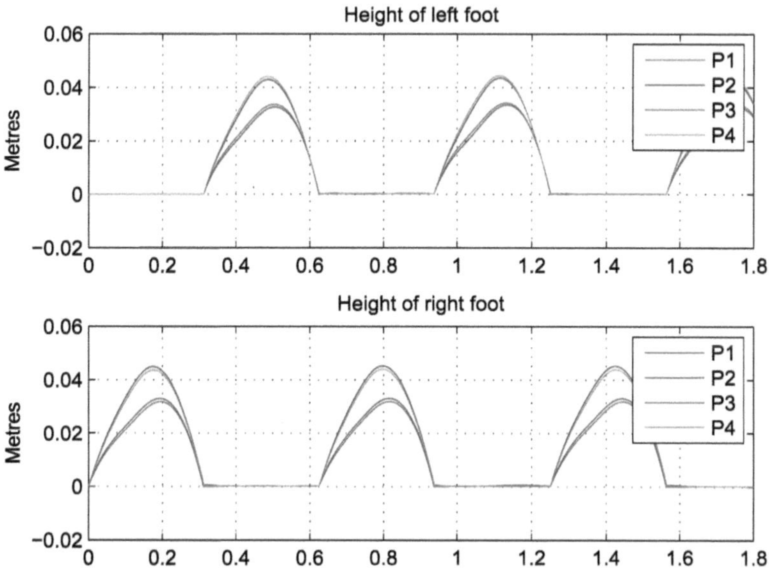

Fig. 15.7 Feet heights for 6 steps for Romeo, representing a stable motion. P1–P4 represent the heights of the corners of each foot

of this disturbance (both ankles, knees, and hip joints). This is depicted in Fig. 15.8, where the error is small and bounded, and the robot maintains a stable walking gait. The torques for these joints are shown in Fig. 15.9, where they stay between the boundaries of ± 150 Nm. Despite the disturbances, good performance of the closed-loop error dynamics, driven by the proposed nonlinear $\mathcal{H}_\infty$ state feedback, is still achieved.

Finally, the performance of the $\mathcal{H}_\infty$ controller was compared against the performance of a PD controller. In order to do a fair comparison, the same pre-feedback (15.38) was used, and just the disturbance attenuator (15.27) was replaced by the PD controller $\mathbf{u} = -\mathbf{K_p}\mathbf{x_1} - \mathbf{K_v}\mathbf{x_2}$, with the constant matrices $[\mathbf{K_p}, \mathbf{K_v}] = \mathbf{B_2^\top P_\varepsilon}$ where $\mathbf{P}_\varepsilon$ is the solution to the algebraic version of the Riccati equation (15.25) with $\dot{\mathbf{P}}_\varepsilon = 0$.

The comparison results with the time-varying disturbance force $10\sin(t) + 10$ N, applied to the hip are shown in Fig. 15.10, where it can be seen that after 6.13 s, the cumulative position tracking error, generated by the developed nonlinear periodic $\mathcal{H}_\infty$ tracking controller, is approximately 26% less than that generated by the PD controller. Thus, a better performance of the proposed synthesis is concluded in comparison to the standard linear $\mathcal{H}_\infty$ PD design.

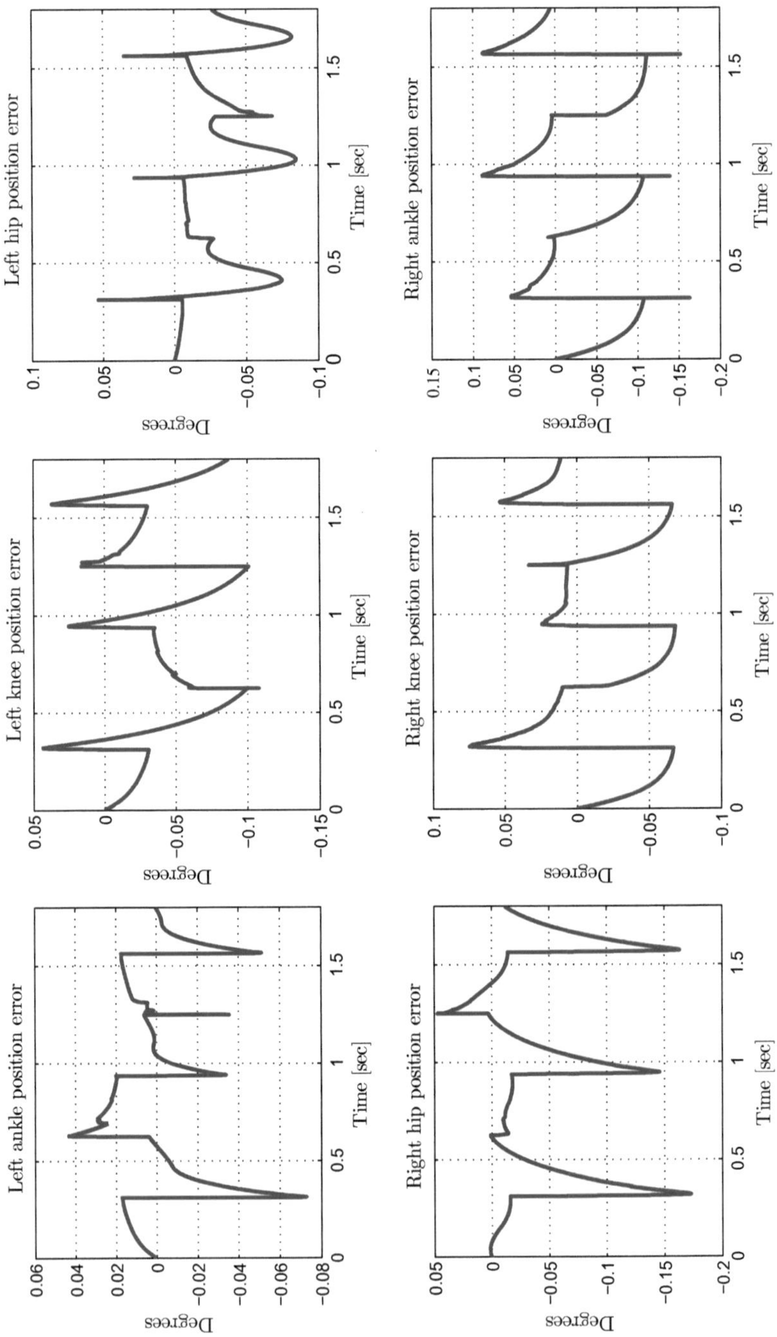

Fig. 15.8 Joints errors for *left* and *right* hips, knees, and ankles, under a persistent continuous disturbance ($10\sin(t)$ Nm) applied on the hip

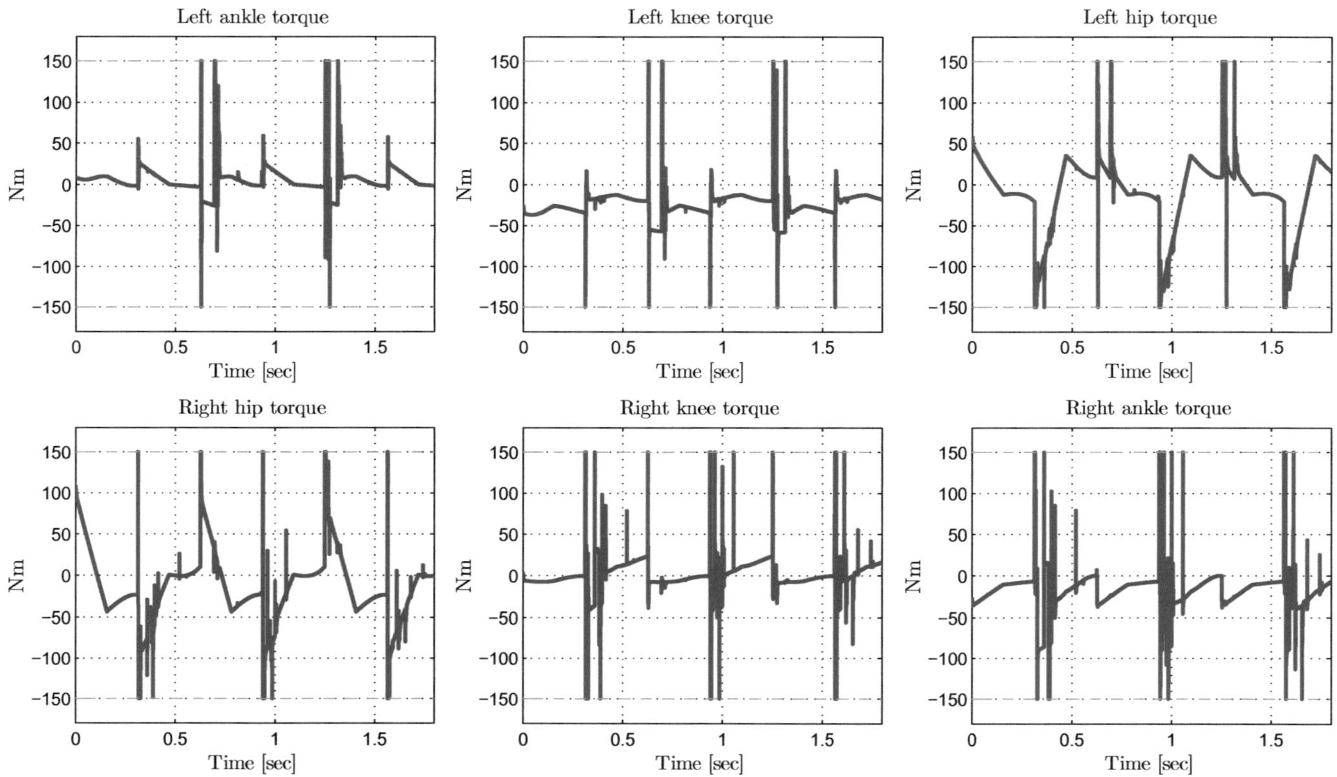

Fig. 15.9 Torques for *left* and *right* hips, knees, and ankles, under a persistent continuous disturbance ($10\sin(t)$ Nm) applied on the hip

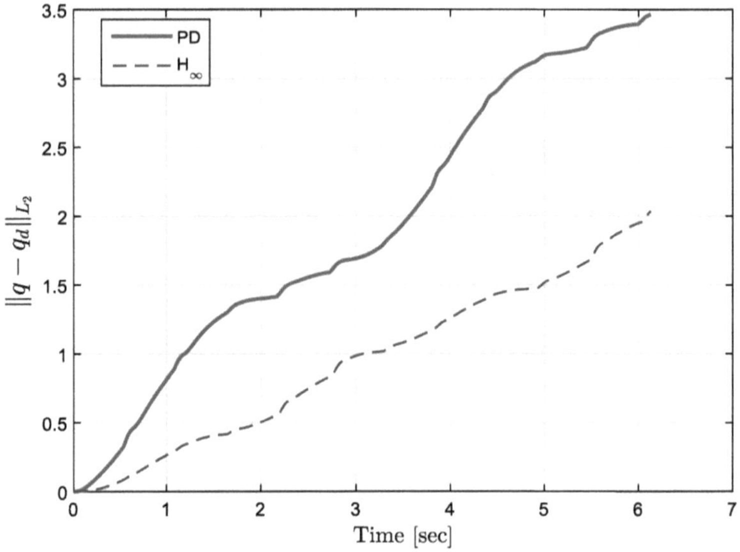

Fig. 15.10 Cumulative error comparison of the nonlinear $\mathcal{H}_\infty$ controller (*dashed lines*) versus the linear $\mathcal{H}_\infty$ PD controller (*solid lines*)

15.5 Conclusions

In this chapter, sufficient conditions for a local solution to the $\mathcal{H}_\infty$ state feedback tracking problem to exist, are presented in terms of the appropriate solvability of an independent inequality on a discrete disturbance factor that occurs in the restitution rule, and two coupled inequalities, involving a Hamilton–Jacobi–Isaacs inequality. The former inequality ensures that the closed-loop impulse dynamics (when the state trajectory hits the unilateral constraint) are ISS whereas the latter inequalities, arising in the continuous-time state feedback, should impose the desired iISS property on the continuous closed-loop dynamics between impacts.

The effectiveness of this synthesis procedure, based on solving a disturbed differential Riccati equation, corresponding to the linearized system, is supported by a numerical study, made for the state feedback design of a stable walking gait of a 32-DOF fully actuated biped robot. The desired disturbance attenuation is satisfactorily achieved under external disturbances during the free-motion phase, and in the presence of uncertainty in the transition phase.

Acknowledgements The authors acknowledge the financial support of ANR Chaslim grant and CONACYT grant no.165958.

References

1. Aoustin, Y., Formalsky, A.: Control design for a biped: reference trajectory based on driven angles as functions of the undriven angle. J. Comput. Syst. Sci. Int. **42**(4), 645–662 (2003)
2. Baras, J., James, M.: Robust output feedback control for discrete-time nonlinear systems: the finite-time case. In: Proceedings of the 32nd IEEE Conference on Decision and Control, 1993, pp. 51–55. IEEE (1993)
3. Biemond, J., van de Wouw, N., Heemels, W., Nijmeijer, H.: Tracking control for hybrid systems with state-triggered jumps. IEEE Trans. Automat. Control **58**(4), 876–890 (2013)
4. Brogliato, B., Niculescu, S., Orhant, P.: On the control of finite-dimensional mechanical systems with unilateral constraints. IEEE Trans. Automat. Control **42**(2), 200–215 (1997)
5. Dai, H., Tedrake, R.: $\mathcal{L}_2$-gain optimization for robust bipedal walking on unknown terrain. In: IEEE International Conference on Robotics and Automation (ICRA), 2013, pp. 3116–3123. IEEE (2013)
6. Del Prete, A., Mansard, N.: Addressing constraint robustness to torque errors in task-space inverse dynamics. In: Robotics, Sciences and Systems 2015 (2015)
7. Goebel, R., Sanfelice, R., Teel, A.: Hybrid dynamical systems. IEEE Control Syst. **29**(2), 28–93 (2009)
8. Grishin, A., Formalsky, A., Lensky, A., Zhitomirsky, S.: Dynamic walking of a vehicle with two telescopic legs controlled by two drives. Int. J. Robot. Res. **13**(2), 137–147 (1994)
9. Haddad, W., Chellaboina, V., Nersesov, S.: Impulsive and Hybrid Dynamical Systems: Stability, Dissipativity, and Control. Princeton University Press (2006)
10. Haddad, W., Kablar, N., Chellaboina, V., Nersesov, S.: Optimal disturbance rejection control for nonlinear impulsive dynamical systems. Nonlinear Anal. Theory, Methods Appl. **62**(8), 1466–1489 (2005)
11. Hamed, K., Grizzle, J.: Robust event-based stabilization of periodic orbits for hybrid systems: application to an underactuated 3D bipedal robot. In: Proceedings of the 2013 American Control Conference (2013)
12. Heemels, W., Brogliato, B.: The complementarity class of hybrid dynamical systems. Eur. J. Control **9**(2), 322–360 (2003)
13. Hespanha, J.P., Liberzon, D., Teel, A.R.: Lyapunov conditions for input-to-state stability of impulsive systems. Automatica **44**(11), 2735–2744 (2008)
14. Hill, D., Moylan, P.: Connections between finite-gain and asymptotic stability. IEEE Trans. Autom. Control **25**(5), 931–936 (1980)
15. Khalil, W., Kleinfinger, J.: A new geometric notation for open and closed-loop robots. In: Proceedings of the 1986 IEEE International Conference on Robotics and Automation 1986, vol. 3, pp. 1174–1179. IEEE (1986)
16. Lin, W., Byrnes, C.: $\mathcal{H}_\infty$-control of discrete-time nonlinear systems. IEEE Trans. Autom. Control **41**(4), 494–510 (1996)
17. Luh, J., Walker, M., Paul, R.: Resolved-acceleration control of mechanical manipulators. IEEE Trans. Autom. Control **25**(3), 468–474 (1980)
18. Miossec, S., Aoustin, Y.: A simple stability study for a biped walk with under and over actuated phases. In: Proceedings of the Fourth International Workshop on Robot Motion and Control, 2004, RoMoCo 2004, pp. 53–59. IEEE (2004)
19. Montano, O., Orlov, Y., Aoustin, Y.: Nonlinear $\mathcal{H}_\infty$-control of mechanical systems under unilateral constraints. In: Proceedings of the 9th World Congress of the International Federation of Automatic Control (IFAC 2014) (2014)
20. Montano, O., Orlov, Y., Aoustin, Y.: Nonlinear $h\infty$-control under unilateral constraints. Int. J. Control **89**(12), 1–23 (2016)
21. Montano, O., Orlov, Y., Aoustin, Y., Chevallereau, C.: Nonlinear orbital $\mathcal{H}_\infty$-stabilization of underactuated mechanical systems with unilateral constraints. In: Proceedings of the 14th European Control Conference, pp. 800–805 (2015)

22. Montano, O., Orlov, Y., Aoustin, Y., Chevallereau, C.: Orbital stabilization of an underactuated bipedal gait via nonlinear $\mathcal{H}_\infty$-control using measurement feedback. Auton. Robot. 1–19 (2016). doi:10.1007/s10514-015-9543-z
23. Montano, O., Orlov, Y., Aoustin, Y., Chevallereau, C.: Robust stabilization of a fully actuated 3D bipedal locomotion via nonlinear $\mathcal{H}\infty$-control under unilateral constraints. In: Proceedings of the IEEE-RAS 16th International Conference on Humanoid Robots. IEEE (2016)
24. Nesic, D., Zaccarian, L., Teel, A.: Stability properties of reset systems. Automatica **44**(8), 2019–2026 (2008)
25. Nikkhah, M., Ashrafiuon, H., Fahimi, F.: Robust control of underactuated bipeds using sliding modes. Robotica **25**(03), 367–374 (2007)
26. Orlov, Y.: Discontinuous Systems–Lyapunov Analysis and Robust Synthesis Under Uncertainty Conditions. Springer (2009)
27. Orlov, Y., Aguilar, L.: Advanced $\mathcal{H}_\infty$ Control-Towards Nonsmooth Theory and Applications. Birkhauser, Boston (2014)
28. Oza, H., Orlov, Y., Spurgeon, S., Aoustin, Y., Chevallereau, C.: Finite time tracking of a fully actuated biped robot with pre-specified settling time: a second order sliding mode synthesis. In: 2014 IEEE International Conference on Robotics and Automation (ICRA), pp. 2570–2575. IEEE (2014)
29. Posa, M., Tobenkin, M., Tedrake, R.: Stability analysis and control of rigid-body systems with impacts and friction. IEEE Trans. Autom. Control **61**(6), 1423–1437 (2016)
30. Prete, A.D., Mansard, N.: Robustness to joint-torque-tracking errors in task-space inverse dynamics. IEEE Trans. Robot. **32**(5), 1091–1105 (2016)
31. Rengifo, C., Aoustin, Y., Plestan, F., Chevallereau, C., et al.: Contact forces computation in a 3D bipedal robot using constrained-based and penalty-based approaches. In: Proceedings of Multibody Dynamics (2011)
32. Robotics, A.: Project romeo. www.ald.softbankrobotics.com/fr/cool-robots/romeo (2016). Accessed 01 July 2016
33. Tlalolini, D., Aoustin, Y., Chevallereau, C.: Design of a walking cyclic gait with single support phases and impacts for the locomotor system of a thirteen-link 3D biped using the parametric optimization. Multibody Sys. Dyn. **23**(1), 33–56 (2010)
34. Van Zutven, P., Kostic, D., Nijmeijer, H.: On the stability of bipedal walking. In: Simulation, Modeling, and Programming for Autonomous Robots, p. 521 (2010)
35. Walker, M., Orin, D.: Efficient dynamic computer simulation of robotic mechanisms. J. Dyn. Syst. Meas. Control **104**(3), 205–211 (1982)
36. Westervelt, E., Grizzle, J., Chevallereau, C., Choi, J., Morris, B.: Feedback Control of Dynamic Bipedal Robot Locomotion. CRC press, Boca Raton (2007)
37. Willems, J.: Dissipative dynamical systems part 1: general theory. Arch. Ration. Mech. Anal. **45**(5), 321–351 (1972)
38. Yuliar, S., James, M., Helton, J.: Dissipative control systems synthesis with full state feedback. Math. Control Signals Syst. **11**(4), 335–356 (1998)
39. Yunt, K., Glocker, C.: Trajectory optimization of mechanical hybrid systems using sumt. In: 9th IEEE International Workshop on Advanced Motion Control, pp. 665–671. IEEE (2005)

Chapter 16
Event-Triggered Sliding Mode Control Strategies for a Class of Nonlinear Uncertain Systems

Antonella Ferrara and Michele Cucuzzella

Abstract This chapter presents novel Sliding Mode Control (SMC) strategies of Event-Triggered (ET) type for a class of nonlinear systems affected by uncertainties and external disturbances. By virtue of its ET nature, the proposed control strategies are particularly appropriate for Networked Control Systems (NCSs), i.e., feedback systems including communication networks. The objective of the proposed control schemes is indeed to reduce the number of data transmissions over the communication network, in order to avoid problems typically due to the network congestion such as jitter and packet loss. In particular, an ET-SMC scheme and an ET Second Order SMC (ET-SOSMC) scheme are designed for a class of nonlinear uncertain NCSs, guaranteeing satisfactory performance of the controlled system even in presence of delayed transmissions. The proposed control schemes are theoretically analyzed in this chapter, showing their capability of enforcing the robust ultimate boundedness of the sliding variable associated with the controlled system, and also of its first time derivative in case of ET-SOSMC. Moreover, in order to guarantee the avoidance of the notorious Zeno behaviour, the existence of a lower bound for the time elapsed between consecutive triggering events is proven.

16.1 Introduction

Sliding Mode Control (SMC) is considered a powerful strategy, able to guarantee satisfactory performance in terms of robustness of the controlled system, even in the presence of unavoidable modelling uncertainties and external disturbances [27, 57–59]. The same robustness property holds for Second Order SMC (SOSMC)

A. Ferrara · M. Cucuzzella (✉)
Dipartimento di Ingegneria Industriale e dell'Informazione,
University of Pavia, Via Ferrata 5, 27100 Pavia, Italy
e-mail: michele.cucuzzella@gmail.com

A. Ferrara
e-mail: antonella.ferrara@unipv.it

J.B. Clempner and W. Yu (eds.), *New Perspectives and Applications of Modern Control Theory*, https://doi.org/10.1007/978-3-319-62464-8_16

methodology [4–8, 25, 26, 34, 40, 52, 55], in which not only the sliding variable, but also its first time derivative are steered to zero in a finite time. This is confirmed by the numerous applications described in the literature (see, for instance, [16–18, 21–23, 42, 56]). Moreover, by virtue of its low complexity, SMC methodology represents a very easy to implement solution adequate also for Networked Control Systems (NCSs), i.e., feedback systems including communication networks [1, 36, 39, 47, 60, 62].

In Networked Control Systems (NCSs), critical problems such as jitter, packet loss and delayed transmissions can occur specially when the network is congested [15, 48, 51], thereby determining the worsening of the performance of the controlled system. For these reasons, the need of designing robust control schemes able to reduce data transmissions over the network, while guaranteeing satisfactory performance of the controlled system even in the presence of uncertainties and delayed transmissions becomes mandatory.

In the literature, a methodology which is very appreciated in designing control schemes capable of reducing data transmission effort over communication networks is the so-called Event-Triggered (ET) control [3, 37, 38, 44, 49, 53, 54, 61]. Differently from traditional time-triggered control implementation, where periodic data transmissions occur, ET control schemes enable the communication between the plant and the controller (feedback path), and between the controller and the actuator (direct path) only when some triggering condition is satisfied (or violated depending on the adopted logic). For this reason the ET control approach can significantly reduce the number of data transmissions, avoiding the congestion of the network and its possible unavailability. Obviously, there exists a trade-off between the performance of the controlled system and the communication rate [24]. However, in spite of aperiodic data transmissions, satisfactory stability properties of the controlled system have been studied in the literature. Specifically, in [53], it was proven that in case of nonlinear systems, relying on threshold-based ET algorithm, the Input-to-State Stability of the controlled system can be guaranteed by ensuring a certain decrease in a suitable Lyapunov function.

Recently, in the literature, the basic ET approach has been developed so as to take into account the knowledge of the nominal model of the plant. This has given rise to the so-called Model-Based ET (MB-ET) control [35, 50]. This methodology has been also exploited together with SMC, and Model Predictive Control (MPC), [28, 32, 41], even in case of Mixed Logical Dynamical (MLD) systems [29–31]. In particular, the use of SMC in conjunction with ET implementation is justified by the necessity of robustness to face modelling uncertainties and external disturbances which can naturally affect the system [10–12, 19, 20].

In the present chapter, an ET-SMC scheme and an ET-SOSMC scheme are presented for nonlinear uncertain systems including communication networks that can be unavailable. The proposed ET-SMC strategy is based on a triggering condition that depends on the sliding variable associated with the controlled system and on the size of a pre-specified boundary layer of the sliding manifold. The proposed control scheme is theoretically analyzed in the chapter, proving that the sliding variable is ultimate bounded in the desired boundary layer, even in presence of delayed transmissions. The ET-SOSMC scheme is based on two triggering conditions and two control laws that depend not only on the sliding variable, but also on its first time derivative. The stability properties of this control strategy are theoretically analyzed proving that the sliding variable and its first time derivative are ultimately bounded in a desired vicinity of the origin, even in presence of delayed transmissions. These results imply the ultimate boundedness of the state of the original uncertain nonlinear system as well. Moreover, in order to avoid the notorious Zeno behaviour [2, 43], the existence of a lower bound for the time elapsed between consecutive triggering events is proven.

The present chapter is organized as follows. In Sect. 16.2.1, a first problem formulation is presented. The proposed ET-SMC strategy is described in Sect. 16.2.2, while in Sect. 16.2.3, the stability properties of the control scheme are theoretically analyzed. In order to demonstrate the efficiency of the proposal, an illustrative example is provided in Sect. 16.2.4. In Sect. 16.3.1, a second problem formulation is introduced, and in Sect. 16.3.2 the proposed ET-SOSMC strategy is described. The stability analysis of the control scheme is formally discussed in Sect. 16.3.3, and an illustrative example is provided in Sect. 16.3.4. Some conclusions are finally gathered in Sect. 16.4.

16.2 Event-Triggered Sliding Mode Control

16.2.1 Problem Formulation

In the following sections, given the set $\mathcal{D} \subset \mathbb{R}$, we denote $\mathcal{D}^{\mathrm{sup}} \triangleq \sup_{d \in \mathcal{D}}\{|d|\}$. Moreover, given the set $\mathcal{B}$, $\partial\mathcal{B}$ will denote the boundary of $\mathcal{B}$. The relative degree of the system, i.e., the minimum order r of the time derivative $\sigma^{(r)}$ of the sliding variable in which the control u explicitly appears, is considered well-defined, uniform and time invariant. Moreover, for the sake of simplicity, the dependence of all the variables on time t is omitted when obvious.

Consider a plant (process and actuator) which can be modelled as

$$\dot{x} = a(x) + b(x)u + d_{\mathrm{m}}(x), \tag{16.1}$$

where $x \in \Omega$ ($\Omega \subset \mathbb{R}^n$ bounded) is the state vector, the value of which at the initial time instant t_0 is $x(t_0) = x_0$, and $u \in \mathbb{R}$ is the control variable, while $a : \Omega \to \mathbb{R}^n$ and $b : \Omega \to \mathbb{R}^n$ are uncertain functions of class C^0. Moreover, system (16.1) is subject to the external disturbance $d_{\mathrm{m}} : \Omega \times \mathbb{R} \to \mathbb{R}^n$. To permit the controller design in the next sections, the following assumption is made on d_{m}:

Assumption 16.1 The external disturbance d_{m} is matched, i.e.,

$$d_{\mathrm{m}}(x) = b(x)d,$$

where d is unknown, but bounded as

$$d \in \mathcal{D} \subset \mathbb{R},$$

$\mathcal{D}^{\text{sup}}$ being a known positive constant.

Define now a suitable output function: the so-called *sliding variable*.

Definition 16.1 $\sigma : \Omega \to \mathbb{R}$ of class C^1 is a sliding variable for system (16.1) provided that the pair (σ, u) has the following property: if u in (16.1) is designed so that, in a finite time $t_r^\star \geq t_0$, $\forall\, x_0 \in \Omega$, $\sigma = 0\ \ \forall t \geq t_r^\star$, then $\forall t \geq t_r^\star$ the origin is an asymptotically stable equilibrium point of (16.1) constrained to the sliding manifold $\sigma = 0$.

Now, regarding the sliding variabile σ as the controlled variable associated with system (16.1), assume that system (16.1) is complete in Ω and has a uniform relative degree equal to 1. The following definitions are introduced.

Definition 16.2 Given $t_r^\star \geq t_0$ (ideal reaching time), if $\forall\, x_0 \in \Omega, \sigma = 0\ \ \forall t \geq t_r^\star$, then an *ideal sliding mode* of system (16.1) is enforced on the sliding manifold $\sigma = 0$.

Definition 16.3 Given $t_r \geq t_0$ (practical reaching time), if $\forall\, x_0 \in \Omega,\ |\sigma| \leq \delta\ \forall t \geq t_r$, then a *practical sliding mode* of system (16.1) is enforced in a vicinity of the sliding manifold $\sigma = 0$.

Moreover, assume that system (16.1) admits a global normal form in Ω, i.e., there exists a global diffeomorphism (see [45]) of the form $\Phi = [\Psi, \sigma]^T = [x_r, \xi]^T$, with $\Phi : \Omega \to \Phi_\Omega$ ($\Phi_\Omega \subset \mathbb{R}^n$ bounded), and $\Psi : \Omega \to \mathbb{R}^{n-1},\ x_r \in \mathbb{R}^{n-1}, \xi \in \mathbb{R}$, such that

$$\begin{cases} \dot{x}_r = a_r(x_r, \xi) & (16.2a) \\ \dot{\xi} = f(x_r, \xi) + g(x_r, \xi)(u + d), & (16.2b) \end{cases}$$

with

$$a_r(x_r, \xi) = \tfrac{\partial \Psi}{\partial x}\left(\Phi^{-1}(x_r, \xi)\right) a\left(\Phi^{-1}(x_r, \xi)\right)$$

$$f(x_r, \xi) = a\left(\Phi^{-1}(x_r, \xi)\right) \cdot \nabla\sigma\left(\Phi^{-1}(x_r, \xi)\right)$$

$$g(x_r, \xi) = b\left(\Phi^{-1}(x_r, \xi)\right) \cdot \nabla\sigma\left(\Phi^{-1}(x_r, \xi)\right).$$

Note that, as a consequence of the uniform relative degree assumption, one has that $g \neq 0$. In the literature, (16.2b) is the so-called *auxiliary system*. Since a_r, f, g are continuous functions and Φ_Ω is a bounded set, one has also that

$$\begin{aligned} |f(x_r, \xi)| &\leq F \\ g(x_r, \xi) &\leq G_{\max}, \end{aligned} \qquad (16.3)$$

F and $G_{\max}$ being known positive constants. Moreover, the following assumption is made on the uncertain function g.

Assumption 16.2 The uncertain function g can be lower bounded as

$$g(x_r, \xi) \geq G_{\min}, \tag{16.4}$$

$G_{\min}$ being a known positive constant.

Now, a preliminary control problem can be formulated.

Problem 16.1 Let Assumptions 16.1 and 16.2 hold. Relying on (16.1)–(16.4), design a feedback control law

$$u^\star = \kappa(\sigma), \tag{16.5}$$

with the following property: $\forall\, x_0 \in \Omega,\ \exists\ t_r^\star \geq t_0$ such that $\sigma = 0\ \ \forall\, t \geq t_r^\star$, in spite of the uncertainties.

Remark 16.1 The solution to Problem 16.1 is a control law capable of robustly enforcing an *ideal sliding mode* of system (16.1) in a finite time, according to Definition 16.2.

In practical implementation the state is sampled at certain time instants $t = \{t_0, t_1, \ldots, t_k, \ldots\}$, $k \in \mathbb{N}$, and the control law, computed as $u(t_k) = \kappa(\sigma(t_k))$, is held constant between two successive samplings. This kind of implementation, called sample-and-hold, can be expressed as

$$u = u(t_k) \quad \forall\, t \in [t_k,\ t_{k+1}[\,,\ k \in \mathbb{N}, \tag{16.6}$$

where $t_k,\ t_{k+1} \in \mathcal{T}$, $\mathcal{T}$ being the set of the triggering time instants. In conventional implementation, the sequence $\{t_k\}_{k\in\mathbb{N}}$ is typically periodic and the time interval $t_{k+1} - t_k$, is a priori fixed. The control approach, in that case, is classified as *time-triggered*.

In the present chapter, instead of relying on time-triggered executions, we introduce a triggering condition which depends on the sliding variable, so that the state of the auxiliary system is transmitted over the communication network only when such a condition is verified. This implies that also the control law is updated and sent to the actuator of the plant only at the triggering time instants. In the literature, this control approach is known as Event-Triggered (ET) control approach. Note that, in this chapter, we do not adopt a mathematical model of the network, but we design the control strategy aiming at reducing data transmission. This, in order to avoid the network congestion and limit its negative consequences, such as packets drop. However, in order to take into account possible malfunctions, we suppose that the presence of communication networks can cause delayed transmissions due to the network unavailability. More precisely, we suppose that data transmissions could occur with time-varying delay Δ_{direct} and Δ_{feedback} in the direct and feedback path, respectively (see Fig. 16.1). Let the overall time delay be $\Delta = \Delta_{\text{direct}} + \Delta_{\text{feedback}}$. The following assumption is made on Δ.

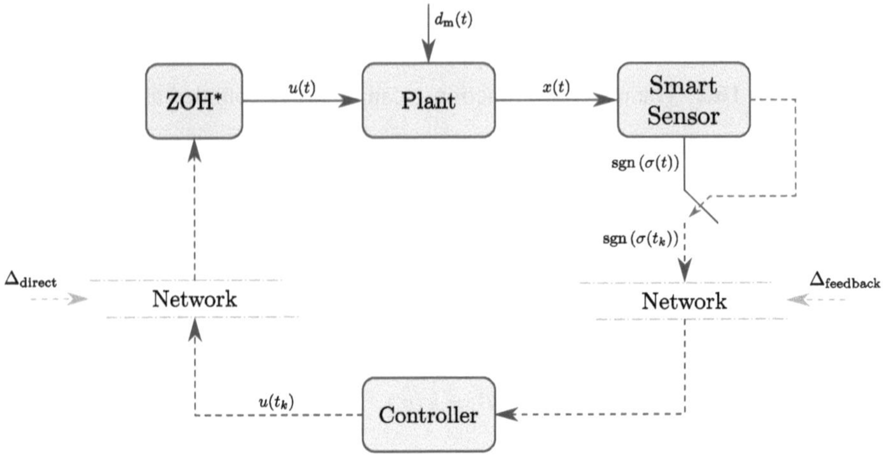

Fig. 16.1 The proposed Event-Triggered Sliding Mode Control (ET-SMC) scheme

Assumption 16.3 The overall time-varying delay Δ can be bounded as

$$\Delta \leq \Delta_{\max}, \tag{16.7}$$

$\Delta_{\max}$ being a known positive constant.

Moreover, we suppose that the plant is equipped with a particular zero-order-hold (indicated in Fig. 16.1 with ZOH*), such that the control variable computed at the last triggering time instant t_k is held constant $\forall\, t \in [t_k,\ t_{k+1}[$. Then, this approach tends to reduce the transmissions over the network both on the direct path (from the controller to the plant) and on the feedback path (from the sensor to the controller).

Taking into account the previous considerations, we can now move from Problem 16.1 and formulate the problem which will be actually solved in this chapter.

Problem 16.2 Let Assumptions 16.1–16.3 hold. Relying on (16.1)–(16.4), design a feedback control law

$$u = u(t_k) = \kappa(\sigma(t_k)) \quad \forall\, t \in [t_k,\ t_{k+1}[,\ k \in \mathbb{N}, \tag{16.8}$$

with the following property: $\forall\, x_0 \in \Omega,\ \exists\ t_r \geq t_0$ such that $|\sigma| \leq \delta\ \ \forall\, t \geq t_r$, in spite of the uncertainties, δ being a positive constant arbitrary set.

Remark 16.2 The solution to Problem 16.2 is an ET control law capable of enforcing a *practical sliding mode* of system (16.1) in a finite time, according to Definition 16.3.

Before illustrating the features of the proposed control scheme, relying on Problem 16.2 let us introduce the following definition.

Definition 16.4 The *boundary layer* for the sliding variable is

$$\mathcal{B}_\delta \triangleq \{\sigma \in \mathbb{R} : |\sigma| \leq \delta\}, \tag{16.9}$$

δ being a positive constant arbitrarily set.

16.2.2 The Proposed Control Scheme

The control scheme proposed to solve Problem 16.2 is reported in Fig. 16.1, where the dashed path means that the corresponding signals are transmitted over the network only at the triggering time instants t_k. The control scheme contains two key blocks: the *Smart Sensor* and the *Controller*.

The Smart Sensor

The considered sensor is smart in the sense that it has some computation capability, i.e., it is able to compute the sliding variable and verify the following triggering condition.

Triggering Condition

$$|\sigma| = \delta, \tag{16.10}$$

δ being a positive constant arbitrarily set (see Definition 16.4).

The switch in Fig. 16.1 is closed only when (16.10) holds, i.e., when the Smart Sensor generates the triggering signal. More precisely, when (16.10) is verified, the sign of the sliding variable is transmitted by the Smart Sensor to the Controller (through the communication network).

The Controller.

Relying on (16.8)–(16.10), the control law we propose to solve Problem 16.2 can be expressed as

$$u = u(t_k) = -U_{\max} \operatorname{sign}(\sigma(t_k)) \quad \forall t \in [t_k, t_{k+1}[, \; k \in \mathbb{N}, \tag{16.11}$$

where

$$U_{\max} > \frac{F}{G_{\min}} + \mathcal{D}^{\text{sup}} \tag{16.12}$$

is a positive value suitably selected in order to enforce a sliding mode (see Fig. 16.2).

Note that the control signal is transmitted by the Controller to the plant (through the communication network) only when (16.10) holds.

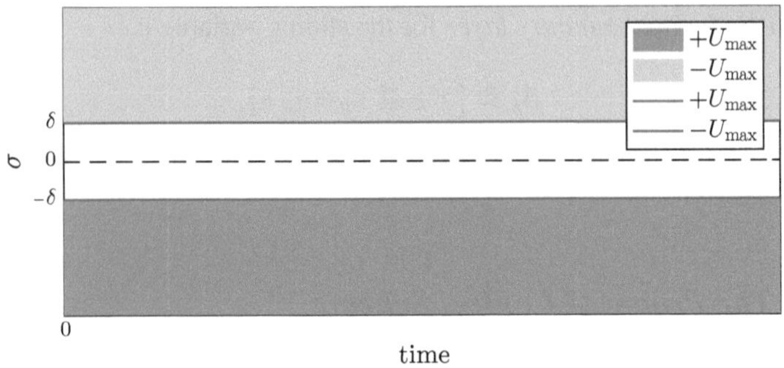

Fig. 16.2 Representation of the proposed ET-SMC strategy

16.2.3 Stability Analysis

In this section, the stability properties of system (16.1) controlled via the proposed ET-SMC strategy are analyzed. To this end, it is convenient to introduce the following definitions.

Definition 16.5 The set $\mathcal{B}_\delta$ is said to be attractive if the solution to the auxiliary system (16.2b), $\forall\, \sigma \in \mathbb{R} \setminus \mathcal{B}_\delta$, satisfies the so-called η-reachability condition [27, 58]

$$\sigma\dot{\sigma} \leq -\eta|\sigma|\,. \tag{16.13}$$

Definition 16.6 The solution σ to the auxiliary system (16.2b) is said to be ultimately bounded with respect to the set $\mathcal{B}_\delta$ if

$$\forall\, x_0 \in \Omega,\ \exists\, t_\mathrm{r} \geq t_0 : \sigma \in \mathcal{B}_\delta \qquad \forall\, t \geq t_\mathrm{r}\,. \tag{16.14}$$

Definition 16.7 Let σ be the solution to the auxiliary system (16.2) starting from the initial condition $\sigma(t_0)$. The set $\mathcal{B}_\delta$ is said to be positively invariant if

$$\sigma(t_0) \in \mathcal{B}_\delta \Rightarrow \sigma \in \mathcal{B}_\delta \qquad \forall\, t \geq t_0\,. \tag{16.15}$$

Definition 16.8 In analogy with [46], given the bounded sets $\Omega,\ \Omega_\delta \subset \Omega$, then, the origin of system (16.1) is said to be practically stable with respect to $(t_\mathrm{r}^\star, t_{\mathrm{r},x}, \Omega, \Omega_\delta, \mathcal{D})$ if

$$\forall\, t_\mathrm{r}^\star \geq t_0,\ \forall\, d \in \mathcal{D},\ \forall\, x_0 \in \Omega,\ \exists\, t_{\mathrm{r},x} \geq t_\mathrm{r}^\star : x \in \Omega_\delta \quad \forall\, t \geq t_{\mathrm{r},x}\,. \tag{16.16}$$

Before showing the theoretical results, the following assumption is made on the initial condition of the auxiliary system (16.2b).

Assumption 16.4 Given the auxiliary system (16.2b), let the sign of the initial condition $\sigma(t_0)$ be known.

Now, making reference to the auxiliary system (16.2b) the following results can be proved.

Lemma 16.1 *Let Assumptions 16.1, 16.2 and 16.4 hold with* $|\sigma(t_0)| > \delta$, δ *being a positive constant arbitrarily set. Given the auxiliary system (16.2b) controlled by* (16.11), (16.12) *with the triggering condition* (16.10)*, then, the boundary layer* $\mathcal{B}_\delta$ *is attractive for the solution* σ *to (16.2b).*

Proof Consider the η-reachability condition (16.13). Since Assumption 16.4 holds, one has that $\text{sign}(\sigma) = \text{sign}(\sigma(t_0)), \forall t \in [t_0, t_1]$, t_1 being the first triggering time instant. Making reference to system (16.2), since $\sigma \, \text{sign}(\sigma) = |\sigma|$, it yields

$$\begin{aligned} \sigma\dot{\sigma} &= \sigma\,(f - gU_{\max}\,\text{sign}(\sigma) + gd) \\ &\leq (F - G_{\min}(U_{\max} - \mathcal{D}^{\text{sup}}))\,|\sigma|, \end{aligned} \tag{16.17}$$

By virtue of inequality (16.12), one can easily verify that (16.13) holds with $\eta = -\,(F - G_{\min}(U_{\max} - \mathcal{D}^{\text{sup}})) > 0$. Then, integrating the inequality $\sigma\dot{\sigma} \leq -\eta|\sigma|$ from $t_0 = 0$ to t_{r}, one has

$$t_{\text{r}} \leq \frac{|\sigma(0)| - \delta}{\eta}, \tag{16.18}$$

implying the finite time convergence of the sliding variable to $\mathcal{B}_\delta$. Moreover, one can also conclude that the first transmission over the network is executed at the triggering time instant $t_1 = t_{\text{r}}$. □

Remark 16.3 Note that, by virtue of Lemma 16.1, the proposed control solution avoids to transmit the value of $\text{sign}(\sigma)$ and u over the network during the entire reaching phase, i.e., till the sliding variable enters the boundary layer $\mathcal{B}_\delta$ at the time instant $t_1 = t_{\text{r}}$.

Lemma 16.2 *Let Assumptions 16.1, 16.2 and 16.4 hold with* $|\sigma(t_0)| \leq \delta$, δ *being a positive constant arbitrarily set. Given the auxiliary system (16.2b) controlled by* (16.11), (16.12) *with the triggering condition* (16.10)*, then, the boundary layer* $\mathcal{B}_\delta$ *is a positively invariant set for the solution* σ *to (16.2b).*

Proof Consider two different cases in order to prove the result.

Case 1 ($|\sigma| < \delta$). In this case, according to the proposed ET-SMC strategy, $\forall t \in [t_k, t_{k+1}[$ the control law is not updated, i.e., its sign does not change. This implies that the sliding variable evolves in the boundary layer $\mathcal{B}_\delta$ until it reaches its border, so that Case 2 occurs.

Case 2 ($|\sigma| = \delta$). In this second case, the triggering condition is verified. Then, $\text{sign}(\sigma)$ is sent to the controller and the control law is updated. In particular, the sign of the control law changes, and the sliding variable is steered towards the interior of $\mathcal{B}_\delta$, so that Case 1 occurs again. This implies that $\forall \sigma(t_0) \in \mathcal{B}_\delta$, then, $\forall t \geq t_0$, $\sigma \in \mathcal{B}_\delta$, i.e., $\mathcal{B}_\delta$ is a positively invariant set, according to Definition 16.7. □

Now, relying on Lemmas 16.1 and 16.2, one can prove the major result concerning the evolution of the auxiliary system (16.2b) controlled via the proposed strategy.

Theorem 16.1 *Let Assumptions 16.1, 16.2 and 16.4 hold, and let δ be a positive constant arbitrarily set. Given the auxiliary system (16.2b) controlled by (16.11),* (16.12) *with the triggering condition* (16.10)*, then the solution σ to (16.2b) is ultimately bounded with respect to $\mathcal{B}_\delta$.*

Proof The proof is a straightforward consequence of Lemmas 16.1 and 16.2. In fact, by virtue of Lemma 16.1, applying the control law (16.11), (16.12) there exists a time instant t_r when σ enters $\mathcal{B}_\delta$, i.e., the triggering condition (16.10) is verified. Then, by virtue of Lemma 16.2, $\forall t \geq t_r$, σ remains in $\mathcal{B}_\delta$, which implies that it is ultimately bounded with respect to $\mathcal{B}_\delta$. □

Remark 16.4 Note that the proposed control scheme, because of its ET nature, cannot generate an ideal sliding mode (see Definition 16.2), but only a practical sliding mode (see Definition 16.3). However, by virtue of the Regularization Theorem [58, Chap. 2], it can be proven that also the state of system (16.1) is ultimately bounded. This implies that Problem 16.2 is equivalent to the problem of designing a bounded control such that, according to Definition 16.8, the origin of system (16.1) is practically stable.

Now, since the triggering time instants are implicitly defined and only known at the execution times, we prove the existence of a lower bound for the so-called *inter-execution* or *inter-event* times [53]. More specifically, let $\tau_{\min}$ be the minimum inter-event time, such that $t_{k+1} - t_k \geq \tau_{\min}$ for any $k \in \mathbb{N}^+$.

Theorem 16.2 *Let Assumptions 16.1, 16.2 and 16.4 hold, and let δ be a positive constant arbitrarily set. Given the auxiliary system (16.2b) controlled by* (16.11), (16.12) *with the triggering condition* (16.10)*, then, $\forall t > t_r$ the inter-event times are lower bounded by*

$$\tau_{\min} = \frac{2\delta}{F + G_{\max}(U_{\max} + \mathcal{D}^{\sup})}.$$

Proof Since $\text{sign}(\sigma)$ and u are transmitted over the network only when the triggering condition (16.10) is verified, the theorem will be proved by computing the time interval $t_{k+1} - t_k$ that σ takes to evolve from $-\delta$ to δ with the maximum velocity, i.e., $\dot{\sigma}_{\max} = F + G_{\max}(U_{\max} + \mathcal{D}^{\sup})$. Then, one has

$$\begin{aligned} \sigma(t_{k+1}) - \sigma(t_k) &= \int_{t_k}^{t_{k+1}} \dot{\sigma}_{\max}\, d\tau \\ \delta - (-\delta) &= \dot{\sigma}_{\max}(t_{k+1} - t_k) \\ 2\delta &= (F + G_{\max}(U_{\max} + \mathcal{D}^{\sup}))\tau_{\min}, \end{aligned} \tag{16.19}$$

where the equality $t_{k+1} - t_k = \tau_{\min}$ follows from the assumption that σ evolves with constant maximum velocity $\dot{\sigma}_{\max}$. Finally, from (16.19) one obtains

$$\tau_{\min} = \frac{2\delta}{F + G_{\max}(U_{\max} + \mathcal{D}^{\text{sup}})}, \tag{16.20}$$

which proves the theorem. □

Remark 16.5 Note that Theorem 16.2 guarantees that the time elapsed between consecutive triggering events does not become arbitrarily small, avoiding the notorious Zeno behaviour [2, 43]. In practical cases, this result is very useful to assess the feasibility of the proposed scheduling policy.

Taking now into account the possible occurrence of delayed data transmissions, due to the presence of communication networks, the following result can be proven. More precisely, relying on Problem 16.2 we will prove that by modifying the triggering condition (16.10), the auxiliary state-space trajectory is ultimately bounded with respect to the desired boundary layer $\mathcal{B}_\delta$.

Theorem 16.3 *Let Assumptions 16.1–16.4 hold. Given the auxiliary system (16.2b) controlled by* (16.11), (16.12) *with the triggering condition* (16.10), *then, for any desired*

$$\delta > (F + G_{\max}(U_{\max} + \mathcal{D}^{\text{sup}}))\Delta_{\max},$$

the triggering condition

$$|\sigma| = \delta',$$

with

$$\delta' = \delta - (F + G_{\max}(U_{\max} + \mathcal{D}^{\text{sup}}))\Delta_{\max},$$

enforces the following inequality

$$|\sigma| \leq \delta \quad \forall t \geq t_{\text{r}}', \tag{16.21}$$

t_{r}' *being the reaching time instant of the boundary layer*

$$\mathcal{B}_{\delta'} \triangleq \{\sigma \in \mathbb{R} : |\sigma| \leq \delta'\}.$$

Proof In analogy with Lemma 16.1, one can easily prove that there exists a time instant t_{r}' when σ enters the inner boundary layer $\mathcal{B}_{\delta'}$. Suppose now that the transmission of $\sigma(t_k) = \delta'$ occurs with maximum time delay $\Delta_{\max}$. Moreover, assume that the sliding variable evolves with constant maximum velocity $\dot{\sigma}_{\max} = F + G_{\max}(U_{\max} + \mathcal{D}^{\text{sup}})$. In order to enforce inequality (16.21), we impose that $\sigma(t_k + \Delta_{\max}) = \delta$. Then, one has

$$\begin{aligned} \sigma(t_k + \Delta_{\max}) - \sigma(t_k) &= \int_{t_k}^{t_k+\Delta_{\max}} \dot{\sigma}_{\max}\, d\tau \\ \delta - \delta' &= \dot{\sigma}_{\max} \Delta_{\max} \\ \delta' &= \delta - (F + G_{\max}(U_{\max} + \mathcal{D}^{\sup}))\Delta_{\max}, \end{aligned} \tag{16.22}$$

which concludes the proof. □

Remark 16.6 Note that in case of delayed transmissions, the lower bound $\tau_{\min}$ can be obtained by using δ' instead of δ in Theorem 16.2.

16.2.4 Illustrative Example

In this section, an illustrative example is briefly discussed. Consider the perturbed double integrator

$$\begin{cases} \dot{x}_1 = x_2 \\ \dot{x}_2 = u + d\,. \end{cases} \tag{16.23}$$

with $d = \mathcal{D}^{\sup}\sin(t)$, $\mathcal{D}^{\sup} = 3$. Let the initial condition be $x(0) = \begin{bmatrix}1 & 1\end{bmatrix}^T$, and the sliding variable be $\sigma = x_1 + x_2$. In the triggering condition (16.10) the threshold is $\delta = 0.2$. Then, the control amplitude $U_{\max}$ is selected equal to 5. In Fig. 16.3 the time evolution of the sliding variable σ is shown. Note that, according to Theorem 16.1, σ is ultimately bounded with respect to the boundary layer $\mathcal{B}_\delta$, the size of which is δ. In Fig. 16.3 the time evolution of the system states x_1, x_2, and the inter-event times $\tau_k = t_{k+1} - t_k$, are also shown. In particular, one can appreciate that the minimum inter-event time $\tau_{\min} = 0.0404$ s, according to Theorem 16.2, is a lower bound for the inter-event times. Moreover, considering a sampling time T_s = 1e-4 s, and a simulation time $T = 10$ s, the number of transmissions is 101, i.e., 99.9% less than the number required by the conventional (i.e., time-driven) SMC implementation. Obviously, reducing the size of the boundary layer implies the improvement of the convergence accuracy. Yet, a larger number of transmissions could be required. The correct balance between convergence accuracy and transmission load has to be searched depending on the specific application. Finally, in Fig. 16.3 the time evolution of the sliding variable σ in presence of transmissions with maximum time delay $\Delta_{\max} = 0.005$ s acting from $t = 1$ s to $t = 4$ s is shown. Note that, by selecting $\delta' = 0.151$ (see Theorem 16.3), even in presence of maximum time delay, σ is ultimately bounded with respect to the desired boundary layer $\mathcal{B}_\delta$. In this case, the number of transmissions is 127.

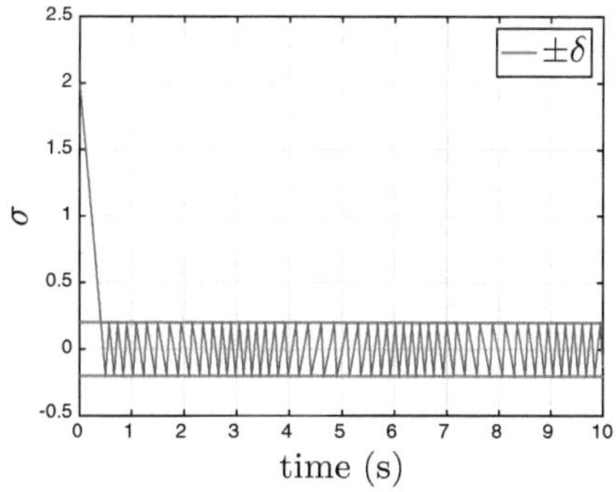

(a) Sliding variable

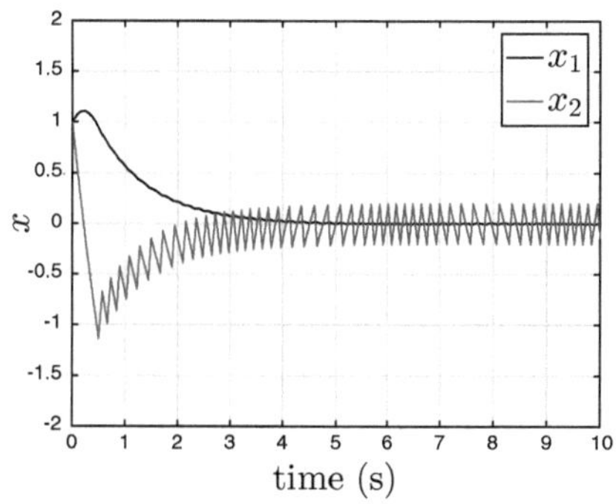

(b) System state

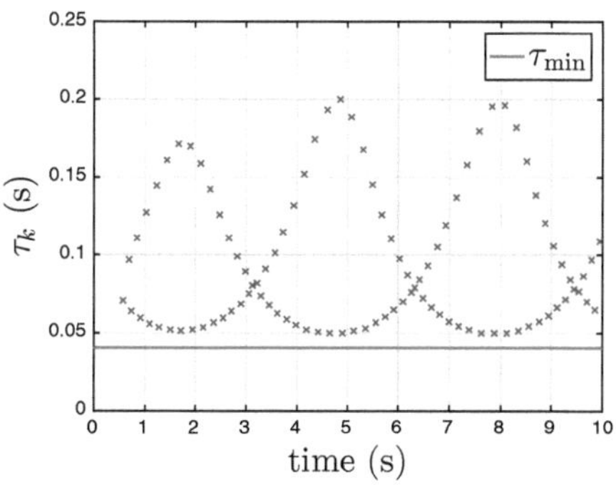

(c) Inter-event times

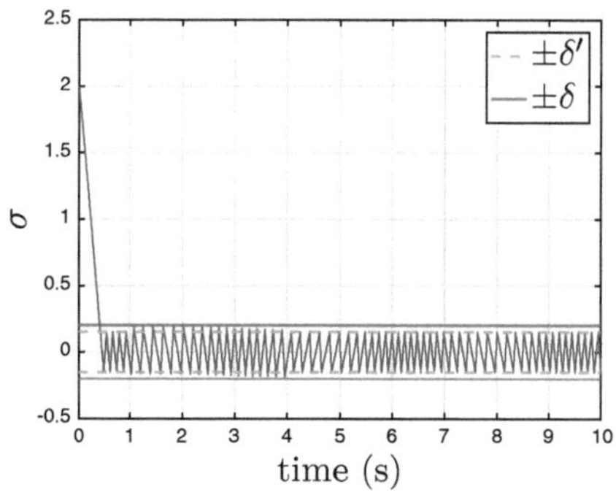

(d) Sliding variable

Fig. 16.3 Event-Triggered Sliding Mode Control. **a** Time evolution of the sliding variable, with visualization of the boundary layer $\mathcal{B}_\delta$ of size δ. **b** Time evolution of the system states x_1 and x_2. **c** Inter-event times $\tau_k = t_{k+1} - t_k$, with visualization of the minimum inter-event time $\tau_{\min}$. **d** Time evolution of the sliding variable in presence of delayed transmissions, with zoom and visualization of both the desired boundary layer $\mathcal{B}_\delta$ of size δ, and the boundary layer $\mathcal{B}_{\delta'}$ of size δ' adopted for the triggering condition

16.3 Event-Triggered Second Order Sliding Mode Control

16.3.1 Problem Formulation

Consider the uncertain nonlinear system (16.1), where $x \in \Omega$ ($\Omega \subset \mathbb{R}^n$, bounded) is the state vector, the value of which at the initial time instant t_0 is $x(t_0) = x_0$, and $u \in \mathbb{R}$ is the control variable, while $a : \Omega \to \mathbb{R}^n$ and $b : \Omega \to \mathbb{R}^n$ are uncertain functions of class C^1. Let Assumption 16.1 hold. We now define the sliding variable as follows.

Definition 16.9 $\sigma : \Omega \to \mathbb{R}$ of class C^2 is a sliding variable for system (16.1) provided that the pair (σ, u) has the following property: if u in (16.1) is designed so that, in a finite time $t_\mathrm{r}^\star \geq t_0, \forall x_0 \in \Omega,\ \sigma = \dot{\sigma} = 0\ \forall t \geq t_\mathrm{r}^\star$, then $\forall t \geq t_\mathrm{r}^\star$ the origin is an asymptotically stable equilibrium point of (16.1) constrained to the sliding manifold $\sigma = \dot{\sigma} = 0$.

Now, regarding the sliding variabile σ as the controlled variable associated with system (16.1), assume that the system (16.1) is complete in Ω and has a uniform relative degree equal to 2. The following definitions are introduced.

Definition 16.10 Given $t_\mathrm{r}^\star \geq t_0$ (ideal reaching time), if $\forall x_0 \in \Omega,\ \sigma = \dot{\sigma} = 0\ \forall t \geq t_\mathrm{r}^\star$, then an *ideal second order sliding mode* of system (16.1) is enforced on the sliding manifold $\sigma = \dot{\sigma} = 0$.

Definition 16.11 Given $t_\mathrm{r} \geq t_0$ (practical reaching time), if $\forall x_0 \in \Omega,\ |\sigma| \leq \delta_1$, $|\dot{\sigma}| \leq \delta_2\ \ \forall t \geq t_\mathrm{r}$, then a *practical second order sliding mode* of the system (16.1) is enforced in a vicinity of the sliding manifold $\sigma = \dot{\sigma} = 0$.

Moreover, assume that system (16.1) admits a global normal form in Ω, i.e., there exists a global diffeomorphism of the form $\Phi = [\Psi,\ \sigma,\ a \cdot \nabla\sigma]^T = [x_\mathrm{r},\ \xi]^T$, with $\Phi : \Omega \to \Phi_\Omega$ ($\Phi_\Omega \subset \mathbb{R}^n$ bounded), and $\Psi : \Omega \to \mathbb{R}^{n-2}$, $x_\mathrm{r} \in \mathbb{R}^{n-2}$, $\xi = [\sigma,\ \dot{\sigma}]^T \in \mathbb{R}^2$, such that

$$\begin{cases} \dot{x}_\mathrm{r} = a_\mathrm{r}(x_\mathrm{r}, \xi) & (16.24\mathrm{a}) \\ \dot{\xi}_1 = \xi_2 & (16.24\mathrm{b}) \\ \dot{\xi}_2 = f(x_\mathrm{r}, \xi) + g(x_\mathrm{r}, \xi)(u + d), & (16.24\mathrm{c}) \end{cases}$$

with

$$a_\mathrm{r}(x_\mathrm{r}, \xi) = \tfrac{\partial \Psi}{\partial x}\left(\Phi^{-1}(x_\mathrm{r}, \xi)\right) a\left(\Phi^{-1}(x_\mathrm{r}, \xi)\right)$$

$$f(x_\mathrm{r}, \xi) = a\left(\Phi^{-1}(x_\mathrm{r}, \xi)\right) \cdot \nabla\left(a\left(\Phi^{-1}(x_\mathrm{r}, \xi)\right) \cdot \nabla\sigma\left(\Phi^{-1}(x_\mathrm{r}, \xi)\right)\right)$$
$$g(x_\mathrm{r}, \xi) = b\left(\Phi^{-1}(x_\mathrm{r}, \xi)\right) \cdot \nabla\left(a\left(\Phi^{-1}(x_\mathrm{r}, \xi)\right) \cdot \nabla\sigma\left(\Phi^{-1}(x_\mathrm{r}, \xi)\right)\right).$$

Note that, as a consequence of the uniform relative degree assumption, one has that $g \neq 0$. In the literature, see for instance [6], subsystem (16.24b), (16.24c) is the so-called *auxiliary system*. Since a_r, f, g are continuous functions and Φ_Ω is a bounded set, one has also that

$$\begin{aligned} |f(x_\mathrm{r}, \xi)| &\leq \mathcal{F} \\ g(x_\mathrm{r}, \xi) &\leq \mathcal{G}_{\max}, \end{aligned} \tag{16.25}$$

$\mathcal{F}$ and $\mathcal{G}_{\max}$ being known positive constants. Moreover, the following assumption is made on the uncertain function g.

Assumption 16.5 The uncertain function g can be lower bounded as

$$g(x_\mathrm{r}, \xi) \geq \mathcal{G}_{\min}, \tag{16.26}$$

$\mathcal{G}_{\min}$ being a known positive constant.

Now, a preliminary control problem can be formulated.

Problem 16.3 Let Assumptions 16.1 and 16.5 hold. Relying on (16.1) and (16.24)–(16.26), design a feedback control law

$$u^\star = \kappa(\sigma, \dot{\sigma}), \tag{16.27}$$

with the following property: $\forall\, x_0 \in \Omega$, $\exists\, t_\mathrm{r}^\star \geq t_0$ such that $\sigma = \dot{\sigma} = 0$, $\forall\, t \geq t_\mathrm{r}^\star$, in spite of the uncertainties.

Remark 16.7 The solution to Problem 16.3 is a control law capable of robustly enforcing an *ideal second order sliding mode* of system (16.1) in a finite time, according to Definition 16.10.

Taking into account the considerations made in Sect. 16.2.1, in the present chapter, instead of relying on time-triggered executions, we will introduce two different triggering conditions, transmitting data over the network only when such conditions are verified (Event-Triggered implementation).

Now, we can move from Problem 16.3 and formulate the problem which will be actually solved in this chapter.

Problem 16.4 Let Assumptions 16.1, 16.3 and 16.5 hold. Relying on (16.1) and (16.24)–(16.26), design a feedback control law

$$u = u(t_k) = \kappa(\sigma(t_k), \dot{\sigma}(t_k)) \quad \forall t \in [t_k,\, t_{k+1}[,\ k \in \mathbb{N}, \tag{16.28}$$

with the following property: $\forall\, x_0 \in \Omega$, $\exists\, t_\mathrm{r} \geq t_0$ such that $|\sigma| \leq \delta_1$, and $|\dot{\sigma}| \leq \delta_2$, $\forall\, t \geq t_\mathrm{r}$, in spite of the uncertainties, δ_1 and δ_2 being positive constants arbitrarily set.

Remark 16.8 The solution to Problem 16.4 is an ET control law capable of enforcing a *practical second order sliding mode* of system (16.1) in a finite time, according to Definition 16.11.

Before illustrating the features of the proposed control scheme, relying on Problem 16.4 let us introduce the following definition.

Definition 16.12 The convergence set for the solution $(\sigma, \dot{\sigma})$ to (16.24b), (16.24c) is

$$\mathcal{B} \triangleq \mathbb{R}^2 \setminus \{S_1 \cup S_2 \cup S_3 \cup S_4 \cup S_5\}, \tag{16.29}$$

where

$$\begin{aligned}
S_1 &\triangleq \left\{(\sigma, \dot{\sigma}) : \ |\dot{\sigma}| \geq \delta_2\right\} \\
S_2 &\triangleq \left\{(\sigma, \dot{\sigma}) : \ \sigma \geq \delta_1 , \ -\delta_2 < \dot{\sigma} \leq 0\right\} \\
S_3 &\triangleq \left\{(\sigma, \dot{\sigma}) : \ \sigma \leq -\delta_1 , \ 0 \leq \dot{\sigma} < \delta_2\right\} \\
S_4 &\triangleq \left\{(\sigma, \dot{\sigma}) : \ \sigma \geq -\frac{\dot{\sigma}\,|\dot{\sigma}|}{2\alpha_{\mathrm{r}}} + \delta_1 , \ 0 < \dot{\sigma} < \delta_2\right\} \\
S_5 &\triangleq \left\{(\sigma, \dot{\sigma}) : \ \sigma \leq -\frac{\dot{\sigma}\,|\dot{\sigma}|}{2\alpha_{\mathrm{r}}} - \delta_1 , \ -\delta_2 < \dot{\sigma} < 0\right\},
\end{aligned}$$

δ_1, δ_2 being positive constants arbitrarily set, and α_{r} being a positive constant defined as

$$\alpha_{\mathrm{r}} \triangleq \mathcal{G}_{\min}(U_{\max} - \mathcal{D}^{\sup}) - \mathcal{F} > 0, \tag{16.30}$$

where $U_{\max}$ is the control amplitude (see Fig. 16.4).

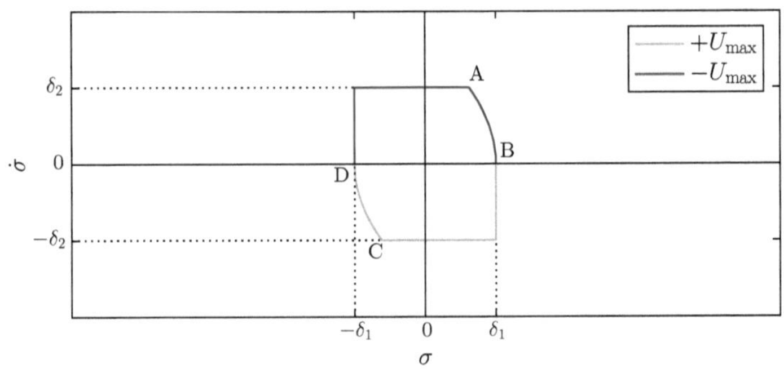

Fig. 16.4 Representation of the convergence set $\mathcal{B}$

16.3.2 The Proposed Control Scheme

The control scheme proposed to solve Problem 16.4 is analogous to the scheme reported in Fig. 16.1, where the dashed path means that the corresponding signals are transmitted over the network only at the triggering time instants t_k. The control scheme contains two key blocks: the *Smart Sensor* and the *Controller*. The signals transmitted by the Smart Sensor to the Controller (through the communication network) are specified in the following.

The Smart Sensor

The considered sensor is smart in the sense that it has some computation capability, i.e., it is able to compute σ, $\dot{\sigma}$, and verify two different triggering conditions. The first one being used only during the reaching of the convergence set (16.29), the second one being used for the rest of the control interval.

Triggering Condition 1

For any $(\sigma, \dot{\sigma}) \notin \{\mathcal{B} \cup \partial\mathcal{B}\}$ the adopted triggering condition is

$$\text{sign}\left(\sigma(t) - \frac{1}{2}\sigma_{\text{M}_i}\right) \neq \text{sign}\left(\sigma(t_k) - \frac{1}{2}\sigma_{\text{M}_i}\right), \tag{16.31}$$

$\{\sigma_{\text{M}_i}\}$ denoting the sequence of the extremal values of the sliding variable, i.e., $\sigma_{\text{M}_i} = \sigma(t_{\text{M}_i})$ such that $\dot{\sigma}(t_{\text{M}_i}) = 0$.

The switch in Fig. 16.1 is closed only when (16.31) holds, i.e., when the Smart Sensor generates the triggering signal. More precisely, when (16.31) is verified, $\text{sign}(\sigma - \frac{1}{2}\sigma_{\text{M}_i})$ is transmitted by the Smart Sensor to the Controller (through the communication network). Note that the Smart Sensor checks the Triggering Condition 1 only during the reaching phase, i.e., before $(\sigma, \dot{\sigma})$ reaches $\partial\mathcal{B}$. For the rest of the control interval a second triggering condition is adopted.

Triggering Condition 2

For any $(\sigma, \dot{\sigma}) \in \{\mathcal{B} \cup \partial\mathcal{B}\}$ the adopted triggering condition is

$$(\sigma, \dot{\sigma}) \in \partial\mathcal{B}, \tag{16.32}$$

$\mathcal{B}$ being the desired convergence set for the solution $(\sigma, \dot{\sigma})$ (see Definition 16.12).

When (16.31) holds, $\text{sign}(\dot{\sigma})$ is transmitted by the Smart Sensor to the Controller (through the communication network).

The Controller

The proposed control strategy is based on two different control laws. The first one is used together with (16.31) only during the reaching phase, while the second one is used together with (16.32) for the rest of the control interval.

Control Law 1

In analogy with the Suboptimal SOSM control algorithm in [6], and relaying on (16.28)–(16.31), for any $(\sigma, \dot{\sigma}) \notin \{\mathcal{B} \cup \partial\mathcal{B}\}$, the control law we propose to solve Problem 16.4 can be expressed as

$$u = u(t_k) = -U_{\max}\,\text{sign}\left(\sigma(t_k) - \frac{1}{2}\sigma_{\mathrm{M}_i}\right) \qquad \forall\, t \in [t_k, t_{k+1}[,\, k \in \mathbb{N}\,. \tag{16.33}$$

According to [6], in order to enforce a SOSM, it is necessary to select the control amplitude as

$$U_{\max} > \max\left(\frac{\mathcal{F}}{\alpha^* \mathcal{G}_{\min}} + \frac{\mathcal{D}^{\text{sup}}}{\alpha^*}; \frac{4\mathcal{F} + (3\mathcal{G}_{\min} + \mathcal{G}_{\max})\mathcal{D}^{\text{sup}}}{3\mathcal{G}_{\min} - \alpha^* \mathcal{G}_{\max}}\right) \tag{16.34}$$

with

$$\alpha^* \in (0, 1] \cap \left(0, \frac{3\mathcal{G}_{\min}}{\mathcal{G}_{\max}}\right)$$

Note that the control signal is transmitted by the Controller to the plant (through the communication network) only when (16.31) holds.

When $(\sigma, \dot{\sigma})$ reaches (for the first time) the boundary $\partial\mathcal{B}$, the Smart Sensor communicates (only one time) to the Controller that a second control law is applied for the rest of the control interval.

Control Law 2

Relaying on (16.28)–(16.30) and (16.32), for any $(\sigma, \dot{\sigma}) \in \{\mathcal{B} \cup \partial\mathcal{B}\}$, the control law we propose to solve Problem 16.4 can be expressed as

$$u = u(t_k) = -U_{\max}\,\text{sign}(\dot{\sigma}(t_k)) \qquad \forall\, t \in [t_k, t_{k+1}[,\, k \in \mathbb{N}\,, \tag{16.35}$$

with $U_{\max}$ as in (16.34).

Note that the control signal is transmitted by the Controller to the plant (through the communication network) only when (16.32) holds.

Remark 16.9 Note that, when the system (16.1) has unitary relative degree, in order to perform the so-called “chattering alleviation” [9, 13, 14, 33], the foregoing control solution can be analogously applied by artificially increasing the relative degree of the system.

16.3.3 Stability Analysis

In this section, the stability properties of system (16.1) controlled via the proposed ET-SOSMC strategy are analyzed. To this end, it is convenient to introduce the following definitions.

Definition 16.13 The solution $(\sigma, \dot{\sigma})$ to the auxiliary system (16.24b), (16.24c) is said to be ultimately bounded with respect to the convergence set $\{\mathcal{B} \cup \partial\mathcal{B}\}$ if

$$\forall x_0 \in \Omega, \ \exists\, t_r \geq t_0 : (\sigma, \dot{\sigma}) \in \{\mathcal{B} \cup \partial\mathcal{B}\} \quad \forall t \geq t_r . \tag{16.36}$$

Definition 16.14 Let $(\sigma, \dot{\sigma})$ be the solution to the auxiliary system (16.24b), (16.24c) starting from the initial condition $(\sigma(t_0), \dot{\sigma}(t_0))$. The set $\{\mathcal{B} \cup \partial\mathcal{B}\}$ is said to be positively invariant if

$$(\sigma(t_0),\ \dot{\sigma}(t_0)) \in \{\mathcal{B} \cup \partial\mathcal{B}\} \Rightarrow (\sigma, \dot{\sigma}) \in \{\mathcal{B} \cup \partial\mathcal{B}\} \ \forall t \geq t_0 . \tag{16.37}$$

Before showing the theoretical results, the following assumption is made on the initial conditions of the auxiliary system (16.24b), (16.24c).

Assumption 16.6 Given the auxiliary system (16.24b), (16.24c), let the sign of the initial conditions $(\sigma(t_0), \dot{\sigma}(t_0))$ be known.

Now, making reference to the auxiliary system (16.24b), (16.24c) the following results can be proven.

Lemma 16.3 *Let Assumptions 16.1, 16.5 and 16.6 hold with* $(\sigma(t_0), \dot{\sigma}(t_0)) \notin \{\mathcal{B} \cup \partial\mathcal{B}\}$, δ_1 *and* δ_2 *in* (16.29) *being positive constants arbitrarily set. Given the auxiliary system (16.24b), (16.24c) controlled by* (16.33), (16.34) *with the triggering condition* (16.31)*, then, the solution* $(\sigma, \dot{\sigma})$ *to (16.24b), (16.24c) is steered to the convergence set* $\{\mathcal{B} \cup \partial\mathcal{B}\}$ *in a finite time.*

Proof For the proof of this Lemma we refer to [6], where it is proved that the Suboptimal control law generates the contraction of the extremal values of the sliding variable. In this chapter, the event-triggered implementation of the Suboptimal control law does not disrupt the contraction property, since the sign changes are performed exactly at the same time instants when they occur in the non event-triggered implementation. In the other time instants the sign of the control is kept constant by the ZOH*. □

Lemma 16.4 *Let Assumptions 16.1, 16.5 and 16.6 hold with* $(\sigma(t_0), \dot{\sigma}(t_0)) \in \{\mathcal{B} \cup \partial\mathcal{B}\}$, δ_1 *and* δ_2 *in* (16.29) *being positive constants arbitrarily set. Given the auxiliary system (16.24b), (16.24c) controlled by* (16.35)*, with the triggering condition* (16.32)*, then, the convergence set* $\{\mathcal{B} \cup \partial\mathcal{B}\}$ *is a positively invariant set.*

Proof Since $\text{sign}(\dot{\sigma})$ and u are updated only when (16.32) holds, i.e., when $(\sigma, \dot{\sigma}) \in \partial\mathcal{B}$, the Lemma will be proven showing that for any $(\sigma(t_0), \dot{\sigma}(t_0)) \in \partial\mathcal{B}$, the vector field $(\dot{\sigma}, \ddot{\sigma})$ never points outside $\mathcal{B}$. Let $\partial\mathcal{B}^+$ denote $(\sigma, \dot{\sigma}) \in \partial\mathcal{B} : \dot{\sigma} > 0$, and $\partial\mathcal{B}^-$ denote $(\sigma, \dot{\sigma}) \in \partial\mathcal{B} : \dot{\sigma} < 0$ (in Fig. 16.4, $\partial\mathcal{B}^+$ is blue and $\partial\mathcal{B}^-$ is red). Assume that $(\sigma(t_0), \dot{\sigma}(t_0)) \in \partial\mathcal{B}^-$. The vector field is $(\dot{\sigma}, f + g(u + d))$ with $\dot{\sigma} < 0$ and, according to (16.35), $u = U_{\max}$. Then, $\ddot{\sigma} \geq \alpha_r > 0$, so that the vector field points up-left, that is inside $\mathcal{B}$. Note that, if $(\sigma(t_0), \dot{\sigma}(t_0)) \in \overline{\mathrm{CD}}$ (all the points on this curve verify $\sigma = -\frac{\dot{\sigma}|\dot{\sigma}|}{2\alpha_r} - \delta_1$), then the vector field can be, at most, tangent to $\overline{\mathrm{CD}}$, never pointing outside $\mathcal{B}$. Analogous considerations can be done if $(\sigma(t_0), \dot{\sigma}(t_0)) \in \partial\mathcal{B}^+$. □

Relying now on Lemmas 16.3 and 16.4, one can prove the major result concerning the evolution of the auxiliary system (16.24b), (16.24c) controlled via the proposed strategy.

Theorem 16.4 *Let Assumptions 16.1, 16.5 and 16.6 hold. Given the auxiliary system (16.24b), (16.24c) controlled by* (16.33), (16.34) *with the triggering condition* (16.31) *when* $(\sigma, \dot{\sigma}) \notin \{\mathcal{B} \cup \partial\mathcal{B}\}$, *and by* (16.35) *with the triggering condition* (16.32) *when* $(\sigma, \dot{\sigma}) \in \{\mathcal{B} \cup \partial\mathcal{B}\}$, *then, the solution* $(\sigma, \dot{\sigma})$ *to (16.24b), (16.24c) is ultimately bounded with respect to the desired convergence set* $\{\mathcal{B} \cup \partial\mathcal{B}\}$, δ_1 *and* δ_2 *in* (16.29) *being arbitrarily set positive constants.*

Proof The proof is a straightforward consequence of Lemmas 16.3 and 16.4. By virtue of Lemma 16.3, there exists a time instant t_r when the trajectory $(\sigma, \dot{\sigma})$ enters $\{\mathcal{B} \cup \partial\mathcal{B}\}$. Then, by virtue of Lemma 16.4, $\forall t \geq t_r$, $(\sigma, \dot{\sigma})$ is ultimately bounded with respect to the convergence set $\{\mathcal{B} \cup \partial\mathcal{B}\}$. □

Remark 16.10 Note that the proposed control scheme, because of its ET nature, cannot generate an ideal second order sliding mode (see Definition 16.10), but only a practical second order sliding mode (see Definition 16.11). However, in analogy with the Regularization Theorem [58], it can be proven that, also in case of second order sliding mode, the state of system (16.1) is ultimately bounded. This implies that Problem 16.4 is equivalent to the problem of designing a bounded control such that, according to Definition 16.8, the origin of system (16.1) is practically stable.

Now, since the triggering time instants are implicitly defined and only known at the execution times, we prove the existence of lower-bounds for the *inter-event* times [53]. Let $\tau_{\min}$ be the minimum inter-event time when $(\sigma, \dot{\sigma}) \in \{\mathcal{B} \cup \partial\mathcal{B}\}$, the following result can be proved.

Theorem 16.5 *Let Assumptions 16.1, 16.5 and 16.6 hold, and let* δ_1 *and* δ_2 *in* (16.29) *be positive constants arbitrarily set. Given the auxiliary system (16.24b), (16.24c) controlled by* (16.35) *with the triggering condition* (16.32) *when* $(\sigma, \dot{\sigma}) \in \{\mathcal{B} \cup \partial\mathcal{B}\}$, *then, the inter-event times are lower bounded by*

$$\tau_{\min} = \frac{\delta_2}{\mathcal{F} + \mathcal{G}_{\max}(U_{\max} + \mathcal{D}^{\sup})}$$

Proof Since $\text{sign}(\dot{\sigma})$ and u are transmitted over the network only when the triggering condition (16.32) is verified, the theorem will be proven by computing the time interval $t_{k+1} - t_k$ that $\dot{\sigma}$ takes to evolve from 0 to δ_2 with the maximum acceleration $\alpha_R \triangleq \mathcal{G}_{\max}(U_{\max} + \mathcal{D}^{\text{sup}}) + \mathcal{F}$. Then, it yields

$$\begin{aligned}\dot{\sigma}(t_{k+1}) - \dot{\sigma}(t_k) &= \int_{t_k}^{t_{k+1}} \ddot{\sigma}_{\max}\, d\tau \\ \delta_2 - 0 &= \alpha_R\ (t_{k+1} - t_k) \\ \delta_2 &= (\mathcal{F} + \mathcal{G}_{\max}(U_{\max} + \mathcal{D}^{\text{sup}}))\tau_{\min}\ .\end{aligned} \tag{16.38}$$

Analogous considerations can be used if we consider the evolution of $\dot{\sigma}$ from 0 to $-\delta_2$. □

Remark 16.11 Note that Theorem 16.5 guarantees that the time elapsed between consecutive triggering events does not become arbitrarily small, avoiding the notorious Zeno behaviour [2, 43]. In practical cases, this result is very useful to assess the feasibility of the proposed scheduling policy.

Remark 16.12 Note also that, by virtue of the contraction property of the Suboptimal control law [6], there exists a sequence of time instants when an extremal value of the sliding variable occurs. Then, one can conclude that also during the reaching of the desired convergence set $\{\mathcal{B} \cup \partial\mathcal{B}\}$, the notorious Zeno behaviour is avoided.

Taking now into account the possible occurrence of delayed data transmissions, due to the presence of the communication network, the following result can be proved. More precisely, relying on Problem 16.4 we will prove that by modifying the triggering condition (16.32), the auxiliary state-space trajectory is ultimately bounded with respect to the desired convergence set $\{\mathcal{B} \cup \partial\mathcal{B}\}$.

Theorem 16.6 *Let Assumptions 16.1, 16.3, 16.5 and 16.6 hold. Given the auxiliary system (16.24b), (16.24c) controlled by* (16.35) *when* $(\sigma, \dot{\sigma}) \in \{\mathcal{B} \cup \partial\mathcal{B}\}$, *then, for any desired*

$$\begin{aligned}\delta_2 &> (\mathcal{F} + \mathcal{G}_{\max}(U_{\max} + \mathcal{D}^{\text{sup}}))\Delta_{\max} \\ \delta_1 &> \left(\frac{\delta_2^2}{2} - \frac{(\delta_2 - (\mathcal{F} + \mathcal{G}_{\max}(U_{\max} + \mathcal{D}^{\text{sup}}))\Delta_{\max})^2}{2}\right)\left(\frac{1}{\alpha_R} + \frac{1}{\alpha_r}\right),\end{aligned}$$

the triggering condition

$$(\sigma, \dot{\sigma}) \in \partial\mathcal{B}'\ ,$$

with

$$\mathcal{B}' \triangleq \mathbb{R}^2 \setminus \{S'_1 \cup S'_2 \cup S'_3 \cup S'_4 \cup S'_5\}$$

and

$$
\begin{aligned}
S'_1 &\triangleq \left\{(\sigma,\, \dot\sigma) : \ |\dot\sigma| \geq \delta'_2\right\} \\
S'_2 &\triangleq \left\{(\sigma,\, \dot\sigma) : \ \sigma \geq \delta_1\,,\ -\delta'_2 < \dot\sigma \leq 0\right\} \\
S'_3 &\triangleq \left\{(\sigma,\, \dot\sigma) : \ \sigma \leq -\delta_1\,,\ 0 \leq \dot\sigma < \delta'_2\right\} \\
S'_4 &\triangleq \left\{(\sigma,\, \dot\sigma) : \ \sigma \geq -\frac{\dot\sigma|\dot\sigma|}{2\alpha_\mathrm{r}} + \delta'_1\,,\ 0 < \dot\sigma < \delta'_2\right\} \\
S'_5 &\triangleq \left\{(\sigma,\, \dot\sigma) : \ \sigma \leq -\frac{\dot\sigma|\dot\sigma|}{2\alpha_\mathrm{r}} - \delta'_1\,,\ -\delta'_2 < \dot\sigma < 0\right\} \\
\delta'_2 &\triangleq \delta_2 - (\mathcal{F} + \mathcal{G}_{\max}(U_{\max} + \mathcal{D}^{\mathrm{sup}}))\Delta_{\max} > 0 \\
\delta'_1 &\triangleq \delta_1 - \left(\frac{\delta_2^2}{2} - \frac{(\delta'_2)^2}{2}\right)\left(\frac{1}{\alpha_\mathrm{R}} + \frac{1}{\alpha_\mathrm{r}}\right) > 0
\end{aligned}
$$

enforces the following relation

$$
(\sigma, \dot\sigma) \in \{\mathcal{B} \cup \partial\mathcal{B}\} \quad \forall t \geq t'_\mathrm{r}, \tag{16.39}
$$

t'_r being the reaching time instant of $\left\{\mathcal{B}' \cup \partial\mathcal{B}'\right\}$.

Proof In analogy with Lemma 16.3, it follows that there exists a time instant t'_r when $(\sigma, \dot\sigma)$ enters the inner convergence set $\left\{\mathcal{B}' \cup \partial\mathcal{B}'\right\}$. Suppose now that the transmission of $\dot\sigma(t_k) = \delta'_2$ occurs with maximum time delay $\Delta_{\max}$. Moreover, assume that the sliding variable evolves with constant maximum acceleration $\alpha_\mathrm{R} \triangleq \mathcal{G}_{\max}(U_{\max} + \mathcal{D}^{\mathrm{sup}}) + \mathcal{F}$. In order to enforce relation (16.39), we impose that $\dot\sigma(t_k + \Delta_{\max}) = \delta_2$. Then, one has that

$$
\begin{aligned}
\dot\sigma(t_k + \Delta_{\max}) - \dot\sigma(t_k) &= \int_{t_k}^{t_k+\Delta_{\max}} \ddot\sigma_{\max}\, d\tau \\
\delta_2 - \delta'_2 &= \alpha_\mathrm{R}\Delta_{\max} \\
\delta'_2 &= \delta_2 - \alpha_\mathrm{R}\Delta_{\max}.
\end{aligned} \tag{16.40}
$$

Now, one can easily verify that the parabolic auxiliary state-space trajectory passing through point A in Fig. 16.4 with acceleration α_R is

$$
\sigma = \frac{\dot\sigma|\dot\sigma|}{2\alpha_\mathrm{R}} + \delta_1 - \frac{\delta_2^2}{2}\left(\frac{1}{\alpha_\mathrm{R}} + \frac{1}{\alpha_\mathrm{r}}\right).
$$

Then, the intersection point of this curve with $\dot\sigma = \delta'_2$ is

$$
\left(\sigma = \delta_1 + \frac{(\delta'_2)^2}{2\alpha_\mathrm{R}} - \frac{\delta_2^2}{2}\left(\frac{1}{\alpha_\mathrm{R}} + \frac{1}{\alpha_\mathrm{r}}\right),\ \dot\sigma = \delta'_2\right) \tag{16.41}
$$

Finally, by imposing that the curve $\sigma = -\frac{\dot{\sigma}|\dot{\sigma}|}{2\alpha_r} + \delta_1'$ passes through the point (16.41), one can compute the value of δ_1'. □

Remark 16.13 Note that in case of delayed transmissions, the lower bound $\tau_{\min}$ can be obtained by using δ_2' instead of δ_2 in Theorem 16.5.

16.3.4 Illustrative Example

In this section, an illustrative example is briefly discussed. Consider the perturbed double integrator (16.23) with $d = \mathcal{D}^{\text{sup}} \cos(2t)$, $\mathcal{D}^{\text{sup}} = 2.4$. Let the initial condition be $x(0) = \begin{bmatrix} 1 & 0 \end{bmatrix}^T$, and the sliding variable be $\sigma = x_1$. Relying on system (16.23) it is possible to set the bounds in (16.25) and (16.26) equal to $\mathcal{F} = 0$, $\mathcal{G}_{\min} = \mathcal{G}_{\max} = 1$. Then, according to (16.34), the control amplitude $U_{\max}$ is selected equal to 5, with $\alpha^* = 1$. Moreover, the convergence set $\mathcal{B}$ (see Definition 16.12) is chosen by selecting $\delta_1 = 0.005$, $\delta_2 = 0.1$, and, according to (16.30), $\alpha_r = 2.6$.

In Fig. 16.5 the auxiliary state-space trajectory is shown. Note that, according to Theorem 16.4, $(\sigma, \dot{\sigma})$ is ultimately bounded with respect to the convergence set $\mathcal{B}$. Moreover, in Fig. 16.5 the time evolution of the sliding variable and the time evolution of its first time derivative are reported, showing that they are ultimately bounded with respect to the boundary layer of size δ_1 and δ_2, respectively. Finally, the inter-event times $\tau_k = t_{k+1} - t_k$ are also shown. In particular, one can appreciate that the minimum inter-event time $\tau_{\min} = 0.0135$ s, according to Theorem 16.5, is a lower bound for the inter-event times. Note that, considering a sampling time $T_s = 1\text{e-}4$ s, and a simulation time $T = 10$ s, the number of transmissions is 203, i.e., 99.8% less than the number required by the conventional (i.e., time-driven) implementation. Obviously, reducing the size of the convergence set implies the improvement of the convergence accuracy. Yet, a larger number of transmissions could be required. The correct balance between convergence accuracy and transmission load has to be searched depending on the specific application.

In Fig. 16.6 the auxiliary state-space trajectory in presence of transmissions with maximum time delay $\Delta_{\max} = 0.005$ s acting from $t = 2$ s to $t = 4$ s is shown. Note that, by selecting $\delta_1' = 0.0034$ and $\delta_2' = 0.063$ (see Theorem 16.6), even in presence of maximum time delay, $(\sigma, \dot{\sigma})$ is ultimately bounded with respect to the desired convergence set $\mathcal{B}$. Moreover, in Fig. 16.6 are reported the time evolution of the sliding variable, the time evolution of its first time derivative and the inter-event times, with $\tau_{\min} = 0.0085$ s. In this case, the number of transmissions is 300.

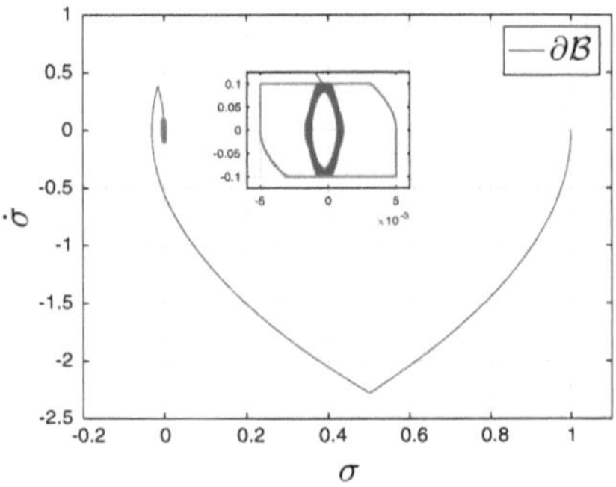

(a) Auxiliary state space trajectory

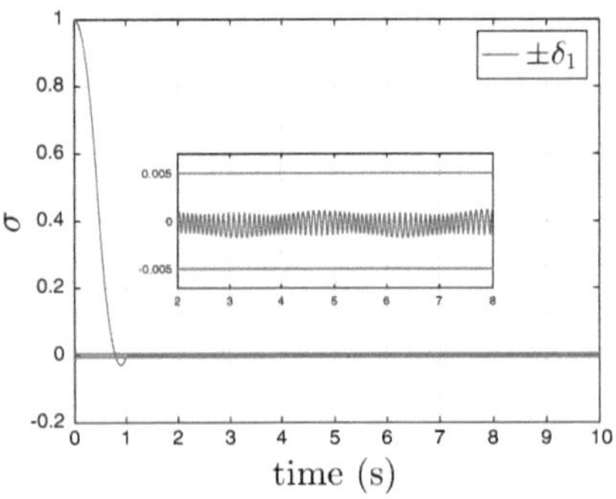

(b) Sliding variable

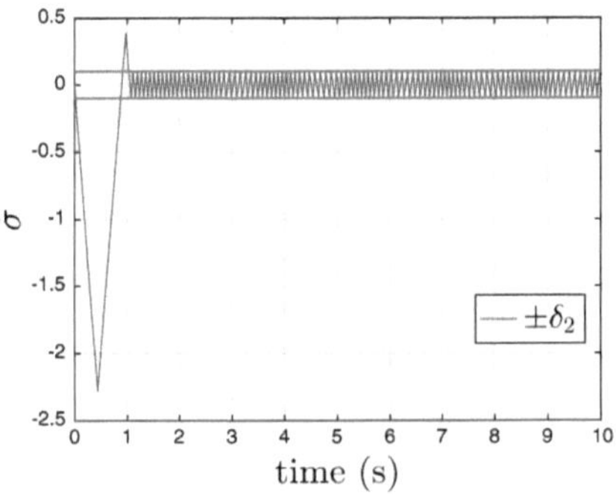

(c) First time derivative of the sliding variable

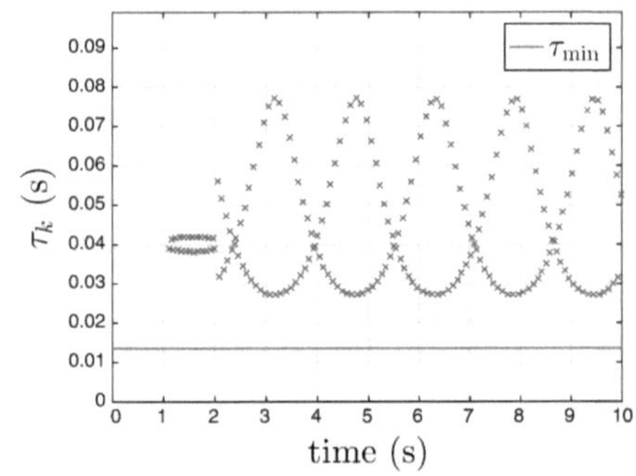

(d) Inter-event times

Fig. 16.5 Event-Triggered Second Order Sliding Mode Control in absence of delayed transmissions. **a** Auxiliary state-space trajectory, with zoom and visualization of the convergence set $\mathcal{B}$. **b** Time evolution of the sliding variable, with zoom and visualization of the boundary layer of size δ_1. **c** Time evolution of the first time derivative of the sliding variable, with visualization of the boundary layer of size δ_2. **d** Inter-event times $\tau_k = t_{k+1} - t_k$, with visualization of the minimum inter-event time $\tau_{\min}$

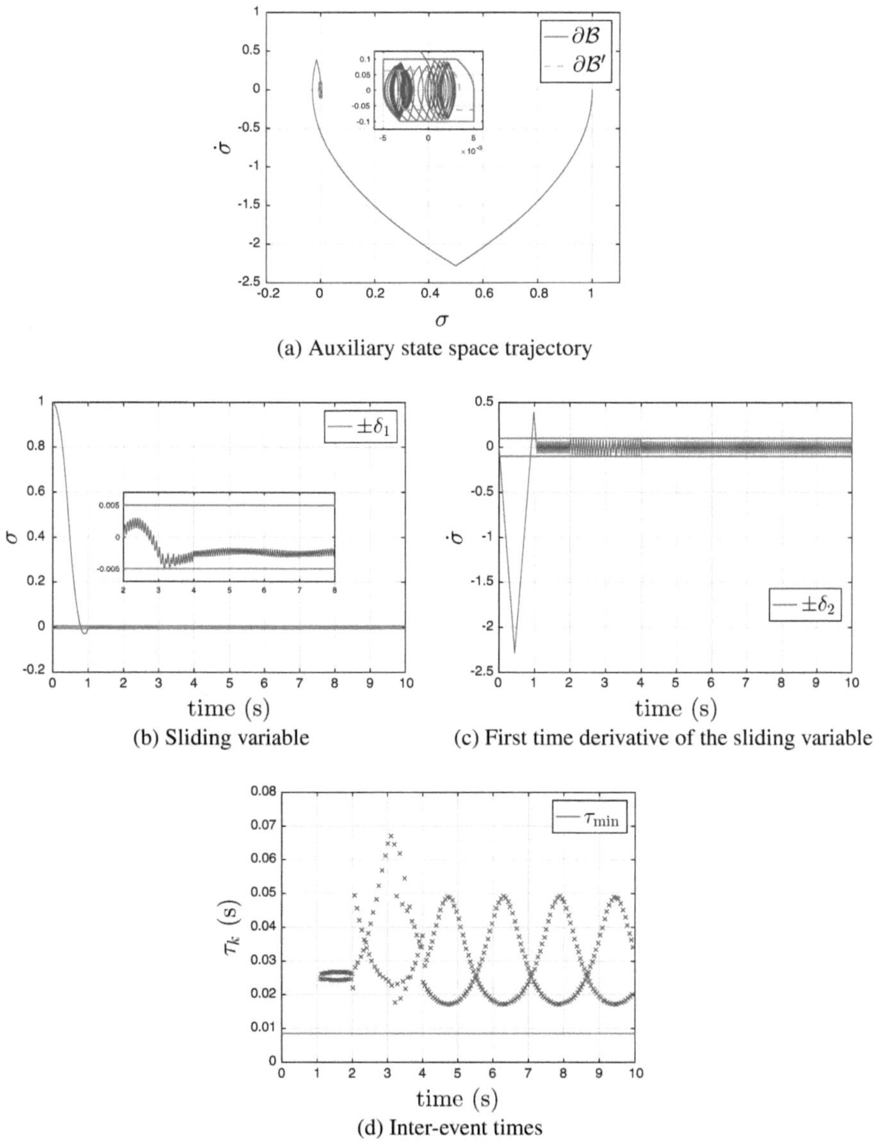

Fig. 16.6 Event-Triggered Second Order Sliding Mode Control in presence of delayed transmissions. **a** Auxiliary state-space trajectory, with zoom and visualization of both the desired convergence set $\mathcal{B}$, and the convergence set $\mathcal{B}'$ adopted for the triggering condition. **b** Time evolution of the sliding variable, with zoom and visualization of the boundary layer of size δ_1. **c** Time evolution of the first time derivative of the sliding variable, with visualization of the boundary layer of size δ_2. **d** Inter-event times $\tau_k = t_{k+1} - t_k$, with visualization of the minimum inter-event time $\tau_{\min}$

16.4 Conclusions

In this chapter, the Event-Triggered control approach and the Sliding Mode control methodology are coupled to design robust control schemes for nonlinear uncertain systems including communication networks. The main objective is indeed to reduce the data transmission effort, while guaranteeing satisfactory stability properties of the controlled system even in presence of modelling uncertainties and delayed transmissions due to the network unavailability. Both the Event-Triggered Sliding Mode Control and the Event-Triggered Second Order Sliding Mode Control schemes are based on suitable triggering conditions. The stability properties of the proposed control schemes are theoretically analyzed, proving the ultimate boundedness of the auxiliary system state, which implies the ultimate boundedness of the solution of the controlled system. Moreover, lower bounds for the time elapsed between consecutive triggering events are provided, in order to guarantee the avoidance of the notorious Zeno behaviour.

References

1. Alur, R., Arzen, K.E., Baillieul, J., Henzinger, T., Hristu-Varsakelis, D., Levine, W.S.: Handbook of Networked and Embedded Control Systems. Springer Science & Business Media. Birkhäuser, Boston (2007)
2. Ames, A.D., Tabuada, P., Sastry, S.: On the stability of zeno equilibria. In: International Workshop on Hybrid Systems: Computation and Control, pp. 34–48. Springer, Berlin (2006)
3. Aström, K.J.: Event based control. In: Analysis and Design of Nonlinear Control Systems, pp. 127–147. Springer, Berlin (2008)
4. Bartolini, G., Ferrara, A., Usai, E.: Output tracking control of uncertain nonlinear second-order systems. Automatica **33**(12), 2203–2212 (1997)
5. Bartolini, G., Ferrara, A., Pisano, A., Usai, E.: Adaptive reduction of the control effort in chattering-free sliding-mode control of uncertain nonlinear systems. Appl. Math. Comput. Sci. **8**(1), 51–71 (1998)
6. Bartolini, G., Ferrara, A., Usai, E.: Chattering avoidance by second-order sliding mode control. IEEE Trans. Autom. Control **43**(2), 241–246 (1998)
7. Bartolini, G., Ferrara, A., Usai, E.: On boundary layer dimension reduction in sliding mode control of siso uncertain nonlinear systems. In: Proceedings of the 1998 IEEE International Conference on Control Applications, 1998, vol. 1, pp. 242–247. IEEE (1998)
8. Bartolini, G., Ferrara, A., Levant, A., Usai, E.: On second order sliding mode controllers. Variable Structure Systems. Sliding Mode and Nonlinear Control, pp. 329–350. Springer, Berlin (1999)
9. Bartolini, G., Ferrara, A., Usai, E., Utkin, V.I.: On multi-input chattering-free second-order sliding mode control. IEEE Trans. Autom. Control **45**(9), 1711–1717 (2000)
10. Behera, A.K., Bandyopadhyay, B.: Event based sliding mode control with quantized measurement. In: 2015 International Workshop on Recent Advances in Sliding Modes (RASM), pp. 1–6. IEEE (2015)
11. Behera, A.K., Bandyopadhyay, B.: Event-triggered sliding mode control for a class of nonlinear systems. Int. J. Control **89**(9), 1916–1931 (2016)
12. Behera, A.K., Bandyopadhyay, B., Xavier, N., Kamal, S.: Event-triggered sliding mode control for robust stabilization of linear multivariable systems. In: Recent Advances in Sliding Modes: From Control to Intelligent Mechatronics, pp. 155–175. Springer, Berlin (2015)

13. Boiko, I.M.: Analysis of chattering in sliding mode control systems with continuous boundary layer approximation of discontinuous control. In: American Control Conference (ACC), 2011, pp. 757–762. IEEE (2011)
14. Boiko, I., Fridman, L., Pisano, A., Usai, E.: Analysis of chattering in systems with second-order sliding modes. IEEE Trans. Autom. Control **52**(11), 2085–2102 (2007)
15. Buttazzo, G., Cervin, A.: Comparative assessment and evaluation of jitter control methods. In: Proceedings of the 15th conference on Real-Time and Network Systems, pp. 163–172 (2007)
16. Cucuzzella, M., Incremona, G.P., Ferrara, A.: Design of robust higher order sliding mode control for microgrids. IEEE J. Emerg. Sel. Top. Circuits Syst. **5**(3), 393–401 (2015)
17. Cucuzzella, M., Incremona, G.P., Ferrara, A.: Master-slave second order sliding mode control for microgrids. In: American Control Conference (ACC), 2015, pp. 5188–5193. IEEE (2015)
18. Cucuzzella, M., Incremona, G.P., Ferrara, A.: Third order sliding mode voltage control in microgrids. In: 2015 European Control Conference (ECC), pp. 2384–2389. IEEE (2015)
19. Cucuzzella, M., Ferrara, A.: Event-triggered second order sliding mode control of nonlinear uncertain systems. In: 2016 European Control Conference (ECC), pp. 295–300. IEEE (2016)
20. Cucuzzella, M., Incremona, G.P., Ferrara, A.: Event-triggered sliding mode control algorithms for a class of uncertain nonlinear systems: experimental assessment. In: American Control Conference (ACC), 2016, pp. 6549–6554. IEEE (2016)
21. Cucuzzella, M., Incremona, G.P., Guastalli, M., Ferrara, A.: Sliding mode control for maximum power point tracking of photovoltaic inverters in microgrids. In: 2016 IEEE 55th Conference on Decision and Control (CDC), pp. 7294–7299. IEEE (2016)
22. Cucuzzella, M., Rosti, S., Cavallo, A., Ferrara, A.: Decentralized sliding mode voltage control in DC microgrids. In: Proceedings of the American Control Conference (2017)
23. Cucuzzella, M., Trip, S., De Persis, C., Ferrara, A.: Distributed second order sliding modes for optimal load frequency control. In: Proceeding of the American Control Conference (2017)
24. Demirel, B., Gupta, V., Johansson, M.: On the trade-off between control performance and communication cost for event-triggered control over lossy networks. In: 2013 European Control Conference (ECC), pp. 1168–1174. IEEE (2013)
25. Dinuzzo, F., Ferrara, A.: Finite-time output stabilization with second order sliding modes. Automatica **45**(9), 2169–2171 (2009)
26. Dinuzzo, F., Ferrara, A.: Higher order sliding mode controllers with optimal reaching. IEEE Trans. Autom. Control **54**(9), 2126–2136 (2009)
27. Edwards, C., Spurgeon, S.: Sliding Mode Control: Theory and Applications. CRC Press, Boca Raton (1998)
28. Ferrara, A., Incremona, G.P., Magni, L.: Model-based event-triggered robust MPC/ISM. In: 2014 European Control Conference (ECC), pp. 2931–2936. IEEE (2014)
29. Ferrara, A., Sacone, S., Siri, S.: Event-triggered strategies for the networked control of freeway traffic systems. In: 2014 European Control Conference (ECC), pp. 2594–2599. IEEE (2014)
30. Ferrara, A., Sacone, S., Siri, S.: Design of networked freeway traffic controllers based on event-triggered control concepts. International Journal of Robust and Nonlinear Control (2015)
31. Ferrara, A., Sacone, S., Siri, S.: Event-triggered model predictive schemes for freeway traffic control. Trans. Res. Part C: Emerg. Technol. **58**, 554–567 (2015)
32. Ferrara, A., Sacone, S., Siri, S.: Model-based event-triggered control for freeway traffic systems. In: 2015 International Conference on Event-based Control, Communication, and Signal Processing (EBCCSP), pp. 1–6. IEEE (2015)
33. Fridman, L.: The problem of Chattering: an Averaging Approach. Variable Structure Systems, Sliding Mode and Nonlinear Control, pp. 363–386. Berlin, Berlin (1999)
34. Fridman, L., Levant, A.: Higher order sliding modes. Sliding Mode Control Eng. **11**, 53–102 (2002)
35. Garcia, E., Antsaklis, P.J.: Model-based event-triggered control with time-varying network delays. In: 2011 50th IEEE Conference on Decision and Control and European Control Conference (CDC-ECC), pp. 1650–1655. IEEE (2011)
36. Gupta, R.A., Chow, M.Y.: Networked control system: overview and research trends. IEEE Trans. Ind. Electron. **57**(7), 2527–2535 (2010)

37. Heemels, W., Sandee, J., Van Den Bosch, P.: Analysis of event-driven controllers for linear systems. Int. J. Control **81**(4), 571–590 (2008)
38. Heemels, W., Johansson, K.H., Tabuada, P.: An introduction to event-triggered and self-triggered control. In: 2012 IEEE 51st Annual Conference on Decision and Control (CDC), pp. 3270–3285. IEEE (2012)
39. Hespanha, J.P., Naghshtabrizi, P., Xu, Y.: A survey of recent results in networked control systems. Proc. IEEE **95**(1), 138–162 (2007)
40. Incremona, G.P., Cucuzzella, M., Ferrara, A.: Adaptive suboptimal second-order sliding mode control for microgrids. Int. J. Control **89**(9), 1849–1867 (2016)
41. Incremona, G.P., Ferrara, A.: Adaptive model-based event-triggered sliding mode control. Int. J. Adapt. Control Signal Process. **30**(8–10), 1099–1117 (2016)
42. Incremona, G.P., Cucuzzella, M., Magni, L., Ferrara, A.: MPC with sliding mode control for the energy management system of microgrids. In: Proceedings of the 20th IFAC World Congress IFAC, 7658–7663 (2017)
43. Johansson, K.H., Lygeros, J., Sastry, S., Egerstedt, M.: Simulation of zeno hybrid automata. In: Proceedings of the 38th IEEE Conference on Decision and Control, 1999, vol. 4, pp. 3538–3543. IEEE (1999)
44. Ke-You, Y., Li-Hua, X.: Survey of recent progress in networked control systems. Acta Autom. Sin. **39**(2), 101–117 (2013)
45. Khalil, H.K.: Nonlinear Systems. Prentice-Hall, Upper Saddle River (1996)
46. La Salle, J., Lefschetz, S.: Stability by Liapunov's Direct Method with Applications by Joseph L Salle and Solomon Lefschetz, vol. 4. Elsevier, Academic Press, New York (2012)
47. Liu, G.P., Xia, Y., Chen, J., Rees, D., Hu, W.: Networked predictive control of systems with random network delays in both forward and feedback channels. IEEE Trans. Ind. Electron. **54**(3), 1282–1297 (2007)
48. Luck, R., Ray, A.: An observer-based compensator for distributed delays. Automatica **26**(5), 903–908 (1990)
49. Mazo, M., Tabuada, P.: On event-triggered and self-triggered control over sensor/actuator networks. In: CDC 2008. 47th IEEE Conference on Decision and Control, 2008, pp. 435–440. IEEE (2008)
50. Montestruque, L.A., Antsaklis, P.: Stability of model-based networked control systems with time-varying transmission times. IEEE Trans. Autom. Control **49**(9), 1562–1572 (2004)
51. Nilsson, J., Bernhardsson, B.: Analysis of real-time control systems with time delays. In: Proceedings of the 35th IEEE Conference on Decision and Control, 1996, vol. 3, pp. 3173–3178. IEEE (1996)
52. Rubagotti, M., Ferrara, A.: Second order sliding mode control of a perturbed double integrator with state constraints. In: American Control Conference (ACC), 2010, pp. 985–990. IEEE (2010)
53. Tabuada, P.: Event-triggered real-time scheduling of stabilizing control tasks. IEEE Trans. Autom. Control **52**(9), 1680–1685 (2007)
54. Tallapragada, P., Chopra, N.: Decentralized event-triggering for control of nonlinear systems. IEEE Trans. Autom. Control **59**(12), 3312–3324 (2014)
55. Tanelli, M., Ferrara, A.: Enhancing robustness and performance via switched second order sliding mode control. IEEE Trans. Autom. Control **58**(4), 962–974 (2013)
56. Trip, S., Cucuzzella, M., Ferrara, A., De Persis, C.: An energy function based design of second order sliding modes for automatic generation control. In: Proceedings of the 20th IFAC World Congress IFAC, 1211–12123 (2017)
57. Utkin, V.: Sliding mode control in discrete-time and difference systems. Variable Structure and Lyapunov Control, pp. 87–107. Springer, UK (1994)
58. Utkin, V.I.: Sliding Modes in Control and Optimization. Springer Science & Business Media, Berlin (2013)
59. Utkin, V., Guldner, J., Shi, J.: Sliding Mode Control in Electro-mechanical Systems, vol. 34. CRC Press, London (2009)

60. Wang, F.-Y., Liu, D.: Networked Control Systems: Theory and Applications. Springer, London (2008)
61. Yu, H., Antsaklis, P.J.: Event-triggered real-time scheduling for stabilization of passive and output feedback passive systems. In: American Control Conference (ACC), 2011, pp. 1674–1679. IEEE (2011)
62. Zhang, W., Branicky, M.S., Phillips, S.M.: Stability of networked control systems. IEEE Control Syst. **21**(1), 84–99 (2001)

Chapter 17
Hybrid-Impulsive Higher Order Sliding Mode Control

Yuri B. Shtessel, Fathi M. Aldukali and Frank Plestan

Abstract A hybrid-impulsive second order/higher order sliding mode (2-SMC/HOSM) control is explored in order to reduce dramatically the convergence time practically to zero, achieving instantaneous (or short time) convergence and uniformity. For systems of relative degree 2, the impulsive portion of the control function drives the system's output (the sliding variable) and it's derivative to zero instantaneously (or in short time) achieving a uniform convergence. Then the discontinuous state or output feedback stabilizes system's trajectory at the origin (or its close vicinity), while achieving the ideal or real second order sliding mode (2-SM). The Lyapunov analysis of the considered hybrid-impulsive-discontinuous systems is performed. Hybrid-impulsive continuous HOSM (CHOSM) control is studied in systems of *arbitrary relative degree* with impulsive action that achieves almost instantaneous convergence and uniformity. This approach allows reducing the CHOSM amplitude, since the task of compensating the initial conditions is addressed by the impulsive action. Two hybrid-impulsive 2-SMCs are studied in systems of *arbitrary relative degree* in a *reduced information environment*. Only "snap" knowledge of the all states is required to facilitate the impulsive action. The efficacy of studied hybrid-impulsive control algorithms is illustrated via simulations.

Y.B. Shtessel (✉) · F.M. Aldukali
Department of Electrical and Computer Engineering, The University of Alabama in Huntsville, Huntsville, AL 35899, USA
e-mail: shtessy@uah.edu

F.M. Aldukali
e-mail: fma0001@uah.edu

F. Plestan
Ecole Centrale de Nantes, Institut de Recherche en Communications et Cybernétique de Nantes - IRCCyN, Nantes, France
e-mail: Franck.Plestan@irccyn.ec-nantes.fr

J.B. Clempner and W. Yu (eds.), *New Perspectives and Applications of Modern Control Theory*, https://doi.org/10.1007/978-3-319-62464-8_17

17.1 Introduction

The insensitivity and finite time convergence properties enjoyed by sliding mode controllers (SMC) [25], make them a useful approach for systems with bounded perturbations. The additional properties of higher order sliding mode (HOSM) control with respect to SMC can be listed as [15, 17, 22, 25]:

1. the capability to handle systems with arbitrary relative degree,
2. generate the continuous control function by a price of artificial increase of relative degree,
3. the enhanced accuracy of the sliding variable stabilization, while the HOSM controller is implemented in discrete time,
4. and make HOSM control an attractive technique for a theoretical study and a practical implementation.

Providing finite time convergence, HOSM control cannot guaranty its uniformity. In other words, the finite convergence time depends on the initial conditions that very often are unknown. In the work [3], the authors propose the finite time convergent dynamics for the HOSM differentiator in order to provide the uniformity, i.e. to make the convergence time from an arbitrary initial condition to be uniformly bounded. The uniformity is achieved by a price of a very aggressive feedback when the norm of the initial conditions is large. The interest in impulsive control as a control solution that allows driving the system's states to the origin in a short time has increased over the past few years [2, 3, 5, 8–12, 18, 19, 21, 23, 24, 26, 27, 30–32]. There exist numerous practical tasks for which impulsive control is not just an option, but the only solution to achieve the required performance, since a large deviation from equilibrium often need to be corrected in very short time. Reaction control systems (RCS) for quick steering and attitude control of aerospace vehicles can serve as a good example of utility of the impulsive control [5, 11, 18, 24]. In addition, the impulsive control was effectively applied for synchronization of the networks with time delay [32]. For this kind of application, the goal is to drive some variable to zero in a short time. System is assumed exhibiting the impulsive effects on a countable set of the time instants $t_0 < t_1 < t_2 < t_3 < \cdots$ when the states of the control system change instantaneously in accordance with restitution rules [21]. Therefore, such systems can be also treated as discrete-continuous or hybrid-impulsive systems [19, 21]. For practical applications, the problem is in designing the impulsive control by taking into account the non-zero sampling intervals. The results presented in this chapter are inspired by pioneering work [10, 23, 31], where the second order sliding mode control and relay output control were combined with the impulsive actions achieving uniformity and short convergence time.

In this chapter, we explore a hybrid-impulsive HOSM control in order to reduce dramatically the convergence time practically to zero, achieving instantaneous (or short time) convergence and uniformity [8, 9, 23, 26, 27, 30]. Specifically, the robustness and uniform convergence in systems with relative degree $r \geq 2$ is studied using the state and output feedback discontinuous-hybrid-impulsive algorithms. The impulsive portion of the control function drives the system's output (the sliding variable) and its derivative to zero instantaneously (or in short time) achieving a uniform convergence. Then, the discontinuous state or output feedback stabilizes system's trajectory at the origin (or its close vicinity), while achieving the ideal or real second order sliding mode (2-SM). The restitution rules are enforced using the Dirac delta function and its derivative.

In order to avoid the sliding variable differentiation in HOSM control algorithms the use of the discontinuous output feedback in 2-SMC in a concert with the hybrid-impulsive control is explored. It is known that such systems are only stable in the absence of the disturbance, and may be unstable in its presence. It is shown in [8, 9, 23, 26, 27, 30] that the hybrid-impulsive-discontinuous output feedback control uniformly stabilizes the output (the sliding variable) and its derivative at zero (in a non-perturbed case) and in a small domain (in a perturbed case). Therefore, the ideal or real 2-SM is uniformly reached via discontinuous-impulsive output feedback. The stability of considered hybrid-impulsive system is studied using the Lyapunov Functions approaches [13, 21] for the ideal and practically implemented impulsive control laws as in the work [8, 9, 21, 23, 26]. The studied hybrid-impulsive discontinuous algorithms were implemented in [2, 8, 26] using a practical realization of the delta function, which is introduced on a compact time set [26, 30]. The implementation of delta function was studied in [8, 26] while taking into account the sampling rate and the limitations to the amplitude of the control functions. In this chapter, we also explore a *continuous* HOSM (CHOSM) of *arbitrary relative degree* with impulsive action [1] in order to reduce dramatically the convergence time practically to zero, achieving almost instantaneous convergence and uniformity. After a short convergence time due to the impulsive action, CHOSM takes over and compensates for the disturbance, while keeping the system states in the origin. This approach allows reducing the CHOSM amplitude since the task of compensating the initial conditions is addressed by the impulsive action. By approximating the Dirac delta function and its derivatives, we determine practical impulsive input signals that bring the states of the system to the origin in a short time.

Robust control of perturbed dynamic systems in a *reduced information environment* [2] that requires a short convergence time is another important practical task [5, 11] that is studied in this chapter. In many cases, only one or two last variables of the system's dynamic equations are available for measurement. For instance, a speed of a car usually is measured by a tachometer, while the position is not directly measured. Another example includes a launch vehicle attitude control [5, 11, 14] when only angular accelerations (and, sometimes, angular velocities) are directly measured by the measuring devices located in the Inertial Measurement Unit (IMU) [14]. The controlled attitude angles may not be directly measured. In this chapter only the *last one or two* states of the perturbed system presented in a format of consecutive r integrators are assumed measured. The control problem that consists in driving *all* state variables to zero instantaneously or in a short time and keep them in the origin thereafter in the presence of the bounded perturbations is addressed via impulsive control [2, 8, 19, 21, 26, 30] employed in a concert with second order sliding mode control (2-SMC) [16, 25]. Two 2-SMC algorithms are employed in a concert with the impulsive action in the reduced information environment, specifically [2, 5, 25]:

1. Super-twisting (STW) control algorithm that drives the *last* state variable (assumed measurable) and its derivative to zero in finite time in the presence of the smooth disturbance, which derivative is bounded.
2. Twisting control (TW) that drives the *two last* state variables (assumed measurable) to zero in finite time in the presence of the bounded disturbance.

Impulsive control, applied in a concert with super-twisting and twisting control functions, drives the rest of the variables to zero instantaneously or in a very short time. Only "snap" knowledge (or knowledge in a single time instant) of the all states is required in order to facilitate the effectiveness of the impulsive action. The efficacy of the studied hybrid-impulsive 2-SMC/HOSM algorithms is verified via simulations.

A structure of the chapter is as follows: the mathematical background is presented in Sects. 17.2 and 17.3 contains the problems description; hybrid-impulsive control in 2-SMC systems is studied in Sect. 17.4; second order systems with hybrid-impulsive discontinuous output feedback are discussed in Sect. 17.5; simplified discontinuous-hybrid-impulsive feedback is presented in Sects. 17.6 and 17.7 is dedicated to continuous HOSM with impulsive actions; hybrid-impulsive 2-SMC in reduced information environment is studied in Sect. 17.8; case studies are discussed in Sect. 17.9; conclusions are presented in Sect. 17.10 that is followed by the references section.

17.2 Mathematical Background

17.2.1 System Dynamics

Consider SISO input–output dynamic system of the form

$$\begin{aligned} \dot{z} &= a(t,z) + b(t,z)u \\ \sigma &= \sigma(t,z) \end{aligned} \tag{17.1}$$

where $z \in \mathrm{R}^n$ is a state vector, $u \in \mathrm{R}$ is a control function, $a(t,z) \in \mathrm{R}^n$ is a smooth enough partially known vector-field, $b(t,z) \in \mathrm{R}^n$ is a smooth enough known vector-field, and $\sigma \in \mathrm{R}$ is a smooth enough output (a sliding variable).

The sliding variable input–output dynamics can be presented in a form of rth order differential equation

$$\sigma^{(r)} = f(t,z) + \Delta f(t) + \underbrace{g(t,z)u}_{v} \tag{17.2}$$

that can be presented as

$$\sigma^{(r)} = \Delta f(t) + v_1 \tag{17.3}$$

where $f(t, z) \in \mathrm{R}$ is a known drift term, $\Delta f(t) \in \mathrm{R}$ is the perturbation, r is relative degree, and $v = -f(t, z) + v_1$.

The sliding variable dynamics (17.3) can be also presented in a state variable format

$$\dot{x} = Ax + B(\Delta f + v_1), \quad x(0) = x_0 \tag{17.4}$$

where

$$x = [x_1, x_2, \ldots, x_r]^T = \left[\sigma, \dot{\sigma}, \ldots, \sigma^{(r-1)}\right]^T, A = \begin{bmatrix} 0\ 1 \ldots 0 \\ \cdot\ \cdot\ \cdot\ \cdot\ \cdot \\ 0\ 0 \ldots 1 \\ 0\ 0 \ldots 0 \end{bmatrix}, \quad B = \begin{bmatrix} 0 \\ 0 \\ \ldots \\ 1 \end{bmatrix} \tag{17.5}$$

Assumptions:
(A1) Relative degree r is assumed to be constant and known;
(A2) The zero dynamics of system (17.1) are stable;
(A3) The derivative of the perturbation $\Delta f(t)$ is bounded with a known boundary, i.e. $\left|\Delta \dot{f}\right| \le L_1, \ \ L_1 > 0$, or/and $\Delta f(t)$ is bounded by itself, when required, i.e. $|\Delta f| \le L_2, \ \ L_2 > 0$;
(A4) The function $g(t, z) \neq 0, \ \ \forall z \in \mathrm{R}, \ \ \forall t \in \mathrm{R}^+$ is known.

17.2.2 *The Impulsive Input Approach*

The control function in (17.3) is augmented as

$$v_1 = v_2 + v_{imp} \tag{17.6}$$

where v_2 is HOSM control to be designed later on, and v_{imp} is the impulsive input that is considered in the form [8, 26, 30]

$$v_{imp} = \sum_{k=0}^{r-1} q_k \delta_\varepsilon^{(k)}(t) \tag{17.7}$$

where $\delta_\varepsilon^{(k)}$ are the generalized derivatives of the Dirac delta distribution centered in $\varepsilon > 0$ defined as [7, 29]

$$\int \delta_\varepsilon^{(k)}(t)\phi(t)dt = (-1)^k \phi^{(k)}(\varepsilon) \tag{17.8}$$

for any smooth enough test function $\phi(t)$, and q_k are scalar to be determined. In order to identify q_k, firstly, a solution of system (17.4), (17.7) is determined as:

$$x(t) = exp(At)x_0 + \int_0^t expA(t-\tau)B\left(q_k \sum_{k=0}^{r-1} \delta_\varepsilon^{(k)}(\tau) + v_2(\tau) + \Delta f(\tau)\right)d\tau \tag{17.9}$$

Consider unperturbed system (17.4) with $\Delta f(t) \equiv 0$ and $v_2 \equiv 0$, then (17.9) becomes

$$x(t) = exp(At)x_0 + \sum_{k=0}^{r-1} exp\left(A(t-\varepsilon)\right) A^k B q_k \tag{17.10}$$

for $t \geq \varepsilon$, and $\varepsilon > 0$ can be infinitesimal. Requiring $x(\varepsilon) = 0$, the coefficients q_k can be computed

$$\left[q_0,\ q_1,\ \ldots,\ q_{r-1}\right]^T = -\left[B,\ AB,\ \ldots,\ A^{r-1}B\right]^{-1} exp\left(A\varepsilon\right) x_0 \tag{17.11}$$

For system (17.4), (17.5), (17.10), (17.11) the impulsive control (17.7) becomes

$$v_{imp} = -\sum_{k=0}^{r-1} x_{r-k}(0)\delta_\varepsilon^{(k)}(t) \tag{17.12}$$

Remark 17.1 Apparently, being applied at $t = 0$ (or for $\varepsilon = 0^+$) the impulsive control (17.12) drives $x(t) \to 0$ (or $\sigma, \dot{\sigma}, \ldots, \sigma^{(r-1)} \to 0$) instantaneously.

17.2.3 Uniformity

The definition of uniformity follows the one given in [3].

Definition 17.1 System (17.1) or (17.4), (17.5) is said to be

1. uniformly exact finite time convergent if there exists $T \geq 0$ such that $\forall\ \ x(t_0) \in \mathrm{R}^r$, $x(t) \equiv 0\ \ \forall t \geq T$
2. uniformly exact convergent if for any $r_0 > 0$ there exists $T_r \geq 0$ such that $\forall\, x(t_0) \in \mathrm{R}^r\ \|x(t)\| \leq r_0\ \ \forall t \geq T_r$

In the other words, the convergence is uniform if the convergence time does not depend on initial conditions.

17.2.4 Approximation of Delta Function and Its Derivative

The approximation of the impulsive input by the Gaussian function was proposed and studied in quite a few references (see, for instance, [4, 7])

$$\Phi_h(t) = \frac{1}{\sqrt{2\pi}h} \exp\left(-t^2/(2h^2)\right) \tag{17.13}$$

The Gaussian function (17.13) does not have compact support that makes the approximation less accurate. The following kernel function with finite time support is proposed [29]

$$\omega_h(t) = \begin{cases} \frac{1}{\kappa h} \exp\left(\frac{t^2}{t^2-h^2}\right), & if \ |t| < h \\ 0, & if \ |t| \geq h \end{cases} \tag{17.14}$$

where $\kappa = \int_{-1}^{1} \exp\left(\frac{t^2}{t^2-1}\right) dt$ is a normalized factor, and $h > 0$ is an arbitrary (usually small) constant. It is well known that the functions ω_h are C^∞ smooth and, as $h \to 0$, these functions approximate the Dirac delta distribution [29].

Another kernel function with finite time support that is more practical for the implementation is proposed [8, 26, 30] for piece-wise constant approximation of delta function:

$$\delta_h(t) = \begin{cases} \frac{1}{h}, & if \ 0 \leq t \leq h \\ 0, & \text{otherwise} \end{cases} \tag{17.15}$$

The first derivative of the delta function can be approximated as

$$\dot{\delta}_h(t) = \frac{\delta_{h/2}^{(0)}(t+h/2) - \delta_{h/2}^{(0)}(t)}{h/2} = \begin{cases} 4/h^2, & 0 \leq t \leq h/2 \\ -4/h^2, & h/2 < t \leq h \\ 0, & \text{otherwise} \end{cases} \tag{17.16}$$

The higher order derivatives can be approximated iteratively

$$\delta_h^{(k)}(t) = \frac{\delta_{h/2}^{(k-1)}(t+h/2) - \delta_{h/2}^{(k-1)}(t)}{h/2}, \quad k \geq 1 \tag{17.17}$$

The coefficients $q_0, q_1, \ldots, q_{r-1}$ in impulsive control (17.7) are to be calculated taking into account the piece-wise constant *approximation* of the δ-function and it derivatives in (17.15)–(17.17). Recall that, the goal of the impulsive control (17.7) is to drive $x(t) \to 0$ by the time $t = h$ while the δ-function and its derivatives are approximated as in (17.15)–(17.17).

Therefore, the coefficients $q_0, q_1, \ldots, q_{r-1}$ are to satisfy the following condition, while v_2 and $\Delta f(t)$ are assumed disabled for $t \in [0, h]$

$$0 = x_0 + \int_0^h exp(-A\tau)B\left(q_k \sum_{k=0}^{r-1} \delta_h^{(k)}(\tau)\right) d\tau \tag{17.18}$$

Specifically for $r = 3$ the coefficients are calculated as

$$q_0 = -x_3(0), \quad q_1 = -4\left(x_2(0) + \frac{h}{2}x_3(0)\right), \quad q_2 = -8\left(x_1(0) - x_2(0)h - x_3(0)\frac{h^2}{2}\right) \tag{17.19}$$

17.2.5 Continuous HOSM Control

Consider controlling SISO input–output dynamics of the form (17.3). Then the following result is available

Theorem 17.1 ([4]) *Consider the unperturbed system* (17.3) *with* $\Delta f(t) \equiv 0$. *Let* $\gamma_i > 0$ *be such that the polynomial* $s^r + \gamma_r s^{r-1} + \cdots + \gamma_2 s + \gamma_1$ *is Hurwitz, then there exists* $\varepsilon \in (0, 1)$ *such that for every* $\eta \in (1 - \varepsilon, \varepsilon)$ *the origin of the system* (17.3), $\sigma = \dot{\sigma} = \cdots = \sigma^{(r-1)} = 0$ *is a globally finite time stable equilibrium under the feedback control*

$$v_1 = -\gamma_r \left|\sigma^{(r-1)}\right|^{\eta_r} \operatorname{sign}\left(\sigma^{(r-1)}\right) - \gamma_{r-1}\left|\sigma^{(r-2)}\right|^{\eta_{r-1}} \operatorname{sign}\left(\sigma^{(r-2)}\right) - \cdots - \gamma_1 |\sigma|^{\eta_1} \operatorname{sign}(\sigma) \tag{17.20}$$

where the coefficients $\eta_1, \eta_2, \ldots, \eta_r$ *satisfy*

$$\eta_{i-1} = \frac{\eta_i \eta_{i+1}}{2\eta_{i+1} - \eta_i}, \quad \eta_{r+1} = 1, \quad \eta_r = \eta \quad i = 2, \ldots, r \tag{17.21}$$

Remark 17.2 Note that the control (17.20), (17.21) is continuous.

Remark 17.3 The statement of Theorem 17.1 gives the existence conditions for the finite convergence time controller (17.20), (17.21).

These conditions can be easily transformed into the design algorithm. For instance, considering the roots of the Hurwitz polynomial $s^r + \gamma_r s^{r-1} + \cdots + \gamma_2 s + \gamma_1$ to be equal to -2 the control (17.20) can be parameterized for $r \leq 4$ as

$$\begin{aligned}
&r = 1,\ \eta = 1,\ v_1 = -2|\sigma|^{5/10}\operatorname{sign}(\sigma)\\
&r = 2,\ \eta = 0.6,\ v_1 = -4|\sigma|^{6/14}\operatorname{sign}(\sigma) - 4|\sigma|^{6/10}\operatorname{sign}(\dot{\sigma})\\
&r = 3,\ \eta = 0.7,\ v_1 = -8|\sigma|^{7/16}\operatorname{sign}(\sigma) - 12|\dot{\sigma}|^{7/13}\operatorname{sign}(\dot{\sigma}) - 6|\ddot{\sigma}|^{7/10}\operatorname{sign}(\ddot{\sigma})\\
&r = 4,\ \eta = 0.8,\ v_1 = -16|\sigma|^{8/16}\operatorname{sign}(\sigma) - 32|\dot{\sigma}|^{8/14}\operatorname{sign}(\dot{\sigma}) - 24|\ddot{\sigma}|^{8/12}\operatorname{sign}(\ddot{\sigma})\\
&-8\left|\dddot{\sigma}\right|^{8/10}\operatorname{sign}(\dddot{\sigma})
\end{aligned} \tag{17.22}$$

The theoretical result that allows designing the finite time converging continuous control for the perturbed system (17.3) is formulated in the following theorem.

Theorem 17.2 ([6]) *Consider the perturbed system (17.3)–(17.5) with a smooth perturbation $\Delta f \neq 0$, and its derivative is bounded $\left|\Delta \dot{f}\right| \leq L_1$. Let $\gamma_i > 0$ be such that the polynomial $s^r + \gamma_r s^{r-1} + \cdots + \gamma_2 s + \gamma_1$ is Hurwitz, and $\varepsilon \in (0, 1)$ is identified so that unperturbed system (17.3) is a finite time convergent with the control feedback (17.20), (17.21). Then the origin of the perturbed system (17.3), $\sigma = \dot{\sigma} = \cdots = \sigma^{(r-1)} = 0$ is a globally finite time stable equilibrium under the continuous control feedback*

$$v_1 = -\gamma_r \left|\sigma^{(r-1)}\right|^{\eta_r} \operatorname{sign}\left(\sigma^{(r-1)}\right) - \gamma_{r-1} \left|\sigma^{(r-2)}\right|^{\eta_{r-1}} \operatorname{sign}\left(\sigma^{(r-2)}\right) - \cdots - \gamma_1 \left|\sigma\right|^{\eta_1} \operatorname{sign}(\sigma) - \varpi \tag{17.23}$$

where

$$\begin{aligned} \varpi &= \mu_1 \left|s\right|^{1/2} \operatorname{sign}(s) + \xi \\ \dot{\xi} &= \mu_2 \operatorname{sign}(s) \end{aligned} \tag{17.24}$$

with

$$\begin{aligned} s &= \sigma^{(r-1)} + z, \quad \dot{z} = u - \varpi \\ \mu_1 &= 1.5 L_1^{1/2}, \quad \mu_2 = 1.1 L_1 \end{aligned} \tag{17.25}$$

Remark 17.4 It is worth noting to mention that Eqs. (17.23)–(17.25) represent super-twisting disturbance observer, and the term ϖ becomes exactly equal to the perturbation Δf in finite time.

Remark 17.5 The continuous controller (17.23)–(17.25) can be claimed to be a continuous HOSM (CHOSM) controller for the system in (17.1), since it drives $\sigma, \dot{\sigma}, \ldots, \sigma^{(r-1)} \to 0$ in finite time in the presence of the smooth disturbance Δf with known bounded derivative $\left|\Delta \dot{f}\right| \leq L_1$.

17.2.6 Lyapunov Analysis of 2-SMC Control Algorithms

17.2.6.1 Lyapunov Analysis of Super-Twisting Control

The following result is available:

Theorem 17.3 ([20]) *Consider system driven by super-twisting control [16],*

$$\begin{aligned} \dot{x}_1 &= -\alpha \left|x_1\right|^{1/2} \operatorname{sign}(x_1) + x_2 \\ \dot{x}_2 &= -\beta \operatorname{sign}(x_1) + \varphi(t) \end{aligned} \tag{17.26}$$

where $x_1, x_2 \in \mathrm{R}$, $|\varphi(t)| \leq L_1$, $L_1 > 0$; $\alpha, \beta > 0$. Then there exists a strict Lyapunov function

$$\begin{aligned} V(x_1, x_2) &= (\lambda + 4\mu^2)\left|x_1\right| + x_2^2 - 4\mu x_2 \left|x_1\right| \operatorname{sign}(x_1), \quad \lambda, \mu > 0 \\ \dot{V}(x_1, x_2) &\leq -r_0 V^{1/2}(x_1, x_2), \quad r_0 = const, \quad r_0 > 0 \end{aligned} \tag{17.27}$$

with

$$\alpha > \frac{\left(2L_1 + \lambda + 4\mu^2\right)^2}{12\lambda\mu} - \frac{\mu\left(4L_1 + 1\right)}{\lambda}, \quad \beta = 2\mu\alpha \tag{17.28}$$

17.2.6.2 Lyapunov Analysis of Twisting Control

The following result is available:

Theorem 17.4 ([28]) *Consider system driven by twisting control [16]*

$$\begin{aligned} \dot{x}_1 &= x_2 \\ \dot{x}_2 &= -\bar{\alpha}\,\mathrm{sign}(x_1) - \bar{\beta}\,\mathrm{sign}(x_2) + \varphi(t) \end{aligned} \tag{17.29}$$

where $x_1,\ x_2 \in \mathrm{R}$, $|\varphi(t)| \leq L_2,\ \ L_2 > 0$. *Then there exists a strict Lyapunov function*

$$\begin{aligned} V(x_1, x_2) &= \bar{\alpha}x_1^2 + \gamma\, x_2\, |x_1|^{3/2}\ \mathrm{sign}(x_1) + \bar{\alpha}x_2^2\,|x_1| + \tfrac{1}{4}x_2^4 \\ \dot{V}(x_1, x_2) &\leq -r_1\,[V(x_1, x_2)]^{3/4},\ \ r_1 = const,\ \ r_1 > 0 \end{aligned} \tag{17.30}$$

with

$$\bar{\alpha} - L_2 > \bar{\beta},\ \ \bar{\beta} > L_2,\ \ \bar{\alpha},\ \bar{\beta} > 0 \tag{17.31}$$

Corollary 17.1 *In systems* (17.26), (17.28) *and* (17.29), (17.31) *the following conditions hold:*

1. *the variables* $x_1,\ x_2 \to 0$ *in finite time,*
2. $x_1 = x_2 = 0\ \forall t \geq 0$ *if* $x_1(0) = x_2(0) = 0$.

Note that the conditions (17.28) and (17.31) may be very conservative, since they are derived using Lyapunov function techniques, and the coefficients $\bar{\alpha},\ \bar{\beta} > 0$ can be easily identified by tuning in the simulations.

17.3 Studied Problems

The problem discussed in this chapter is three-fold:

1. Study the robustness and uniform convergence of systems (17.3)–(17.5) with relative degree $r = 2$ using the state and *output feedback* discontinuous-hybrid-impulsive algorithms.
2. Study a *continuous* HOSM (CHOSM) control in perturbed systems (17.3)–(17.5) of *arbitrary relative degree* with impulsive action in order to reduce dramatically the convergence time practically to zero, achieving almost instantaneous convergence and uniformity.

3. Study a hybrid-impulsive second order sliding mode control (2-SMC) in perturbed dynamic systems (17.3)–(17.5) of *arbitrary relative degree* in a *reduced information environment*. Only last *one or two* variables in system (17.3)–(17.5) are assumed available.

17.4 Hybrid-Impulsive Control in 2-SMC Systems

Consider a single-input-single output (SISO) uncertain nonlinear system (17.3) with $r = 2$. The system (17.3)–(17.5) can be presented as

$$\begin{cases} \dot{x}_1 = x_2 \\ \dot{x}_2 = \Delta f(t) + v_1, \quad x_1(t_0) = x_{10}, \quad x_2(t_0) = x_{20} \end{cases} \tag{17.32}$$

where $x_1(t) = \sigma(t)$ and $x_2(t) = \dot{\sigma}(t)$.

17.4.1 Hybrid-Impulsive Effects

It is assumed that system (17.32) exhibits impulse effects on a countable set of time instants $t_1 < t_2 < t_3 \dots$ when the states of the system (17.32) are instantaneously changed in accordance with the restitution rules:

$$\begin{aligned} x_1(t_i^+) &= U_1(x_1(t_i^-), t_i) \\ x_2(t_i^+) &= U_2(x_2(t_i^-), t_i), \quad i = 0, 1, 2, \dots \end{aligned} \tag{17.33}$$

where the restitution rules $U_1(x_1(t_i^-), t_i)$, $U_2(x_2(t_i^-), t_i)$ are defined as:

$$\begin{aligned} U_1(x_1(t_i^-), t_i) &= x_1(t_i^-) + \gamma_{1i} \\ U_2(x_2(t_i^-), t_i) &= x_2(t_i^-) + \gamma_{2i}, \quad \gamma_{1i}, \gamma_{2i} \in \mathrm{R} \end{aligned} \tag{17.34}$$

The following assumptions are made [21]:

(A5) Discrete-continuous system (17.32)–(17.34) possess a globally defined, Lipschitz continuous, positive definite, decrescent function $V(t, x) : 0 < V_0(x) \leq V(t, x) \leq V_1(x)$ such that $\dot{V}(t, x) \leq 0$, computed along the trajectory $x(t) = [x_1(t), x_2(t)]^T$ of Eq. (17.32), almost everywhere (excluding, possibly, the impulsive effects (17.33), (17.34)).

(A6) At the time instants $t_i, \quad i = 0, 1, 2, 3, \dots$ the function $V(t, x)$, computed on the trajectory $x(t)$ of Eq. (17.32) taking into account the restitution rule (17.33), takes the values $V(t_i, x_1(t_i^+), x_2(t_i^+))$ so that

$$\lim_{i \to \infty} V(t_i, x_1(t_i^+), x_2(t_i^+)) = 0 \tag{17.35}$$

17.4.2 Stability of Hybrid-Impulsive Systems

Recall the results of stability analysis of system (17.32)–(17.35).

Theorem 17.5 ([21]) *Suppose that the assumption A5 and A6 hold, then system* (17.32) *with the hybrid-impulsive effects* (17.33), (17.34) *is globally asymptotically stable.*

Corollary 17.2 *If* $V(t_i, x_1(t_i^+), x_2(t_i^+)) = 0$ *holds for all* $i \geq k$ *where k is a finite integer number, then system* (17.5) *is practically asymptotically stable with finite convergence time.*

Fulfillment of the restitution rule
Consider system (17.5) with

$$v_1 = v_2 + v_3 \tag{17.36}$$

where $|v_2| \leq \varpi, \quad \varpi > 0$, and v_3 is impulsive control

$$v_3 = -x_{20}\delta(t - t_0) - x_{10}\dot{\delta}(t - t_0) \tag{17.37}$$

with $\delta(\tau), \quad \dot{\delta}(\tau)$ are Dirac function (distribution) and its derivative [7, 29]. The following lemma constitutes the fulfillment of the restitution rule [8, 21, 26].

Lemma 17.1 ([21]) *Impulsive control* (17.37) *fulfills the restitution rule* (17.33), (17.34) *changing the values* $x_1(t), \quad x_2(t)$ *at* $t = t_0$ *instantaneously to* $x_1(t_0^+) = x_2(t_0^+) = 0$.

17.4.3 Twisting-Hybrid-Impulsive Control

17.4.3.1 On Uniformity of Convergence of Twisting Control

It is known that the perturbed system's dynamics (17.32) controlled by the twisting controller

$$v_1 = -K(\operatorname{sign}(x_1) + 0.5\operatorname{sign}(x_2)), \quad K > 2L_2 \tag{17.38}$$

is globally finite time convergent [15, 25]. The following estimation of finite convergence time starting with the time moment, when $x_1 = 0$ is available:

$$T \leq \frac{|\tilde{x}_2|}{(1 - \bar{q})(0.5K - L_2)} \tag{17.39}$$

where $\tilde{x}_2 = x_2$ at the time moments when $x_1 = 0$. Also, it is assumed the successive crossings of the axis $x_1 = 0$ satisfy the inequality $\frac{|x_{2,i+1}|}{|x_{2,i}|} \leq \bar{q} < 1$ (Fig. 17.1). Apparently, the convergence is not uniform, since it depends on the initial conditions.

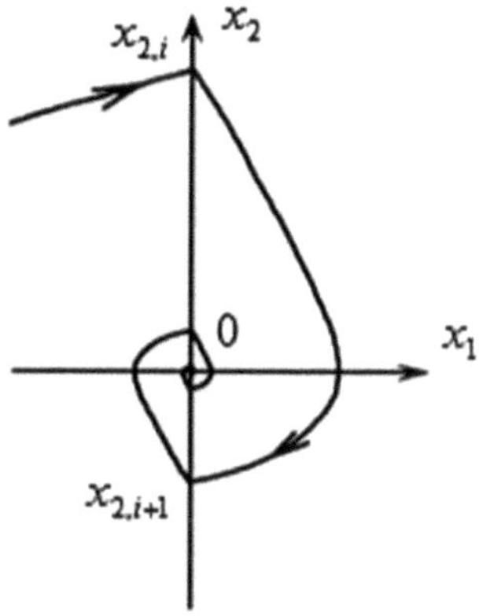

Fig. 17.1 Twisting controller trajectory

17.4.3.2 Stability and Uniformity of Twisting-Hybrid-Impulsive Control

Consider system (17.32) controlled by twisting-hybrid-impulsive control given in a form

$$v_1 = -K(\operatorname{sign}(x_1) + 0.5\operatorname{sign}(x_2)) - x_{20}\delta(t - t_0) - x_{10}\dot{\delta}(t - t_0), \quad K > 2L_2 \tag{17.40}$$

The results of stability and uniformity analysis of system (17.32), (17.40) are presented in the following Theorem (see [8, 26]).

Theorem 17.6 *The twisting-hybrid-impulsive controller* (17.40) *instantaneously drives the system* (17.32) *to 2-SM making the system* (17.32), (17.38) *uniformly exact finite time convergent.*

Remark 17.6 Note that if the initial states x_{10} and x_{20} are uncertain, then the impulsive control (17.37) will drive instantaneously the state variables x_1 and x_2 not to the origin, but to a domain $\Omega(x_1, x_2) : \{|x_1| \le \bar{\varepsilon}_1,\ |x_2| \le \bar{\varepsilon}_2\}$, where $\bar{\varepsilon}_i > 0$ is the amplitude of the measurement/estimation uncertainty of the initial states x_{i0} $i = 1, 2$. Then the twisting control takes over, given the states x_1, x_2 measured with noise (the variable x_2 may contain noise also after the differentiation of noisy x_1 using the higher order sliding mode differentiator [16]), i.e. $x_i^m - x_i = \tilde{\omega}_i, |\tilde{\omega}_i| \le \tilde{\varepsilon}_i,\ \tilde{\varepsilon}_i > 0$ $i = 1, 2$, where x_i^m is the measured value of the state x_i. It is known [16] that in the system run by 2-SMC (specifically by twisting control) the stabilization accuracy of the variable x_1 is proportional to $O(\tilde{\varepsilon})$ and of the variable x_2 is proportional to $O(\tilde{\varepsilon}^{1/2})$, where $\tilde{\varepsilon} = \max(\tilde{\varepsilon}_1, \tilde{\varepsilon}_2),\ \ \tilde{\varepsilon}_1, \tilde{\varepsilon}_2 > 0$, after a finite time convergence.

17.4.3.3 Twisting Control with Simplified Hybrid-Impulsive Action

In this section, the hybrid-impulsive-twisting control law is studied assuming that either $\dot{\delta}(t - t_0)$ or $\delta(t - t_0)$ is not available. Next, system (17.32) is considered under either one of the following simplified twisting-hybrid-impulsive control:

$$v_1 = -K\,(\,\text{sign}(x_1) + 0.5\,\text{sign}(x_2)) - x_{20}\delta(x_1), \quad K > 2L_2 \tag{17.41}$$

or

$$v_1 = -K\,(\,\text{sign}(x_1) + 0.5\,\text{sign}(x_2)) - x_{10}\dot{\delta}(x_2), \quad K > 2L_2 \tag{17.42}$$

The main results of a study of the stability, dynamics and uniformity of system's (17.32), (17.42) are presented in the following theorems [8, 26].

Theorem 17.7 *Assume that* $x_{10} = x_1(t_0) = 0$ *and* $x_{20} = x_2(t_0^-) \neq 0$, *then system* (17.32), (17.41) *is uniformly exact finite time convergent to 2-SM* $x_1, x_2 \to 0$ *with zero convergence time.*

Theorem 17.8 *Assume that* $x_{20} = x_2(t_0^-) = 0$ *and* $x_{10} = x_1(t_0^-) \neq 0$, *then system* (17.32), (17.42) *is uniformly exact finite time convergent to 2-SM* $x_1(t),\ x_2(t) \to 0$ *with zero convergence time.*

Remark 17.7 Uniformity of convergence in system (17.32), (17.41) will be compromised if the initial condition $x_{10} = x_1(t_0) \neq 0$. In this case, system (17.32) will be controlled by the controller (17.41) with mute impulsive action until the state $x_1(t)$ becomes equal to zero in *finite time* $t = t_1$ (see Fig. 17.1) that depends on the initial condition x_{10}. Then, as a result of the impulsive action and in accordance with Theorem 17.7, the 2-SM will be reached in system (17.32), (17.41) instantaneously (uniformly). The same comments are valid about the uniformity of convergence of system (17.32), (17.42), when $x_{20} = x_2(t_0^-) \neq 0$.

17.5 Discontinuous Output Feedback-Hybrid-Impulsive Control

In order to avoid the sliding variable differentiation in the twisting control algorithm (17.38) the use of the discontinuous output feedback in a concert with the hybrid-impulsive control is proposed and explored in this section.

Consider system (17.32) controlled by output feedback discontinuous-hybrid-impulsive control

$$v_1 = -K\,\text{sign}(x_1) - \sum_{i=0,1,2,\ldots} x_2(t_i^-)\delta(t - t_i) - \sum_{i=0,1,2,\ldots} x_1(t_i^-)\dot{\delta}(t - t_i), \ \ K > 0 \tag{17.43}$$

Remark 17.8 It is assumed that the value x_1 is measured continuously. As to x_2, a "snap" knowledge of $x_2(t_i^-)$, $i = 1, 2, 3, \ldots$ is required for implementing the output feedback control law (17.43) (by "snap" knowledge we mean that the variable x_2 is accessible in isolated time instants only).

17.5.1 Unperturbed Case

The results of analysis of unperturbed system (17.5), (17.43) are presented in the following Theorem [8, 26].

Theorem 17.9 *Assume the perturbation* $\Delta f(t) \equiv 0$ *in system* (17.32)*, then system* (17.32), (17.43) *with non-zero initial conditions* $x_1(t_0^-) = x_{10}, \quad x_2(t_0^-) = x_{20}$ *is uniformly exact finite time convergent (to 2-SM) for* $i = 0$ *with zero convergence time.*

17.5.2 Perturbed Case

Consider perturbed system (17.32), (17.43) with $\Delta f(t) \neq 0, \quad |\Delta f(t)| \leq L_2, \quad L_2 > 0$ and the non-zero initial conditions $x_1(t_0^-) = x_{10}, \quad x_2(t_0^-) = x_{20}$. Then after application of the impulsive action at $t = t_0$ system (17.32), (17.43) becomes

$$\begin{cases} \dot{x}_1 = x_2 \\ \dot{x}_2 = \Delta f(t) - K \operatorname{sign}(x_1), \quad K > L_2 \end{cases} \tag{17.44}$$

with $x_1(t_0^+) = x_2(t_0^+) = 0$ for all $t \in [t_0^+, t_1]$.

Verifying a validity of the assumption (A5), a candidate to a Lyapunov function $V_1(x_1, x_2) = V(x_1, x_2) = K|x_1| + \frac{1}{2}x_2^2$ is introduced, and its derivative is computed as:

$$\dot{V}(x_1, x_2) = K\dot{x}_1 \operatorname{sign}(x_1) + x_2\dot{x}_2 = Kx_2 \operatorname{sign}(x_1) - x_2 K \operatorname{sign}(x_1) + x_2 \Delta f(.) = x_2 \Delta f(.) \leq |x_2| L_2 \tag{17.45}$$

Apparently, $\dot{V}(x_1, x_2)$ in (17.45) is sign indefinite, and the assumption (A5) in Theorem 17.5 is not met. It is difficult to expect that there exists any other positive definite function that has negative definite (negative semi-definite) derivative in a case of perturbed system (17.44).

Therefore, the application of Theorem 17.5 for analysis of the stability of the perturbed output feedback-hybrid-impulsive system (17.32), (17.43) is questionable. Prior to studying the stability of the perturbed system (17.5), (17.16) the stability of the perturbed system (17.32)–(17.34) is studied upon the following assumption.
(A7) Discrete-continuous system (17.32)–(17.34) possess a globally defined, Lipschitz continuous, positive definite, decrescent function $0 \leq V_0(x) \leq V(t, x) \leq$

$V_1(x)$, and there exist $\beta_0 > 0$ and $c > 0$ so that $\forall x \in W, W : \{V_1(x) \leq \beta_0\}$ inequality $\dot{V}_1(x) \leq c$ computed along the trajectory $x(t) = [x_1(t), x_2(t)]^T$ of (17.32), holds on any finite time interval $t \in [t_m, t_M]$, $0 \leq t_m < t_M$ (excluding, possibly, the time instants with impulsive effects (17.33), (17.34)).

The results of analysis of the discontinuous output feedback-hybrid-impulsive perturbed system (17.32)–(17.34) that satisfies the assumption (A7) are presented in the following theorem (see [8, 26]).

Theorem 17.10 *Assume that the assumption (A7) holds for system* (17.32)*, then there exist the restitution rules* (17.33)*,* (17.34) *so that the trajectory* $x(t) = [x_1(t), x_2(t)]^T$ *stays in the domain* $V_1(x_1(t), x_2(t)) \leq \beta_0$ *for all* $t \geq t_0^+$, $\beta_0 > 0$.

Next, the stability and uniformity in the perturbed system (17.32), (17.43) is studied, and the results are formulated in the following Theorem [8, 26].

Theorem 17.11 *Given* $\beta_0 > 0$*, then the perturbed system* (17.32) *with the output feedback discontinuous-hybrid-impulsive control* (17.43) *(where* t_{i+1}, $i=0, 1, 2, \ldots$ *are defined from* $V_1(x_1(t_{i+1}^-), x_2(t_{i+1}^-)) = K\left|x_1(t_{i+1}^-)\right| + \frac{1}{2}x_2^2(t_{i+1}^-) = \beta_0$, $i = 0, 1, 2, \ldots$*) is uniformly convergent with zero convergence time to real 2-SM in the domain* $V_1(x_1(t), x_2(t)) \leq \beta_0$, $\forall t \geq t_0$.

Remark 17.9 It is worth noting that the value of $\beta_0 > 0$ can be taken arbitrary small, which yields an arbitrary small size of the domain of convergence $V_1(x_1(t), x_2(t)) \leq \beta_0$, $\forall t \geq t_0$.

17.6 Simplified Discontinuous Output Feedback-Hybrid-Impulsive Control

In this section, we consider *perturbed* system (17.32) controlled by discontinuous output feedback with a simplified hybrid-impulsive control given in a form

$$v_1 = -K\,\mathrm{sign}(x_1) - \sum_{i=0,1,2,3,\ldots} x_2(t_i^-)\delta(t - t_i), \quad K > D \tag{17.46}$$

The advantage of such simplified hybrid-impulsive discontinuous output feedback is in the use of only delta function (the implementation of the derivative of the delta function requires larger control amplitude and can be more challenging in real applications) for implementing the restitution rules (17.33), (17.34). The time instants t_i in Eq. (17.46) correspond to the cases when $x_1(t_i) = 0$. Then the restitution rules in (17.33), (17.34) take a form

$$x_1(t_i^+) = x_1(t_i^-) = 0, \quad x_2(t_i^+) = x_2(t_i^-) - x_2(t_i^-) = 0, \quad i = 0, 1, 2, .. \tag{17.47}$$

It is assumed that only x_1 measurement is available continuously and x_2 measurement is available when $x_1 = 0$ ("snap" measurement).

17.6.1 Unperturbed System

The results of stability analysis of unperturbed system (17.5) with a simplified discontinuous output feedback-hybrid-impulsive control (17.19) is presented in the following theorem [8, 26].

Theorem 17.12 *Assume that in system* (17.32) $x_{10} = x_1(t_0) = 0$ *and* $\Delta f(t) \equiv 0$, *then unperturbed system* (17.32), (17.46) *for* $i = 0$ *is uniformly exact finite time convergent (to 2-SM) with zero convergence time (see [8, 26]).*

Discussion. It is worth noting that the uniformity of exact finite time convergence is compromised as soon as the initial condition $x_{10} = x_1(t_0) \neq 0$. The discontinuous output feedback controller $v_1 = -K\ \mathrm{sign}(x_1)$ makes the condition of Theorem 17.12 valid as soon as $x_1$is driven to zero in *finite time*.

17.6.2 Perturbed System

Theorem 17.11 is applied for the analysis of perturbed system (17.32) with the hybrid-impulsive discontinuous output feedback (17.46) and a simplified restitution rule (17.47). The results of the analysis are presented in the following theorem (see [8, 26]).

Theorem 17.13 *Assume that* $x_{10} = x_1(t_0) = 0,\ x_{20} = x_2(t_0) \neq 0,$ *and* $x_1(t_i^+) = \varepsilon_{1i},\ x_2(t_i^+) = \varepsilon_{2i}$ *(due to the imperfect snap knowledge of* x_2*), where* $|\varepsilon_{1i}|, |\varepsilon_{2i}| \in \mathrm{R}^+$ *are small real numbers, then the perturbed system* (17.5), (17.19) *is uniformly convergent with zero convergence time to real 2-SM in the domain*

$$\begin{aligned}
&V_1(x_1(t), x_2(t)) \leq \beta_0 \\
&\beta_0 = \max_{i=1,2,3,\ldots} \beta_i, \\
&\beta_i = K\left(|\varepsilon_{1i}| + |\varepsilon_{2i}|\,\tfrac{|\varepsilon_{2i}|+\sqrt{\varepsilon_{2i}^2+2|\varepsilon_{1i}|(K-D)}}{K-D} + \tfrac{(D+K)}{2}\left(\tfrac{|\varepsilon_{2i}|+\sqrt{\varepsilon_{2i}^2+2|\varepsilon_{1i}|(K-D)}}{K-D}\right)^2\right) + \\
&\tfrac{1}{2}\left(|\varepsilon_{2i}|\,\tfrac{(D+K)}{K-D}\left(|\varepsilon_{2i}| + \sqrt{\varepsilon_{2i}^2 + 2\,|\varepsilon_{1i}|\,(K-D)}\right)\right)
\end{aligned} \tag{17.48}$$

Discussion. The results of the Theorem 17.13 are illustrated in Fig. 17.2, where the state $x_2(t)$ is moved instantaneously to $x_2(t_0^+) \in \varepsilon_{20}$ and $x_1(t_0^+) \to \varepsilon_{10} > 0$ by the impulsive action (17.46), (17.47) at $t = t_0$. The output feedback control $v_1 = -K sign(x_1)$ overcomes the bounded perturbation $\Delta f(t)$ and drives the states until $x_1(t)$ becomes equal to zero at $t = t_i$ $(i = 1)$. Then the impulsive action (17.46), (17.47) drives $x_2(t)$ instantaneously to $x_2(t_i^+) \in \varepsilon_{2i}$ and $x_1(t_i^+) \in \varepsilon_{1i} < 0$, and so on. The states $x_1(t),\ x_2(t)$ are uniformly exact convergent with zero convergence time to real 2-SM in the domain $x_1(t),\ x_2(t) \in \{\, V_1(x_1(t), x_2(t)) \leq \beta_0\}\ \forall t \geq t_0^+$.

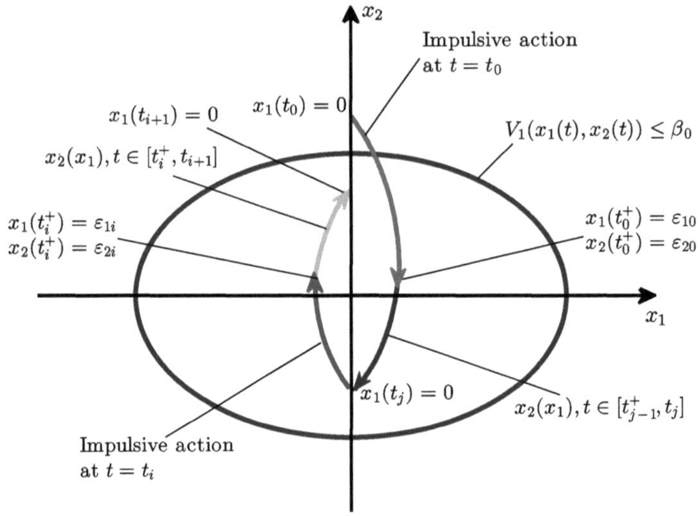

Fig. 17.2 Simplified hybrid-impulsive effects in perturbed system with the discontinuous output feedback

Corollary 17.3 *If ε_{1i}, ε_{2i} that may characterize the implementation error of the restitution rules* (17.47) *are equal to zero, then $\beta_0 = 0$ and the system* (17.32), (17.46) *is uniformly exact convergent with zero convergence time to 2-SM $x_1(t) = x_2(t) = 0$ $\forall t \geq t_0^+$ as soon as $x_1(t)$ becomes equal to zero.*

The study of the implementation issues of delta function $\delta_\varepsilon(t)$ and its derivatives $\delta_\varepsilon^{(k)}(t)$, $k = 1, 2, \ldots, r-1$ as in (17.15)–(17.17) are available in [8, 26]. Also, this study is presented in the Sects. 17.7 and 17.8, and in examples discussed in Sect. 17.9.

17.7 Continuous Higher Order Sliding Mode Control with Impulsive Action

In this section, we explore a *continuous* HOSM (CHOSM) control of *arbitrary relative degree* with impulsive action in order to reduce dramatically the convergence time practically to zero, achieving almost instantaneous convergence and uniformity. This approach allows reducing the CHOSM amplitude, since the task of compensating the initial conditions is addressed by the impulsive action.

Consider SISO input–output perturbed dynamics in Eqs. (17.3)–(17.5) with the smooth perturbation $\Delta f(t)$ whose derivative is bounded $\left|\Delta \dot{f}\right| \leq L_1$, $L_1 > 0$. Next, in accordance with Theorem 17.2, the CHOSM control (17.23)–(17.25) drives $\sigma, \dot{\sigma}, \ldots, \sigma^{(r-1)} \to 0$ in finite time and holds $\sigma, \dot{\sigma}, \ldots, \sigma^{(r-1)}$ in the origin all consecutive time.

It is worth noting that the equilibrium point $\sigma = \dot{\sigma} = \cdots = \sigma^{(r-1)} = 0$ is stabilized by CHOSM control (17.23)–(17.25) in finite time but not uniformly.

In this section, we address the following tasks:

1. Design the hybrid-impulsive CHOSM control that drives $\sigma, \dot{\sigma}, \ldots, \sigma^{(r-1)} \to 0$ instantaneously (uniformly) and holds $\sigma = \dot{\sigma} = \cdots = \sigma^{(r-1)} = 0$ for all consecutive time.
2. Implement the delta function and its derivatives and study the system's (17.3)–(17.5) controlled by hybrid-impulsive CHOSM control with the practically implemented impulsive action.

17.7.1 Hybrid-Impulsive CHOSM Control with Available Delta Function and Its Derivatives

The first main result is formulated in the following theorem.

Theorem 17.14 *Assume that:*

1. *The initial condition vector $x(0) = x_0 \in R^r$ is available,*
2. *the delta function $\delta_\varepsilon(t)$ and its consecutive derivatives $\delta_\varepsilon^{(k)}(t), \;\; k = 1, 2, \ldots, r-1$ are exactly implementable*

then the hybrid-impulsive CHOSM control law (17.6)*: $v_1 = v_2 + v_{imp}$, where v_2 is the CHOSM control function computed in terms of v_1 in Eqs.* (17.23)–(17.25)*, and v_{imp} is the impulsive input that is considered in the form* (17.7)*, or in the form* (17.12)*, (the initial values of the states are assumed available), drives $x(t) \to 0$ instantaneously and keeps it there all consecutive time after a finite time transient response.*

Proof The impulsive control v_{imp} in (17.7), (17.12) drives vector $x(t) \to 0$ instantaneously due to the analysis given in Sect. 17.2.2. Then it requires *finite time* for the super-twisting disturbance observer in Eqs. (17.24) and (17.25) to reconstruct the perturbation Δf, i.e. $\varpi \equiv \Delta f \;\; \forall t \geq t_f$, where t_f is the convergence time of the super-twisting disturbance observer in (17.24), (17.25). Next, the control (17.23)–(17.25) drives $x(t) \to 0$ in *finite time* $\bar{t}_f > t_f$ in accordance with Theorem 17.2. Then $x(t) \equiv 0 \, \forall t \geq \bar{t}_f$. Theorem is proven.

17.7.2 Hybrid-Impulsive CHOSM Control with Impulsive Function Piecewise-Constant Approximation

In this subsection, the Hybrid-Impulsive CHOSM control with delta function and its derivatives being *approximated* as in Eqs. (17.15)–(17.17) is explored. In Eq. (17.7) for impulsive control in the approximation case the coefficients $q_0, q_1, \ldots, q_{r-1}$

are to be calculated using Eq. (17.18). Specifically, for $r = 3$ the coefficients are calculated as in Eq. (17.19).

The second main result is formulated in the following proposition

Theorem 17.15 *Assume that:*

1. *The initial condition vector* $x(0) = x_0 \in R^r$ *is available,*
2. *The delta function and its* $r - 1$ *derivatives are approximated in accordance with Eqs.* (17.15)–(17.17),
3. *The CHOSM control* (17.23)–(17.25) *is mute and* $\Delta f(t) \equiv 0$ *for* $t \in [0, h]$.

then the CHOSM-impulsive control law in Eqs. (17.3)–(17.7), (17.23)–(17.25) *drives* $x(t) \to 0$ *by time* $t = h$ *and keeps it there all consecutive time after a finite time transient response.*

Proof The impulsive control v_{imp} in (17.7), (17.15)–(17.17) drives vector $x(t) \to 0$ by time $t = h$ due to a special selection of the coefficients $q_0, q_1, \ldots, q_{r-1}$ in accordance with Eq. (17.18). The rest of the proof is the same as in Theorem 17.14.

17.8 Impulsive-Second Order Sliding Mode Control in Reduced Information Environment

In many cases, only one or two last variables of the system's dynamic equations are available for measurement. For instance, a speed of a car usually is measured by a tachometer, while the position is not directly measured. Another example includes a launch vehicle attitude control, when only angular accelerations (and, sometimes, angular velocities) are directly measured by the measuring devices located in the Inertial Measurement Unit (IMU) [14]. The controlled attitude angles may not be directly measured. In this subsection, it is assumed that

(A8) Only last *one (in a case of super-twisting control) or two* (in a case of twisting control) variables in system (17.4) are available.

The control problem that consists in driving *all* state variables to zero instantaneously or in a short time and keep them in the origin thereafter in the presence of the bounded perturbations presents a challenge. Here we propose to employ the impulsive control [19, 21, 30] in a concert with second order sliding mode control (2-SMC) [16, 25] to address the discussed challenge.

17.8.1 Motivation Examples

Consider perturbed system (17.4) *with relative degree* $r = 2$ *and the only state* $x_2(t)$ *is available*

In accordance with Theorem 17.1 the super-twisting control [1, 18] $v_1 = v_{STW}$

$$\begin{aligned} v_{STW} &= -\alpha \left|x_2\right|^{1/2} \operatorname{sign}(x_2) + w \\ \dot{w} &= -\beta \operatorname{sign}(x_2) \end{aligned} \tag{17.49}$$

with $\alpha,\ \beta$ selected to meet (17.28) will drive the sliding set $x_2, \dot{x}_2 \to 0$ in finite time $t = t_f$. In 2-SM the system's (17.4), (17.49) dynamics become $\dot{x}_1 = x_2,\ \ x_2 = 0$, which means that $x_1 = x_1(t_f) = const$ in 2-SM, and the goal $x_1 = 0,\ \ x_2 = 0\ \forall t \geq t_f$ is not achieved.

Consider perturbed system (17.4) *with relative degree* $r = 3$, *and the only state* $x_3(t)$ *is available*

The super-twisting control $v_1 = v_{STW}$

$$\begin{aligned} v_{STW} &= -\alpha \left|x_3\right|^{1/2} \operatorname{sign}(x_3) + w \\ \dot{w} &= -\beta \operatorname{sign}(x_3) \end{aligned} \tag{17.50}$$

with $\alpha,\ \beta > 0$ selected to meet (17.28) will drive the sliding set $x_3, \dot{x}_3 \to 0$ in finite time $t = t_f$. In 2-SM the system's (17.4), (17.50) dynamics become $\dot{x}_1 = x_2,\ \ \dot{x}_2 = x_3,\ \ x_3 = 0$, which means $x_2 = x_2(t_f),\ \ x_1 = x_1(t_f) + x_2(t_f)\left(t - t_f\right) \forall t \geq t_f$ in 2-SM, and the goal $x_1 = x_2 = x_3 = 0\ \forall t \geq t_f$ is not achieved. Furthermore, the variable $x_1 \to \infty$ as time increases in 2-SM. Therefore, the assumption (A8) appeared to be too restrictive for the super-twisting controller design in the perturbed system (17.4) with relative degree $r \geq 2$.

Consider perturbed system (17.4) *with relative degree* $r = 3$, *and the state* $x_2(t),\ x_3(t)$ *are available.*

The twisting control [22, 25] $v_1 = v_{TW}$

$$v_{TW} = -\bar{\alpha} \operatorname{sign}(x_2) - \bar{\beta} \operatorname{sign}(x_3) \tag{17.51}$$

with $\bar{\alpha},\ \bar{\beta} > 0$ selected to meet (17.31) will drive the sliding set $x_2, x_3 \to 0$ in finite time $t = t_f$. In 2-SM the system's (17.4), (17.51) dynamics become $\dot{x}_1 = x_2,\ \ x_2 = 0,\ \ x_3 = 0$, which means that $x_1 = x_1(t_f) = const$ in 2-SM, and the goal $x_1 = x_2 = x_3 = 0\ \forall t \geq t_f$ is not achieved.

17.8.2 Problem Statement and Plan of Attack

The problem is in designing the impulsive/2-SMC control function v_1 for the perturbed system (17.3) or (17.4) in the reduced information environment that drives the state vector x (or sliding set $\sigma,\ \dot{\sigma},\ \ldots, \sigma^{(r-1)}$) to zero instantaneously or in a short time and keep it there thereafter in the presence of the bounded perturbation (in a case of using impulsive control in the concert with twisting control), and the smooth perturbation with the bounded derivative (in a case of using impulsive control in the concert with super-twisting control), assuming (A1)–(A8) hold.

We will seek the control function of a form

$$v_1 = v_{2-SMC} + v_{imp} \tag{17.52}$$

The impulsive control (17.7), (17.12) is applied at $t = \varepsilon, \quad \varepsilon = 0$ assuming that the perturbation $\Delta f(0) \equiv 0$, and the snap knowledge of the state $x \in R^r$ at $t = 0$ is available (in other words, the initial vector $x(0) = x_0$ is assumed measured). Therefore, the impulsive control (17.7), (17.12) drives $x(t) \to 0$ instantaneously, which yields $x(0^+) = 0$ is Eq. (17.4).

1. Plan of attack for $v_{2-SMC} = v_{STW}$ *in Eq.* (17.50)

Here, it is assumed that $\Delta f(t)$ is a smooth function with the bounded derivative, and only x_r is measured $\forall t \geq 0^+$. Next, as soon as the impulsive control (17.7), (17.12) is applied, then the super-twisting control $v_{2-SMC} = v_{STW}$

$$\begin{aligned} v_{STW} &= -\alpha\,|x_r|^{1/2}\,\text{sign}(x_r) + w \\ \dot{w} &= -\beta\,\text{sign}(x_r) \end{aligned} \tag{17.53}$$

is to be activated in system (17.4) at $\forall t = 0^+$ to keep $x_r, \dot{x}_r = 0$ (or $\sigma^{(r-1)}, \sigma^{(r)} = 0$) $\forall t \geq 0^+$ in the presence of the disturbance with the bounded derivative.

2. Plan of attack for $v_{2-SMC} = v_{TW}$ *in Eq.* (17.51)

Here, it is assumed that $\Delta f(t)$ is a bounded function, and only x_{r-1} and x_r are measured $\forall t \geq 0^+$. Next, as soon as the impulsive control (17.7), (17.12) is applied, the twisting control $v_{2-SMC} = v_{TW}$

$$v_{TW} = -\bar{\alpha}\,\text{sign}(x_{r-1}) - \bar{\beta}\,\text{sign}(x_r) \tag{17.54}$$

is to be activated in system (17.4) at $\forall t = 0^+$ to keep $x_{r-1}, x_r = 0$ (or $\sigma^{(r-2)}, \sigma^{(r-1)} = 0$) $\forall t \geq 0^+$ in the presence of the bounded disturbance.

It is worth noting that the impulsive control (17.7), (17.12) ($v_1 = v_{imp}$) drives the state vector $x \to 0$ of the unperturbed ($\Delta f \equiv 0$) system (17.4) instantaneously and keeps it there for all consecutive time.

Remark 17.10 Note that for the perturbed system (17.4) with the perturbation

$$\Delta f = \begin{cases} 0, & t \in [0, \varepsilon] \\ \Delta\bar{f}, & t > \varepsilon \end{cases} \tag{17.55}$$

the impulsive control (17.7), (17.12) cannot stabilize the state vector x at zero, since $x(\varepsilon^+) = 0$ and $\forall t \geq \varepsilon$ the state vector dynamics are described by system (17.4) with $v_1 \equiv 0$ and $\Delta f \neq 0$. In this case, a state vector time history can be defined

$$x(t) = \int_{\varepsilon}^{t} \exp\left[A(t-\tau)\right] B\left(\Delta\bar{f}\right) d\tau \;\; \forall t \geq \varepsilon \tag{17.56}$$

and $x(t)$ deviates from the origin in a perturbed system controlled by the impulsive action only. Therefore, based on the analysis presented in Sects. 17.2 and 17.3 it can be concluded that neither 2-SMC nor $v_{imp}(t)$ individually can drive the system's state vector to zero instantaneously or in a short time and keep it there thereafter in the reduced information environment in the presence of bounded perturbations.

17.8.3 Hybrid-Impulsive 2-SMC with Ideal Delta Function and Its Derivative

Assuming availability of the ideal delta function and its derivative, the results are summarized in the following theorems

Theorem 17.16 *Assume that*

1. *Assumptions A1–A8 hold*
2. *The delta function and its* $(r-1)$ *consecutive derivatives are available,*

then the impulsive-super-twisting control (17.52), (17.53) *drives the state vector* $x \to 0$ *instantaneously by the time* $t = 0^+$ *and keeps it there for all consecutive time.*

Proof As soon as the impulsive control (17.7), (17.12) is applied at $t = 0$, it drives $x \to 0$ instantaneously, i.e. $x(0^+) = 0$. Then the system's (17.4) dynamics become

$$\dot{x} = Ax + B(\Delta f + v_{STW}), \quad x(0^+) = 0 \tag{17.57}$$

where v_{STW} is defined by Eq. (17.53) and the coefficients $\alpha, \beta > 0$ are selected in accordance with Eq. (17.28). In accordance with Theorem 17.1 $x_r(t) = \dot{x}_r(t) = 0 \;\; \forall t \geq 0^+$ in second order sliding mode (2-SM). Then system's (17.4) dynamics in 2-SM become $\forall t \geq 0^+$

$$\begin{aligned} &\dot{\bar{x}} = \bar{A}\bar{x}, \quad x_r = 0, \\ &\bar{x} = \left[x_1, x_2, \ldots, x_{r-1}\right]^T, \quad \bar{x}(0^+) = 0, \\ &\bar{A} = \begin{bmatrix} 0\;1\;\ldots\;0 \\ \cdot\;\;\cdot\;\;\cdot\;\;\cdot \\ 0\;0\;\ldots\;1 \\ 0\;0\;\ldots\;0 \end{bmatrix} \in \mathrm{R}^{(r-1)\times(r-1)} \end{aligned} \tag{17.58}$$

which yields $\bar{x}(t) = 0, \;\; x_r = 0 \;\; \forall t \geq 0^+$. Theorem is proven.

Theorem 17.17 *Assume that:*

1. *Assumptions A1–A8 hold*
2. *The delta function and its* $(r-1)$ *consecutive derivatives are available,*

then the impulsive-twisting control (17.52), (17.54) *drives the state vector* $x \to 0$ *instantaneously i.e.* $x(0^+) = 0$ *and keeps it there for all consecutive time.*

Proof As soon as the impulsive control (17.7), (17.12) is applied at $t = 0$, it drives $x \to 0$ instantaneously, i.e. $x(0^+) = 0$. Then the system's (17.4) dynamics become

$$\dot{x} = Ax + B(\Delta f + v_{STW}), \quad x(0^+) = 0 \tag{17.59}$$

where v_{TW} is defined by Eq. (17.54) and $\bar{\alpha}, \ \bar{\beta} > 0$ are selected in accordance with Eq. (17.31). In accordance with Theorem 17.2 $x_{r-1}(t) = x_r(t) = 0 \ \ \forall t \geq 0^+$ in 2-SM. Then, system's (17.4) dynamics in 2-SM become $\forall t \geq 0^+$

$$\begin{gathered} \dot{\tilde{x}} = \tilde{A}\tilde{x}, \quad x_{r-1}(t) = x_r(t) = 0, \\ \tilde{x} = \left[x_1, x_2, \ldots, x_{r-2}\right]^T, \quad \tilde{x}(0^+) = 0, \\ \tilde{A} = \begin{bmatrix} 0 \ 1 \ \ldots \ 0 \\ \cdot \ \cdot \ \cdot \ \cdot \\ 0 \ 0 \ \ldots \ 1 \\ 0 \ 0 \ \ldots \ 0 \end{bmatrix} \in \mathrm{R}^{(r-2)\times(r-2)} \end{gathered} \tag{17.60}$$

which yields $\tilde{x}(t) = 0, \quad x_{r-1}(t) = x_r(t) = 0 \ \ \forall t \geq 0^+$. Theorem is proven.

17.8.4 *Hybrid-Impulsive-2-SMC Control with Approximation of the Delta Function and Its Derivatives*

In this subsection, the *practical* impulsive/2-SMC control with delta function and its derivatives being *approximated* as in (17.15)–(17.17) is explored. For delta function centered at zero the *practical* impulsive action in Eq. (17.7) can be rewritten as

$$v_{imp\_h} = \sum_{k=0}^{r-1} q_k \delta_h^{(k)}(t) \tag{17.61}$$

The coefficients $q_0, \ q_1, \ \ldots, q_{r-1}$ are to be identified so that $x(h) = 0$ as soon as the *practical* impulsive control (17.61) is applied.

The following assumptions are made for identification of the coefficients q_0, $q_1, \ \ldots, \ q_{r-1}$ in Eq. (17.18) with approximated delta function and its derivatives:
(A9) The perturbation $\Delta f(t)$ satisfies the condition (17.55), where $\Delta f(t)$ is bounded (in a case of using impulsive control in the concert with twisting control), and is smooth with the bounded derivative (in a case of using impulsive control in the concert with super-twisting control) for all $t > h$.
(A10) The 2-SMC, v_{2-SMC}, is disabled/mute while the *practical* impulsive control v_{imp_h} in (17.61) is applied.

Specifically, for $r = 3$ the coefficients are calculated as in Eq. (17.19). Next, we will seek the control function in the system (17.4) of a form

$$v_1 = v_{2-SMC} + v_{imp\_h} \tag{17.62}$$

The obtained results are summarized in the following two theorems.

Theorem 17.18 *Assume that the assumptions (A1)–(A10) hold, then the practical impulsive-super-twisting control* (17.61) *with* $v_{2-SMC} = v_{STW}$, *where* v_{STW} *defined by Eqs.* (17.52) *and* (17.53) *that is active in system* (17.4) $\forall t \geq h$ *and impulsive action* v_{imp_h} *in* (17.18), (17.61) *drive the state vector* $x \to 0$ *in a short time by* $t = h$ *and keeps it there for all consecutive time.*

Proof As soon as the *practical* impulsive control (17.18), (17.61) is applied at $t = 0$ (with disabled v_{STW} $\forall t \in [0, h]$) it makes $x(h) = 0$. Then the system (17.4) becomes

$$\dot{x} = Ax + B(\Delta f + v_{STW}), \quad x(h) = 0 \tag{17.63}$$

where v_{STW} that is active $\forall t \geq h$ is defined by Eq. (17.53) and $\alpha, \beta > 0$ are selected in accordance with Eq. (17.28). In accordance with Theorem 17.1 $x_r(t) = \dot{x}_r(t) = 0$ $\forall t \geq h$ in 2-SM. Then system (17.4) in 2-SM become $\forall t \geq h$

$$\begin{gathered} \dot{\bar{x}} = \bar{A}\bar{x}, \quad x_r = 0, \\ \bar{x} = \left[x_1, x_2, \ldots, x_{r-1}\right]^T, \quad \bar{x}(h) = 0, \\ \bar{A} = \begin{bmatrix} 0 & 1 & \ldots & 0 \\ \cdot & \cdot & \cdot & \cdot \\ 0 & 0 & \ldots & 1 \\ 0 & 0 & \ldots & 0 \end{bmatrix} \in \mathrm{R}^{(r-1)\times(r-1)} \end{gathered} \tag{17.64}$$

which yields $\bar{x}(t) = 0, \quad x_r = 0 \;\; \forall t \geq h$. Theorem is proven.

Theorem 17.19 *Assume that the assumptions (A1)–(A10) hold, then the practical impulsive-twisting control* (17.62) *with* $v_{2-SMC} = v_{TW}$, *where* v_{TW} *defined by Eqs.* (17.52), (17.54) $\forall t \geq h$ *and the impulsive action* v_{imp_h} *defined by Eqs.* (17.15)–(17.18), (17.61), *makes the state vector* $x(h) = 0$ *and keeps it there for all consecutive time.*

Proof As soon as the practical impulsive control (17.15)–(17.18), (17.61), is applied at $t = 0$ (with disabled v_{TW} $\forall t \in [0, h]$) it makes $x(h) = 0$. Then the system (17.4) becomes

$$\dot{x} = Ax + B(\Delta f + v_{TW}), \quad x(h) = 0 \tag{17.65}$$

where v_{TW} that is active $\forall t \geq h$ is defined by Eq. (17.54) and $\bar{\alpha}, \bar{\beta} > 0$ are selected in accordance with Eq. (17.31). In accordance with Theorem 17.2 $x_{r-1}(t) = x_r(t) = 0$ $\forall t \geq h$ in 2-SM. Then, system (17.4) in 2-SM become $\forall t \geq h$

$$
\begin{aligned}
&\dot{\tilde{x}} = \tilde{A}\tilde{x}, \quad x_{r-1}(t) = x_r(t) = 0, \\
&\tilde{x} = \left[x_1, x_2, \ldots, x_{r-2}\right]^T, \quad \tilde{x}(h) = 0, \\
&\tilde{A} = \begin{bmatrix} 0\ 1 \ldots 0 \\ \cdot\ \ \cdot\ \ \cdot\ \ \cdot \\ 0\ 0 \ldots 1 \\ 0\ 0 \ldots 0 \end{bmatrix} \in R^{(r-2)\times(r-2)}
\end{aligned}
\tag{17.66}
$$

which yields $\tilde{x}(t) = 0, \quad x_{r-1}(t) = x_r(t) = 0 \ \ \forall t \geq h$. Theorem is proven.

17.8.5 Robustness of Impulsive/2-SMC Control in Reduced Information Environment with Respect to Measurement Uncertainty and Chattering

17.8.5.1 Robustness to Uncertainties in Measurement of $x(0)$

If $x(0)$ is measured inaccurately, then the impulsive control v_{imp} in (17.12) or the *practical* impulsive control v_{imp_h} in (17.15)–(17.18), (17.61) will not move $x(t) \to 0$ instantaneously or in short time but to a domain $x(0^+) \in \Omega * (x)$ or $x(h) \in \Omega * (x)$. The problem is that system's (17.4) dynamics in 2-SM in (17.64) or (17.66) are unstable. The instability is due to multiple zero eigenvalues of matrices $\bar{A}$ and $\tilde{A}$. This instability is polynomial, but not exponential, which limits the rate of divergence, but it is still instability. The way in managing instability is inspired by the practice of using Global Positioning System (GPS) in correcting initial conditions of IMU integrators that integrate the measured acceleration while obtaining the velocity and the position during the flight of aerospace vehicles. These GPS measurements (snap knowledge) of the vector $x(t)$ are available in isolated time instants $t_k, \quad k = 1, 2, \ldots$ with a certain time increment Δ.

The following assumption is made about the uncertain initial condition $x(0)$ and, possibly, uncertain snap values $x(t_k)$, $k = 1, 2, 3, \ldots$

(A11) The uncertain values $x(0)$ and $x(t_k)$, $k = 1, 2, 3, \ldots$ yield such coefficients in (17.18) that the practical impulsive control (17.61) being applied at t_k, $k = 1, 2, 3, \ldots$ drives state vector $x(t_k + h)$ to the domain $\Omega * (x)$.

The following corrected impulsive-2-SMC algorithm in a reduced information environment in the presence of uncertain initial conditions $x(0)$ and uncertain $x(t_k)$, $k = 1, 2, 3, \ldots$ is proposed [17]:

(a) The domain $x(t) \in \Omega^*$ that provides a desired accuracy of x-stabilization is introduced

(b) make $k = 0$

(c) if $x(t_k) \in \Omega^*$ then

(d) make $k = k + 1$ and go to (c), otherwise

(e) treat $x(t_k)$ as a snap/initial value of the state vector and apply the *practical* v_{imp} impulsive control v_{imp_h} in (17.15)–(17.18), (17.61), where t_k is the initial time, make $k = k + 1$ and go to (c).

The accuracy of stabilization of the state vector $x(t)$ that is provided by the corrected practical impulsive-2-SMC algorithm in (a)–(c) in the presence of uncertain initial conditions is estimated upon the following assumptions.
(A12) Given $\Omega^* : \left\{\|x\|^2 \le \varepsilon\right\}$ then for $x(t_k) \in \Omega^*,\ \ x(t_{k+1}) \notin \Omega^* \to \|x(t_k)\|^2 \le \varepsilon,\ \ \|x(t_{k+1})\|^2 > \varepsilon$ holds.
(A13) System (17.4), (17.5) is controlled by *practical impulsive-super-twisting* control
(A14) The snap time intervals $\Delta = t\_{k+1} - t_k = const, h << \Delta$.

Theorem 17.20 *Assume that the assumptions (A11)–(A14) hold, then the accuracy* $\Xi = \|x(t_{k+1})\|^2 - \varepsilon$ *of reaching the domain* $\Omega^* : \left\{\|x\|^2 \le \varepsilon\right\}$ *is estimated as*

$$\Xi \le \Delta \sum_{i=1}^{r-1} |a_i| \left(|a_i|\, \Delta + 2\, |x_i(t_k)|\right) \tag{17.67}$$

where

$$a_i = x_{i+1}(t_k) + \frac{1}{2} x_{i+2}(t_k)\Delta + \cdots + \frac{1}{(r-i)!} x_{r-i}(t_k)\Delta^{r-i-1} \tag{17.68}$$

Proof At the time instant t^+_{k+1} the impulsive control v_{imp_h} is applied and drives the state vector $x(t_{k+1})$ to the domain Ω^*, i.e. $\|x(t_{k+1} + h)\|^2 \le \varepsilon$. Assume that the assumptions (A11)–(A13) hold. Then, having a snap knowledge $x(t_k)$ and bearing in mind that $x_r(t_k) = 0$, the system's (17.4), (17.5) dynamics $\forall t \in \left[t_k, t\_{k+1}\right],\ t_k > h$ are described by

$$\begin{aligned} &\dot{\bar{x}} = \bar{A}\bar{x},\ \ x_r = 0, \\ &\bar{x} = \left[x_1, x_2, \ldots, x_{r-1}\right]^T,\ \ \bar{x}(t_k) = \bar{x}_k, \\ &\bar{A} = \begin{bmatrix} 0\ 1\ \ldots\ 0 \\ .\ \ .\ \ .\ \ .\ \ . \\ 0\ 0\ \ldots\ 1 \\ 0\ 0\ \ldots\ 0 \end{bmatrix} \in \mathrm{R}^{(r-1)\times(r-1)} \end{aligned} \tag{17.69}$$

Integrating Eq. (17.69) on the time interval $t \in \left[t_k, t\_{k+1}\right]$ we obtain

$$x_i(t_{k+1}) = x_i(t_k) + a_i \Delta \tag{17.70}$$

where a_i are defined by Eq. (17.68). Next computing $\Xi = \|x(t_{k+1})\|^2 - \varepsilon$ we obtain

$$\Xi = \|x(t_{k+1})\|^2 - \varepsilon \le \|x(t_{k+1})\|^2 - \|x(t_k)\|^2 = \Delta \sum_{i=1}^{r-1} |a_i| \left(|a_i|\, \Delta + 2\, |x_i(t_k)|\right) \tag{17.71}$$

The theorem is proven.

Remark 17.11 Note that the proposed corrective algorithm in (a)–(e) guarantees a certain stabilization accuracy in the reduced information environment, while avoiding integrating of IMU-measured (snap knowledge) states in order to obtain the non-measurable states for closing a full state control feedback.

17.8.5.2 Chattering Effect

Assume that x_r (in a case of super-twisting control) exhibits chattering that can be modeled as a high frequency periodical signal with a frequency ω and zero mean in real 2-SM. Then, as a result of integration of the system (17.63) or (17.65), the amplitudes of the remaining variables $|x_{r-i}| \sim 1/\omega^i \ \forall i = 1, 2, \ldots, r-1$. In other words, the effect of chattering is reduced in the remaining variables, since the amplitude of chattering is attenuated reciprocally to ω^i. The same logic can be applied to the twisting controller that operates in the reduced information environment.

17.9 Case Studies

17.9.1 Output Feedback-Hybrid-Impulsive Control of Perturbed Double Integrator: DC-Motor Velocity Stabilization

Consider the following mathematical model of a DC-motor [33]:

$$\dot{\omega}(t) = \frac{1}{J}\left(-b\omega(t) + K_m i_a(t) - T_L\right)$$

$\frac{d\, i_a(t)}{dt} = \frac{1}{L_a}\left(-R_a i_a(t) - K_a\omega(t) + V_a(t)\right)$ where $\omega(t)\,[1/s]$ is the angular velocity, $i_a(t)$ [A] is the armature current, $V_a(t)$ [V] is the armature voltage (control law input), J $[\mathrm{kg}\cdot\mathrm{m}^2]$ is the rotor's moment of inertia, K_m $[\frac{\mathrm{N\cdot m}}{\mathrm{A}}]$ is the motor constant, K_a $[\frac{\mathrm{A}}{\mathrm{H}}]$ is the feedback electromotive force constant, R_a $[\Omega]$ is the armature resistance, L_a [H] is the armature inductance, b $[\frac{\mathrm{kg\cdot m^2}}{\mathrm{s}}]$ is the viscous friction coefficient, and T_L $[\mathrm{N}\cdot\mathrm{m}]$ is the bounded smooth disturbance. It is assumed that only $\omega(t)$ is measured.

The objective is to design the simplified discontinuous output feedback-hybrid-impulsive control $V_a(t)$ in a form of Eq. (17.43) that drives $\omega(t)$, $\dot{\omega}(t) \to 0$ in the presence of the smooth disturbance T_L.

The input–output dynamics of relative degree 2 are obtained:

$$\ddot{\omega}(t) = \left(\frac{b^2}{J^2} - \frac{R_a K_m}{J L_a}\right)\omega(t) - \left(\frac{bK_m}{J^2} + \frac{R_a K_m}{J L_a}\right) i_a(t) + \frac{K_m}{J L_a} V_a(t) - \frac{1}{J}\dot{T}_L + \frac{b}{J^2} T_L$$

The control input $V_a(t)$ is represented as $V_a(t) = V_{a1}(t) + V_{a2}(t)$, where: $V_{a1}(t) = \frac{JL_a}{K_m}\left\{-\left(\frac{b^2}{J^2} - \frac{K_a K_m}{JL_a}\right)\omega(t)\right\}$ and $V_{a2}(t) = \frac{JL_a}{K_m}[v_1(t)]$,$\varphi(t) = -\left(\frac{bK_m}{J^2} + \frac{R_a K_m}{JL_a}\right) i_a(t) - \frac{1}{J}\dot{T}_L + \frac{b}{J^2}T_L$. Then the DC-motor dynamics is reduced to a form (17.5) or (17.32) $\ddot{\omega}(t) = u(t) + \varphi(t)$. The values of the DC-motor parameter used for simulation are taken as [29]: $K_m = K_a = 0.005$, $L_a = 0.02$ [H], $R = 1$ [Ω], $b = 0.35$, $J = 0.01$ [kg $\cdot$ m^2], and $T_L = 0.2\ \sin(t) + 0.1\ \cos(2t)$ [Nm]. The initial conditions are taken as: $\omega(0) = 20$ 1/s, $i(0) = 95$ A, $h = 0.005$ s, and $K = 255$. It is assumed that a "snap" knowledge of $\dot{\omega}$ (when $\omega = 0$) is available. The results of the simulations are presented in Figs. 17.3, 17.4, 17.5, 17.6 and 17.7.

The presented simulation plots (Figs. 17.3, 17.4, 17.5, 17.6 and 17.7) confirm the efficacy of the proposed discontinuous output feedback-hybrid-impulsive control on a practical example. The impulsive control drives $\dot{\omega}(t)$ close to the origin in a very short time as soon as $\omega(t)$ becomes equal to zero at $t = 0.76$ s. Then the discontinuous output feedback in terms of the armature voltage takes over (Fig. 17.5). Next, the impulsive control compensates the effect of the disturbance, etc. It is clear from Figs. 17.3, 17.4, 17.5, 17.6 and 17.7 that the simplified discontinuous output feedback-hybrid-impulsive control (17.43) drives $\omega(t)$, $\dot{\omega}(t) \to 0$ of the *perturbed* DC-motor with a high accuracy.

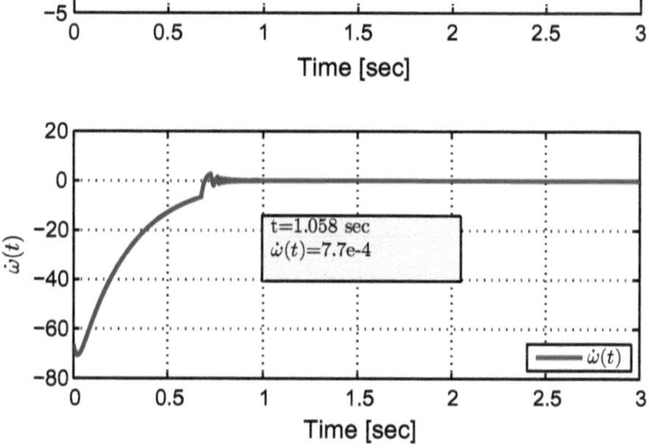

Fig. 17.3 Time history of the states $\dot{\omega}[1/s^2]$, $\omega[1/s]$ with $h = 0.05$[s]

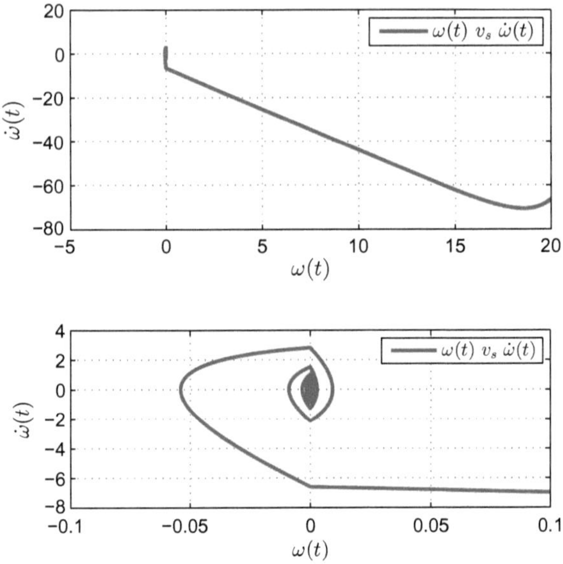

Fig. 17.4 From the *top* to the *bottom*: the phase portrait of the states $\dot{\omega}[1/\text{s}^2]$, $\omega[1/\text{s}]$ and its zoomed portion

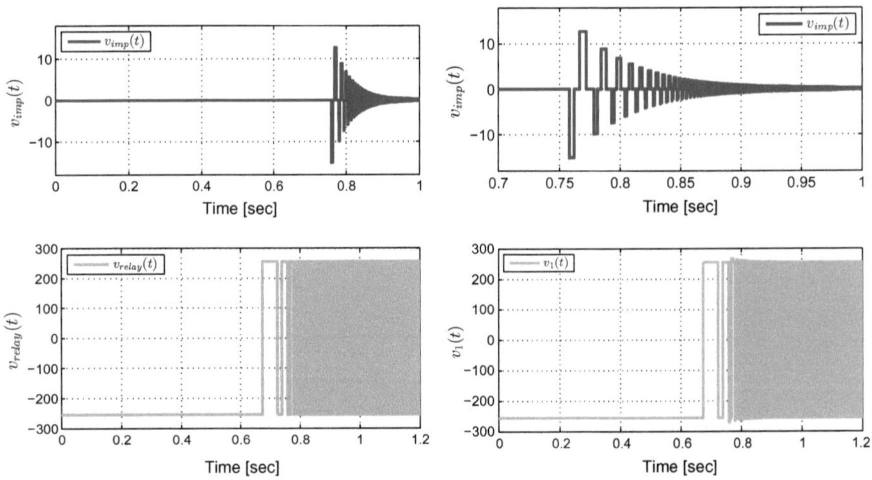

Fig. 17.5 From the *top* to the *bottom*: impulsive control $v_{imp}(t) = -\sum_{i=1,2,3,...} \dot{\omega}(t_i^-)\delta(t - t_i)$[V], its zoomed portion, output feedback relay control $v_{relay}(t) = -K\,\text{sign}(\omega)$[V], and the simplified discontinuous output feedback-hybrid-impulsive control (17.43): $v_1(t) = v_{imp}(t) + v_{relay}(t)$[V]

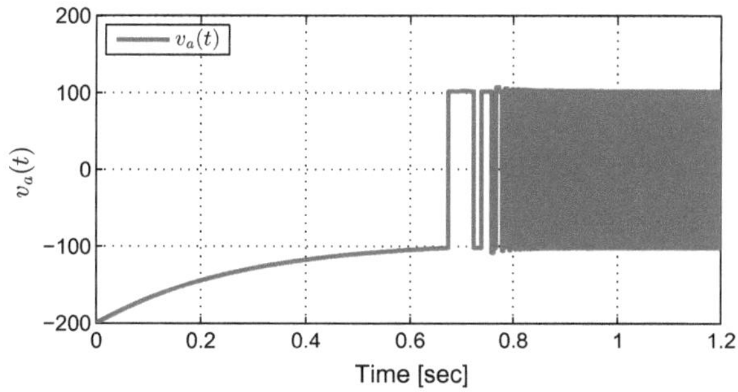

Fig. 17.6 Armature voltage $V_a(t)$[V]

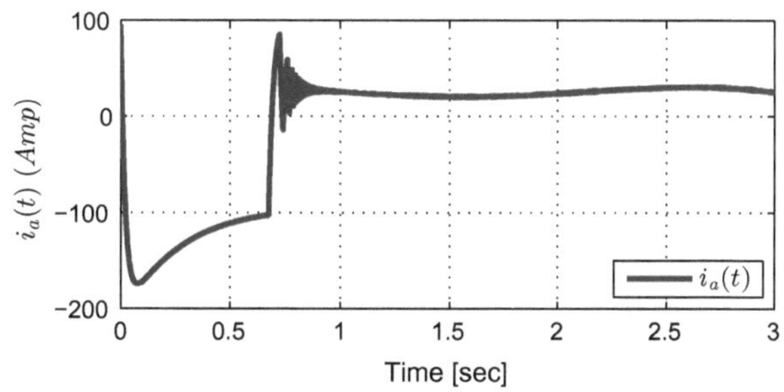

Fig. 17.7 Armature current $i_a(t)$[A]

17.9.2 Third Order Hybrid-Impulsive Continuous HOSM Control

Simulation set up. The system in Eqs. (17.3) or (17.4), (17.5) is studied with $r = 3$, $\Delta f(t) = \sin(t) + \cos(2t)$, and initial conditions $x_1(0) = 1,\ x_2(0) = -1,\ x_3(0) = 1$. The width of the delta impulse is taken as $h = 0.01$ s, and coefficients $q_0,\ q_1,\ q_2$ are calculated as: $q_0 = -1,\ q_1 = 3.98,\ q_2 = -8.01$. Also, $\gamma_1 = 8$, $\gamma_2 = 12$, $\gamma_3 = 6$, $\eta_1 = 0.56$, $\eta_2 = 0.66$, $\eta_3 = 0.8$.

The impulsive control took a form (17.72)

$$v_{imp\_h}(t) = q_0\delta(t) + q_1\dot{\delta}(t) + q_2\ddot{\delta}(t) \tag{17.72}$$

The following cases have been considered:
Case 1. The system was simulated with control (17.23)–(17.25) without impulsive control. The results of the simulations are shown in Fig. 17.8a–c.
Case 2. The unperturbed system was simulated with control (17.6), (17.7), (17.23)–(17.25). The control function (17.23)–(17.25) was applied with a time delay $\delta = h = 0.01$ s. The results of the simulations are shown in Fig. 17.9a–f. It can be seen from Fig. 17.9a–c that the finite time convergence of $x_1, x_2, x_3 \to 0$ is achieved by time $t = h = 0.01$ s. The components of the impulsive control are shown in Fig. 17.9d–f.
Case 3. The perturbed system was simulated with control (17.6), (17.7), (17.23). The control function (17.23)–(17.25) and the disturbance were applied with a time

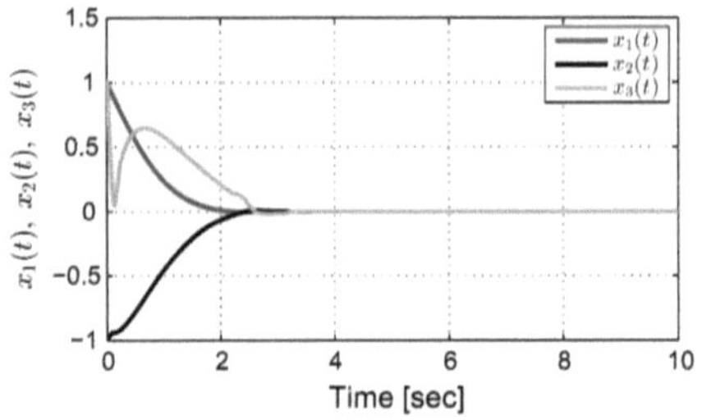

(a) Finite time outputs stabilization of the states.

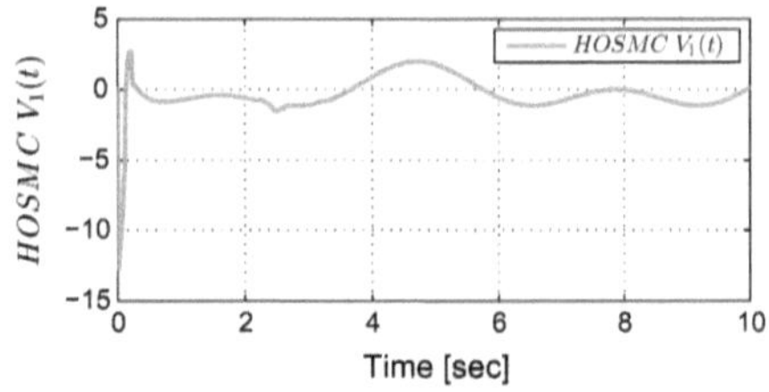

(b) The continuous HOSMC control.

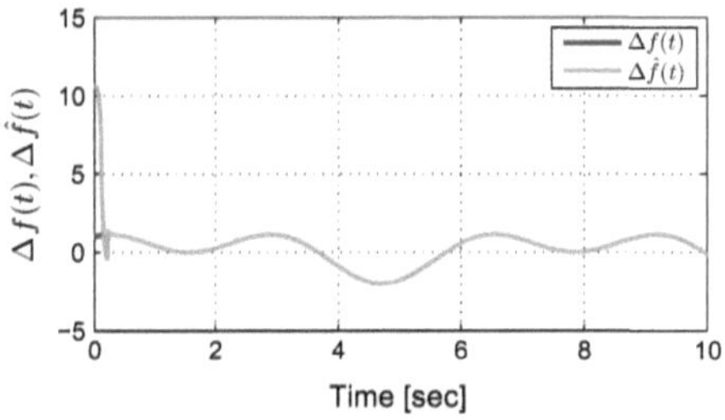

(c) A The continuous HOSMC disturbance observer with $\Delta f(t)$.

Fig. 17.8 It is clear that the finite time convergence of $x_1, x_2, x_3 \to 0$ has been achieved by the continuous control in the presence of the bounded disturbance

delay $\delta = h = 0.01$ s. The results of the simulations are shown in Fig. 17.10a–c. Apparently, the convergence time in the perturbed system) with impulsive action (see Fig. 17.10a–c) is shorter than the convergence time in the same system without impulsive action (see Fig. 17.8a).

17.9.3 *Hybrid-Impulsive 2-SMC in Reduced Information Environment: Perturbed System with Practical Impulsive-Super-Twisting Control and Uncertain Measurement of the Initial Conditions*

Simulation set up. Consider perturbed system (17.4) with relative degree $r = 3$

$$\dot{x}_1 = x_2, \quad \dot{x}_2 = x_3, \quad \dot{x}_3 = v + \Delta f(t) \tag{17.73}$$

and only state x_3 is measured in the case of super-twisting control, and the states x_2, x_3 are measured in the case of twisting control, while a snap knowledge of $x_1(0)$, $x_2(0)$ is required. The system (17.73) is studied with

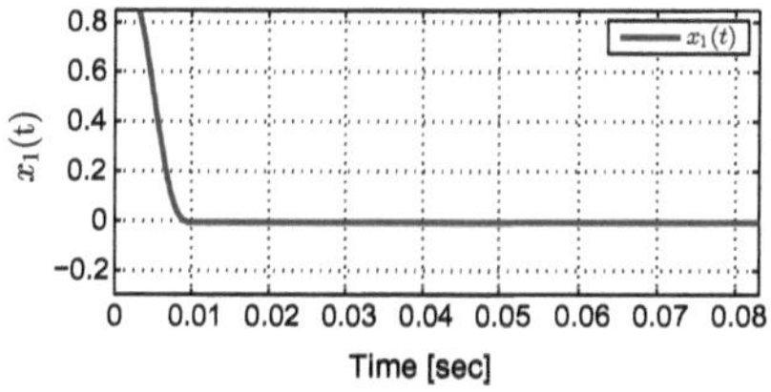

(a) Time history of $x_1(t)$ finite time convergence with impulsive control only.

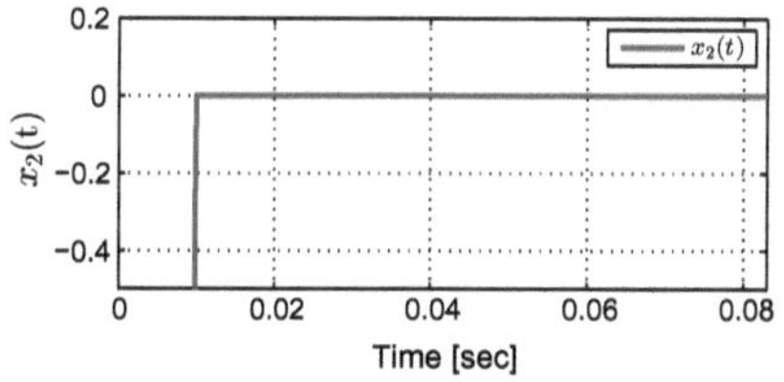

(b) Time history of $x_2(t)$ finite time convergence with impulsive control only.

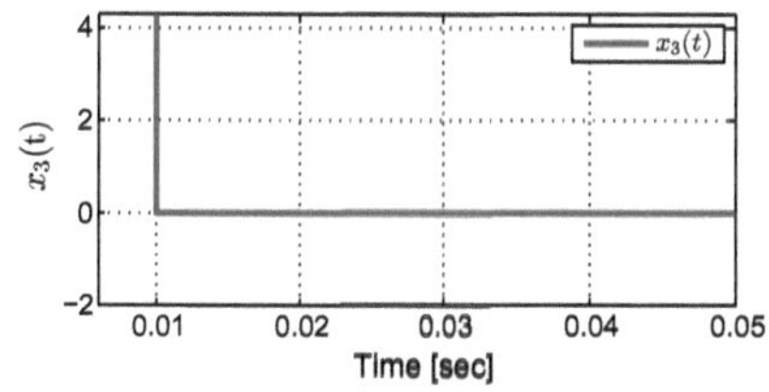

(c) Time history of $x_3(t)$ finite time convergence with impulsive control only.

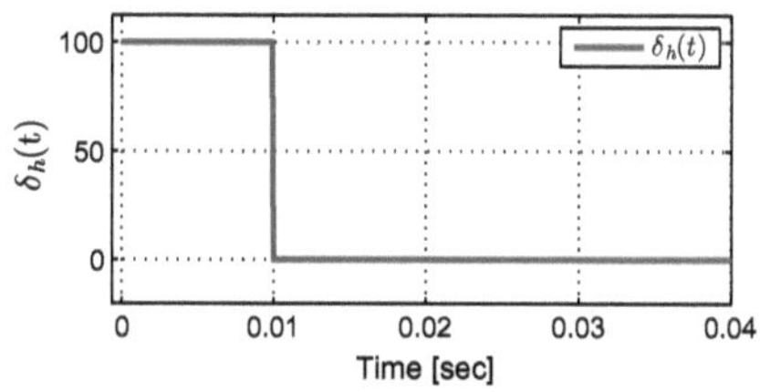

(d) Time history of $\delta(t)$ impulsive control.

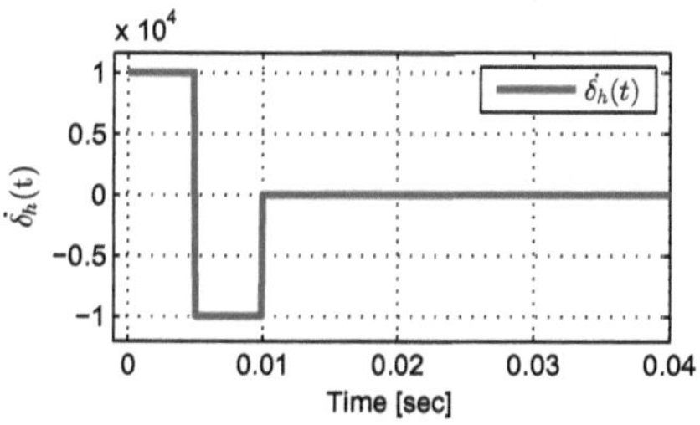

(e) Time history of $\dot{\delta}(t)$ impulsive control.

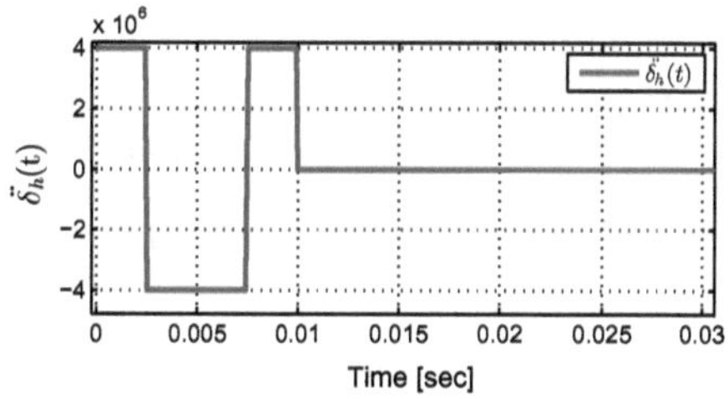

(f) Time history of $\ddot{\delta}(t)$ impulsive control.

Fig. 17.9 Case 2 of study

$$\Delta f(t) = \begin{cases} 0, & if\ t \in [0, h] \\ \sin(t), & if\ t \geq h \end{cases} \tag{17.74}$$

and initial conditions $x_{10} = 1, \quad x_{20} = 1, \quad x_{30} = 1$. The perturbation (17.74) and its derivative are bounded: $|\Delta f(t)| \leq 1, \quad \left|\Delta \dot{f}(t)\right| \leq 1$. The width of the *practical* delta impulse and its derivatives is taken as $h = 0.1$ s.

Taking into account Eqs. (17.18) and (17.19), the impulsive control takes a form

$$v_{imp\_h}(t) = -\delta_h(t) - 1.1\dot{\delta}_h(t) - 1.105\ddot{\delta}_h(t), \quad t \in [0,\ h] \tag{17.75}$$

It is assumed that last variable $x_3(t)$ being measured exactly in continuous time, and impulsive-super-twisting control is given by

$$v_1(t) = v_{STW}(t) + v_{imp\_h}(t) \tag{17.76}$$

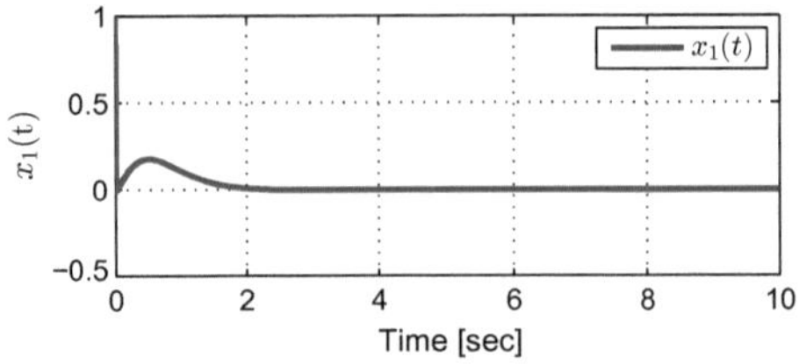

(a) Time history of $x_1(t)$ finite time convergence.

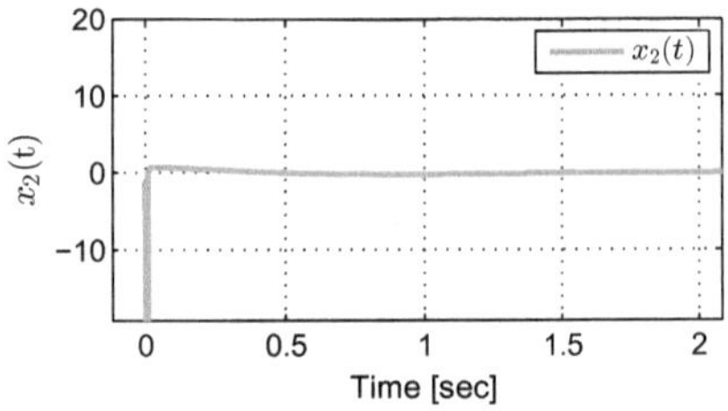

(b) Time history of $x_2(t)$ finite time convergence.

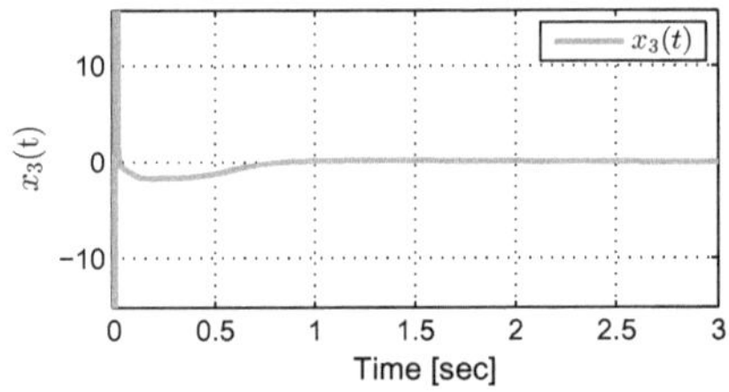

(c) Time history of $x_3(t)$ finite time convergence.

Fig. 17.10 It becomes clear from Fig. 17.10a–c that the convergence of $x_1, x_2, x_3 \to 0$ is achieved by time $t = h = 0.01$[s]. However, the finite time transient response follows due to some finite time required for the super-twisting disturbance observer to reconstruct the disturbance and a consecutive finite time convergence that is enforced by the control (17.4)

with $v_{imp\_h}(t)$ given by Eq. (17.75), where $x_3(0)$ is measure exactly and $x_3(0) = 1.0$. The initial values $x_1(0)$ and $x_2(0)$ that both equal to 1.0 are measured with 2% error and the available values are $x_{1m}(0) = 0.98$ and $x_{1m}(0) = 1.02$, where $x_{1m}(0),\ x_{1m}(0)$ are measured values of $x_1(0)$ and $x_2(0)$. These errors will affect the impulsive control (17.75). The super-twisting control is

$$\begin{aligned} v_{STW} &= -\alpha\,|x_3|^{1/2}\,\mathrm{sign}(x_3) + w \\ \dot{w} &= -\beta\,\mathrm{sign}(x_3), \qquad \alpha = 2, \quad \beta = 1, \quad t \geq h \end{aligned} \tag{17.77}$$

A required stabilization accuracy is: $x \in \Omega^*(x),\quad \Omega^*(x) : \left\{\|x\|^2 \leq 0.01\right\}$. The "snap" values $x_1(t_k), x_2(t_k),\quad k = 1, 2, 3, \ldots$ are assumed to be obtained exactly by an external measurement device (for instance it can be GPS) with a time increment $\Delta = 1$[s], i.e. at time instances $t_1 = 1.0,\quad t_2 = 2.0,\quad t_3 = 3.0, \ldots$. The variable $x_3(t)$ is exactly measured continuously. The control strategy is as described in Sect. 17.8.

The time histories of $\|x\|^2$ and x_1, x_2, x_3 are shown in Fig. 17.11a and b–d respectively. The plots of *practical* impulses $\delta_h(t),\ \dot{\delta}_h(t)$, and $\ddot{\delta}_h(t)$ applied at time $t = 0$ s and at time $t = 6$ s are shown in Fig. 17.11e–g. The impulsive control $v_{imp\_h}(t)$ applied at time $t = 0$ s, and at time $t = 6$ s is shown in Fig. 17.11h. Plots of $\Delta f(t) = \sin(t)$ and $v_{STW}(t)$ are presented in Fig. 17.11i, j respectively. Simulations

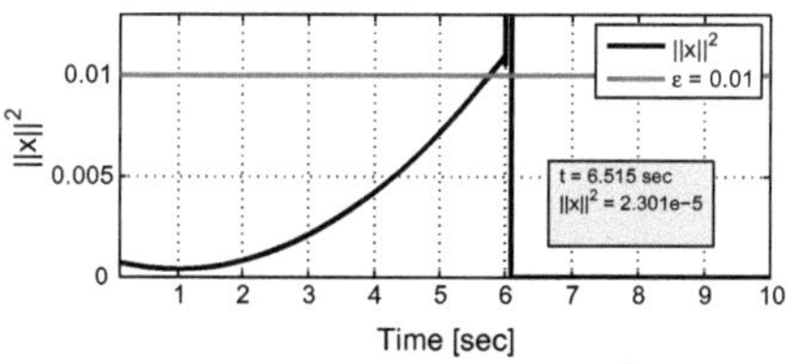

(a) Time history of $\|x\|^2$.

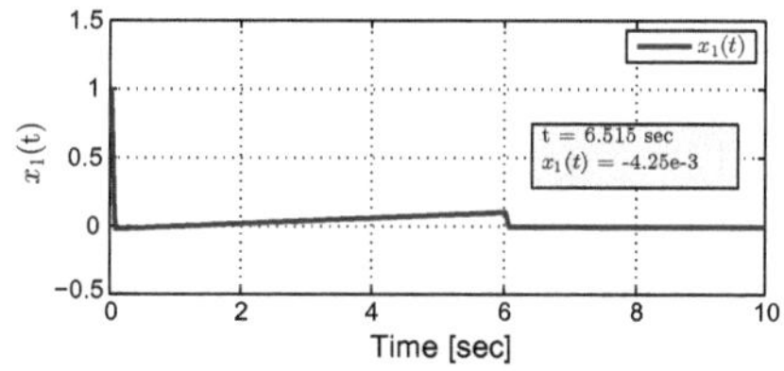

(b) Time history of $x_1(t)$ with impulsive-super-twisting control.

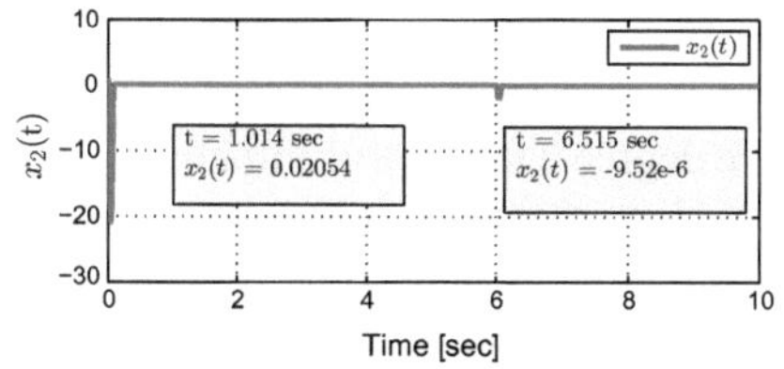

(c) Time history of $x_2(t)$ with impulsive-super-twisting control.

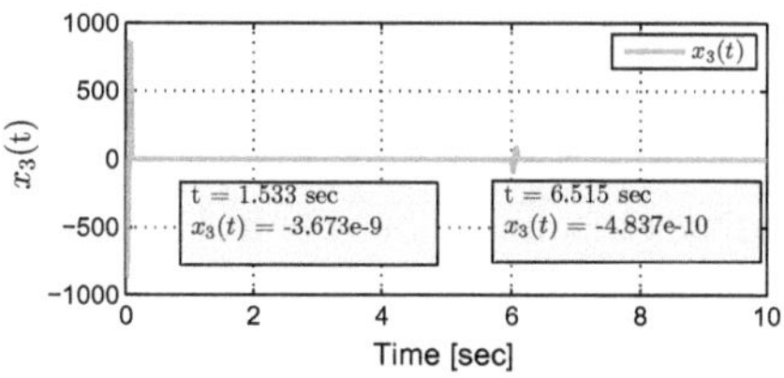

(d) Time history of $x_3(t)$ with impulsive-super-twisting control.

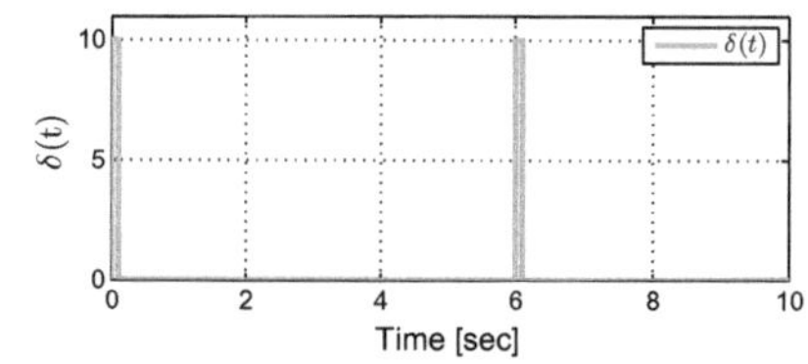

(e) Time history of $\delta(t)$ practical impulse.

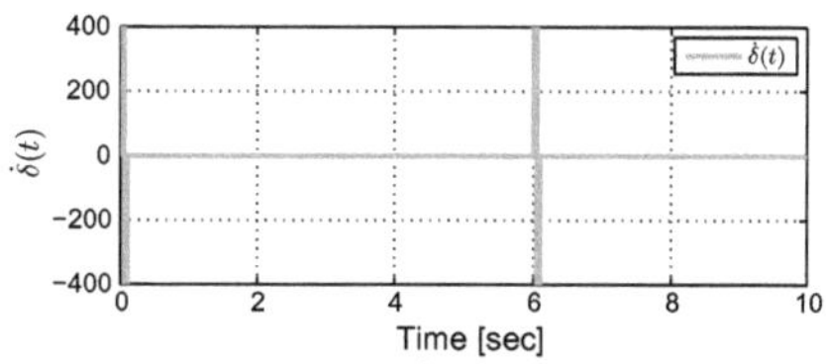

(f) Time history of $\dot{\delta}(t)$ practical impulse.

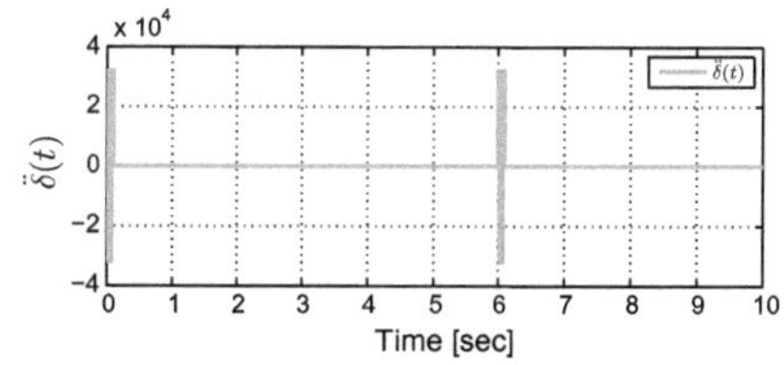

(g) Time history of $\ddot{\delta}(t)$ practical impulse.

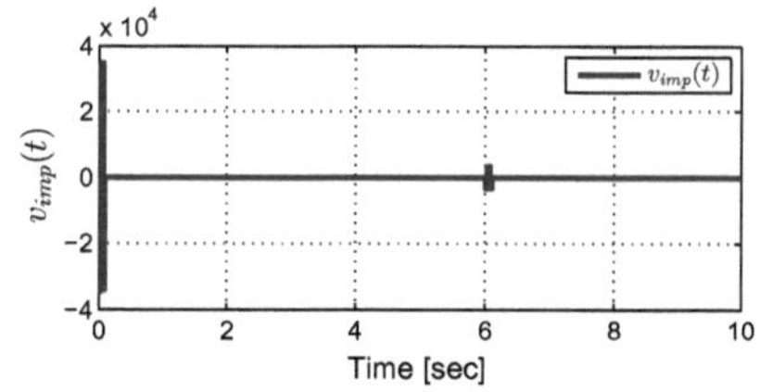

(h) The impulsive control $v_{imp}(t)$.

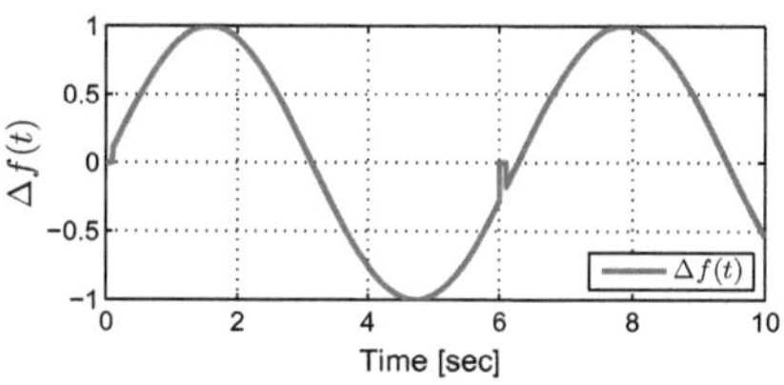

(i) Time history of the disturbance $\Delta f(t)$.

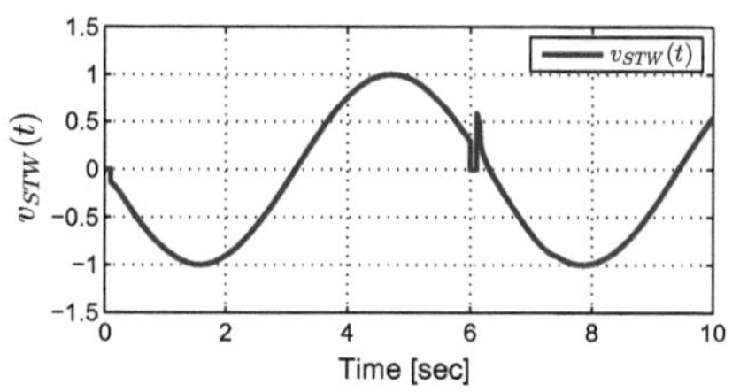

(j) Time history of $v_{STW}(t)$ super-twisting control.

Fig. 17.11 Case 2 of study

confirmed the theoretical expectations for the case with uncertain initial conditions and exact consecutive snap knowledge $x_1(t_k)$, $x_2(t_k)$ in the isolated time instants $t = 1.0,\ 2.0,\ 3.0, \ldots$ s.

17.10 Conclusion

A hybrid-impulsive second order/higher order sliding mode SMC/HOSM control approach that allows reducing dramatically the convergence time practically to zero, and achieving instantaneous (or short time) convergence and uniformity is studied in this chapter. Specifically, for systems of relative degree 2 the hybrid-impulsive output feedback discontinuous control achieves almost instantaneous uniform convergence to the origin and stabilization of the trajectory tracking error in a close vicinity of the origin. The restitution rules are enforced using the Dirac delta function and its derivative. The discontinuous-hybrid-impulsive *output feedback* control with *practically implemented* impulsive actions guarantees the asymptotic stability of the system's origin. The Lyapunov analysis of the considered hybrid-impulsive-discontinuous systems is performed. In systems of *arbitrary relative degree* hybrid-impulsive continuous HOSM (CHOSM) control reduces dramatically the convergence time practically to zero. This approach allows reducing the CHOSM amplitude in the sliding mode, since the task of compensating the initial conditions is addressed by the impulsive action, which is implemented in a piece-wise constant format. In systems of *arbitrary relative degree* in a *reduced information environment* two hybrid-impulsive 2-SMCs are studied. Only last *one or two* system variables are assumed available. Only "snap" knowledge (or knowledge in a single time instant) of the all states is required in order to facilitate the effective impulsive action. The efficacy of studied hybrid-impulsive control algorithms are illustrated via simulations.

References

1. Aldukali, F.M., Shtessel, Y.B.: Continuous higher order sliding mode control with impulsive action. In: 2015 IEEE 54th Annual Conference on Decision and Control (CDC), pp. 5420–5425. IEEE (2015)
2. Aldukali, F.M., Shtessel, Y.B., Glumineau, A., Plestan, F.: Impulsive-super-twisting control in reduced information environments. In: American Control Conference (ACC), pp. 7207–7212. IEEE (2016)
3. Angulo, M.T., Moreno, J.A., Fridman, L.: An exact and uniformly convergent arbitrary order differentiator. In: 2011 50th IEEE Conference on Decision and Control and European Control Conference (CDC-ECC), pp. 7629–7634. IEEE (2011)
4. Bhat, S.P., Bernstein, D.S.: Geometric homogeneity with applications to finite-time stability. Math. Control Signals Syst. (MCSS) **17**(2), 101–127 (2005)
5. Butt, A., Popp, C., Pitts, H., Sharp, D.: NASA ares i launch vehicle roll and reaction control systems design status. In: 45th AIAA/ASME/SAE/ASEE Joint Propulsion Conference and Exhibit, p. 5130 (2009)
6. Edwards, C., Shtessel, Y.: Adaptive continuous higher order sliding mode control. IFAC Proc. Vol. **47**(3), 10826–10831 (2014)

7. Gelfand, I.M., Shilov, G.E.: Generalized Functions. Applications of Harmonic Analysis, vol. 4. Academic Press, New York (1964)
8. Glumineau, A., Shtessel, Y., Plestan, F.: Impulsive-sliding mode adaptive control of second order system. IFAC Proc. Vol. **44**(1), 5389–5394 (2011)
9. Glumineau, A., Shtessel, Y., Plestan, F.: Lyapunov stability of a hybrid impulsive-sliding mode adaptive controller for second order system. In: 2012 IEEE 51st Annual Conference on Decision and Control (CDC), pp. 5477–5481. IEEE (2012)
10. Guan, Z.H., Hill, D.J., Shen, X.: On hybrid impulsive and switching systems and application to nonlinear control. IEEE Trans. Autom. Control **50**(7), 1058–1062 (2005)
11. Honeywell: reaction control system (2010). http://www51.honeywell.com/aero/common/documents/myaerospacecatalog-documents/Missiles-Munitions/$Reaction_Control_Systems_(Tactical).pdf$
12. Karageorgos, A.D., Pantelous, A.A., Kalogeropoulos, G.I.: Transferring instantly the state of higher-order linear descriptor (regular) differential systems using impulsive inputs. J. Control Sci. Eng. **2009**, 6 (2009)
13. Khalil, H.K.: Noninear Systems. Prentice-Hall, New Jersey (2002)
14. King, A.: Inertial navigation-forty years of evolution. GEC Rev. **13**(3), 140–149 (1998)
15. Levant, A.: Sliding order and sliding accuracy in sliding mode control. Int. J. Control **58**(6), 1247–1263 (1993)
16. Levant, A.: Higher-order sliding modes, differentiation and output-feedback control. Int. J. Control **76**(9–10), 924–941 (2003)
17. Levant, A.: Principles of 2-sliding mode design. Automatica **43**(4), 576–586 (2007)
18. Li, H.Y., Luo, Y.Z., Tang, G.J., et al.: Optimal multi-objective linearized impulsive rendezvous under uncertainty. Acta Astronaut. **66**(3), 439–445 (2010)
19. Miller, B.M., Rubinovich, E.Y.: Impulsive Control in Continuous and Discrete-Continuous Systems. Springer Science & Business Media, New York (2012)
20. Moreno, J.A., Osorio, M.: Strict Lyapunov functions for the super-twisting algorithm. IEEE Trans. Autom. Control **57**(4), 1035–1040 (2012)
21. Orlov, Y.V.: Discontinuous Systems: Lyapunov Analysis and Robust Synthesis Under Uncertainty Conditions. Springer Science & Business Media, New York (2008)
22. Pisano, A., Usai, E.: Sliding mode control: a survey with applications in math. Math. Comput. Simul. **81**(5), 954–979 (2011)
23. Plestan, F., Moulay, E., Glumineau, A., Cheviron, T.: Robust output feedback sampling control based on second-order sliding mode. Automatica **46**(6), 1096–1100 (2010)
24. Rosello, A.D.: A vehicle health monitoring system for the space shuttle reaction control system during reentry. Ph.D. thesis, Draper Laboratory (1995)
25. Shtessel, Y., Edwards, C., Fridman, L., Levant, A.: Sliding Mode Control and Observation. Springer, Berlin (2014)
26. Shtessel, Y., Glumineau, A., Plestan, F., Aldukali, F.M.: Hybrid-impulsive second-order sliding mode control: Lyapunov approach. Int. J. Robust Nonlinear Control (2016)
27. Shtessel, Y., Glumineau, A., Plestan, F., Aldukali, F.M.: Hybrid-impulsive second-order sliding mode control: Lyapunov approach. Int. J. Robust Nonlinear Control **27**(7), 1064–1093 (2017)
28. Shtessel, Y.B., Moreno, J.A., Fridman, L.M.: Twisting sliding mode control with adaptation: Lyapunov design, methodology and application. Automatica **75**, 229–235 (2017)
29. Sobolev, S.L., Browder, F.E.: Applications of Functional Analysis in Mathematical Physics. American Mathematical Society, Dunod (1962)
30. Weiss, M., Shtessel, Y.: An impulsive input approach to short time convergent control for linear systems. Advances in Aerospace Guidance, Navigation and Control, pp. 99–119. Springer, Berlin (2013)
31. Yang, T., Chua, L.O.: Impulsive stabilization for control and synchronization of chaotic systems: theory and application to secure communication. IEEE Trans. Circuits Syst. I: Fundam. Theory Appl. **44**(10), 976–988 (1997)
32. Yang, X., Cao, J., Ho, D.W.: Exponential synchronization of discontinuous neural networks with time-varying mixed delays via state feedback and impulsive control. Cogn. Neurodynamics **9**(2), 113–128 (2015)

Chapter 18
Sliding Mode Control Design Procedure for Power Electronic Converters Used in Energy Conversion Systems

Yazan M. Alsmadi, Vadim Utkin and Longya Xu

18.1 Introduction

Due to its order reduction property, good dynamic performance and low sensitivity to disturbances and plant parameter variations, sliding mode control (SMC) has been the method of choice for handling nonlinear systems with uncertain dynamics and disturbances. Moreover, this control methodology reduces the complexity of feedback control design since the system is decoupled into independent lower dimensional subsystems [38, 39]. Because of these properties, sliding mode control has a wide range of applications in the areas of electric motors, power systems, power electronics, robotics, aviation and automotive control [1, 38–40].

Power electronic converters belong to the category of circuits controlled by switching devices with "ON/OFF" as the only admissible operating states [22, 28]. Therefore, control variables can take only values from a discrete set. For this type of circuits, SMC is preferred not only from a theoretical view but also a technological one. In the past, State Space Averaging approach was the method of choice to analyze some of the power conversion circuits [26, 35]. Using this technique, state space models and equations are obtained for each possible switch's position. The equations are then averaged over a switching cycle to obtain a low frequency averaged model. Linear

Y.M. Alsmadi (✉)
Department of Electrical Engineering, Jordan University of Science and Technology, Irbid 22110, Jordan
e-mail: ymalsmadi@just.edu.jo

V. Utkin · L. Xu
Department of Electrical and Computer Engineering, The Ohio State University, Columbus, OH 43212, USA
e-mail: utkin.2@osu.edu

L. Xu
e-mail: xu.12@osu.edu

J.B. Clempner and W. Yu (eds.), *New Perspectives and Applications of Modern Control Theory*, https://doi.org/10.1007/978-3-319-62464-8_18

(after linearization if needed) or nonlinear control theory are then applied to design a feedback compensator. SMC provides a powerful tool to control power converters such that linearization is not needed and analysis and control design methods are developed in the framework of the nonlinear models [26, 33, 35, 39].

The main drawback of any system with pulse-width modulation (PWM) control approaches, including SMC, is the appearance of undesirable oscillations having finite amplitude and frequency due to the presence of unmodeled dynamics or discrete time implementation. This phenomenon, called "Ripple" in power electronics literature or "Chattering" in control theory, may lower control accuracy or incur unwanted wear of mechanical components [24, 39, 40]. An additional obstruction of SMC is the fact that it results in frequency variations, which is not acceptable in many applications [24].

Although the methods outlined in literature [4, 9, 10, 12–14, 36] showed efficient suppression of chattering, these methods are disadvantageous or not even applicable when dealing with power electronics systems controlled by switches with "ON/OFF" as the only admissible operating states. For example, gain-dependent and equivalent-control based methods are not applicable to reduce chattering since the output of power converters can only take two (or finite number of) values. Equivalent control method of SMC also yields the continuous motion equation directly. This motion, called "slow motion," exits along with high frequency components [40]. The boundary layer approach, on the other hand, avoids generating sliding mode by replacing the discontinuous control action with a continuous saturation function [36, 38–40]. However, control discontinuities are inherent to these power electronics systems. When implementing a continuous controller, techniques such as PWM have to be exploited to adopt the continuous control law and feed it to the discontinuous system inputs. Therefore, it seems unjustified to bypass the inherent discontinuities in the system by converting a continuous control law to a discontinuous one by means of PWM. As an alternative, discontinuous control inputs should be used directly in control, and new methods should be investigated to reduce chattering under these operating conditions [38–40].

The most natural and straightforward way to reduce chattering in power converters is to increase the switching frequency. As technology progresses, switching devices are manufactured with enhanced switching frequency (up to 100s KHz) and high power rating. However, heat power losses enforce a new restriction. Even though switching is possible with high switching frequency; it is limited to the maximum allowable heat losses caused by switching.

The objective of this chapter is to provide a comprehensive sliding mode control design procedure for power electronic converters that are used in energy conversion systems. This includes the DC/DC, AC/DC and DC/AC power electronic converters.

First, a comprehensive control design procedure for DC/DC power converters based on sliding mode control methodology. The input inductor current control and subsequently output capacitor voltage regulation for different types of DC/DC are investigated. Sliding mode control algorithms for Buck and Boost converters with incomplete information about system states are developed by designing state observers. Chattering issues are also discussed and a chattering reduction design approach for power converters is proposed.

Second, a new sliding model control strategy for AC/DC power converters is presented. The basic idea is to apply a feedback implementation of Pulsewidth Modulation (PWM). The proposed control algorithm exhibits low sensitivity to disturbances and fast dynamic performance in addition to the main converter properties. This includes unity power factor, sinusoidal input currents, and low level of DC output voltage ripple.

Third, Sliding Mode Pulsewidth Modulation (SMPWM) control methodologies for current-controlled inverters are proposed. Two novel approaches based on the sliding mode concept are presented to make the system tracking reference inputs. The proposed approaches control the phase currents and the neutral point voltage simultaneously. Optimization of different operational criteria is also presented.

18.2 Sliding Mode Control of DC/DC Power Converters

The input inductor current and the output capacitor voltage are normally selected as state variables for DC/DC converters. For most converters used in practice, the motion rate of the current is much faster than the motion rate of the output voltage. Calling upon the Singular Perturbation theory [19, 20],[1] the control problems of the DC/DC converters can be solved by using a cascaded control structure with two control loops: an inner current loop and an outer voltage loop. The current control loop is based on PWM that can be implemented using SMC or hysteresis control while output voltage regulation is normally a result of the current control loop such that output voltage converges to the desired reference value if designing the ideal current tracking system is accomplished [32, 34, 39].

18.2.1 Direct SMC of DC/DC Power Converters

(A) Buck-Type DC/DC Converters

The circuit structure of a Buck-type DC/DC converter is shown in Fig. 18.1, where E is the source voltage, C is the storage capacitor, L is the loop inductor, R is the load resistance, I is the input current and V is the output voltage.

Based on circuit analysis, the dynamic model of the Buck converter is given by:

$$\dot{x}_1 = -\frac{1}{L}x_2 + u\frac{E}{L} \tag{18.1}$$

[1]Formally, conventional Singular Perturbation Theory is not applicable for differential equations with discontinuous right-hand side. But, if sliding mode appears, sliding mode equations are in compliance with the conditions of the theory.

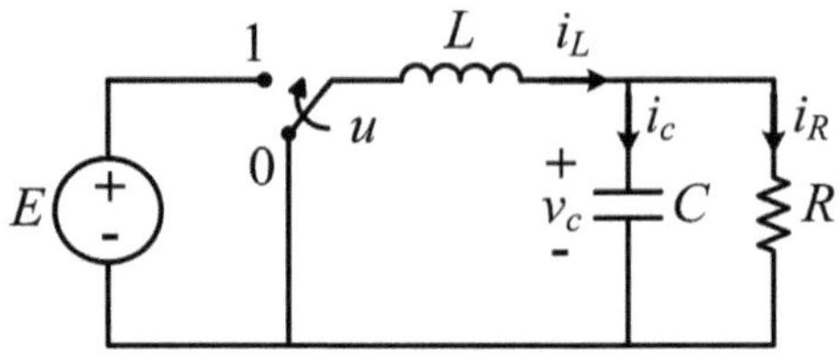

Fig. 18.1 Buck-type DC/DC converter

$$\dot{x}_2 = -\frac{1}{C}x_1 + \frac{1}{RC}x_2 \tag{18.2}$$

where $x_1 = I$ and $x_2 = V$.

The switching function of the switch can be defined as:

$$u = \begin{cases} 1 & switch\ is\ closed \\ 0 & switch\ is\ open \end{cases} \tag{18.3}$$

The control objective is to achieve a constant output voltage denoted by V_d. In other words, the steady state behavior of the Buck converter should be given by:

$$x_2 = V_d \tag{18.4}$$

$$\dot{x}_2 = \dot{V}_d = 0 \tag{18.5}$$

The proposed design methodology follows a two-step procedure known as integrator backstepping or regular form control [23, 25, 43]. First, it is assumed that x_1 in Eq. (18.2) can be handled as a control input. However, since x_1 is also the output of the current loop, given by Eq. (18.1), this first design step leads to the desired current $x^*{}_1$. After substituting Eqs. (18.4) and (18.5) into (18.2), the desired current is given by:

$$x^*{}_1 = \frac{V_d}{R} \tag{18.6}$$

The goal of the second step of the proposed design procedure is to ensure that the actual current x_1 tracks the desired current $x^*{}_1$ exactly. Due to its ideal tracking properties, sliding mode approach is an efficient tool for this task. If sliding mode is enforced in:

$$s = x_1 - x^*{}_1 = 0 \tag{18.7}$$

then, $x_1 = V_d/R$. In order to enforce sliding mode in the manifold $s = 0$, control u is defined as:

$$u = \frac{1}{2}(1 - sign(s)) \tag{18.8}$$

The condition for sliding mode to exist is derived from ($\lim_{t\to+0} \dot{s} < 0$, $\lim_{t\to-0} \dot{s} > 0$). As a result, sliding mode is enforced if:

$$0 < x_2 < E \tag{18.9}$$

Equation (18.9) defines an attraction domain of the sliding manifold. Since the control u, given by Eq. (18.8), does not contain a control gain that needs to be adjusted, the domain of attraction, given by Eq. (18.9) is predetermined by the system architecture. In steady state conditions, Eq. (18.9) is fulfilled by the definition of a Buck converter where the output voltage is smaller than the source voltage.

After the state of the inner current loop has reached the sliding manifold, i.e. converged to $s = 0$ at time $t = t_h$, $x_1 = x^*{}_1' = V_d/R$ holds for $t > t_h$ and the outer voltage loop is governed by:

$$\dot{x}_2 = -\frac{1}{RC}x_2 + \frac{1}{RC}V_d \tag{18.10}$$

The solution of Eq. (18.10) is given by:

$$x_2 = V_d + (x_2(t_h) - V_d)\ e^{\frac{-1}{RC}(1-t_h)} \tag{18.11}$$

where x_2 tends to V_d exponentially. Hence, the goal of the proposed control design is achieved.

(B) Boost-Type DC/DC Converters

The circuit structure of a Boost-type DC/DC converter is shown in Fig. 18.2, where circuit variables are similar to those defined for the Buck converter. Based on circuit analysis, the dynamic model of the Boost converter is given by:

$$\dot{x}_1 = -(1-u)\frac{1}{L}x_2 + \frac{E}{L} \tag{18.12}$$

$$\dot{x}_2 = (1-u)\frac{1}{C}x_1 - \frac{1}{RC}x_2 \tag{18.13}$$

where $x_1 = I$ and $x_2 = V$ and switching function u is given by Eq. (18.3).

Boost converters are more difficult to control than Buck converters. This is because of the fact that the control u appears in the voltage and current equations, and both in a bilinear fashion. Such a configuration implies a highly nonlinear system with

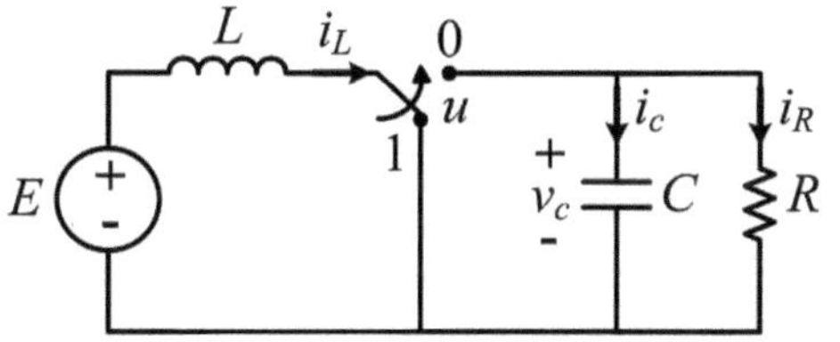

Fig. 18.2 Boost-type DC/DC converter

a difficult control design. It also leads to unstable dynamics with respect to current when voltage is the only variable to be controlled [32, 34].

Since Boost converters satisfy the Motion-Separation principle that is based on the *Singular Perturbation theory* [19, 20], their control problems can also be solved by using two cascaded control loops: an inner current and an outer voltage control loops. Sliding Mode control techniques will be used again to design the current control loop.

Similar to the proposed control design for the Buck converter, the desired current is obtained from the outer voltage loop as:

$$x^*{}_1 = \frac{V_d{}^2}{RE} \tag{18.14}$$

where V_d is the desired output capacitor voltage. The switching function of the inner current control is defined as:

$$s = x_1 - x^*{}_1 \tag{18.15}$$

In order to enforce the current x_1 to track the desired current $x^*{}_1$, control u can be designed as:

$$u = \frac{1}{2}\left(1 - sign\left(s\right)\right) \tag{18.16}$$

Under the above control scheme, the equivalent control is formally derived by substituting $\dot{s} = \dot{x_1} = 0$ in (18.12) and solve for the control input u as:

$$u_{eq} = 1 - \frac{E}{x_2} \tag{18.17}$$

where x_2 is the output voltage of the slow voltage loop. The motion equation of the outer voltage loop during sliding mode is obtained by substituting the equivalent control, given by Eq. (18.17), into (18.13).

$$\dot{x}_2 = -\frac{1}{RC}\left(x_2 - \frac{V_d{}^2}{x_2}\right) \tag{18.18}$$

Equation (18.18) can be solved explicitly as:

$$x_2 = \sqrt{V_d{}^2 + \left(x_2{}^2\left(t_h\right) - V_d{}^2\right)\, e^{\frac{-2}{RC}\left(t - t_h\right)}} \tag{18.19}$$

where t_h represents the reaching instant of the sliding manifold $s = 0$ and $x_2\left(t_h\right)$ is the output voltage at time t_h. Apparently, x_2 tends to V_d asymptotically as time goes to infinity.

The attraction domain of the sliding manifold $s = 0$ is found by applying the convergence condition $s\dot{s} < 0$ to the Eqs. (18.12) and (18.13), yielding:

$$x_2 > E \; or \; 0 < u_{eq} = 1 - \frac{E}{x_2} < 1 \tag{18.20}$$

Condition (18.20) implies that sliding mode can be enforced as long as the output voltage is higher than the source voltage. This requirement is essential for a Boost-type DC/DC converter. A careful consideration of the initial conditions is also required to guarantee the convergence to $s = 0$.

18.2.2 Observer-Based Sliding Mode Control of DC/DC Converters

In recent years, research on sliding mode control theory has revealed great advantages of incorporating certain dynamics into the sliding mode controllers [38–40]. This new approach, called *Observer-based Sliding Mode Control*, fall into the category of dynamic feedback control where the sliding mode controller simulates the ideal plant model in parallel with the real one [25, 38, 42]. This helps to reduce the number of states that need to be measured since the real plant states are substituted by observer states. The mismatch between the measurable state(s) and the observed one is the 'bridge' that establishes coupling between the plant system space and the controller space and keeps both systems operating close to each other. It can also be been utilized in different ways to improve the control performance. Overall, an observer can be viewed as an artificially introduced auxiliary dynamic system in order to improve the control performance [11, 16].

This subsection investigates the input inductor current control and subsequently output capacitor voltage regulation for different types of DC/DC. It develops sliding mode control algorithms for Buck and Boost converters with incomplete information about system states by designing state observers. For Buck converters governed by linear equations, conventional observer design methods could be applied directly while non-linear observers should be designed for Boost converters. It is assumed that the variable under control is measured only for both converters.

(A) Observer-Based Control of Buck-Type DC/DC Converters

The observer equations, or the auxiliary system, of the buck converter are designed as:

$$\widehat{\dot{x}}_1 = -\frac{1}{L}\hat{x}_2 + u\frac{E}{L} - l_1\left(\hat{x}_1 - x_1\right) \tag{18.21}$$

$$\widehat{\dot{x}}_2 = \frac{1}{C}\hat{x}_1 - \frac{1}{RC}\hat{x}_2 - l_2\left(\hat{x}_1 - x_1\right) \tag{18.22}$$

where l_1, l_2 are constant scalar observer gains and $\hat{x}_1$, $\hat{x}_2$ are the estimates of the inductor current and output voltage. Note that it is only required to measure the inductor current x_1.

Equations with respect to the mismatches $\overline{x}_1,\ \overline{x}_2$ are defined as:

$$\overline{x}_1 = \hat{x}_1 - x_1 \tag{18.23}$$

$$\overline{x}_2 = \hat{x}_2 - x_2 \tag{18.24}$$

The error dynamics can be derived by subtracting Eqs. (18.1) and (18.2) from (18.21) and (18.22):

$$\dot{\overline{x}}_1 = -\frac{1}{L}\overline{x}_2 - l_1\overline{x}_1 \tag{18.25}$$

$$\dot{\overline{x}}_2 = \frac{1}{C}\overline{x}_1 - \frac{1}{RC}\overline{x}_2 - l_2\overline{x}_1 \tag{18.26}$$

The characteristic polynomial of the above linear system is defined as:

$$P^2 + \left(l_1 - \frac{1}{RC}\right)P + \left(-\frac{l_1}{RC} - \frac{1}{LC} - \frac{l_2}{L}\right) = 0 \tag{18.27}$$

For the second order error system given in Eq. (18.27), stability and desired rate of convergence can be provided by the proper choice of l_1, l_2. This implies that the observer errors $\overline{x}_1$ and $\overline{x}_2$ tend to zero asymptotically.

The switching function for the sliding mode current controller will be designed based on the observed current $\hat{x}_1$ instead of the measured current x_1 as done in Eq. (18.7):

$$\hat{s} = \hat{x}_1 - \frac{V_d}{R} \tag{18.28}$$

The control u that is applied to the real plant and the observer system is of the same form as in the case of the control scheme without an observer.

$$u = \frac{1}{2}\left(1 - sign\left(\hat{s}\right)\right) \tag{18.29}$$

Assuming that sliding mode can be enforced in the vicinity of the sliding manifold $\hat{s} = 0$, $\hat{x}_1$can be calculated as:

$$\hat{x}_1 \equiv \frac{V_d}{R}\ with\ \forall t > t_h \tag{18.30}$$

where t_h denotes the reaching instant of the sliding manifold $\hat{s}$. The equivalent control of u can be obtained by solving $\dot{\hat{s}} = 0$ with respect to control:

$$u_{eq} = \hat{x}_2\left(\frac{1}{L}\hat{x}_2 - l_1\left(\hat{x}_1 - x_1\right)\right)\frac{L}{E} \tag{18.31}$$

After substituting u_{eq} into the real plant model, given by Eqs. (18.1) and (18.2), and considering the observer model, given by Eqs. (18.21) and (18.22), the motion in sliding mode can be represented as a reduced order system, comprising the motion of the real plant and the slow dynamics (output voltage loop) of the observer, as shown in the following equations:

$$\dot{x}_1 = -\frac{1}{L}\overline{x_2} + l_1\left(\frac{V_d}{R} - x_1\right) \tag{18.32}$$

$$\dot{x}_2 = \frac{1}{C}x_1 - \frac{1}{RC}x_2 \tag{18.33}$$

$$\dot{\hat{x}}_2 = \frac{1}{C}\frac{V_d}{R} - \frac{1}{RC}\hat{x}_2 - l_2\left(\hat{x}_1 - x_1\right) \tag{18.34}$$

Further analysis of Eqs. (18.32)–(18.34) results in:

$$\lim_{t\to\infty} x_1(t) = \hat{x}_1(t)\Big|_{t\geq t_h} = \frac{V_d}{R} \tag{18.35}$$

$$\lim_{t\to\infty} x_2(t) = \lim_{n\to\infty} \hat{x}_1(t) = V_d \tag{18.36}$$

The remaining task is to find the condition under which the occurrence of the sliding mode can be guaranteed. Applying the existence condition of sliding mode to Eq. (18.21) along with the substitution of Eqs. (18.28) and (18.29) yields:

$$-Ll_1\left(\hat{x}_1 - x_1\right) < \hat{x}_2 < E - Ll_1(\hat{x}_1 - x_1) \tag{18.37}$$

$$0 < u_{eq} = \left(\frac{1}{L}\hat{x}_2 + l_1\left(\hat{x}_1 - x_1\right)\right)\frac{L}{E} < 1 \tag{18.38}$$

Since $s(0) = 0$ with zero initial conditions for the power converter and E is high enough, sliding mode existence condition (18.38) is fulfilled.

(B) Observer-Based Control of Boost-Type DC/DC Converters

In order to simplify the derivation of the observer-based control scheme of Boost converters, a new control input is introduced and defined as $v = (1 - u)$. The observer dynamics that are designed for a Boost converter are governed by:

$$\dot{\hat{x}}_1 = -v\frac{1}{L}\hat{x}_2 + \frac{E}{L} - l_1\left(\hat{x}_1 - x_1\right) \tag{18.39}$$

$$\dot{\hat{x}}_2 = v\frac{1}{C}\hat{x}_1 - \frac{1}{RC}\hat{x}_2 - l_2\left(\hat{x}_1 - x_1\right) \tag{18.40}$$

where $\hat{x}_1, \hat{x}_2$ are the observed inductor current and output voltage, i.e. outputs of the observer; and l_1, l_2 are positive scalar observer gains.

Equations with respect to the mismatches $\overline{x}_1$, $\overline{x}_2$ are defined as:

$$\overline{x}_1 = \hat{x}_1 - x_1 \tag{18.41}$$

$$\overline{x}_2 = \hat{x}_2 - x_2 \tag{18.42}$$

The error dynamics can be derived by subtracting Eqs. (18.12) and (18.13) from (18.41) and (18.42):

$$\dot{\overline{x}}_1 = -v\frac{1}{L}\overline{x}_2 - l_1\overline{x}_1 \tag{18.43}$$

$$\dot{\overline{x}}_2 = v\frac{1}{C}\overline{x}_1 - \frac{1}{RC}\overline{x}_2 - l_2\overline{x}_1 \tag{18.44}$$

Equations (18.43) and (18.44) are nonlinear, since the system states are multiplied by the control input v. For the convergence proof, select a *Lyapunov* function candidate as:

$$V = \frac{1}{2}(L\overline{x}_1^2 + C\overline{x}_2^2) > 0 \tag{18.45}$$

The time derivative $\dot{V}$ on the system trajectory can be easily calculated as:

$$\dot{V} = -Ll_1\overline{x}_1^2 - \frac{1}{R}\overline{x}_2^2 - Cl_2\overline{x}_1\overline{x}_2 \tag{18.46}$$

Therefore, the convergence rate can be selected can be selected by proper choice of l_1, l_2 under Sylvester condition $\left(\frac{Ll_1}{R} > \frac{(Cl_2)^2}{4}\right)$ for quadratic form (18.46). As a result, observer errors $\overline{x}_1$ and $\overline{x}_2$ tend to zero asymptotically. The convergence rate of the inductor current estimation can be adjusted by the observer gains l_1, l_2.

The switching function of the sliding mode current controller will be designed based on the observed current $\hat{x}_1$ instead of measured current x_1 as done in (18.15):

$$\hat{s} = \hat{x}_1 - \frac{V_d^2}{RE} \tag{18.47}$$

The control u that is applied to the real plant and the observer system is of the same form as in the case of the control scheme without an observer.

$$u = \frac{1}{2}\left(1 - sign\left(\hat{s}\right)\right) \tag{18.48}$$

Equation (18.48) can be represented in terms of the new control input $v = (1 - u)$ as:

$$v = \frac{1}{2}\left(1 - sign\left(\hat{s}\right)\right) \tag{18.49}$$

Assuming that sliding mode can be enforced in the vicinity of the sliding manifold $\hat{s} = 0$, $\hat{x}_1$can be calculated as:

$$\hat{x}_1 \equiv \frac{V_d^2}{RE} \; with \; \forall t > t_h \tag{18.50}$$

where t_h denotes the reaching instant of the sliding manifold $\hat{s}$. The equivalent control of u can be obtained by solving $\dot{\hat{s}} = 0$:

$$v_{eq} = \frac{E - Ll_1(\frac{V_d^2}{RE} - x_1)}{\hat{x}_2} \tag{18.51}$$

After substituting v_{eq} into the real plant model, given by Eqs. (18.12) and (18.13), and considering the observer model, given by Eqs. (18.39) and (18.40), the motion in sliding mode can be represented as a reduced order system, comprising the motion of the real plant and the slow dynamics (output voltage loop) of the observer.

$$\dot{x}_1 = -v_{eq}\frac{1}{L}x_2 + \frac{E}{L} \tag{18.52}$$

$$\dot{x}_2 = v_{eq}\frac{1}{C}x_1 - \frac{1}{RC}x_2 \tag{18.53}$$

$$\dot{\hat{x}}_2 = v_{eq}\frac{1}{C}\frac{V_d^2}{RE} - \frac{1}{RC}\hat{x}_2 - l_2\left(\frac{V_d^2}{RE} - x_1\right) \tag{18.54}$$

By defining the errors $\overline{x}_1^* = \frac{V_d^2}{RE} - x_1$ and $\overline{x}_2 = \hat{x}_2 - x_2$ and since $\overline{x}_1$ tends to zero at the desired rate, the above equations can be transformed into a second order error system as:

$$\dot{\overline{x}}_1^* = -\frac{x_2}{\hat{x}_2}l_1\overline{x}_1^* + \frac{E}{L}\frac{x_2}{\hat{x}_2} - \frac{E}{L} \tag{18.55}$$

$$\dot{\overline{x}}_2 = \frac{E}{C}\frac{1}{\hat{x}_2}\overline{x}_1^* - \frac{Ll_1}{C\hat{x}_2}(\overline{x}_1^*)^2 - \frac{1}{RC}\overline{x}_2 - l_2\overline{x}_1^* \tag{18.56}$$

Further analysis of Eqs. (18.55) and (18.56) results in:

$$\lim_{t\to\infty} \hat{x}_2 = x_2 \tag{18.57}$$

$$\lim_{t\to\infty} \left(\frac{E}{L}\frac{x_2}{\hat{x}_2} - \frac{E}{L}\right) = 0 \tag{18.58}$$

This means that $\overline{x}_1^*$ converges to zero asymptotically at a rate that depends on l_1. As a result, $\overline{x}_2$ and x_2 tend respectively to zero and V_d. The remaining task is to find the condition under which the occurrence of the sliding mode can be guaranteed. Applying the existence condition of sliding mode to (18.47) along with the substitution of Eqs. (18.39), (18.40) and (18.49) yields:

$$0 < E - Ll_1\left(\hat{x}_1 - x_1\right) < \hat{x}_2 \tag{18.59}$$

$$0 < v_{eq} = \frac{E - Ll_1(\hat{x}_1 - x_1)}{\hat{x}_2} < 1 \tag{18.60}$$

Equation (18.60) can be represented in terms of the new control input $v = (1 - u)$ as:

$$0 < u_{eq} < 1 \tag{18.61}$$

Since x_1 is measured and $\hat{x}_1$, $\hat{x}_2$ are state variables in the controller space, the initial conditions of the observer, $\hat{x}_1(0)$ and $\hat{x}_2(0)$, can be designed such that the occurrence of the sliding mode is always guaranteed.

18.2.3 Simulation Results

In order to evaluate the proposed sliding mode control strategies for DC/DC power converters, several computer simulations have been conducted using MATLAB/Simulink software.

(A) Direct Sliding Mode Control of Boost Converters

Parameters of the simulated Buck converter are listed in Table 18.1. Figures 18.3 and 18.4 show the simulation results of the proposed control algorithm for Boost converters. As it can be seen from the figures, both the inductor current and the output capacitor voltage converge rapidly to their reference values.

(B) Direct Sliding Mode Control of Boost Converters

Parameters of the simulated Boost converter are listed in Table 18.2. Figures 18.5 and 18.6 show the simulation results of the proposed control algorithm for Boost converters. As it can be seen from the figures, both the inductor current and the output capacitor voltage converge rapidly to their reference values.

(C) Observer-Based Sliding Mode Control of Buck Converters

Parameters of the simulated Buck converter are listed in Table 18.1. The observer gain is selected as $l = 200$. Two sets of the observer initial conditions, listed in Table 18.3, were selected in order to compare the influence of observer initial conditions on the system response.

Table 18.1 Simulation parameters of the Buck-type DC/DC converter

Parameter	L (mH)	C (μf)	V_d (V)	R (Ω)	E (V)
Value	40	4	7	40	20

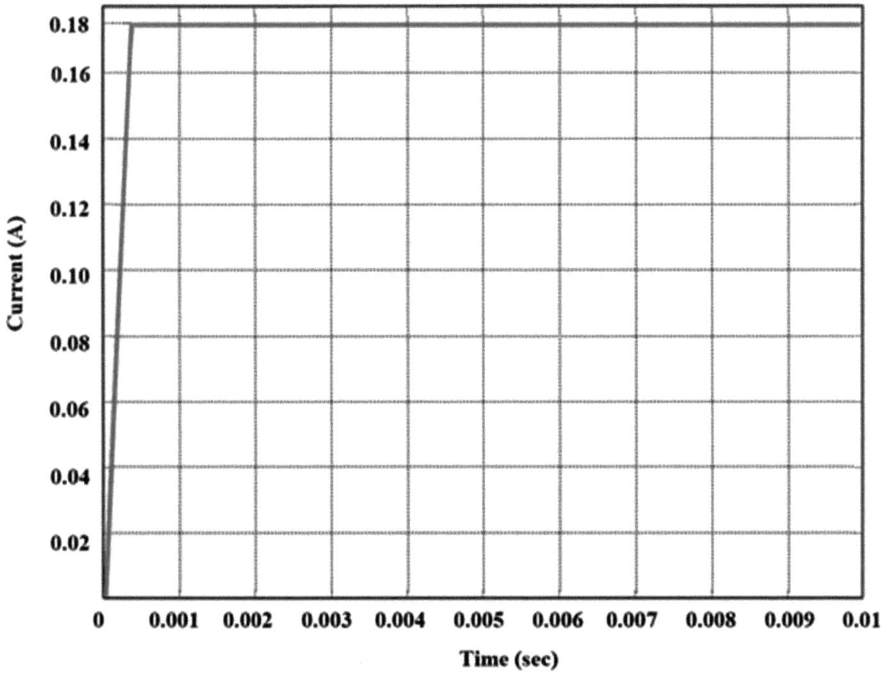

Fig. 18.3 Current response of a sliding mode-controlled Buck DC/DC power converter

Figures 18.7 and 18.8 show the response of real and estimated inductor current and output capacitor voltage under different initial conditions. Note that both the inductor current and the output capacitor voltage converge rapidly to their reference values and the system response can be influenced by the design of the observer initial conditions.

(D) Observer-Based Sliding Mode Control of Boost Converters

Parameters of the simulated Boost converter are listed in Table 18.2. The observer gain is selected as $l = 200$. Two sets of the observer initial conditions, listed in Table 18.4, were selected in order to compare the influence of observer initial conditions on the system response.

Figures 18.9 and 18.10 show the response of real and estimated inductor current and output capacitor voltage under different initial conditions. Note that both the inductor current and the output capacitor voltage converge rapidly to their reference values and the system response can be influenced by the design of the observer initial conditions.

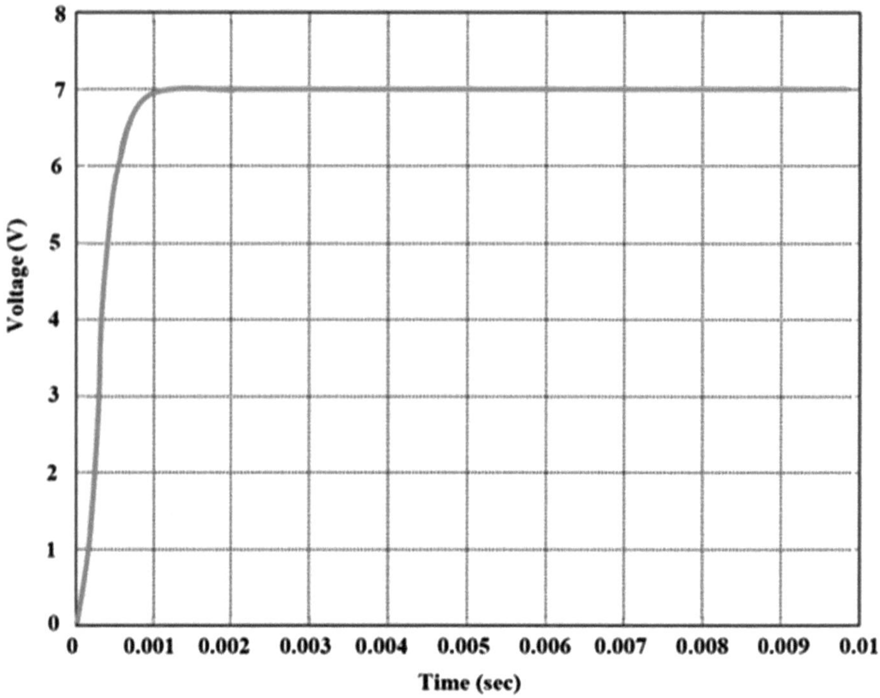

Fig. 18.4 Voltage response of a sliding mode-controlled Buck DC/DC power converter

Table 18.2 Simulation parameters of the Boost-type DC/DC converter

Parameter	L (mH)	C (μf)	V_d (V)	R (Ω)	E (V)
Value	40	4	40	40	20

18.2.4 Sliding Mode Control of Multi-phase DC/DC Power Converters

This section presents a new chattering reduction approach for power converters. The proposed scheme is based on the idea of designing a multi-phase converter system that operates at a given and fixed switching frequency without any additional dynamic elements. This may be possible by providing a desired phase shift between phases for any loads or frequencies in order to implement the "*Harmonic Cancellation*" method [24]. Several attempts have been made to apply this idea to PWM such that phase shifts are interconnected and can be controlled using a transformer with primary and secondary coils in different phases. On the other hand, the phase shift was obtained using delays, filters, or set of triangular inputs with selected delays [27, 45, 48].

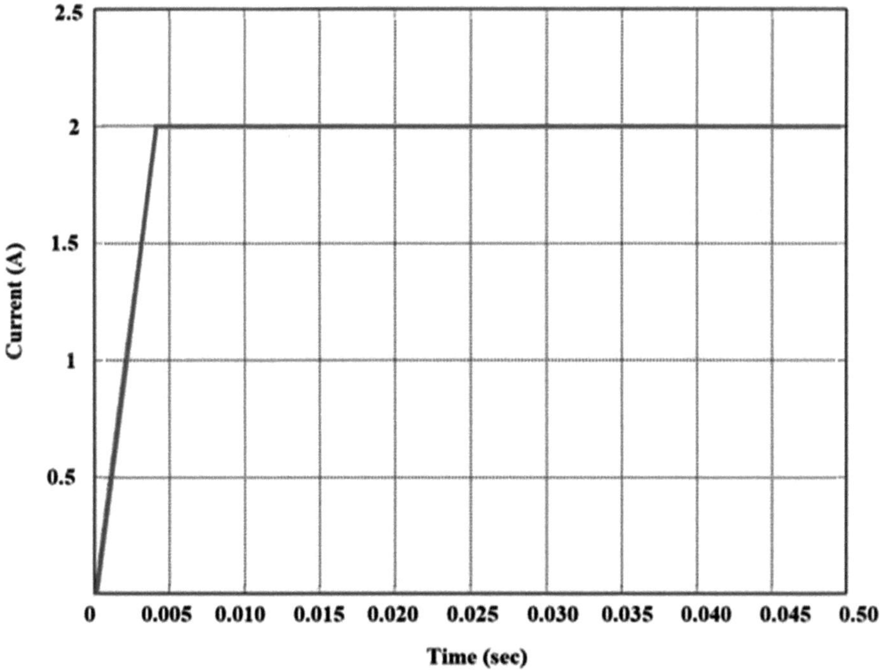

Fig. 18.5 Current response of a sliding mode-controlled Boost DC/DC power converter

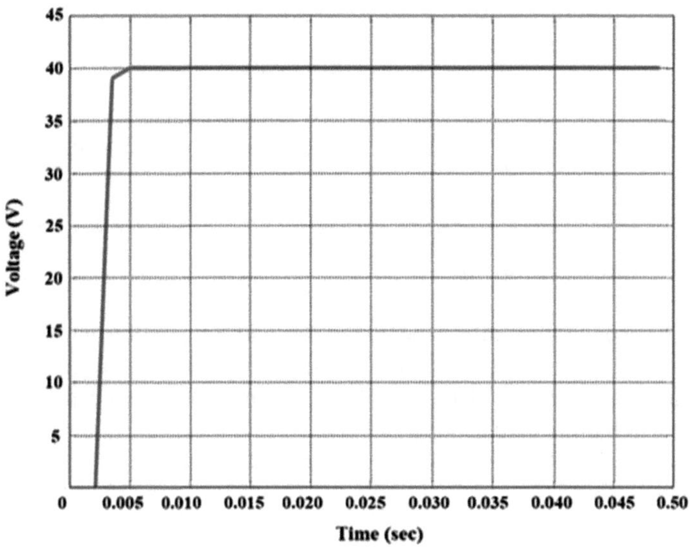

Fig. 18.6 Voltage response of a sliding mode-controlled Boost DC/DC power converter

Table 18.3 Observer initial conditions for the simulated Buck-type DC/DC converter

Initial condition	$\hat{x}_1$ (0) (A)	$\hat{x}_2$ (0) (V)
Set #1	0.12	5
Set #2	0.07	2.5

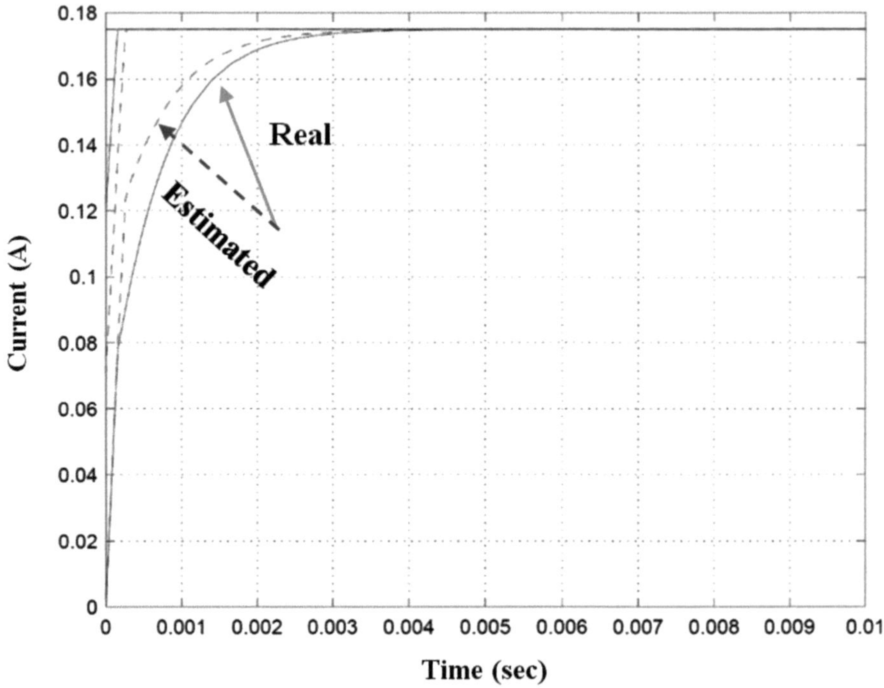

Fig. 18.7 Response of real (curves start from zero) and estimated current of a buck converter under different initial conditions. *Solid line* $\hat{x}_1(0) = 0.12$, $\hat{x}_2(0) = 5.0$; *dashed line* $\hat{x}_1(0) = 0.07$, $\hat{x}_2(0) = 2.5$

18.2.4.1 Frequency Control

Chattering frequency normally depends on the width of the hysteresis loop in the switching element which is usually selected such that the frequency does not exceed the maximum admissible level f_{des} for all operation modes. This can be achieved by varying the width of hysteresis loop [9, 29].

It order to formulate the problem statement, the following system with scalar control is considered.

$$\dot{x} = f(x, t) + b(x, t)\,u \qquad (x, f, b \in \mathcal{R}^n) \tag{18.62}$$

It is assumed that control u should be designed as a continuous function of state variables ($u_0(x)$). This situation is usually assumed with the "*Cascade Con-*

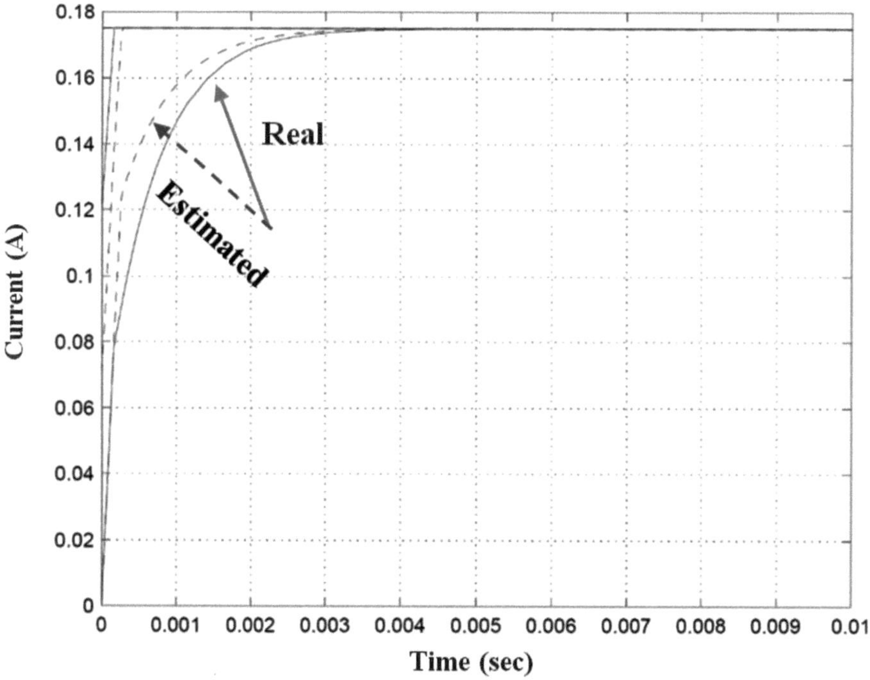

Fig. 18.8 Response of real (curves start from zero) and estimated voltage of a buck converter under different initial conditions. *Solid line* $\hat{x}_1(0) = 0.12$, $\hat{x}_2(0) = 5.0$; *dashed line* $\hat{x}_1(0) = 0.07$, $\hat{x}_2(0) = 2.5$

Table 18.4 Observer initial conditions for the simulated Boost-type DC/DC converter

Initial condition	$\hat{x}_1(0)$ (A)	$\hat{x}_2(0)$ (V)
Set #1	0	0
Set #2	1.95	38.5

trol" approach that is used for electric motors with the current as a control input [23, 39]. Power converters often utilize PWM as a principle operation mode in order to implement the desired control algorithm [22, 28]. Sliding mode is one of the tools to implement this operation mode based on the feedback approach as shown in Fig. 18.11, which illustrates that output u tracks the reference input $u_0(x)$ in sliding mode.

Select the sliding surface as:

$$s = u_0(x) - u, \;\; \dot{u} = v = M sign(s), \;\; M > 0 \tag{18.63}$$

$$\dot{s} = g(x) - M sign(s), \; g(x) = [grad(u_0)]^T (f + bu) \tag{18.64}$$

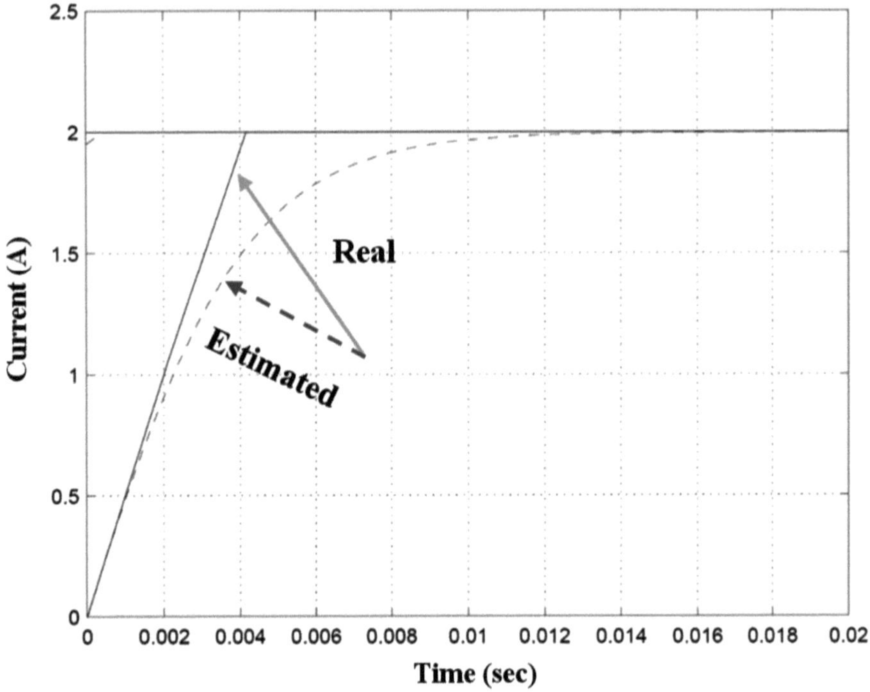

Fig. 18.9 Response of real (curves start from zero) and estimated current of a Boost converter under different initial conditions. *Solid line* $\hat{x}_1(0) = 0, \hat{x}_2(0) = 0$; *dashed line* $\hat{x}_1(0) = 1.95, \hat{x}_2(0) = 38.5$

It is evident that sliding mode, in the surface $s = 0$ with $u \equiv u_0$, exists if $M > |g(x)|$. If the control is implemented with a hysteresis loop, the switching frequency can be maintained at the desired level f_{des} by varying the width of the hysteresis loop Δ [24].

$$\Delta = \frac{M^2 - g(x)^2}{2M}\frac{1}{f_{des}} \tag{18.65}$$

18.2.4.2 Chattering Suppression

Suppose that the desired control is implemented by m power converters with $s_i = \frac{u_0}{m} - u_i$, $(i = 1, 2, \ldots m)$ and $\frac{u_0}{m}$ as reference inputs as shown in Fig. 18.12. If each power converter operates properly, the output is equal to the desired control $u_0(x)$. Amplitude and frequency of chattering in each converter can be found as:

$$A = \frac{\Delta}{2},\ f = \frac{M^2 - \left(\frac{g(x)}{m}\right)^2}{2M\Delta} \tag{18.66}$$

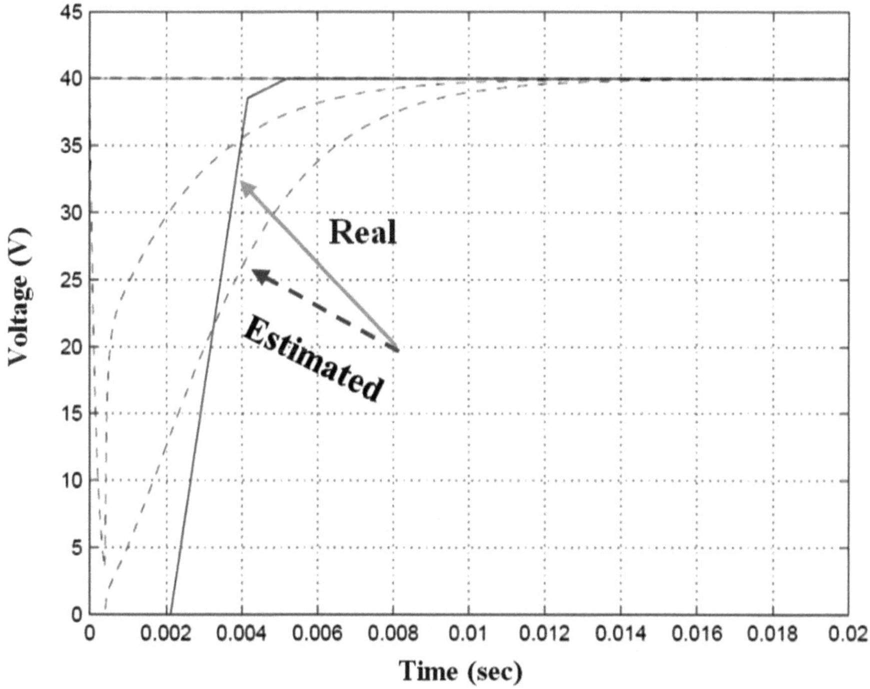

Fig. 18.10 Response of real (curves start from zero) and estimated voltage of a Boost converter under different initial conditions. *Solid line* $\hat{x}_1(0) = 0, \hat{x}_2(0) = 0$; *dashed line* $\hat{x}_1(0) = 1.95$, $\hat{x}_2(0) = 38.5$

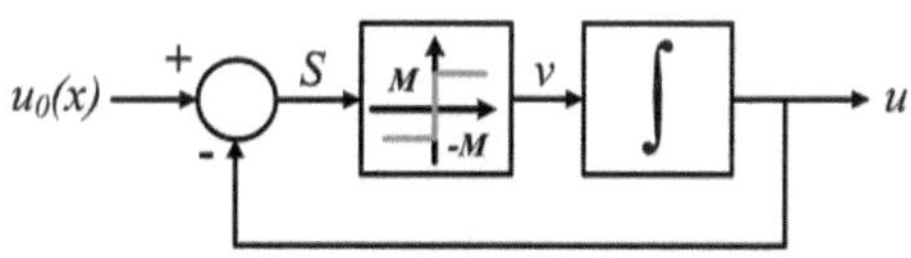

Fig. 18.11 Sliding mode control for a simple power converter model structure

The chattering amplitude of output u depends on the oscillation in each phase. Its maximum value is m times higher than that of each converter. For the system shown in Fig. 18.12, phases depend on initial conditions and cannot be controlled.

First, it should be demonstrated that, by varying phases, the output oscillation amplitude can be controlled. Suppose that a multiphase converter with m phases is to be designed such that the period of chattering T is the same in each phase, and two subsequent phases have phase shift T/m. Since chattering is a periodic time function, it can be represented as Fourier series with frequencies:

$$\omega_k = \omega.k, \omega = \frac{2\pi}{T} \; (k = 1.2\ldots..\infty) \tag{18.67}$$

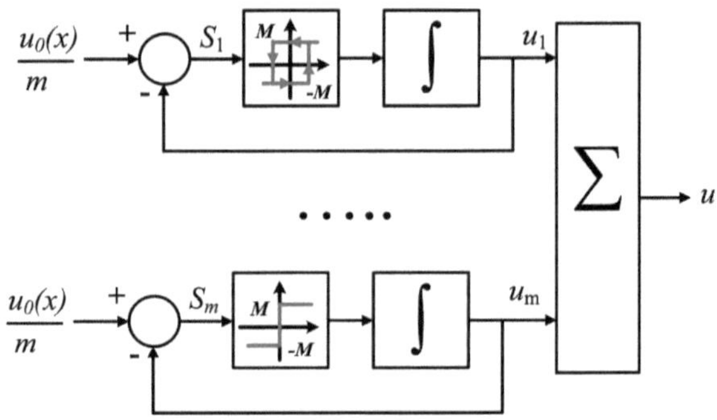

Fig. 18.12 m-phase converter with evenly distributed reference

The effect of the k-th harmonic in the output signal is the sum of individual outputs from all phases and can be easily calculated as:

$$\begin{aligned} \sum_{i=0}^{m-1} \sin\left(\omega_k\left(t - \frac{2\pi}{\omega m} i\right)\right) &= \sum_{i=0}^{m-1} Im[e^{j\omega_{kt} - \frac{2\pi k}{m} i}] \\ &= Im\left(e^{j\omega_k t} Z\right), Z = \sum_{i=0}^{m-1} e^{-j\frac{2\pi k}{m} i} \end{aligned} \tag{18.68}$$

To find Z, consider the following equation:

$$Ze^{-j\frac{2\pi k}{m}} = \sum_{i=0}^{m-1} e^{-j\frac{2\pi k}{m}(i+1)} = \sum_{i'=0}^{m-1} e^{-j\frac{2\pi k}{m} i'} \tag{18.69}$$

Since

$$Ze^{-j\frac{2\pi k}{m}}\Big|_{i=m} = e^{-j\frac{2\pi k}{m} i}\Big|_{i=0} \tag{18.70}$$

then,

$$Ze^{-j\frac{2\pi k}{m}} = Z \tag{18.71}$$

The function $e^{-j\frac{2\pi k}{m}}$ is equal to 1 only if k/m is an integer or $k = m, 2m, \ldots$, which means that $Z = 0$ for all other cases.

The above analysis shows that all harmonics except for lm ($l = 1, 2, \ldots$) are suppressed in the output signal. As a result, the amplitude of chattering can be reduced to the desired level by increasing the number of phases and providing a desired phase shift between each two subsequent phases.

18.2.4.3 Design Principle

In order to describe the design idea of the proposed chattering suppression approach, suppose that two power converters are implemented as shown in Fig. 18.13.

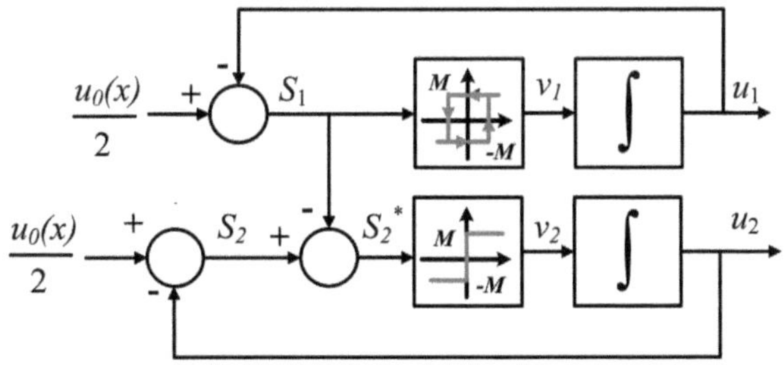

Fig. 18.13 A controller model with two interconnected phases

The switching function for the second converter is selected as:

$$s_2^* = s_2 - s_1 \tag{18.72}$$

where,

$$s_1 = u_0/2 - u_1 \tag{18.73}$$

$$s_2 = \frac{u_0}{2} - u_2 \tag{18.74}$$

$$v_1 = Msign(s_1)v_2 = Msign(s_2^*) \tag{18.75}$$

The time derivatives of $s_1,\ s_2^*$ are:

$$\dot{s}_1 = a - Msign\,(s_1) \qquad \left(a = \frac{g\,(x)}{2}\right) \tag{18.76}$$

$$\dot{s}_2^* = Msign\,(s_1) - Msign\,(s_2) \tag{18.77}$$

Figures 18.14 and 18.15 analyze the system behavior in the plane s_1 and s_2^* with $a = 0$, $\propto = 1$. Δ and α are the widths of the hysteresis loops for the two sliding surfaces. T can be easily calculated as:

$$T = \frac{\Delta}{M - a} + \frac{\Delta}{M + a} = \frac{2M\Delta}{M^2 - a^2} \tag{18.78}$$

By analyzing Fig. 18.14, it can be seen that the phase shift becomes:

$$\Phi = \frac{\alpha\Delta}{2M} \tag{18.79}$$

where Φ is equal to the time for changing s_2^* from $\frac{\alpha\Delta}{2}$ to $-\frac{\alpha\Delta}{2}$ or vice versa.

The previous analysis demonstrates that the phase shift between oscillations can be selected by proper choice of $\propto$ for any switching frequency without using dynamic elements. The value $\propto$ is calculated to provide the desired phase shift. Since Φ must be equal to *T/m,* $\propto$can be found as:

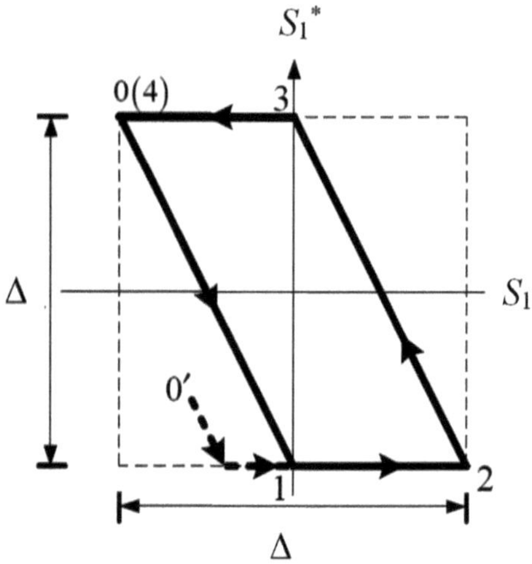

Fig. 18.14 The system behavior in s-plane

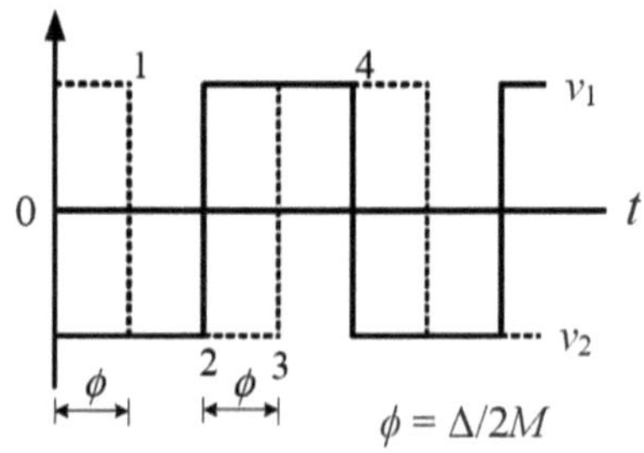

Fig. 18.15 Control of the phase between v_1, v_2

$$\propto = \frac{4M^2}{m(M^2 - a^2)} \tag{18.80}$$

For $a > 0$, $\propto$ can be selected such that phase shift is equal to what we need. Finally, the proposed chattering suppression design procedure for multiphase converters can be summarized as follows:

1. Select the width of the hysteresis loop as a state function such that the switching frequency in the first phase is maintained at the desired level.
2. Determine the number of phases for a given range of α variation.
3. Find the parameter $\propto$ as a function of α such that the phase shift between two subsequent phases is equal to $\frac{1}{m}$ of the oscillation period of the first phase.

Finally, another approach, called "master-slave" method of multiphase converters can be used for chattering reduction even if for a given number of phases m, parameter a is beyond the admissible domain and the desired phase shift cannot be guaranteed by varying the width of the hysteresis loop. For the master-slave implementation, each phase can be complimented by several sequentially connected "slaves", as illustrated

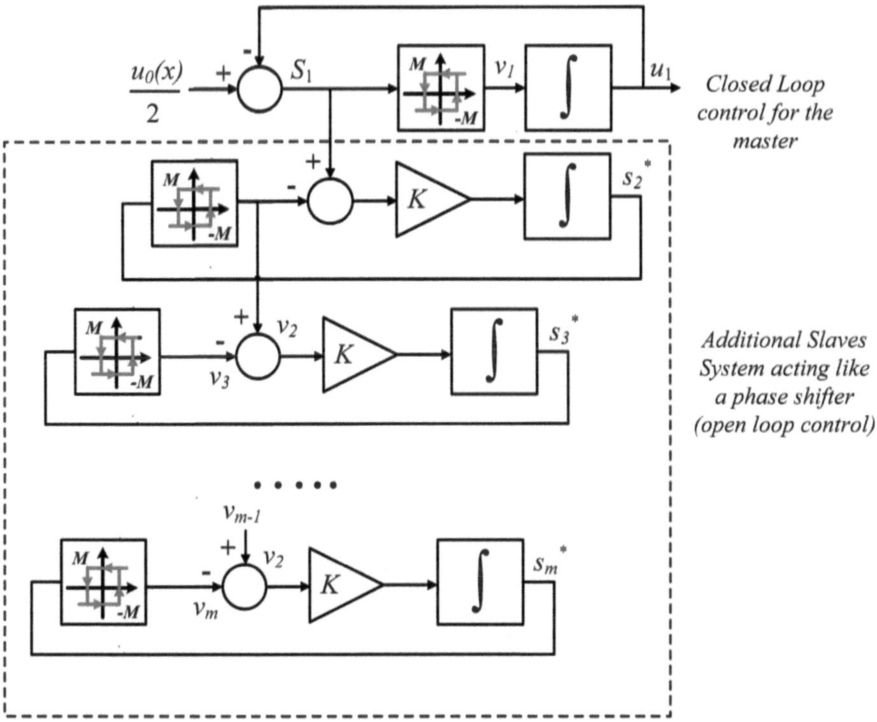

Fig. 18.16 A modified master-slave mode schematic with two more additional systems. $v_{2,3}$ is the switching command for the second channel

in Fig. 18.16 for a two-phase converter, such that the total phase shift is equal to the desired value.

18.2.4.4 Simulation Results

In order to evaluate the proposed sliding mode chattering suppression control strategies for multi-phase DC/DC power converters, computer simulations have been conducted using MATLAB/Simulink software.

Multi-phase DC/DC Converter Topology

The chattering suppression approach developed in the previous section is applied to DC/DC multi-phase buck converters. The objective is to demonstrate, via simulations, the effectiveness of the proposed design methodology in multiphase power converters. Simulations also check the range of the function "***a***" for which the chattering suppression takes place. The "*master-slave*" method is utilized in all simulations. A two-phase DC/DC converter is shown in Fig. 18.17. Simulation parameters are listed in Table 18.5.

Fig. 18.17 A two-phase DC/DC converter

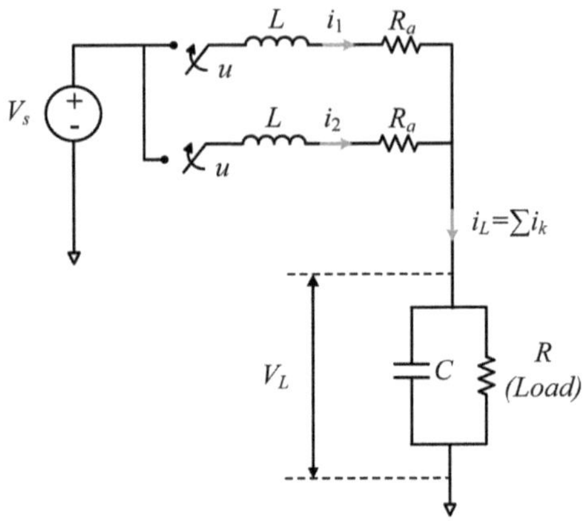

Table 18.5 Simulation parameters of the two-phase DC/DC converter

Parameter	L (H)	C (F)	R_a (Ω)	R_L (Ω)	V_s (V)
Value	$5 * 10^{-8}$	$1 * 10^{-3}$	$3 * 10^{-4}$	$1 * 10^{-2}$	12

For simulation purposes, the governing equations of the m-phase converter are assumed as follows:

$$\dot{I}_k = \frac{1}{L}(-I_k R_a + u_k - V_L),\ k = 1, 2, \ldots, m \tag{18.81}$$

$$V_L = \frac{1}{C}\left(\sum_{k=1}^{m} I_k - \frac{V_L}{R_L}\right) \tag{18.82}$$

The following control law is used for the two-phase power converter ($m = 2$) represented in Eqs. (18.81) and (18.82):

$$s_1 = I_1 - \frac{I_{ref}}{m},\ I_{ref} = \frac{V_{ref}}{R_L} \tag{18.83}$$

$$u_1 = V_s\left(\frac{1 - sign\,(s_1)}{2}\right) \tag{18.84}$$

$$u_2 = V_s\left(\frac{1 - sign\,(s_3^*)}{2}\right) \tag{18.85}$$

$$\dot{s}_1 = \frac{1}{L}\left(-R_a s_1 - \frac{V_s}{2} sign\,(s_1)\right) + \left(\frac{V_s}{2} - \frac{I_{ref} R_a}{m} - V_L\right) = -M sign\,(s_1) - b s_1 + a \tag{18.86}$$

$$\dot{s}_2^* = M\left(sign\left(s_1\right) - sign\left(s_2^*\right)\right) \tag{18.87}$$

$$\dot{s}_3^* = M\left(sign\left(s_2^*\right) - sign\left(s_3^*\right)\right) \tag{18.88}$$

where V_{ref} and I_{ref} are the reference input voltage and the corresponding reference load current, respectively. $a = \frac{V_s}{2} - \frac{I_{ref} R_a}{m} - V_L$, $M = \frac{V_s}{2L}$ and $b = \frac{R_a}{L}$. The desired phase shift $T/2$ is obtained by using two additional blocks with a phase shift of $T/4$ in each of them. The four-phase converter ($M = 4$) is simulated with switching frequency control of the first phase by appropriate choice of hysteresis width or hysteresis loop gain K_h in order to maintain the switching frequency at 50 Hz. The selected function $K_h\left(V_{ref}\right) = -0.0013V_{ref}^2 + 0.0127V_{ref} - 0.0007$ is shown in Fig. 18.18.

Figures 18.19, 18.20 and 18.21 demonstrates the effectiveness of the proposed chattering suppression algorithm on a 4-phase DC/DC power converter with three different reference inputs, $v_{ref1} = 3$ V, $v_{ref2} = 6$ V, and $v_{ref3} = 8$ V, respectively. Note that the inductance is relatively small in order to have fast converter dynamics. This leads to a high level of chattering in each phase, but it is practically suppressed in the output signal.

Finally, it is demonstrated that chattering can be reduced considerably following the "master-slave method" even if for a given number of phases m, parameter a is beyond the admissible domain and the desired phase shift cannot be guaranteed by

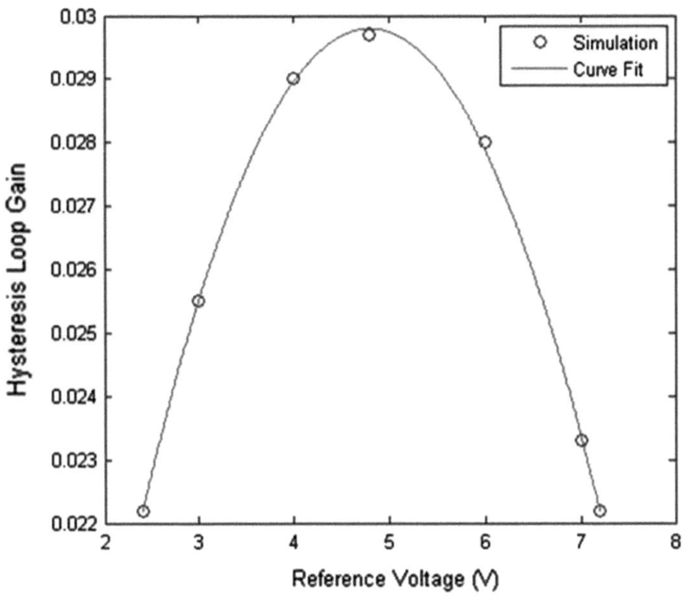

Fig. 18.18 Hysteresis loop gain K_h

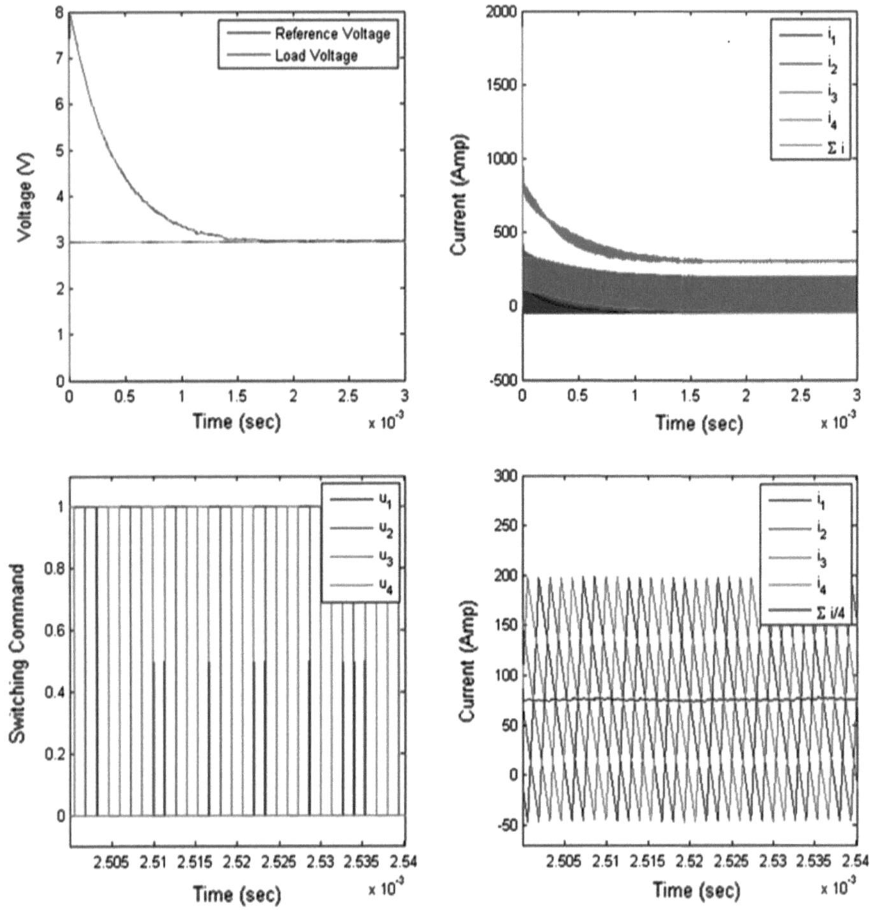

Fig. 18.19 Simulation results for the four-phase DC/DC converter, $V_{ref} = 3\,\text{V}$

varying the width of the hysteresis loop. For the master-slave implementation, each phase can be complimented by several sequentially connected "slaves", as illustrated in Fig. 18.16 for two-phase converter, such that the total phase shift is equal to the desired value. Of course, chattering can be suppressed by increasing the number of phases preserving the "*one slave in one phase*" approach, as shown in Fig. 18.22 for an 8-phase converter.

18.2.4.5 Experimental Results

Experimental results were conducted and discussed in [2] in order to demonstrate the effectiveness of the proposed sliding mode chattering suppression approach for multi-phase converters and also verify the simulation results.

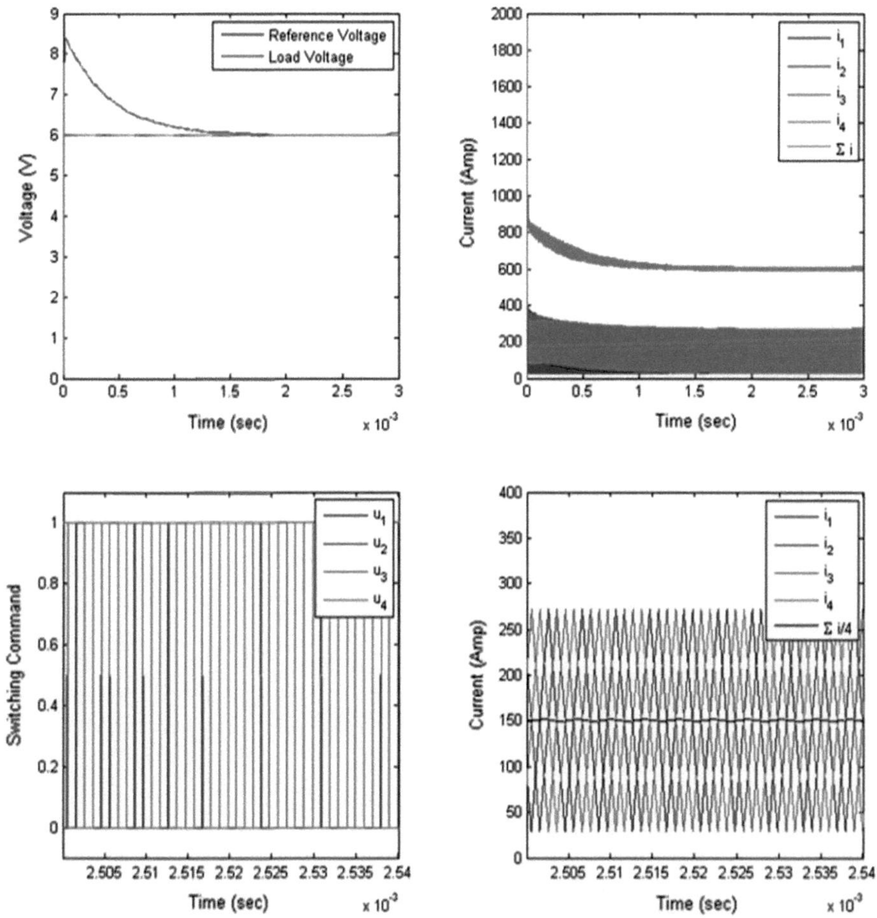

Fig. 18.20 Simulation results for the four-phase DC/DC converter, $V_{ref} = 6\,\mathrm{V}$

Figure 18.23 shows the experimental setup. The system consists of a 4-phase DC/DC buck converter that is controlled using two loops: A current inner loop that is sliding mode controlled with a hysteresis band, and a voltage outer loop that defines the current reference through a PI controller. Table 18.6 shows the converter parameters.

Figure 18.24 shows the experimental waveforms of the input current, output voltage and control command for a half- bridge, 1-phase DC/DC converter. The reference output voltage is 5 V. The average input current value is 2.44 A and the chattering amplitude is 0.47 A. The average output voltage is 4.96 V.

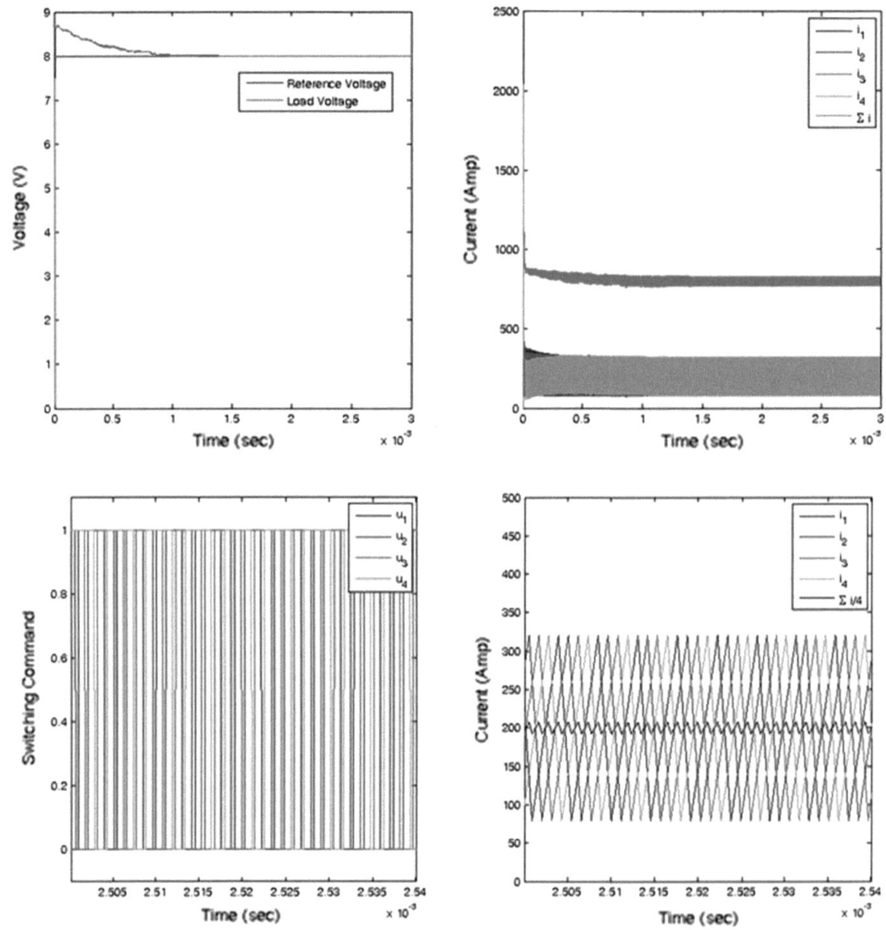

Fig. 18.21 Simulation results for the four-phase DC/DC converter, $V_{ref} = 8\,\text{V}$

Figures 18.25 and 18.26 show the four shifted input currents and the MOSFET drain voltages of a 4-phases DC/DC converter, respectively. Both Figures demonstrate that the master-slave algorithm works properly, with respect to the current shifting. The current values of each phase are $i_{L1} = 0.62 \mp 0.43\,\text{A}$, $i_{L2} = 0.6 \mp 0.42\,\text{A}$, $i_{L3} = 0.63 \mp 0.43\,\text{A}$, and $i_{L4} = 0.66 \mp 0.42\,\text{A}$.

The effectiveness of the proposed chattering suppression algorithm is demonstrated in Fig. 18.27, which represents the input current, output voltage and control command of a 4-phase DC/DC converter, respectively. The reference output voltage is 5 V. As it can be seen, the current chattering amplitude is 0.095, which is five times less than the current chattering amplitude of the 1-phase DC/DC converter.

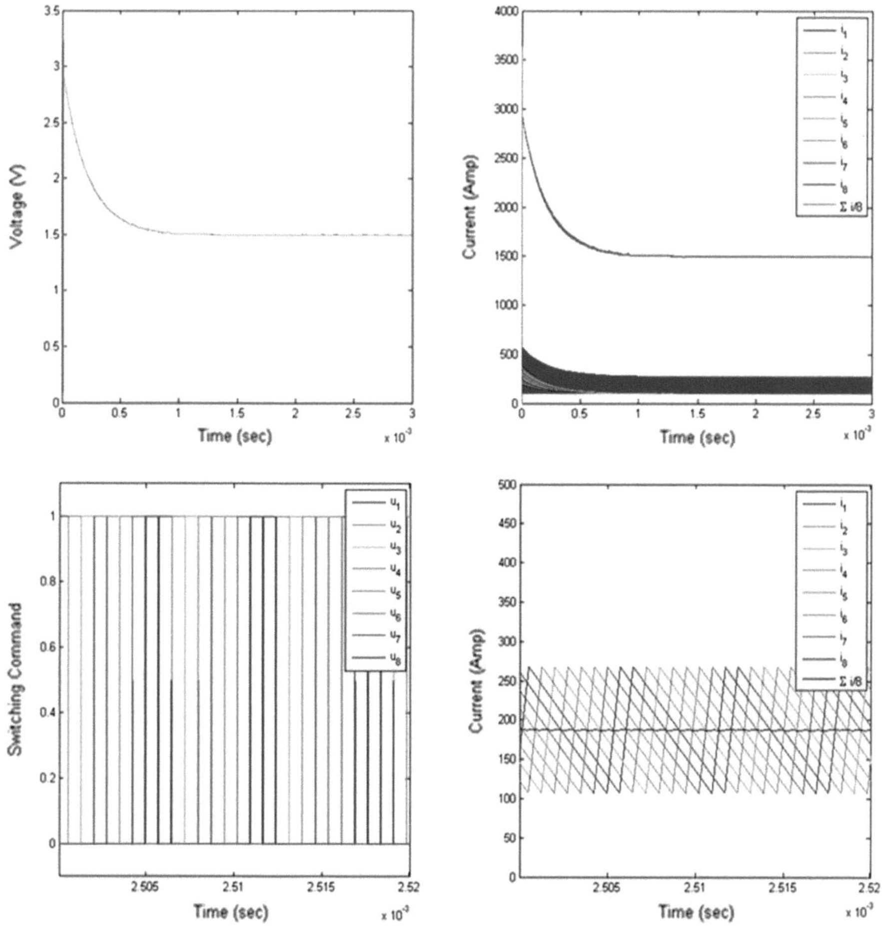

Fig. 18.22 Simulation results for the eight-phase DC/DC converter, $V_{ref} = 1.5\,\mathrm{V}$

Fig. 18.23 Experimental Setup of a 4-phase DC/DC buck converter [2]

Table 18.6 Parameters of the experimental setup (4-phase DC/DC buck converter) [2]

L (H)	C (F)	R_L (Ω)	E (V)	*Switching Frequency*
22 μ	10 μ	2	10	100 KHz

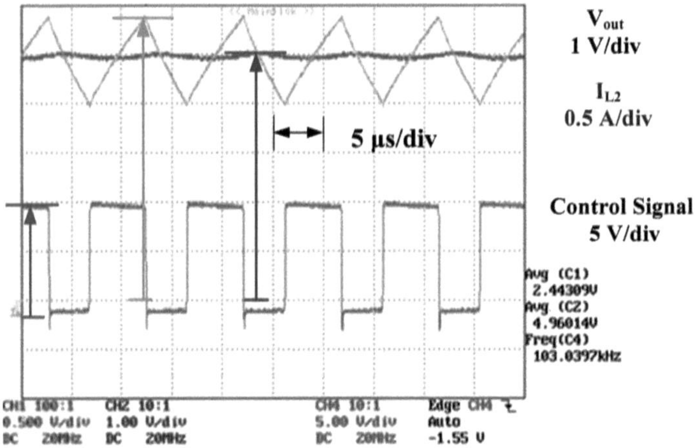

Fig. 18.24 Input current, output voltage and command control of a 1-phase DC/DC converter, $v_{dc\_ref} = 5\,\mathrm{V}$

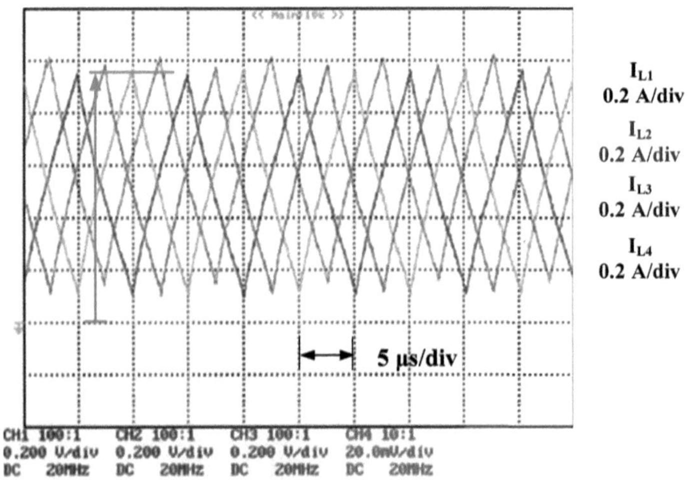

Fig. 18.25 Input phase currents a 4-phase DC/DC converter

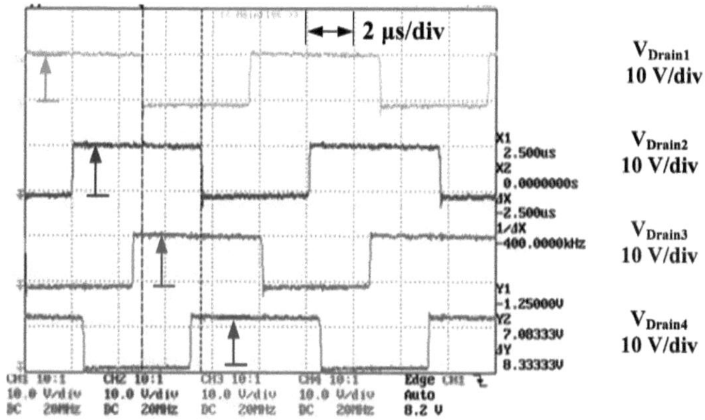

Fig. 18.26 MOSFET drain voltages of a 4-phases DC/DC converter

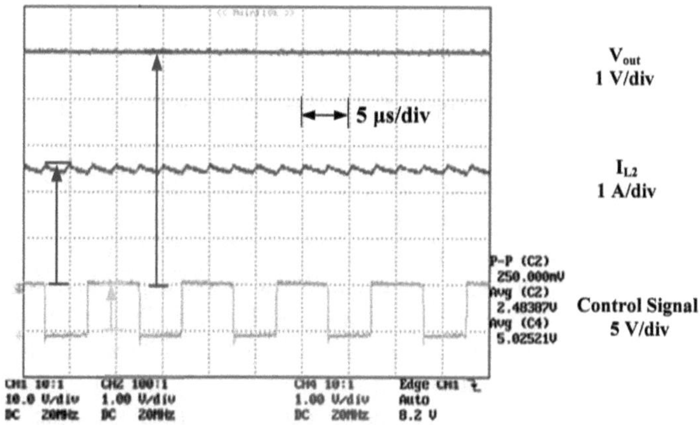

Fig. 18.27 Input current, output voltage and command control of a 4-phase DC/DC converter, $v_{dc\_ref} = 5\,\text{V}$

Figure 18.28, shows the input current, output voltage and control command of a 4-phase DC/DC converted, where the reference output voltage is 4.59 V. It can be noticed that the algorithm almost cancelled the chattering completely. The average value of the output current is 2.28 A, with a chattering width of 0.033, which is about 14 times less than the current chattering amplitude of the 1-phase DC/DC converter. This is due to the fact that the four duty cycles are equal to 0.5, and the corresponding Fourier harmonic coefficients cancel each other.

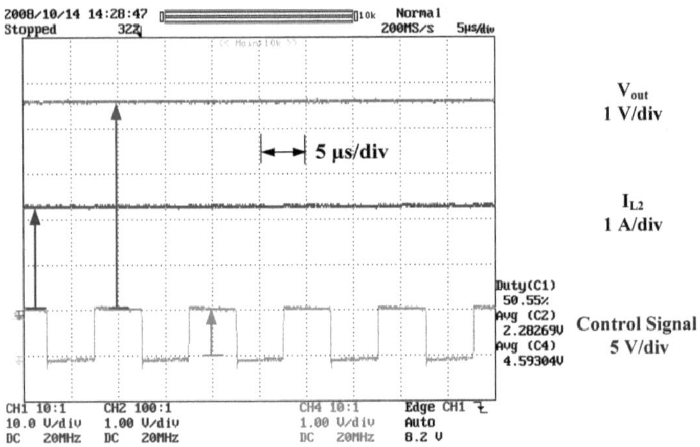

Fig. 18.28 Input current, output voltage and command control of a 4-phase DC/DC converter, $v_{dc\_ref} = 4.59\,\mathrm{V}$

18.3 Sliding Mode Control of AC/DC Power Converters

Recently, a variety of power electronic devices, including three-phase AC/DC power converters, has been widely used in energy conversion systems [28]. One of their disadvantages is that they cause inherent problems of reactive power generation and higher harmonic content in the input current. These practical problems have become more serious as the AC/DC converter capacity becomes much larger [8, 30, 44].

The ideal AC/DC converter has a constant dc output voltage (or current) and a pure sinusoidal input current at unity power factor from the ac line. Conventional thyristor phase-controlled converters have an inherent drawback which is that the power factor decreases as the firing angle increases and that harmonics of the line current are relatively high [21, 46]. In recent years, there has been a tendency to operate AC/DC converters with PWM switching patterns to improve the input and output performance of the converter. PWM AC/DC converters offer distinct advantages over the conventional rectifiers [15, 47]. These advantages include unity power factor, capability of bidirectional power flow, low harmonic components in input current and low ripple in output voltage. All of these features simplify filtering problems on both ac and dc sides of the converter [8, 39, 47].

The objective of this section is to develop a feedback control algorithm for AC/DC power converters such that the output voltage is maintained at the desired level with zero steady state error, input currents are free of higher harmonics and the reactive power is equal to zero. This mission will be accomplished using the framework of SMC methodology implying that the order of the motion is reduced with state trajectories in the pre-selected manifold with the system state space. Furthermore, the control system is designed in two reference frames: the *abc* natural reference frame, and the *dq* synchronous reference frame.

18.3.1 Circuit Model and Design Methodology

The circuit model and analysis of the three-phase PWM AC/DC voltage source converter in the natural *abc* and the rotating *dq* reference frames is presented next. First, consider the three-phase PWM AC/DC voltage source converter shown in Fig. 18.29, where e_a, e_b, e_c are the balanced three-phase AC voltages representing the infinite bus, $R_g$and L_g represent the grid-side resistance and inductance, respectively, C_{dc} is the dc-link capacitance, i_a, i_b, i_c are the three-phase AC input currents, i_{dc} is dc-link current, v_{dc} is dc-link voltage and i_L is the load current.

Based on circuit analysis, the AC input current equations are given by:

$$L_g\frac{di_a}{dt} = e_a - R_g i_a - v_{an} \tag{18.89}$$

$$L_g\frac{di_b}{dt} = e_b - R_g i_a - v_{bn} \tag{18.90}$$

$$L_g\frac{di_c}{dt} = e_c - R_g i_c - v_{cn} \tag{18.91}$$

where v_{an}, v_{bn}, v_{cn} are the voltages from the AC side of the converter to the power neutral point *n*. The balanced three-phase AC voltages are expressed as:

$$e_a = E_0 \sin(\omega t) \tag{18.92}$$

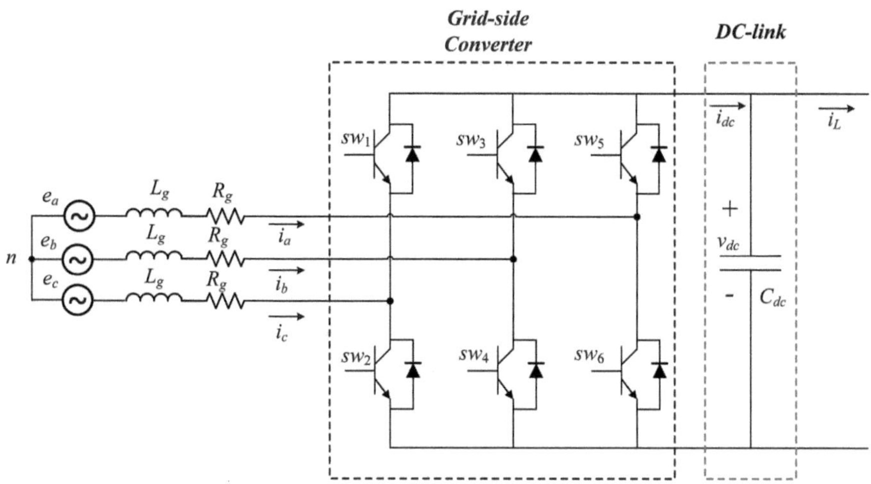

Fig. 18.29 Three-phase PWM AC/DC voltage source converter scheme

$$e_b = E_0 \sin\left(\omega t - \frac{2\pi}{3}\right) \tag{18.93}$$

$$e_c = E_0 \sin\left(\omega t + \frac{2\pi}{3}\right) \tag{18.94}$$

where E_0 is the amplitude of the phase voltage and ω is the ac power source angular frequency. If we assume that $i = \begin{bmatrix} i_a \\ i_b \\ i_c \end{bmatrix}$, $e = \begin{bmatrix} e_a \\ e_b \\ e_c \end{bmatrix}$, $v_s = \begin{bmatrix} v_{an} \\ v_{bn} \\ v_{cn} \end{bmatrix}$, then equations in (18.89)–(18.91) can be re-written in a compact form as:

$$L_g \frac{di}{dt} = e - R_g i - v_s \tag{18.95}$$

The switching function S of each switch can be defined as:

$$S_j = \begin{cases} 1 & S_j \quad is\ close \\ -1 & S_j \quad is\ open \end{cases} \quad j = a, b, c \tag{18.96}$$

The voltage vector v_s be expressed in terms of the switching functions $S = \begin{bmatrix} S_a \\ S_b \\ S_c \end{bmatrix}$ as:

$$v_s = \frac{1}{3} v_{dc} \begin{bmatrix} 2 & -1 & -1 \\ -1 & 2 & -1 \\ -1 & -1 & 2 \end{bmatrix} S \tag{18.97}$$

By substituting Eq. (18.97) into (18.95), the AC input current equations can be expressed as:

$$L_g \frac{di}{dt} = e - R_g i - \frac{1}{3} v_{dc} \begin{bmatrix} 2 & -1 & -1 \\ -1 & 2 & -1 \\ -1 & -1 & 2 \end{bmatrix} S \tag{18.98}$$

Finally, the output voltage equation can be given by:

$$C \frac{dv_{dc}}{dt} = -i_L + i^T S \tag{18.99}$$

The operating principle of the three phase PWM grid-side voltage source converter can also be analyzed by the classical rotating field theory with the well-known *Park and Clarke* transformations. The dynamic model of three-phase AC/DC voltage source converter in the *d-q* rotating reference frame can be described as [28]:

$$L_g \frac{di_d}{dt} = e_d - R_g i_d + \omega_g L_g i_q - v_d \tag{18.100}$$

$$L_g \frac{di_q}{dt} = e_q - R_g i_q - \omega_g L_g i_d - v_q \tag{18.101}$$

$$C \frac{dv_{dc}}{dt} = -i_L + \frac{1}{2}(i_d s_d + i_q s_q) \tag{18.102}$$

where, $v_d = v_{dc} s_d$ and $v_q = v_{dc} s_q$. s_d and s_q are the switching functions in the *d-q* synchronous reference frame. i_d, i_q, e_d and e_q are the input currents of the converter and grid voltages in the synchronous *d-q* reference frame, respectively. Note that $v_{dc} = v_{dc\_d}$ due to the field orientation.

The design methodology proposed in this chapter is different compared to the one usually used for AC/DC power converters. It is common to start with solving the main problem of maintaining the DC output voltage at the desired level and then minimizing high order harmonics and reactive power at the input. In this chapter, the inverse sequence of actions is offered. First, a tracking system is designed with sinusoidal current references proportional to the input AC voltages. This means that the reactive power is automatically equal to zero. If the tracking system is ideal, the input current will be free of higher harmonics. The second phase of the research work is the output voltage control. It will be shown that the output voltage will be constant and depends only on the amplitude of the reference input if designing the ideal tracking system is accomplished.

Furthermore, the proposed strategy does not utilize conventional SMC. As previously mentioned, the objective is to control three variables: two phase current and output voltage, using three dimensional control vectors. Generally, this can be done by enforcing sliding mode at the intersection of three sliding surfaces that depend on the mismatches between the reference and real values. However, the following analysis will show that matrix multiplying control is singular and conventional SMC methodology is not applicable.

18.3.2 Proposed Control Scheme in the abc Natural Reference Frame

(A) Sliding Mode Current Tracking Control

According to the proposed design procedure, phase currents should track sinusoidal reference inputs proportional to input ac voltages. The tracking problem is solved in the framework of sliding mode methodology. Substituting Eq. (18.100) into (18.99) results in:

$$L \frac{di_{abc}}{dt} = e_{abc} + v_{dc} \Gamma_0 S \tag{18.103}$$

where, matrix Γ_0 is given by:

$$\Gamma_0 = \begin{bmatrix} \frac{-2}{3} & \frac{1}{3} & \frac{1}{3} \\ \frac{1}{3} & \frac{-2}{3} & \frac{1}{3} \\ \frac{1}{3} & \frac{1}{3} & \frac{-2}{3} \end{bmatrix}, \quad det\Gamma_0 = 0 \tag{18.104}$$

Since the sum of the three-phase currents is equal to zero, three state variables should only be controlled: output voltage and two phase currents in a system with a three dimensional control vector S. But, unfortunately the conventional sliding mode method cannot be applied directly since matrix Γ_0 is singular. Because of that, we will start with designing a tracking system for two phase currents only. It means that sliding mode should be enforced on the intersection of two surfaces $\sigma_a = L\left(i_{aref} - i_a\right)$ and $\sigma_b = L\left(i_{bref} - i_b\right)$ or in vector form:

$$\sigma_{ab} = L\left(i_{abref} - i_{ab}\right) \tag{18.105}$$

Excluding the phase current $i_c = -i_a - i_b$ yields to:

$$L\frac{di_{ab}}{dt} = e_{ab} + u_c \Gamma S \tag{18.106}$$

$$C\frac{dv_{dc}}{dt} = -\frac{v_{dc}}{R} + (i_a, i_b, -i_a - i_b)S \tag{18.107}$$

where Γ is a 2 × 3 matrix given by:

$$\Gamma = \begin{bmatrix} \frac{-2}{3} & \frac{1}{3} & \frac{1}{3} \\ \frac{1}{3} & \frac{-2}{3} & \frac{1}{3} \end{bmatrix} \tag{18.108}$$

The ideal tracking system is based on Lyapunov function:

$$V = \frac{1}{2}\,{\sigma_{ab}}^T\;\sigma_{ab} \tag{18.109}$$

Discontinuous control should be selected such that time derivative of V is negative definite. The time derivative $\dot{V}$ on the system trajectory can be easily calculated as:

$$\dot{V} = {\sigma_{ab}}^T F\,(.) - \;u_c\,(\alpha, \beta, \gamma)\,S \tag{18.110}$$

where $F\,(.)$ is state function, which does not depend on control. α, β and γ are given by:

$$\begin{cases} \alpha = \left(\frac{-2}{3}\sigma_a + \frac{1}{3}\sigma_b\right) \\ \beta = \left(\frac{1}{3}\sigma_a + \frac{-2}{3}\sigma_b\right) \\ \gamma = \left(\frac{1}{3}\sigma_a + \frac{-2}{3}\sigma_b\right) \end{cases} \tag{18.111}$$

If v_{dc}/L is large enough, $F(.)$ can be suppressed with control S given by:

$$S = \begin{cases} S_a = sign\ (\alpha) \\ S_b = sign\ (\beta) \\ S_c = sign\ (\gamma) \end{cases} \tag{18.112}$$

where α, β and γ are selected such that $\dot{V} < 0$. This means that σ_{ab} tends to zero. Even more, σ_{ab} becomes zero after a finite time interval t_s [10]. As a result, sliding mode occurs with $i_{ab} = i_{abref}$. If a tracking system is designed with sinusoidal current references proportional to the input AC voltages as:

$$i_{ab} = K * e_{ab} \tag{18.113}$$

where K is constant. Then, the first problem of having zero reactive power and phase currents free of higher harmonics is solved.

(B) Sliding Mode Output Voltage Control

The main objective is to maintain the DC output voltage at the desired level. The first question to be answered is what is the output voltage after sliding mode occurs? Sliding mode equation can be derived using the so called "equilibrium control" procedure [39]. Step 1: time derivative of vector σ_{ab} on the state trajectory is made equal to zero. Step 2: the obtained algebraic equation should be solved with respect to control S and then substituted in the original system.

The equivalent control:

$$(\Gamma S)_{eq} = \left(L\frac{di_{abref}}{dt} - e_{ab}\right)\frac{1}{u_c} \tag{18.114}$$

is the solution of the following equation:

$$\frac{d\sigma_{ab}}{dt} = L\frac{di_{abref}}{dt} - e_{ab} - v_{dc}(\Gamma S)_{eq} = 0 \tag{18.115}$$

The three phase input currents can also be written as:

$$\left(i_{a,}\ i_{b,}\ -i_a - i_b\right) = -(i_a - i_c,\ i_b - i_c)\ \Gamma \tag{18.116}$$

After substituting Eqs. (18.114) and (18.115) into (18.107), the sliding mode equation can be obtained as:

$$C\left(\frac{dv_{dc}}{dt}\right)=\left(-\frac{v_{dc}}{R}\right)-\frac{L}{u_c}\left(i_a\frac{di_a}{dt}+i_b\frac{di_b}{dt}+ic\frac{di_c}{dt}\right)+\frac{1}{v_{dc}}\left(i_ae_a+i_bi_b+i_ce_c\right) \tag{18.117}$$

For the given balanced three phase input voltages given by Eqs. (18.88)–(18.91), and after sliding mode occurs ($i_{ab}=Ke_a$), it can be noticed that:

$$\begin{cases} i_a\frac{di_a}{dt}+i_b\frac{di_b}{dt}+i_c\frac{di_c}{dt}=0 \\ i_ae_a+i_be_b+i_ce_c=\frac{3}{2}KE_0{}^2=costant \end{cases} \tag{18.118}$$

After substituting Eq. (18.118) into (18.117), the sliding mode equation is given by:

$$C\frac{dv_{dc}}{dt}=-\frac{v_{dc}}{R_L}+\frac{3}{2}KE_0{}^2\frac{1}{u_c} \tag{18.119}$$

The last equation has only one asymptotically stable equilibrium point that can be described by:

$$v_{dcss}=\sqrt{\frac{3}{2}KE_0{}^2R_L} \tag{18.120}$$

In order to prove the stability of the equilibrium point, represent the right hand side of Eq. (18.119) in the form:

$$\begin{cases} f\left(v_{dc}\right)=-\frac{v_{dc}}{R_L}+\frac{3}{2}KE_0{}^2\frac{1}{u_c} \\ f\left(v_{dc}\right)=-g\left(v_{dc}\right)\left(v_{dc}-v_{dcss}\right)\ ,\ g\left(v_{dc}\right)>0 \\ \Delta v_{dc}=\left(v_{dc}-v_{dcss}\right) \end{cases} \tag{18.121}$$

The derivative of $f\left(v_{dc}\right)$ is negative and is given by:

$$f'\left(v_{dc}\right)=-\frac{1}{R_L}-\frac{3}{2}KE_0{}^2\frac{1}{v_{dc}^2}<0 \tag{18.122}$$

Therefore, $g\left(u_c\right)>0$ and $C\frac{d\Delta u_c}{dt}=-g\Delta u_c$, $\Delta u_c\rightarrow 0$. This proves that the equilibrium point of (18.119) is asymptotically stable. It can be seen that the output voltage tends to a constant value which depends on the value of constant K in the reference current input equation. If R_L is known, K can be easily assigned. Otherwise, K should be varied such that the equilibrium point is equal to desired value u_{cref}. As in any boost converter, the output voltage should be greater than some minimum value. Correspondingly, $K>K_{min}$ and $K_{min}>0$. Selected K as an integral function of the mismatch as:

$$\dot{K} = \alpha \left(u_{cref} - u_c\right) + M\, sg\,(K_{min} - K)\,,\ \alpha > 0,\ M > \alpha \left|v_{dc\_ref} - v_{dc}\right| \tag{18.123}$$

where α is constant. The last term of (18.123) is added to prevent K from being less than K_{min}. The only steady state of the system is $u_c = u_{cref}$, and the value of K is given by:

$$K = \frac{2}{3}\frac{v_{dc\_ref}^2}{{E_0}^2 R_L} \tag{18.124}$$

If $v_{dc\_ref} > 0$, $K\,(0) > 0$, and $v_{dc}\,(0) > 0$, then, according to (18.120) and (18.123), $K\,(t) > 0$ and $v_{dc}\,(t) > 0$ for any $t > 0$. Substituting $u_c = u_{cref} + x > 0$ in Eq. (18.120) results in:

$$\frac{dx}{dt} = -Av_{dc_{ref}} - Ax + \frac{BK}{v_{dc_{ref}} + x} \tag{18.125}$$

Differentiating Eq. (18.125) results in:

$$\ddot{x} = -\,(t)\,\dot{x} - f\,(x)\,,\ (t) > 0,\ x\, f\,(x) > 0 \tag{18.126}$$

Select the Lyapunov function:

$$V = \frac{\dot{x}^2}{2} + \int_0^x f(\gamma)d\gamma > 0 \tag{18.127}$$

The time derivative $\dot{V}$ on the system trajectory can be easily calculated as:

$$\dot{V} = -(t)\dot{x}^2 \tag{18.128}$$

Since $\dot{x}\,(t) \equiv 0$, this means that $x\,(t) \equiv 0$ and the equilibrium point is globally asymptotically stable.

18.3.3 Proposed Control Scheme in the dq Synchronous Reference Frame

(A) Sliding Mode Current Tracking Control

Similar to the control design for electric motors, the current control of a *Boost*-type AC/DC power converter can be designed either in phase coordinates or in the *dq* coordinate frame. Since the control criteria (as listed in the performance characteristics) are normally given in the *dq* coordinate frame, it is more convenient to design the current control in the *dq* coordinate frame than in phase coordinates. Represent the current equations in (18.100) and (18.101) in a vector form as:

$$\dot{I}_{dq} = f_{dq}(I_{dq}, e_{dq}, \omega_g) - bs_{dq} \tag{18.129}$$

where,

$$\begin{cases} I_{dq} = \begin{bmatrix} i_d \\ i_q \end{bmatrix} \\ s_{dq} = \begin{bmatrix} s_d \\ s_q \end{bmatrix} \\ f_{dq}\left(I_{dq}, e_{dq}, \omega_g\right) = \begin{bmatrix} \frac{R_g}{L_g} i_d + \frac{e_d}{L_g} + \omega_g i_q \\ -\frac{R_g}{L_g} i_q + \frac{e_q}{L_g} - \omega_g i_d \end{bmatrix} \\ b = \frac{v_{dc}}{2L} \end{cases} \tag{18.130}$$

The switching functions for the current control are designed as:

$$\sigma_d = i_d^* - i_d \tag{18.131}$$

$$\sigma_q = i_q^* - i_q \tag{18.132}$$

where i_d^* and i_q^* are the desired values of the currents in the (*d*, *q*) coordinate frame. The next task is to find the condition under which sliding mode can be enforced. It can be noticed that no control gain can be adjusted for the control design of AC/DC converters. The solution is to find a domain in the system space from which any state trajectory converges to the sliding manifold defined by $\sigma_d = 0$, $\sigma_q = 0$. Defining $\sigma_{dq} = [\sigma_d, \sigma_q]^T$ and taking the time derivative of σ_{dq} results in:

$$\dot{\sigma}_{dq} = \dot{I}_{dq}^* - f_{dq}\left(I_{dq}, e_{dq}, \omega_g\right) + b\, s_{dq} = F_{dq} + DS \tag{18.133}$$

where $I_{dq}^* = [i_d^*, i_q^*]^T$, $F_{dq} = \dot{I}_{dq}^* - f_{dq}\left(I_{dq}, e_{dq}, \omega_g\right)$, $D = bA_{d,q}^{\alpha,\beta} A_{\alpha,\beta}^{a,b,c}$, $A_{d,q}^{\alpha,\beta} A_{\alpha,\beta}^{a,b,c}$ are the Park and Clarke transformation matrices, respectively, S is the vector of transformed switching functions defined in (18.8).

Design the matrix of switching functions S as:

$$S = -sign(S^*) \tag{18.134}$$

where $S^* = [S_a^*, S_b^*, S_c^*]^T$ is a vector of transformed switching functions and $sign\,(S^*) = [sign(S_a^*), sign(S_b^*), sign(S_c^*)]^T$. The transformed vector S^* should be designed such that, s_{fd} and s_{fq} disappear in finite time. Vector S^* is selected as:

$$S^* = \frac{3}{2b^2} D^T \sigma_{dq} \tag{18.135}$$

Design a *Lyapunov* function candidate as:

$$V = \sigma_{dq}^{T}\sigma_{dq} \tag{18.136}$$

The time derivative of the *Lyapunov* function on the system trajectory can be easily calculated as:

$$\dot{V} = (S^{*})^{T}F^{*} + (S^{*})^{T}D^{T}DS \tag{18.137}$$

where $F^{*} = [\,F_{a}^{*}\ F_{b}^{*}\ F_{c}^{*}\,]^{T} = D^{T}F_{dq}$. Substituting control equation (18.134) into (18.137) results in:

$$\dot{V} = (S^{*})^{T}F^{*} - \left(S^{*}\right)^{T}D^{T}D\,sign(S^{*}) \tag{18.138}$$

where $D^{T}D$ is a singular matrix that can be calculated as:

$$D^{T}D = b^{2}\frac{4}{9}\begin{bmatrix} 1 & -\frac{1}{2} & -\frac{1}{2} \\ -\frac{1}{2} & 1 & -\frac{1}{2} \\ -\frac{1}{2} & -\frac{1}{2} & 1 \end{bmatrix} \tag{18.139}$$

Equation (18.138) can be expanded as:

$$\dot{V} = \left(S_{a}^{*}F_{a}^{*} + S_{b}^{*}F_{b}^{*} + S_{c}^{*}F_{c}^{*}\right) - \left(\frac{2}{3}\right)^{2}b^{2}\left(2\left|S_{l}^{*}\right| + \left|S_{m}^{*}\right| + \left|S_{n}^{*}\right|\right) \tag{18.140}$$

where $l \neq m \neq n\ and\ l, m, n\ \in \{a, b, c\}$. Equation (18.140) can be further represented as:

$$\dot{V} = \left(S_{a}^{*}F_{a}^{*} + S_{b}^{*}F_{b}^{*} + S_{c}^{*}F_{c}^{*}\right) - \left(\frac{2}{3}\right)^{2}b^{2}\left(\left|S_{l}^{*}\right| + \left|S_{m}^{*}\right| + \left|S_{n}^{*}\right|\right) - \left(\frac{2}{3}\right)^{2}b^{2}\left|S_{l}^{*}\right| \tag{18.141}$$

This means that the sufficient condition for $\dot{V} < 0$ can be given by:

$$\left(\frac{2}{3}\right)^{2}b^{2} > \max(F_{a}^{*}, F_{b}^{*}, F_{c}^{*}) \tag{18.142}$$

Equation (18.142) defines a subspace in the system space in which the state trajectories converges to the sliding manifold defined by $\sigma_{dq} = 0$ in finite time. This is to show that the attraction domain in the sliding manifold is bounded in the state space. It is also important to notice that parameter $b = \frac{v_{dc}}{2L}$ should be high enough at the initial time instant. The output voltage is not zero at the initial time instant since $v_{dc\_d} = v_{dc}$. In critical applications, this can be achieved by starting the converter operation with an open-loop control. The last step of the current control design is that the resulting controls S_{a}, S_{b} and S_{c} should be mapped into the switching patterns that can be applied to the power converter. This can be done using the following system of equations:

$$\begin{cases} s_{w1} = \frac{1}{2}\left(1 + S_a\right), s_{w4} = 1 - s_{w1} \\ s_{w2} = \frac{1}{2}\left(1 + S_b\right), s_{w5} = 1 - s_{w2} \\ s_{w3} = \frac{1}{2}\left(1 + S_c\right), s_{w6} = 1 - s_{w3} \end{cases} \quad (18.143)$$

(B) Sliding Mode Output Voltage Control

It is necessary to determine the reference currents feeding to the current controller, i_d^* and i_q^*, in order to ensure asymptotic stability of the output voltage regulation. Neglecting the voltage drop over the phase resistance R_g, Eqs. (18.100)–(18.102) can be simplified to:

$$L_g \frac{di_d}{dt} = e_d + \omega_g L_g i_q - v_d \quad (18.144)$$

$$L_g \frac{di_q}{dt} = e_q - \omega_g L_g i_d - v_q \quad (18.145)$$

$$C \frac{dv_{dc}}{dt} = -i_L + \frac{1}{2}\left(i_d s_d + i_q s_q\right) \quad (18.146)$$

In general, the value of the inductance satisfies $L \ll 1$, while the right-hand sides of equations in (18.144)–(18.146) have the values of the same order. Therefore, $di_d/dt, di_q/dt \gg du_{cd}/dt$ indicating that the dynamics of i_d and i_q are much faster than those of v_{dc}. In case that the fast dynamics are stable, the output voltage control can be simplified considerably. Based on the *Singular Perturbation Theory* [19, 20], the left-hand sides of Eqs. (18.144)–(18.146) can be formally equal to zero, and the algebraic equations for s_d and s_q can be solved. Therefore, the following equation system is valid for control design of the slow-manifold:

$$s_d = \frac{2\left(e_d + \omega_g L i_q^*\right)}{v_{dc}} \quad (18.147)$$

$$s_q = \frac{2\left(e_q + \omega_g L i_d^*\right)}{v_{dc}} \quad (18.148)$$

$$\frac{dv_{dc}}{dt} = -\frac{i_L}{C} + \frac{i_d^* s_d + i_q^* s_q}{2C} \quad (18.149)$$

where i_d^* and i_q^* are the reference values of i_d and i_q, respectively. The real currents are replaced with their reference values in Eqs. (18.147) and (18.149), since we assume that the inner current control loop is in sliding mode with $\sigma_d = i_d^* - i_d, \sigma_q = i_q^* - i_q$.

As mentioned in the introduction, the objective is to control the three-phase AC/DC power converter such that the output voltage is maintained at the desired level with zero steady state error, input currents are free of higher harmonics and the reactive power is equal to zero. The demand of sinusoidal input currents has been fulfilled automatically by involving the (*d,q*) transformation. The following characteristics will cover the remaining major requirements of a well-controlled three-phase AC/DC power converter:

1. The output voltage should converge to its reference value u_c^*.
2. The input current phase-angle, $\rho^* = \arctan(i_q^*/i_d^*)$, should trace its reference value.
3. The power-balance condition should be satisfied, i.e. $e_d i_d^* + e_q i_q^* = u_c^* i_l = v_{dc}^* i_L$.

The reference currents i_d^* and i_q^* should be calculated satisfying these requirements. Substitution of (18.147) and (18.148) into (18.149) yields:

$$\frac{dv_{dc}}{dt} = -\frac{i_L}{C} + \frac{i_d^* s_d + i_q^* s_q}{C v_{dc}} \tag{18.150}$$

Taking into account the power-balance condition, the above equation can be simplified to:

$$\frac{dv_{dc}}{dt} = -\frac{i_L}{C} + \frac{v_{dc}^* i_L}{C v_{dc}} \tag{18.151}$$

For a system with a pure load resistance R_L, $i_L = v_{dc}/R_L$. As a result, linear dynamics for output voltage u_{cd} can be calculated as:

$$\frac{dv_{dc}}{dt} = \frac{v_{dc}^* - v_{dc}}{C R_L} \tag{18.152}$$

Define the voltage regulation error as $\overline{v}_{dc} = v_{dc}^* - v_{dc}$ with a constant desired voltage $\dot{v}_{dc}^* = 0$ such that:

$$\overline{v}_{dc} + R_L C \dot{\overline{v}}_{dc} = 0 \tag{18.153}$$

Equation (18.153) shows that the voltage error tends to zero asymptotically with the time-constant $R_L C$. It also means that the output voltage converges to its reference value automatically if the power-balance condition is fulfilled. Finally, i_d^* and i_q^* can be calculated as:

$$i_d^* = \frac{v_{dc}^* i_L}{e_d + e_d tan\rho^*} \tag{18.154}$$

$$i_q^* = \frac{v_{dc}^* i_L \tan\rho^*}{e_d + e_d \tan\rho^*} \tag{18.155}$$

where ρ^* is the input reference current phase angle. It is usually determined by the control designer.

18.3.4 Simulation Results

In order to evaluate the proposed sliding mode control strategies for AC/DC power converters, several computer simulations have been conducted using MATLAB/Simulink software. Parameters of the simulated three-phase AC/DC converter are listed in Table 18.7. All of the reported results were obtained under the assumption that ideal sliding modes are enforced in the systems. Figure 18.30 shows the DC output voltage using the proposed SMC algorithm. As it can be seen, the output voltage is maintained at the desired level with zero steady state error. $v_{dc\_ref} = 250\,\mathrm{V}$ in this case,

Results of the current tracking control system are shown in Fig. 18.31. After the transient stage, phase currents track references very well. This proves that the proposed sliding mode control strategy ensures that the input currents are free of higher order harmonics.

Figure 18.32 shows the source voltage and current. It illustrates the ability of the proposed control algorithm to reduce the reactive power to zero and maintain a unity power factor.

Figure 18.33 illustrates the ability of the input current and output voltage control systems under the proposed strategy to quickly follow a sudden change in the dc link voltage reference ($v_{dc\_ref1} = 300\,\mathrm{V}$, $v_{dc\_ref2} = 250\,\mathrm{V}$). In this case, K should be

Table 18.7 Simulation parameters of the three-phase AC/DC converter

Parameter	L (mH)	C (μf)	f (Hz)	R_L (Ω)	E_0 (V)
Value	7.5	820	60	100	120

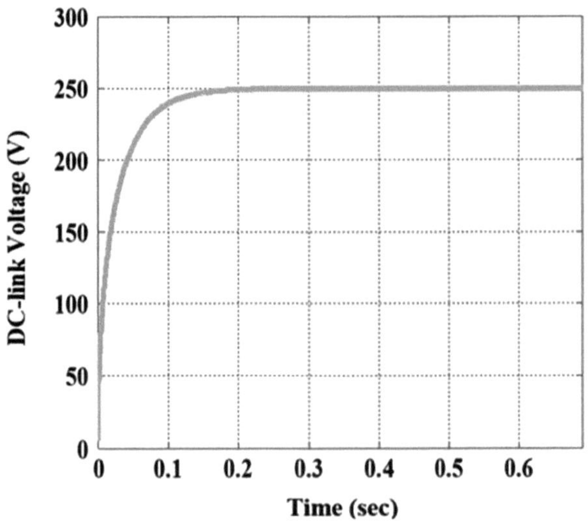

Fig. 18.30 DC-link output voltage of the AC/DC power converter

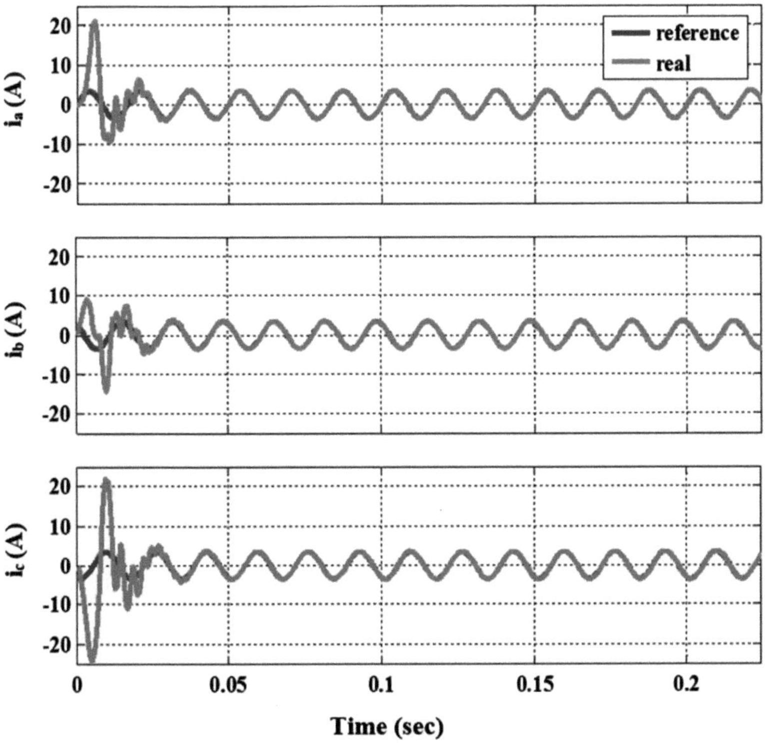

Fig. 18.31 Three-phase current tracking using the proposed SMC strategy

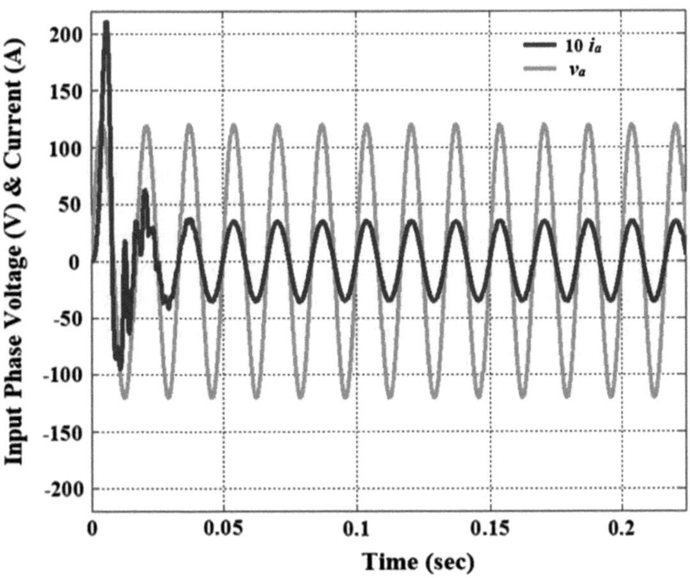

Fig. 18.32 Input phase voltage and phase current

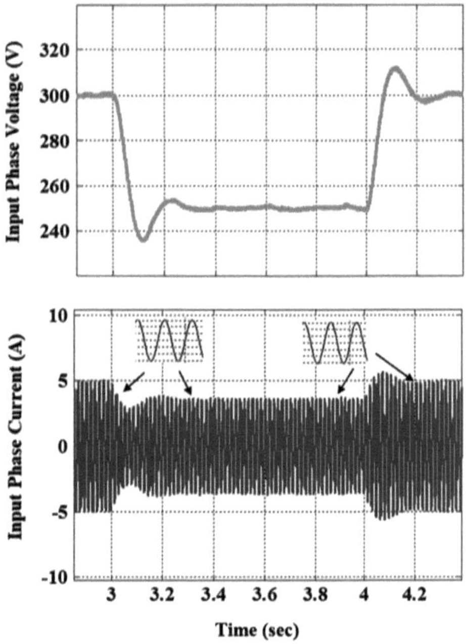

Fig. 18.33 System response to a sudden change of the dc link voltage and input current

varied such that the equilibrium point is equal to desired value u_{cref}. For this purpose, K is selected as an integral function of the mismatch $v_{dc\_ref} - v_{dc}$.

18.4 Sliding Mode Control of DC/AC Power Electronic Inverters

PWM originally refers to an approach used to realize or amplify a continuous function by using on/off implementation. In power electronics industry, the term PWM also refers to the control of semiconductor switches in converter circuits. With the development of new control algorithms for power converters, several PWM techniques have been proposed and implemented in the control system of DC/AC inverters, like harmonic elimination PWM [5, 6] space vector PWM [17, 18, 41], sine-triangle PWM [28], hybrid PWM [3], hysteresis band control, etc.

PWM control approaches can be classified into two main categories: open-loop control using a pre-calculated switching pattern, such as harmonic elimination PWM, and close-loop control whose switching actions depend on feedback information, such as hysteresis band PWM approach. Some of them can be used in both open-loop and close-loop schemes, such as space vector PWM [41]. Comparative studies of different PWM techniques can be found in [31, 37], and other literature.

Performance of power electronic inverters is evaluated in different aspects, for instance harmonic loss factor, reference tracking ability and robustness against variation of circuit parameters and disturbances. Therefore, Sliding mode control (SMC) is very natural to be used in the control of power inverters [1, 38–40], since the switching between two discrete values can be used directly as gating signals to the semiconductor switching devices in power converter circuits. SMC has excellent reference tracking ability and robust against parameter variations of the inverter.

This section presents a new PWM approach for power electronic inverters called Sliding Mode PWM (SMPWM). It is a close-loop control method based on the sliding mode concept. The control objective is to track proper reference inputs using a three-phase full-bridge inverter. It considers the fact that three arbitrary phase current references cannot be tracked for three-phase three-wire systems.

18.4.1 Circuit Model and Design Methodology

Consider the system in Fig. 18.34. The three-phase full-bridge inverter under control is to provide desired currents to the load, taking into account that they can be dependent.

Based on circuit analysis, the system equations are:

$$\begin{cases} L\frac{di_a}{dt} + Ri_a + E_a = s_a\, U_{dc} - v_n \\ L\frac{di_b}{dt} + Ri_b + E_b = s_b U_{dc} - v_n \\ L\frac{di_c}{dt} + Ri_c + E_c = s_c\, U_{dc} - v_n \end{cases} \tag{18.156}$$

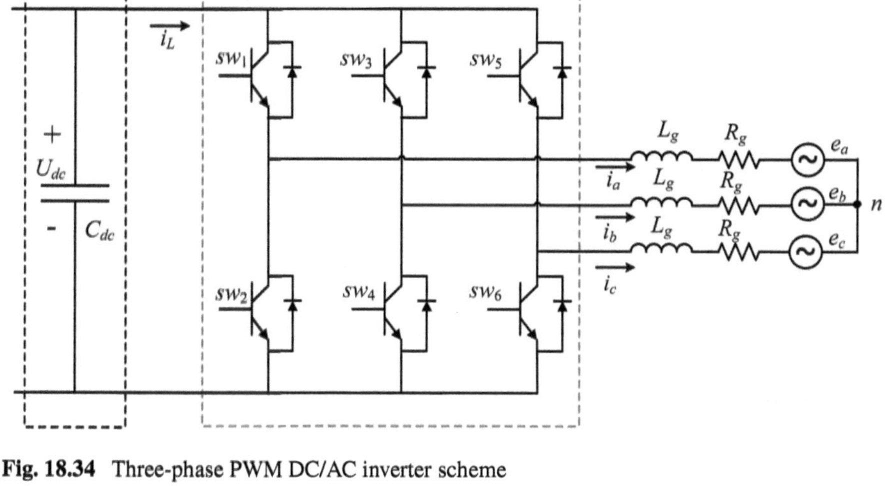

Fig. 18.34 Three-phase PWM DC/AC inverter scheme

where v_n is the voltage at neutral point. $E_a,\ E_b,\ E_c$ are balanced three phase voltage sources which represent three back EMF in ac motors. $s_a,\ s_b,\ s_c \in \{\pm 1\}$ represent the on/off control signals for six switching devices of the three-phase full-bridge converter. If a switching device is conducting, value "1" is assigned to it, otherwise "-1" is assigned. Since sum of the currents is equal to zero, v_n of the three-phase system can be found as:

$$v_n = \frac{U_{dc}}{3}(s_a + s_b + s_c) \tag{18.157}$$

It looks natural that v_n is discontinuous, but its average value changes continuously.

18.4.2 Sliding Mode Pulse Width Modulation (SMPWM)

The objective of the inverter's control system is to track proper selected voltage or current references. Some existing PWM techniques use three-phase current errors as inputs to its controller, for instance hysteresis band PWM. It controls the three phases independently based on corresponding current error. However, as it was mentioned, the three currents cannot be independent.

Denote

$$\begin{cases} J_a = -\left(\frac{R}{L}\right) i_a - \frac{E_a}{L} \\ J_b = -\left(\frac{R}{L}\right) i_b - \frac{E_b}{L} \\ J_c = -\left(\frac{R}{L}\right) i_c - \frac{E_c}{L} \end{cases} \tag{18.158}$$

Substituting (18.157) into (18.156) results in:

$$\frac{di}{dt} = J + \frac{U_{dc}}{3L} DS \tag{18.159}$$

where,

$$i = \begin{bmatrix} i_a \\ i_b \\ i_c \end{bmatrix},\ D = \begin{bmatrix} 2 & -1 & -1 \\ -1 & 2 & -1 \\ -1 & -1 & 2 \end{bmatrix},\ S = \begin{bmatrix} s_a \\ s_b \\ s_c \end{bmatrix},\ J = \begin{bmatrix} J_a \\ J_b \\ J_c \end{bmatrix}$$

Notice that matrix D is singular, therefore among i_a, i_b, i_c, only two phase currents can be controlled independently. If the sum of reference currents $\sum i_z^* \neq 0$, where $z = a, b, c$, then problem can not be solved. If $\sum i_z^* = 0$, only two of the reference inputs can be given independently, then we have one superfluous degree of freedom since the dimension of control equals to three. As result, although three currents are equal to the desired values, the motion of system is not unique. For example the voltage v_n in (18.157) can be equal to different values. In future, we assume that reference inputs and are given and discuss how to utilize the additional degree of freedom.

It should be noted from (18.156) that, since the equation for each phase current depends on all three discontinuous functions s_a, s_b, s_c, the frequency analysis cannot be performed independently. Thus, although hysteresis band PWM works in practical implementation, its analysis needs to be carefully checked. Control algorithm should be developed to control three-phase currents while only two of them can be controlled independently. The extra one degree of freedom can be used to control one more variable.

The proposed control algorithm is based on complementing the original system by the first order equation:

$$\dot{s}_3 = v_n^* - v_n \tag{18.160}$$

where v_n is defined in Eq. (18.157).

The sliding surfaces $s = [s_a, s_b, s_3]^T$ can be defined as:

$$\begin{cases} s_a = i_a^* - i_a \\ s_b = i_b^* - i_b \\ s_3 = \int (v_n^* - v_n)\, d\tau \end{cases} \tag{18.161}$$

where i_a^*, i_b^*, v_n^* are reference signals. The derivative of vector s can be easily derived as:

$$\begin{cases} \dot{s}_a = \dot{i}_a^* - J_a - \frac{2}{3}\frac{U_{dc}}{L} s_a + \frac{1}{3}\frac{U_{dc}}{L} s_b + \frac{1}{3}\frac{U_{dc}}{L} s_c \\ \dot{s}_b = \dot{i}_b^* - J_b - \frac{2}{3}\frac{U_{dc}}{L} s_b + \frac{1}{3}\frac{U_{dc}}{L} s_a + \frac{1}{3}\frac{U_{dc}}{L} s_c \\ \dot{s}_3 = v_n^* - (U_{dc}/3)\,(s_a + s_b + s_c) \end{cases} \tag{18.162}$$

Three-dimensional discontinuous control J can be designed such that sliding mode is enforced on $s = 0$. When sliding mode occurs on $s = 0$, all three components of are equal to zero and the tracking problem is solved. Since motion equation in sliding mode depends on parameter v_n^*, it can be selected in correspondence with some performance criterion. Different control methods can be used to enforce sliding mode.

Having proposed sliding surface s, the proposed control algorithm should be designed such that vector s is reduced to zero after finite time. Select the Lyapunov function:

$$V = \frac{1}{2}\, s^T s \tag{18.163}$$

Its time derivative is given by:

$$\dot{V} = F(.) - \frac{U_{dc}}{3L}[\alpha s_a + \beta s_b + \gamma s_c] \tag{18.164}$$

where,

$$\begin{cases} F(.) = s_a\left(i_a^* - J_a\right) + s_b\left(i_b^* - J_b\right) + s_3 v_n^* \\ \alpha = (2s_a - s_b + s_3 L) \\ \beta = (2s_b - s_a + s_3 L) \\ \gamma = (-s_a - s_b + s_3 L) \end{cases} \tag{18.165}$$

If u_c/L is large enough, $F(.)$ can be suppressed with control S given by:

$$S = \begin{cases} s_a = sign\ (\alpha) \\ s_b = sign\ (\beta) \\ s_c = sign\ (\gamma) \end{cases} \tag{18.166}$$

The proposed SMPWM control design for AC/DC inverters has the following advantages:

1. SMPWM with Lyapunov approach takes into account all sliding surfaces directly for each phase's control.
2. SMPWM does not require the dc voltage source to have constant value.
3. SMPWM can tolerate significant amount of disturbances or fluctuations from voltage source. This is an attractive advantage of SMPWM.

18.4.3 SMPWM Using Sliding Surface Decoupling Approach

Although the SMPWM method proposed in Sect. 4.2 is simple, it does not provide the capability to analyze the motion in the vicinity of the sliding manifold, since the three equations in (18.162) are interconnected. Therefore, this section presents another SMPWM technique that takes into account all sliding surfaces indirectly for each phase's switching control.

Equation (18.162) can be written in matrix form

$$\dot{s} = f_2 + \frac{U_{dc}}{3L} BJ \tag{18.167}$$

where the vector $f_2 = \left[\frac{di_a^*}{dt} - K_a, \frac{di_b^*}{dt} - K_b, v_n^*\right]$ includes terms without control variables, and $B = \begin{bmatrix} 2 & -1 & -1 \\ -1 & 2 & -1 \\ L & L & L \end{bmatrix}$. Since B is a non-singular matrix, it can be transformed into a diagonal matrix, and control variables can be decoupled for each sliding surface. Introduce new switching manifold $\overline{s} = B^{-1}s = 0$. In this case,

$$B^{-1} = \frac{1}{9L}\begin{bmatrix} 3L & 0 & 3 \\ 0 & 3L & 3 \\ -3L & -3L & 3 \end{bmatrix} \tag{18.168}$$

$$\overline{s} = \frac{1}{9L}\begin{bmatrix} 3Ls_a + 3s_3 \\ 3Ls_b + 3s_3 \\ -3Ls_a - 3Ls_b + 3s_3 \end{bmatrix} = \begin{bmatrix} \overline{s}_1 \\ \overline{s}_2 \\ \overline{s}_3 \end{bmatrix} \tag{18.169}$$

Differentiation of (18.169) gives $\dot{\overline{s}}_1 = B^{-1}\dot{s}$ and $\dot{s}$ can be as:

$$\dot{\overline{s}} = B^{-1}\dot{s} = f_3 - \frac{U_{dc}}{3L}\begin{bmatrix} s_a \\ s_b \\ s_c \end{bmatrix} \tag{18.170}$$

where $f_3 = [f_{31}, f_{32}, f_{33}]^T$ is a 3 * 1 vector whose elements have bounded values as:

$$\begin{cases} f_{31} = \frac{1}{3L}(L\frac{di_a^*}{dt} - LK_a + v_n^*) \\ f_{32} = \frac{1}{3L}(L\frac{di_b^*}{dt} - LK_b + v_n^*) \\ f_{33} = \frac{1}{3L}(-L\frac{di_a^*}{dt} + LK_a - \frac{di_b^*}{dt} + LK_b + v_n^*) \end{cases} \tag{18.171}$$

f_3 does not include control variables, s_a, s_b and s_c. Equation (18.170) shows that the overall motion can be decoupled into three individual ones (with respect to control). Select the control logic for inverter switches as follows:

$$\begin{cases} s_a = sign\ (\overline{s}_1) \\ s_b = sign\ (\overline{s}_2) \\ s_c = sign\ (\overline{s}_3) \end{cases} \tag{18.172}$$

If $\frac{U_{dc}}{3L} > \sup \|f_3\|$, the sliding mode existence conditions $\overline{s}_1\dot{\overline{s}}_1 < 0, \overline{s}_2\dot{\overline{s}}_2 < 0, \overline{s}_3\dot{\overline{s}}_3 < 0$ hold. Sliding manifold is reached after finite time. Reference input v_n^* can be selected depending on some operation criteria: control of switching frequency, minimization of this frequency with given accuracy, and so on.

18.4.4 Sliding Mode Control of Neutral Point Voltage (v_n)

This section explains the physical meaning of the proposed control methodology. As it follows from (18.162), the time derivative of s_3 depends on varying at high frequency in sliding mode. It means high frequency will be rejected on sliding surface and will depend on average value of v_n.

Section 4.3 shows that SMPWM can track reference currents and control the average value of v_n at the same time. As a result, proper v_n^* can be found to minimize some performance criterion. As an example, the switching frequency is to be minimized in order to reduce switching losses with a given accuracy. Ideally, sliding mode is a mathematical abstraction where the sliding motion trajectories are strictly on discontinuous surfaces. However, sliding mode in a real life system occurs not strictly on its discontinuous surfaces. Instead, it appears within some boundary layer. Analysis of sliding motion in boundary layer can be found in [40]. Assume that switching devices have hysteresis loop Δ. Then Δ defines the accuracy of the system, as shown in Fig. 18.35. In our system, the switching frequency depends on state velocities. Moreover, the switching frequency is not a constant value. We consider the system using surface decoupling approach shown in Sect. 4.3. Then each phase can be handled independently. The switching frequency is determined by two time intervals with $\dot{s} > 0$ and $\dot{s} < 0$. Consider the switching behavior of the i-th motion in (18.170) where $i \in \{1, 2, 3\}$. For each motion in (18.170), the time duration of "switch on" and the time duration of "switch off" can be written as functions of v_n^*.

The function f_3 in (18.170) depends on v_n^*, as shown in (18.171). For each motion in (18.170), the time duration of "switch on" and the time duration of "switch off" can be written as functions of v_n^*

$$\begin{cases} t_{oni}(v_n^*) = \dfrac{\Delta}{\left\| f_{3i}(v_n^*) - \frac{U_{dc}}{3L} \right\|} \\ \\ t_{offi}(v_n^*) = \dfrac{\Delta}{\left\| f_{3i}(v_n^*) + \frac{U_{dc}}{3L} \right\|} \end{cases} \tag{18.173}$$

It means that the switching frequency of the i-th motion is a function of v_n^* as well.

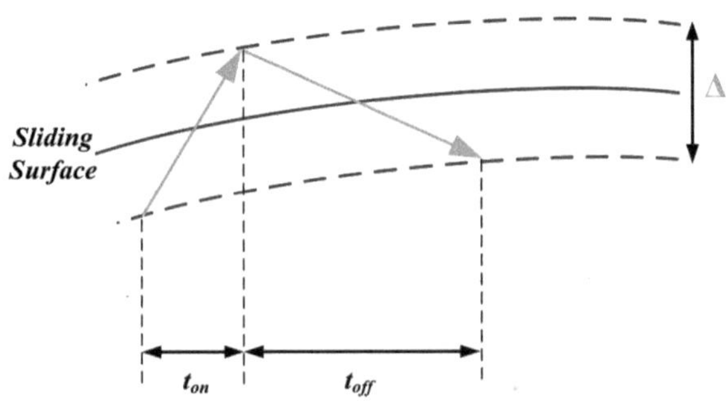

Fig. 18.35 Sliding manifold of SMPWM

$$f_{switch\_i}(v_n^*) = \frac{1}{t_{oni}(v_n^*) + t_{offi}(v_n^*)} \tag{18.174}$$

Taking into account all three motions in (18.170), the variable f_{switch} defined in (18.87) can be used as a measurement of the overall switching frequency of the system:

$$f_{switch\_i}(v_n^*) = \sum_{i=1}^{3} f_{switch\_i}(v_n^*) \tag{18.175}$$

Let $f_{switch\_i}(v_n^*)$ be the function to be minimized under the constraint $\frac{U_{dc}}{3L} > sup \| f_3 \|$. Then, the optimal v_n^* can be found from (18.172) and (18.173), such that $f_{switch\_i}(v_n^*)$ is minimized. In this particular example, the performance criterion is a function of v_n^*. In other problems, criterion could be a function of v_n. Then let v_n^* equal to the optimal value of v_n, and the tracking $v_n = v_n^*$ is provided by SMPWM.

Table 18.8 Simulation parameters of the three-phase AC/DC inverter

Parameter	U_{dc} (V)	$E_{abc,\ rms}$ (V)	$i^*_{abc,\ peak}$ (A)	R (Ω)	L (mH)	Step (μs)	Freq. (Hz)
Value	650	200	18	0.06	12	10	60

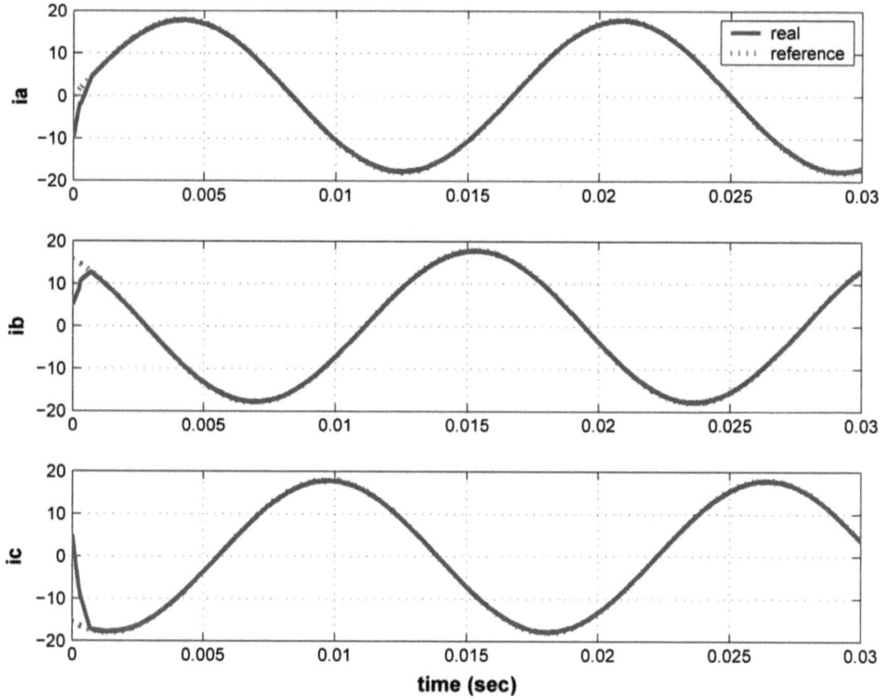

Fig. 18.36 Current tracking using SMPWM

18.4.5 Simulation Results

In order to evaluate the proposed control algorithm, computer simulations have been conducted using MATLAB/SIMULAINK software. Parameters used in simulation are listed in Table 18.8. This simulation used sliding surface decoupling approach. Results of three-phase current tracking (zoomed-in) are shown in Fig. 18.35. After the transient stage, phase currents track the references very well. Discontinuous output voltages v_a, v_b, v_c are shown in Fig. 18.36. In Fig. 18.37, the average value of v_n tracks a time varying reference. This average value is obtained by a first order dynamic $\mu \dot{x} = -x + v_n$ and $\mu = 0.001$. This low pass filter is used to calculate the average value of v_n in order to illustrate how close is this value to the v_n^*. Of course this filter is not needed for implementation of the system. In this simulation, the reference v_n^* is a randomly selected time varying function. It is not optimized according to some criterion. The only objective of Fig. 18.37 is to show the ability to track a time varying.

It is worth mentioning that SMPWM (see Fig. 18.38) is suitable for both analog and digital implementations. This due to the fact that SMPWM can be implemented with or without timing functionality, while other PWM approaches like Space Vector PWM require the DSP controller to have the ability to handle timing [7, 41].

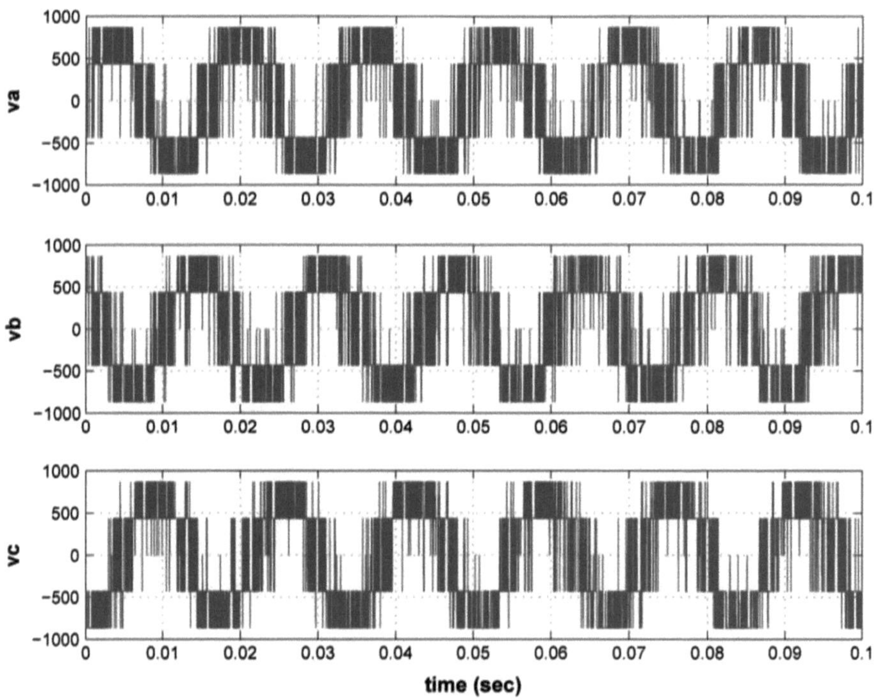

Fig. 18.37 Three-phase output voltages using SMPWM

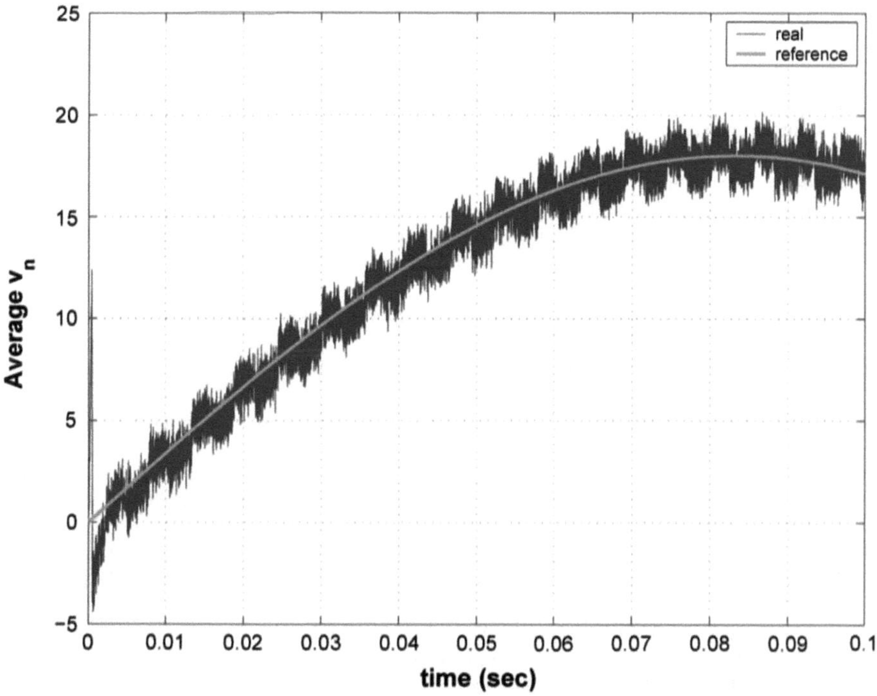

Fig. 18.38 Tracking a time varying $v_n^*(t)$ using SMPWM

References

1. Bartolini, G., Fridman, L., Pisano, A., Usai, E.: Modern Sliding Mode Control Theory: New Perspectives and Applications. Lecture Notes in Control and Information Sciences, vol. 375 (2008)
2. Biel, D., Fossas, E.: Some experiments on chattering suppression in power converters. In: Control Applications, (CCA) & Intelligent Control, (ISIC), 2009 IEEE, pp. 1523–1528. IEEE (2009)
3. Blasko, V.: Analysis of a hybrid pwm based on modified space-vector and triangle-comparison methods. IEEE Trans. Ind. Appl. **33**(3), 756–764 (1997)
4. Bondarev, A., Bondarev, S., Kostyleva, N., Utkin, V.I.: Sliding modes in systems with asymptotic state observers. Avtomatika i Telemekhanika **6**, 5–11 (1985)
5. Bowes, S.R., Clark, P.R.: Transputer-based harmonic-elimination pwm control of inverter drives. IEEE Trans. Ind. Appl. **28**(1), 72–80 (1992)
6. Bowes, S.R., Clark, P.R.: Regular-sampled harmonic-elimination pwm control of inverter drives. IEEE Trans. Power Electron. **10**(5), 521–531 (1995)
7. Chassaing, R.: DSP Applications Using C and the TMS320C6x DSK, vol. 13. Wiley, New York (2003)
8. Chen, L., Blaabjerg, F., Frederiksen, P.: An improved predictive control for three-phase pwm ac/dc converter with low sampling frequency. In: 20th International Conference on Industrial Electronics, Control and Instrumentation, 1994. IECON'94, vol. 1, pp. 399–404. IEEE (1994)

9. Cortes, D., Alvarez, J.: Robust sliding mode control for the boost converter. In: Power Electronics Congress, 2002. Technical Proceedings. CIEP 2002. VIII IEEE International, pp. 208–212. IEEE (2002)
10. DeCarlo, R.A., Zak, S.H., Matthews, G.P.: Variable structure control of nonlinear multivariable systems: a tutorial. Proc. IEEE **76**(3), 212–232 (1988)
11. Drakunov, S., Utkin, V., Zarei, S., Miller, J.: Sliding mode observers for automotive applications. In: Proceedings of the 1996 IEEE International Conference on Control Applications, 1996, pp. 344–346. IEEE (1996)
12. Drakunov, S.V., Izosimov, D., Lukyanov, A., Utkin, V., Utkin, V.: A hierarchical principle of the control systems decomposition based on motion separation. In: 9th IFAC Congress, vol. V, pp. 134–139 (1984)
13. Drakunov, S.V., Izosimov, D., Lukyanov, A., Utkin, V., Utkin, V.: Block control principle. ii. Autom. Remote Control **51**(6), 737–746 (1990)
14. Emelyanov, S., Utkin, V., Taran, V., Kostyleva, N., Shubladze, A., Ezerov, V., et al.: Theory of Variable Structure Systems. Nauka, Moscow (1970)
15. Habetler, T.G.: A space vector-based rectifier regulator for ac/dc/ac converters. IEEE Trans. Power Electron. **8**(1), 30–36 (1993)
16. Hashimoto, H., Utkin, V., Xu, J.X., Suzuki, H., Harashima, F.: Vss observer for linear time varying system. In: Industrial Electronics Society, 1990. IECON'90, 16th Annual Conference of IEEE, pp. 34–39. IEEE (1990)
17. Holtz, J., Lotzkat, W., Khambadkone, A.M.: On continuous control of pwm inverters in the overmodulation range including the six-step mode. IEEE Trans. Power Electron. **8**(4), 546–553 (1993)
18. Kazmierkowski, M.P., Dzieniakowski, M.A., Sulkowski, W.: Novel space vector based current controllers for pwm-inverters. IEEE Trans. Power Electron. **6**(1), 158–166 (1991)
19. Kokotović, P.V.: Applications of singular perturbation techniques to control problems. SIAM Rev. **26**(4), 501–550 (1984)
20. Kokotovic, P.V., O'malley, R., Sannuti, P.: Singular perturbations and order reduction in control theoryan overview. Automatica **12**(2), 123–132 (1976)
21. Komurcugil, H., Kukrer, O.: A novel current-control method for three-phase pwm ac/dc voltage-source converters. IEEE Trans. Ind. Electron. **46**(3), 544–553 (1999)
22. Krein, P.T.: Elements of Power Electronics, vol. 126. Oxford University Press, New York (1998)
23. Krstic, M., Kanellakopoulos, I., Kokotovic, P.V.: Nonlinear and Adaptive Control Design. Wiley, New York (1995)
24. Lee, H., Utkin, V.: Chattering analysis. Advances in Variable Structure and Sliding Mode Control. Lecture Notes in Control and Information Science, vol. 334, pp. 107–121. Springer, Berlin (2006)
25. Lukyanov, A., Utkin, V.: Methods of reducing equations for dynamic-systems to a regular form. Autom. Remote Control **42**(4), 413–420 (1981)
26. Martinez-Salamero, L., Cid-Pastor, A., Giral, R., Calvente, J., Utkin, V.: Why is sliding mode control methodology needed for power converters? In: Power Electronics and Motion Control Conference (EPE/PEMC), 2010 14th International, pp. S9–25. IEEE (2010)
27. Miwa, B.A., Otten, D.M., Schlecht, M.: High efficiency power factor correction using interleaving techniques. In: Applied Power Electronics Conference and Exposition, 1992. APEC'92. Conference Proceedings 1992., Seventh Annual, pp. 557–568. IEEE (1992)
28. Mohan, N., Undeland, T.M.: Power Electronics: Converters, Applications, and Design. Wiley, New York (2007)
29. Nguyen, V.M., Lee, C.: Tracking control of buck converter using sliding-mode with adaptive hysteresis. In: Power Electronics Specialists Conference, 1995. PESC'95 Record., 26th Annual IEEE, vol. 2, pp. 1086–1093. IEEE (1995)
30. Ooi, B.T., Dixon, J.W., Kulkarni, A.B., Nishimoto, M.: An integrated ac drive system using a controlled-current pwm rectifier/inverter link. In: Power Electronics Specialists Conference, 1986 17th Annual IEEE, pp. 494–501. IEEE (1986)
31. Rashid Muhammad, H.: Power Electronics Handbook. Academic Press, San Diego (2001)

32. Sira-Ramirez, H.: Sliding-mode control on slow manifolds of dc-to-dc power converters. Int. J. Control **47**(5), 1323–1340 (1988)
33. Sira-Ramírez, H.: Sliding Mode Control: The Delta-Sigma Modulation Approach. Birkhäuser (2015)
34. Sira-Ramirez, H., Escobar, G., Ortega, R.: On passivity-based sliding mode control of switched dc-to-dc power converters. In: IEEE Conference on Decision and Control, vol. 3, pp. 2525–2526. Institute Of Electrical Engineers Inc (IEE) (1996)
35. Sira-Ramírez, H., Silva-Ortigoza, R.: Control Design Techniques in Power Electronics Devices. Springer Science & Business Media, London (2006)
36. Slotine, J.J.E.: Sliding controller design for non-linear systems. Int. J. Control **40**(2), 421–434 (1984)
37. Trzynadlowski, A.M., Wang, Z., Nagashima, J.M., Stancu, C., Zelechowski, M.H.: Comparative investigation of pwm techniques for a new drive for electric vehicles. IEEE Trans. Ind. Appl. **39**(5), 1396–1403 (2003)
38. Utkin, V.: Sliding Modes in Control and Optimization. Springer, Berlin (1992)
39. Utkin, V., Guldner, J., Shi, J.: Sliding Mode Control in Electro-mechanical Systems, vol. 34. CRC Press, Boca Raton (2009)
40. Utkin, V.I.: Sliding Modes and Their Application in Variable Structure Systems. Mir publishers, Moscow (1978)
41. van Der Broeck, H.W., Skudelny, H.C., Stanke, G.V.: Analysis and realization of a pulsewidth modulator based on voltage space vectors. IEEE Trans. Ind. Appl. **24**(1), 142–150 (1988)
42. Utkin, V.I., Chen, D.S., Zarei, S., Miller, J.: Discrete time sliding mode observer for automotive alternator. In: Control Conference (ECC), 1997 European, pp. 3025–3030. IEEE (1997)
43. Venkataramanan, R., Sabanovic, A., Cuk, S.: Sliding mode control of dc-to-dc converters. In: Proceedings, IEEE Conference on Industrial Electronics, Control and Instrumentations (IECON), pp. 251–258 (1985)
44. Wong, C., Mohan, N., He, J.: Adaptive phase control for three phase pwm ac-to-dc converters with constant switching frequency. In: Conference Record of the Power Conversion Conference, 1993. Yokohama 1993, pp. 73–78. IEEE (1993)
45. Woo, W., Lee, N., Schuellein, G.: Multi-phase buck converter design with two-phase coupled inductors, applied power electronics conference and exposition, 2006. In: APEC, vol. 6, p. 6 (2006)
46. Wu, R., Dewan, S.B., Slemon, G.R.: A pwm ac-to-dc converter with fixed switching frequency. IEEE Trans. Ind. Appl. **26**(5), 880–885 (1990)
47. Wu, R., Dewan, S.B., Slemon, G.R.: Analysis of an ac-to-dc voltage source converter using pwm with phase and amplitude control. IEEE Trans. Ind. Appl. **27**(2), 355–364 (1991)
48. Xu, P., Wei, J., Lee, F.C.: Multiphase coupled-buck converter-a novel high efficient 12 v voltage regulator module. IEEE Trans. Power Electron. **18**(1), 74–82 (2003)

Chapter 19
An Adaptive Finite Time Sliding Mode Observer

Dongya Zhao, Sarah K. Spurgeon and Xinggang Yan

Abstract This chapter develops a novel adaptive finite time observer which can achieve finite time unmatched parameter estimation and finite time system state observation. The proposed approach has strong robustness and rapid convergence. A step by step proof is given which employs finite time stability and sliding mode principles. It is seen that the method also enables lumped matched uncertainty to be estimated. An illustrative example is used to validate the effectiveness of the proposed approach.

19.1 Introduction

In the system and control area, numerous control methodologies use a state feedback approach to guarantee stability. Such methods require that all the system states are available for controller implementation [5, 13]. However, some states may be difficult, if not impossible or prohibitively expensive, to measure in practice. For example, the velocity in a mechanical system or the reactant concentration in a chemical system may not be routinely measured [3]. In the case, the system states cannot be measured, state feedback based control will not work or not work well. Hence, to acquire all of the state information is a very important problem for both controller design and implementation in practical applications. One of the effective ways to estimate the system states is to use a state observer [16].

D. Zhao (✉)
Department of Chemical Equipment and Control Engineering,
China University of Petroleum, No. 66, Changjiang West Road,
Huangdao District 266580, Qingdao, China
e-mail: dongyazhao@139.com; dyzhao@upc.edu.cn
URL: http://www.pmc.upc.edu.cn

S.K. Spurgeon
Department of Electronic and Electrical Engineering, University College London,
Torrington Place WC1E 7JE, UK
e-mail: s.spurgeon@ucl.ac.uk

X. Yan
School of Engineering and Digital Arts, University of Kent, Canterbury, UK
e-mail: x.yan@kent.ac.uk

J.B. Clempner and W. Yu (eds.), *New Perspectives and Applications of Modern Control Theory*, https://doi.org/10.1007/978-3-319-62464-8_19

Observer design is also important in the domain of fault detection and isolation as well as condition monitoring [12]. Here the observer is a dynamical system representing nominal system behavior and some appropriate measure of the deviation between the dynamical behavior of the observer and the dynamical behavior of the plant or process is used to monitor the condition of the plant or process and detect the presence of faults or failures.

There are many approaches and algorithms to estimate the system states, such as the Luenberger observer [19], the high gain observer [8], the H_∞ observer [1], sliding mode observers [15] and the Kalman filter based observer [17]. For many observer design paradigms, the presence of unmatched uncertainty poses problems. Problems with unmatched uncertainty frequently occur in chemical and biological process control, as seen for example in the continuously stirred tank reactor, crude oil unit and fermentation processes [9]. If there is unmatched uncertainty present in the system, the observer accuracy may degrade and the observer may become unstable.

The sliding mode observer is known to possess strong robustness and appropriate design can eliminate the effects of system uncertainty and external disturbances. Recently, a class of step by step sliding mode observers has been developed for a plant consisting of a series of integrators. For systems with this structure, the observer error will converge to zero in finite time [6, 14, 20]. However the approach cannot deal with unmatched uncertainty effectively.

A novel finite time dynamic parameter estimator has been presented which uses low pass filters [11]. However this requires that all of the system states can be measured. It should also be mentioned that there is a finite time disturbance observer which also requires that all of the states can be measured [10].

Besides strong robustness, a finite time observer has faster convergence speed than an asymptotic observer, which is desirable in practice [4]. Note that if the designer wants to guarantee finite time observation, the observer has to eliminate the uncertainties and external disturbances in finite time. It is very challenging to design a finite time observer where the system contains unmatched and matched uncertainties. To address this problem, this chapter proposes a new sliding mode observer for a class of dynamical systems which contain unmatched uncertain parameters and matched lumped uncertainty. The uncertain parameters can be estimated in finite time by appealing to the adaptive principle. The system states can then be observed step by step in finite time by using the sliding mode principle and appealing to the principle of the equivalent injection. It should be mentioned that the proposed observer can also estimate the matched lumped system uncertainty in finite time. The proposed observer is thus not only useful for the implementation of state feedback controllers, but also provides a candidate observer for fault detection and monitoring as both the unknown parameters as well as the matched uncertainty, which can be thought of as an unknown input, are estimated. By appealing to the Lyapunov method, the system stability will be proved step by step. In comparison with existing sliding mode observers [6, 14, 20], the proposed approach can deal with unmatched uncertain dynamical parameters effectively. In comparison with the existing finite time parameter estimator [11], the proposed approach can estimate the dynamical parameters without requiring all of the system states to be measured. It should be mentioned that an adaptive output feedback finite time control has been designed for the case of a second order nonlinear system, which can also estimate the system parameters and states in finite time [21]. However, this chapter extends the approach to a class of higher order systems and a more general design procedure is developed.

This chapter is organized as follows: In Sect. 19.2, the problem is formulated and preliminary results and assumptions will be presented. In Sect. 19.3, an adaptive finite time observer will be designed and the stability analysis presented. In Sect. 19.4, an illustrative example will be used to validate the proposed approach. Section 19.5 will present concluding remarks.

19.2 Problem Formulation and Preliminaries

Consider the following system which is subject to unmatched uncertain parameters and matched lumped uncertainty:

$$\begin{aligned}
\dot{x}_1 &= x_2 + d_1(\bar{x}_1)\theta_1 \\
&\vdots \\
\dot{x}_i &= x_{i+1} + d_i(\bar{x}_i)\theta_i \\
&\vdots \\
\dot{x}_{n-1} &= x_n + d_{n-1}(\bar{x}_{n-1})\theta_{n-1} \\
\dot{x}_n &= u + \rho \\
y &= x_1
\end{aligned} \tag{19.1}$$

where $x = [x_1, x_2, \ldots, x_n]^T \in R^n$ is the system state vector, $\bar{x}_i = [x_1, x_2, \ldots, x_i]^T$, $y \in R$ is the system output, $u \in R$ is the control input, $\theta_i \in R^{l_i}$ is the unmatched uncertain parameter vector, $d_i(\bar{x}_i) \in R^{1\times l_i}$ is a known function with appropriate dimension and $\rho \in R$ denotes the matched uncertainty.

Remark 19.1 Note that (19.1) includes unmatched uncertain parameter vectors θ_i $(i = 1, \ldots, n-1)$ and lumped matched uncertainty ρ. Assuming θ_i is a constant vector which is unknown, the parameter vector should be estimated in finite time online if the observer is to be finite time stable.

Definition 19.1 A vector or matrix function $\phi(x)$ is persistently excited (PE) if there exist $t > 0$ and $\varepsilon > 0$ such that $\int_t^{t+\Delta t} \phi(r)\phi^T(r)\,dr \geq \varepsilon I$, $\forall t \geq 0$.

Assumption 1 $|d_i(\bar{x}_i)\theta_i| \leq \bar{d}_i, i = 1, \ldots, n-1, \bar{d}_i > 0$ is known and $d_i(\bar{x}_i)$ is PE.

Assumption 2 $|\rho| \leq \bar{\rho}, \bar{\rho} > 0$ is known.

The objective of this chapter is to design a finite time stable observer by using the adaptive method and the sliding mode principle in order to compensate for the presence of unmatched uncertain parameters and the matched lumped uncertainty. It will be seen that an added advantage of the proposed observer is the ability to estimate in finite time the matched lumped uncertainty, which can be considered an unknown input to the system. The designed observer thus belongs to the class of so called unknown input observers.

19.3 Adaptive Finite Time State Observer

In this section, a finite time parameter estimator will be developed firstly for (19.1), then an adaptive finite time state observer will be designed in light of the parameter estimator.

19.3.1 Finite Time Parameter Estimator

In this section, all of the system states are assumed to be measurable. A finite time parameter estimator is first designed by using filters as demonstrated in [2, 11].

For $\dot{x}_i = x_{i+1} + d_i(\bar{x}_i)\theta_i$ $(i = 1, \ldots, n-1)$, the following filters are chosen:

$$\begin{aligned} k\dot{x}_{i\_f} + x_{i\_f} &= x_i \\ k\dot{x}_{(i+1)\_f} + x_{(i+1)\_f} &= x_{(i+1)} \\ k\dot{d}_{i\_f} + d_{i\_f} &= d_i \end{aligned} \tag{19.2}$$

where $k > 0$ is a constant and the initial conditions of the three filters are given by $x_{i\_f}(0) = 0$, $x_{(i+1)\_f}(0) = 0$ and $d_{i\_f}(0) = 0$, respectively.

According to (19.2):

$$\dot{x}_{i\_f} = \frac{x_i - x_{i\_f}}{k} = x_{(i+1)\_f} + d_{i\_f}\theta_i \tag{19.3}$$

Define an auxiliary filter as:

$$\begin{aligned} \dot{p}_i &= -l p_i + d_{i\_f}^T d_{i\_f} \\ \dot{q}_i &= -l q_i + d_{i\_f}^T \left(\frac{x_i - x_{i\_f}}{k} - x_{(i+1)\_f} \right) \end{aligned} \tag{19.4}$$

where $l > 0$ is a design parameter.

The solution of (19.4) is:

$$p_i(t) = \int_0^t \exp(-l(t-r))\, d_{i\_f}^T(r)\, d_{i\_f}(r)\, dr$$
$$q_i(t) = \int_0^t \exp(-l(t-r)) d_{i\_f}^T(r) \left[\left(x_i(r) - x_{i\_f}(r) \right)/k - x_{(i+1)\_f}(r) \right] dr \tag{19.5}$$

It is obvious that:

$$\theta_i = p_i^{-1}(t)\, q_i(t) \tag{19.6}$$

Define an auxiliary vector as:

$$w_i(t) = p_i(t)\, \hat{\theta}_i(t) - q_i(t) \tag{19.7}$$

Then design the following adaptive parameter estimator for θ_i:

$$\dot{\hat{\theta}}_i = -\Gamma p_i^T(t)\, \mathrm{sgn}(w_i(t)) \tag{19.8}$$

where $\Gamma_i \in R^{l_i \times l_i}$ is a constant positive definite matrix, for a vector $z \in R^n$, $\mathrm{sgn}(z(t)) = [sgn(z_1(t)), \ldots, sgn(z_n(t))]^T$, z_i denotes ith element of vector z.

Remark 19.2 In the following state observer design, estimation of ρ is not required. Its effect can be eliminated by robust injection, as explained in the next section.

Lemma 19.1 ([11]) *If the regressor matrix $d_i(\bar{x}_i)$ is PE, the matrix $p_i(t)$ is positive definite and satisfies $\lambda_{\min}(p_i(t)) > \sigma_i$ for $t > T_i$ and $\sigma_i > 0$, $T_i > 0$, in which $\lambda_{\min}(\cdot)$ denotes the minimum eigenvalue of a positive matrix.*

Lemma 19.2 ([18]) *If $a_1, a_2, \ldots, a_n$ are all positive numbers and $0 < p < 2$, then the following inequality holds: $\left(a_1^2 + a_2^2 + \cdots + a_n^2\right)^p \le \left(a_1^p + a_2^p + \cdots a_n^p\right)^2$.*

Define the estimation error as:

$$\tilde{\theta}_i = \theta_i - \hat{\theta}_i \tag{19.9}$$

Proposition 19.1 *Consider Assumptions 1 and 2. If the parameter estimator is designed as in (19.8), the estimation error will converge to 0 in finite time.*

Proof Select a Lyapunov function as:

$$V_i^e = \frac{1}{2\Gamma} \tilde{\theta}_i^T \tilde{\theta}_i \tag{19.10}$$

Differentiate (19.10) with respect to time:

$$\dot{V}_i^e = -\frac{1}{\Gamma}\tilde{\theta}_i^T\dot{\hat{\theta}}_i \tag{19.11}$$

$$= \tilde{\theta}_i^T p_i^T(t)\,\mathrm{sgn}(w_i(t)) \tag{19.12}$$

$$= \tilde{\theta}_i^T p_i^T(t)\left[\mathrm{sgn}\left(-\left(p_i(t)\tilde{\theta}_i\right)_1\right),\ldots,sgn\left(-\left(p_i(t)\tilde{\theta}_i\right)_n\right)\right]^T \tag{19.13}$$

According to Lemmas 19.1 and 19.2, it follows that:

$$\dot{V}_i^e = -\sum_{i=1}^{n}\left(p_i(t)\tilde{\theta}_i\right)_i$$

$$\leq -\left\|p_i(t)\tilde{\theta}_i\right\| \tag{19.14}$$

$$\leq -\mu_i\sqrt{V_i^e} \tag{19.15}$$

where $\mu_i = \sigma_i\sqrt{2/\lambda_{\max}\left(\Gamma^{-1}\right)}$, and $\tilde{\theta}_i = 0$ as $t \geq t_i^e$, $t_i^e \leq 2\sqrt{V_i^e(0)}/\mu_i$.

Remark 19.3 The estimation law (19.8) requires that all of the states are measurable. However, it can be very difficult to measure all of the states in practice. For example, the velocity and acceleration can be difficult to measure in mechanical problems. Inspired by (19.8), an adaptive state observer will now be developed.

19.3.2 Adaptive Finite Time System State Observer

Before designing the observer, a further assumption is necessary.

Assumption 3 $\left|\left(p_i(t)\zeta_i(t)\right)_j\right| < \left|\left(p_i(t)\tilde{\theta}_i(t)\right)_j\right|$, $i = 1,\ldots,n-1$ and $j = 1,\ldots,l_i$, in which $\zeta_i(t)$ is caused by the observer error. This means the effect of the deviation $\zeta_i(t)$ caused by $\hat{x}_{i+1}$ is smaller than that due to the parameter estimation error $\tilde{\theta}_i(t)$.

Remark 19.4 Assumption 3 means that the observer may cause a deviation in the low pass filter which may in turn affect the parameter estimation. If the observer's effect is small enough, it will not affect the finite time parameter estimation. It is not a difficult situation to deal with in practice with appropriate selection of design parameters.

An adaptive finite time state observer is now designed for $i = 1,\ldots,n-1$. The filters defined in (19.2) are now defined in terms of the state estimates as follows

$$\begin{aligned} k\dot{\hat{x}}_{i\_f} + \hat{x}_{i\_f} &= \hat{x}_i \\ k\dot{\hat{x}}_{(i+1)\_f} + \hat{x}_{(i+1)\_f} &= \hat{x}_{(i+1)\_f} \\ k\dot{d}_{i\_f} + d_{i\_f} &= d_i \end{aligned} \tag{19.16}$$

with initial conditions $\hat{x}_{i\_f}(0) = 0$, $\hat{x}_{(i+1)\_f}(0) = 0$ and $d_{i\_f}(0) = 0$ respectively where

$$\begin{aligned} \dot{\hat{x}}_{i\_f} &= \frac{\hat{x}_i - \hat{x}_{i\_f}}{k} \\ &= \hat{x}_{(i+1)\_f} + d_{i\_f}\theta_{i\_f} - \zeta_i(t) \end{aligned} \tag{19.17}$$

and $\zeta_i(t)$ defined in Assumption 3 denotes the mismatch resulting from the use of the observer state $\hat{x}(t)$ rather than the state $x(t)$. The corresponding auxiliary filters are defined in terms of the estimated state by

$$\begin{aligned} \dot{p}_i &= -l p_i + d_{i\_f}^T d_{i\_f} \\ \dot{q}_i &= -l q_i + d_{i\_f}^T \left(\frac{\hat{x}_i - \hat{x}_{i\_f}}{k} - \hat{x}_{(i+1)\_f} \right) \end{aligned} \tag{19.18}$$

with corresponding solutions

$$\begin{aligned} p_i(t) &= \int_0^t e^{-l(t-r)} d_{i\_f}^T(r)\, d_{i\_f}(r)\, dr \\ q_i(t) &= \int_0^t e^{-l(t-r)} d_{i\_f}^T(r) \left[\left(\hat{x}_i(r) - \hat{x}_{i\_f}(r) \right)/k - \hat{x}_{(i+1)\_f}(r) \right] dr \end{aligned} \tag{19.19}$$

Then

$$\theta_i = p_i^{-1}(t)\, q_i(t) + \zeta_i(t) \tag{19.20}$$

with

$$\widehat{x}_i = \left[\alpha_{(i-1)} \mathrm{sgn}\left(\tilde{x}_{(i-1)} \right) \right]_{eq} + \hat{x}_i \tag{19.21}$$

where $[.]_{eq}$ denotes the equivalent injection signal as defined in the domain of sliding mode control [7]. This equivalent injection signal is not the signal applied in practice but represents the average behavior of the corresponding discontinuous signal required to maintain a sliding condition. This equivalent injection may be obtained by passing the applied discontinuous signal through a low pass filter.

The adaptive observer for x_i is designed as:

$$\dot{\hat{x}}_i = \hat{x}_{(i+1)} + d_i\left(\hat{\widehat{x}}_i \right) \hat{\theta}_i + \alpha_i \mathrm{sgn}\left(\widehat{x}_i - \hat{x}_i \right) \tag{19.22}$$

$$\dot{\hat{\theta}}_i = -\Gamma_i \left(p_i^T(t)\,\mathrm{sgn}\,(w_i(t)) - d_i^T\left(\hat{\bar{x}}_i\right)\tilde{x}_i \right) \tag{19.23}$$

If $i = n$, the state observer can be designed as:

$$\dot{\hat{x}}_n = u + \alpha_n \mathrm{sgn}(\widehat{x}_n - \hat{x}_n) \tag{19.24}$$

where $\alpha > \bar{\rho}$, $\widehat{x}_n = \left[\alpha_{(n-1)}\mathrm{sgn}\left(\tilde{x}_{(n-1)}\right)\right]_{eq} + \hat{x}_n$.

Theorem 19.1 *Consider the system (19.1) which is assumed to satisfy Assumptions 1–3. If the observer is designed as (19.16)–(19.24) under the conditions Γ_i is positive and α_i is large enough, the system states and dynamic parameters can be estimated in finite time. Further, the matched lumped system uncertainty can be estimated in finite time by using the sliding mode injection as:*

$$\rho = \left[\alpha_n \mathrm{sgn}\left(\widehat{x}_n - \hat{x}_n\right)\right]_{eq} \tag{19.25}$$

Proof To prove the theorem, the observer error is defined as:

$$\tilde{x}_i = x_i - \hat{x}_i \tag{19.26}$$

A step by step proof will be used.

Step 1

At the first step, the state observer and parameter adaptation law are designed as follows:

$$\dot{\hat{x}}_1 = \hat{x}_2 + d_1\left(\hat{\bar{x}}_1\right)\hat{\theta}_1 + \alpha_1 \mathrm{sgn}\,(\tilde{x}_1) \tag{19.27}$$

$$\dot{\hat{\theta}}_1 = -\Gamma_1 \left(p_1^T(t)\,\mathrm{sgn}\,(w_1(t)) - d_1^T\left(\hat{\bar{x}}_1\right)\tilde{x}_1 \right) \tag{19.28}$$

The observer error dynamic equation is given by:

$$\dot{\tilde{x}}_1 = \tilde{x}_2 + d_1\left(\hat{\bar{x}}_1\right)\tilde{\theta}_1 + \Delta d_1(\bar{x}_1)\,\theta_1 - \alpha_1 \mathrm{sgn}\,(\tilde{x}_1) \tag{19.29}$$

where $\Delta d_1(\bar{x}_1) = d_1(\bar{x}_1) - d_1\left(\hat{\bar{x}}_1\right)$.

Choose a Lyapunov function as:

$$V_1^o = \frac{1}{2}\tilde{x}_1^2 + \frac{1}{2\Gamma_1}\tilde{\theta}_1^T\tilde{\theta}_1 \tag{19.30}$$

Differentiate (19.30) with respect to time along (19.29):

$$\dot{V}_1^o = \tilde{x}_1\dot{\tilde{x}}_1 + \frac{1}{\Gamma_1}\tilde{\theta}_1^T\dot{\tilde{\theta}}_1 \tag{19.31}$$

Substitute (19.27)–(19.29) into (19.31):

$$\begin{aligned}\dot{V}_1^o &= \tilde{x}_1\left(\tilde{x}_2 + d_1\left(\hat{\bar{x}}_1\right)\tilde{\theta}_1 + \Delta d_1\left(\bar{x}_1\right)\theta_1 - \alpha_1 \mathrm{sgn}\left(\tilde{x}_1\right)\right) + \\ &\quad \tilde{\theta}_1^T\left(p_1^T(t)\,\mathrm{sgn}\left(w_1(t)\right) - d_1\left(\hat{\bar{x}}_1\right)\tilde{x}_1\right) \qquad (19.32)\\ &= -\alpha_1\left|\tilde{x}_1\right| + \tilde{x}_1\left(\tilde{x}_2 + \Delta d_1\left(\bar{x}_1\right)\theta_1\right) + \\ &\quad \tilde{\theta}_1^T p_1^T(t)\,\mathrm{sgn}\left(-p_1(t)\,\tilde{\theta}_1 + p_1\zeta_1(t)\right) \qquad (19.33)\end{aligned}$$

According to Lemma 19.2:

$$\dot{V}_1^o \le -\left|\tilde{x}_1\right|\left(\alpha_1 - \left|\tilde{x}_2 + \Delta d_1\left(\bar{x}_1\right)\theta_1\right|\right) - \left\|p_1(t)\,\tilde{\theta}_1\right\| \qquad (19.34)$$

If α_1 is large enough, there is a positive number $\eta_1 > 0$ such that:

$$\alpha_1 - \left|\tilde{x}_2 + \Delta d_1\left(\bar{x}_1\right)\theta_1\right| \ge \eta_1$$

Equation (19.34) then becomes:

$$\begin{aligned}\dot{V}_1^o &\le -\eta_1\left|\tilde{x}_1\right| - \left\|p_1(t)\,\theta_1\right\| \qquad (19.35)\\ &\le -c_{11}\sqrt{\frac{1}{2}\tilde{x}_1^2} - c_{12}\sqrt{\frac{1}{2}\tilde{\theta}_1^T\tilde{\theta}_1} \qquad (19.36)\end{aligned}$$

where $c_{11} = \sqrt{2}\eta_1$, $c_{12} = \sigma_1\sqrt{2/\lambda_{\max}\left(\Gamma_1^{-1}\right)}$. Let $c_1 = \min\{c_{11}, c_{12}\}$. Then, according to Lemma 19.2, the following inequality holds:

$$\dot{V}_1^o \le -c_1 V_1^o \qquad (19.37)$$

According to the finite time stability principle, $\tilde{x}_1$ and $\tilde{\theta}_1$ will be 0 after $t \ge t_1^o$, $t_1^o = \frac{\sqrt{V_1^o(0)}}{2c_1}$.

After $t \ge t_1^o$, $\tilde{\theta}_1 = 0$ and $\Delta d_1\left(\bar{x}_1\right) = 0$. According to (19.29), the following equation holds:

$$\tilde{x}_2 = \left[\alpha_1 \mathrm{sgn}\left(\tilde{x}_1\right)\right]_{eq} \qquad (19.38)$$

where $\left[\alpha_1 \mathrm{sgn}\left(\tilde{x}_1\right)\right]_{eq}$ is the equivalent injection corresponding to $\alpha_1 \mathrm{sgn}\left(\tilde{x}_1\right)$.

Step 2

The state observer and parameter adaptive law are designed as follows:

$$\begin{aligned}\dot{\hat{x}}_2 &= \hat{x}_3 + d_2(\hat{\bar{x}}_2)\hat{\theta}_2 + \alpha_2 \mathrm{sgn}(\widehat{x}_2 - \hat{x}_2) \qquad (19.39)\\ \dot{\hat{\theta}}_2 &= -\Gamma_2\left(p_2^T(t)\,\mathrm{sgn}\left(w_2(t)\right) - d_2^T\left(\hat{\bar{x}}_2\right)\tilde{x}_2\right) \qquad (19.40)\end{aligned}$$

where $\widehat{x}_2 = \left[\alpha_1 \mathrm{sgn}\left(\tilde{x}_1\right)\right]_{eq} + \hat{x}_2$ and $\tilde{x}_2$ is given in Eq. (19.38).

The corresponding observer error dynamic equation is given by:

$$\dot{\tilde{x}}_2 = \tilde{x}_3 + d_2\left(\hat{\bar{x}}_2\right)\tilde{\theta}_2 + \Delta d_2\left(\bar{x}_2\right)\theta_2 - \alpha_2 \text{sgn}\left(\tilde{x}_2\right) \tag{19.41}$$

where $\Delta d_2\left(\bar{x}_2\right) = d_2\left(\bar{x}_2\right) - d_2\left(\hat{\bar{x}}_2\right)$.

Choose a Lyapunov function as:

$$V_2^o = \frac{1}{2}\tilde{x}_2^2 + \frac{1}{2\Gamma_2}\tilde{\theta}_2^T\tilde{\theta}_2 \tag{19.42}$$

By using the procedure as employed for *Step 1* and described in Eqs. (19.30)–(19.37), it can be proved that $\tilde{x}_2 = 0$ and $\tilde{\theta}_2 = 0$ in finite time as $t \geq t_2^o$, $t_2^o = \frac{\sqrt{V_2^o(0)}}{2c_2}$, $c_{21} = \sqrt{2}\eta_2$, $c_{22} = \sigma_2\sqrt{2/\lambda_{\max}\left(\Gamma_2^{-1}\right)}$, $c_2 = \min\{c_{21}, c_{22}\}$.

Step i

The state observer and parameter adaptive law are designed as follows:

$$\dot{\hat{x}}_i = \hat{x}_{(i+1)} + d_i\left(\hat{\bar{x}}_i\right)\hat{\theta}_i + \alpha_i \text{sgn}(\widehat{x}_i - \hat{x}_i) \tag{19.43}$$

$$\dot{\hat{\theta}}_i = -\Gamma_i\left(p_i^T(t)\,\text{sgn}\left(w_i(t)\right) - d_i^T\left(\hat{\bar{x}}_i\right)\tilde{x}_i\right) \tag{19.44}$$

where $\widehat{x}_i = \left[\alpha_{(i-1)}\text{sgn}\left(\tilde{x}_{(i-1)}\right)\right]_{eq} + \hat{x}_i$ and $\tilde{x}_i = \left[\alpha_{(i-1)}\text{sgn}\left(\tilde{x}_{(i-1)}\right)\right]_{eq}$.

The corresponding observer error dynamic equation becomes:

$$\dot{\tilde{x}}_i = \tilde{x}_{(i+1)} + d_i\left(\hat{\bar{x}}_i\right)\tilde{\theta}_i + \Delta d_i\left(\bar{x}_i\right)\theta_i - \alpha_i \text{sgn}\left(\tilde{x}_i\right) \tag{19.45}$$

where $\Delta d_i\left(\bar{x}_i\right) = d_i\left(\bar{x}_i\right) - d_i\left(\hat{\bar{x}}_i\right)$.

Choose a Lyapunov function as:

$$V_i^o = \frac{1}{2}\tilde{x}_i^2 + \frac{1}{2\Gamma_i}\tilde{\theta}_i^T\tilde{\theta}_i \tag{19.46}$$

By using the same approach as in *Step 1* and *Step 2*, it can be proved that $\tilde{x}_i = 0$ and $\tilde{\theta}_i = 0$ as $t \geq t_i^o$, $t_i^o = \frac{\sqrt{V_i^o(0)}}{2c_i}$, $c_{i1} = \sqrt{2}\eta_i$, $c_{i2} = \sigma_i\sqrt{2/\lambda_{\max}\left(\Gamma_i^{-1}\right)}$, $c_i = \min\{c_{i1}, c_{i2}\}$.

Step n

In this final step, it is not required to compute the equivalent injection as the system output $y = x_n$ and the observer error $\tilde{x}_n$ are known. Hence the adaptive law is not required. The observer is designed as:

$$\dot{\hat{x}}_n = u + \alpha_n \text{sgn}(\widehat{x}_n - \hat{x}_n) \tag{19.47}$$

where $\widehat{x}_n = \left[\alpha_{(n-1)}\text{sgn}\left(\tilde{x}_{(n-1)}\right)\right]_{eq} + \hat{x}_n$, $\tilde{x}_n = \left[\alpha_{(n-1)}\text{sgn}\left(\tilde{x}_{(n-1)}\right)\right]_{eq}$.
The observer error dynamic equation will be:

$$\dot{\tilde{x}}_n = -\alpha_n \text{sgn}\left(\tilde{x}_n\right) + \rho \tag{19.48}$$

Choose a Lyapunov function as:

$$V_n^o = \frac{1}{2}\tilde{x}_n^2 \tag{19.49}$$

Differentiate (19.49) with respect to time along (19.48):

$$\dot{V}_n^o = \tilde{x}_n\left(-\alpha_n \text{sgn}\left(\tilde{x}_n\right) + \rho\right) \tag{19.50}$$

$$= -\alpha_n\left|\tilde{x}_n\right| + \tilde{x}_n\rho \le -\alpha_n\left|\tilde{x}_n\right| + \left|\tilde{x}_n\right|\left|\rho\right| \le -\left|\tilde{x}_n\right|\left(\alpha_n - \bar{\rho}\right) \tag{19.51}$$

From Assumptions 1–3, if α_n is large enough such that $\alpha_n - \bar{\rho} > \eta_n > 0$, Eq. (19.51) becomes:

$$\dot{V}_n^o \le -\eta_n\left|\tilde{x}_n\right| = -c_n\sqrt{\frac{1}{2}\tilde{x}_n^2} = -c_n V_n^o \tag{19.52}$$

where $c_n = \sqrt{2}\eta_n$. $\tilde{x}_n = 0$ as $t \ge t_n^o$, $t_n^o = \frac{\sqrt{V_n^o(0)}}{2c_n}$.
Further, if $\tilde{x}_n = 0$, according to (19.48), it follows that:

$$\rho = \alpha_n \text{sgn}\left(\tilde{x}_n\right) \tag{19.53}$$

In light of the sliding principle, the lumped matched uncertainty can be estimated as follows from the corresponding equivalent injection signal:

$$\rho = \left[\alpha_n \text{sgn}\left(\tilde{x}_n\right)\right]_{eq} \tag{19.54}$$

This concludes the proof.

Remark 19.5 An adaptive finite time state observer has been developed. The results relating to the finite stability time can be used to determine the observer parameters. However, given the inherent conservatism in the Lyapunov method of proof, some parameter tuning via simulation may improve the performance. The proposed observer not only estimates the unmatched uncertain parameters and system states, but also estimates the lumped matched uncertainty present in the system. The observer framework can be used in fault detection and condition monitoring problems as well as being useful for state measurement and uncertainty compensation for controller design and implementation.

19.4 Illustrative Example

In this section, an illustrative example will be used to validate the proposed approach. The following nonlinear dynamic system will be considered:

$$\dot{x}_1 = x_2 + d_1(\bar{x}_1)\theta_1 \tag{19.55}$$

$$\dot{x}_2 = x_3 + d_2(\bar{x}_2)\theta_2 \tag{19.56}$$

$$\dot{x}_3 = u + \rho \tag{19.57}$$

where $d_1(\bar{x}_1) = x_1$ and $d_2(\bar{x}_2) = [x_1, x_2]$ are regressor matrices. $\theta_1 = 0.5$ and $\theta_2 = [0.5, 0.7]^T$ are the unmatched uncertain parameter vectors. $\rho = 3sin(t)$ denotes the lumped matched uncertainty.

For the simulation experiments the following parameters are chosen. In the filters defined in Eq. (19.16) the free parameter $k = 10$ and for the auxiliary filters in Eq. (19.18), $l = 10$. Within the state observer defined in Eqs. (19.22) and (19.24) the parameters α_i are selected as $\alpha_1 = 5$, $\alpha_2 = 3$, $\alpha_3 = 3$. In the Eq. (19.23), the parameters Γ_i are selected as $\Gamma_1 = 10$, $\Gamma_2 = 0.5$, $\Gamma_3 = 0.5$.

Figures 19.1, 19.2 and 19.3 show the system states and the corresponding state estimates; the solid lines represent the system states and the dashed lines the estimates. From these simulation results, one can see the proposed adaptive finite time observer can estimate the system states accurately and in finite time.

Figures 19.4, 19.5 and 19.6 show the corresponding parameter estimates. From these figures, it can be seen that the proposed adaptive parameter estimator can accurately estimate the parameters in finite time. It should be noted that to achieve finite time state estimation by using a step by step sliding mode observer, it is necessary to estimate the unmatched uncertain parameters in finite time. Otherwise, finite time observation cannot be achieved. This is readily seen from the proof of the theorem. During the step by step sliding mode observer proof, a corresponding equivalent injection is used to estimate the observer error. If there is unmatched uncertainty

Fig. 19.1 System state x_1 and its estimate $\hat{x}_1$

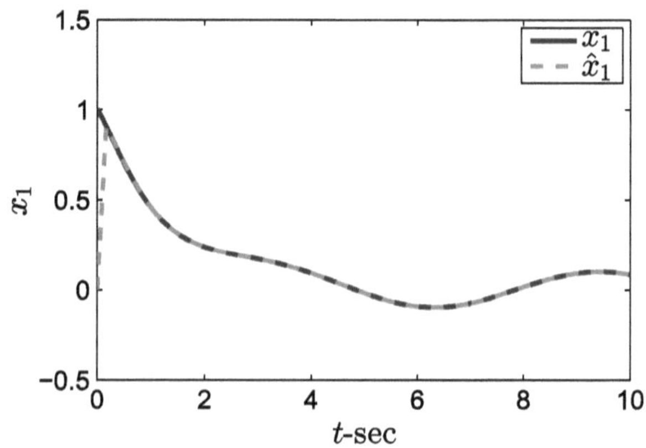

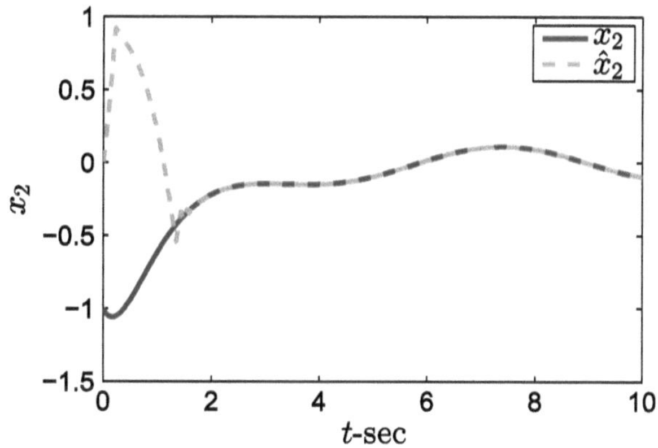

Fig. 19.2 System state x_2 and its estimate $\hat{x}_2$

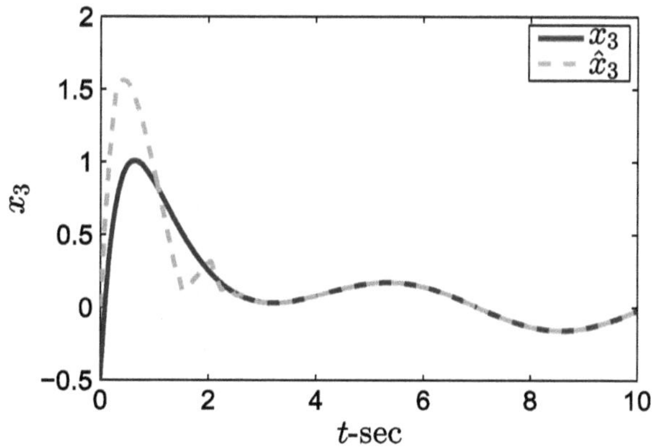

Fig. 19.3 System state x_3 and its estimate $\hat{x}_3$

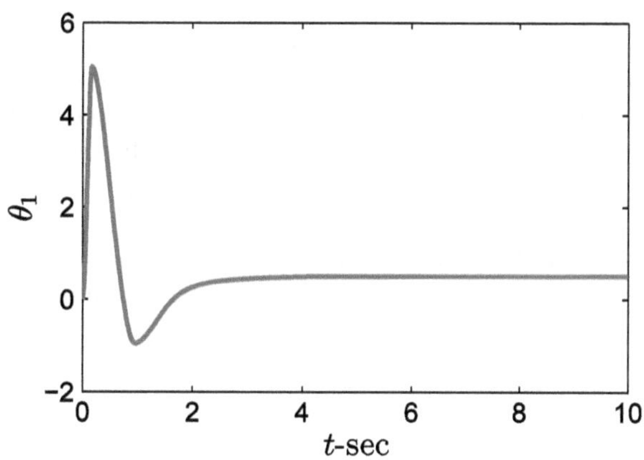

Fig. 19.4 Estimation of parameter θ_1

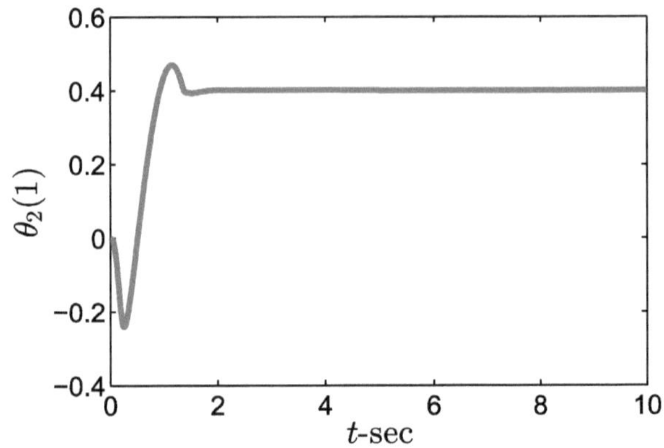

Fig. 19.5 Estimation of parameter $\theta_1(1)$

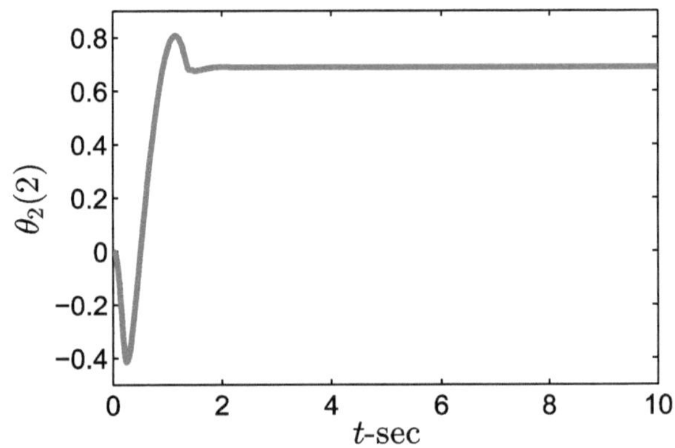

Fig. 19.6 Estimation of parameter $\theta_2(2)$

present in the error dynamic equations, this would contribute to the corresponding equivalent injection signals at every step. It follows that the observer error would not be accurately estimated. Hence, the unmatched uncertainty must be compensated. If the parameters are estimated asymptotically, the observer would not be finite time stable. This is the reason why the parameter estimator is required to be finite time stable.

Figure 19.7 shows the estimate of the lumped matched uncertainty obtained from Eq. (19.25). In the simulation results, the discontinuous injection signals were implemented using the unit vector approximation

$$\mathrm{sgn}(.) \approx \frac{(.)}{\mathrm{abs}(.) + \delta} \tag{19.58}$$

where $\delta > 0$ is selected to be small. For the simulations, $\delta = 0.01$ which was sufficient to smooth the corresponding injection signal. It is seen that a high-quality estimate of the lumped matched uncertainty is obtained.

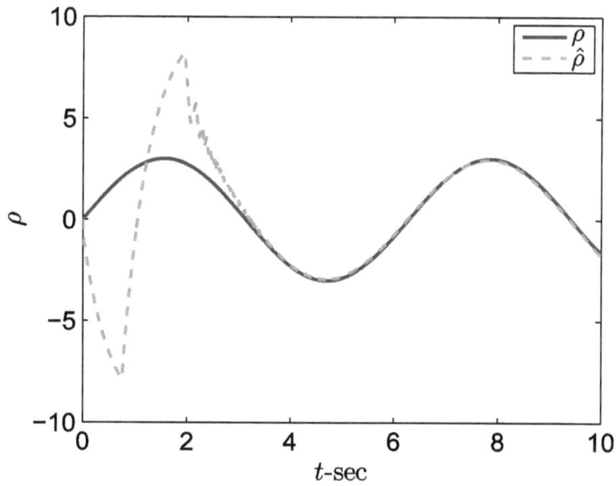

Fig. 19.7 Estimate of the disturbance from the observer equivalent injection given from Eq. (19.25)

19.5 Conclusion

In practical systems, such as chemical processes and mechanical systems, it can be very challenging to design a feedback control strategy based only on partial state measurement. In such cases, a state observer is very beneficial for controller implementation. In addition state and unknown input observers can be very useful tools in the development of fault detection and isolation systems and for condition monitoring. This chapter presents a novel step by step sliding mode observer including a finite time parameter estimator. The proposed approach can achieve good performance in the presence of unmatched uncertain dynamic parameters as well as lumped matched uncertainty. A constructive proof is developed, which is very helpful in understanding the proposed approach. The proof employs the Lyapunov approach as well as the principle of the equivalent injection from the domain of sliding mode control and observation. As well as estimating the system states and the unknown parameters, it is also possible to estimate the lumped matched uncertainty in the system. An illustrative example is used to demonstrate the observer performance. Feedback controller design in the light of the proposed observer is interesting and is a future topic of research.

References

1. Abbaszadeh, M., Marquez, H.J.: Robust $H\infty$ observer design for sampled-data Lipschitz nonlinear systems with exact and Euler approximate models. Automatica **44**(3), 799–806 (2008)
2. Adetola, V., Guay, M.: Finite-time parameter estimation in adaptive control of nonlinear systems. IEEE Trans. Autom. Control **53**(3), 807–811 (2008)
3. Besancon, G.: Nonlinear Observers and Applications. Springer, Berlin (2007)

4. Bhat, S.P., Bernstein, D.S.: Finite-time stability of continuous autonomous systems. SIAM J. Control Optim. **38**(3), 751–766 (2006)
5. Chen, C.T.: Linear System Theory and Design. Holt, Rinehart, and Winston, New York (1984)
6. Daly, J.M., Wang, D.W.L.: Output feedback sliding mode control in the presence of unknown disturbances. Syst. Control Lett. **58**(3), 188–193 (2009)
7. Edwards, C., Spurgeon, S.K.: Sliding Mode Control: Theory and Applications. Taylor and Francis, London (1998)
8. Khalil, H.K., Laurent, P.: High-gain observers in nonlinear feedback control. Int. J. Robust Nonlinear Control **24**(6), 993–1015 (2014)
9. Lee, P.L.: Nonlinear Process Control. Springer, London (1993)
10. Li, S., Sun, H., Yang, J., Yu, X.: Continuous finite-time output regulation for disturbed systems under mismatching condition. IEEE Trans. Autom. Control **60**(1), 277–282 (2015)
11. Na, J., Mahyuddin, M.N., Herrmann, G., Ren, X., Barber, P.: Robust adaptive finite - time parameter estimation and control for robotic systems. Int. J. Robust Nonlinear Control **25**(16), 3045–3071 (2014)
12. Patton, R., Chen, J.: Observer-based fault detection and isolation: robustness and applications. Control Eng. Pract. **5**(5), 671–682 (1997)
13. Radke, A., Gao, Z.: A survey of state and disturbance observers for practitioners. In: American Control Conference, pp. 5183–5188 (2006)
14. Slotine, J.J.E., Hedrick, J.K., Misawa, E.A.: On sliding observers for nonlinear systems. In: American Control Conference, pp. 1794–1800 (1986)
15. Spurgeon, S.K.: Sliding mode observers: a survey. Int. J. Syst. Sci. **39**(8), 751–764 (2008)
16. Tsui, C.C.: Observer design survey. Int. J. Autom. Comput. **12**(1), 50–61 (2015)
17. Velardi, S.A., Hammouri, H., Barresi, A.A.: In-line monitoring of the primary drying phase of the freeze-drying process in vial by means of a Kalman filter based observer. Chem. Eng. Res. Des. **87**(10A), 1409–1419 (2009)
18. Yu, S., Yu, X., Shirinzadeh, B., Man, Z.: Continuous finite-time control for robotic manipulators with terminal sliding mode. Automatica **41**(11), 1957–1964 (2005)
19. Zeitz, M.: The extended Luenberger observer for nonlinear systems. Syst. Control Lett. **9**(2), 149–156 (1987)
20. Zhao, D., Li, S., Zhu, Q.: Output feedback terminal sliding mode control for a class of second order nonlinear systems. Asian J. Control **15**(1), 237–247 (2013)
21. Zhao, D., Spurgeon, S.K., Yan, X.: Adaptive output feedback finite time control for a class of second order nonlinear systems. In: International Workshop on Variable Structure Systems, pp. 53–58 (2016)

MIX
Papier aus verantwortungsvollen Quellen
Paper from responsible sources
FSC® C105338

If you have any concerns about our products,
you can contact us on
ProductSafety@springernature.com

In case Publisher is established outside the EU,
the EU authorized representative is:
Springer Nature Customer Service Center GmbH
Europaplatz 3, 69115 Heidelberg, Germany

Printed by Libri Plureos GmbH
in Hamburg, Germany